Merrill GEOMETRY

FOSTER • CUMMINS • YUNKER

CHARLES E. MERRILL PUBLISHING CO.
A Bell & Howell Company
Columbus, Ohio

Toronto • London • Sydney

ISBN 0-675-05839-2

Published by
CHARLES E. MERRILL PUBLISHING CO.
A Bell & Howell Company
Columbus, Ohio 43216

Printed in the United States of America

Merrill
GEOMETRY

AUTHORS

Senior Author, **Alan G. Foster,** is head of the Mathematics Department at Addison Trail High School, Addison, Illinois. He has taught mathematics courses at every level of the high school curriculum. Mr. Foster obtained his B. S. from Illinois State University and his M. A. in mathematics from the University of Illinois, with additional work at University of Colorado, University of Wyoming, Northern Illinois University, Purdue, Northwestern, Princeton, Illinois Institute of Technology, and National College of Education. Mr. Foster is active in professional organizations at local, state, and national levels, frequently speaking or conducting workshops. He is a past president of the Illinois Council of Teachers of Mathematics.

Jerry J. Cummins is head of the Mathematics Department at Proviso West High School, Hillside, Illinois. In addition to teaching experience at a junior high school and community college, he has taught mathematics courses at every level of the high school curriculum. Mr. Cummins obtained a B. S. in mathematics education and an M. S. in educational administration and supervision from Southern Illinois University. He also holds an M. S. in mathematics education from the University of Oregon. He has done additional graduate work at the Illinois Institute of Technology. Mr. Cummins has spoken on mathematics at the local, state, and national levels. He is past president of the Metropolitan Mathematics Club of Chicago and a past president of the Illinois Council of Teachers of Mathematics.

Lee E. Yunker is head of the Mathematics Department at West Chicago Community High School, West Chicago, Illinois. He has taught mathematics courses at every level of the high school curriculum. Mr. Yunker obtained his B. S. from Elmhurst College and his M. Ed. in mathematics from the University of Illinois, with additional graduate work at the University of Illinois, Northwestern University, University of Montana, Northern Illinois University, and National College of Education. Mr. Yunker is very active in professional mathematics organizations at the local, state, and national levels frequently speaking or conducting workshops on a variety of topics. Mr. Yunker has had articles published in *The Illinois Mathematics Teacher* and the *New Mexico Council of Teachers of Mathematics Journal.*

REVIEWERS

David E. Sagan
Head, Mathematics Department
Pickerington High School
Pickerington, Ohio

Richard Galbraith
Head, Mathematics Department
Bryant High School
Bryant, Arkansas

Jerry Williams
Head, Mathematics Department
Carmi Community High School
Carmi, Illinois

Barbara Rugys
Mathematics Resource Consultant
Kennedy High School
Patterson, New Jersey

Vincent F. O'Connor
Curriculum Specialist, Mathematics
Milwaukee, Wisconsin

Paul S. Miller
Head, Mathematics Department
Coshocton High School
Coshocton, Ohio

Ms. Bobbie Robicheaux
Mathematics Teacher
North Richland Junior High School
Fort Worth, Texas

Lynda Ward Wagster
Mathematics Teacher
Antioch High School
Antioch, Tennessee

John M. Lockhart
Mathematics Consultant
Oklahoma City, Oklahoma

Richard Hansen
Mathematics Teacher
Rio Americano High School
Sacramento, California

STAFF

Editorial Staff

Project Editors: Elaine C. Murphy, Arthur L. Block
Managing Editor: Donald W. Collins
Assistant Editors: Jack Witherspoon, Ann Kinkley Stollenwerk, Cynthia Rockey Zengler, Darlene Diller Lewis, Rita Mudd
Photo Editor: Sue Marquart

Art Staff

Project Artist: Dotte Turner Russell
Art Director: Lester Shumaker
Book Designer: Lester Shumaker

PHOTO CREDITS

Cover Photo: Hedrich Blessing
1, Advertising Image Photography; **2,** Hickson-Bender Photography; **8,** Reprinted with permission of Ralston Purina Co., St. Louis, Missouri; **15,** Advertising Image Photography; **16,** Photo supplied by Dick Ruhl-(APRO) Aerial Phenomena Research Org. Inc., Tucson, Arizona/N. Y.; **33,** Gale Constable/Duomo; **38,** Advertising Image Photography; **57,** Hickson-Bender Photography; **63,** Photography Dept. Ohio State University; **64, 82, 87,** Advertising Image Photography; **89,** Radio Shack; **95, 96,** Advertising Image Photography; **103,** Mayflower Corp.; **114,** General Motors Research Laboratories; **131,** Cheryl Blair; **142,** U.S. Department of Energy; **151,** Black Dot; **159,** Tracy I. Borland; **160,** Cheryl Blair; **170,** Advertising Image Photography; **174,** U. S. Bureau of Reclamation; **180,** Courtesy of John Hancock Mutual Life Insurance Co.; **184,** Jerry Bowling; **193,** Ray Manley; **202, 206,** Larry Hamill; **213,** Grumman Corp.; **213,** Smithsonian Institution photo by Dane A. Penland; **214,** Kevin Roche, John Dinkeloo and Associates; **229,** Jerry Bowling; **230,** Bill Barr for the Kansas City Kings; **236,** Graig Kramer; **238,** Eric Hoffhines; **242,** Department of Defense; **242,** A. Ottolenghi, Ohio State University; **242,** Karen Sass; **247,** Ohio Department of Natural Resources; **249,** Larry Hamill; **263,** Advertising Image Photography; **279,** Eric Hoffhines; **287,** U.S.D.A. Forest Service; **293,** William C. Maddox; **294,** Photo courtesy of DEKALB AgResearch, Inc.; **298,** Russ Lappa; **304,** Photography Dept. Ohio State University; **311,** University of Virginia; **320,** George Anderson/ICOM; **327,** Studio Ten; **328,** Advertising Image Photography; **354,** EPA Documerica; **360,** Jerry Bowling; **367,** Hickson-Bender Photography; **385,** Mountain Meadows Ski Touring Center; **389,** Paul Brown; **405,** Larry Hamill; **419,** Linda Young; **431,** Keith Turpie; **439,** Frank Lerner; **446,** Department of Energy; **447,** Larry Hamill; **453,** Larry Hamill

PREFACE

Merrill Geometry is a text developed by experienced high school teachers for the classroom. Its goals are to develop proficiency with geometric skills, to expand understanding of geometric concepts, to improve logical reasoning, to provide a complete course in high school geometry, and to promote success. To achieve these goals the following strategies are used.

Build a Solid Foundation. Geometric concepts are introduced intuitively, drawing upon students' past experience with geometry in real life. Students learn to organize their ideas and gradually are led to the concept of geometric proof. In this way, the needed background is built. Students can thus progress successfully.

Utilize Sound Pedagogy. *Merrill Geometry* covers all topics generally presented at this level, in logical sequence. Each concept presented is then *used* within that lesson and in later lessons. Ideas are introduced when they are needed.

Facilitate Learning. The reading level has been carefully controlled to ease the special problems that occur with geometry. Furthermore, many appealing photographs, illustrations, charts, graphs, and tables make it easy for the students to visualize the ideas presented. As a result, students can both read and understand *Merrill Geometry*.

Use Relevant, Real-Life Applications. Applications provide motivation by showing how concepts have practical value and how geometry can help prepare students for the future.

Merrill Geometry offers a variety of useful aids for the student studying geometry.

Student Annotations	Help students to identify important concepts as they study.
Selected Answers	Allow students to check their progress as they work. These answers are provided at the back of the text.
Vocabulary	Provides students with a list of important mathematical words.
Chapter Summary	Provides students with a compact listing of major concepts presented within each chapter.
Chapter Review	Permits students to review each chapter by working sample problems from each lesson.
Chapter Test	Enables students to check their progress.

The following special features which appear periodically throughout the text, provide interesting and useful extra topics.

Careers	Depict a variety of people in different careers using mathematics. These careers are typical of careers that students may pursue.
Using Money	Provides insights into how mathematics can be used in making consumer decisions.
Using Geometry	Illustrates how geometry can be and is used in everyday life.
Reading Geometry	Gives students special instructions needed to read and interpret mathematical symbols, especially those found in geometry.
Excursions in Geometry	Enliven and help maintain student interest by providing interesting side trips. Topics are varied and include glimpses into the development and uses of geometry, as well as puzzles, games, and history.
Using Computers	Shows students the many uses of computers in society and alerts them to what computers can and cannot do.

Students will find the practical, straightforward approach of *Merrill Geometry* both interesting and easy to understand. Teachers will find that the careful sequencing of topics and thorough treatment of essential ideas provide an effective course in high school geometry.

CONTENTS

Coordinates — 367

Loci and Constructions — 405

Transformations — 431

Chapter

1

Points, Lines, and Planes

In geometry, terms are described in a mathematical way. These terms, however, are used as models for real life phenomena. For example, a point is a model for the position of a source of light.

1–1 Points, Lines, and Planes

Points, lines, and **planes** are used in many ways. They are the basic terms of geometry.

A point has no dimension.

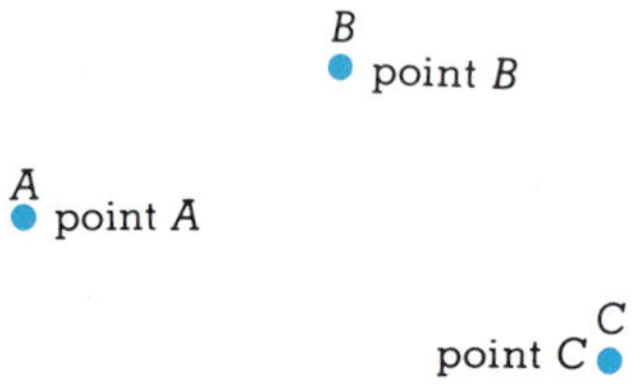

A location or a pinhole suggest the idea of a point. Points have no size. Points are represented by dots and named by capital letters.

A line has one dimension.

A line, in geometry, means a straight line.

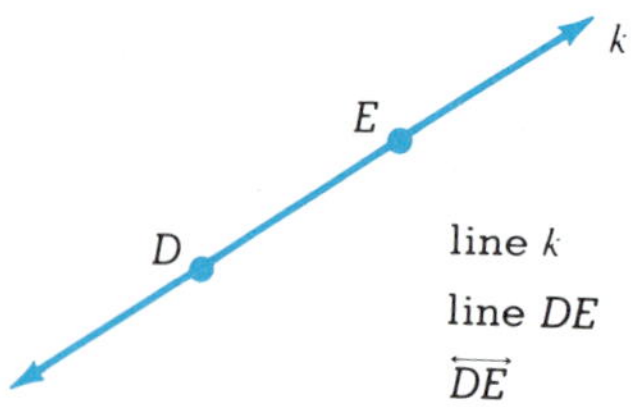

A flight path or a taut wire suggest the idea of a line. Lines extend indefinitely and have no thickness or width. Lines are represented by double arrows and named by lowercase letters. A line also can be named by two points of the line.

A plane has two dimensions.

plane $\mathscr{R}$

A flat map or a pane of glass in a window suggest the idea of a plane. A plane extends indefinitely in two directions and has no thickness. Planes are represented by four-sided figures and named by capital script letters. Planes also can be named by three points of the plane that are *not* on the same line.

Two lines, or lines and planes **intersect** if they have points in common. For example, the two lines below intersect at point E.

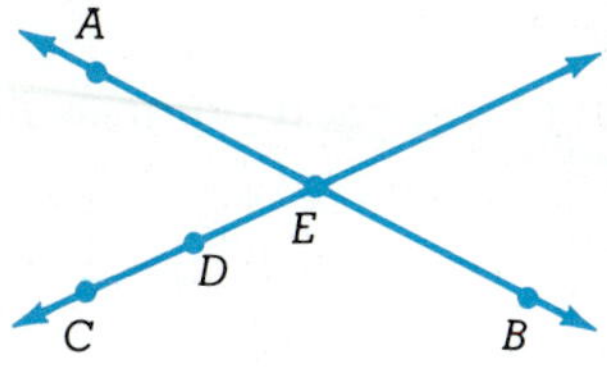

$\overleftrightarrow{AB}$ intersects $\overleftrightarrow{CD}$ at E.
The intersection of $\overleftrightarrow{AB}$ and $\overleftrightarrow{CD}$ is E.

examples

1 **Find the intersection of plane $\mathscr{A}$ and plane $\mathscr{B}$.**

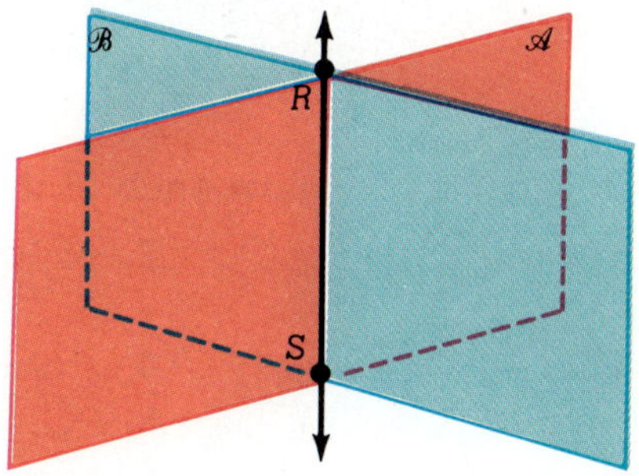

Plane $\mathscr{A}$ and plane $\mathscr{B}$ have $\overleftrightarrow{RS}$ in common. Their intersection is $\overleftrightarrow{RS}$.

2 **Find the intersection of plane $\mathscr{A}$ and $\overleftrightarrow{TR}$.**

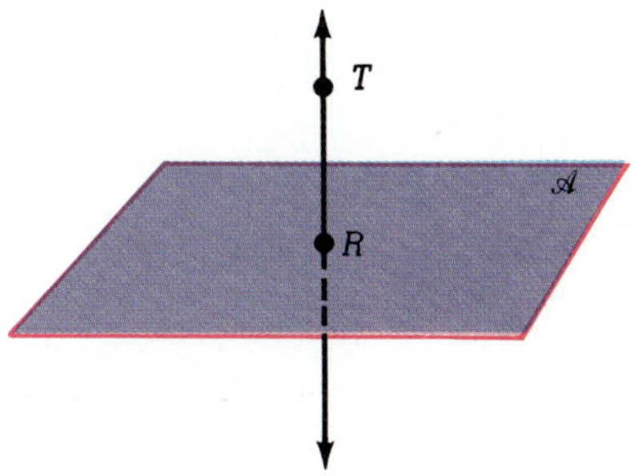

Plane $\mathscr{A}$ and $\overleftrightarrow{TR}$ have R in common. Their intersection is R.

In the figure on the right, all the points of $\overleftrightarrow{AB}$ are also points of $\mathscr{M}$. We say that $\overleftrightarrow{AB}$ **lies on** $\mathscr{M}$, or $\mathscr{M}$ **contains** $\overleftrightarrow{AB}$. Also, A *lies on* $\overleftrightarrow{AB}$ and *lies on* $\mathscr{M}$. Alternatively, both $\overleftrightarrow{AB}$ and $\mathscr{M}$ contain A.

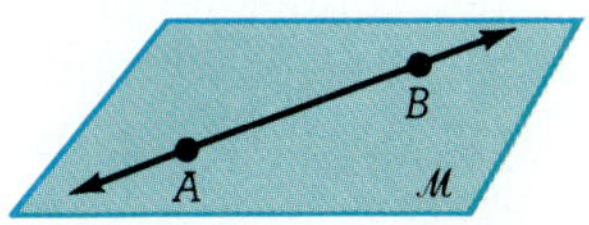

Lines contain points.
Planes contain points and lines.

Points lie on lines and planes.
Lines lie on planes.

exercises

Exploratory State whether each of the following suggest a *point*, a *line*, or a *plane*.

1. corner of a box
2. side of a box
3. edge of a box
4. wall of a room
5. straw
6. parking lot
7. telephone wire
8. grain of salt
9. ceiling of a room
10. clothesline
11. hole in the ground
12. star in the sky
13. straight path on a map
14. city on a map

Written Draw and label a diagram to show each of the following.

1. plane $\mathscr{L}$ intersects plane $\mathscr{M}$ at $\overleftrightarrow{PQ}$
2. plane $\mathscr{A}$, plane $\mathscr{B}$, and plane $\mathscr{C}$ intersect at $\overleftrightarrow{ST}$
3. $\overleftrightarrow{AB}$ and $\overleftrightarrow{CD}$ intersect at P
4. ℓ and m intersect at R
5. $\overleftrightarrow{RS}$ and plane $\mathscr{M}$ do *not* intersect
6. $\overleftrightarrow{PQ}$, $\overleftrightarrow{RS}$, and plane $\mathscr{N}$ intersect at T
7. plane $\mathscr{L}$, plane $\mathscr{M}$, and plane $\mathscr{N}$ intersect at P
8. plane $\mathscr{M}$, plane $\mathscr{N}$, and plane $\mathscr{L}$ do *not* intersect
9. P lies on $\overleftrightarrow{AB}$
10. $\overleftrightarrow{AB}$ contains R
11. t contains Q and R, but does *not* contain P and S
12. A, B, and C lie on m, but D does *not* lie on m
13. plane $\mathscr{M}$ contains n and R
14. t contains P and lies on plane $\mathscr{N}$

For each of the following, use the figure on the right.

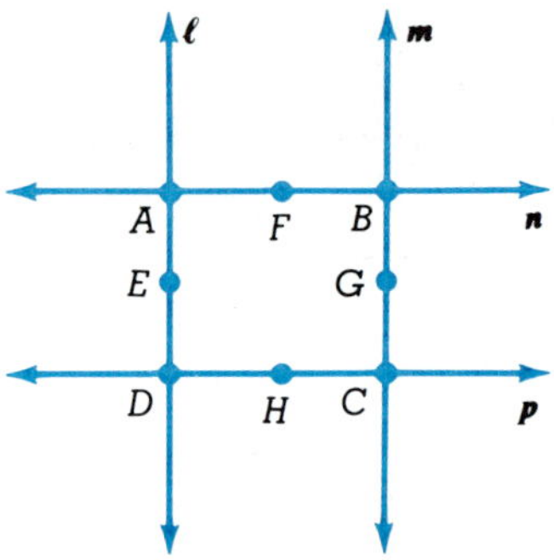

15. Write another name for $\overleftrightarrow{AB}$.
16. Write another name for ℓ.
17. What points do $\overleftrightarrow{AD}$ and $\overleftrightarrow{BC}$ have in common?
18. What points do $\overleftrightarrow{AB}$ and m have in common?
19. Name two points that lie on $\overleftrightarrow{AD}$.
20. Name two points that lie on p.
21. Write three other names for n.
22. Write three other names for $\overleftrightarrow{BC}$.

Solve each problem.

23. Suppose P and Q are two points. How many lines lie on P and Q?
24. Suppose P and Q are two points. How many planes contain P and Q?
25. Suppose P, Q, and R are three points *not* on the same line. How many lines can be drawn through at least two of the points?
26. Think about a board or a piece of stiff cardboard. What is the *minimum* number of legs needed to support it in a fixed position?

1–2 Definitions

To define a word, you use other words that are left **undefined.** The simplest undefined terms of geometry are **point, line,** and **plane.** They are used to define other terms of geometry.

Undefined terms are words whose meaning is readily understood.

The following definitions are based on points, lines, and planes.

Points are collinear if and only if they lie on the same line.

Definition of Collinear Points

Points are coplanar if and only if they lie on the same plane.

Definition of Coplanar Points

examples

1 **According to the figure on the right, are points A and D collinear?**

Points A and D both lie on $\overleftrightarrow{AD}$.
Yes, they are collinear.

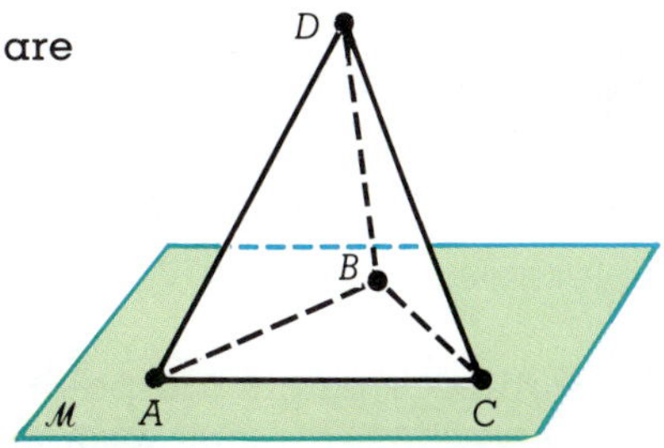

2 **According to the figure above, are points A, B, C, and D coplanar?**

Points A, B, and C lie in plane $\mathcal{M}$.
Point D does *not* lie in plane $\mathcal{M}$.
No, all four points are *not* coplanar.

Good definitions must be written very carefully. The following guidelines will help you write good definitions.

1. Name the term being defined.
2. Use only undefined terms or already defined terms.
3. Identify the set to which the term belongs.
4. State the properties which distinguish the term from others in its set.
5. Make it reversible.

Guidelines for a Good Definition

Reread the definition of collinear points. Notice that it satisfies all the guidelines for a good definition.

Name the term being defined. The term being defined is *collinear*.

Use only undefined terms or already defined terms. The definition uses *points* and *line*. Both these terms have already been used.

Identify the set to which the term belongs. Collinear points are *points*.

State the properties which distinguish the term from others in its set. Collinear points *lie on the same line*. Some sets of points do *not* lie on the same line.

Make it reversible. Points are collinear if they lie on the same line. If points lie on the same line, they are collinear. Using *if and only if* makes a definition reversible.

examples

3 **The statement *they lie in the same plane* is *not* a good definition. Tell why. Assume the terms in the statement are previously defined.**

The statement does *not* satisfy guidelines **1**, **3**, and **5**. It does *not* name the term being defined, does *not* identify the set to which the term belongs, and is *not* reversible.

4 **Noncollinear points are points that are *not* collinear. Write a good definition for *noncollinear points* without using the word *collinear*.**

Name the term being defined. The term being defined is noncollinear points.

Use only undefined terms or already defined terms. Use only *points* and *line*.

Identify the set to which the term belongs. Noncollinear points are *points*.

State the properties which distinguish the term from others in its set. Collinear points lie on the same line. Some sets of points do *not* lie on the same line.

Make it reversible. Points are noncollinear if they do *not* lie on the same line. If points do *not* lie on the same line, they are noncollinear. Put these two sentences together using *if and only if*.

Points are noncollinear if and only if they do *not* lie on the same line.

exercises

Exploratory Use the figure to state whether each of the following is true or false.

1. *A, B,* and *C* are collinear.
2. *A, B,* and *C* are coplanar.
3. *E, F, G,* and *H* are coplanar.
4. *A, F, E, G* are coplanar.
5. *B, J,* and *K* are coplanar.
6. *C* and *G* are collinear.
7. *H, G,* and *D* are collinear.
8. *B, J,* and *C* are collinear.
9. *B, J, H,* and *G* are coplanar.
10. *B, J, H,* and *G* are collinear.

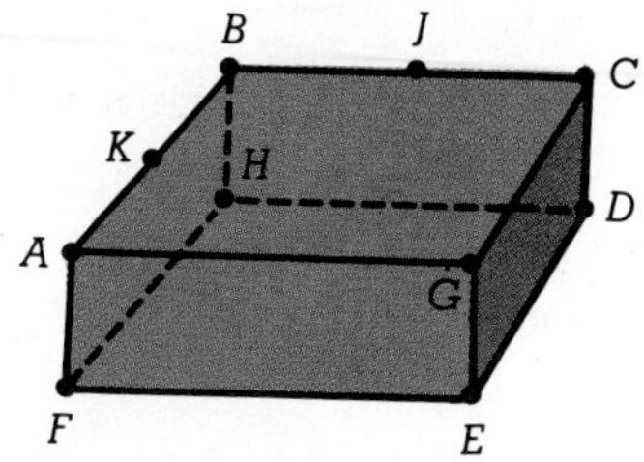

Written Points are *noncollinear* if and only if they do *not* lie on the same line. Use the figure to identify each of the following as *collinear* or *noncollinear*.

1. *F, G, E*
2. *A, J, E*
3. *B, F, A*
4. *H, B, A*
5. *C, J, E*
6. *K, G, C*

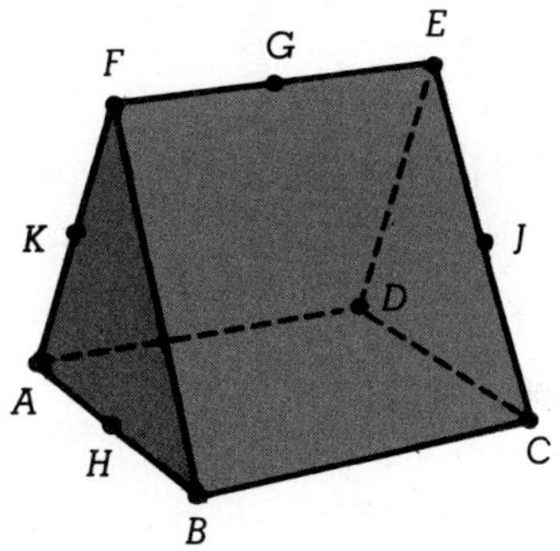

Points are *noncoplanar* if and only if they do *not* lie on the same plane. Use the figure to identify each of the following as *coplanar* or *noncoplanar*.

7. *F, G, E, K*
8. *A, J, E, D*
9. *B, F, A, G*
10. *H, B, A, C*
11. *C, J, E, B*
12. *K, G, C, A*

State whether or *not* each of the following statements is a good definition. If *not*, tell why. Assume the terms in the statement are previously defined.

13. Skew lines are lines which are not coplanar.
14. A bedroom is a room where people sleep.
15. A person who lives in Detroit lives in Michigan.
16. A triangle is a polygon having three sides and three angles.
17. A diagonal of a quadrilateral divides the quadrilateral into two triangles.
18. Concurrent lines are lines which have all their points in common.
19. Points are noncoplanar points if and only if they do *not* lie on the same plane.
20. Points are collinear points if and only if they lie on the same line.

21. Write your own definition of pencil. State how it satisfies the qualities of a good definition.

1–3 Conditional Statements

If you want a healthy cat, then start with a healthy kitten.

Conditional statements have two parts. One part can begin with *if*, and the other part can begin with *then*. The statement below is a conditional statement with *then* omitted.

Conditional statements are written in many different ways, but all conditional statements can be written in *if-then* form. For example, the following statements could *all* be written as, *If you are a genius, then you are left-handed.*

All geniuses are left-handed.
You are left-handed if you are a genius.
You are a genius only if you are left-handed.
Being a genius implies you are left-handed.
You are left-handed when you are a genius.

example 1

Rewrite the following conditional statement in *if-then* form.

There is at least one plane that contains a line and a point *not* on the line.

If there is a line and a point *not* on the line, *then* there is at least one plane that contains them.

The part following *if* in a conditional statement is called the **hypothesis.** It tells what is to be *assumed*. The part following *then* in a conditional statement is called the **conclusion.** It tells the result, what is believed to follow from the assumption.

Conditional	If there is a line and a point *not* on the line, then there is at least one plane that contains them.
Hypothesis	there is a line and a point *not* on the line
Conclusion	there is at least one plane that contains them

A conditional statement is false only when the hypothesis is true and the conclusion is false.

Conditional statements may be true or false. Questions and exclamations are *not* conditional statements because usually they are *neither* true *nor* false.

Question	Who wrote *I Know Why the Caged Bird Sings?*
Exclamation	Be careful!

Given a conditional statement, a new statement can be formed by interchanging the hypothesis and conclusion. This new statement is called the **converse** of the conditional.

Statement	If it is raining, then there are clouds in the sky.
Converse	If there are clouds in the sky, then it is raining.

As the above example shows, the converse of a true statement may *not* be true.

example 2

Write the converse of the following true conditional statement. Is the converse a true statement?

If two points lie on the same line,
then they are collinear.

Converse If two points are collinear, then they lie on the same line.

In this case, the converse is a true statement.

A statement in *if and only if* form can be written as two conditional statements.

Two planes are parallel if and only if they do *not* intersect.

means

If two planes are parallel, then they do *not* intersect.
and If two planes do *not* intersect, then they are parallel.

Not all reversible statements are definitions.

Statements that can be written in *if and only if* form are called **biconditional statements.** All definitions are biconditional statements because they are reversible.

example 3

Rewrite the following pair of statements as a single statement in *if and only if* form.

> **If two lines are skew, then they do *not* intersect and are *not* in the same plane.**
>
> **If two lines do *not* intersect and are *not* in the same plane, then they are skew.**

Two lines are skew if and only if they do *not* intersect and are *not* in the same plane.

exercises

Exploratory **Identify the hypothesis and conclusion for each of the following statements.**

1. If it rains, then the grass gets wet.
2. If a triangle has a right angle, then it is a right triangle.
3. If you live in Texas, then you are an American.
4. If you ride a bicycle, then you have strong legs.
5. If two lines are perpendicular, then the two lines intersect.
6. If $3x = 12$, then $x = 4$.
7. If n is even, then n^2 is even.
8. If a number is rational, then it is real.
9. If an animal is a chimpanzee, then it is a mammal.
10. If you practice hard, then you will make the team.

Written **Rewrite each of the following conditional statements in *if-then* form.**

1. Parallel lines do *not* intersect.
2. A square is *not* a triangle.
3. A square is a rectangle.
4. All integers are real numbers.
5. $x < 0$ implies $5x > 6x$.
6. $2n + 1 = 5$ implies $n = 2$.
7. When it rains it pours.
8. If you want a new account, phone for it.
9. Two intersecting lines are contained in exactly one plane.
10. Two angles are congruent when they are vertical angles.

11–20. Write the converse of each conditional statement in problems **1–10.** Use *if-then* form.

Tell whether or *not* each of the following is a conditional statement. Write *yes* or *no*. Then explain your answer.

21. If it is blue, then it is *not* green.
22. If it is *not* green, then it is blue.
23. When animals are hungry, they eat.
24. A person who is crying is *not* happy.
25. Be home by twelve o'clock.
26. Define collinear points.
27. Every line contains at least two points.
28. Sugar is *not* always white.
29. How are points and lines different?
30. Why are some people afraid of high places?

31–40. For each of the statements in problems 21–30, write *true*, *false*, or *neither*.

Give an example for each of the following.

41. A conditional and its converse that are both true.
42. A conditional and its converse that are both false.
43. A true conditional statement whose converse is false.
44. A true conditional statement whose converse is true.

Rewrite each of the following statements as two *if-then* statements, one of which is the converse of the other.

45. Points are collinear if and only if they lie on the same line.
46. Points are coplanar if and only if they lie on the same plane.
47. Two line segments are congruent if and only if they have the same length.
48. Two angles are congruent if and only if they have the same measure.
49. Two angles are supplementary if and only if the sum of their measures is 180.
50. Two angles are complementary if and only if the sum of their measures is 90.

Rewrite the following pairs of statements as single statements using *if and only if*.

51. If two lines are perpendicular, then they intersect at right angles.

    If two lines intersect at right angles, then they are perpendicular.

52. In a plane, if two lines do *not* intersect, then they are parallel.

    In a plane, if two lines are parallel, then they do *not* intersect.

53. If two lines are intersected by a transversal so that alternate interior angles are congruent, then the lines are parallel.

    If two lines are parallel and intersected by a transversal, then the alternate interior angles are congruent.

54. If two lines are parallel and intersected by a transversal, then the corresponding angles are congruent.

    If two lines are intersected by a transversal so that corresponding angles are congruent, then the lines are parallel.

Postulates are statements that are accepted as true.

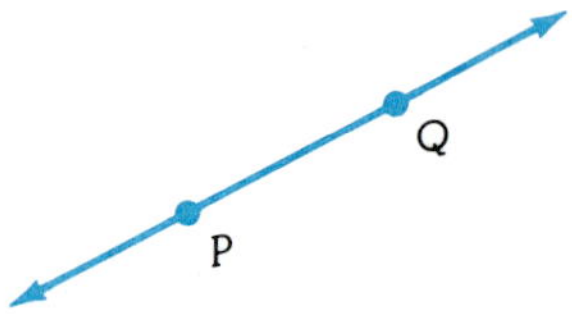

1–4 Postulates

In order to play a game, you must know the rules. In order to study geometry, you must know the **postulates.** The rules, or postulates, of geometry tell how different sets of points are related.

Postulate **1–1** relates points and lines. It guarantees that *P* and Q, for example, always determine a line.

Postulate 1–1 **Through any two points there is exactly one line.**

Postulate **1–2** relates points and planes.

Postulate 1–2 **Through any three points *not* on the same line there is exactly one plane.**

examples

1 **How many lines are determined by four points no three of which are collinear?**

For every two points there is exactly one line. Thus, for four points there are *at most* six lines that can be drawn.

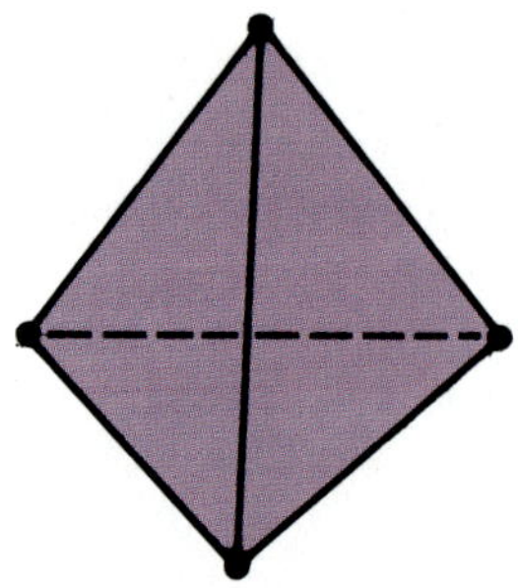

2 **How many planes are determined by four points no three of which are collinear?**

For every three points *not* on the same line there is exactly one plane. Each face of the pyramid in the figure represents a plane containing three of the four points. The figure shows that there are at most four planes determined by four points.

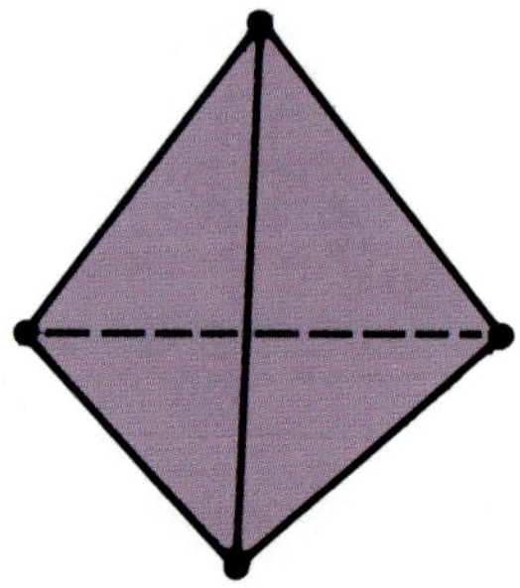

Postulates **1–3** and **1–4** state the minimum numbers of points on a line and on a plane.

A line contains at least two points. — Postulate 1–3

A plane contains at least three points *not* on the same line. — Postulate 1–4

Postulate **1–5** relates points, lines, and planes.

If two points lie in a plane, then the entire line containing those two points lies in that plane. — Postulate 1–5

example 3

Suppose three noncollinear points, *A*, *B*, and *C*, lie in the same plane. Name the lines containing *A*, *B*, and *C* which lie in that plane.

$\overleftrightarrow{AB}$, $\overleftrightarrow{BC}$, and $\overleftrightarrow{CA}$ are in the plane.

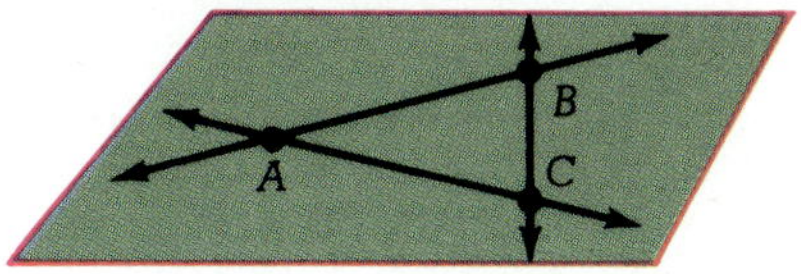

Postulate **1–6** relates lines and planes.

If two planes intersect, then their intersection is a line. — Postulate 1–6

example 4

Suppose plane $\mathcal{M}$ contains noncollinear points *A*, *C*, and *E*. Suppose plane $\mathcal{N}$ contains noncollinear points *A*, *C*, and *D*. Name the intersection of the two planes.

The two planes intersect in $\overleftrightarrow{AC}$.

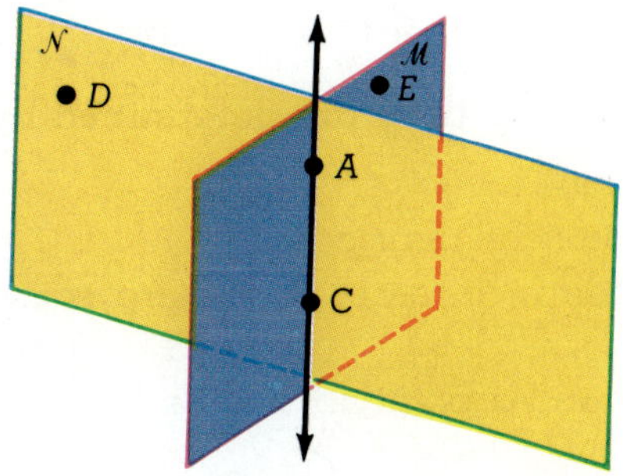

exercises

Exploratory For each of the following statements, name the postulate that explains the relationship. Use the figure below.

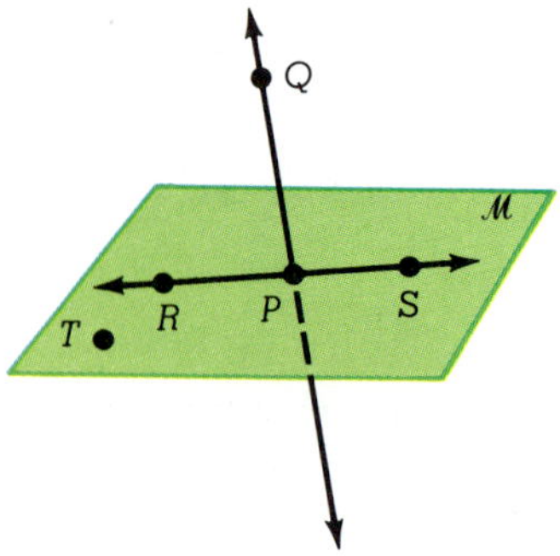

1. Exactly one line contains Q and P.
2. The entire line containing R and S lies in plane $\mathcal{M}$.
3. There are at least three points in plane $\mathcal{M}$.
4. There are at least two points on $\overleftrightarrow{QP}$.
5. There is at least one point in plane $\mathcal{M}$ not on $\overleftrightarrow{RS}$.
6. Exactly one line contains R and S.
7. P and S lie on exactly one line.
8. Plane $\mathcal{M}$ is the only plane containing R, S, and T.
9. The intersection of plane $\mathcal{M}$ and the plane determined by Q, P, and R is $\overleftrightarrow{RP}$.
10. The intersection of plane $\mathcal{M}$ and the plane determined by Q, P, and T is a line through P and T.

Written In the figure below, P, Q, and R are collinear. Points P and S lie in plane $\mathcal{M}$. Points R and T lie in plane $\mathcal{N}$. State whether each of the following statements is *true* or *false*.

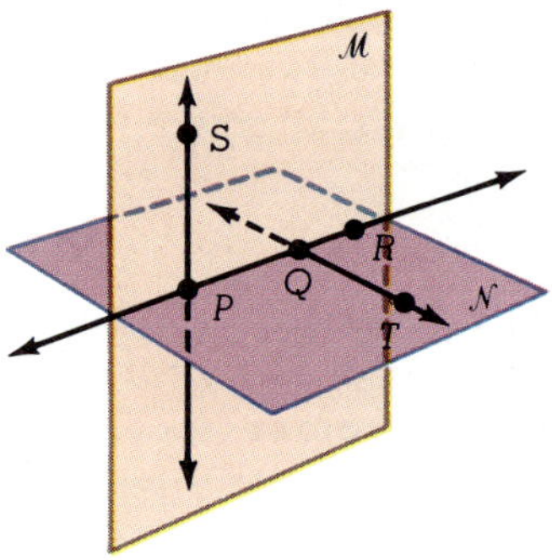

1. P, Q, and R lie in plane $\mathcal{N}$.
2. P, Q, and R lie in plane $\mathcal{M}$.
3. $\overleftrightarrow{PS}$ does *not* lie in plane $\mathcal{M}$.
4. Q lies in plane $\mathcal{M}$.
5. P, Q, R, and S are coplanar.
6. Q does *not* lie in plane $\mathcal{M}$.
7. Q, R, and T are collinear.
8. P, Q, S, and T are coplanar.
9. Plane $\mathcal{M}$ and plane $\mathcal{N}$ intersect in $\overleftrightarrow{PR}$.
10. $\overleftrightarrow{QT}$ lies in plane $\mathcal{N}$.

State how many lines are determined by each of the following.

11. one point
12. two points
13. three noncollinear points
14. three collinear points
15. four points, three of which are collinear
16. five points, no three of which are collinear
17. a line and a point *not* on the line
18. the intersection of two planes

19–26. State how many planes are determined by each of the sets in problems **11–18.**

CAREERS

Paralegal Aide

April Barlow is a paralegal aide for the Tucson City prosecuting attorney's office. She is gathering information regarding a shoplifting case. She must find evidence to prove that E. Val Deeds did shoplift at Levy's Department Store.

From the police report and witnesses, she has the following statements.

A. E. Val Deeds was in the store at the time of the shoplifting.
B. E. Val Deeds is 16 years old.
C. E. Val Deeds has no receipt for the merchandise.
D. E. Val Deeds attends high school.
E. E. Val Deeds has blue eyes.
F. E. Val Deeds was in the parking lot with the merchandise.

Suppose the definition of shoplifting is, *the removal of merchandise from a store by an individual(s) without proof of payment.* Then the information in statements **A**, **C**, and **F** can be used as evidence of shoplifting.

Exercises For each of the following, tell which statements can be used as evidence.

1. A *speeding violation* occurs when a motorist drives at a speed over the posted limit.

   A. Bell Wilkins was driving 30 mph in a 35 mph zone.

   B. Jon Bunch was found driving 45 mph in a 20 mph zone.

2. An *isosceles triangle* is a triangle with at least two sides congruent.

   A. Triangle *ABC* has three congruent sides.

   B. Triangle *RST* has three congruent angles.

3. A *good swimmer* is a person who can do the 100 meter backstroke in less than 01:02.00.

   A. Kornelia Ender can do the 100 meter freestyle in 00:55.65.

   B. Ulriche Richter can do the 100 meter backstroke in 01:01.51.

4. *Supplementary angles* are two angles whose degree measures total 180.

   A. The degree measure of angle C is 50, and the degree measure of angle D is 40.

   B. The degree measure of angle R is 125, and the degree measure of angle S is 55.

1–5 Theorems

Are you convinced? Does this photograph prove there are flying saucers?

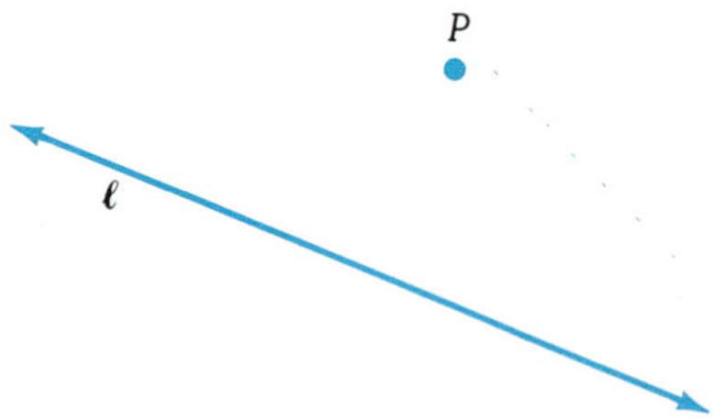

This figure shows a line ℓ and a point P *not* on the line.

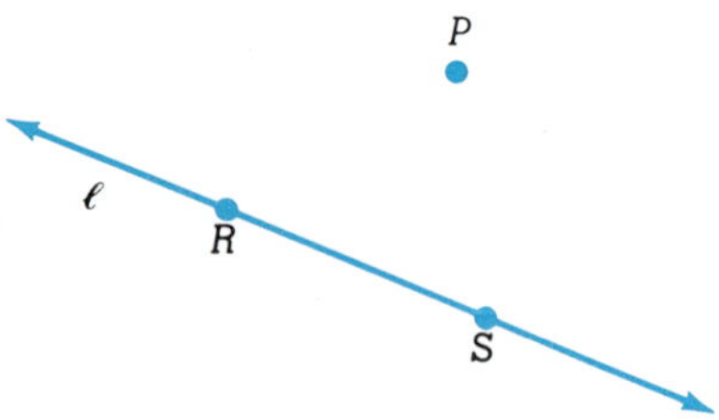

According to Postulate 1–3, a line contains at least two points. The points are labeled R and S. So, there are at least three points, P, R, and S, *not* on the same line.

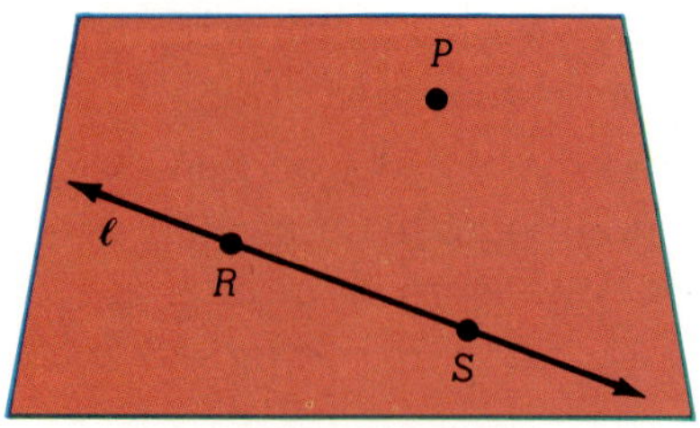

According to Postulate 1–2, through any three points *not* on the same line there is exactly one plane.

According to Postulate 1–5, if two points lie in a plane, then the entire line containing them lies in that plane. So there is at least one plane containing P and ℓ.

Are you convinced?

Not everyone has the same idea of what is convincing. Some arguments are more convincing than others. Geometry uses a system of reasoning that many people find convincing.

First, certain information is taken for granted. In geometry, undefined terms, definitions, and postulates are taken for granted.

This information is *__not__ proved.*

Undefined terms are basic terms from which other terms are defined.

Definitions are explanations of how words are to be used.

Postulates are statements that describe fundamental properties of the basic terms.

The information from undefined terms, definitions, and postulates is used to find new information, written as a **theorem.**

Theorems are statements that must be proved.

For example, the argument on page 16 leads to the following theorem.

Theorem 1–1

If there is a line and a point *not* on the line, then there is exactly one plane that contains them.

To prove a theorem, start with the hypothesis. Then use the information from postulates and definitions to reach the conclusion.

Different postulates can lead to different conclusions.

Formal proofs have five main parts, theorem, given, prove, a diagram, and the proof with statements and reasons. Formal proofs are written in the form shown below.

Theorem: *Write out the theorem to be proved. Sometimes this part is omitted.*

Given: *Write the information given in the hypothesis of the theorem. Give names for the terms of the diagram.*

Prove: *Write the conclusion of the theorem. Use the names from the diagram.*

A diagram may be provided for the given statements. It should picture the information in the hypothesis of the theorem. Label the diagram with names from the given statement.

Proof:

STATEMENTS	REASONS
1. *Write a list of numbered statements arranged in order so*	1. *Write a list of numbered reasons which go with the statements on*
2. *that they lead to the conclusion.*	2. *the left. Reasons used can be given information, definitions,*
3. *Each statement must be justified by a reason.*	3. *postulates, or theorems that have already been proved.*

example 1

State the *given*, then draw and label a diagram to illustrate the *given* for Theorem 1–1.

Theorem **1–1** states, *if there is a line and a point* not *on the line, then there is exactly one plane that contains them.* The hypothesis, the part following *if*, states there is a line and a point *not* on the line. Name the line ℓ and the point P.

Given: P *not* on ℓ

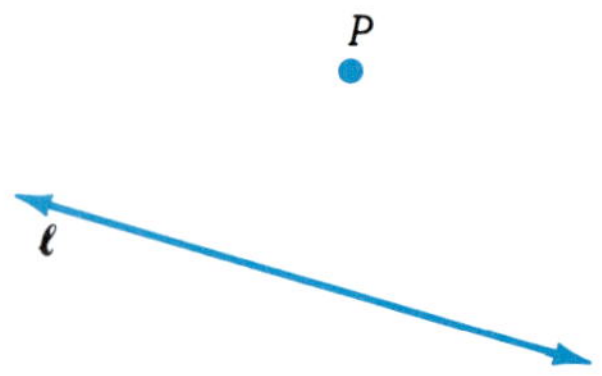

exercises

Exploratory **Solve the following problems about formal proofs.**

1. Name the five main parts.
2. What information is in *given*?
3. What information is in *prove*?
4. What does the diagram illustrate?
5. Must a theorem be stated? Write *yes* or *no*.
6. Must a diagram be shown? Write *yes* or *no*.
7. Must the *given* be stated? Write *yes* or *no*.
8. Must the *prove* be stated? Write *yes* or *no*.
9. Name the headings for the two columns in *proof*.
10. What is the purpose of the statements?
11. What is the purpose of the reasons?
12. What types of reasons can be used?

State the hypothesis for each of the following theorems.

13. If two lines intersect, then exactly one plane contains both lines.
14. If two lines intersect, then they intersect in exactly one point.
15. If a line and a plane intersect, and the plane does *not* contain the line, then their intersection is a point.
16. If four points are collinear, then they are coplanar.

17–20. State the conclusion for the theorems in problems **13–16**.

Written State the *given* for each of the following theorems.

1. If two lines intersect, then exactly one plane contains both lines.
2. If two lines intersect, then they intersect in exactly one point.
3. If a line and a plane intersect, and the plane does *not* contain the line, then their intersection is a point.
4. If four points are collinear, then they are coplanar.

5–8. State the *prove* for each of the theorems in problems **1–4**.

Draw and label a diagram to illustrate each of the following *given* statements.

9. **Given:** line ℓ and point P *not* on ℓ
10. **Given:** points A, B, and C *not* on the same line
11. **Given:** line m intersecting line n
12. **Given:** line m contained in plane $\mathscr{P}$
13. **Given:** point R *not* on plane $\mathscr{M}$
14. **Given:** coplanar lines m and n point P on line m

Draw and label a diagram to illustrate the hypotheses for each of the following theorems.

15. If two lines intersect, then exactly one plane contains both lines.
16. If two lines intersect, then they intersect in exactly one point.
17. If a line and a plane intersect, and the plane does *not* contain the line, then their intersection is a point.
18. If four points are collinear, then they are coplanar.

The following diagrams are taken from formal proofs. State the *given* for each diagram.

19.

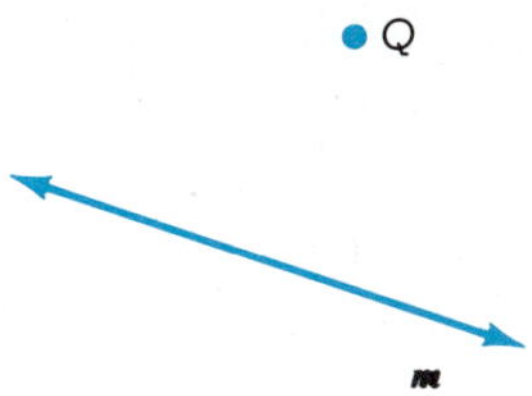

20.

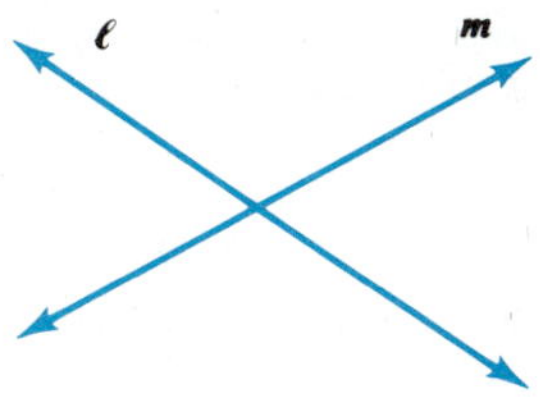

21.

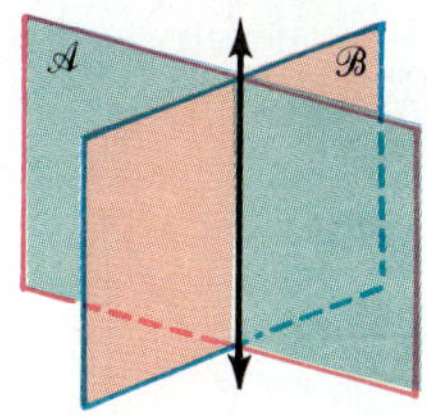

22.

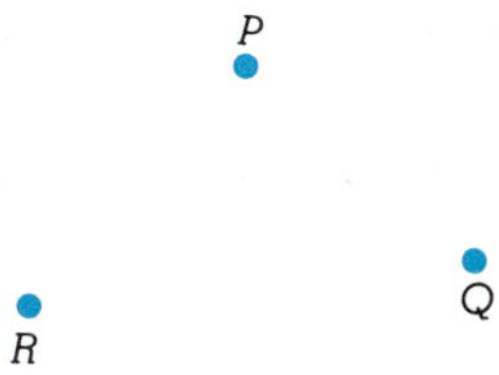

23.

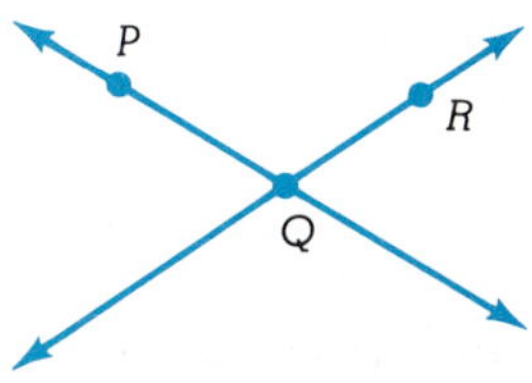

24.

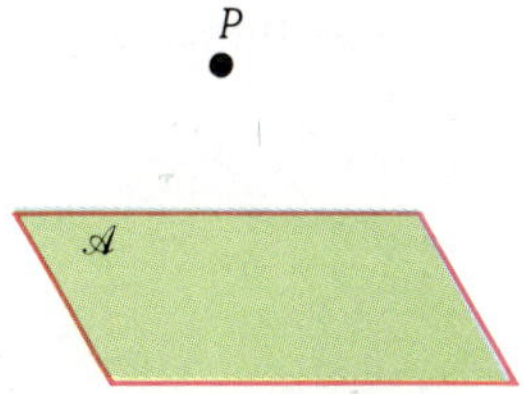

25.

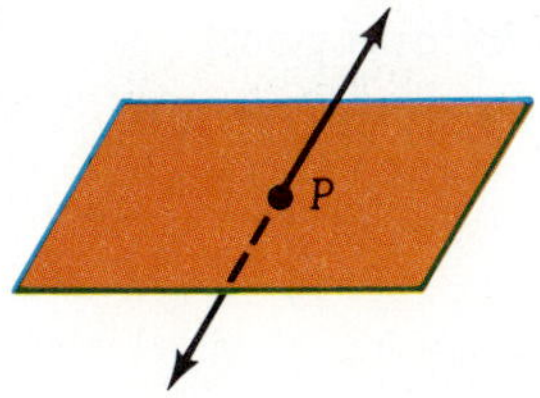

26.

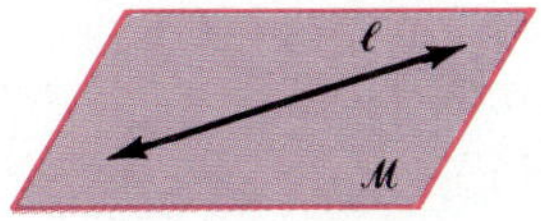

27.

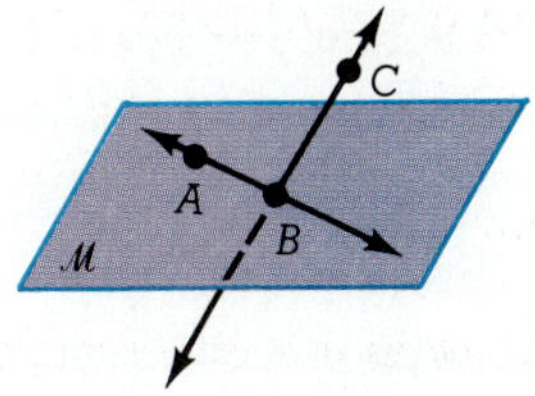

Using Geometry

Point, line and *plane* are undefined terms of geometry. The postulates of geometry describe the fundamental properties of these terms. Different geometries can be made by changing the undefined terms and postulates.

A miniature geometry can be created by changing the interpretation of *point, line* and *plane*.

Postulate 1 For every two points, there is exactly one line that contains them.

Postulate 2 If a line contains two distinct points of a plane, it is contained in the plane.

Postulate 3 For any three points, there is at least one plane, containing them. For any three non-collinear points, there is exactly one plane containing them.

Postulate 4 If two distinct planes intersect, then their intersection is a line.

Postulate 5 Every line contains at least two points, every plane contains at least three non-collinear points, and space contains exactly four non-coplanar points.

The following three-dimensional figure is one model for the miniature geometry.

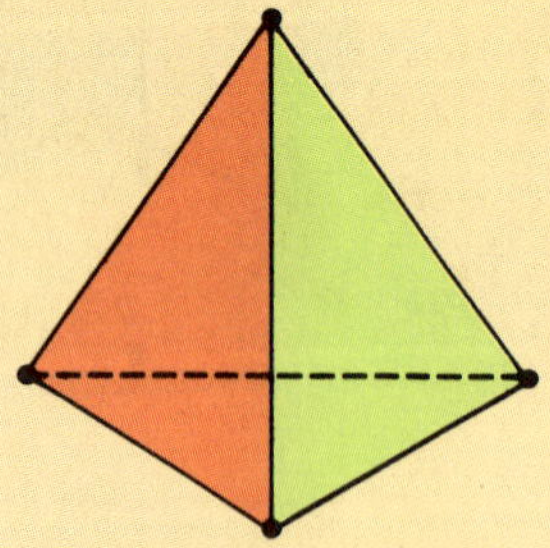

Each dot represents a point.
Each edge represents a line.
Each surface represents a plane.

Check this model against each postulate.

Using Postulates 1–5, the following theorems can be proved.

Theorem 1 If two lines intersect, the intersection contains exactly one point.

Theorem 2 A line and a point *not* on that line are contained in exactly one plane.

Theorem 3 Every line contains exactly two points.

Theorem 4 Every plane contains exactly three points.

Theorem 5 Every plane contains exactly three lines.

Geometries can be used to describe a variety of situations. Telephone trunk lines, electrical circuits, and utility lines are some examples. Another example is a commission of people that has several committees. The undefined terms are *person* and *committee*. The postulates are the rules of the organization.

Rule 1 There is at least one committee.

Rule 2 Every committee contains at least three different people.

Rule 3 Any two committees have at least one person in common.

Rule 4 Any two people are on exactly one committee together.

Rule 5 *Not* all people are on the same committee, and each person is on at least one committee.

Replacing the names of the terms in a system does *not* change the system itself. For example, some companies have collaborated to place several satellites in orbit. Call the set of all satellites that a given company helped to place in orbit a *network*. Replace *person* with *satellite*, and *committee* with *network* in the above rules to see another situation it describes. For another example, try replacing *person* with *point*, and *committee* with *line*.

Exercises Solve each problem. Problems **1–4** refer to the first miniature geometry.

1. How many points are in the geometry?
2. How many lines are in the geometry?
3. How many planes are in the geometry?
4. Name the postulates or theorems that justify the conclusions of Theorem 5.

Problems **5–9** refer to the commission of people. Write *true* or *false* for each of statements **5–8**. Then, solve problem **9**. (Hint: Draw a picture to represent the geometry about the commission.)

5. There are at least four people on the commission.
6. Any two committees have exactly one person in common.
7. If Ann Salvatore is on the commission, there is at least one committee to which she does *not* belong.
8. Each person is in at least three committees.
9. Suppose Ann Salvatore, Lebaron Dudley, Debra White, Janet Rosenfeld, Arthur Lebowitz, Keith St. Pierre, and Karen Elliott are on the commission. Make a list that shows all the possible committees, and names the people on each committee. Label the committees Committee 1, Committee 2, and so on.

1–6 Writing Proofs

A formal proof really is a summary. It does *not* show the thinking and planning done to make a logical argument. The following suggestions will help you write a formal proof.

Write out the theorem.

Theorem: If there is a line and a point *not* on the line, then there is at least one plane that contains them.

Write the given and prove.

Given: line ℓ and point P *not* on ℓ

Prove: at least one plane contains ℓ and P

Draw a diagram.

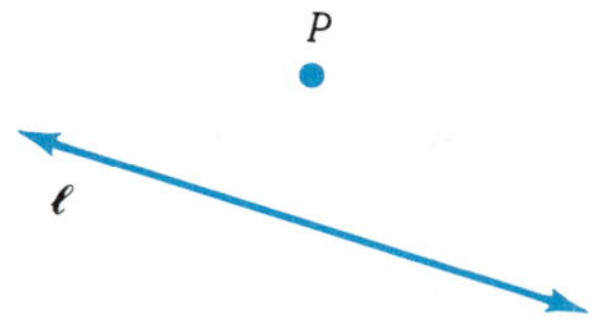

Plan the proof.

Review the definitions, postulates, and theorems already proved that relate to the theorem.

Postulate 1–2: Through any three points *not* on the same line, there is exactly one plane.

Postulate 1–3: A line contains at least two points.

Postulate 1–5: If two points lie in a plane, then the entire line containing those points lies in that plane.

Arrange the ideas in logical order. Start with the *given* and end with the *prove*.

line ℓ and point P *not* on ℓ
ℓ contains 2 points, say R and S
there are 3 points, namely R, S, and P
exactly one plane contains R, S, and P
R and S lie in a plane, so ℓ lies in the plane
ℓ and P lie in at least one plane

Write the proof.

List the statements and reasons.

Procedure for Writing Formal Proofs

1. Write out the theorem.
2. Write the given and prove.
3. Draw a diagram.
4. Plan the proof.
5. Write the proof.

example 1

Write a formal proof for the following theorem.

Theorem: If there is a line and a point *not* on the line, then there is at least one plane that contains them.

Given: line ℓ and point P *not* on ℓ

Prove: at least one plane contains ℓ and P

Proof:

STATEMENTS	REASONS
1. line ℓ and point P *not* on ℓ	1. Given
2. points R and S on ℓ	2. Postulate 1–3: A line contains at least two points.
3. exactly one plane contains R, S, and P	3. Postulate 1–2: Through any three points *not* on the same line there is exactly one plane.
4. at least one plane contains ℓ and P	4. Postulate 1–5: If two points lie in a plane, then the entire line containing those two points lies in that plane.

The following theorem relates lines and planes.

If two lines intersect, then exactly one plane contains both lines.

Theorem 1–2

example 2

Plan a proof to show that at least one plane contains two intersecting lines.

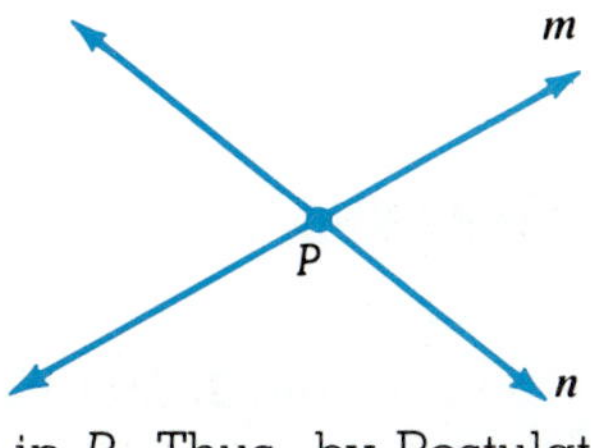

The lines m and n intersect in P. Thus, by Postulate 1–3, m and n both contain two points. By Postulate 1–3, there are three points, namely P, R, and S. Then, by Postulate 1–2, exactly one plane contains P, R, and S. Finally, Postulate 1–5 says that the plane contains both m and n.

exercises

Exploratory For each of the following, name the postulates or theorems you have learned that have the same hypothesis, or write *none*.

1. If two lines intersect, then exactly one plane contains both lines.
2. If two planes intersect, then their intersection contains at least two points.
3. A plane contains at least three lines.
4. If two lines intersect, then their intersection contains one point.
5. Through any three points *not* on the same line, there are exactly three lines.
6. Through four collinear points, there is exactly one line.

7–12. For each of the theorems in problems 1–6, name the postulates or theorems that have the same conclusion, or write *none*.

Written For each of the following, fill in the blank with a reason. Use a definition, a postulate, or a proved theorem.

1. *S* and *Q* are points — Given
   There is exactly one line through *S* and *Q* — ______
2. *Q* and *R* are points — Given
   There is exactly one line containing *Q* and *R* — ______
3. *Q*, *S*, and *R* are noncollinear points — Given
   There is exactly one plane containing *Q*, *S*, and *R* — ______
4. *P*, *Q*, and *R* are noncollinear — Given
   $\mathscr{A}$ is the only plane containing *P*, *Q*, and *R* — ______
5. *Q* and *R* lie in $\mathscr{L}$ — Given
   $\overleftrightarrow{QR}$ lies in $\mathscr{L}$ — ______
6. *P* is *not* on *m* — Given
   $\mathscr{L}$ is the only plane containing *P* and *m* — ______
7. *S* is *not* on *m* — Given
   Exactly one plane contains *S* and *m* — ______
8. *S*, *P*, and *Q* lie on ℓ — Given
   S, *P*, and *Q* are collinear — ______
9. *S* and *R* lie on *m* — Postulate 1–1
   P does *not* lie on *m* — Given
   S, *R*, and *P* are noncollinear — ______
10. *P* does *not* lie on ℓ — Given
    A and *B* lie on ℓ — Given
    A, *B*, and *P* are coplanar — ______

For each of the following, fill in the blank with a statement that follows from the reason.

11. ℓ and *m* intersect — Given
    ______ — Theorem 1–2: If two lines intersect, then exactly one plane contains both lines.
12. *P*, *Q*, and *R* are noncollinear — Given
    ______ — Postulate 1–2: Through any three points *not* on the same line, there is exactly one plane.

13. $\mathscr{M}$ and $\mathscr{N}$ intersect — Given

______________ — Postulate **1–6:** If two planes intersect, then their intersection is a line.

14. $\mathscr{L}$ is a plane — Given

______________ — Postulate **1–4:** A plane contains at least three points *not* on the same line.

15. ℓ is a line — Given

______________ — Postulate **1–3:** A line contains at least two points.

16. P and Q are two points — Given

______________ — Postulate **1–1:** Through any two points there is exactly one line.

17. ℓ is a line and P is a point *not* on ℓ — Given

______________ — Theorem **1–1:** If there is a line and a point *not* on the line, then there is exactly one plane that contains them.

18. R and S lie in $\mathscr{B}$ — Given

______________ — Postulate **1–5:** If two points lie in a plane, then the entire line containing those two points lies in that plane.

For each of the following theorems, name the *given*, *prove*, and draw a diagram you would use in a formal proof.

19. If two planes intersect, then their intersection contains at least two points.

20. If two lines intersect, then at least one plane contains both lines.

21. Two intersecting lines contain at least three points.

22. A plane contains at least three lines.

23–26. Write a formal proof for each of the theorems in problems **19–22**.

Apples and Oranges

Excursions in Geometry

The three boxes shown below are labeled *apples*, *oranges*, and *apples and oranges*. Each label is incorrect. Suppose you may pull out one piece of fruit from one box. You are *not* allowed to feel around in the box or to peek. Tell from which box you would choose. Then, explain how to label the boxes correctly.

1–7 Inverses and Contrapositives

A statement formed by denying another statement is called a *negation*.

A statement and its negation cannot both be true, nor can they both be false.

Sentence Today is Sunday.

Negation Today is not Sunday.

SUN	MON	TUE	WED	THU	FRI	SAT
		1	2	3	4	5
6	7	8	9	10	11	12
13	14	15	16	17	18	19
20	21	22	23	24	25	26
27	28	29	30			

Sentence Three points are noncollinear.

Negation Three points are collinear.

If a statement is true, then its negation is false. If a statement is false, then its negation is true.

The negation of a statement may be written several ways. Some times a statement may *appear* to be the negation of another, when it is *not*.

example 1

Name which statements are negations of the following statement.

The point is on the line.

a. **No point is on the line.**
b. **The point is *not* on the line.**
c. **The line is *not* the point.**
d. **The line does *not* contain the point.**
e. **The point is never on the line.**
f. **No line contains the point.**
g. **The point is *not* the line.**

Only **b** and **d** are negations of the given statement.

Given a conditional statement, its **inverse** can be formed by *negating* both the hypothesis and conclusion.

Statement If a figure is a square, then the figure is a rectangle.

Inverse If a figure is *not* a square, then the figure is *not* a rectangle.

As the example shows, the inverse of a true statement may *not* be true.

Conditional statements also have **contrapositives.** A contrapositive statement is formed by interchanging the hypothesis and conclusion, and negating both.

Statement	If a figure is a square, then the figure is a rectangle.
Contrapositive	If a figure is *not* a rectangle, then the figure is *not* a square.

The contrapositive of a true statement is *always* a true statement. The contrapositive of a false statement is *always* a false statement.

example 2 **Write the inverse and contrapositive of the following true statement. State whether or *not* they are true.**

If two points lie in a plane, then the entire line containing the points lies in that plane.

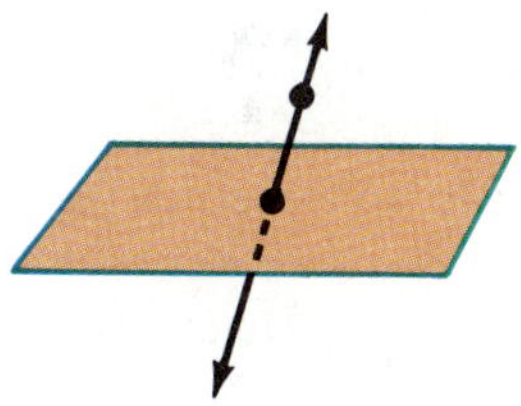

Inverse If two points do *not* lie in a plane, then the entire line containing the points does *not* lie in the plane.

In this case, the inverse is true.

Contrapositive If one entire line containing two points does *not* lie in a plane, then the two points do *not* both lie in the plane.

The contrapositive of a true statement is *always* true.

exercises

Exploratory **State the negation for each of the following.**

1. a figure is a triangle
2. two points lie in a plane
3. three points are collinear
4. a figure is a polygon
5. $\mathcal{M}$ intersects $\mathcal{N}$
6. $\overleftrightarrow{AB}$ does *not* intersect $\mathcal{M}$
7. ℓ does *not* lie in $\mathcal{M}$
8. ℓ intersects t
9. P, Q, and R are noncoplanar
10. A, B, and C are noncollinear
11. two points lie in different planes
12. two lines are in distinct planes

Written Tell whether or *not* each sentence is a negation of the given sentence. Write *yes* or *no*.

$^{-}1$ is a real number.

1. $^{-}1$ is an imaginary number.
2. $^{-}1$ is really a number.
3. 1 is a real number.
4. $^{-}1$ is *not* really a number.
5. $^{-}1$ is *not* a real number.
6. $^{-}1$ is a rational number.

A triangle has four sides.

7. A triangle has three sides.
8. A triangle does *not* have sides.
9. A triangle has no sides.
10. A triangle does *not* have four sides.
11. Four triangles have sides.
12. A triangle has foresight.

Tomorrow it will rain.

13. It will *not* rain tomorrow.
14. Tomorrow it won't rain.
15. Tomorrow the sun will shine.
16. It will *not* rain today.
17. Tomorrow it will *not* rain.
18. Today it will rain.

A room is in a house.

19. A house is *not* a room.
20. A house does *not* contain a room.
21. No house contains a room.
22. A house is *not* in a room.
23. A room is *not* in a house.
24. No room is in a house.

Write the inverse for each of the following statements. If there is no inverse, write *none*.

25. If it rains, then the grass gets wet.
26. If a triangle has a right angle, then it is a right triangle.
27. If you live in Texas, then you are an American.
28. If you ride a bicycle, then you have strong legs.
29. If two lines are perpendicular, then the two lines intersect.
30. If $3x = 12$, then $x = 4$.
31. If n is even, then n^2 is even.
32. If a number is rational, then it is real.
33. If an animal is a chimpanzee, then it is a mammal.
34. If you practice hard, then you will make the team.
35. Parallel lines do *not* intersect.
36. A square is *not* a triangle.
37. A square is a rectangle.
38. Integers are real numbers.
39. $x < 0$ implies $5x > 6x$.
40. $2n + 1 = 5$ implies $n = 2$.
41. When it rains it pours.
42. If you want a new account, phone for it.
43. At least one plane is determined by two intersecting lines.
44. Two angles are congruent when they are vertical angles.

45–64. Write the contrapositive for each statement in problems **25–44.** If there is no contrapositive, write *none*.

Vocabulary

point (1)
line (1)
plane (1)
intersect (2)
undefined terms (5)
collinear points (5)
coplanar points (5)
noncollinear points (6)
noncoplanar points (7)
conditional statements (8)
if-then form (8)
hypothesis (9)
conclusion (9)
converse (9)
if and only if form (9)
biconditional statement (10)
postulates (12)
theorem (17)
formal proof (17)
negation (26)
inverse (26)
contrapositive (27)

Chapter Summary

1. Points have no size. (2)
2. Lines extend indefinitely and have no thickness or width. (2)
3. A plane extends indefinitely in two directions and has no thickness. (2)
4. Two lines, or lines and planes intersect if they have points in common. (3)
5. Guidelines for Good Definition: (5)
   1. Name the term being defined.
   2. Use only undefined terms or already defined terms.
   3. Identify the set to which the term belongs.
   4. State the properties which distinguish the term from others in its set.
   5. Make it reversible.
6. All conditional statements can be written in *if-then* form and may be true or false. (8)
7. The converse of a conditional statement is formed by interchanging the hypothesis and conclusion. (9)
8. All definitions can be written in *if and only if* form. (10)
9. Postulate 1–1: Through any two points there is exactly one line. (12)
10. Postulate 1–2: Through any three points *not* on the same line there is exactly one plane. (12)
11. Postulate 1–3: A line contains at least two points. (13)
12. Postulate 1–4: A plane contains at least three points *not* on the same line. (13)

13. Postulate 1–5: If two points lie in a plane, then the entire line containing those two points lies in that plane. (13)
14. Postulate 1–6: If two planes intersect, then their intersection is a line. (13)
15. Theorem 1–1: If there is a line and a point *not* on the line, then there is exactly one plane that contains them. (17)
16. Formal proofs have five main parts: theorem, given, prove, a diagram, and the proof with statements and reasons. (17)
17. Procedure for Writing Formal Proofs: (22)
    1. Write out the theorem.
    2. Write the given and prove.
    3. Draw a diagram.
    4. Plan the proof.
    5. Write the proof.
18. Theorem 1–2: If two lines intersect, then exactly one plane contains both lines. (23)
19. The inverse of a conditional statement is formed by negating both the hypothesis and conclusion. (26)
20. The contrapositive of a conditional statement is formed by interchanging the hypothesis and conclusion, and negating both. (27)

Chapter Review

1–1 Draw and label a diagram to show each of the following.

1. ℓ intersects $\mathcal{M}$ at A
2. A and B lie on ℓ, but C does *not*
3. P does *not* lie on m
4. P does *not* lie on $\mathcal{M}$

1–2 Solve each problem.

5. Noncoplanar points are points that are *not* coplanar. Write a good definition for *noncoplanar points* without using the word *coplanar*.
6. *They do not intersect* is *not* a good definition. Tell why.

1–3 Rewrite each of the following conditional statements in *if-then* form. Identify the hypothesis and the conclusion.

7. Through any two points there is exactly one line.
8. Through any three points *not* on the same line there is exactly one plane.

9–10. Write the converse for each statement in problems **7** and **8**.

Rewrite each of the following biconditional statements in *if and only if* form.

11. Skew lines are lines which are *not* coplanar.
12. Concurrent lines are lines which have all their points in common.

1–4 **Solve each problem.**

13. Suppose a set consists of a plane and a line *not* in the plane. State the minimum number of points in the set.

14. Suppose two different planes both contain A and B. Describe the intersection of the planes.

1–5 **Draw and label a diagram to illustrate each of the following *given* statements.**

15. Given: ℓ and $\mathscr{A}$ do *not* intersect

16. Given: A, B, and C are noncollinear

17. Given: p and q intersect at A

18. Given: $\mathscr{L}$ and m intersect at R

1–6 **Theorem:** If there is a line and a point *not* on the line, then there is at least one plane that contains them.

Given: ℓ and point P *not* on ℓ

Prove: at least one plane contains ℓ and P

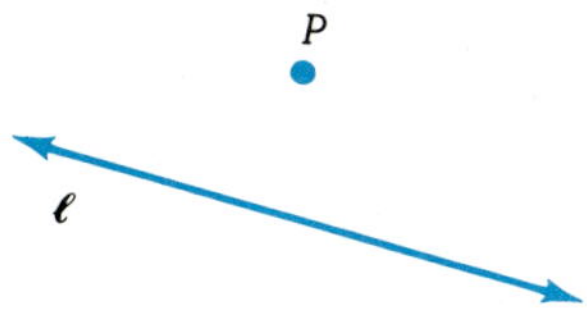

Proof:

STATEMENTS	REASONS
1. ℓ and point P *not* on ℓ	1. Given
2. R and S on ℓ	2. Postulate 1–3: A line contains at least two points.
3. at least one plane contains R, S, and P	3. ______________
4. ______________	4. Postulate 1–5: If two points lie in a plane, then the entire line containing those two points lies in that plane.

19. Complete statement **4** in the above proof.

20. Complete reason **3** in the above proof.

1–7 **Write the inverse for each of the following statements. If there is no inverse, write *none*.**

21. If you want something, ask for it.

22. When the sun shines, the sky is blue.

23. If $2x = 8$, then $x = 4$.

24. A plane contains at least three points.

25–28. Write the contrapositive for each statement in problems **21–24.** If there is no contrapositive, write *none*.

Chapter Test

Draw and label a diagram to illustrate each of the following.

1. P lies on $\mathcal{M}$
2. P, Q, and R are coplanar
3. ℓ and m intersect at S
4. p intersects $\mathcal{M}$ at A
5. $\mathcal{L}$ does *not* contain A or B
6. $\mathcal{M}$ and $\mathcal{N}$ intersect

Rewrite each of the following conditional statements in *if-then* form. Identify the hypothesis and conclusion.

7. A line contains at least two points.
8. A plane contains at least three points *not* on the same line.
9. The intersection of two planes is a line.
10. Two lines intersect in at least one point.

11–14. Write the converse for each statement in problems **7–10.** If there is no converse, write *none*.
15–18. Write the inverse for each statement in problems **7–10.** If there is no inverse, write *none*.
19–22. Write the contrapositive for each statement in problems **7–10.** If there is no contrapositive, write *none*.

Write the *given*, *prove*, and draw a diagram for each of the following.

23. If two lines intersect, then at least one plane contains both lines.
24. Two intersecting lines contain at least three points.

Chapter 2

Measure

Sports competition often uses measurement. A shotputter, for example, wins by putting the shot the greatest distance. The distance is measured from the back of the shot to the outer edge of the shotput ring.

2–1 Number Lines

While a point on a line has *no* size or shape, it does have a position. Numbers can be used to name the positions of different points on a line.

The following figure is a **number line.** Some points on the line and their corresponding numbers are shown. These numbers are called **natural numbers** or counting numbers.

Negative numbers usually are used to name points to the left of zero. The following number line shows the **integers.**

Many numbers can be written as the quotient of two nonzero integers. Any number of this type is called a **rational number.** The following number line shows some rational numbers.

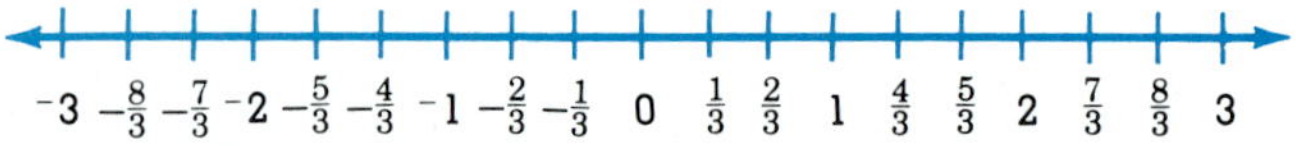

Every rational number can be written as a fraction or a decimal. To change a fraction to a decimal, divide the numerator by the denominator.

example 1

Change $\frac{5}{8}$ to a decimal.

```
   0.625
8)5.000
  4 8
    20
    16
     40
     40
      0 ← remainder
```

The decimal form for $\frac{5}{8}$ is 0.625.

Notice that the division process in Example 1 stops or terminates when a remainder of 0 occurs. The resulting decimal quotient in these cases is called a *terminating decimal.*

A decimal in which a digit or a group of digits repeat is called a *repeating decimal.*

example 2

Change $\frac{9}{27}$ to a decimal.

$$\begin{array}{r} 0.333 \\ 27\overline{)9.000} \\ \underline{8\,1} \\ 90 \\ \underline{81} \\ 90 \\ \underline{81} \\ 9 \end{array} \leftarrow \textit{remainder}$$

Notice that the same remainder, 9, repeats. Thus, the same quotient digit, 3, will keep repeating.

The decimal form for $\frac{9}{27}$ is 0.333 . . .

A repeating decimal can be shown by placing a *bar* over the digit or digits that repeat. Thus, $\frac{9}{27}$ can be written as $0.\overline{3}$.

You can change a repeating decimal to a fraction by the following method.

example 3

Change $0.\overline{345}$ to a fraction.

Let $x = 0.\overline{345}$.

Multiply both sides by 1000. *Why?*

$$1000\,x = 345.\overline{345}$$

$$x = 0.\overline{345} \qquad \textit{Subtract the original equation.}$$

$$999\,x = 345$$

$$\frac{999}{999}x = \frac{345}{999} \qquad \textit{Divide both sides by 999.}$$

$$x = \frac{345}{999} \qquad \textit{Change } \frac{345}{999} \textit{ to simplest form.}$$

$$x = \frac{115}{333}$$

The fraction form of $0.\overline{345}$ is $\frac{115}{333}$.

Some decimals are nonterminating and nonrepeating. For example, the decimal name for π is nonterminating and nonrepeating. Nonterminating, nonrepeating decimals represent **irrational numbers.**

$$\pi \text{ is } 3.14159265358979323846264338\ldots$$

Unless otherwise stated, the word number means real number.

Taken together, the rational numbers and irrational numbers make up the **real numbers.** Each point on a number line can be named by a real number.

Postulate 2–1 Points on the Number Line

Each real number corresponds to exactly one point on a number line. Each point on a number line corresponds to exactly one real number.

A number associated with a point on a number line is called the **coordinate** of the point. On the number line below, the coordinate of A is $^-7$. The coordinate of B is 2.

exercises

Exploratory State the coordinate of each point.

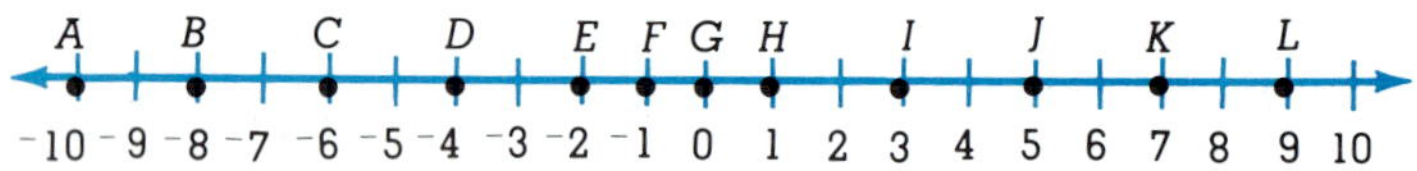

1. H	2. F	3. B	4. E
5. G	6. A	7. J	8. K
9. C	10. I	11. D	12. L

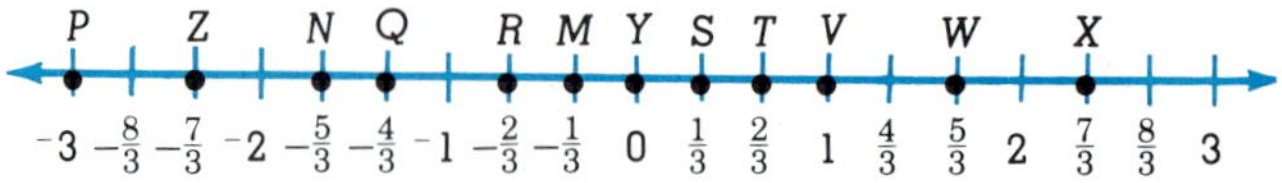

13. T	14. S	15. Q	16. V
17. P	18. R	19. X	20. W
21. M	22. N	23. Y	24. Z

Written Change each fraction to decimal form.

1. $\frac{3}{10}$ 2. $\frac{1}{4}$ 3. $-\frac{4}{5}$ 4. $\frac{7}{8}$

5. $-\frac{3}{4}$ 6. $\frac{5}{8}$ 7. $\frac{2}{3}$ 8. $-\frac{3}{7}$

9. $\frac{4}{12}$ 10. $-\frac{5}{6}$ 11. $\frac{4}{9}$ 12. $\frac{1}{12}$

Change each mixed numeral to decimal form.

> Sample: $^{-}4\frac{1}{4} = {}^{-}4 + -\frac{1}{4}$
> $= {}^{-}4 + {}^{-}0.25$
> $= {}^{-}4.25$

13. $1\frac{1}{5}$ 14. $3\frac{3}{4}$ 15. $^{-}2\frac{7}{10}$ 16. $^{-}4\frac{3}{8}$

17. $2\frac{2}{3}$ 18. $^{-}2\frac{1}{6}$ 19. $^{-}7\frac{1}{7}$ 20. $5\frac{2}{9}$

Change each repeating decimal to a fraction in simplest form.

21. $0.\overline{7}$ 22. $0.\overline{23}$ 23. $0.\overline{5}$ 24. $1.\overline{2}$

25. $4.2\overline{6}$ 26. $0.3\overline{27}$ 27. $1.\overline{26}$ 28. $0.\overline{524}$

29–39. Complete the following chart for natural numbers. Write *yes* or *no*.

Number	$\frac{\pi}{2}$	$^{-}13$	0.07	$^{-}11\frac{2}{3}$	$0.3\overline{3}$	$\frac{1.7}{3}$	π	6.9	$\frac{5}{4}$	$1.74\overline{21}$	$-\frac{35}{3}$
Natural Number											
Integer											
Rational Number											
Irrational Number											
Real Number											

40–50. Complete the chart for integers. Write *yes* or *no*.

62–72. Complete the chart for irrational numbers. Write *yes* or *no*.

51–61. Complete the chart for rational numbers. Write *yes* or *no*.

73–83. Complete the chart for real numbers. Write *yes* or *no*.

2–2 Distance

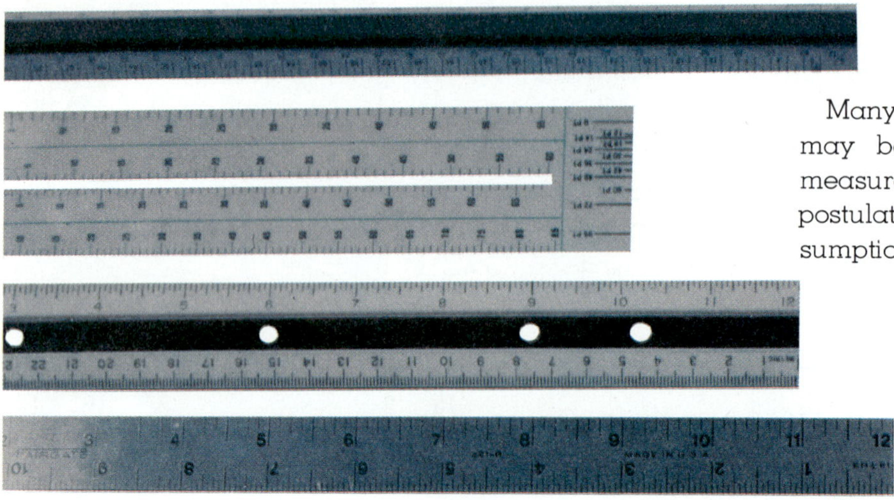

Many different rulers may be used to make measurements. The ruler postulate states this assumption.

Postulate 2–2
Ruler Postulate

The points on a number line can be paired with the real numbers so that, given any two points P and Q on the line, P corresponds to zero, and Q corresponds to a positive number.

When a measurement is made, a number is found. This number is called a **measure.** A measurement also has a unit, called the **unit of measure.** The measure of the distance between two points may differ depending on the unit of measure.

The window is 26 inches wide.
The measure is 26.
The unit of measure is inches.

The window is 57.2 centimeters wide.
The measure is 57.2.
The unit of measure is centimeters.

We do *not* always state the unit of measure. If no units are stated, assume the same unit is used for all measurements.

The Distance Postulate assumes that, given a certain unit of measure, the distance between two points is unique.

Postulate 2–3
Distance Postulate

For any two points on a line and a given unit of measure, there is a unique positive number called the measure of the distance between the two points.

The absolute value of the difference between two numbers represents the measure of the distance between their corresponding points on the number line.

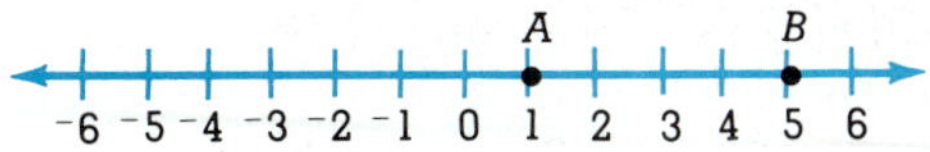

$$|5 - 1| = |4| \text{ or } 4$$

The measure of the distance between A and B is 4.

To indicate the measure of the distance between A and B, write AB. Thus, in the example above, AB is 4.

examples

1 **Find CD, CB, AB, and AC.**

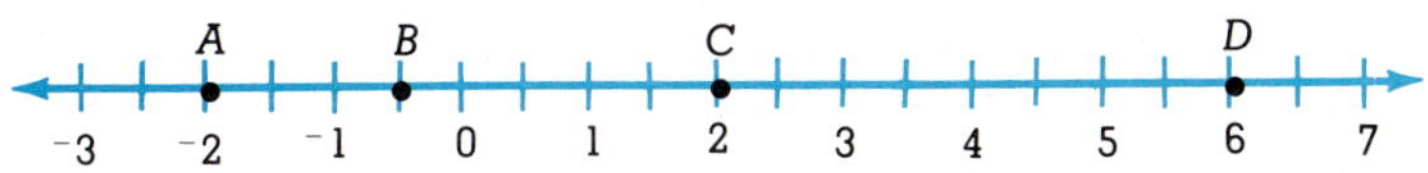

$CD = |6 - 2|$
$= |4|$ or 4

$CB = |-\frac{1}{2} - 2|$
$= |-2\frac{1}{2}|$ or $2\frac{1}{2}$

$AB = |-\frac{1}{2} - {}^-2|$
$= |1\frac{1}{2}|$ or $1\frac{1}{2}$

$AC = |2 - {}^-2|$
$= |4|$ or 4

2 **Find PQ, QR, PS, and RS.**

$PQ = |7.2 - 2.5|$
$= |4.7|$ or 4.7

$QR = |7.7 - 7.2|$
$= |0.5|$ or 0.5

$PS = |8.4 - 2.5|$
$= |5.9|$ or 5.9

$RS = |8.4 - 7.7|$
$= |0.7|$ or 0.7

The symbol AA represents the distance from point A to point A. Thus, $AA = 0$.

exercises

Exploratory State the absolute value for each of the following.

1. 12	2. $^-33$	3. 30	4. 90
5. $^-15$	6. 17	7. 11	8. $^-7$
9. $^-3$	10. $^-61$	11. 89	12. 39
13. $^-44$	14. 97	15. $^-52$	16. $^-58$
17. $^-2.3$	18. $7\frac{1}{2}$	19. $-\frac{3}{4}$	20. $^-0.88$
21. $^-12\frac{7}{8}$	22. $^-0.01$	23. 10.02	24. $\frac{1}{4}$

Written Find each difference.

1. $28 - 19$	2. $4 - 7$	3. $0 - 12$
4. $24 - 42$	5. $^-27 - 42$	6. $^-59 - 38$
7. $^-77 - 76$	8. $^-99 - 29$	9. $^-6 - {^-17}$
10. $^-17 - {^-23}$	11. $^-115 - {^-28}$	12. $^-9 - {^-12}$
13. $^-78 - {^-19}$	14. $0 - {^-35}$	15. $^-41 - {^-36}$

Find each value.

16. $\lvert 2 - {^-2} \rvert$	17. $\lvert {^-2} - {^-5} \rvert$	18. $\lvert 0 - {^-6} \rvert$
19. $\lvert 5 - {^-4} \rvert$	20. $\lvert 3.6 - 9.2 \rvert$	21. $\lvert {^-5.4} - 7.8 \rvert$
22. $\lvert 7.7 - {^-7.2} \rvert$	23. $\lvert {^-5\frac{1}{4}} - {^-2\frac{1}{2}} \rvert$	24. $\lvert 9\frac{3}{5} - {^-6} \rvert$
25. $\lvert 9 - {^-2\frac{3}{4}} \rvert$	26. $\lvert 6\frac{1}{3} - 8\frac{2}{3} \rvert$	27. $\lvert {^-2\frac{2}{3}} - 2\frac{1}{6} \rvert$

Use the number line below to find each measure.

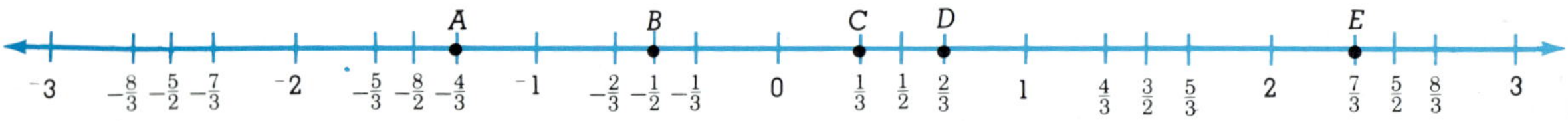

28. *AB*	29. *CD*	30. *DA*	31. *AE*	32. *BE*
33. *BC*	34. *AC*	35. *BD*	36. *AA*	37. *EC*
38. *ED*	39. *AD*	40. *CC*	41. *DC*	42. *CA*

Use the ruler to find each measure.

43. *PQ*	44. *RS*	45. *SP*	46. *PT*	47. *QT*
48. *QR*	49. *PR*	50. *QS*	51. *SS*	52. *TR*
53. *TS*	54. *PS*	55. *RR*	56. *SR*	57. *RP*

Keeping a Car

Did you know that the typical automobile costs about $2000 per year to operate? Car owners not only pay for gas, but pay for insurance, licenses, maintenance and repairs.

Gloria Barnett's car was new three years ago. She has kept a record of expenses.

Item	First Year	Second Year	Third Year
Miles traveled	10,528	12,213	9,760
Loan Payments	$ 1,529.52	$ 1,529.52	$ 1,529.52
Insurance	$ 400.00	$ 415.00	$ 430.00
License and fees	$ 55.00	$ 40.00	$ 40.00
Parking and Tolls	$ 250.00	$ 265.00	$ 280.00
Gas	$ 324.53	$ 399.41	$ 436.42
Maintenance	$ 130.39	$ 185.57	$ 277.65
Tires	—	$ 125.60	—

Loan payments, insurance, license and fees, and parking and tolls are called **fixed costs.** Gas, maintenance and repairs, and tires are called **variable costs.** What were Gloria Barnett's fixed costs and variable costs the first year?

Fixed Costs

$1529.52	*Loan payments*
400.00	*Insurance*
55.00	*License and fees*
+ 250.00	*Parking and tolls*
$2234.52	

Variable Costs

$324.53	*Gas*
130.39	*Maintenance and repairs*
+ 0.00	*Tires*
$454.92	

Altogether, Gloria's operating expenses were $2234.52 + $454.92, or $2689.44, on her car the first year.

Exercises Solve each problem.

1. Find Gloria's fixed costs for the second year.
2. Find Gloria's fixed costs for the third year.
3. Find Gloria's variable costs for the second year.
4. Find Gloria's variable costs for the third year.
5. Find Gloria's operating expenses for the second year.
6. Find Gloria's operating expenses for the third year.

7–9. For each year, find Gloria's operating expenses if she had *no* loan payments.

10. Explain how you think Gloria's operating expenses will change for the fourth year.

Unless stated otherwise, betweenness and collinearity of points may be assumed if they are given in a diagram.

Look at the figure at the right. Point Q is between P and R. Points W, X, Y and Z are *not* between P and R. In geometry, **between** is defined in the following way.

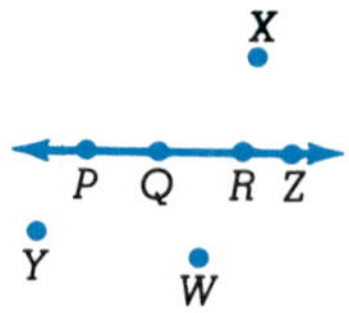

Definition of Between

A point Q is between points P and R if and only if the following holds.
1. P, Q, and R are collinear, and
2. $PQ + QR = PR$.

example 1

Show that Z is *not* between P and R in the figure below.

$$PZ + ZR \stackrel{?}{=} PR$$
$$|{}^{-}2 - 5| + |5 - 2| \stackrel{?}{=} |{}^{-}2 - 2|$$
$$7 + 3 \stackrel{?}{=} 4$$
$$10 \neq 4$$

Thus, $PZ + ZR \neq PR$ *The three points do not satisfy condition 2 of the definition of between.*

The part of the line below that includes points A and B and all the points between them is called a **line segment** or segment. Points A and B are called the *endpoints* of the segment.

line segment AB
$\overline{AB}$

The *length* of a line segment is the distance between the two endpoints of the segment.

The **midpoint** of a line segment separates the segment into two segments of equal measure.

Definition of Midpoint

A point M is the midpoint of a line segment, $\overline{PQ}$, if and only if M is between P and Q, and $PM = MQ$.

example 2

Show that M is the midpoint of $\overline{PQ}$.

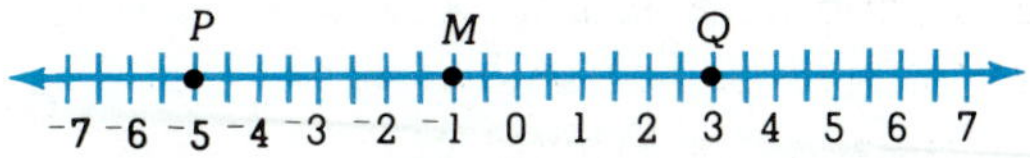

There are two conditions to be met.

1. Show that M is between P and Q.

   The figure shows that all three points are collinear.

$$PM + MQ \stackrel{?}{=} PQ$$
$$|{}^-5 - {}^-1| + |{}^-1 - 3| \stackrel{?}{=} |3 - {}^-5|$$
$$|{}^-4| + |{}^-4| \stackrel{?}{=} |8|$$
$$8 = 8$$

2. Show that $PM = MQ$.

$$PM \stackrel{?}{=} MQ$$
$$|{}^-5 - {}^-1| \stackrel{?}{=} |{}^-1 - 3|$$
$$|{}^-4| \stackrel{?}{=} |{}^-4|$$
$$4 = 4$$

Because distance is defined as unique, the midpoint of a line segment is unique.

A line segment has exactly one midpoint. Theorem 2–1

exercises

Exploratory State the coordinates of the endpoints for each of the following line segments.

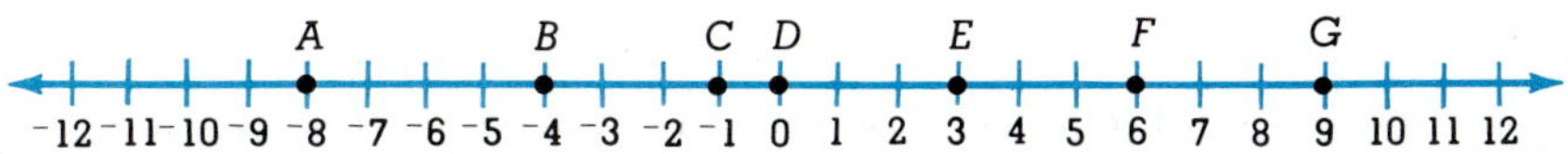

1. $\overline{CD}$
2. $\overline{BF}$
3. $\overline{AG}$
4. $\overline{EB}$
5. $\overline{BA}$
6. $\overline{CG}$
7. $\overline{FE}$
8. $\overline{GA}$
9. $\overline{AD}$
10. $\overline{BE}$
11. $\overline{CF}$
12. $\overline{FG}$
13. $\overline{FA}$
14. $\overline{CB}$
15. $\overline{BD}$
16. $\overline{AE}$
17. $\overline{EF}$
18. $\overline{CE}$
19. $\overline{BG}$
20. $\overline{EG}$

Written Find the measure of the length of each of the following line segments.

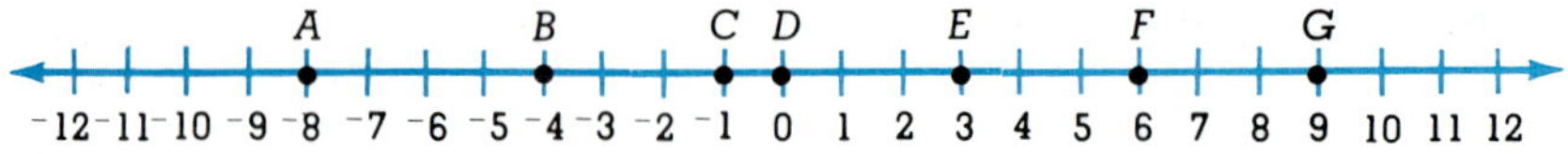

1. $\overline{CD}$
2. $\overline{BF}$
3. $\overline{AG}$
4. $\overline{EB}$
5. $\overline{BA}$
6. $\overline{CG}$
7. $\overline{FE}$
8. $\overline{GA}$
9. $\overline{AD}$
10. $\overline{BE}$
11. $\overline{CF}$
12. $\overline{FG}$
13. $\overline{FA}$
14. $\overline{CB}$
15. $\overline{BD}$
16. $\overline{AE}$
17. $\overline{EF}$
18. $\overline{CE}$
19. $\overline{BG}$
20. $\overline{EG}$

Use the number line below to solve each of the following problems.

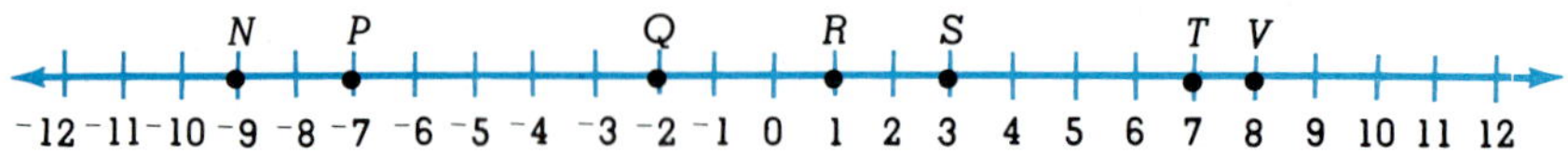

21. Show that T is *not* between R and S.

22. Show that Q is *not* between V and R.

23. Show that N is *not* the midpoint of $\overline{PS}$.

24. Show that R is *not* the midpoint of $\overline{PS}$.

25. Show that Q is the midpoint of $\overline{PS}$.

26. Show that S is the midpoint of $\overline{QV}$.

27. Find the coordinate of the midpoint of $\overline{PT}$.

28. Find the coordinate of the midpoint of $\overline{NS}$.

For each of the following, draw a diagram to illustrate the given.

29. Given: A has coordinate $^{-}2$
B has coordinate 5
M has coordinate 1.5
M is between A and B
Prove: M is the midpoint of $\overline{AB}$

30. Given: T has coordinate $^{-}9$
R has coordinate $^{-}6$
S has coordinate $^{-}3$
R is between T and S
Prove: R is the midpoint of $\overline{TS}$

31. Given: A, B, and C are collinear
$AB = 3$
$AC = 5$
$BC = 2$
Prove: B is between A and C

32. Given: P, Q, and R are collinear
$RQ = 6.5$
$PQ = 2.3$
Q is between P and R
Prove: $PR = 8.8$

33. Given: N is the midpoint of $\overline{AB}$
$AB = 4.5$
Prove: $NB = 2.25$

34. Given: M is the midpoint of $\overline{PQ}$
$MQ = 7\frac{1}{2}$
Prove: $PM = 7\frac{1}{2}$

35–40. List the definitions, postulates, or theorems you would use to complete the proofs in problems **29–34.**

41. The coordinate of B is $^{-}5$ and $AB = 6$. Find the possible coordinates of A.

42. The coordinate of P is 3 and $PQ = 5$. Find the possible coordinates of Q.

43. The coordinate of B is $^{-}5$ and $AB = 6$. Find the possible coordinates of the midpoint of $\overline{AB}$.

44. The coordinate of P is 3 and $PQ = 5$. Find the possible coordinates of the midpoint of $\overline{PQ}$.

Length in the Metric System

The most widely used system of measurement is the **metric system.** It is a decimal system, which means that the units increase or decrease in size by multiples of ten.

Each kind of measurement has a basic unit. Prefixes are attached to the unit to define larger and smaller units. For example, the standard unit of length is the *meter*. Other commonly used units of length are kilo*meter*, centi*meter*, and milli*meter*.

The following table shows some metric prefixes and their meanings.

Prefix	Pronunciation	Symbol	Increase or decrease in unit
mega	MEHG uh	M	1,000,000 (one million)
kilo	KIHL oh	k	1,000 (one thousand)
hecto	HEHK toh	h	100 (one hundred)
deka	DEHK uh	da	10 (ten)
deci	DEHS uh	d	0.1 (one-tenth)
centi	SEHN tuh	c	0.01 (one-hundredth)
milli	MIHL uh	m	0.001 (one-thousandth)
micro	MY kroh	μ	0.000001 (one-millionth)

The unit *kilometer* comes from *kilo* and *meter*. Thus, one kilometer is one thousand meters. Thirty-four kilometers is thirty-four thousand meters. Three-tenths of a kilometer is three hundred meters.

1 km = 1000 m m *is the symbol for meter*
34 km = 34,000 m
0.3 km = 300 m

Exercises Tell which unit is longer.

1. 1 meter or 1 kilometer
2. 1 centimeter or 1 millimeter
3. 1 meter or 1 millimeter
4. 1 centimeter or 1 meter
5. 1 centimeter or 1 kilometer
6. 1 kilometer or 1 millimeter

Are the measurements equivalent? Write *yes* or *no*.

7. 450 cm, 4.5 m
8. 7.2 m, 7200 cm
9. 200 m, 2 cm
10. 54 mm, 0.54 cm
11. 925 mm, 92.5 cm
12. 4.1 cm, 41 mm
13. 36.2 km, 3620 m
14. 1.94 km, 194 m
15. 56 km, 0.56 m
16. 14 m, 140 cm
17. 6.8 mm, 68 cm
18. 9.2 mm, 0.92 cm

2–4 Properties of the Real Number System

The measures of length and distance are real numbers. Thus, the properties of the real number system will be used.

In your arithmetic you have assumed these properties of equality.

Postulate 2–4
Reflexive Property of Equality

For any number a, $a = a$.

Postulate 2–5
Symmetric Property of Equality

For all numbers a and b, if $a = b$, then $b = a$.

Postulate 2–6
Transitive Property of Equality

For all numbers a, b, and c, if $a = b$ and $b = c$, then $a = c$.

example 1

Prove that if P, Q, R, and T lie on a line in order so that $PQ = QR$ and $PQ = RT$, then $RT = QR$.

Given: $PQ = QR$
$PQ = RT$

Prove: $RT = QR$

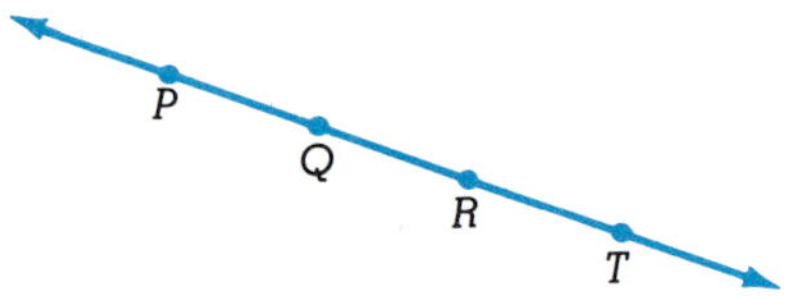

Proof:

STATEMENTS	REASONS
1. $PQ = RT$	1. Given
2. $RT = PQ$	2. Postulate 2–5: Symmetric Property of Equality
3. $PQ = QR$	3. Given
4. $RT = QR$	4. Postulate 2–6: Transitive Property of Equality

To shorten geometric proofs, statements having the reflexive, symmetric, and transitive properties as reasons may be omitted.

Many equations can be solved using the following properties.

Postulate 2–7
Addition and Subtraction Properties of Equality

For any numbers a, b, and c, if $a = b$, then
$a + c = b + c$ and
$a - c = b - c$.

Postulate 2–8
Multiplication and Division Properties of Equality

For any numbers a, b, and c,
if $a = b$, then $a \cdot c = b \cdot c$ and,
if c is *not* zero, then $\frac{a}{c} = \frac{b}{c}$.

For any numbers a and b, if $a = b$, then a may be replaced by b.

Postulate 2–9
Substitution Property of Equality

The following chart summarizes properties of addition and multiplication with real numbers.

For any numbers a, b, and c			
Property	**Addition**	**Multiplication**	
Commutative	$a + b = b + a$	$a \cdot b = b \cdot a$	Postulate 2–10
Associative	$(a + b) + c = a + (b + c)$	$(a \cdot b) \cdot c = a \cdot (b \cdot c)$	Postulate 2–11
Identity	$a + 0 = a = 0 + a$	$a \cdot 1 = a = 1 \cdot a$	Postulate 2–12
Inverse	$a + {}^{-}a = 0 = {}^{-}a + a$	If a is *not* zero, then $a \cdot \frac{1}{a} = 1 = \frac{1}{a} \cdot a$.	Postulate 2–13
Distributive of multiplication over addition: $a(b + c) = ab + ac$ and $(b + c)a = ba + ca$			Postulate 2–14

example 2

Prove that if $5x - 7 = 23$, then $x = 6$.

Given: $5x - 7 = 23$

Prove: $x = 6$

Proof:

STATEMENTS	REASONS
1. $5x - 7 = 23$	1. Given
2. $5x - 7 + 7 = 23 + 7$	2. Postulate 2–7: Addition Property of Equality
3. $5x = 30$	3. Postulate 2–9: Substitution Property of Equality, Postulate 2–12
4. $\frac{5x}{5} = \frac{30}{5}$	4. Postulate 2–8: Division Property of Equality
5. $x = 6$	5. Postulate 2–9: Substitution Property of Equality

In references to algebraic properties, often the postulate numbers are omitted.

In algebra, proofs usually are written in a different way. Statements are shown, although they may be combined. Reasons may or may *not* be shown.

example 3 **Prove that if $6(a + 5) + 10a = {}^-2$, then $a = {}^-2$.**

$6(a + 5) + 10a = {}^-2$	*Given*
$6a + 30 + 10a = {}^-2$	*Distributive and Substitution Properties of Equality*
$16a + 30 = {}^-2$	*Commutative, Distributive, and Substitution Properties of Equality*
$16a = {}^-32$	*Addition and Substitution Properties of Equality*
$a = {}^-2$	*Division and Substitution Properties of Equality*

exercises

Exploratory State the property shown in each of the following.

1. $3 + (2 + 3) = 3 + (2 + 3)$
2. $3 + (2 + 3) = 3 + 5$
3. If $3 = 2 + 1$, then $2 + 1 = 3$.
4. If $5 + 7 = 7 + 5$, then $7 + 5 = 5 + 7$.
5. If $8 = 3 + 5$ and $3 + 5 = \sqrt{64}$, then $8 = \sqrt{64}$.
6. If $8 + 1 = 9$ and $9 = 3 + 6$, then $8 + 1 = 3 + 6$.
7. $12 = 12$
8. $9 + 2 = (6 + 3) + 2$
9. If $2 + 2 = 4$, then $6 + (2 + 2) = 6 + 4$.
10. If $8 = 6 + 2$ and $6 + 2 = 5 + 3$, then $8 = 5 + 3$.
11. $AB = AB$
12. If $PQ = RS$, then $RS = PQ$.
13. If $AB = BC$ and $BC = CD$, then $AB = CD$.
14. If $PQ = RS$, then $PQ + TV = RS + TV$.
15. If $AB = BC$, then $AB + EF = BC + EF$.
16. $PQ = PQ$
17. If $AB = BC$, then $BC = AB$.
18. If $PQ = QR$ and $QR = RS$, then $PQ = RS$.
19. If $AB = CD$, then $AB + TV = CD + TV$.
20. If $PQ = QR$, then $PQ - RS = QR - RS$.

Written State the missing reason for each of the following proofs.

1. **Given:** $9x + 4 = 2\frac{1}{2}$

   Prove: $x = -\frac{1}{6}$

   Proof:

STATEMENTS	REASONS
1. $9x + 4 = 2\frac{1}{2}$	1. Given
2. $9x + 4 - 4 = 2\frac{1}{2} - 4$	2. ________________
3. $9x = -\frac{3}{2}$	3. Postulate 2–9: Substitution Property of Equality
4. $9x \div 9 = -\frac{3}{2} \div 9$	4. Postulate 2–8: Division Property of Equality
5. $x = -\frac{1}{6}$	5. Postulate 2–9: Substitution Property of Equality

2. **Given:** $\frac{3}{4}x = \frac{1}{3}$

Prove: $x = \frac{4}{9}$

Proof:

STATEMENTS	REASONS
1. $\frac{3}{4}x = \frac{1}{3}$	1. Given
2. $\frac{4}{3} \cdot \frac{3}{4}x = \frac{4}{3} \cdot \frac{1}{3}$	2. Postulate 2–8: Multiplication Property of Equality
3. $x = \frac{4}{9}$	3. ______________

3. **Given:** $3x + 8 = 29$

Prove: $x = 7$

Proof:

STATEMENTS	REASONS
1. $3x + 8 = 29$	1. Given
2. $3x + 8 - 8 = 29 - 8$	2. Postulate 2–7: Subtraction Property of Equality
3. $3x = 21$	3. ______________
4. $\frac{3x}{3} = \frac{21}{3}$	4. Postulate 2–8: Division Property of Equality
5. $x = 7$	5. Postulate 2–9: Substitution Property of Equality

4. **Given:** $4 - 7x = 25$

Prove: $x = {}^{-}3$

Proof:

STATEMENTS	REASONS
1. $4 - 7x = 25$	1. Given
2. ${}^{-}4 + 4 - 7x = {}^{-}4 + 25$	2. Postulate 2–7: Addition Property of Equality
3. ${}^{-}7x = 21$	3. Postulate 2–9: Substitution Property of Equality
4. $\frac{{}^{-}7x}{{}^{-}7} = \frac{21}{{}^{-}7}$	4. ______________
5. $x = {}^{-}3$	5. Postulate 2–9: Substitution Property of Equality

5. **Given:** $AB = CD$

Prove: $AC = BD$

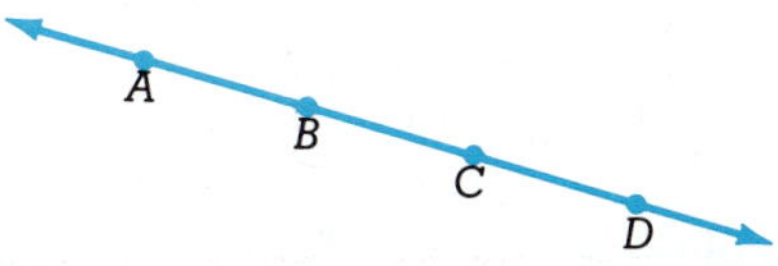

Proof:

STATEMENTS	REASONS
1. $AB = CD$	1. Given
2. $AB + BC = CD + BC$	2. ______________
3. $AC = BD$	3. Postulate 2–9: Substitution Property of Equality and Definition of Between.

6. **Given:** $PM = MS$

M is the midpoint of $\overline{PQ}$

Prove: $MS = MQ$

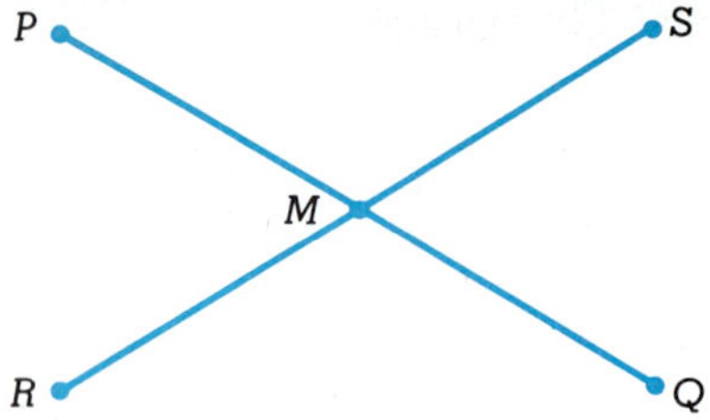

Proof:

STATEMENTS	REASONS
1. $PM = MS$	1. Given
2. M is the midpoint of $\overline{PQ}$	2. Given
3. $PM = MQ$	3. Definition of midpoint
4. $MQ = MS$	4. Postulate 2–9: Substitution Property of Equality
5. $MS = MQ$	5. __________

7. **Given:** $MB = MQ$

Prove: $AB = AM + MQ$

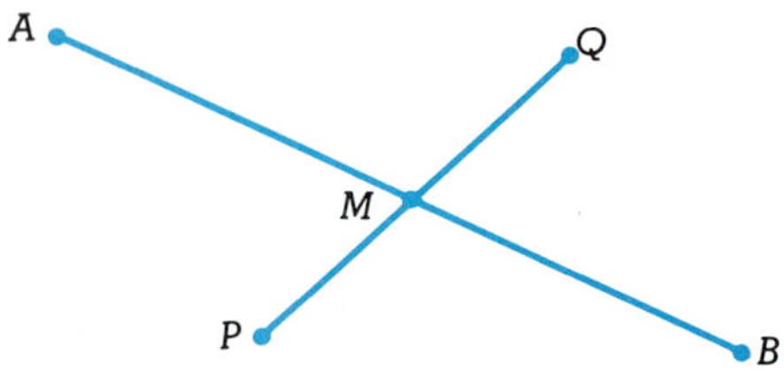

Excursions in Geometry

Significant Digits

All measurements are approximate. No matter how small the unit of measure, the exact measurement cannot be determined. The degree of accuracy of a measurement can be shown by the number of **significant digits.**

45.3 has *three* significant digits.
1.250 has *four* significant digits.

When leading zeros are used to show the position of the decimal point, they are *not* counted as significant digits.

0.071 has *two* significant digits.

Suppose you are told that a metal rod is 75.3 cm long. Then you can be reasonably sure the actual length is between 75.2 cm and 75.4 cm.

Exercises Tell how many significant digits are in each of the following.

1. 19.5	2. 120.0	3. 0.006
4. 212.01	5. 0.090	6. 7.0
7. 3304	8. 0.87	9. 190
10. 0.040	11. 19.0	12. 5400

2–5 Congruent Segments

Segments having exactly the same length are called **congruent segments.** Segments $\overline{PQ}$ and $\overline{RS}$ below are congruent because they are both 4 units long.

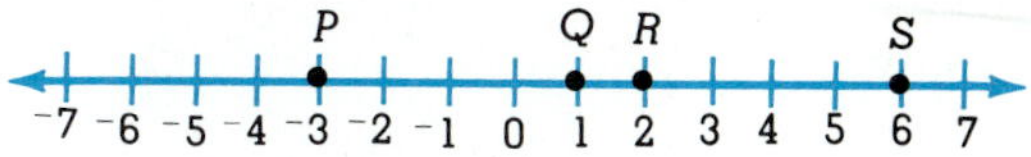

The symbol ≅ means *congruent* or *is congruent to*. In the example above, $\overline{PQ} \cong \overline{RS}$ means $\overline{PQ}$ *is congruent to* $\overline{RS}$.

Since congruence of line segments is related to measure, the following theorems hold.

Congruence of line segments is reflexive. — Theorem 2–2

Congruence of line segments is symmetric. — Theorem 2–3

Congruence of line segments is transitive. — Theorem 2–4

The proofs of these theorems all are similar. One proof is shown below.

example 1

Prove that congruence of line segments is symmetric.

Given: $\overline{PQ} \cong \overline{RS}$

Prove: $\overline{RS} \cong \overline{PQ}$

Proof:

STATEMENTS	REASONS
1. $\overline{PQ} \cong \overline{RS}$	1. Given
2. $PQ = RS$	2. Definition of congruent line segments
3. $RS = PQ$	3. Postulate 2–5: Symmetric property of equality
4. $\overline{RS} \cong \overline{PQ}$	4. Definition of congruent line segments

The proof of the following theorem shows that the midpoint of a line segment separates the segment into two congruent segments.

If M is the midpoint of $\overline{PQ}$, then $\overline{PM} \cong \overline{MQ}$. — Theorem 2–5

example 2

Prove Theorem 2–5.

Given: M is the midpoint of $\overline{PQ}$

Prove: $\overline{PM} \cong \overline{MQ}$

Proof:

STATEMENTS	REASONS
1. M is the midpoint of $\overline{PQ}$	1. Given
2. $PM = MQ$	2. Definition of midpoint
3. $\overline{PM} \cong \overline{MQ}$	3. Definition of congruent line segments

The converse of Theorem 2–5 also holds. Thus, congruence could be used to define midpoint.

If a plane, a line, or a line segment intersects a line segment at its midpoint, it is said to **bisect** the segment. It is called a bisector of the segment.

$\mathcal{N}$ bisects $\overline{PQ}$
$\overline{TM}$ bisects $\overline{PQ}$
$\overleftrightarrow{RM}$ bisects $\overline{PQ}$

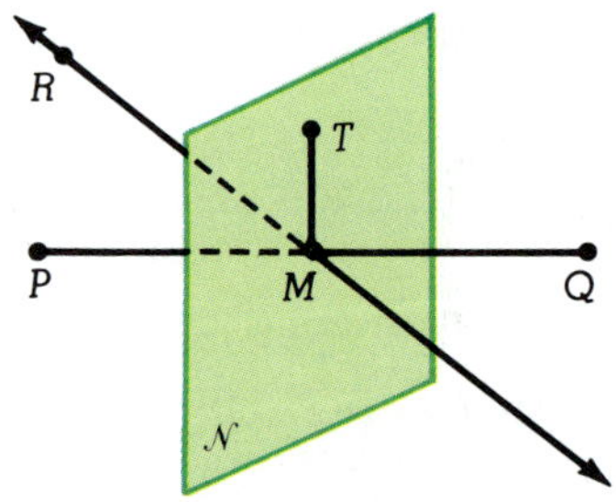

example 3

Prove that if $\overline{AB}$ bisects $\overline{CD}$ at M, then $CM = MD$.

Given: $\overline{AB}$ bisects $\overline{CD}$ at M

Prove: $CM = MD$

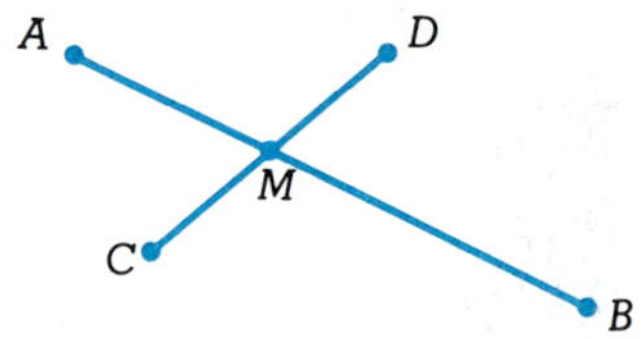

Proof:

STATEMENTS	REASONS
1. $\overline{AB}$ bisects $\overline{CD}$ at M	1. Given
2. M is the midpoint of $\overline{CD}$	2. Definition of bisector
3. $CM = MD$	3. Definition of midpoint

Theorem 2–6 **If $\overline{PQ}$ is bisected at point M, then $\overline{PM} \cong \overline{MQ}$.**

exercises

Exploratory State the property shown in each of the following.

1. $\overline{RS} \cong \overline{RS}$
2. If $\overline{RS} \cong \overline{PQ}$ and $\overline{PQ} \cong \overline{TV}$, then $\overline{RS} \cong \overline{TV}$.
3. If $\overline{PQ} \cong \overline{RS}$, then $\overline{RS} \cong \overline{PQ}$.
4. $\overline{PQ} \cong \overline{PQ}$
5. If $\overline{AB} \cong \overline{CD}$ and $\overline{CD} \cong \overline{EF}$, then $\overline{AB} \cong \overline{EF}$.
6. If $\overline{AB} \cong \overline{BC}$ and $\overline{BC} \cong \overline{CD}$, then $\overline{AB} \cong \overline{CD}$.
7. If $\overline{RX} \cong \overline{PQ}$ and $\overline{PQ} \cong \overline{TR}$, then $\overline{RX} \cong \overline{TR}$.
8. If $\overline{AB} \cong \overline{BC}$, then $\overline{BC} \cong \overline{AB}$.

Written For each of the following statements, either write *true* or *false*.

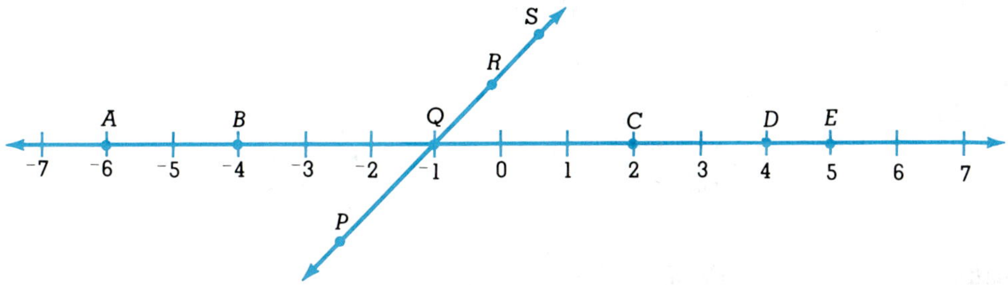

1. $\overline{PS}$ bisects $\overline{BC}$
2. D is the midpoint of $\overline{CE}$
3. $\overline{AB} \cong \overline{CD}$
4. $\overline{BQ} \cong \overline{CE}$
5. $\overline{AQ} \cong \overline{QD}$
6. $\overline{PR}$ bisects $\overline{AD}$
7. Q is the midpoint of $\overline{BC}$
8. $\overline{BC} \cong \overline{QE}$
9. $\overline{EC} \cong \overline{QC}$
10. $\overline{BQ} \cong \overline{QC}$

Solve each problem.

11. State the hypothesis of Theorem 2–5.
12. State the conclusion of Theorem 2–5.
13. Write the converse of Theorem 2–5.
14. Write a definition of midpoint based on congruence.
15. State the hypothesis of Theorem 2–6.
16. Draw and label a diagram to illustrate the hypothesis of Theorem 2–6.
17. State the *given* for a proof of Theorem 2–6.
18. State the conclusion of Theorem 2–6.
19. State the *prove* for a proof of Theorem 2–6.
20. Give a definition for *bisects*.
21. Give a definition for *midpoint*.
22. State the theorem in this lesson that is about midpoints and congruent line segments.
23. Give a plan for a proof of Theorem 2–6.
24. Write a formal proof for Theorem 2–6.

Prove each of the following statements.

25. Congruence of line segments is reflexive.
26. Congruence of line segments is transitive.
27. If M is between P and Q, and $\overline{PM} \cong \overline{MQ}$, then M is the midpoint of $\overline{PQ}$.
28. If $\overline{LP}$ bisects $\overline{RS}$ at M, then $\overline{RM} \cong \overline{MS}$.

2–6 Order

Line segments may have the same length, but there are other possibilities.

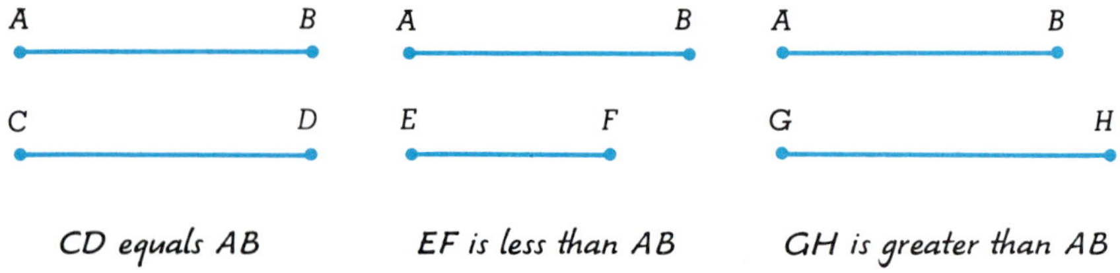

Inequalities and equations are used to describe these situations.

$CD = AB$ $\qquad$ $EF < AB$ $\qquad$ $GH > AB$

The Comparison Property states that these are the only relationships that exist among measures.

Postulate 2–15
Comparison Property

For any two numbers a and b, exactly one of the following statements is true.

$$a < b \qquad a = b \qquad a > b$$

The symbols $\neq$, $\leq$, and $\geq$ also can be used when comparing numbers. For example, $x \leq y$ means $x < y$ or $x = y$.

Operating with the same number on both sides of an inequality may or may *not* affect its truth value. The following properties state when the truth value is unchanged.

Postulate 2–16
Addition and Subtraction Properties of Inequality

For any numbers a, b, and c,

1. **if $a > b$, then $a + c > b + c$ and $a - c > b - c$.**
2. **if $a < b$, then $a + c < b + c$ and $a - c < b - c$.**

Postulate 2–17
Multiplication and Division Properties of Inequality

For any numbers a, b, and c,

1. **if $c > 0$ and $a < b$, then $ac < bc$ and $\frac{a}{c} < \frac{b}{c}$.**
2. **if $c > 0$ and $a > b$, then $ac > bc$ and $\frac{a}{c} > \frac{b}{c}$.**
3. **if $c < 0$ and $a < b$, then $ac > bc$ and $\frac{a}{c} > \frac{b}{c}$.**
4. **if $c < 0$ and $a > b$, then $ac < bc$ and $\frac{a}{c} < \frac{b}{c}$.**

examples

1

Prove that if 3x > 9x + 6, then x < ⁻1.

Given: $3x > 9x + 6$

Prove: $x < {}^{-}1$

Proof:

STATEMENTS	REASONS
1. $3x > 9x + 6$	1. Given
2. $^{-}6x > 6$	2. Postulate 2–16: Subtraction Property of Inequality and Postulate 2–9: Substitution
3. $x < {}^{-}1$	3. Postulate 2–17: Division Property of Inequality and Postulate 2–9: Substitution

2

Two line segments, $\overline{PQ}$ and $\overline{RS}$, intersect at point T so that $\overline{RT}$ is congruent to $\overline{TQ}$. PQ is greater than RS. Write an equation or inequality that relates PT and TS.

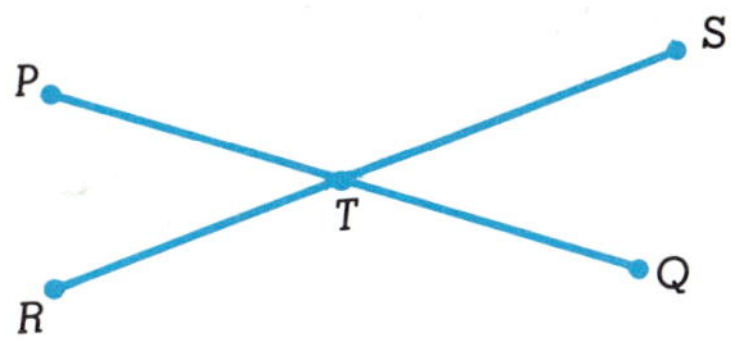

$PQ > RS$	*Given*
$PQ - RT > RS - RT$	*Subtraction Property of Inequality*
$\overline{RT} \cong \overline{TQ}$	*Given*
$RT = TQ$	*Definition of congruent segments*
$PQ - TQ > RS - RT$	*Substitution*
$PT > TS$	*Substitution*

exercises

Exploratory Replace each ■ with <, >, or = to make each sentence true.

1. 2 ■ 6
2. 7 ■ 15
3. 6 ■ 2
4. ⁻5 ■ ⁻1
5. ⁻1 ■ ⁻6
6. 0.01 ■ 0.001
7. 2 + 3 ■ 4 + 1
8. $3\frac{1}{2} + 1$ ■ $3\frac{1}{2}$
9. 1.25 ■ 1.253
10. ⁻5 + 4 ■ 2 + ⁻3
11. $2\frac{1}{4} - 3$ ■ $2\frac{1}{4}$
12. $2\frac{1}{4} - 3$ ■ 3
13. $1\frac{7}{8}$ ■ $1\frac{3}{4}$
14. $^{-}2\frac{1}{2}$ ■ $^{-}2\frac{3}{4}$
15. $\frac{1}{2}$ ■ $\frac{3}{6}$

Written Replace each ■ with <, >, or = to make each sentence true. Refer to the number line below.

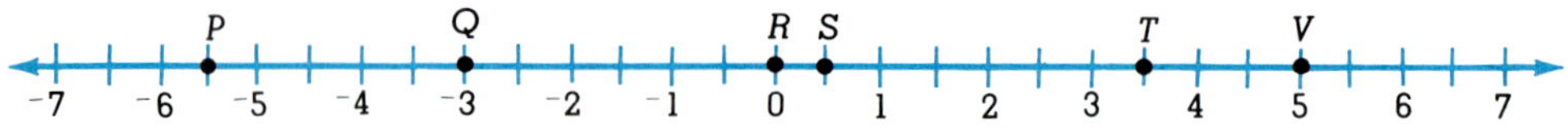

1. *QR* ■ *ST*
2. *PS* ■ *QV*
3. *PQ* ■ *TV*
4. *PS* ■ *RV*
5. *RT* ■ *QS*
6. *PV* ■ *RS*
7. *PT* ■ *QV*
8. *PR* ■ *RV*
9. *PQ* ■ *QS*
10. *RS* ■ *TV*
11. *SV* ■ *PR*
12. *PS* ■ *QT*

Prove each of the following statements.

13. If $3x + 7 > 43$, then $x > 12$.
14. If $2t - 9 < 21$, then $t < 15$.
15. If $11 - 5y < {}^-77$, then $y > 17.6$.
16. If $8 - 3x < 44$, then $x > {}^-12$.
17. If $9(x + 2) < 72$, then $x < 6$.
18. If $5(5z - 3) > 60$, then $z > 3$.

19. Suppose *P*, *Q*, *R*, and *S* are collinear points with *Q* between *P* and *R*, and *R* between *Q* and *S*. Also, $\overline{PQ}$ is congruent to $\overline{RS}$, and *QR* is less than *RS*. Write an equation or inequality that relates *PQ* and *QR*.
20. Suppose *A*, *B*, *C*, and *D* are collinear points with *B* between *A* and *C*, and *C* between *B* and *D*. Also, $\overline{AB}$ is congruent to $\overline{CD}$, and *BC* is greater than *CD*. Write an equation or inequality that relates *BD* and *AC*.
21. If $AB < CD$ and $CD < EF$, how are *AB* and *EF* related?
22. If $PQ > RS$ and $RS > TV$, how are *PQ* and *TV* related?

Using Computers

History

If you don't know Japanese, then you won't understand instructions that are given in Japanese. If a computer does *not* have a compiler that translates BASIC, then the computer won't understand programs that are written in BASIC.

Most computers have a compiler which enables the computer to translate incoming programs into machine language. For example, the BASIC compiler for a PDP 11 computer translates a complete BASIC program into a list of number codes in the machine language of that computer. The compiler does *not* do the calculations of the program. It only translates the program.

The first compiler was developed by Dr. Grace M. Hopper in 1952 for UNIVAC. Hopper studied and taught mathematics at Vassar College. She has worked for several computer corporations, lectured at many colleges and universities, and worked for the information systems division of the U. S. Navy. She has won several awards and published in professional journals. Her contribution is significant because it allows people to design programming languages to meet the needs of certain kinds of problems without having to consider the details of computer hardware.

CAREERS

Chemist

Jennings Lee is a chemist at Juniper Lane Chemicals. He tests products to be sure they are safe for consumers. Product testing helps protect consumers from harmful or poisonous substances. Even some of the smallest particles of products, called atoms, are examined. The atoms often are bonded together. Chemists sometimes must estimate the distance between bonded atoms.

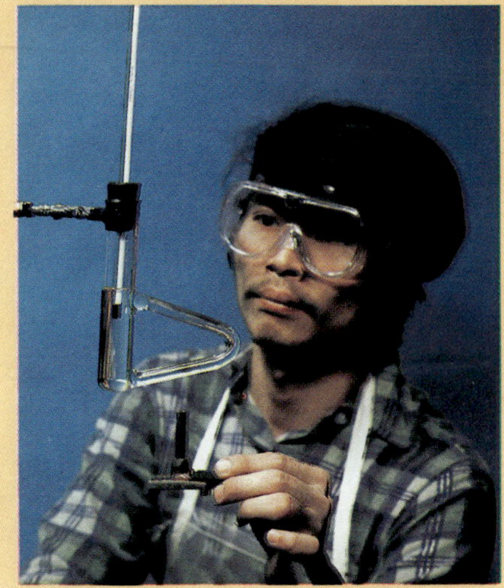

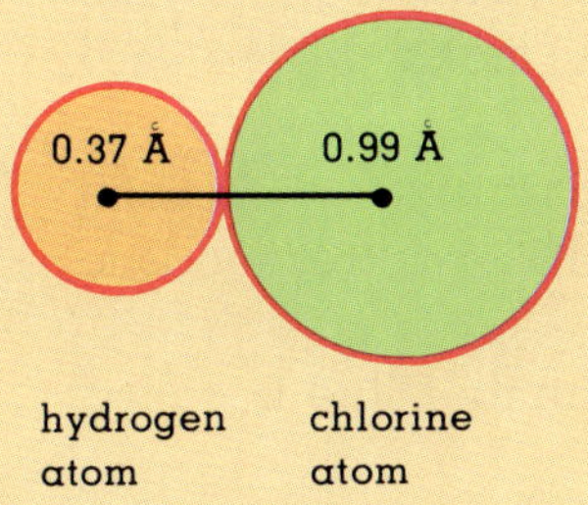

The figure at the left represents an atom of hydrogen and an atom of chlorine. The radius of a hydrogen atom measures 0.37 Å and the radius of a chlorine atom measures 0.99 Å. The bond distance between the two atoms is estimated by adding the two measures. *Å, or angstrom, is a very small unit of metric measure.*

$$\text{bond distance} = \text{radius of hydrogen} + \text{radius of chlorine}$$
$$B = 0.37 + 0.99$$
$$B = 1.36$$

The bond distance between the hydrogen and chlorine atoms is 1.36 Å.

Exercises 1–5. Complete the following chart.

Atom	Radius	Atom	Radius	Bond Distance
boron B	0.88 Å	fluorine F	0.64 Å	
beryllium Be	0.89 Å	chlorine Cl	0.99 Å	
carbon C	0.77 Å	silicon Si	1.17 Å	
aluminum Al	1.25 Å	iodine I	1.33 Å	
nitrogen N	0.70 Å	hydrogen H	0.37 Å	

Vocabulary

number line (34)
natural number (34)
integer (34)
rational number (34)
terminating decimal (35)
repeating decimal (35)
irrational number (36)
real number (36)
coordinate (36)
measure (38)
unit of measure (38)
distance (38)
between (42)
line segment (42)
length (42)
midpoint (42)
congruent segments (51)
bisect (52)

Chapter Summary

1. Counting numbers are called *natural numbers*. Positive counting numbers, negative counting numbers, and zero are called *integers*. Numbers that can be written as the quotient of two integers are called *rational numbers*. Nonterminating, nonrepeating decimals represent *irrational numbers*. Taken together, the rational numbers and irrational numbers make up the *real numbers*. (34)
2. Postulate 2–1: Each real number corresponds to exactly one point on a number line. Each point on a number line corresponds to exactly one real number. (36)
3. A number associated with a point on a number line is called the coordinate of the point. (36)
4. Postulate 2–2: The points on a number line can be paired with the real numbers so that, given any two points P and Q on the line, P corresponds to zero, and Q corresponds to a positive number. (38)
5. Postulate 2–3: For any two points on a line and a given unit of measure, there is a unique positive number called the measure of the distance between the two points. (38)
6. The absolute value of the difference between two numbers represents the measure of the distance between their corresponding points on the number line. (39)
7. Definition of Between: A point Q is between points P and R if and only if the following holds. **1.** P, Q, and R are collinear, and **2.** $PQ + QR = PR$. (42)

8. Definition of Midpoint: A point M is the midpoint of a line segment, $\overline{PQ}$, if and only if M is between P and Q, and $PM = MQ$. (42)
9. Theorem 2–1: A line segment has exactly one midpoint. (43)
10. Properties of equality: (47)

	For any numbers a, b, and c	
Postulate 2–4	reflexive	$a = a$
Postulate 2–5	symmetric	If $a = b$, then $b = a$.
Postulate 2–6	transitive	If $a = b$ and $b = c$, then $a = c$.
Postulate 2–7	addition and subtraction	If $a = b$, then $a + c = b + c$, and $a - c = b - c$.
Postulate 2–8	multiplication and division	If $a = b$, then $a \cdot c = b \cdot c$, and if c is *not* zero, $\frac{a}{c} = \frac{b}{c}$.
Postulate 2–9	substitution	If $a = b$, then a may be replaced by b.

11. Properties of operations: (47)

	For any numbers a, b, and c		
		addition	multiplication
Postulate 2–10	commutative	$a + b = b + a$	$a \cdot b = b \cdot a$
Postulate 2–11	associative	$(a + b) + c =$ $a + (b + c)$	$(a \cdot b) \cdot c =$ $a \cdot (b \cdot c)$
Postulate 2–12	identity	$a + 0 = a = 0 + a$	$a \cdot 1 = a = 1 \cdot a$
Postulate 2–13	inverse	$a + {}^{-}a = 0 = {}^{-}a + a$	If a is *not* zero, then $a \cdot \frac{1}{a} = 1 = \frac{1}{a} \cdot a$.
Postulate 2–14	distributive of multiplication over addition: $a(b + c) = ab + ac$ and $(b + c)a = ba + ca$		

12. Segments having exactly the same length are called congruent segments. (51)
13. Theorem 2–2: Congruence of line segments is reflexive. (51)
14. Theorem 2–3: Congruence of line segments is symmetric. (51)
15. Theorem 2–4: Congruence of line segments is transitive. (51)
16. Theorem 2–5: If M is the midpoint of $\overline{PQ}$, then $\overline{PM} \cong \overline{MQ}$. (51)
17. Theorem 2–6: If $\overline{PQ}$ is bisected at point M, then $\overline{PM} \cong \overline{MQ}$. (52)

18. Properties of inequality: (54)

	For any number a, b, and c	
Postulate 2–15	comparison	$a < b$, or $a = b$, or $a > b$
Postulate 2–16	addition and subtraction	1. If $a > b$, then $a + c > b + c$ and $a - c > b - c$. 2. If $a < b$, then $a + c < b + c$ and $a - c < b - c$.
Postulate 2–17	multiplication and division	1. If $c > 0$ and $a < b$, then $ac < bc$ and $\frac{a}{c} < \frac{b}{c}$. 2. If $c > 0$ and $a > b$, then $ac > bc$ and $\frac{a}{c} > \frac{b}{c}$. 3. If $c < 0$ and $a < b$, then $ac > bc$ and $\frac{a}{c} > \frac{b}{c}$. 4. If $c < 0$ and $a > b$, then $ac < bc$ and $\frac{a}{c} < \frac{b}{c}$.

Chapter Review

2–1 **State the coordinate of each point.**

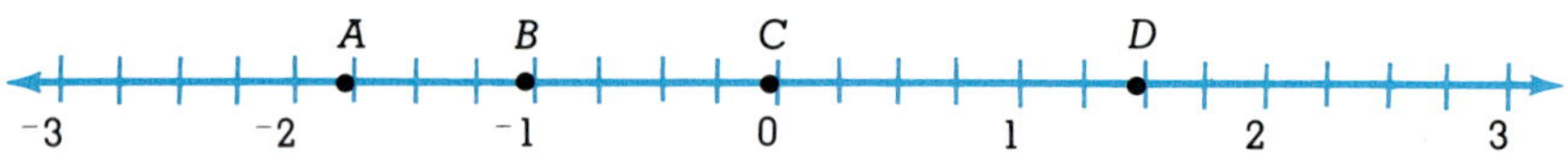

1. *A* **2.** *B* **3.** *C* **4.** *D*

Change each fraction or mixed numeral to decimal form.

5. $\frac{7}{10}$ **6.** $^{-}1\frac{2}{5}$ **7.** $-\frac{2}{3}$ **8.** $2\frac{1}{6}$

Change each decimal to a fraction in simplest form.

9. $0.\overline{3}$ **10.** $1.\overline{4}$ **11.** $0.\overline{12}$ **12.** $6.2\overline{5}$

2–2 **Use the number line to find each measure.**

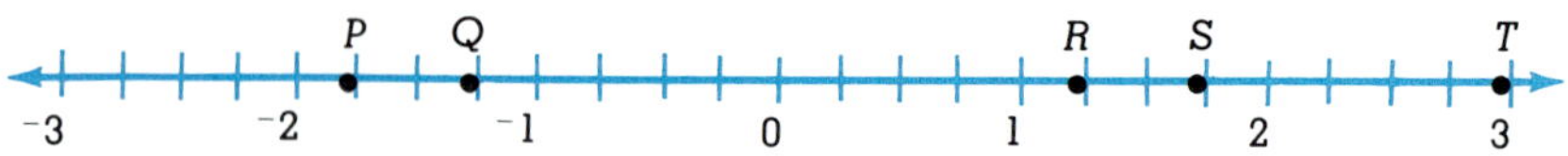

13. *PQ* **14.** *QR* **15.** *PT* **16.** *SQ*
17. *PS* **18.** *QT* **19.** *SR* **20.** *RT*

2–3 **Use the number line to solve each of the following problems.**

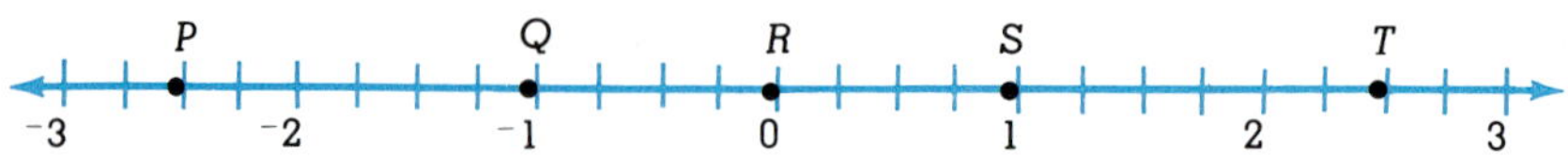

21. Show that *Q* is between *P* and *R*.

22. Show that *Q* is *not* between *R* and *T*.

23. Show that *S* is *not* the midpoint of *Q* and *T*.

24. Show that *R* is the midpoint of *P* and *T*.

2–4 **25.** Complete the reasons for the following proof.

Given: $3(6 - 4x) = 2 + 4x$

Prove: $x = 1$

Proof:

STATEMENTS	REASONS
1. $3(6 - 4x) = 2 + 4x$	1. ________
2. $18 - 12x = 2 + 4x$	2. ________
3. $18 - 12x - 4x = 2 + 4x - 4x$	3. ________
4. $18 - 16x = 2$	4. ________
5. $^{-}18 + 18 - 16x = {}^{-}18 + 2$	5. ________
6. $^{-}16x = {}^{-}16$	6. ________
7. $\frac{^{-}16x}{^{-}16} = \frac{^{-}16}{^{-}16}$	7. ________
8. $x = 1$	8. ________

2–5 **26.** Complete the reasons for the following proof.

Given: M is between P and Q
$\overline{PM} \cong \overline{MQ}$

Prove: M is the midpoint of $\overline{PQ}$

Proof:

STATEMENTS	REASONS
1. M is between P and Q	1. ________
2. $\overline{PM} \cong \overline{MQ}$	2. ________
3. $PM = MQ$	3. ________
4. M is the midpoint of $\overline{PQ}$	4. ________

2–6 **27.** Complete the reasons for the following proof.

Given: $9(x + 2) < 72$

Prove: $x < 6$

Proof:

STATEMENTS	REASONS
1. $9(x + 2) < 72$	1. ________
2. $9x + 18 < 72$	2. ________
3. $9x + 18 - 18 < 72 - 18$	3. ________
4. $9x < 54$	4. ________
5. $\frac{9x}{9} < \frac{54}{9}$	5. ________
6. $x < 6$	6. ________

Chapter Test

Change each fraction or mixed numeral to decimal form.

1. $\frac{3}{100}$ 2. $-\frac{5}{8}$ 3. $^{-}3\frac{1}{3}$

Change each decimal to a fraction in simplest form.

4. $1.\overline{6}$ 5. $0.0\overline{3}$ 6. $1.\overline{45}$

Use the following number line to solve problems 7–16.

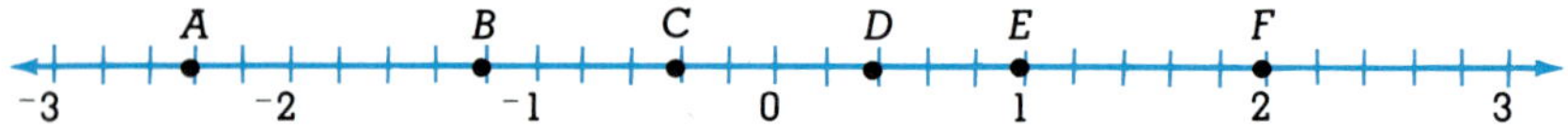

State the coordinate of each point.

7. *B* 8. *C* 9. *D* 10. *F*

Find each measure.

11. *AC* 12. *BD* 13. *CF* 14. *AF*

15. Show that *B* is between *A* and *E*.

16. Show that *C* is the midpoint of $\overline{BD}$.

17. Complete the reasons for the following proof.

Given: $6(3 - 2x) \geq 2 + 4x$

Prove: $x \leq 1$

Proof:

STATEMENTS	REASONS
1. $6(3 - 2x) \geq 2 + 4x$	1. ________________
2. $18 - 12x \geq 2 + 4x$	2. ________________
3. $18 - 12x - 4x \geq 2 + 4x - 4x$	3. ________________
4. $18 - 16x \geq 2$	4. ________________
5. $^{-}18 + 18 - 16x \geq {}^{-}18 + 2$	5. ________________
6. $^{-}16x \geq {}^{-}16$	6. ________________
7. $\frac{^{-}16x}{^{-}16} \leq \frac{^{-}16}{^{-}16}$	7. ________________
8. $x \leq 1$	8. ________________

18. Complete the following proof.

Given: *P*, *M*, and *Q* are collinear
$PM + MQ = PQ$
$\overline{PM} \cong \overline{MQ}$

Prove: *M* is the midpoint of $\overline{PQ}$

Chapter 3

Angles and Perpendiculars

To keep a player from scoring, a hockey goalie tries to reduce the angle from which the player shoots.

3–1 Rays and Angles

A lasar beam looks like a **ray.** A never ending straight path in one direction is represented by a ray.

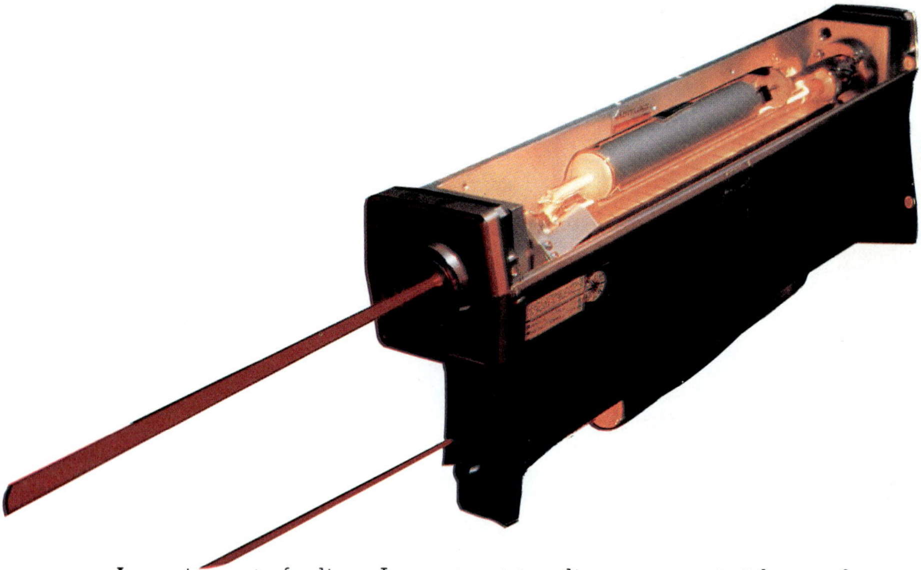

A ray is part of a line. In contrast to a line segment, it has only one endpoint. To name a ray, name the endpoint first. Then name any other point on the ray. $\overrightarrow{PQ}$ is shown below.

Definition of Ray

$\overrightarrow{PQ}$ is a ray if and only if it is the set of points $\overline{PQ}$ and all points S for which Q is between P and S.

Opposite rays have a common endpoint.

Any point on a line determines two rays on that line. The point is the endpoint of the two rays that head in opposite directions.

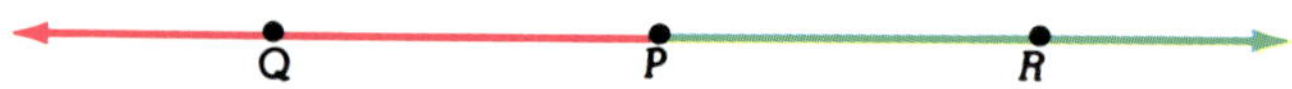

$\overrightarrow{PQ}$ and $\overrightarrow{PR}$ are **opposite rays.**

Definition of Opposite Rays

$\overrightarrow{PQ}$ and $\overrightarrow{PR}$ are opposite rays if and only if P, Q, and R are collinear and P is between Q and R.

An **angle** is a pair of rays that have the same endpoint but do *not* lie on the same line. The rays are called the **sides** of the angle. The common endpoint is called the **vertex** of the angle. A **straight angle** is a pair of opposite rays.

The pictures below can be described using angles.

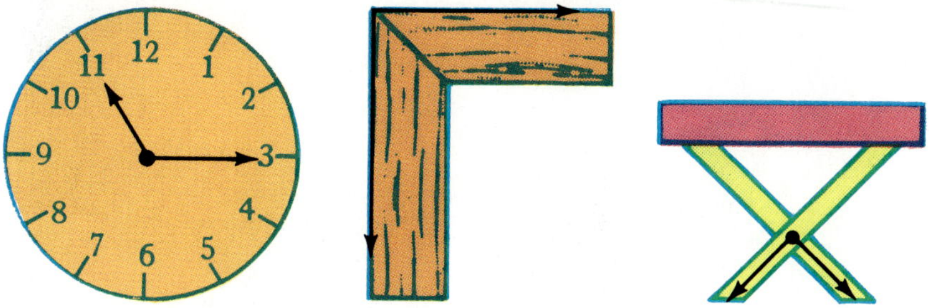

Angles can be named in several ways, with letters or numbers. When three letters are used to name an angle, the letter naming the vertex is between the other two letters.

angle P	angle RPQ	angle 1
$\angle P$	$\angle RPQ$	$\angle 1$

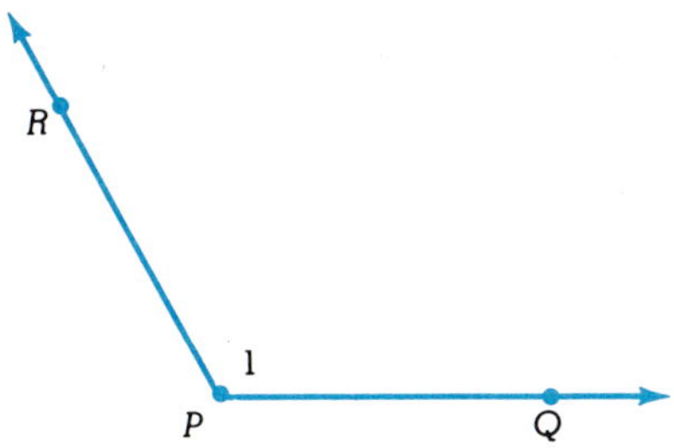

example 1

Draw two angles which have a common side. Label the angles and name the common side.

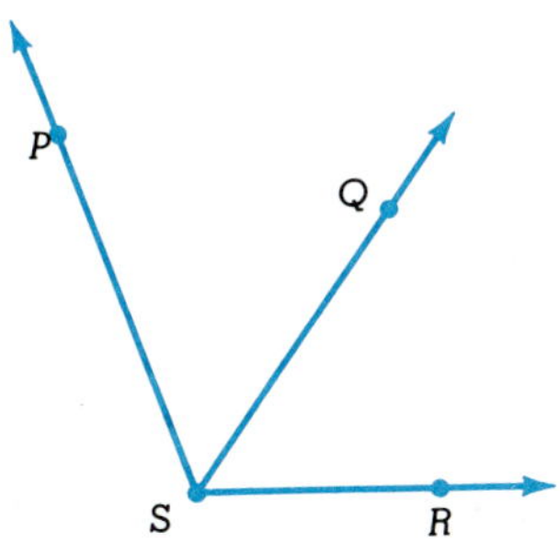

$\angle PSQ$ and $\angle QSR$ have side $\overrightarrow{SQ}$ in common.

Neither the interior nor the exterior of an angle contain the angle.

An angle separates a plane into three parts. The parts are the *interior*, the *exterior*, and the angle itself.

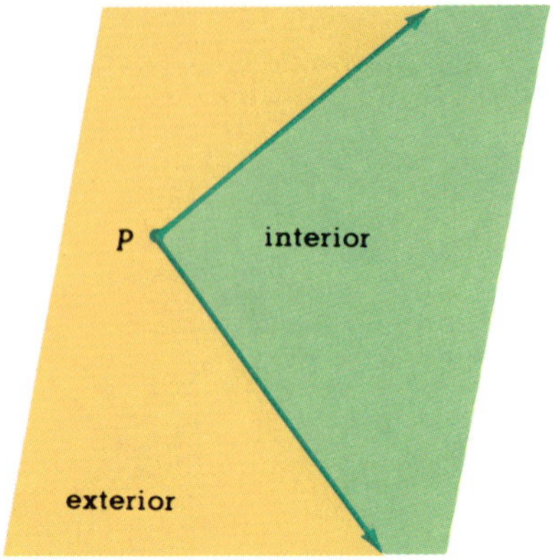

Any point in the green part of the plane is in the interior of $\angle P$. Any point in the yellow part of the plane is in the exterior of $\angle P$.

Any point in the interior of an angle is between two points other than the vertex, one on each side of the angle. In the diagram at the right, R is in the interior of $\angle P$ because R lies between Q and S.

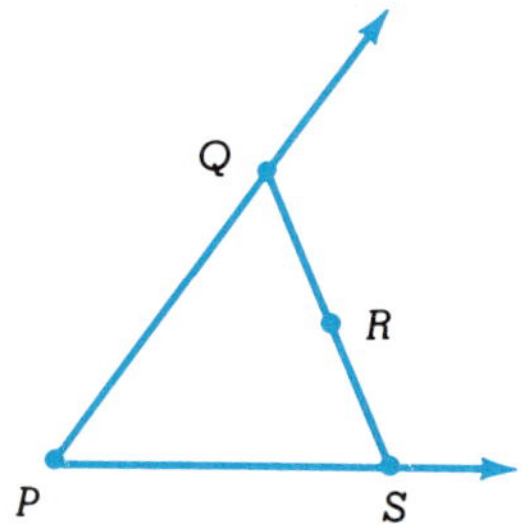

exercises

Exploratory For each of the following statements, write *yes* or *no*.

1. $\overrightarrow{RT}$ and $\overrightarrow{RS}$ name the same ray.
2. $\overrightarrow{QT}$ and $\overrightarrow{TQ}$ name the same ray.
3. $\overrightarrow{PQ}$ is another name for $\overrightarrow{QP}$.
4. $\overrightarrow{SR}$ is another name for $\overrightarrow{SP}$.
5. T is the endpoint of $\overrightarrow{TR}$.
6. S is the endpoint of $\overrightarrow{SV}$.
7. P is the endpoint of $\overrightarrow{QP}$.
8. R is the endpoint of $\overrightarrow{TR}$.
9. $\overrightarrow{QS}$ and $\overrightarrow{SQ}$ have the same endpoint.
10. $\overrightarrow{RT}$ and $\overrightarrow{ST}$ have the same endpoint.
11. R lies on $\overrightarrow{SV}$.
12. V lies on $\overrightarrow{ST}$.
13. P lies on $\overrightarrow{PQ}$.
14. S lies on $\overrightarrow{QS}$.
15. T lies on $\overrightarrow{VS}$.
16. Q lies on $\overrightarrow{PR}$.
17. $\overrightarrow{QR}$ and $\overrightarrow{RQ}$ are opposite rays.
18. $\overrightarrow{RQ}$ and $\overrightarrow{RS}$ are opposite rays.
19. $\overrightarrow{TR}$ and $\overrightarrow{TV}$ are opposite rays.
20. $\overrightarrow{QP}$ and $\overrightarrow{TV}$ are opposite rays.

Written Solve each problem.

1. How many endpoints does a ray have?
2. Which point is the endpoint of $\overrightarrow{PR}$, point P or point R?
3. What is the intersection of two opposite rays?
4. If two rays have the same endpoint, are they opposite rays?
5. Suppose $\overrightarrow{PQ}$ and $\overrightarrow{PR}$ are opposite rays on a number line. The coordinate of P is 0. The coordinate of R is 10. Is the coordinate of Q positive or negative?
6. Suppose $\overrightarrow{AB}$ and $\overrightarrow{AC}$ are opposite rays on a number line. The coordinate of A is 0. The coordinate of C is $^{-}4$. Is the coordinate of B positive or negative?

For each of the following, draw and label two rays $\overrightarrow{PQ}$ and $\overrightarrow{RS}$ so that the given conditions are satisfied.

7. $\overleftrightarrow{PQ}$ and $\overleftrightarrow{RS}$ intersect, but $\overline{PQ}$ and $\overline{RS}$ do *not* intersect.
8. $\overrightarrow{PQ}$ and $\overrightarrow{RS}$ intersect in point P, but $\overline{PQ}$ and $\overline{RS}$ do *not* intersect.
9. $\overrightarrow{PQ}$ and $\overrightarrow{RS}$ intersect in point Q, but $\overline{PQ}$ and $\overline{RS}$ do *not* intersect.
10. P, Q, R, and S are collinear, but $\overrightarrow{PQ}$ and $\overrightarrow{RS}$ do *not* intersect.

Use the figure at the right to solve each of the following.

11. $\angle 1$ and $\angle PQS$ name the same angle. Write *yes* or *no*.
12. $\angle TQV$ and $\angle VQT$ name the same angle. Write *yes* or *no*.
13. Name the vertex of $\angle TQR$.
14. Name the vertex of $\angle 3$.
15. Name the sides of $\angle PQT$.
16. Name the sides of $\angle 2$.

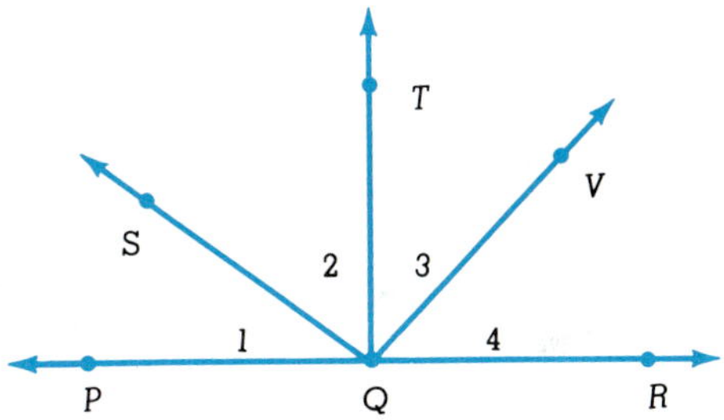

17. Name the common side of $\angle SQT$ and $\angle TQR$.
18. Name the common side of $\angle 1$ and $\angle 2$.
19. Name all angles with $\overrightarrow{QV}$ as a side.
20. Name all angles with $\overrightarrow{QP}$ as a side.
21. Name a point that lies on $\angle 4$.
22. Name a point that lies on $\angle 2$.
23. Name a point interior to $\angle TQR$.
24. Name a point interior to $\angle VQS$.
25. Name a point exterior to $\angle 2$.
26. Name a point exterior to $\angle 1$.

For each of the following, draw two angles so that the given conditions are satisfied.

27. Two angles intersect in a point.
28. Two angles intersect in a line segment.
29. Two angles intersect in a ray.
30. Two angles intersect in two points.
31. Two angles intersect in three points.
32. Two angles intersect in four points.
33. The vertex of one angle is interior to the vertex of the other and vice versa.
34. Two angles have the same vertex.

3–2 Measuring Angles

Angles often are measured in units called **degrees.** The Angle Measure Postulate guarantees a degree measure for every angle.

Postulate 3–1
Angle Measure Postulate

For every angle there is a unique positive number between 0 and 180 called the degree measure of the angle.

A protractor is used to find the *degree measure* of a given angle.

Place the point of the center arrow of the protractor on the vertex of the angle. Line up the 0 mark on the inner scale with one ray of the angle. Read the lower scale where it falls on the other ray of the angle.

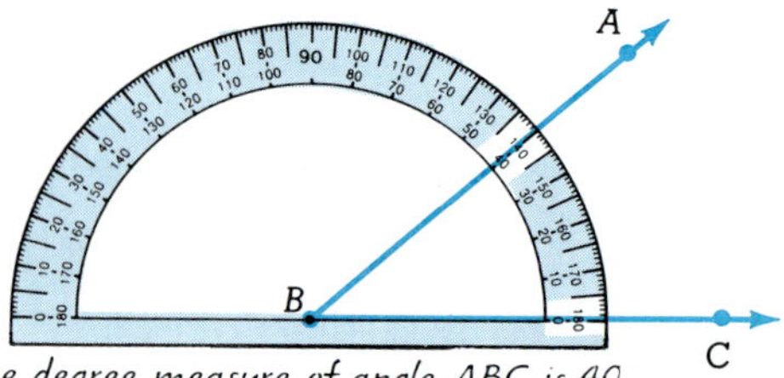

The degree measure of angle ABC is 40.
$m\angle ABC = 40$

$\overleftrightarrow{QT}$ is the edge of each half plane.

In the diagram below, $\overleftrightarrow{QT}$ separates the plane into two **half planes.** The ray, $\overrightarrow{QT}$, is the side of two angles, each measuring 40°. The Protractor Postulate guarantees that in each half plane there is only one 40° angle having $\overrightarrow{QT}$ as a side.

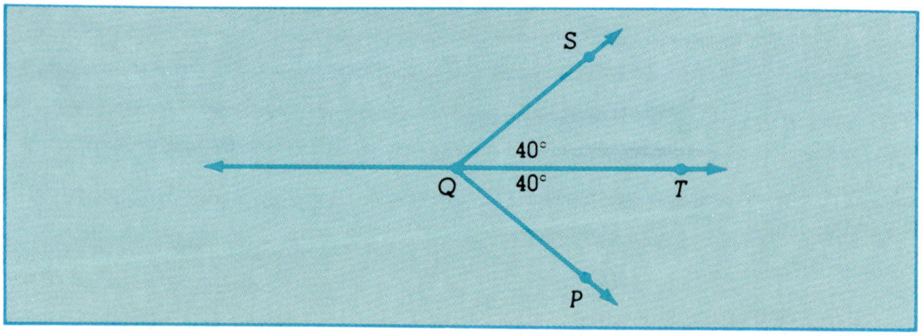

Postulate 3–2
Protractor Postulate

Given a ray QT on the edge of a half plane, for every positive number r between 0 and 180 there is exactly one ray $\overrightarrow{QP}$ in the half plane so that $m\angle PQT = r$.

example 1

Find the angle measurements for each sector of the circle graph below.

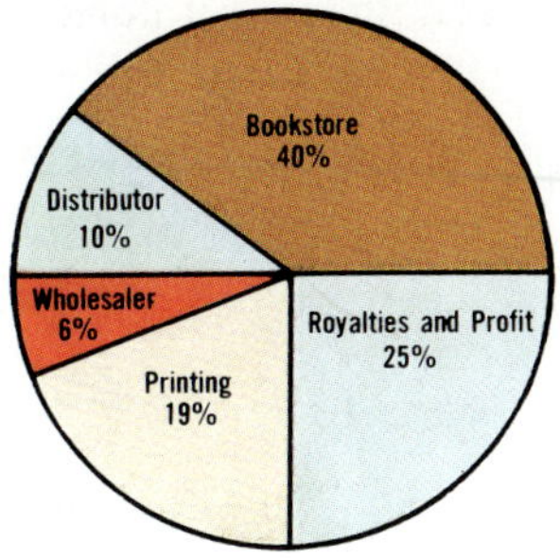

Bookstore	144°
Royalties and Profit	90°
Printing	68°
Wholesaler	22°
Distributor	36°

Measure each angle with a protractor. The center of the circle is the vertex of each angle.

Angles having the same measure are called **congruent angles.** In the diagram below, $\angle PQR$ and $\angle SQV$ are congruent because they both are 50° angles.

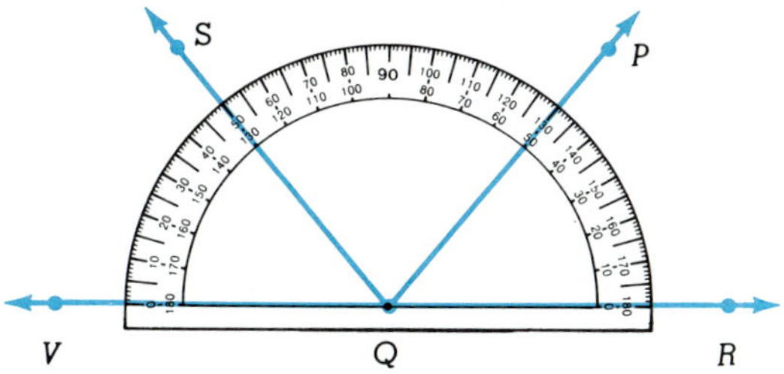

angle PQR is congruent to angle SQV

$\angle PQR \cong \angle SQV$

Congruence of angles is related to equality of numbers. Therefore, the following theorem holds.

Congruence of angles is reflexive, symmetric, and transitive.

Theorem 3–1

The proof of this theorem parallels the proofs of the similar theorems about congruence of line segments.

Two angles are **adjacent** when they have a common side and vertex, but *no* points in the interior of one angle are in the interior of the other. In the diagram below, $\angle PQR$ and $\angle RQS$ are adjacent angles. The angles, $\angle PQS$ and $\angle RQS$, are *not* adjacent angles.

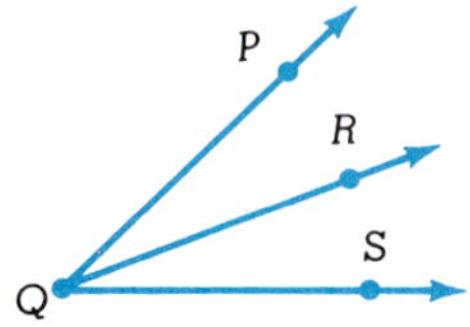

Measures of adjacent angles can be added to find measures of other angles. For example $m\angle DAB$ is 110 and $m\angle DAC + m\angle CAB$ is 80 + 30 or 110.

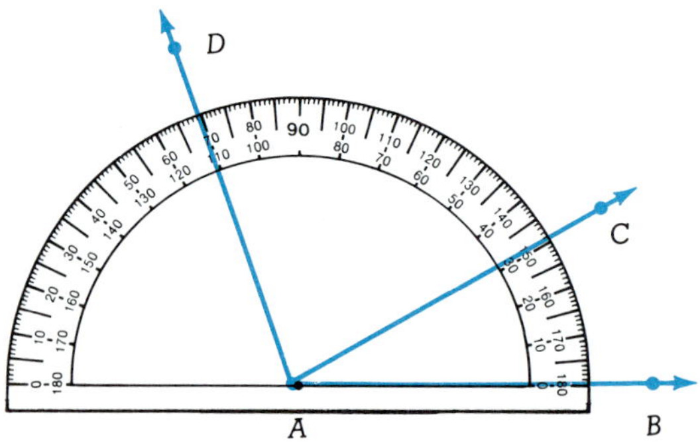

Postulate 3–3
Angle Addition Postulate

If R is in the interior of $\angle PQS$, then $m\angle PQS = m\angle PQR + m\angle RQS$.

example 2

In the figure below, $m\angle PTR = 130$ and $m\angle PTQ = 40$. Find $m\angle QTR$.

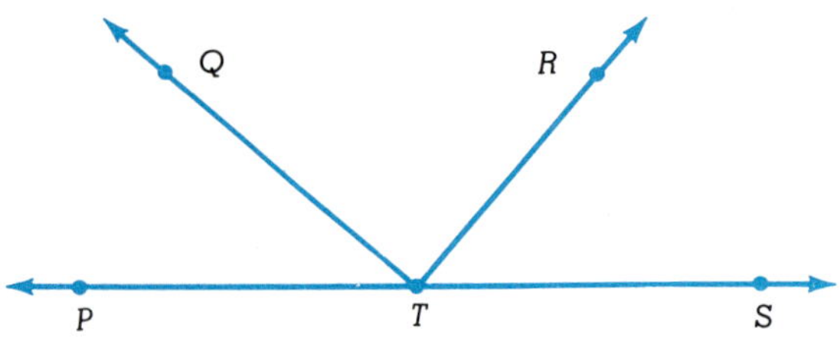

$m\angle PTQ + m\angle QTR = m\angle PTR$	*Angle Addition Postulate*
$40 + m\angle QTR = 130$	*Substitution*
$m\angle QTR = 90$	*Subtraction Property of Equality*

$m\angle QTR$ is 90.

exercises

Exploratory Find the degree measure for each of the following. Use the figure below.

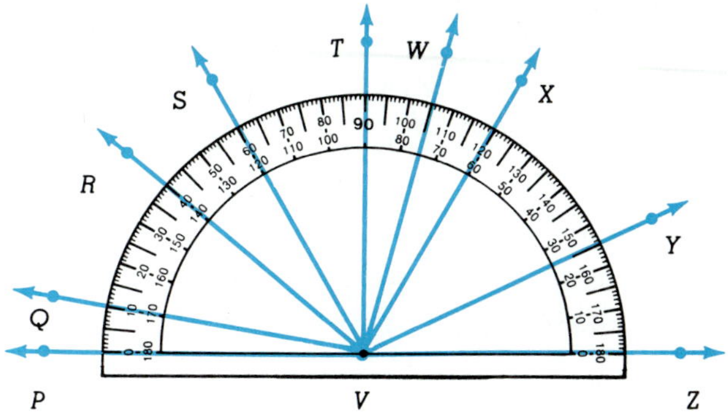

1. $\angle QVZ$	2. $\angle RVZ$	3. $\angle SVZ$	4. $\angle TVZ$
5. $\angle WVZ$	6. $\angle XVZ$	7. $\angle YVZ$	8. $\angle RVY$
9. $\angle QVT$	10. $\angle QVP$	11. $\angle SVP$	12. $\angle SVX$
13. $\angle TVY$	14. $\angle WVX$	15. $\angle RVW$	16. $\angle SVY$

Written Use a protractor to draw angles having the following measure.

1. 45°	2. 60°	3. 144°	4. 135°
5. 75°	6. 29°	7. 179°	8. 120°
9. 57°	10. 85°	11. 163°	12. 170°
13. 15°	14. 67°	15. 25°	16. 99°

Use a protractor to find the angle measurements for the given sectors of each circle graph.

17. **Energy Reserves**

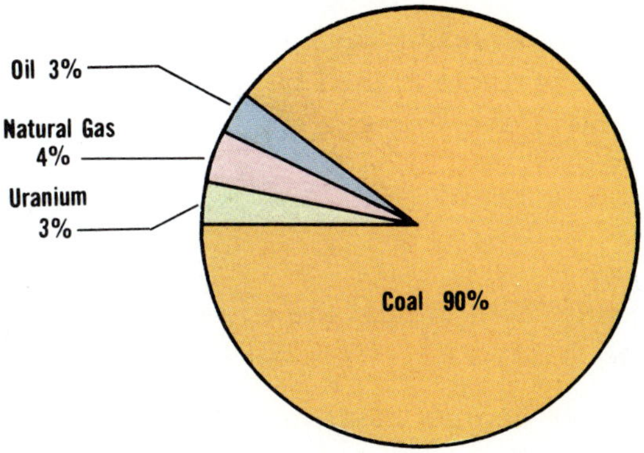

18. **Energy Use**

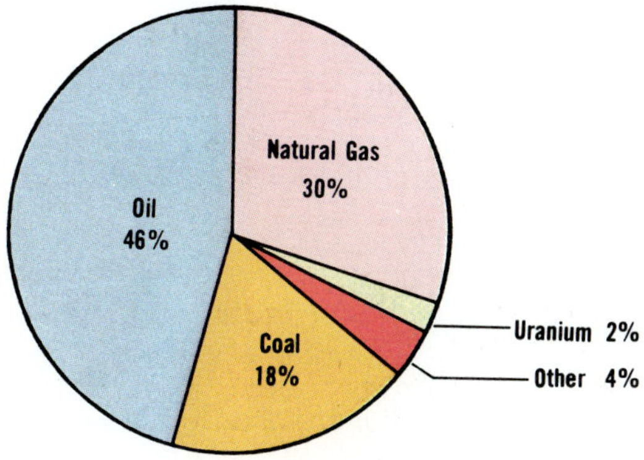

For each of the following write *true* or *false*.

19. $\angle A \cong \angle A$.
20. If $\angle A \cong \angle B$, then $m\angle A = m\angle B$.
21. If $\angle B \cong \angle C$ and $\angle C \cong \angle A$, then $\angle B \cong \angle A$.
22. If $\angle P \cong \angle Q$ and $m\angle P = 30$, then $m\angle Q = 30$.
23. If $m\angle ABC = 50$ and $m\angle CBD = 50$, then $\angle ABC \cong \angle CBD$.
24. Angle congruence is transitive.
25. If $\angle BMA \cong \angle XYZ$, then $\angle XYZ \cong \angle BMA$.
26. Angle congruence is reflexive.
27. If $\angle A \cong \angle P$ and $\angle P \cong \angle Z$ and $m\angle A = 40$, then $m\angle Z = 80$.
28. Angle congruence is *not* symmetric.

29. Prove that angle congruence is reflexive.
30. Prove that angle congruence is symmetric.
31. Prove that angle congruence is transitive.

The sum of the degree measures for $\angle 1$, $\angle 2$, $\angle 3$, $\angle 4$, $\angle 5$, $\angle 6$, and $\angle 7$ is 360. Also, $\angle 1 \cong \angle 5$, $\angle 6 \cong \angle 2$, $\angle 2 \cong \angle 3$. If $m\angle 7 = m\angle 3 + m\angle 4$, $m\angle 3 = 30$, and $m\angle 1 = m\angle 4$, find the degree measures for each of the following angles.

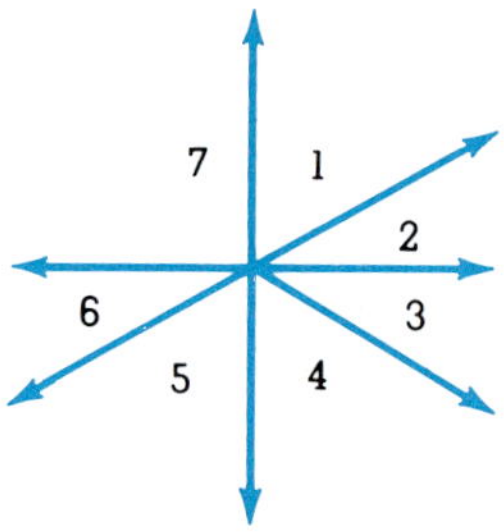

32. $\angle 2$
33. $\angle 6$
34. $\angle 4$
35. $\angle 1$
36. $\angle 5$
37. $\angle 7$

Excursions in Geometry

Radian Measure

One unit of angle measure is the degree. Another common unit of angle measure is the radian. A radian is defined using a circle.

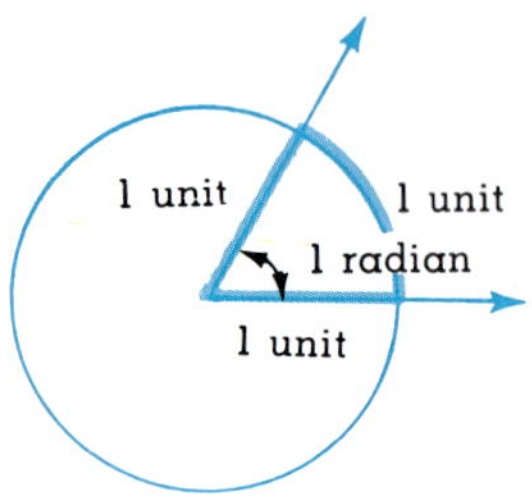

Cut a piece of string that is the same length as the radius of a circle. Stretch the string along the edge of the circle. Mark the beginning point and endpoint of the string on the circle. The angle formed by drawing two rays from the center, one through each point, measures 1 radian.

Exercises If a circle has a 1 unit radius, then the measurement around the edge of the circle is 2π units. Use this information to solve each problem.

1. Find the radian measure of a right angle.
2. If the degree measures of two angles totals 180, find their total radian measure.

3–3 Pairs of Angles

Pairs of angles can be related by the sum of their measures.

Definition of Supplementary Angles

Two angles are supplementary if and only if the sum of their degree measures is 180.

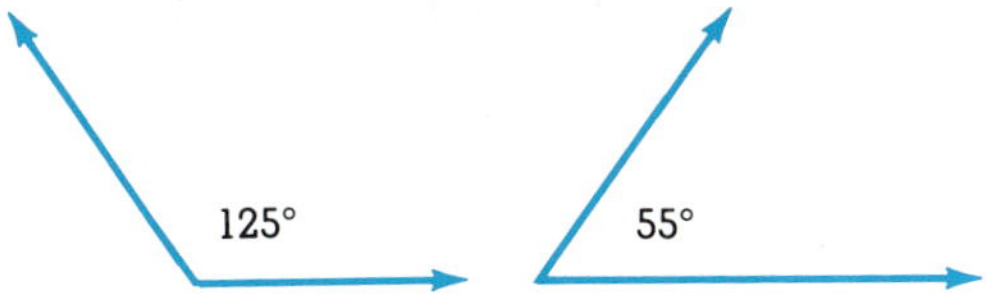

$125 + 55 = 180$

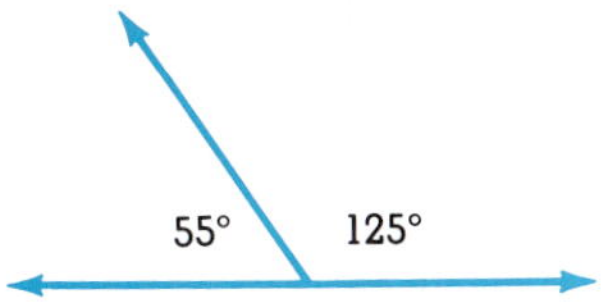

$55 + 125 = 180$

Each angle is called a supplement of the other.

Definition of Complementary Angles

Two angles are complementary if and only if the sum of their degree measures is 90.

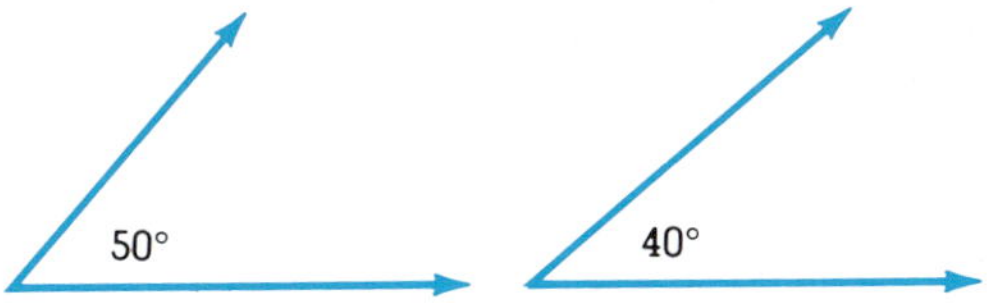

$50 + 40 = 90$

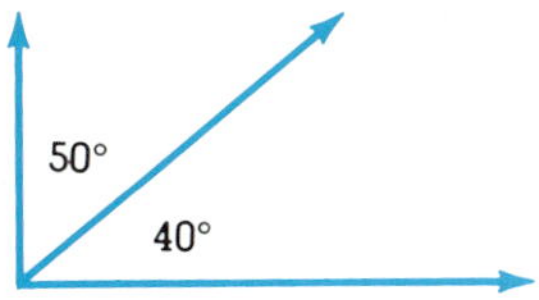

$50 + 40 = 90$

Each angle is called a complement of the other.

If the noncommon sides of adjacent angles together form a straight angle, the angles are called a **linear pair.** In general, all linear pairs are supplementary angles. The sum of their degree measures is 180.

$\angle 1$ and $\angle 2$ are a linear pair because $\overrightarrow{YX}$ and $\overrightarrow{YZ}$ form a straight angle in the figure at the right.

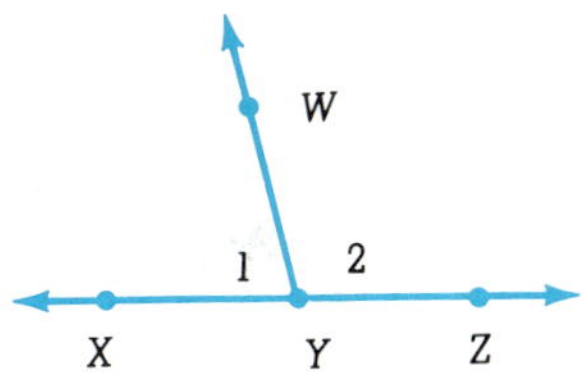

Postulate 3–4 Supplement Postulate

The angles in a linear pair are supplementary angles.

Suppose $\angle A$ is the supplement of $\angle B$. Also, suppose $\angle C$ is the supplement of $\angle B$. If $\angle B$ measures 40, what can you say about the degree measures of $\angle A$ and $\angle C$?

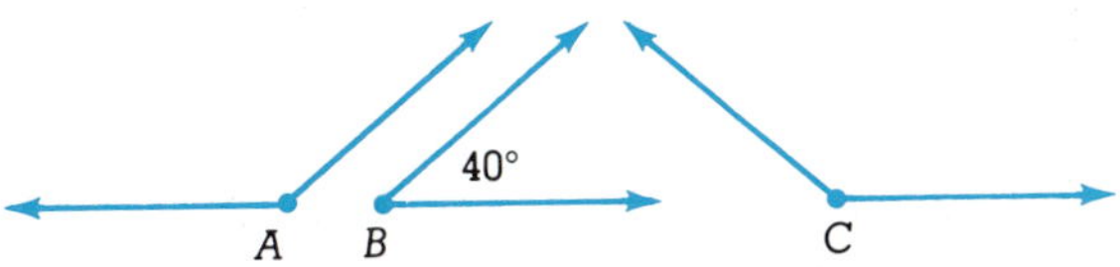

Theorem 3–2

If two angles are supplementary to the same angle, then they are congruent.

examples

1 **Suppose $\angle A$ and $\angle B$ are supplementary to $\angle C$. If $\angle C$ measures 25, find the degree measures of $\angle A$ and $\angle B$.**

$m\angle A + m\angle C = 180$	*$\angle A$ and $\angle C$ are supplementary.*
$m\angle A + 25 = 180$	*Substitute 25 for $m\angle C$.*
$m\angle A = 180 - 25$ or 155	*Subtraction Property of Equality*
$m\angle B + m\angle C = 180$	*$\angle B$ and $\angle C$ are supplementary.*
$m\angle B + 25 = 180$	*Substitute 25 for $m\angle C$.*
$m\angle B = 180 - 25$ or 155	*Subtraction Property of Equality*

Both $\angle A$ and $\angle B$ measure 155.

2 **Prove Theorem 3–2.**

Given: $\angle A$ is the supplement of $\angle B$
$\angle C$ is the supplement of $\angle B$

Prove: $\angle A \cong \angle C$

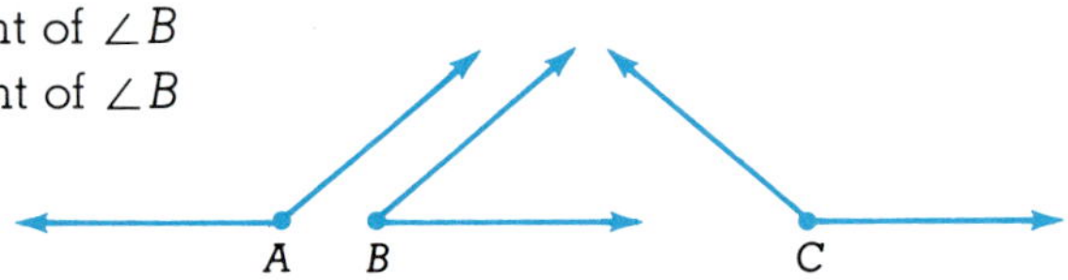

Proof:

STATEMENTS	REASONS
1. $\angle A$ is the supplement of $\angle B$ $\angle C$ is the supplement of $\angle B$	1. Given
2. $m\angle A + m\angle B = 180$ $m\angle C + m\angle B = 180$	2. Definition of supplementary angles
3. $m\angle A + m\angle B =$ $m\angle C + m\angle B$	3. Postulate 2–9: Substitution
4. $m\angle A = m\angle C$	4. Postulate 2–7: Subtraction Property of Equality
5. $\angle A \cong \angle C$	5. Definition of congruent angles

Suppose two angles are supplementary to congruent angles. The diagram below shows two linear pairs. The 60-degree angles are congruent. What can you say about the other two angles?

If two angles are supplementary to two congruent angles, then the two angles are congruent to each other.

Theorem 3–3

example 3

Prove Theorem 3–3.

Given: $\angle A$ is the supplement of $\angle B$
$\angle C$ is the supplement of $\angle D$
$\angle B \cong \angle D$

Prove: $\angle A \cong \angle C$

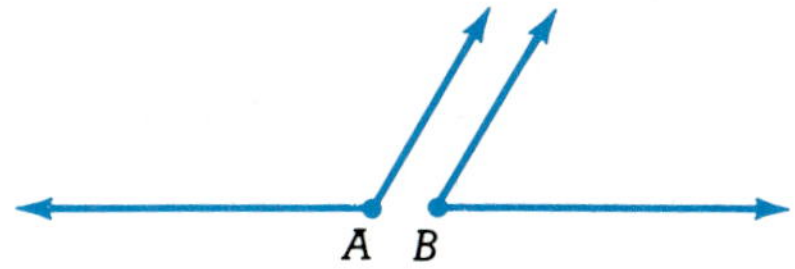

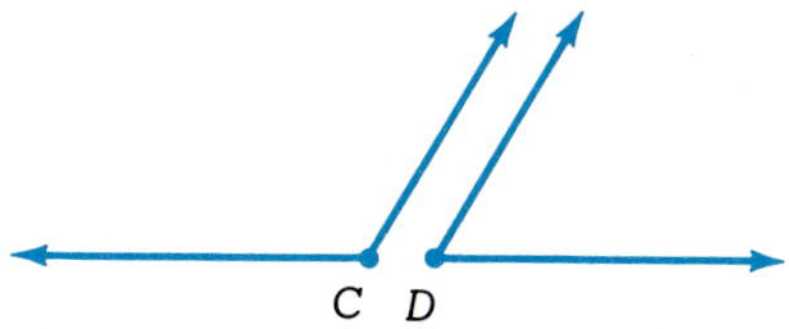

Proof:

STATEMENTS	REASONS
1. $\angle A$ is the supplement of $\angle B$ $\angle C$ is the supplement of $\angle D$	1. Given
2. $m\angle A + m\angle B = 180$ $m\angle C + m\angle D = 180$	2. Definition of supplementary angles
3. $m\angle A + m\angle B =$ $m\angle C + m\angle D$	3. Postulate 2–9: Substitution
4. $\angle B \cong \angle D$	4. Given
5. $m\angle B = m\angle D$	5. Definition of congruent angles
6. $m\angle A + m\angle B =$ $m\angle C + m\angle B$	6. Postulate 2–9: Substitution *Steps 3 and 5*
7. $m\angle A = m\angle C$	7. Postulate 2–7: Subtraction Property of Equality
8. $\angle A \cong \angle C$	8. Definition of congruent angles

Theorems similar to those about supplementary angles can be proved for complementary angles.

If two angles are complementary to the same angle, then they are congruent to each other.

Theorem 3–4

If two angles are complementary to two congruent angles, then the two angles are congruent to each other.

Theorem 3–5

exercises

Exploratory Use the following theorem to solve problems 1–8.

If two sides of adjacent angles lie in a line, then the angles are supplementary.

1. State the hypothesis of the theorem.
2. Draw and label a diagram to illustrate the hypothesis of the theorem.
3. State the *given* that goes with your diagram for a proof of the theorem.
4. State the conclusion of the theorem.
5. State the *prove* that goes with your diagram for a proof of the theorem.
6. Give another name for two adjacent angles that lie in a line.
7. State the theorem or postulate from this lesson that is about adjacent angles and lines.
8. Give a plan for the proof of the theorem.

Written **In the diagram at the right the following pairs of angles are complementary, $\angle 1$ and $\angle 2$, $\angle 3$ and $\angle 4$, $\angle 5$ and $\angle 6$, $\angle 7$ and $\angle 8$. If $\angle 1$ measures 45° and $\angle 3$ measures 30°, $\angle 5 \cong \angle 6$, and $\angle 8$ measures 10°, find the degree measures for each of the following.**

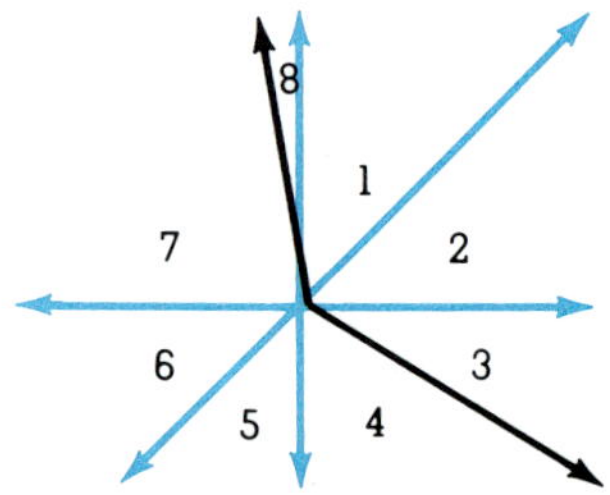

1. $\angle 2$
2. $\angle 4$
3. $\angle 7$
4. $\angle 6$
5. $\angle 5$

In the diagram at the right, $\angle QVR \cong \angle RVS$. Also, $m\angle PVQ = 72$ and $m\angle TVS = 70$. P, V, and T are collinear.

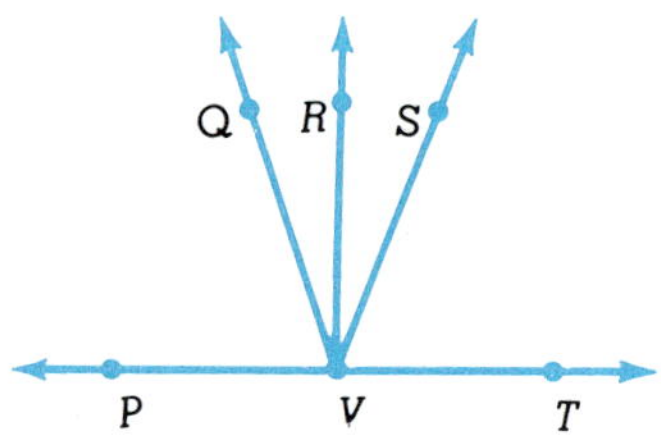

6. Find $m\angle QVS$.
7. Find $m\angle QVR$.
8. Find $m\angle RVS$.
9. Find $m\angle PVR$.
10. Find $m\angle TVR$.
11. Find $m\angle PVS$.
12. Find $m\angle TVQ$.

13–19. Solve problems **6–12** if $m\angle PVQ = 65$ and $m\angle TVS = 60$.

20. Complete the statements and reasons for the following.

Theorem: If two angles are congruent and supplementary, then they each measure 90.

Given: $\angle 1 \cong \angle 2$
$\angle 1$ and $\angle 2$ are supplementary

Prove: $\angle 1$ and $\angle 2$ each measure 90

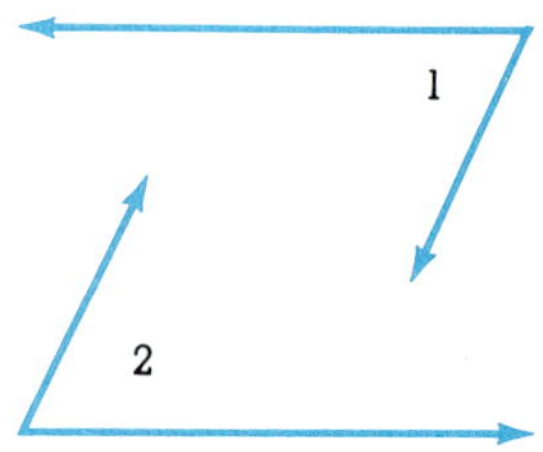

21. Prove Theorem 3–4.
22. Prove Theorem 3–5.

Flow Proofs

One way to present a proof is to write a **flow proof.** The following example gives a flow proof for Theorem 3–2.

Given: $\angle A$ is the supplement of $\angle B$
$\angle C$ is the supplement of $\angle B$

Prove: $\angle A \cong \angle C$

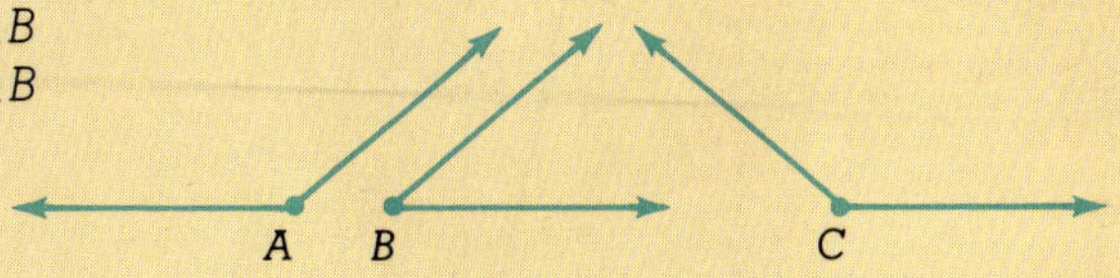

Flow Proof:

$\left.\begin{array}{l}\angle A \text{ is supplement of } \angle B \xrightarrow{1} m\angle A + m\angle B = 180 \\ \angle C \text{ is supplement of } \angle B \xrightarrow{1} m\angle C + m\angle B = 180\end{array}\right\} \xrightarrow{2} m\angle A + m\angle B = m\angle C + m\angle B \xrightarrow{3}$

$\xrightarrow{3} m\angle A = m\angle C \xrightarrow{4} \angle A \cong \angle C$

1. Definition of supplementary angles
2. Substitution
3. Subtraction Property of Equality
4. Definition of congruent angles

A flow proof organizes a series of statements in logical order, starting with given statements. The arrows show the order the statements should follow.

The numbers above the arrows refer to the reasons that allow the statements to be made. The reasons are written below the proof. For example, if $\angle A$ is the supplement of $\angle B$, then the definition of supplementary angles implies that $m\angle A + m\angle B = 180$.

Given: $\angle A$ is the supplement of $\angle B$, $\angle B \cong \angle D$
$\angle C$ is the supplement of $\angle D$

Prove: $\angle A \cong \angle C$

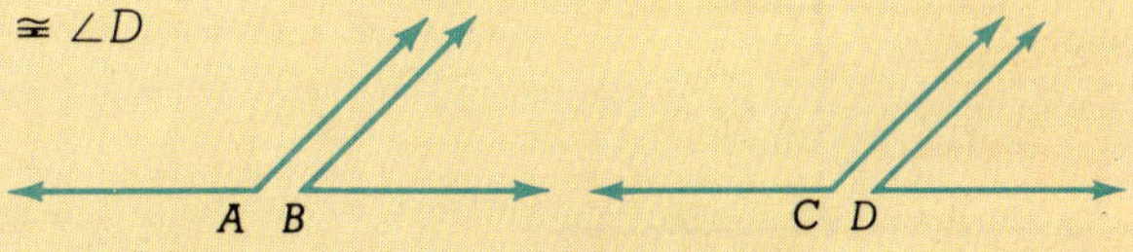

Flow Proof:

$\left.\begin{array}{l}\left.\begin{array}{l}\angle A \text{ is supplement of } \angle B \xrightarrow{1} m\angle A + m\angle B = 180 \\ \angle C \text{ is supplement of } \angle D \xrightarrow{1} m\angle C + m\angle D = 180\end{array}\right\} \xrightarrow{2} m\angle A + m\angle B = m\angle C + m\angle D \\ \qquad\qquad\qquad\qquad \angle B \cong \angle D \xrightarrow{3} m\angle B = m\angle D\end{array}\right\} \xrightarrow{4}$

$\xrightarrow{4} m\angle A + m\angle B = m\angle C + m\angle B \xrightarrow{5} m\angle A = m\angle C \xrightarrow{6} \angle A \cong \angle C$

1. Draw the symbol that is used to show that one statement follows from another.
2. Draw the symbol used to show that two or more statements imply another.
3. Write the statements that imply $m\angle A + m\angle B = m\angle C + m\angle B$.
4. Write the reason for arrow 1.
5. Write the reason for arrow 2.
6. Write the reason for arrow 3.
7. Write the reason for arrow 4.
8. Write the reason for arrow 5.
9. Write the reason for arrow 6.
10. Write a flow proof for Theorem 3–4.

3–4 Right Angles

The symbol ⌝ is used to indicate right angles.

Angles can be classified according to their measures.

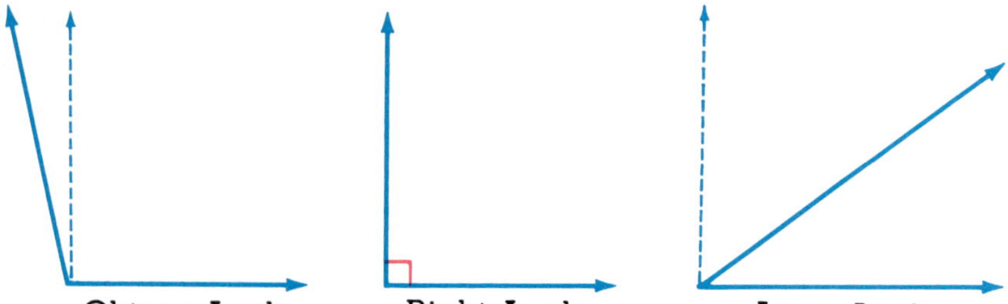

Definition of Right, Acute, and Obtuse Angles

A right angle is an angle whose degree measure is 90. An acute angle is one whose degree measure is less than 90. An obtuse angle is one whose degree measure is greater than 90.

All right angles are congruent because they have the same measures.

Theorem 3–6

If two angles are right angles, then the angles are congruent.

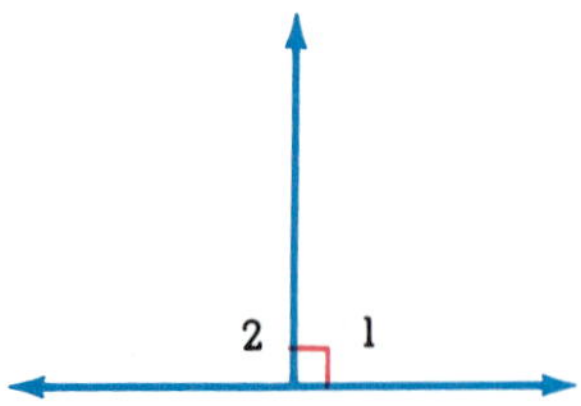

Suppose two angles form a linear pair and one of them is a right angle. In the figure at the left, $\angle 1$ is a right angle.

$m\angle 1 + m\angle 2 = 180$	*The angles in a linear pair are supplementary.*
$m\angle 1 = 90$	*$\angle 1$ is a right angle.*
$90 + m\angle 2 = 180$	*Substitution*
$m\angle 2 = 90$	*Subtraction Property of Equality*

This deduction shows that $\angle 2$ also is a right angle.

Theorem 3–7

If one angle in a linear pair is a right angle, then the other angle also is a right angle.

To prove the following theorem, use substitution and the fact that congruent angles have the same measure.

Theorem 3–8

If two angles are congruent and supplementary, then each is a right angle.

The proof of the following theorem is based on Theorem 3–7.

If two intersecting lines form one right angle, then they form four right angles.

Theorem 3–9

example 1

Prove Theorem 3–9.

Given: Two intersecting lines forming ∠1, ∠2, ∠3, ∠4.
∠1 is a right angle.

Prove: ∠2, ∠3, ∠4 are right angles

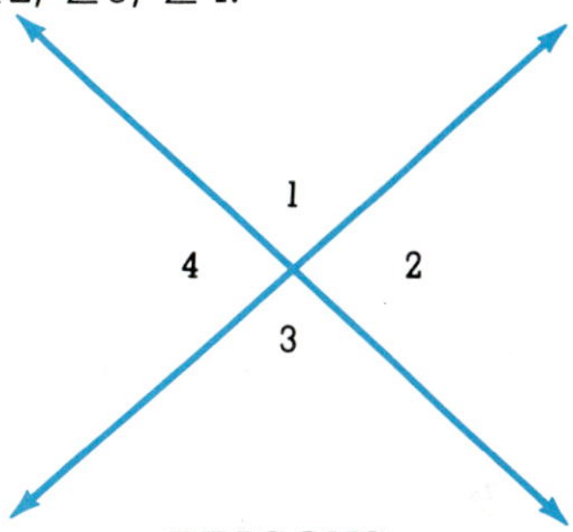

Proof:

STATEMENTS	REASONS
1. Two intersecting lines forming ∠1, ∠2, ∠3, ∠4	1. Given
2. ∠1 and ∠2 form a linear pair ∠1 and ∠4 form a linear pair	2. Definition of linear pair
3. ∠1 is a right angle	3. Given
4. ∠2 is a right angle ∠4 is a right angle	4. Theorem 3–7: If one angle in a linear pair is a right angle, then the other angle also is a right angle.
5. ∠3 and ∠4 form a linear pair	5. Definition of linear pair
6. ∠3 is a right angle	6. Theorem 3–7

Vertical angles are two nonadjacent angles formed by two intersecting lines. In the figure, the angles with yellow interiors are vertical angles. The angles with orange interiors are vertical angles. Notice that an angle with a yellow interior and an angle with an orange interior form a linear pair.

Vertical angles are congruent.

Theorem 3–10

exercises

Exploratory Classify angles with each of the following degree measures as either *acute*, *right*, or *obtuse*.

1. 61	2. 174	3. 3
4. 140	5. 96	6. 31
7. 90	8. 105	9. 89
10. 13	11. 166	12. 94

State the *hypothesis* for a proof of each of the following theorems.

13. An angle with degree measure 90 and a right angle are congruent.
14. If two angles are vertical angles, then the angles are congruent.
15. If two angles are right angles, then the angles are supplementary.
16. If two angles are congruent and supplementary, then each is a right angle.
17. If one angle in a linear pair is a right angle, then the other angle also is a right angle.
18. If two angles are right angles, then the angles are congruent.

19–24. State the *conclusion* for a proof of each of the theorems in problems **13–18**.

Written Solve each problem.

1. Complete the reasons for the following proof.

Theorem: Vertical angles are congruent.

Given: $\angle 1$ and $\angle 2$ are vertical angles

Prove: $\angle 1 \cong \angle 2$

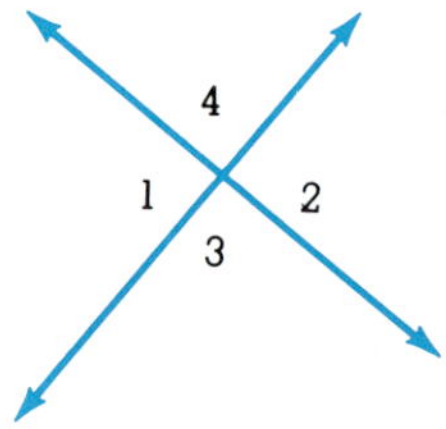

Proof:

STATEMENTS	REASONS
1. $\angle 1$ and $\angle 2$ are vertical angles.	1. ______________
2. $\angle 1$ and $\angle 4$ form a linear pair $\angle 2$ and $\angle 4$ form a linear pair	2. ______________
3. $\angle 1$ and $\angle 4$ are supplementary $\angle 2$ and $\angle 4$ are supplementary	3. ______________
4. $\angle 1 \cong \angle 2$	4. ______________

2. Complete the reasons for the following proof.

Theorem: If two angles are congruent and supplementary, then each is a right angle.

Given: $\angle 1 \cong \angle 2$
$\angle 1$ and $\angle 2$ are supplementary

Prove: $\angle 1$ is a right angle
$\angle 2$ is a right angle

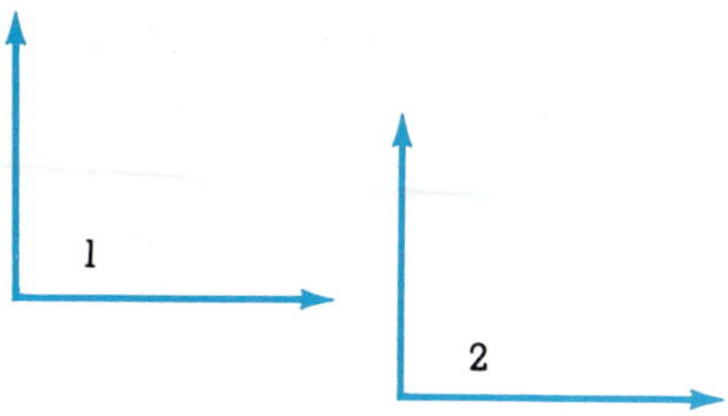

Proof:

STATEMENTS	REASONS
1. $\angle 1$ and $\angle 2$ are supplementary	1. ________________
2. $m\angle 1 + m\angle 2 = 180$	2. ________________
3. $\angle 1 \cong \angle 2$	3. ________________
4. $m\angle 1 = m\angle 2$	4. ________________
5. $m\angle 1 + m\angle 1 = 180$	5. ________________
6. $2\,m\angle 1 = 180$	6. ________________
7. $m\angle 1 = 90$	7. ________________
8. $m\angle 2 = 90$	8. ________________
9. $\angle 1$ is a right angle $\angle 2$ is a right angle	9. ________________

Prove each of the following theorems.

3. An angle with degree measure 90 and a right angle are congruent.
5. If one angle in a linear pair is a right angle, then the other angle also is a right angle.
4. If two angles are right angles, then the angles are supplementary.
6. If two angles are right angles, then the angles are congruent.

Excursions in Geometry

Puzzle

The diagram at the right shows four pieces of chain with three links each. Explain how to construct a necklace by breaking and joining only three links. The necklace must contain all twelve links.

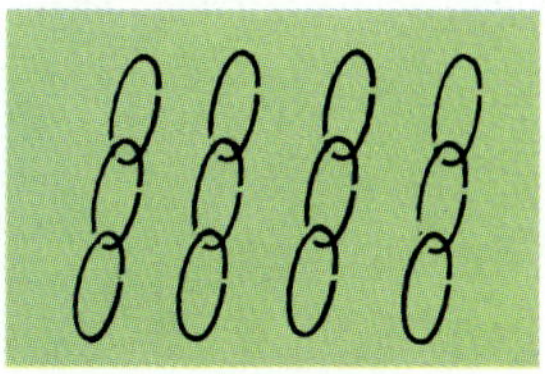

3–5 Perpendiculars

The weather vane in the photograph provides a clear example of **perpendicular lines.** The north-south line is perpendicular to the east-west line.

Definition of Perpendicular Lines

Two lines are perpendicular if and only if they intersect to form a right angle.

The symbol ⊥ means *is perpendicular to*. In the figure below, the two lines are perpendicular.

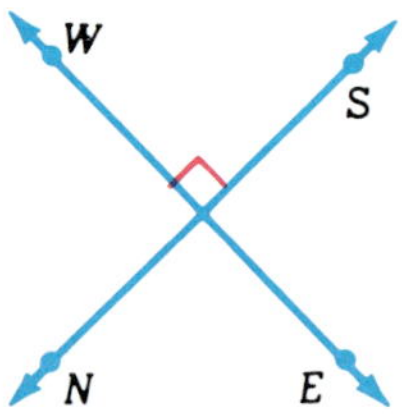

$\overleftrightarrow{NS} \perp \overleftrightarrow{EW}$

$\overleftrightarrow{NS}$ is perpendicular to $\overleftrightarrow{EW}$

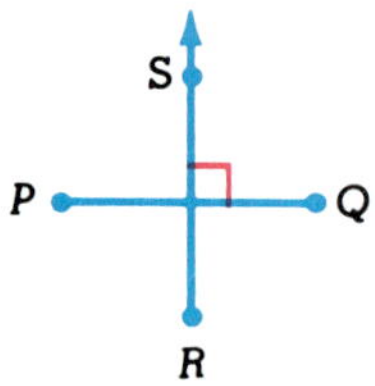

Parts of lines are perpendicular to each other if the lines containing them are perpendicular. For example, a ray can be perpendicular to a line segment. In the figure at the left, $\overrightarrow{RS} \perp \overline{PQ}$.

The same term, perpendicular, and the same notation, ⊥, is used for lines, segments, and rays.

Recall that when two intersecting lines form one right angle, they form four right angles. This information leads to the following theorem.

Theorem 3–11

If two lines are perpendicular, then they form four right angles.

A given line may have many lines perpendicular to it. In a plane, through any given point on the line, there is *exactly one* line perpendicular to the given line. In space, there may be infinitely many lines perpendicular to the line and through the given point. *There is one line perpendicular for each plane that contains the line.*

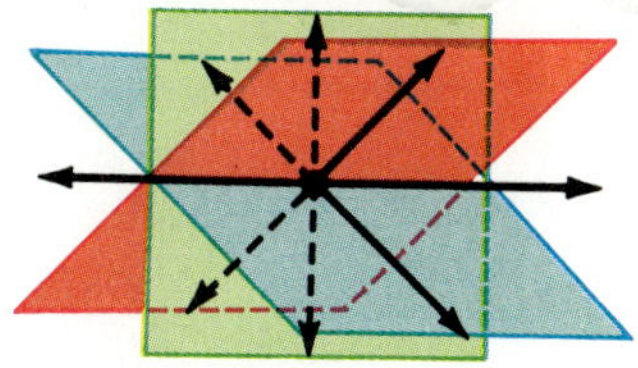

In a plane, through a given point on a line, there is exactly one line perpendicular to the given line. Theorem 3–12

Theorems 3–13 and 3–14 are converses of one another. Thus, one test for perpendicular lines is congruent adjacent angles.

If two lines are perpendicular, then they form congruent adjacent angles. Theorem 3–13

If two intersecting lines form congruent adjacent angles, then they are perpendicular. Theorem 3–14

example 1

Prove Theorem 3–13.

Given: $\overleftrightarrow{PQ} \perp \overleftrightarrow{QR}$

Prove: $\angle PQR \cong \angle RQS$
$\angle PQR \cong \angle PQT$
$\angle TQS \cong \angle RQS$
$\angle TQS \cong \angle PQT$

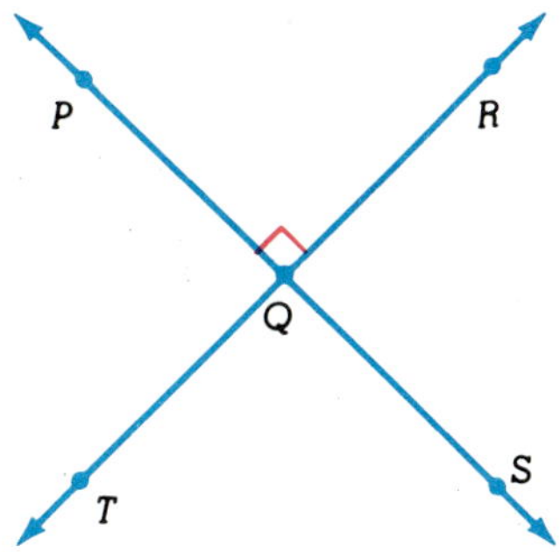

Proof:

STATEMENTS	REASONS
1. $\overleftrightarrow{PQ} \perp \overleftrightarrow{QR}$	1. Given
2. $\overleftrightarrow{PQ}$ and $\overleftrightarrow{QR}$ form a right angle	2. Definition of perpendicular lines
3. $\overleftrightarrow{PQ}$ and $\overleftrightarrow{QR}$ form four right angles *$\angle PQR$, $\angle RQS$, $\angle SQT$, and $\angle TQP$*	3. Theorem 3–9: If two intersecting lines form one right angle, then they form four right angles.
4. $\angle PQR \cong \angle RQS$ $\angle PQR \cong \angle PQT$ $\angle TQS \cong \angle RQS$ $\angle TQS \cong \angle PQT$	4. Theorem 3–6: If two angles are right angles, then the angles are congruent.

exercises

Exploratory In the diagram at the right $\overleftrightarrow{AB} \perp \overleftrightarrow{FE}$, $\overleftrightarrow{AE} \perp \overleftrightarrow{GC}$, and C is the midpoint of $\overline{AE}$. For each of the following statements, write *yes* or *no*.

1. $\overrightarrow{BV} \perp \overrightarrow{VB}$
2. $\overrightarrow{GV} \perp \overleftrightarrow{AE}$
3. $m\angle 1 + m\angle 2 = 90$
4. $\overline{AC} \cong \overline{CE}$
5. $\overline{AC} \perp \overleftrightarrow{GC}$
6. $m\angle BVE = 90$
7. $\angle 3 \cong \angle 4$
8. $m\angle 3 + m\angle 4 = 180$
9. $m\angle AVF = 90$
10. $\overleftrightarrow{VB} \perp \overline{VA}$
11. $m\angle 4 = m\angle 1 + m\angle 2$
12. $\angle 2$ and $\angle 6$ are complementary
13. $\angle FVB$ and $\angle 4$ are complementary
14. $\angle GVA$ is a right angle
15. $\angle 6 \cong \angle 1$
16. $\overrightarrow{AE} \perp \overleftrightarrow{FV}$
17. $\overline{VC} \cong \overline{AC}$
18. $\angle 4 \cong \angle 1$
19. $\angle 6$ and $\angle 3$ are supplementary
20. $\angle AVE$ and $\angle BVF$ are supplementary

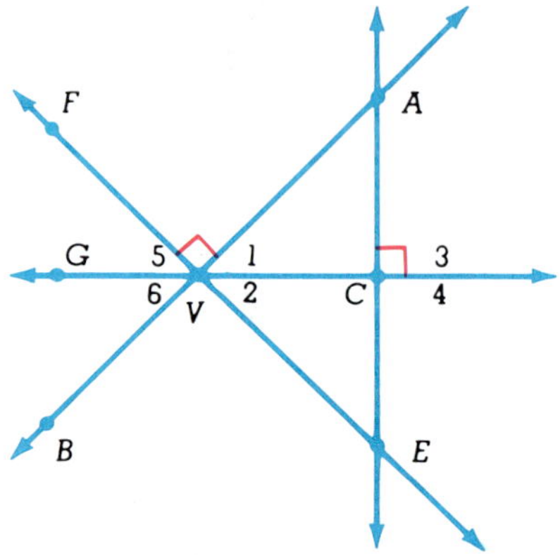

Written 1–9. Complete the reasons for the following proof.

Theorem: If two intersecting lines form two congruent adjacent angles, then they are perpendicular.

Given: ℓ and m intersect
$\angle 1 \cong \angle 2$

Prove: $\ell \perp m$

Proof:

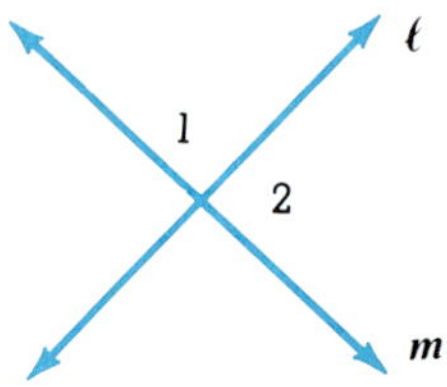

STATEMENTS	REASONS
1. $\angle 1$ and $\angle 2$ form a linear pair	1. ______
2. $\angle 1$ and $\angle 2$ are supplementary	2. ______
3. $m\angle 1 + m\angle 2 = 180$	3. ______
4. $\angle 1 \cong \angle 2$	4. ______
5. $m\angle 1 = m\angle 2$	5. ______
6. $m\angle 1 + m\angle 1 = 180$	6. ______
7. $2\,m\angle 1 = 180$	7. ______
8. $m\angle 1 = 90$, $\angle 1$ is a right angle	8. ______
9. $\ell \perp m$	9. ______

10. Prove Theorem 3–11.

11. Prove that if the sides of two adjacent acute angles are perpendicular, then the angles are complementary.

Using Money

Dial-direct Calls

The cost of a long distance phone call depends on the area being called, the type of call being made, and the time of day the call is made.

The following chart shows some sample rates for dial-direct calls. These are calls made without the help of an operator.

Sample Rates From City of Columbus to:	8:00 A.M.–5:00 P.M. WEEKDAY FULL RATE First Minute	Each Additional Minute	5:00 P.M.–11:00 P.M. EVENING 35% DISCOUNT First Minute	Each Additional Minute	11:00 P.M.–8:00 A.M. NIGHT & WEEKEND 50% DISCOUNT First Minute	Each Additional Minute
Atlanta, Ga.	0.50	0.34	0.32	0.23	0.20	0.14
Boston, Mass.	0.50	0.34	0.32	0.23	0.20	0.14
Chicago, Ill.	0.46	0.32	0.29	0.21	0.18	0.13
Denver, Colo.	0.52	0.36	0.33	0.24	0.20	0.15
Detroit, Mich.	0.44	0.30	0.28	0.20	0.17	0.12
Houston, Tex.	0.52	0.36	0.33	0.24	0.20	0.15
Philadelphia, Pa.	0.48	0.34	0.31	0.23	0.19	0.14
St. Louis, Mo.	0.48	0.34	0.31	0.23	0.19	0.14
Seattle, Wash.	0.54	0.38	0.35	0.25	0.21	0.16

On dial-direct calls, you pay for the minutes you talk. There is a charge for the first minute you talk, and a charge for each minute thereafter. The rates are less during the evening hours, and even less at night and on weekends. Holiday calls are billed at the night rate.

Suppose a dial-direct call is made from Columbus to Denver at 8:00 P.M. for 15 minutes. What is the cost of the call?

The charge is $0.33 for the first minute,
and $0.24 for each of the next fourteen minutes.

$$0.33 + 14(0.24) = 3.69$$

The call cost $3.69.

There is a federal tax on phone calls. The tax rate is 3% of the amount of the call. Thus, the Denver call actually cost $3.69 + $0.11 or $3.80.

Exercises **1–6.** Complete each amount for the phone bill.

CENTRAL PHONE 614 299 2365

ITEMIZED CALLS APR 17 1981

NO.	DATE	TIME	CALLED PLACE	AREA-NUMBER	*	MIN	AMOUNT
1	3-18	1154 PM	SEATTLE, WASH	206 360 5543	C	15	
2	3-26	1130 PM	SEATTLE, WASH	206 360 5543	C	32	
3	3-28	1103 AM	HOUSTON, TX	512 399 1212	A	3	
4	4-2	944 PM	CHICAGO, ILL	312 661 4997	B	12	
082				TOTAL (EXCLUDING U.S. TAX)			
*RATE PERIOD CODE—SEE REVERSE FOR EXPLANATION					U.S. TAX		

3–6 Perpendicular Planes

A line is perpendicular to a plane if and only if it is perpendicular to *every* line in the plane that intersects the given line.

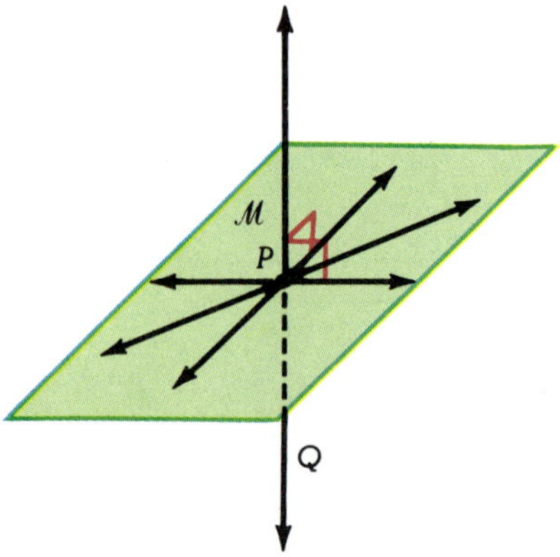

$\overleftrightarrow{PQ} \perp \mathcal{M}$ because it is perpendicular to every line in $\mathcal{M}$ that intersects $\overleftrightarrow{PQ}$. Parts of lines are perpendicular to a plane if and only if they are contained in a line that is perpendicular to the plane. For example, $\overline{PQ} \perp \mathcal{M}$.

A plane is perpendicular to a line if and only if *every* line in the plane that intersects the given line is perpendicular to that line.

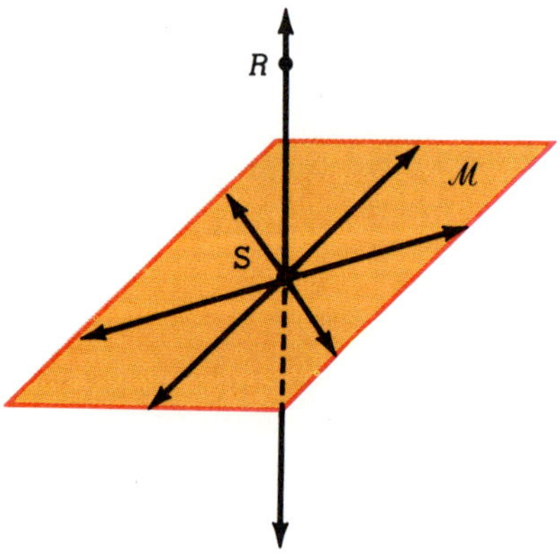

$\mathcal{M}$ is perpendicular to $\overleftrightarrow{RS}$

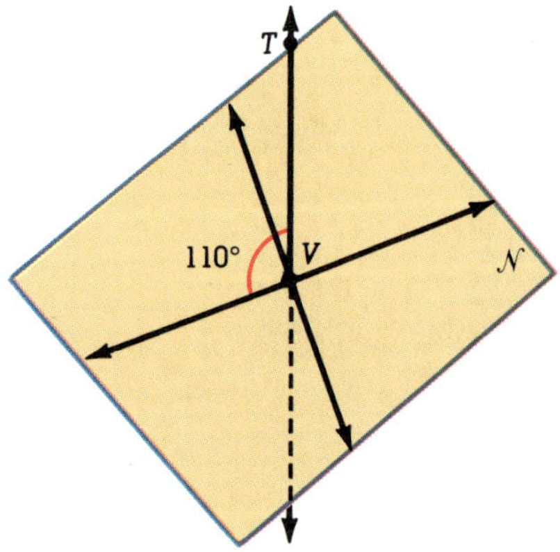

$\mathcal{N}$ is *not* perpendicular to $\overleftrightarrow{TV}$

Just as lines can be perpendicular to lines, planes can be perpendicular to planes. Most walls meet at right angles and provide a model for perpendicular planes.

Definition of Perpendicular Planes

Two planes are perpendicular if and only if any line in one of them that is perpendicular to their line of intersection is perpendicular to the other plane.

When the flaps on the wing of an airplane are up, they form an angle with the wing. Considered in cross section this situation can be described using a geometric angle. Considered in space this situation can be described using another geometric figure called a **dihedral angle.**

Two noncollinear rays with a common endpoint are called an angle. Two noncoplanar half planes with a common edge are called a dihedral angle. Each half plane is called a *face* of the dihedral angle.

Fold a sheet of paper once. The result is a model of a dihedral angle.

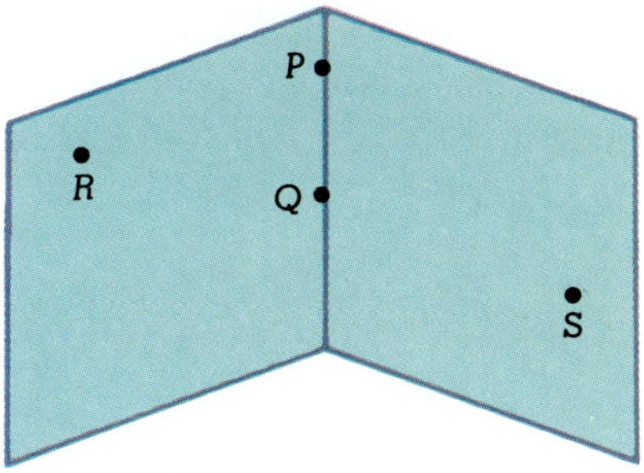

A dihedral angle usually is named by its edge and a point in each face. For example, the dihedral angle at the left is $\angle R\text{-}\overleftrightarrow{PQ}\text{-}S$.

example 1

Name all eight dihedral angles formed by the figure at the right.

angles formed by bottom edges	angles formed by side edges
$\angle T\text{-}\overleftrightarrow{PQ}\text{-}S$	$\angle S\text{-}\overleftrightarrow{TP}\text{-}Q$
$\angle T\text{-}\overleftrightarrow{QR}\text{-}S$	$\angle P\text{-}\overleftrightarrow{TQ}\text{-}R$
$\angle T\text{-}\overleftrightarrow{SR}\text{-}Q$	$\angle Q\text{-}\overleftrightarrow{TR}\text{-}S$
$\angle T\text{-}\overleftrightarrow{SP}\text{-}Q$	$\angle R\text{-}\overleftrightarrow{TS}\text{-}P$

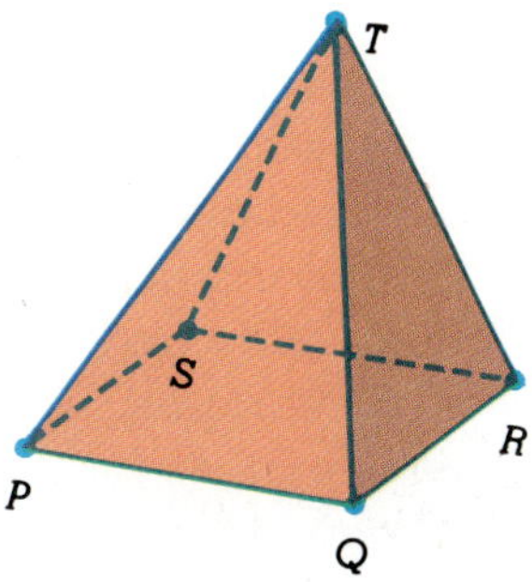

The cross section of a dihedral angle is an angle. Geometrically, the cross section is the intersection of a plane and the dihedral angle. The plane intersects the edge of the dihedral angle in a single point. The following diagrams show two different cross sections.

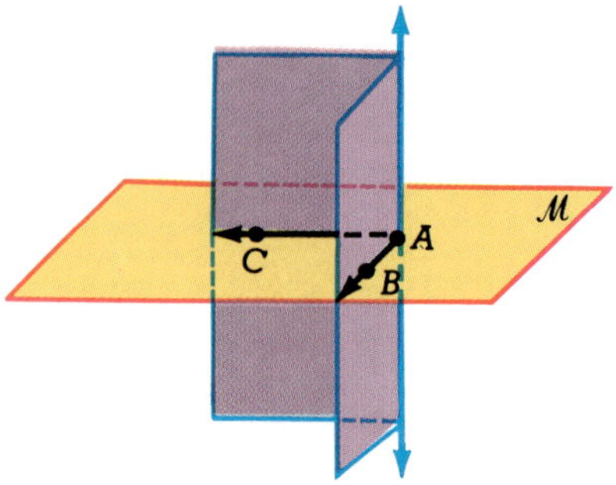

$\mathcal{M}$ is perpendicular to the edge of this dihedral angle.

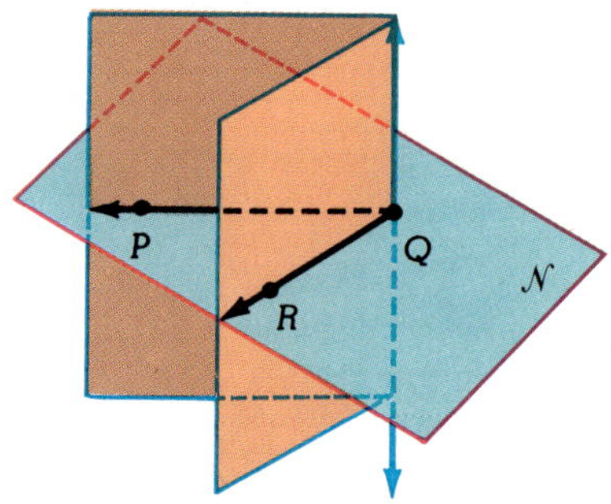

$\mathcal{N}$ is *not* perpendicular to the edge of this dihedral angle.

A *plane angle* of a dihedral angle is the intersection of the dihedral angle and a plane perpendicular to its edge. All the plane angles of a given dihedral angle have the same measure. Thus, the measure of a dihedral angle is defined to be the measure of any of its plane angles.

A dihedral angle is a right dihedral angle if and only if its plane angles are right angles. Just as perpendicular lines form right angles, perpendicular planes form right dihedral angles.

Theorem 3–15 **Two planes are perpendicular if and only if they intersect to form a right dihedral angle.**

exercises

Exploratory In the diagram at the right, $\overline{PQ} \perp \overline{QR}$, $\overline{PQ} \perp \overline{QS}$, and $\overline{QS} \perp \overline{QR}$.

1. Name the dihedral angle with edge $\overleftrightarrow{PQ}$.
2. Name the dihedral angle with edge $\overleftrightarrow{QS}$.
3. Name the dihedral angle with edge $\overleftrightarrow{QR}$.
4. Name the dihedral angle with edge $\overleftrightarrow{PS}$.
5. Name the dihedral angle with edge $\overleftrightarrow{SR}$.
6. Name the dihedral angle with edge $\overleftrightarrow{PR}$.
7. Name a plane angle for $\angle P\text{-}\overleftrightarrow{SQ}\text{-}R$.
8. Name a plane angle for $\angle S\text{-}\overleftrightarrow{PQ}\text{-}R$.
9. Name a plane angle for $\angle P\text{-}\overleftrightarrow{QR}\text{-}S$.

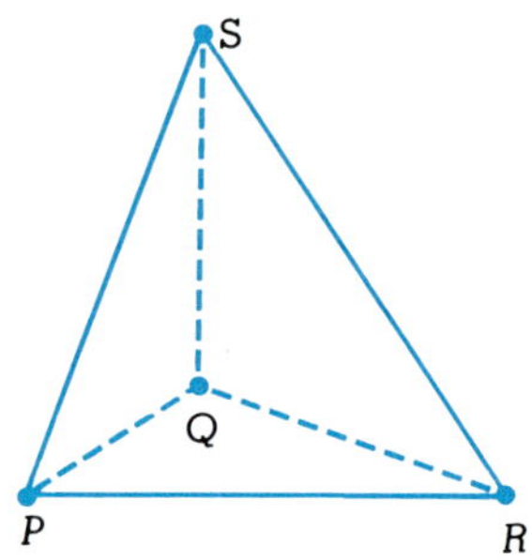

Written In the diagram at the right, $\mathcal{M} \perp \mathcal{L}$ and $\mathcal{L} \perp \mathcal{N}$.

1. Name two dihedral angles.
2. Name a plane angle for $\angle S\text{-}\overleftrightarrow{YQ}\text{-}Z$.
3. Name a plane angle for $\angle R\text{-}\overleftrightarrow{XZ}\text{-}W$.
4. $\overline{SR} \perp \overline{RX}$. Write *yes* or *no*.
5. $\overline{PQ} \perp \overline{QT}$. Write *yes* or *no*.
6. $\overline{RX} \perp \overline{TV}$. Write *yes* or *no*.
7. $\overline{QT} \perp \overline{XW}$. Write *yes* or *no*.

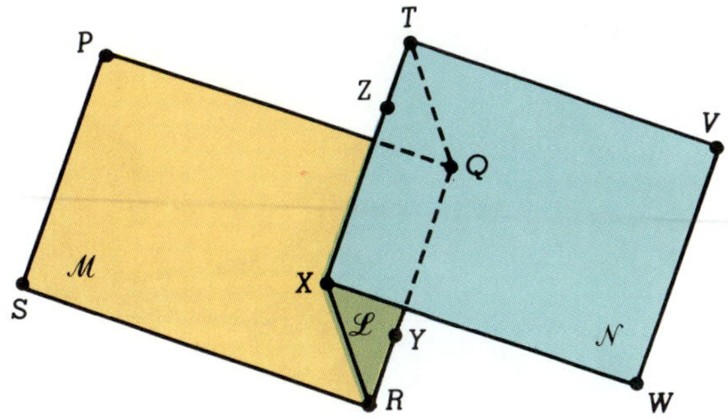

Prove each theorem.

8. If a line is perpendicular to a plane, then any plane containing that line is perpendicular to the plane.
9. If two planes are perpendicular, then they intersect to form a right dihedral angle.
10. If two planes are perpendicular, then they intersect to form four right dihedral angles.
11. If two planes intersect to form a right dihedral angle, then the planes are perpendicular.

Using Computers

Home Computers

By using your telephone, you can make airline reservations, get help with your income tax, or find any UPI news story written in the last eight years. About 2000 kinds of information are available from the computers of Telecomputing Corporation of America. In addition, the computers can be used to send and receive messages across the country.

For a charge of about \$100, a video-display terminal is installed in your home and a special coupler is attached to your telephone. A printer is optional. Then, for about \$20 per hour, you can use the system. In the evening, the hourly fee is less than \$5.

If you don't want to bother with a telephone, you can buy your own home computer. You pay from about \$500 to more than \$3000.

Suppose you can't afford such luxury items. Then, buy your own portable computer, a calculator, for less than \$10.

Using Geometry

Rappelling

Rappelling is a technique used by climbers to make a difficult descent. Climbers back to the edge and spring off.

To avoid slipping, the climber's legs should remain perpendicular to the side of the ledge.

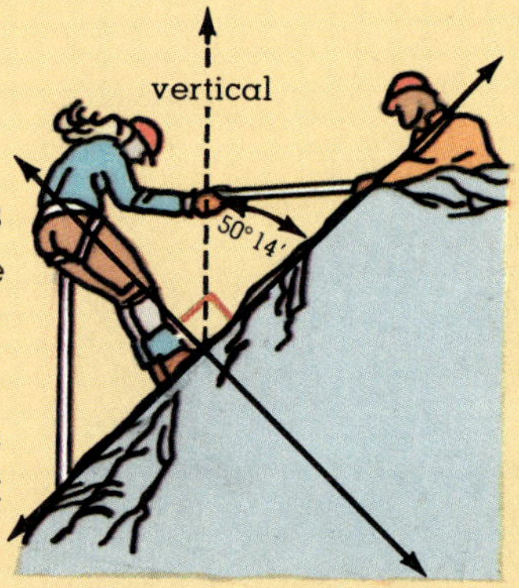

The ledge at the right is 50° 14′ off the vertical. The climber must be perpendicular to the ledge. By instinct, the climber moves to an angle that is the complement of 50° 14′.

The degree measure of an angle plus the degree measure of its complement is 90. Find the complement of 50° 14′.

$$\begin{aligned} \text{complement} + 50° \, 14' &= 90° \, 00' \\ &= 90° \, 00' - 50° \, 14' \\ &\quad \textit{One degree is } 60'. \\ &= 89° \, 60' - 50° \, 14' \\ &= 39° \, 46' \end{aligned}$$

The complement of 50° 14′ is 39° 46′.
The climber's angle measures 39° 46′.

Exercises For each of the following ledge angles, find the climber's angle.

1. 23° **2.** 16° 25′ **3.** 47°

4. 65° 38′ **5.** $18\frac{1}{2}°$ **6.** 36° 50′

Vocabulary

ray (64)
opposite rays (64)
angle (65)
sides of angle (65)
vertex of angle (65)
straight angle (65)
interior of angle (66)
exterior of angle (66)
degrees (68)
protractor (68)
half planes (68)
congruent angles (69)
adjacent angles (70)
supplementary angles (73)
complementary angles (73)
linear pair (73)
acute angle (78)
right angle (78)
obtuse angle (78)
vertical angles (79)
perpendicular lines (82)
perpendicular planes (86)
dihedral angle (87)
plane angle (88)

Chapter Summary

1. A never ending straight path in one direction is represented by a ray. (64)
2. An angle is a pair of rays that have the same endpoint but do *not* lie on the same line. The rays are the sides of the angle, and the common endpoint is the vertex. (65)
3. A point on a line is the endpoint of two rays that head in opposite directions. These two rays are called opposite rays or a straight angle. (65)
4. An angle separates a plane into three parts, the angle, its interior, and its exterior. (66)
5. Angle Measure Postulate: For every angle, there is a unique positive number between 0 and 180 called the degree measure of the angle. (68)
6. Protractor Postulate: Given a ray $\overrightarrow{QT}$ on the edge of a half plane, for every positive number r between 0 and 180 there is exactly one ray, $\overrightarrow{QP}$, in the half plane so that $m\angle PQT = r$. (68)
7. Angles having the same measure are called congruent angles. (69)
8. Theorem 3–1: Congruence of angles is reflexive, symmetric, and transitive. (69)
9. Angle Addition Postulate: If R is in the interior of $\angle PQS$, then $m\angle PQS = m\angle PQR + m\angle RQS$. (70)
10. Definition of Supplementary Angles: Two angles are supplementary if and only if the sum of their degree measures is 180. (73)
11. Definition of Complementary Angles: Two angles are complementary if and only if the sum of their degree measures is 90. (73)

12. Supplement Postulate: The angles in a linear pair are supplementary angles. (73)
13. Theorem 3–2: If two angles are supplementary to the same angle, then they are congruent. (73)
14. Theorem 3–3: If two angles are supplementary to two congruent angles, then the two angles are congruent to each other. (75)
15. Theorem 3–4: If two angles are complementary to the same angle, then they are congruent to each other. (75)
16. Theorem 3–5: If two angles are complementary to two congruent angles, then the two angles are congruent to each other. (75)
17. Definition of Right, Acute, and Obtuse Angles: A right angle is an angle whose degree measure is 90. An acute angle is one whose degree measure is less than 90. An obtuse angle is one whose degree measure is greater than 90. (78)
18. Theorem 3–6: If two angles are right angles, then the angles are congruent. (78)
19. Theorem 3–7: If one angle in a linear pair is a right angle, then the other angle also is a right angle. (78)
20. Theorem 3–8: If two angles are congruent and supplementary, then each is a right angle. (78)
21. Theorem 3–9: If two intersecting lines form one right angle, then they form four right angles. (79)
22. Theorem 3–10: Vertical angles are congruent. (79)
23. Definition of Perpendicular Lines: Two lines are perpendicular if and only if they intersect to form a right angle. (82)
24. Theorem 3–11: If two lines are perpendicular, then they form four right angles. (82)
25. Theorem 3–12: In a plane, through a given point on a line, there is exactly one line perpendicular to the given line. (83)
26. Theorem 3–13: If two lines are perpendicular, then they form congruent adjacent angles. (83)
27. Theorem 3–14: If two intersecting lines form congruent adjacent angles, then they are perpendicular. (83)
28. Definition of Perpendicular Planes: Two planes are perpendicular if and only if any line in one of them that is perpendicular to their line of intersection is perpendicular to the other plane. (86)
29. Two noncoplanar half planes with a common edge are called a dihedral angle. (87)
30. A plane angle of a dihedral angle is the intersection of the dihedral angle and a plane perpendicular to its edge. The measure of a dihedral angle is the measure of any of its plane angles. (88)
31. Theorem 3–15: Two planes are perpendicular if and only if they intersect to form a right dihedral angle. (88)

Chapter Review

3–1 Use the figure at the right to solve each of the following.

1. Name the vertex of $\angle 4$.
2. Name the sides of $\angle 4$.
3. Name a point interior to $\angle 3$.
4. Name a point exterior to $\angle 3$.
5. Name a pair of opposite rays.
6. Write another name for $\angle 2$.

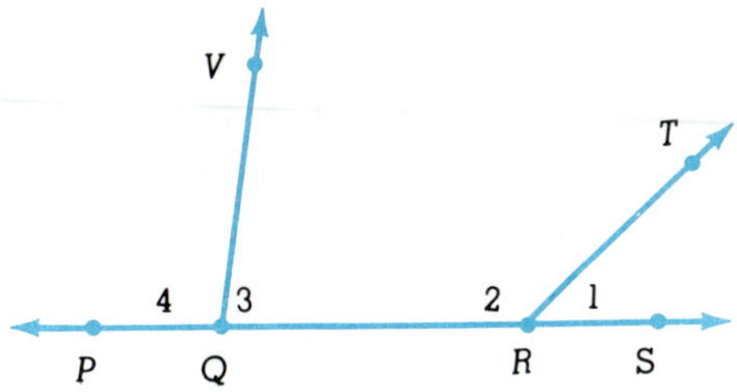

3–2 In the figure at the right, $m\angle QTS = 132$ and $m\angle STR = 37$. Also, $\angle PTS$ is a straight angle.

7. Find $m\angle RTQ$.
8. Find $m\angle QTP$.

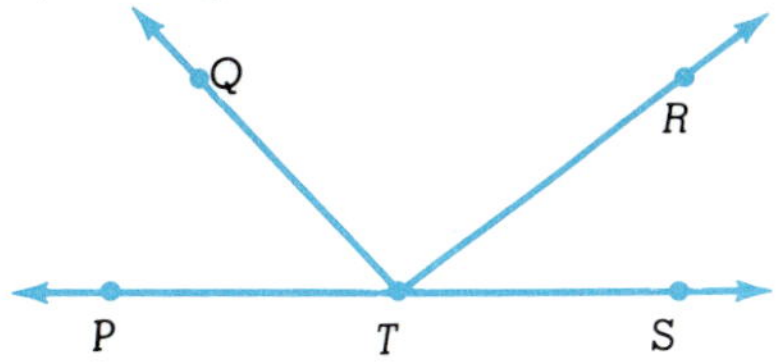

3–3 In the figure at the right, $\angle RPS \cong \angle SPT$. Find the degree measures for each of the following.

9. $\angle TPS$
10. $\angle WPV$
11. $\angle QPW$
12. $\angle RPQ$
13. Prove that if two angles are congruent and supplementary, then they each have degree measure 90.

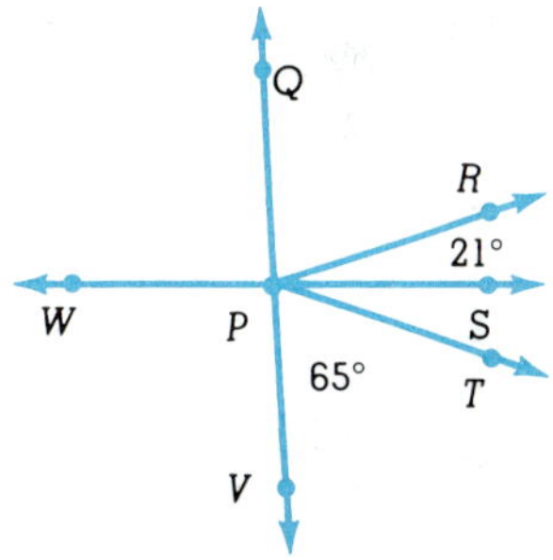

3–4 Prove each of the following.

14. Vertical angles are congruent.
15. If two lines intersect so that one pair of vertical angles is supplementary, then the other pair of vertical angles is supplementary.

3–5 Prove each of the following.

16. If two intersecting lines form two congruent adjacent angles, then they are perpendicular.
17. If the angles in a linear pair are congruent, then the common side is perpendicular to the other two sides.

3–6 In the figure at the right, $\overline{BC} \perp \overline{BE}$ and $\overline{AB} \perp \overline{BE}$. Solve each of the following.

18. Name two dihedral angles.
19. Name a plane angle for $\angle D\text{-}\overleftrightarrow{BE}\text{-}A$.

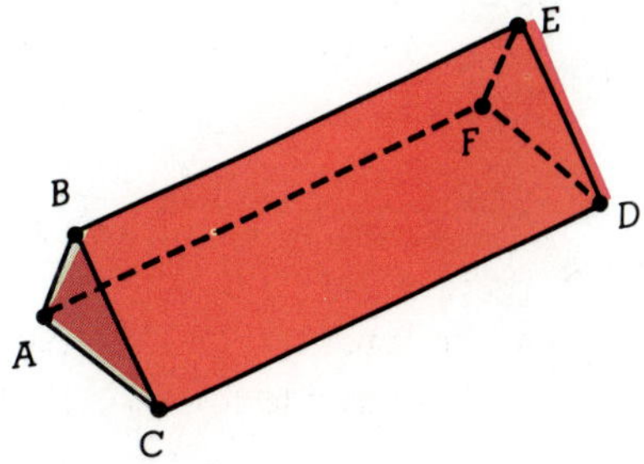

Chapter Test

For each of the following, write *true* or *false*.

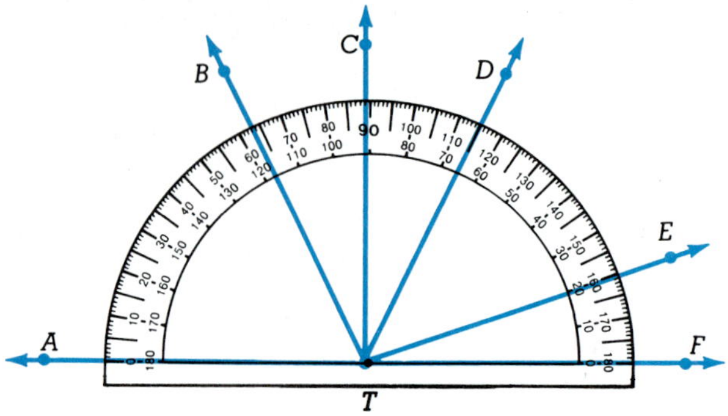

1. $\overrightarrow{TA}$ and $\overrightarrow{TF}$ form a straight angle.
2. D is interior to $\angle BTC$.
3. $\overrightarrow{CT}$ is a side of $\angle FTC$.
4. D is exterior to $\angle BTC$.
5. $\overrightarrow{CT} \perp \overleftrightarrow{AF}$
6. $\angle BTC \cong \angle DTC$
7. $\angle FTD$ and $\angle DTC$ are complementary.
8. $\angle ATB$ and $\angle BTF$ are complementary.
9. $\angle ATB$ and $\angle BTF$ form a linear pair.
10. $\angle BTD$ and $\angle FTE$ are adjacent angles.
11. $m\angle ATB = m\angle DTE + m\angle FTE$
12. $m\angle BTD = m\angle CTB + m\angle DTC$
13. $\angle CTE$ is an acute angle.
14. $\angle ETC$ is an obtuse angle.

In the figure at the right, $\mathscr{A} \perp \mathscr{B}$ and $\mathscr{B} \perp \mathscr{C}$.

15. Name a dihedral angle with edge $\overleftrightarrow{SR}$.
16. Name the faces of $\angle Y\text{-}\overleftrightarrow{VX}\text{-}S$.
17. Find $m\angle TVZ$.
18. Name a plane angle for $\angle P\text{-}\overleftrightarrow{TR}\text{-}W$.

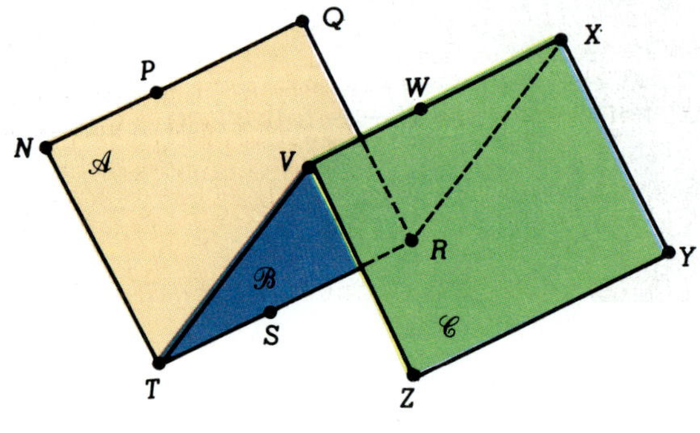

Prove each of the following statements.

19. If two angles are vertical and one angle is a right angle, then the other also is a right angle.
20. If two adjacent angles are congruent and supplementary, then their common side is perpendicular to the other two sides.

Chapter

4

Congruent Triangles

Certain tools, like wrenches, must fit exactly. The wrench opening must have a certain size and shape. Congruent figures have the same size and shape.

4–1 Triangles

Because it is a rigid form, a **triangle** is used in many constructions. Notice the triangles in the electric towers. One is marked.

Three noncollinear line segments connected at their endpoints form a triangle. The line segments are **sides** of the triangle. The endpoints are **vertices** of the triangle. Angles can be formed at each of these vertices. These angles are called angles of the triangle.

At the left, triangle ABC, written $\triangle ABC$, has the following parts.

sides	vertices	angles
$\overline{AB}$	A	$\angle A$
$\overline{BC}$	B	$\angle B$
$\overline{CA}$	C	$\angle C$

Definition of Triangle

A triangle is a figure formed by three noncollinear line segments called sides. Each endpoint of a side is an endpoint of exactly one other side.

Triangles can be classified according to their angles.

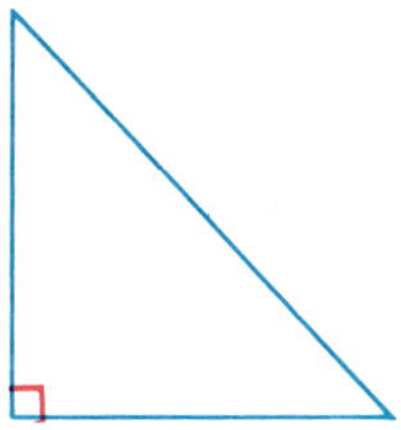

One angle of a *right triangle* is a right angle.

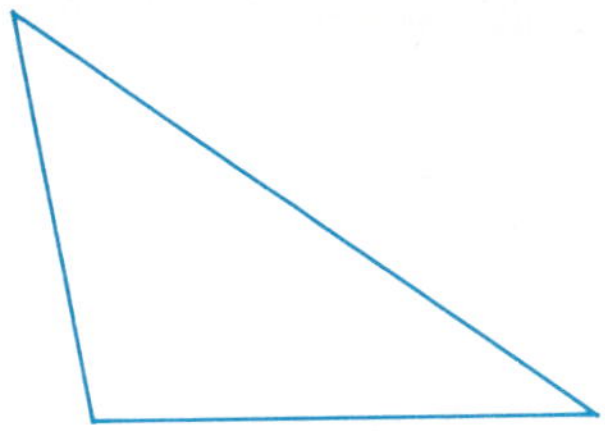

One angle of an *obtuse triangle* is an obtuse angle.

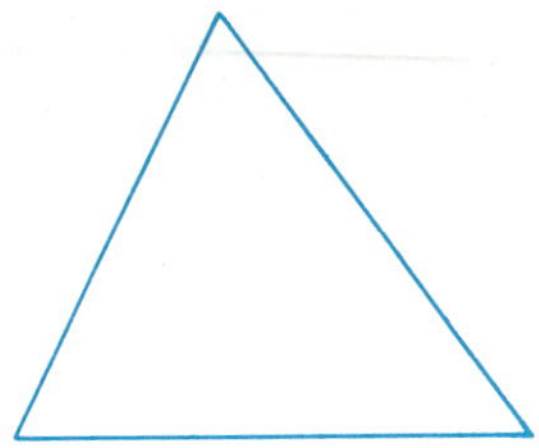

All the angles of an *acute triangle* are acute angles.

If all the angles of a triangle are congruent, then the triangle is *equiangular*.

Triangles also can be classified according to their sides.

Marks on the sides of triangles are used to indicate congruent sides.

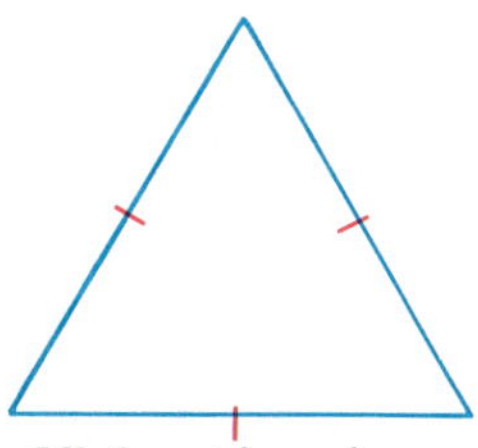

All the sides of an *equilateral triangle* are congruent.

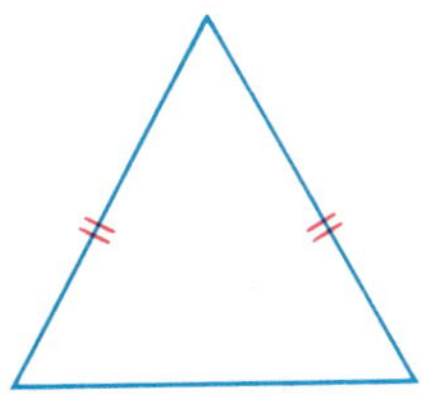

Two sides of an *isosceles triangle* are congruent.

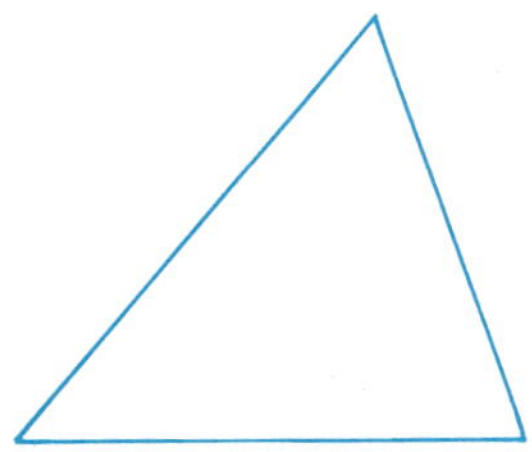

No sides of a *scalene triangle* are congruent.

example 1

Suppose $\triangle PQR$ is an isosceles right triangle as shown. Solve for x and find the measure of each side.

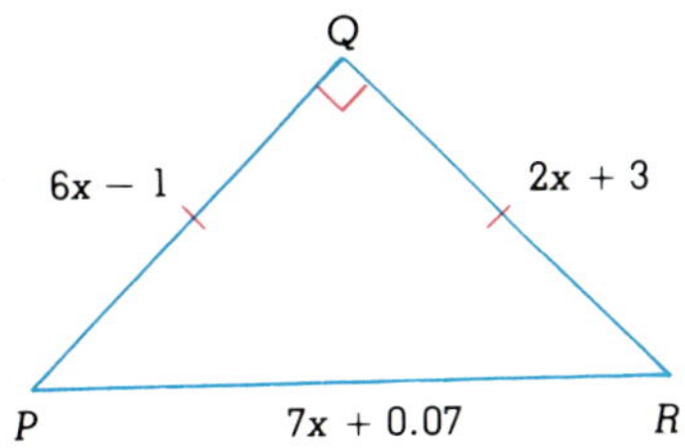

$\overline{PQ} \cong \overline{QR}$	*Given*
$PQ = QR$	*Definition of congruence.*
$6x - 1 = 2x + 3$	*Substitution*
$4x - 1 = 3$	*Subtract 2x from both sides.*
$4x = 4$	*Add 1 to both sides.*
$x = 1$	*Divide both sides by 4.*

$PQ = 6x - 1$
$= 6(1) - 1$
$= 5$

$QR = 2x + 3$
$= 2(1) + 3$
$= 5$

$PR = 7x + 0.07$
$= 7(1) + 0.07$
$= 7.07$

A triangle separates a plane into three parts. The parts are the *interior*, the *exterior*, and the triangle itself.

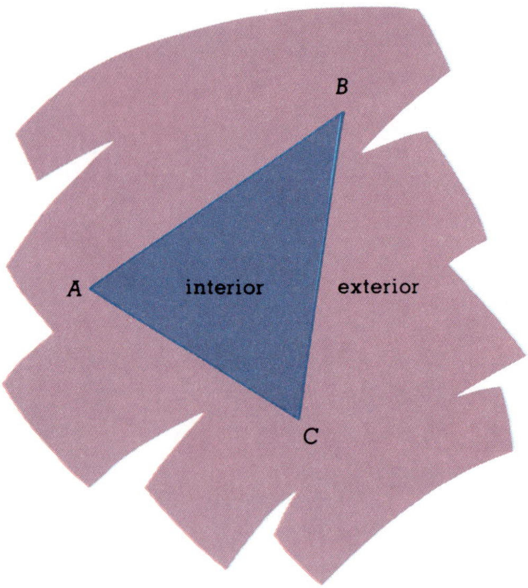

Any point in the blue part of the plane is in the interior of $\triangle ABC$. Any point in the pink part of the plane is in the exterior of $\triangle ABC$.

exercises

Exploratory Use $\triangle DEF$ and $\triangle PQR$ to solve each of the following.

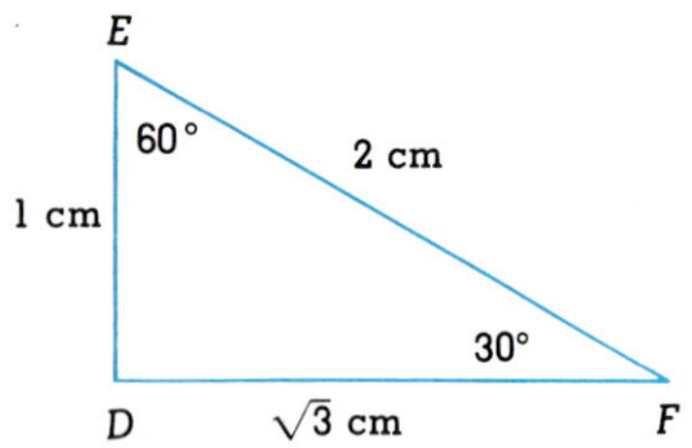

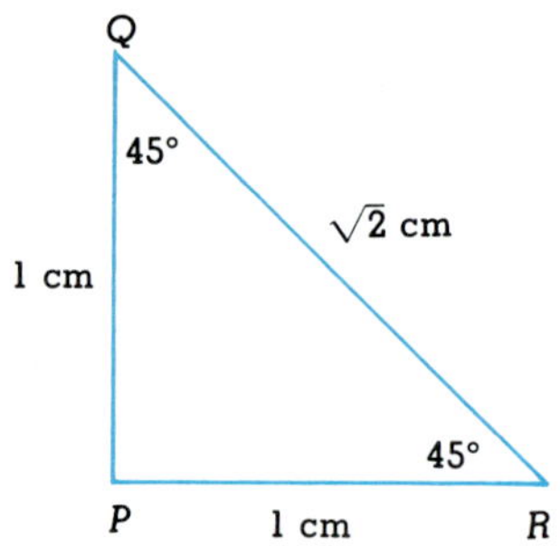

1. Name the vertices of $\triangle DEF$.
2. Name the vertices of $\triangle PQR$.
3. Name the angles of $\triangle DEF$.
4. Name the angles of $\triangle PQR$.
5. Name the sides of $\triangle DEF$.
6. Name the sides of $\triangle PQR$.
7. Classify $\triangle DEF$ by angles.
8. Classify $\triangle PQR$ by angles.
9. Classify $\triangle DEF$ by sides.
10. Classify $\triangle PQR$ by sides.

Written The following figure contains eight triangles.

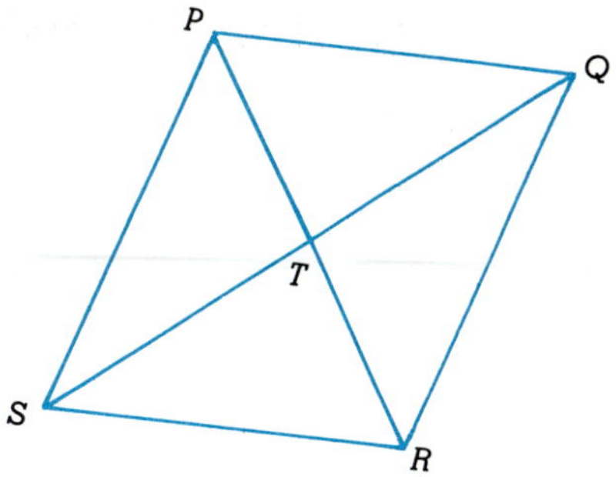

1–8. Name each triangle. Write the letters in alphabetical order.

9. Name the vertices of $\triangle STR$.

10. Name the sides of $\triangle STR$.

11. Name the angles of $\triangle STR$.

12–14. Solve problems 9–11 for $\triangle QSR$.

Give an *if and only if* definition for each of the following.

15. triangle
16. right triangle
17. obtuse triangle
18. acute triangle
19. equiangular triangle
20. equilateral triangle
21. isosceles triangle
22. scalene triangle
23. interior of a triangle
24. exterior of a triangle

For each of the following, draw a triangle that satisfies the given conditions. If no such triangle exists, write *none*.

25. isosceles, right
26. isosceles, acute
27. scalene, acute
28. scalene, obtuse
29. isosceles, obtuse
30. scalene, right
31. isosceles, scalene
32. equilateral, obtuse
33. equilateral, acute
34. equilateral, right
35. scalene, equiangular
36. equilateral, equiangular

Suppose $\triangle PQR$ is an equilateral triangle. Solve for x and find the measure of each side given the following information.

37. $PQ = 3x - 20$
    $QR = x + 16$
    $RP = 2x - 2$

38. $PQ = x + 8$
    $QR = \frac{1}{3}x + 26$
    $RP = 2x - 19$

39. $PQ = \frac{1}{2}x + 9$
    $PR = 2x - 21$

40. $PQ = 0.7x - 20$
    $QR = 0.5x - 12$

Seeing Patterns

Excursions in Geometry

Look at the sequence of figures below. Then, draw the figure that comes next.

4–2 Angle Measures

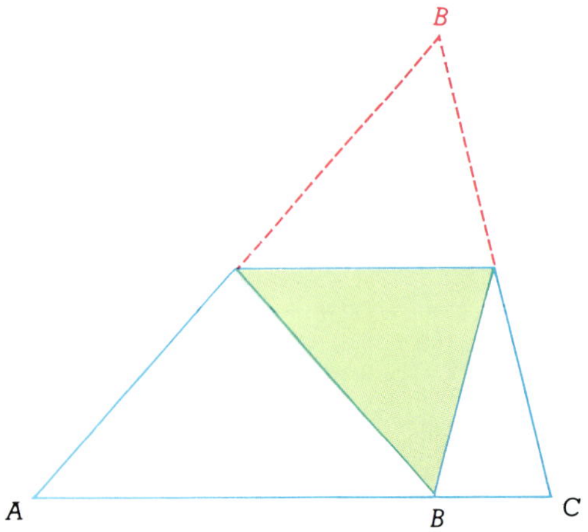

The figure at the left shows a paper cut in the shape of a triangle. The paper is folded so that B lies on $\overline{AC}$.

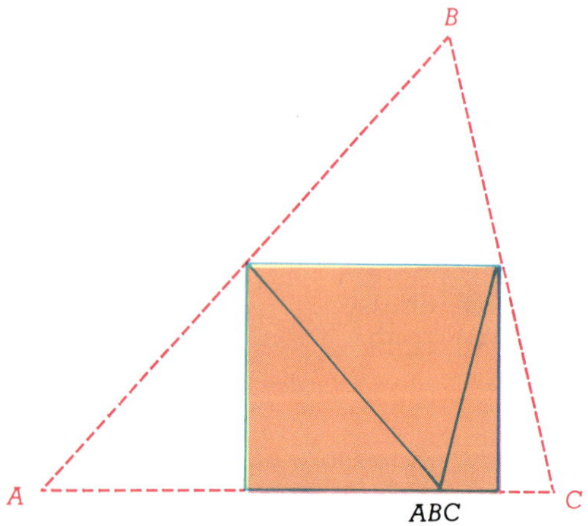

$m\angle A + m\angle B + m\angle C = 180$

The paper is folded in two more places, so that A and C are at B. The folded paper shows that the total degree measures of the angles of the triangle is 180.

Any triangular piece of paper can be folded in the same way. Many experiments such as this one suggest the following theorem.

Theorem 4–1
Angle Sum Theorem

The sum of the degree measures of the angles of a triangle is 180.

All the angles of an equiangular triangle have the same measure. Thus, each angle has degree measure 180 ÷ 3 or 60.

Theorem 4–2

The degree measure of each angle of an equiangular triangle is 60.

If you measure two angles of a triangle, you can find the measure of the third angle.

examples

1 In the triangle at the right, $\angle R$ is a right angle, and $m\angle P = 30$. Find $m\angle Q$.

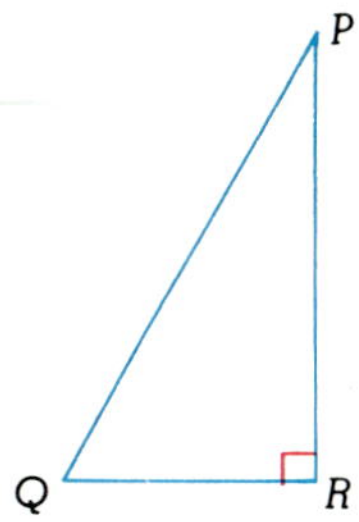

$$m\angle P + m\angle Q + m\angle R = 180 \quad \textit{Theorem 4–1}$$
$$30 + m\angle Q + 90 = 180 \quad \textit{Substitution}$$
$$m\angle Q = 60 \quad \textit{Subtraction Property of Equality}$$

2 In the diagram at the right, $m\angle A = 60$, $m\angle D = 30$, and $m\angle E = 85$. Find $m\angle B$.

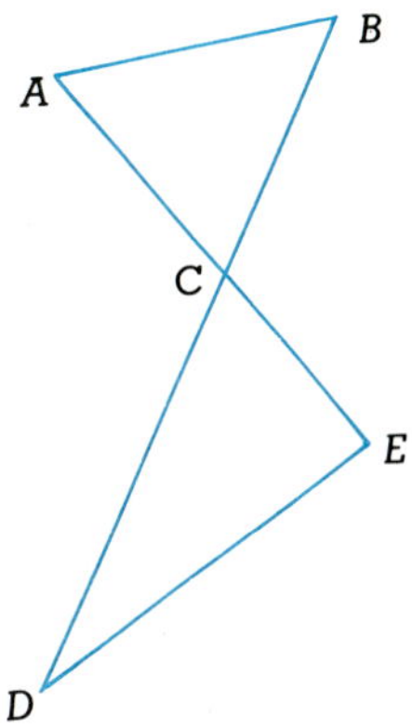

$$m\angle D + m\angle E + m\angle DCE = 180 \quad \textit{Theorem 4–1}$$
$$30 + 85 + m\angle DCE = 180 \quad \textit{Substitution}$$
$$m\angle DCE = 65 \quad \textit{Subtraction Property of Equality}$$
$$m\angle ACB = 65 \quad \angle DCE \textit{ and } \angle ACB \textit{ are vertical angles.}$$
$$m\angle A + m\angle B + m\angle ACB = 180 \quad \textit{Theorem 4–1}$$
$$60 + m\angle B + 65 = 180 \quad \textit{Substitution}$$
$$m\angle B = 55 \quad \textit{Subtraction Property of Equality}$$

The triangle in Example 1 is a right triangle. Notice that the degree measures of the other two angles total 90. The following theorem states this property for all right triangles.

The acute angles of a right triangle are complementary. Theorem 4–3

exercises

Exploratory Use the figure below to solve each of the following.

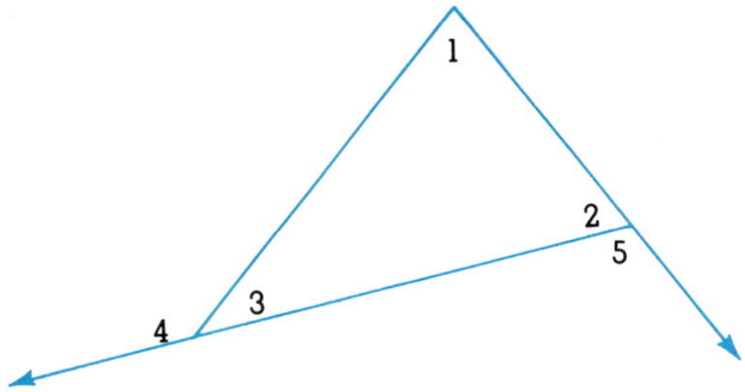

1. Find $m\angle 1$ if $m\angle 2 = 40$ and $m\angle 3 = 55$
2. Find $m\angle 1$ if $m\angle 2 = 60$ and $m\angle 3 = 60$
3. Find $m\angle 1$ if $m\angle 2 = 27$ and $m\angle 3 = 72$
4. Find $m\angle 1$ if $m\angle 2 = 45$ and $m\angle 3 = 45$
5. Find $m\angle 1$ if $m\angle 2 = 81$ and $m\angle 3 = 74$
6. Find $m\angle 1$ if $m\angle 2 = 86$ and $m\angle 3 = 30$
7. Find $m\angle 2$ if $m\angle 1 = 45$ and $m\angle 4 = 105$
8. Find $m\angle 2$ if $m\angle 1 = 59$ and $m\angle 4 = 108$
9. Find $m\angle 2$ if $m\angle 1 = 56$ and $m\angle 4 = 112$
10. Find $m\angle 2$ if $m\angle 1 = 58$ and $m\angle 4 = 125$
11. Find $m\angle 2$ if $m\angle 1 = 47$ and $m\angle 4 = 132$
12. Find $m\angle 2$ if $m\angle 1 = 44$ and $m\angle 4 = 121$
13. Find $m\angle 3$ if $m\angle 1 = 45$ and $m\angle 5 = 98$
14. Find $m\angle 3$ if $m\angle 1 = 42$ and $m\angle 5 = 117$
15. Find $m\angle 3$ if $m\angle 1 = 73$ and $m\angle 5 = 139$
16. Find $m\angle 3$ if $m\angle 1 = 40$ and $m\angle 5 = 123$
17. Find $m\angle 3$ if $m\angle 1 = 60$ and $m\angle 5 = 114$
18. Find $m\angle 3$ if $m\angle 1 = 67$ and $m\angle 5 = 101$

Written Solve each problem.

1. Can a triangle have two right angles? Write *yes* or *no*, then explain your answer.
2. Can a triangle have two obtuse angles? Write *yes* or *no*, then explain your answer.
3. Prove Theorem 4–2.
4. Prove Theorem 4–3.

In the figure at the right, $\overline{AB} \perp \overline{BC}$ and $\overline{CD} \perp \overline{BC}$. Also, $m\angle BEC = 125$, $m\angle A = x$, and $\angle ABE \cong \angle DCE$. Find the degree measure for each of the following.

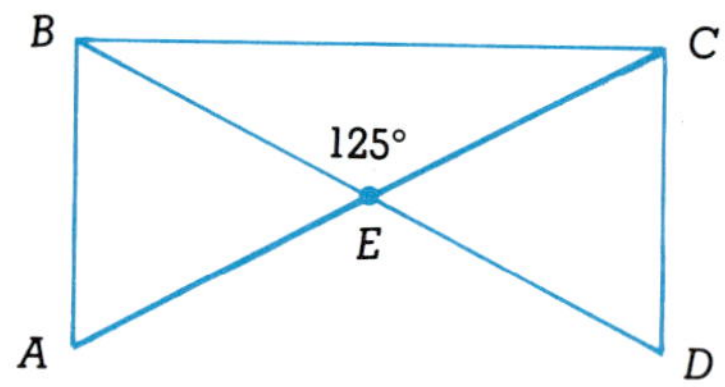

5. $\angle AEB$
6. $\angle EBC$
7. $\angle CED$
8. $\angle ECD$
9. $\angle ECB$
10. $\angle ABE$

11. In the figure below, $\angle Q$ and $\angle T$ are right angles. Prove $m\angle P = m\angle S$.

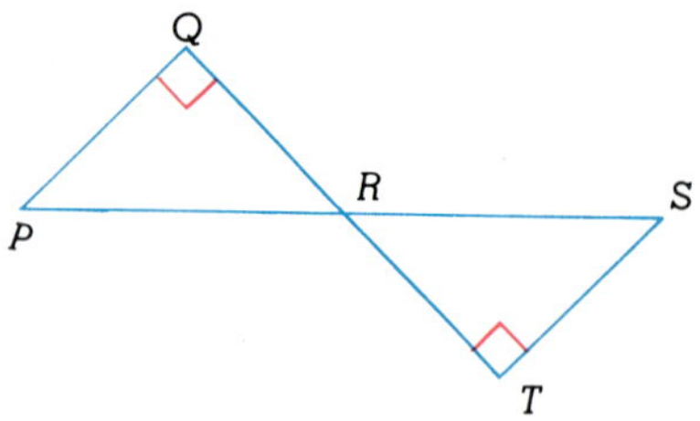

12. In the figure below, $\angle P$ and $\angle T$ are right angles. Also, $m\angle QRS = 70$ and $m\angle SRT = 55$. Prove $m\angle Q = m\angle S$.

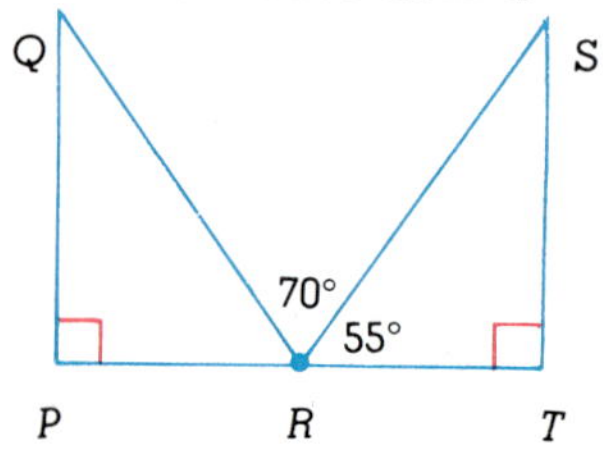

CAREERS

Mover

Before a mover picks up a shipment, a bill of lading is issued as a contract and a receipt for goods shipped. Using the bill of lading, the person who ships goods can plan how much will be paid to unload.

Sally Wilkinson works for A-1 Movers. A family is shipping 4375 pounds of furniture over 373 miles. To complete the bill of lading, Ms. Wilkinson must estimate the charges.

TARIFF	SECT	MILES	WEIGHT	RATE	THIS IS NOT AN ESTIMATE AS DEFINED IN EX PARTE NO. MC-19 SECT 10 EST. TRAN. CHARGES	EST. TOTAL CHARGES
174-A		373	4375	10.97	482.68	587.63

Basic transportation charges are determined by weight. The rate is charged per hundred pounds.

4375 is about 44 hundred
$44 \times 10.97 = 482.68$

Additional service charges are added.

valuation (insurance)	$ 34.50
containers	$ 34.80
packing	$ 30.40
unpacking	$ 5.25
other	$ 0.00
total service	$104.95

The total estimated charges is the sum of all the charges.

transportation charges	$482.68
additional charges	$104.95
total charges	$587.63

The *most* the family will have to pay to have their furniture unloaded is 110% of the total estimated charges, $646.39.

$$587.63 \times 110\% = 587.63 \times 1.10 = 646.39$$

Exercises **1–4.** For each of the following, find the total estimated charges to the nearest cent.

Weight	Rate per 100 lbs	Additional Charges
3590 lbs	$ 16.98	$ 98.60
6902 lbs	$ 14.15	$ 208.15
9450 lbs	$ 5.66	$ 256.38
13020 lbs	$ 12.73	$ 401.56

5–8. Find 110% of the estimate for each shipment in problems **1–4.**

4–3 Congruence

Suppose a piece of paper is cut along a straight line from one corner to the opposite corner. The edges of the piece form two triangles. If one piece is placed on top of the other piece in a certain way, they fit exactly. What must be true about the parts of the two triangles?

The two triangles below represent the pieces cut from the paper. Corresponding parts of the triangles can be used to describe how the papers fit together.

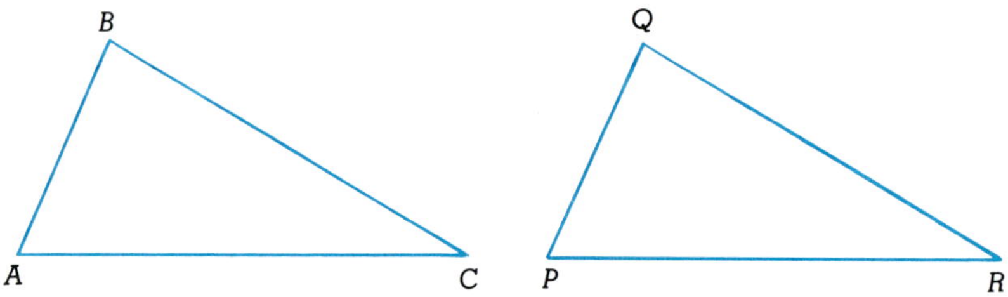

Corresponding angles describe matching the corners of the pieces of paper.

Corresponding sides describe matching the edges of the pieces of paper.

$\angle A$ corresponds to $\angle P$
$\angle B$ corresponds to $\angle Q$
$\angle C$ corresponds to $\angle R$
$\overline{AB}$ corresponds to $\overline{PQ}$
$\overline{BC}$ corresponds to $\overline{QR}$
$\overline{CA}$ corresponds to $\overline{RP}$

The matching parts of the papers fit exactly. Therefore, the corresponding parts of the triangles are congruent and the two triangles are called **congruent triangles.**

Definition of Congruent Triangles

Two triangles are congruent if and only if there is a correspondence such that their corresponding parts are congruent.

Special marks are used to show that certain parts of figures are congruent. Notice how marks are used on the triangles at the right.

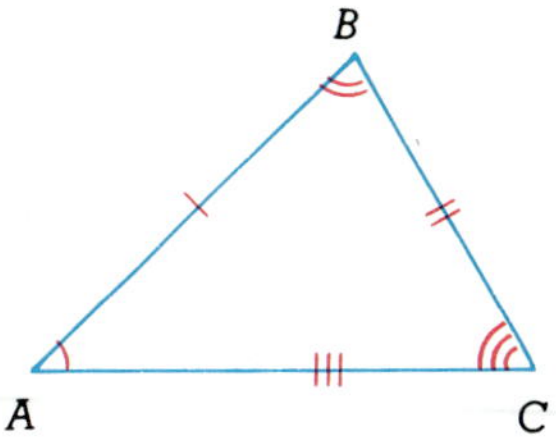

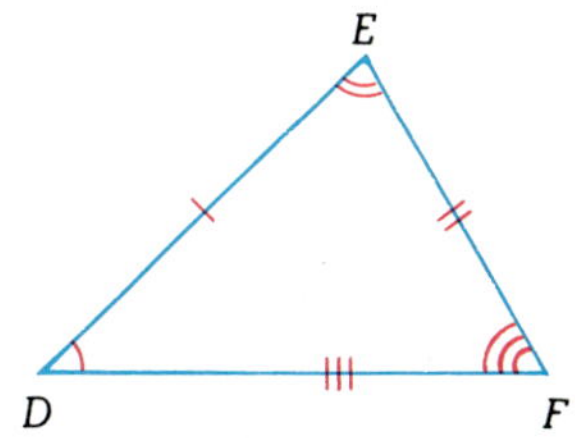

$$\overline{AB} \cong \overline{DE} \qquad \angle A \cong \angle D$$
$$\overline{BC} \cong \overline{EF} \qquad \angle B \cong \angle E$$
$$\overline{CA} \cong \overline{FD} \qquad \angle C \cong \angle F$$

The two triangles are congruent because they have six congruent corresponding parts. To indicate the congruence, write $\triangle ABC \cong \triangle DEF$. The order of letters indicates the correspondence. The first vertices named, A and D, are corresponding vertices. The second vertices named, B and E, are corresponding vertices. The third vertices named, C and F, are corresponding vertices.

When two triangles are congruent, then the corresponding parts are congruent. Often, a geometric problem involves showing that parts of a figure are congruent. One method is to show that the parts are corresponding parts of congruent triangles. Thus, it is important to recognize corresponding parts of congruent triangles.

CPCTC means corresponding parts of congruent triangles are congruent.

example 1

Suppose $\triangle PQS \cong \triangle RQS$. What angle in $\triangle PQS$ is congruent to $\angle R$ in $\triangle RQS$? Which side of $\triangle RQS$ is congruent to $\overline{QS}$ in $\triangle PQS$?

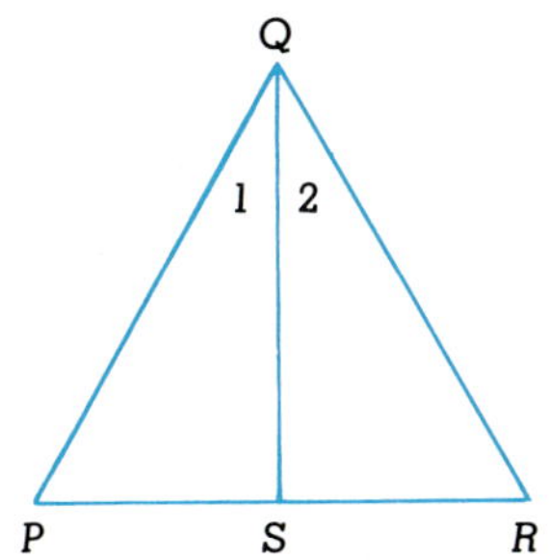

$\angle P$ is congruent to $\angle R$
$\overline{QS}$ is congruent to $\overline{QS}$

Since congruence of line segments and angles is reflexive, symmetric, and transitive, so is congruence of triangles.

Congruence of triangles is reflexive, symmetric, and transitive. **Theorem 4–4**

exercises

Exploratory Suppose $\triangle PQR \cong \triangle STV$. For each of the following, name the corresponding part.

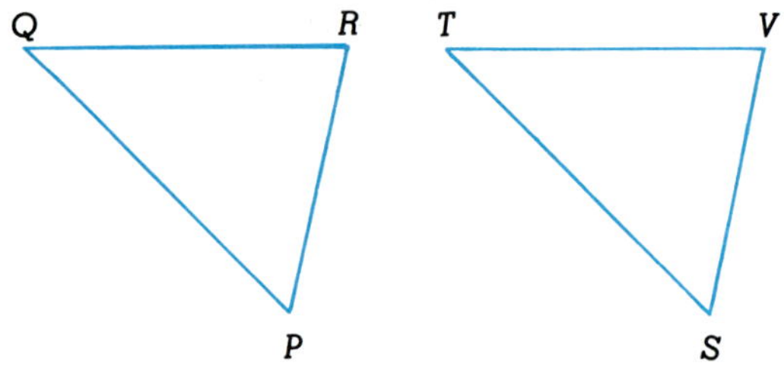

1. $\angle P$
2. $\angle Q$
3. $\angle R$
4. $\angle S$
5. $\angle T$
6. $\angle V$
7. $\overline{PQ}$
8. $\overline{QR}$
9. $\overline{RP}$
10. $\overline{ST}$
11. $\overline{TV}$
12. $\overline{VS}$

13. Suppose $\triangle ABC \cong \triangle FGH$. List the pairs of angles and sides that are congruent.

14. List the information needed to show that $\triangle YZX \cong \triangle RPS$.

Written Tell whether or *not* each of the following statements is correct. Write *yes* or *no*.

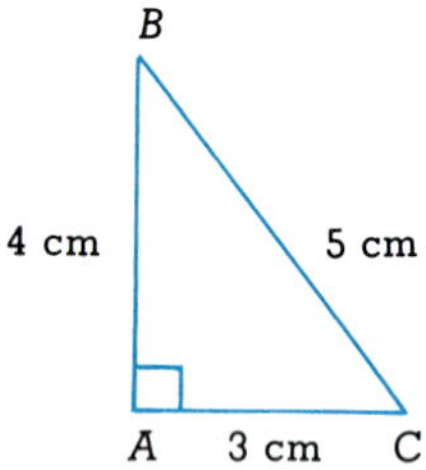

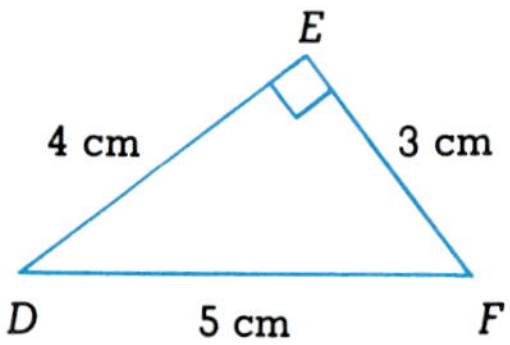

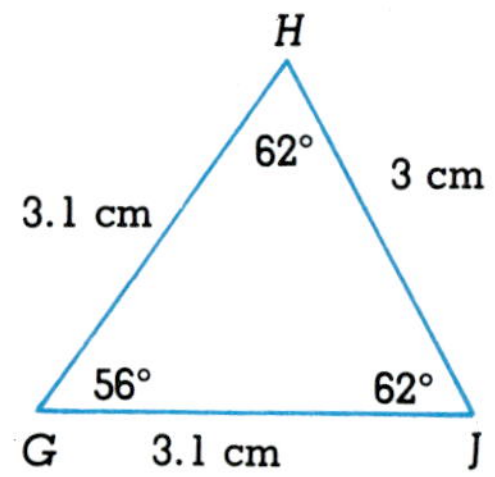

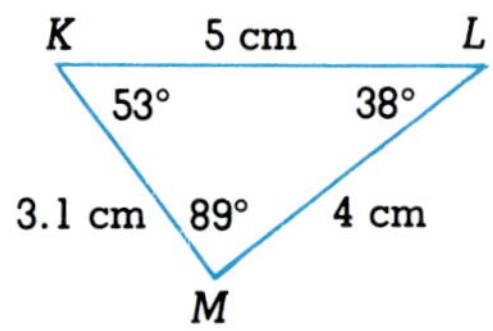

1. $\triangle MLK \cong \triangle ABC$
2. $\triangle DEF \cong \triangle GHJ$
3. $\triangle ACB \cong \triangle EFD$
4. $\triangle CAB \cong \triangle FED$
5. $\triangle KLM \cong \triangle GHJ$
6. $\triangle KML \cong \triangle HGJ$
7. $\triangle JHG \cong \triangle LKM$
8. $\triangle DEF \cong \triangle BAC$
9. $\triangle DFE \cong \triangle BCA$
10. $\triangle BCA \cong \triangle DFE$
11. $\triangle KLM \cong \triangle FDE$
12. $\triangle ABC \cong \triangle MLK$
13. $\triangle HGJ \cong \triangle KML$
14. $\triangle CBA \cong \triangle FDE$
15. $\triangle LKM \cong \triangle FDE$
16. $\triangle LMK \cong \triangle DEF$
17. $\triangle GHJ \cong \triangle CAB$
18. $\triangle FDE \cong \triangle LKM$
19. $\triangle ABC \cong \triangle EDF$
20. $\triangle ABC \cong \triangle FED$
21. $\triangle ABC \cong \triangle DFE$

Suppose $\triangle PQR \cong \triangle STR$ and solve each of the following.

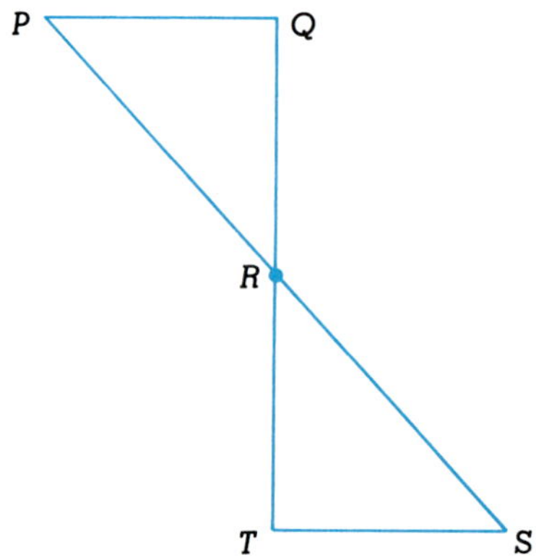

22. What angle in $\triangle PQR$ corresponds to $\angle T$ in $\triangle STR$?
23. What angle in $\triangle STR$ corresponds to $\angle P$ in $\triangle PQR$?
24. Which side of $\triangle PQR$ corresponds to $\overline{RS}$ in $\triangle STR$?
25. Which side of $\triangle PQR$ corresponds to $\overline{TS}$ in $\triangle STR$?
26. Which side of $\triangle STR$ corresponds to $\overline{QR}$ in $\triangle PQR$?
27. Which side of $\triangle STR$ corresponds to $\overline{RP}$ in $\triangle PQR$?
28. Is $\triangle QRP$ congruent to $\triangle TRS$? Write *yes* or *no*.
29. Is $\triangle RQP$ congruent to $\triangle RTS$? Write *yes* or *no*.
30. Is $\triangle RST$ congruent to $\triangle RQP$? Write *yes* or *no*.
31. Is $\triangle TSR$ congruent to $\triangle PQR$? Write *yes* or *no*.

32–34. Prove the three parts of Theorem 4–4.

4–4 Tests for Congruence

To show that two triangles are congruent, it is not necessary to show all six congruent corresponding parts. For example, suppose straws are used to make a model of a triangle. Start with three straws, each having a certain length. Fasten the straws together at the ends. There is only one way to put the straws together. A different-looking triangle cannot be made with the given straws.

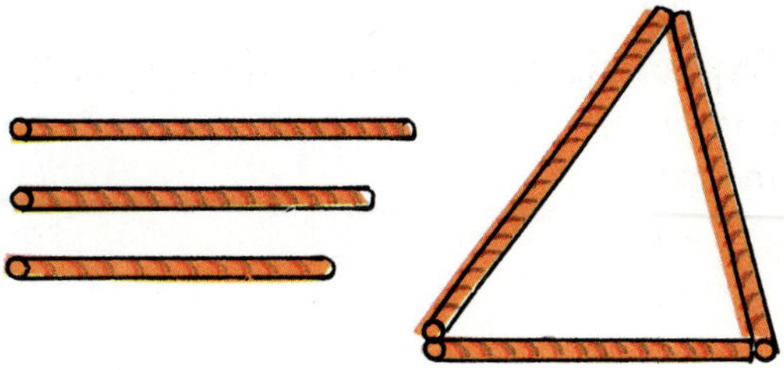

If each side of one triangle is congruent to the corresponding side of another triangle, then the triangles are congruent.

Postulate 4–1
SSS

Suppose two straws are glued together to form a certain angle. There is only one length of straw that can be used to make a triangle with the given straws. The two straws and the angle formed completely determine the triangle.

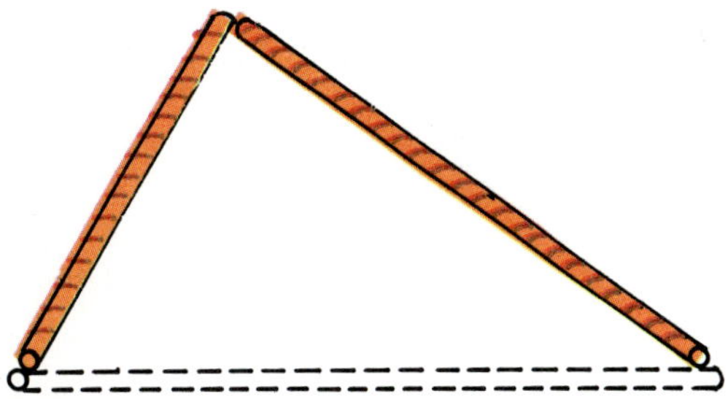

If two sides and the included angle of one triangle are congruent to the corresponding sides and included angle of another triangle, then the triangles are congruent.

Postulate 4–2
SAS

example 1

Use the figure below to tell whether or *not* $\overline{AB} \cong \overline{CD}$.

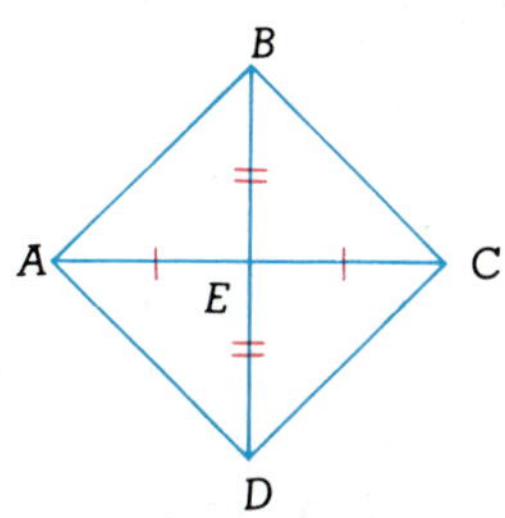

The marks on the figure show $\overline{AE} \cong \overline{CE}$ and $\overline{BE} \cong \overline{DE}$. Since vertical angles are congruent, $\angle AEB \cong \angle CED$. Thus, $\triangle AEB \cong \triangle CED$ by SAS. Since $\overline{AB}$ and $\overline{CD}$ are corresponding parts of the triangles, they are congruent. *CPCTC*

example 2

Given the figure at the right, prove $\angle 1 \cong \angle 2$.

Given: $\overline{PQ} \cong \overline{PS}$
$\overline{QR} \cong \overline{SR}$

Prove: $\angle 1 \cong \angle 2$

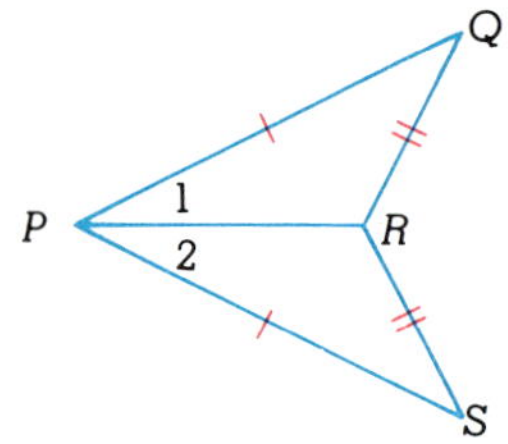

Proof:

STATEMENTS	REASONS
1. $\overline{PQ} \cong \overline{PS}$ $\overline{QR} \cong \overline{SR}$	1. Given
2. $\overline{PR} \cong \overline{PR}$	2. Theorem 2–1: Congruence of line segments is reflexive.
3. $\triangle PQR \cong \triangle PSR$	3. Postulate 4–1: SSS
4. $\angle 1 \cong \angle 2$	4. Definition of congruent triangles *CPCTC*

Postulate 4–3
ASA

If two angles and the included side of one triangle are congruent to the corresponding angles and included side of another triangle, then the triangles are congruent.

example 3

Given the figure at the right, prove $\angle P \cong \angle T$.

Given: $\angle Q$ and $\angle S$ are right angles
$\overline{QR} \cong \overline{SR}$

Prove: $\angle P \cong \angle T$

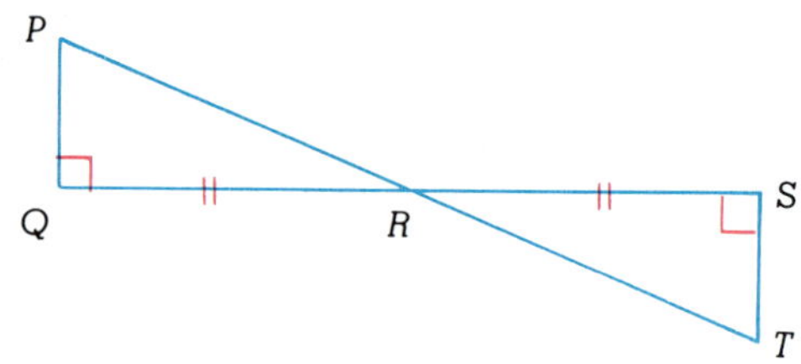

Proof:

STATEMENTS	REASONS
1. $\angle Q$ and $\angle S$ are right angles	1. Given
2. $\angle Q \cong \angle S$	2. Theorem 3–6: If two angles are right angles, then the angles are congruent.
3. $\overline{QR} \cong \overline{SR}$	3. Given
4. $\angle PRQ \cong \angle TRS$	4. Theorem 3–10: Vertical angles are congruent
5. $\triangle PRQ \cong \triangle TRS$	5. Postulate 4–3: ASA
6. $\angle P \cong \angle T$	6. Definition of congruent triangles *CPCTC*

exercises

Exploratory Are the following pairs of triangles congruent by SSS? State *yes* or *no*.

1.
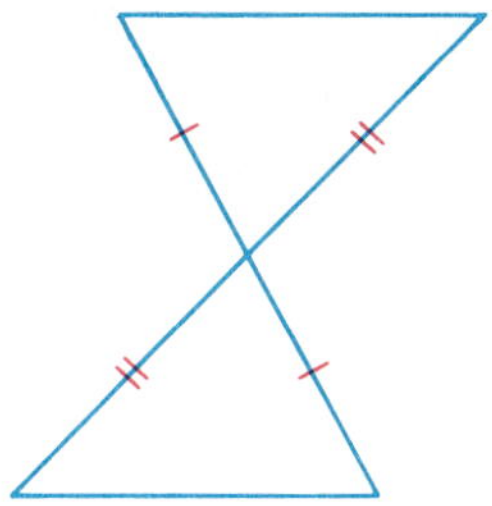

2.
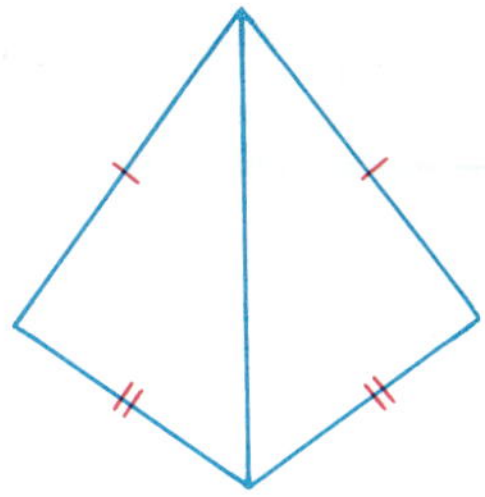

3.
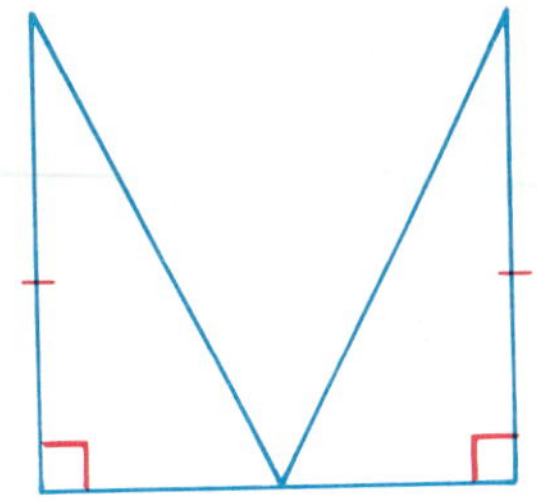

4.
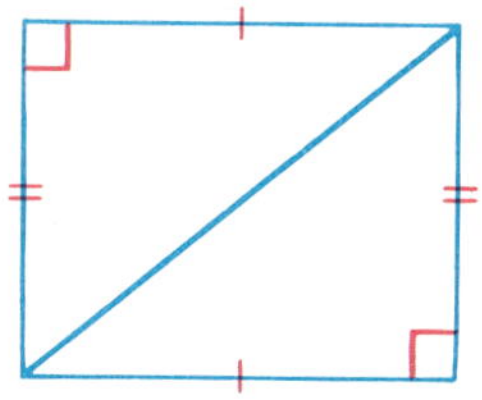

5.
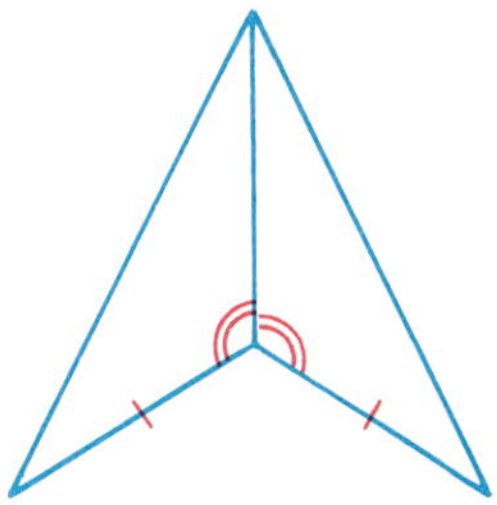

6.
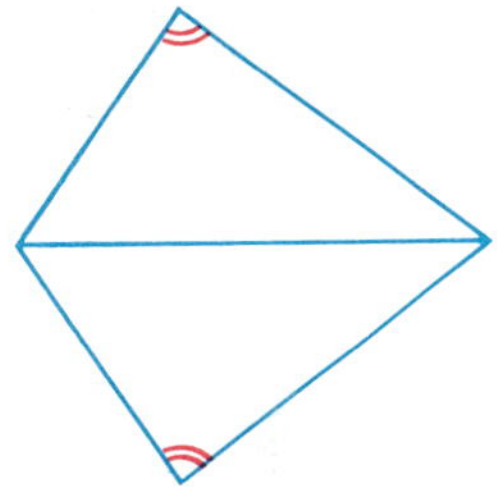

7.
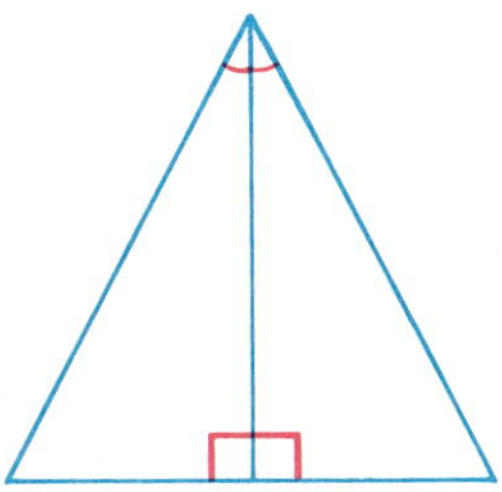

8.
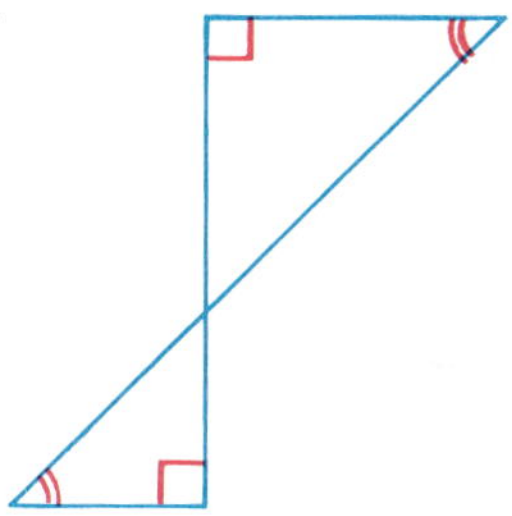

9.
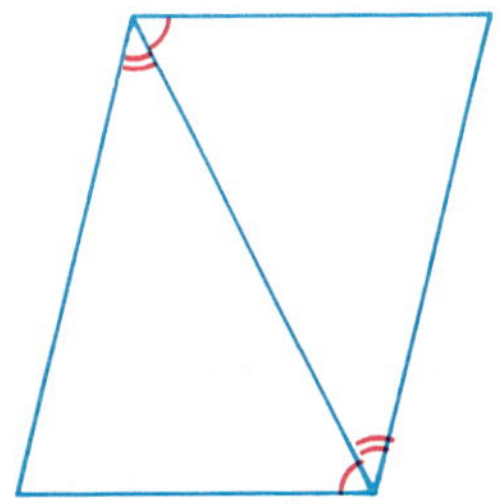

10–18. Are the pairs of triangles in problems 1–9 congruent by SAS? State *yes* or *no*.

19–27. Are the pairs of triangles in problems 1–9 congruent by ASA? State *yes* or *no*.

Written Use the figure at the right to solve each problem. $\angle P$ and $\angle T$ are right angles.

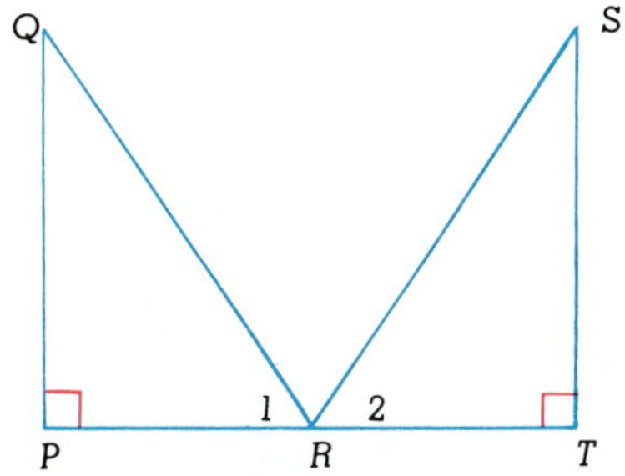

1. **Given:** $\angle 1 \cong \angle 2$
   R is the midpoint of $\overline{PT}$
   Prove: $\overline{QP} \cong \overline{ST}$
2. **Given:** $\angle 1 \cong \angle 2$
   R is the midpoint of $\overline{PT}$
   Prove: $\angle Q \cong \angle S$
3. **Given:** $\overline{QP} \cong \overline{ST}$
   R is the midpoint of $\overline{PT}$
   Prove: $\overline{QR} \cong \overline{SR}$
4. **Given:** $\overline{QP} \cong \overline{ST}$
   R is the midpoint of $\overline{PT}$
   Prove: $\angle 1 \cong \angle 2$
5. **Given:** $\angle 1 \cong \angle 2$
   $\overline{QR} \cong \overline{SR}$
   Prove: $\overline{PR} \cong \overline{TR}$

Use the figure at the right to solve each problem.

6. Given: $\overline{AC} \cong \overline{CE}$
$\angle A \cong \angle E$
Prove: $\overline{AB} \cong \overline{ED}$

7. Given: $\overline{AC} \cong \overline{CE}$
$\angle A \cong \angle E$
Prove: $\overline{BC} \cong \overline{CD}$

8. Given: $\overline{BC} \cong \overline{CD}$
$\angle A \cong \angle E$
Prove: $\overline{AC} \cong \overline{CE}$

9. Given: $\overline{BC} \cong \overline{CD}$
$\angle A \cong \angle E$
Prove: $\overline{AB} \cong \overline{ED}$

10. Given: $\overline{AC} \cong \overline{CE}$
$\overline{BC} \cong \overline{CD}$
Prove: $\angle A \cong \angle E$

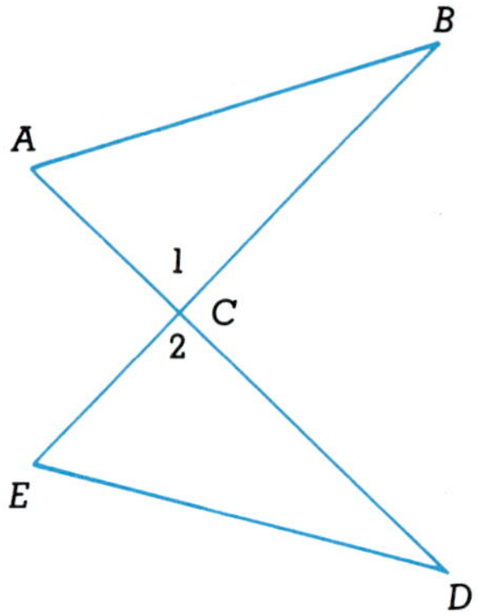

11. Given: $\overline{AC} \cong \overline{CE}$
$\overline{BC} \cong \overline{CD}$
Prove: $\angle B \cong \angle D$

Use the figure to tell whether or *not* each of the following is correct. Write *yes* or *no*.

12. $\triangle PRT$ is isosceles.
13. $\triangle PRT$ is equilateral.
14. $\triangle PVT \cong \triangle TVS$
15. $\triangle QVP \cong \triangle SVT$
16. $\triangle PQT \cong \triangle TSP$
17. $\angle 5 \cong \angle 6$
18. $\angle 3 \cong \angle 4$
19. $\angle 4 \cong \angle 6$
20. $\angle 3 \cong \angle 5$
21. $\angle 1 \cong \angle 2$
22. $\angle 1 \cong \angle 3$
23. $\angle 2 \cong \angle 3$
24. $\angle 1 \cong \angle 6$
25. $\overline{ST} \cong \overline{PQ}$
26. $\overline{RS} \cong \overline{RP}$
27. $\overline{PT} \cong \overline{PQ}$
28. $\overline{RT} \cong \overline{PR}$

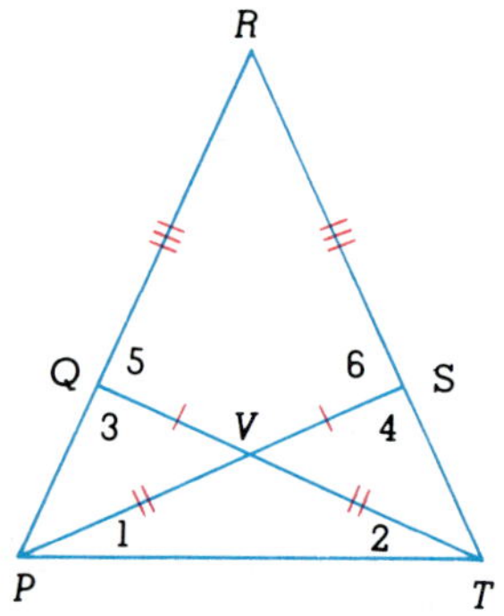

29. $\angle 1$ and $\angle 2$ are complementary.
30. $\angle 3$ and $\angle 6$ are supplementary.
31. $\angle 4$ and $\angle 5$ are supplementary.
32. $\angle 3$ and $\angle 4$ are supplementary.

Challenge Use the figures and information given below to solve each problem.

Given: $\overline{AB} \cong \overline{XY}$
$\angle A \cong \angle X$
$\overline{AC} \cong \overline{XP}$
$\angle ABC \cong \angle XYZ$

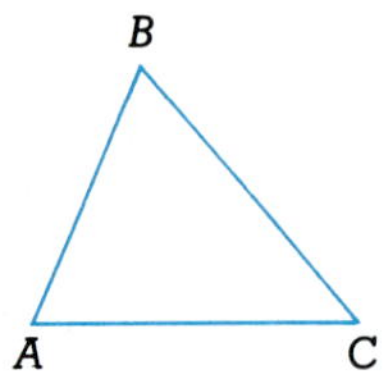

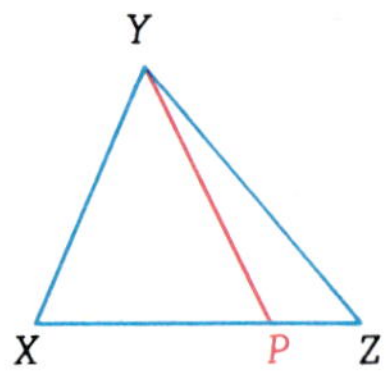

1. Prove that $\triangle ABC \cong \triangle XYP$.
2. Prove that $\angle ABC \cong \angle XYP$.
3. How are $\angle XYP$ and $\angle XYZ$ related? (Hint: Use the results of problem **2.**)
4. How is the figure at the right misleading?

4–5 More Tests for Congruence

To show two triangles congruent, *at least* three congruent corresponding parts are needed. Can any three congruent corresponding parts be used? Three tests already have been postulated, SSS, SAS, and ASA. Another test follows directly from these.

If two angles and a nonincluded side of one triangle are congruent to the corresponding angles and nonincluded side of another triangle, then the triangles are congruent.

Theorem 4–5
AAS

example 1

Prove Theorem 4–5.

Given: $\angle P \cong \angle A$
$\angle Q \cong \angle B$
$\overline{QR} \cong \overline{BC}$

Prove: $\triangle PQR \cong \triangle ABC$

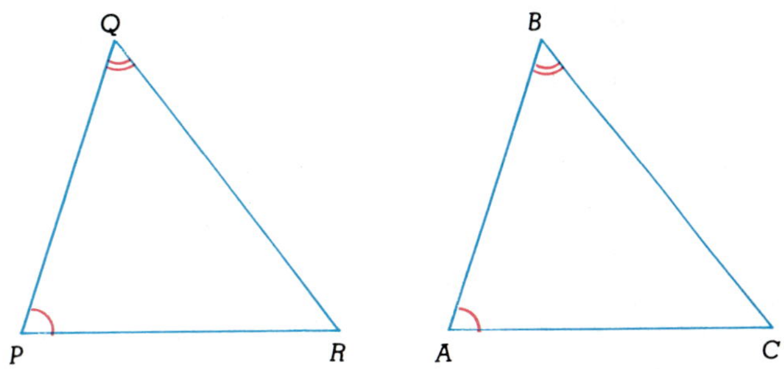

Proof:

STATEMENTS	REASONS
1. $m\angle P + m\angle Q + m\angle R = 180$ $m\angle A + m\angle B + m\angle C = 180$	1. Theorem 4–1: The sum of the degree measures of the angles of a triangle is 180.
2. $\angle P \cong \angle A$ $\angle Q \cong \angle B$	2. Given
3. $m\angle P = m\angle A$ $m\angle Q = m\angle B$	3. Definition of congruent angles
4. $m\angle P + m\angle Q + m\angle R =$ $m\angle P + m\angle Q + m\angle C$	4. Postulate 2–9: Substitution *Step 1 and 3*
5. $m\angle R = m\angle C$	5. Postulate 2–7: Subtraction Property of Equality
6. $\angle R \cong \angle C$	6. Definition of congruent angles
7. $\overline{QR} \cong \overline{BC}$	7. Given
8. $\triangle PQR = \triangle ABC$	8. Postulate 4–3: ASA *Steps 2, 6, and 7*

example 2

In the figure at the right $\angle Q$ and $\angle S$ are right angles. Also, $\angle 1 \cong \angle 3$. Prove $\triangle PQR \cong \triangle RSP$.

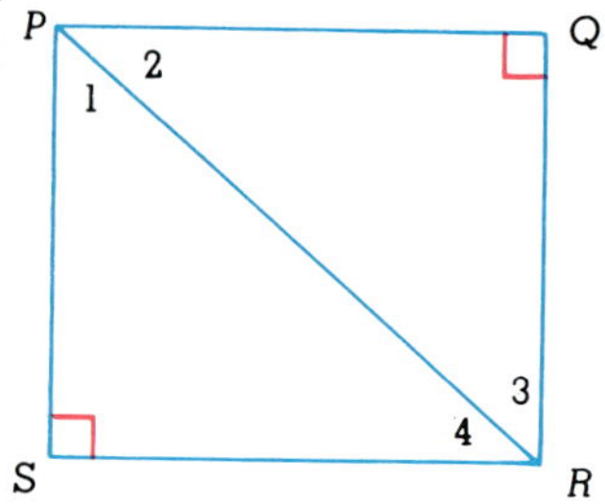

Given: $\angle Q$ and $\angle S$ are right angles
$\angle 1 \cong \angle 3$

Prove: $\triangle PQR \cong \triangle RSP$

Proof:

STATEMENTS	REASONS
1. $\angle Q$ and $\angle S$ are right angles	1. Given
2. $\angle Q \cong \angle S$	2. Theorem 3–6: If two angles are right angles, then the angles are congruent.
3. $\angle 1 \cong \angle 3$	3. Given
4. $\overline{PR} \cong \overline{PR}$	4. Theorem 2–2: Congruence of line segments is reflexive.
5. $\triangle PQR \cong \triangle RSP$	5. Theorem 4–5: AAS *Steps 2, 3, and 4*

Suppose measurements for two sides and a nonincluded angle of a triangle are given. The sides measure 7.3 cm and 5 cm. The angle measures 40°. Can you draw a picture of the triangle?

Two different figures can be drawn using the information given. There is *not* enough information to know which figure is correct.

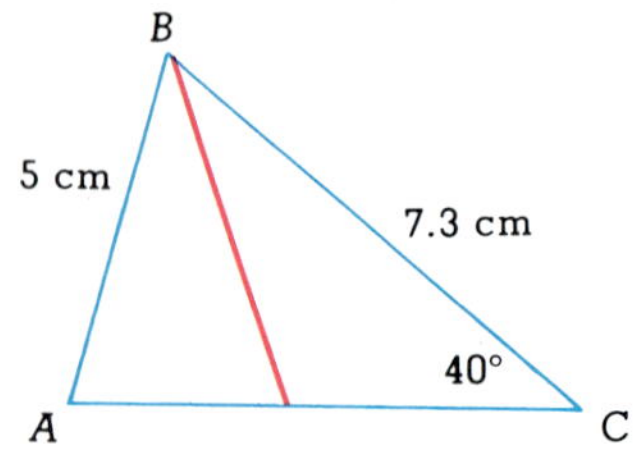

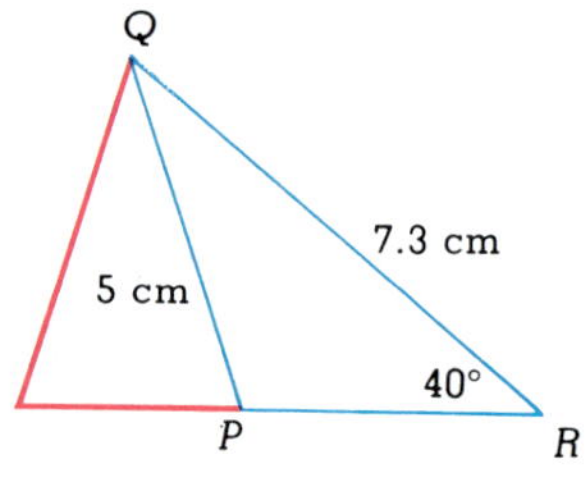

Two sides and a nonincluded angle of $\triangle ABC$ are congruent to two corresponding sides and a nonincluded angle of $\triangle PQR$. The two triangles are *not* congruent. Thus, SSA is *not* a test for congruent triangles. *Is AAA a test for congruent triangles?*

exercises

Exploratory Name the postulate or theorem you would use to prove $\triangle ACB \cong \triangle ZXY$ given the following information. If there is *not* enough information, state none.

1. $\angle B \cong \angle Y$, $\angle A \cong \angle Z$, $\overline{BC} \cong \overline{YX}$
2. $\overline{AB} \cong \overline{ZY}$, $\overline{AC} \cong \overline{ZX}$, $\angle B \cong \angle Y$
3. $\angle C \cong \angle X$, $\angle A \cong \angle Z$, $\overline{CA} \cong \overline{XZ}$
4. $\angle A \cong \angle Z$, $\overline{AC} \cong \overline{ZX}$, $\angle B \cong \angle Y$
5. $\overline{AC} \cong \overline{ZX}$, $\angle B \cong \angle Y$, $\overline{BC} \cong \overline{YX}$
6. $\overline{AB} \cong \overline{ZY}$, $\overline{AC} \cong \overline{ZX}$, $\overline{CB} \cong \overline{XY}$

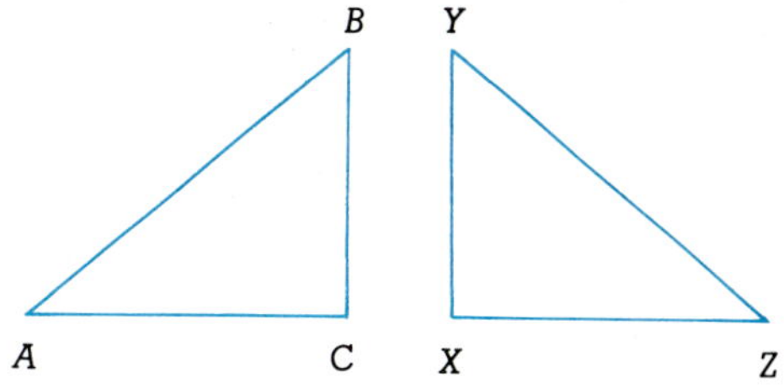

7. $\angle A \cong \angle Z$, $\angle C \cong \angle X$, $\angle B \cong \angle Y$
8. $\angle A \cong \angle Z$, $\angle C \cong \angle X$, $\overline{BC} \cong \overline{YX}$
9. $\overline{AB} \cong \overline{ZY}$, $\angle C \cong \angle X$, $\angle Z \cong \angle A$
10. $\overline{ZY} \cong \overline{AB}$, $\overline{ZX} \cong \overline{AC}$, $\angle X \cong \angle C$
11. $\overline{ZX} \cong \overline{AC}$, $\overline{XY} \cong \overline{CB}$, $\angle X \cong \angle C$
12. $\angle Y \cong \angle B$, $\angle C \cong \angle X$, $\overline{ZX} \cong \overline{AC}$
13. $\angle Y \cong \angle B$, $\overline{ZY} \cong \overline{AB}$, $\angle C \cong \angle X$
14. $\angle A \cong \angle Z$, $\overline{YX} \cong \overline{BC}$, $\overline{ZY} \cong \overline{AB}$
15. $\overline{BC} \cong \overline{YX}$, $\angle X \cong \angle C$, $\overline{AB} \cong \overline{ZY}$
16. $\angle A \cong \angle Z$, $\angle B \cong \angle Y$, $\overline{YX} \cong \overline{AB}$
17. $\overline{AB} \cong \overline{YZ}$, $\angle C \cong \angle Y$, $\angle B \cong \angle X$
18. $\angle A \cong \angle Z$, $\angle B \cong \angle X$, $\overline{AC} \cong \overline{XY}$

Written Use the figure at the right to solve each problem.

1. **Given:** $\angle A \cong \angle C$, $\angle 1 \cong \angle 2$
   Prove: $\overline{AB} \cong \overline{CB}$
2. **Given:** $\angle A \cong \angle C$, $\overline{BD} \perp \overline{AC}$
   Prove: $\overline{AB} \cong \overline{CB}$
3. **Given:** $\angle A \cong \angle C$, $\overline{BD} \perp \overline{AC}$
   Prove: $\triangle ABD \cong \triangle CBD$
4. **Given:** $\angle A \cong \angle C$, $\angle 1 \cong \angle 2$
   Prove: $\overline{AD} \cong \overline{DC}$
5. **Given:** $\angle A \cong \angle C$, $\angle 1 \cong \angle 2$
   Prove: $\overline{BD}$ bisects $\overline{AC}$
6. **Given:** $\angle A \cong \angle C$, $\overline{BD} \perp \overline{AC}$
   Prove: $\overline{BD}$ bisects $\overline{AC}$

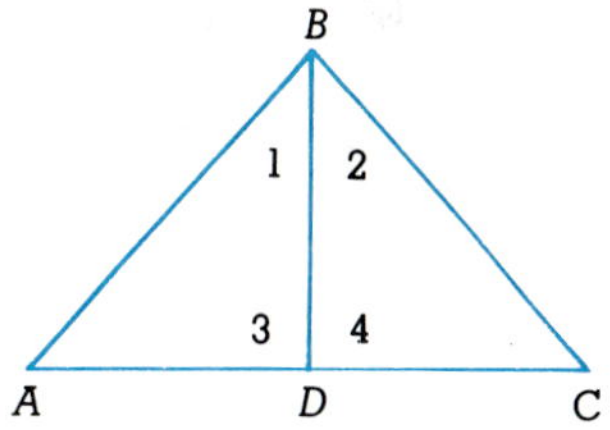

Use the figure at the right to solve the following problem.

7. **Given:** J is the midpoint of $\overline{GK}$
   $\angle H$ is a right angle
   $\angle L$ is a right angle
   Prove: $\overline{HG} \cong \overline{LK}$

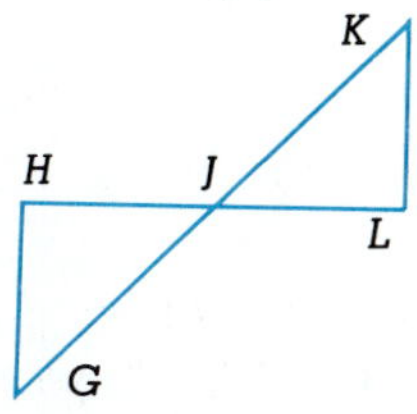

Use the figure at the right to solve each problem.

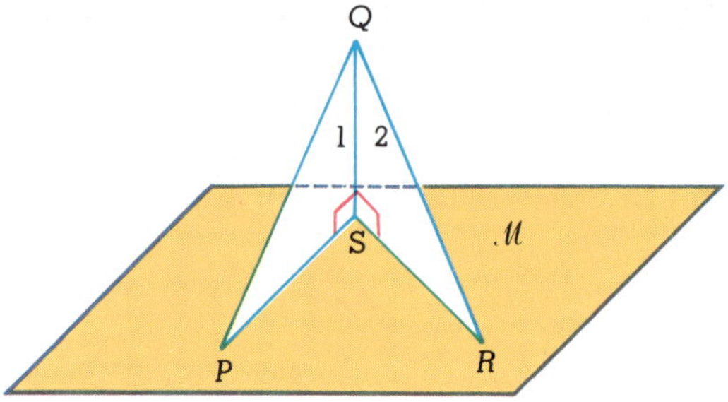

8. **Given:** $\overline{PS}$ and $\overline{SR}$ are in $\mathcal{M}$
$\overline{QS} \perp \mathcal{M}$
$\angle P \cong \angle R$
Prove: $\overline{QP} \cong \overline{QR}$

9. **Given:** $\overline{PS}$ and $\overline{SR}$ are in $\mathcal{M}$
$\overline{QS} \perp \mathcal{M}$
$\angle P \cong \angle R$
Prove: $\overline{PS} \cong \overline{SR}$

10. Draw two triangles that are *not* congruent, but for which SSA holds. Indicate all measurements.
11. Draw two triangles that are *not* congruent, but for which AAA holds. Indicate all measurements.

Using Computers

Studying Traffic

Computer display systems are used to find ways of improving efficiency and safety of traffic. A computer contains a mathematical description of a traffic situation, which appears graphically on a special screen. Researchers type data into the computer. Then, they touch appropriate parts of the screen with a *light pen*. The display changes, showing how the information affects the traffic pattern.

4–6 Medians, Altitudes, and Bisectors

Every triangle has three **medians** and three **altitudes**. They are defined in the following way.

A segment is a median of a triangle if and only if it connects a vertex to the midpoint of the opposite side.

Definition of Median of a Triangle

A segment is an altitude of a triangle if and only if it connects a vertex to the line containing the opposite side so that the segment is perpendicular to that line.

Definition of Altitude of Triangle

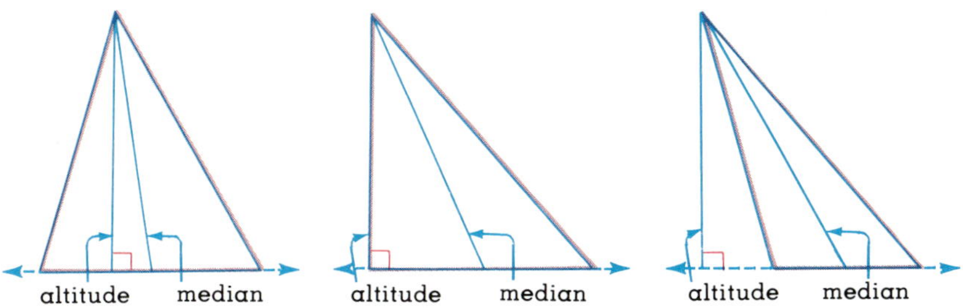

example 1

Complete the following proof.

Given: $\overline{AB} \cong \overline{BC}$

$\overline{BD}$ is a median of $\triangle ABC$

Prove: $\angle ABD \cong \angle CBD$

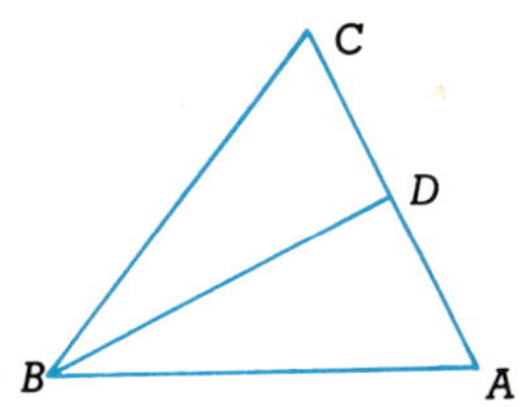

Proof:

STATEMENTS	REASONS
1. $\overline{BD}$ is a median of $\triangle ABC$	1. Given
2. D is a midpoint of $\overline{AC}$	2. Definition of median of triangle
3. $\overline{AD} \cong \overline{DC}$	3. Theorem 2–5: If M is the midpoint of $\overline{PQ}$, then $\overline{PM} \cong \overline{MQ}$.
4. $\overline{AB} \cong \overline{BC}$	4. Given
5. $\overline{BD} \cong \overline{BD}$	5. Theorem 2–2: Congruence of line segments is reflexive.
6. $\triangle ABD \cong \triangle CBD$	6. Postulate 4–1: SSS *Steps 3, 4, 5*
7. $\angle ABD \cong \angle CBD$	7. Definition of congruent triangles *CPCTC*

A ray, $\overrightarrow{QS}$, is called a **bisector of an angle** if and only if it is the common side of two congruent adjacent angles. In the figure at the right, $\angle 1 \cong \angle 2$. Because $\overrightarrow{QS}$ separates $\angle PQR$ into these two congruent angles, it is called the bisector of $\angle PQR$.

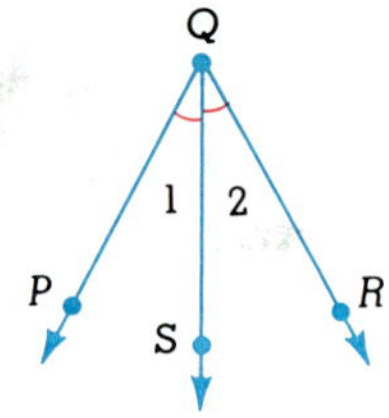

Definition of Bisector of an Angle

A ray, $\overrightarrow{QS}$, is a bisector of $\angle PQR$ if and only if S is in the interior of the angle and $\angle PQS \cong \angle RQS$.

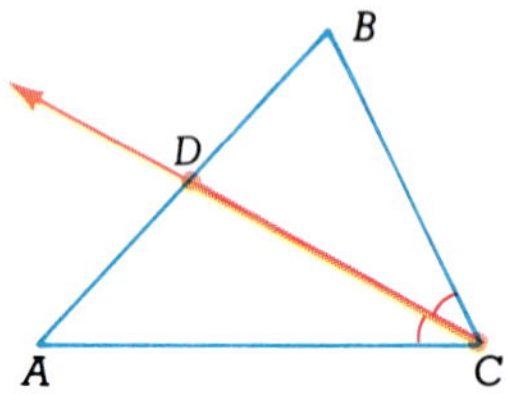

A triangle has three angle bisectors, one for each angle. In the figure at the left, $\overline{CD}$ is an angle bisector of $\triangle ABC$. We say that $\overline{CD}$ bisects $\angle ACB$. *An angle bisector of a triangle is a segment which is contained in a bisector of one of the angles of the triangle, and whose endpoints are on the triangle.*

example 2

Complete the following proof.

Given: $\overline{PR}$ bisects $\angle QPS$
$\overline{PR} \perp \overline{QS}$

Prove: $\triangle PQS$ is isosceles

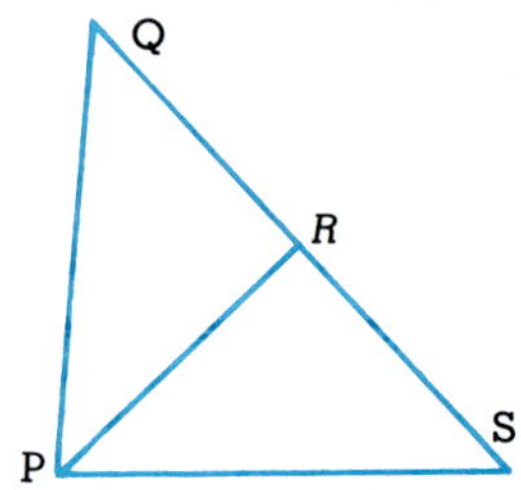

Proof:

STATEMENTS	REASONS
1. $\overline{PR}$ bisects $\angle QPS$	1. Given
2. $\angle QPR \cong \angle SPR$	2. Definition of angle bisector
3. $\overline{PR} \perp \overline{QS}$	3. Given
4. $\angle QRP$ and $\angle SRP$ are right angles	4. Definition of perpendicular lines
5. $\angle QRP \cong \angle SRP$	5. Theorem 3–6: If two angles are right angles, then the angles are congruent.
6. $\overline{PR} \cong \overline{PR}$	6. Theorem 2–2: Congruence of line segments is reflexive.
7. $\triangle PQR \cong \triangle PSR$	7. Postulate 4–3: ASA *Steps 2, 5, 6*
8. $\overline{QP} \cong \overline{SP}$	8. Definition of congruent triangles *CPCTC*
9. $\triangle PQS$ is isosceles	9. Definition of isosceles triangles

exercises

Exploratory In the figure at the right, $\overline{AC}$ bisects $\angle BAD$, and $\overline{CE}$ bisects $\angle ACD$. Also, $\overline{AB}$ is an altitude of $\triangle ABD$, and $\overline{CE}$ is an altitude of $\triangle ACD$. Find the measure for each of the following.

1. $\angle BAD$
2. $\angle CEA$
3. $\angle CED$
4. $\angle BAC$
5. $\angle CAE$
6. $\angle ACE$
7. $\angle DCE$
8. $\angle ECD$
9. $\angle CDE$
10. $\angle DCA$
11. $\angle ACB$
12. $\angle CBA$

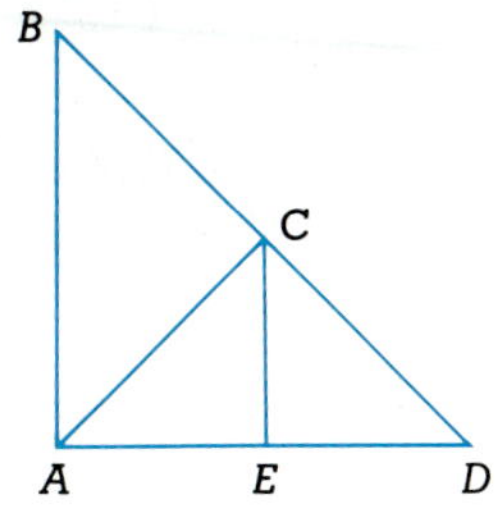

Written Draw diagrams to illustrate each of the following.

1. $\overline{PS}$ is an altitude of $\triangle PQR$, and S is between Q and R.
2. $\overline{PS}$ is a median of $\triangle PQR$, and S is between Q and R.
3. $\overrightarrow{SQ}$ bisects $\angle PSR$.
4. $\overline{HL}$ is an angle bisector of $\triangle HJK$.
5. $\overline{BD}$ is a median of $\triangle ABC$, and bisects $\angle ABC$.
6. $\overline{EC}$ is an altitude of $\triangle ABC$ and bisects $\angle BCA$.
7. $\angle PVR$ and $\angle TVR$ form a linear pair. $\overrightarrow{VQ}$ bisects $\angle PVR$, and $\overrightarrow{VS}$ bisects $\angle TVR$.
8. $\overrightarrow{AP}$ bisects $\angle MLR$ and $\angle NLQ$.
9. $\triangle ABC$ is a right triangle with altitude $\overline{BC}$ and median $\overline{BD}$.
10. $\overline{PT}$ and $\overline{RS}$ are medians of $\triangle PQR$ and intersect at V.

Prove each of the following.

11. **Given:** $\overline{AB}$ is an altitude of $\triangle ABD$
 $\overline{CD}$ is an altitude of $\triangle BCD$
 $\overline{BC} \cong \overline{AD}$, $\overline{CD} \cong \overline{AD}$
 $\overline{AB} \cong \overline{CD}$
 Prove: $\overline{BD}$ bisects $\angle ABC$

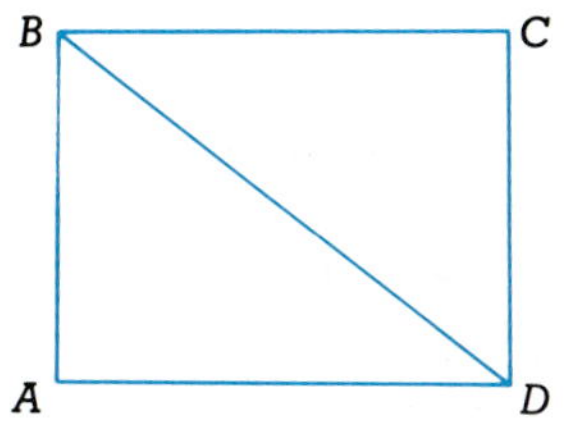

12. **Given:** $\overline{PR}$ is an altitude of $\triangle PQR$
 R is the midpoint of $\overline{PS}$
 R is the midpoint of $\overline{QT}$
 Prove: $\overline{RS}$ is an altitude of $\triangle RST$

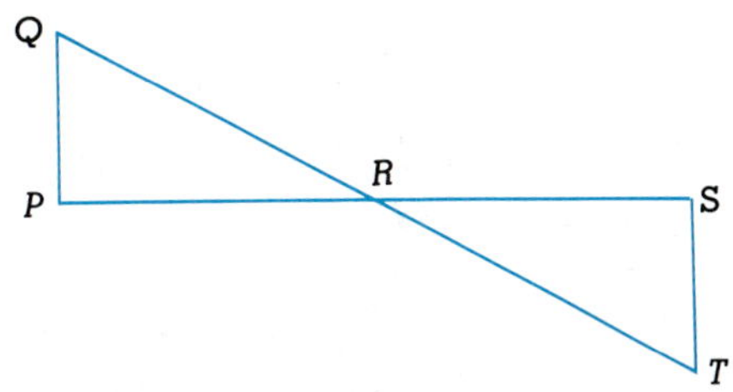

13. The median from the vertex angle of an isosceles triangle bisects the vertex angle.
14. The median from the vertex angle of an isosceles triangle is an altitude of the triangle.
15. If a median of a triangle also is an altitude, then the triangle is isosceles.
16. If a median of a triangle also is an altitude, then it is an angle bisector of the triangle.

4–7 Isosceles Triangles

Some proofs include geometric figures that are *not* given in the theorem. These figures often are lines or parts of lines. They are included on the diagram in color. The proof of the following theorem uses an **auxiliary** line segment.

Theorem 4–6
Isosceles Triangle Theorem

If two sides of a triangle are congruent, then the angles opposite those sides are congruent.

example 1

Prove Theorem 4–6.

Given: $\triangle PQR$
$\overline{PQ} \cong \overline{QR}$

Prove: $\angle P \cong \angle R$

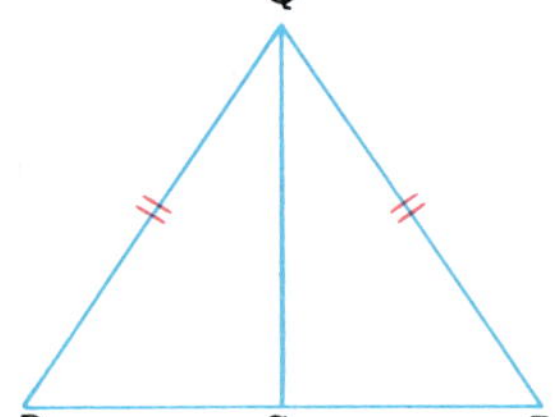

Proof:

STATEMENTS	REASONS
1. Call S the midpoint of $\overline{PR}$ and draw $\overline{QS}$.	1. Theorem 2–1: A line segment has exactly one midpoint, and Postulate 1–1: Through any two points there is exactly one line.
2. $\overline{PS} \cong \overline{SR}$	2. Theorem 2–5: If M is the midpoint of $\overline{PQ}$, then $\overline{PM} \cong \overline{MQ}$.
3. $\overline{PQ} \cong \overline{QR}$	3. Given
4. $\overline{QS} \cong \overline{QS}$	4. Theorem 2–2: Congruence of line segments is reflexive.
5. $\triangle PQS \cong \triangle RQS$	5. Postulate 4–1: SSS
6. $\angle P \cong \angle R$	6. Definition of congruent triangles *CPCTC*

The proofs of the following theorems are based on the Isosceles Triangle Theorem.

Theorem 4–7

If a triangle is equilateral, then the triangle is equiangular.

Theorem 4–8

Each angle of an equilateral triangle has degree measure 60.

It is worthless to introduce an auxiliary figure unless it is useful for proving the given theorem. The line segment in the proof of Theorem 4–6 is useful because it forms two triangles. The proof is based on showing these two triangles are congruent.

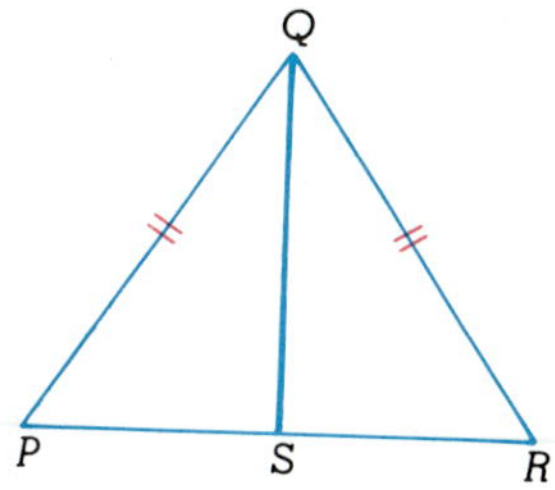

The proof of the following theorem also uses an auxiliary line segment.

If two angles of a triangle are congruent, then the sides opposite those angles are congruent.

Theorem 4–9

According to Theorem 4–7 and the following theorem, a triangle is equilateral if and only if it is equiangular.

If a triangle is equiangular, then the triangle is equilateral.

Theorem 4–10

examples

2 **Plan a proof for Theorem 4–9.**

Draw $\overline{QT}$ perpendicular to $\overline{PR}$. Show $\triangle PQT \cong \triangle RQT$ by AAS. Then $\overline{PQ} \cong \overline{RQ}$ because they are corresponding parts of congruent triangles.

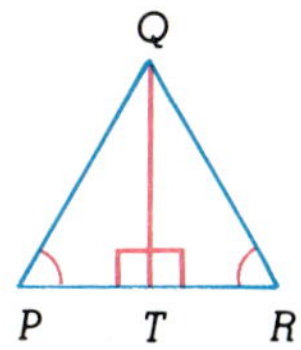

3 **Prove Theorem 4–10.**

Given: $\angle P \cong \angle Q$
$\angle Q \cong \angle R$
$\angle R \cong \angle P$

Prove: $\overline{PQ} \cong \overline{QR}$
$\overline{QR} \cong \overline{PR}$
$\overline{PR} \cong \overline{PQ}$

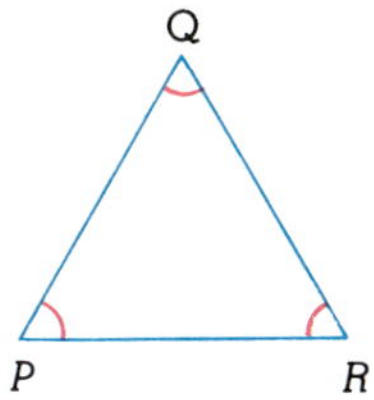

Proof:

STATEMENTS	REASONS
1. $\angle P \cong \angle Q$ $\angle Q \cong \angle R$ $\angle R \cong \angle P$	1. Given
2. $\overline{PQ} \cong \overline{QR}$ $\overline{QR} \cong \overline{PR}$ $\overline{PR} \cong \overline{PQ}$	2. Theorem 4–9: If two angles of a triangle are congruent, then the sides opposite those angles are congruent.

exercises

Exploratory For each of the following will the auxiliary figure be helpful for the proof? State *yes* or *no*.

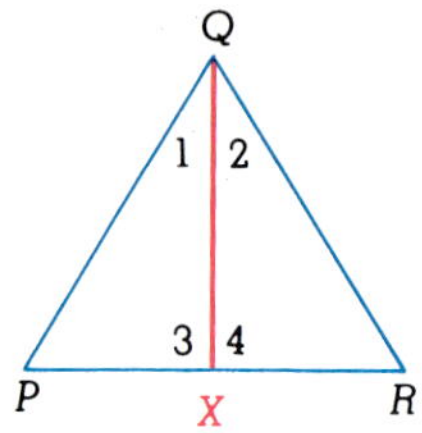

1. **Given:** $\overline{PQ} \cong \overline{QR}$
   Prove: $\angle P \cong \angle R$
   Draw $\overline{QX}$ so it bisects $\angle PQR$
2. **Given:** $\overline{PQ} \cong \overline{QR}$
   Prove: $\angle P \cong \angle R$
   Draw $\overline{QX}$ so it bisects $\overline{PR}$
3. **Given:** $\overline{PQ} \cong \overline{QR}$
   Prove: $\angle P \cong \angle R$
   Draw $\overline{QX}$ perpendicular to $\overline{PR}$
4. **Given:** $\overline{PQ} \cong \overline{QR}$
   Prove: $\angle P \cong \angle R$
   Draw X, the midpoint of $\overline{PR}$
5. **Given:** $\overline{PQ} \cong \overline{PR}$
   Prove: $\angle Q \cong \angle R$
   Draw $\overline{QX}$ so it bisects $\angle PQR$
6. **Given:** $\angle P \cong \angle R$
   Prove: $\overline{PQ} \cong \overline{RQ}$
   Draw $\overline{QX}$ perpendicular to $\overline{PR}$
7. **Given:** $\angle P \cong \angle R$
   Prove: $\overline{PQ} \cong \overline{RQ}$
   Draw $\overline{QX}$ so it bisects $\angle PQR$
8. **Given:** $\angle R \cong \angle P$
   Prove: $\overline{QR} \cong \overline{QP}$
   Draw X, the midpoint of $\overline{PR}$
9. **Given:** $\angle P \cong \angle Q$
   Prove: $\overline{PQ} \cong \overline{PR}$
   Draw median $\overline{QX}$.

Written Solve each problem.

1. Draw $\triangle ABC$ so that $\overline{AD} \perp \overline{BC}$, but $\overline{BD}$ and $\overline{DC}$ are *not* congruent.
2. Draw a triangle, $\triangle ABC$, so that $\overline{BD} \cong \overline{DC}$, but $\overline{AD}$ is *not* perpendicular to $\overline{BC}$.
3. Suppose $\triangle ABC$ is such that $\overline{AD}$ is the perpendicular bisector of $\overline{BC}$. Classify the triangle as *scalene*, *isosceles*, or *equilateral*.
4. Draw $\triangle ABC$ so that $\overline{AD}$ bisects $\angle BAC$, but $\overline{BD}$ and $\overline{DC}$ are *not* congruent.
5. Draw $\triangle ABC$ so that $\overline{BD} \cong \overline{DC}$, but $\overline{AD}$ does *not* bisect $\angle BAC$.
6. Suppose $\triangle ABC$ is such that $\overline{AD}$ bisects both $\angle BAC$ and $\overline{BC}$. Classify the triangle as *scalene*, *isosceles*, or *equilateral*.
7. Draw $\triangle ABC$ so that $\overline{AD}$ bisects $\angle BAC$, but $\overline{AD}$ is *not* perpendicular to $\overline{BC}$.
8. Draw $\triangle ABC$ so that $\overline{AD} \perp \overline{BC}$, but $\overline{AD}$ does *not* bisect $\angle BAC$.
9. Suppose $\triangle ABC$ is such that $\overline{AD}$ bisects $\angle BAC$ and $\overline{AD} \perp \overline{BC}$. Classify the triangle as *scalene*, *isosceles*, or *equilateral*.

Before you introduce an auxiliary figure, be sure it exists and it is *not* a special case. For each of the following, state whether or *not* you can introduce the figure in $\triangle PQR$. Write *yes* or *no*. If your answer is *no*, explain.

Sample **Let $\overline{PX}$ bisect $\angle P$ where X is the midpoint of $\overline{QR}$.**

No. Every angle has a bisector, but it may *not* necessarily intersect the opposite side of the triangle at its midpoint.

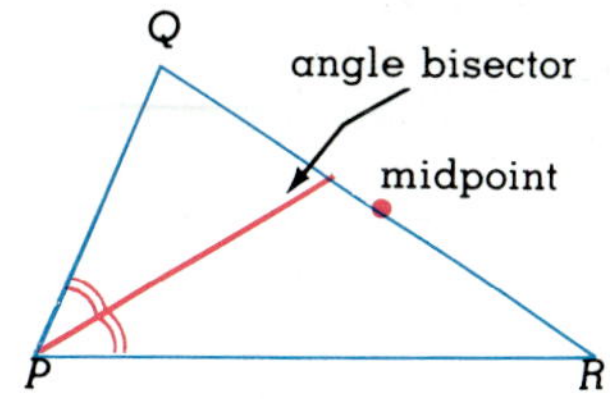

10. Let $\overline{QX}$ bisect $\angle Q$ where X is between P and R.
11. Let $\overline{RX}$ be the perpendicular bisector of $\overline{PQ}$.
12. Let $\overline{RV}$ bisect $\angle R$ and $\overline{PQ}$.
13. Let M be the midpoint of $\overline{QR}$ and draw $\overline{PM}$.
14. Let $\overline{XY}$ bisect $\overline{QP}$ and $\overline{PR}$, where X is on $\overline{QP}$ and Y is on $\overline{PR}$.
15. Let $\overline{XY}$ be perpendicular to $\overline{QP}$ and $\overline{PR}$, where X is on $\overline{QP}$ and Y is on $\overline{PR}$.

Use the figure at the right to solve each problem.

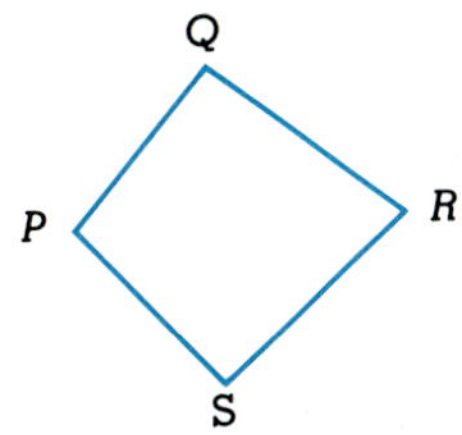

16. **Given:** $\overline{PQ} \cong \overline{PS}$, $\overline{QR} \cong \overline{SR}$
    Prove: $\angle Q \cong \angle S$
17. **Given:** $\overline{PQ} \cong \overline{RS}$, $\overline{QR} \cong \overline{SP}$
    Prove: $\angle Q \cong \angle S$
18. **Given:** $\overline{PQ} \cong \overline{RQ}$, $\overline{PS} \cong \overline{RS}$
    Prove: $\angle P \cong \angle R$
19. **Given:** $\overline{PQ} \cong \overline{RS}$, $\overline{PS} \cong \overline{RQ}$
    Prove: $\angle P \cong \angle R$

Use the figure at the right to solve each problem.

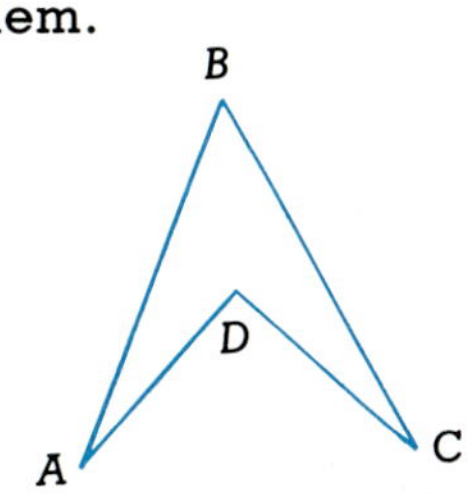

20. **Given:** $\overline{AB} \cong \overline{CB}$, $\overline{AD} \cong \overline{CD}$
    Prove: $\angle A \cong \angle C$
21. **Given:** $\overline{CB} \cong \overline{AD}$, $\overline{CD} \cong \overline{AB}$
    Prove: $\angle A \cong \angle C$
22. **Given:** $\overline{CB} \cong \overline{CD}$, $\overline{CD} \cong \overline{AD}$, $\overline{AD} \cong \overline{AB}$
    Prove: $\angle A \cong \angle C$
23. **Given:** $\overline{BA} \cong \overline{DA}$, $\overline{DC} \cong \overline{BC}$, $\overline{AB} \cong \overline{CB}$
    Prove: $\angle A \cong \angle C$

24. Prove Theorem 4–7.
25. Prove Theorem 4–8.
26. Prove Theorem 4–9.

Reading Geometry

Taking Notes

Taking notes as you move from step to step in a mathematical passage helps your reading comprehension. These notes are *not* summaries, but aids enabling you to continue reading. Such notes usually are thrown out after the passage is completed.

The following pictures show how James Walker took notes as he read page 112.

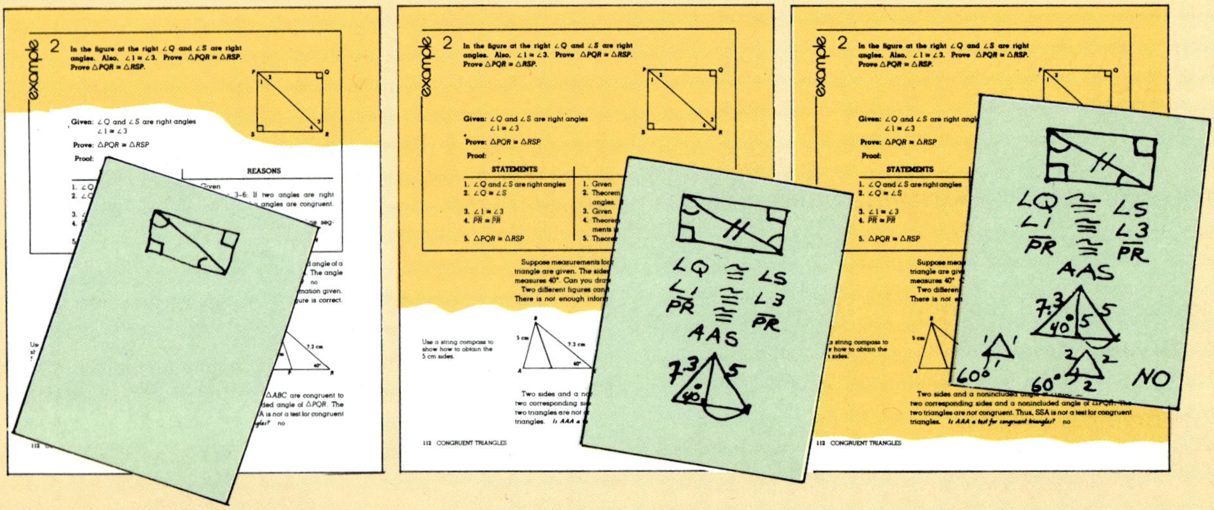

Often, mathematical passages themselves are summaries. They do *not* always show the steps involved in reaching a conclusion. For example, rather than showing several diagrams to indicate the steps followed in finding congruent parts of triangles, a passage will show one completed diagram. In such a case, it is helpful to draw the diagram step-by-step. For another example, sometimes only the results of a computation are shown. In such a case, it is helpful to do the computation.

Exercises For each of the following, suggest ways of taking notes that might be helpful.

1. The segment joining any two points interior to a triangle also is interior to the triangle.
2. The information given about the triangles indicates $\triangle ABC \cong \triangle DEF$. There is *not* enough information to tell whether $\triangle ABC$ is congruent to $\triangle EDF$. What information is needed?
3. Postulate 4–3 in the middle of page 108.
4. Example 1 on page 111.
5. Top half of page 100.
6. Example 2 on page 119.

4-8 Right Triangles

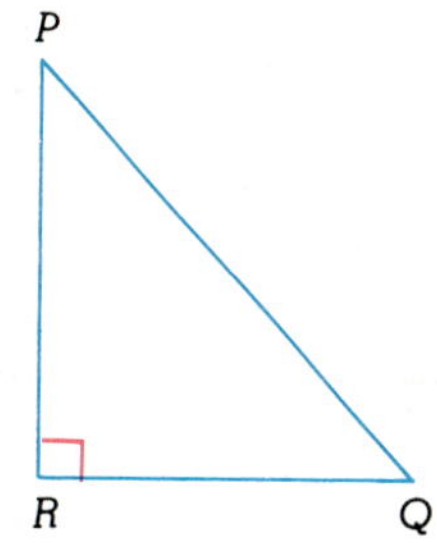

In a right triangle, the side opposite the right angle is called the **hypotenuse.** The other two sides are called the **legs.** The hypotenuse of $\triangle PQR$ is $\overline{PQ}$. The legs are $\overline{PR}$ and $\overline{RQ}$.

All right *angles* are congruent. Thus, any two right triangles automatically have one set of angles congruent. The following theorems state conditions for right triangles to be congruent.

If the hypotenuse and an acute angle of one right triangle are congruent to the corresponding hypotenuse and angle of another right triangle, then the triangles are congruent.

Theorem 4–11
HA

If two legs of one right triangle are congruent to the corresponding legs of another right triangle, then the triangles are congruent.

Theorem 4–12
LL

example 1

In the given figure, $\angle Q$ and $\angle S$ are right angles. Also, $\angle 1 \cong \angle 2$. Prove $\triangle PQR \cong \triangle RSP$.

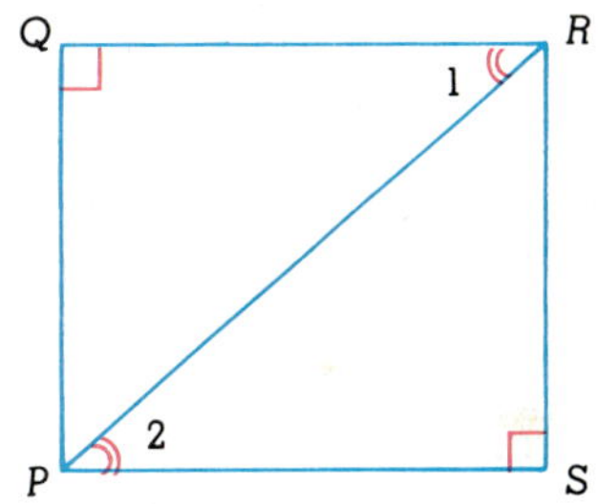

Given: $\angle Q$ and $\angle S$ are right angles
$\angle 1 \cong \angle 2$

Prove: $\triangle PQR \cong \triangle RSP$

Proof:

STATEMENTS	REASONS
1. $\angle Q$ and $\angle S$ are right angles	1. Given
2. $\triangle PQR$ and $\triangle RSP$ are right triangles	2. Definition of right triangles
3. $\angle 1 \cong \angle 2$	3. Given
4. $\overline{PR} \cong \overline{PR}$	4. Theorem 2–2: Congruence of line segments is reflexive.
5. $\triangle PQR \cong \triangle RSP$	5. Theorem 4–11: HA *Steps 3 and 4*

Suppose one leg and an acute angle of one right triangle are congruent to the corresponding leg and angle of another right triangle. Are the triangles congruent? There are two different cases to consider.

CASE 1: The legs are included between the acute angle and the right angles.

CASE 2: The legs are not included between the acute angle and the right angles.

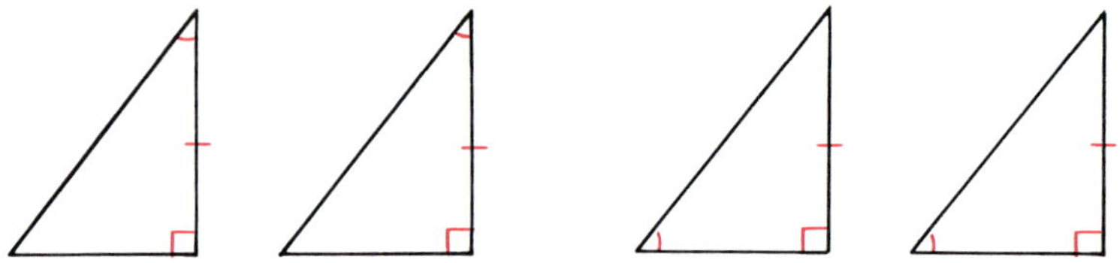

In the first case, the triangles are congruent by ASA. In the second case, the triangles are congruent by AAS.

Theorem 4–13
LA

If one leg and an acute angle of one right triangle are congruent to the corresponding leg and angle of another right triangle, then the triangles are congruent.

example 2

Complete the following proof.

Given: $\overline{BE} \perp \overline{AD}$
C is the midpoint of $\overline{AD}$
$\angle B \cong \angle E$

Prove: $\overline{AB} \cong \overline{DE}$

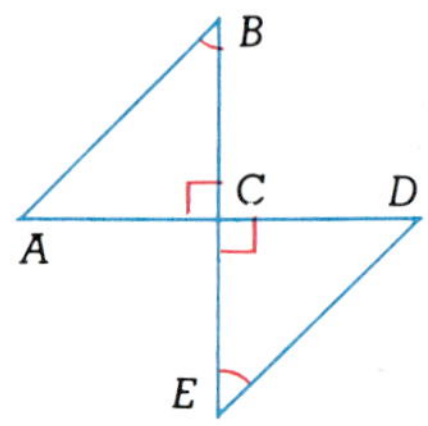

Proof:

STATEMENTS	REASONS
1. $\overline{BE} \perp \overline{AD}$	1. Given
2. $\angle BCA$ and $\angle DCE$ are right angles	2. Theorem 3–11: If two lines are perpendicular, then they form four right angles.
3. $\triangle BCA$ and $\triangle DCE$ are right triangles	3. Definition of right triangles
4. C is the midpoint of $\overline{AD}$	4. Given
5. $\overline{AC} \cong \overline{CD}$	5. Theorem 2–4: The midpoint of a line segment separates the segment into two congruent segments.
6. $\angle B \cong \angle E$	6. Given
7. $\triangle BCA \cong \triangle ECD$	7. Theorem 4–13: LA *Steps 5 and 6*
8. $\overline{AB} \cong \overline{DE}$	8. Definition of congruent triangles *CPCTC*

Viewed from the side, a ladder leaning against a building forms a right triangle. If a ladder is placed against the building so it reaches a certain window, then the foot of the ladder touches the ground at a certain spot. Any other ladder of the same length touches the ground at that same spot.

If the hypotenuse and a leg of one right triangle are congruent to the corresponding sides of another right triangle, then the triangles are congruent.

Postulate 4–4
HL

exercises

Exploratory For each of the following, copy the figure and mark it to describe the problem.

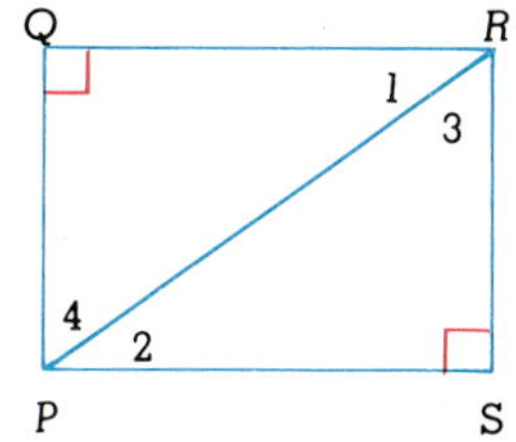

1. **Given:** $\angle Q$ and $\angle S$ are right angles
$\angle 1 \cong \angle 2$
Prove: $\overline{QP} \cong \overline{SR}$
2. **Given:** $\angle Q$ and $\angle S$ are right angles
$\angle 1 \cong \angle 2$
Prove: $\overline{QR} \cong \overline{PS}$
3. **Given:** $\angle Q$ and $\angle S$ are right angles
$\angle 1 \cong \angle 2$
Prove: $\angle QRS$ is a right angle
4. **Given:** $\angle Q$ and $\angle S$ are right angles
$\angle 1 \cong \angle 2$
Prove: $\angle QPS$ is a right angle
5. **Given:** $\angle Q$ and $\angle S$ are right angles
$\overline{QP} \cong \overline{SR}$
Prove: $\overline{QR} \cong \overline{PS}$
6. **Given:** $\angle Q$ and $\angle S$ are right angles
$\overline{QP} \cong \overline{SR}$
Prove: $\angle 4 \cong \angle 3$
7. **Given:** $\angle Q$ and $\angle S$ are right angles
$\overline{QR} \cong \overline{PS}$
Prove: $\angle 1 \cong \angle 2$
8. **Given:** $\angle Q$ and $\angle S$ are right angles
$\overline{QR} \cong \overline{PS}$
Prove: $\angle QRS$ is a right angle

9–16. Name the theorems or postulates you would use to solve problems **1–8**.

Written Solve each problem.

1–8. Write a formal proof for each of problems **1–8** in the Exploratory Exercises.
9. Prove Theorem 4–11.
10. Prove Theorem 4–12.
11. Prove that if one leg included between an acute angle and the right angle of a right triangle are congruent to the corresponding leg and acute angle of another right triangle, then the triangles are congruent.
12. Prove that if one leg *not* included between an acute angle and the right angle of a right triangle are congruent to the corresponding leg and acute angle of another right triangle, then the triangles are congruent.

Vocabulary

triangle (96)
sides (96)
vertices (96)
right triangle (97)
obtuse triangle (97)
acute triangle (97)
equiangular triangle (97)
equilateral triangle (97)
isosceles triangle (97)
scalene triangle (97)
interior (98)
exterior (98)
congruent triangles (104)
SSS (107)
SAS (107)
ASA (108)
AAS (111)
median (115)
altitude (115)
angle bisector (116)
auxiliary figure (118)
hypotenuse (123)
legs (123)
HA (123)
LL (123)
LA (124)
HL (125)

Chapter Summary

1. Three noncollinear line segments connected at their endpoints form a triangle. The line segments are sides and the endpoints are vertices. (96)
2. Triangles classified according to their angles may be right, obtuse, or acute. They also may be equiangular. (97)
3. Triangles classified according to their sides may be equilateral, isosceles, or scalene. (97)
4. A triangle separates a plane into three parts, the interior, the exterior, and the triangle itself. (98)
5. Angle Sum Theorem: The sum of the degree measures of the angles of a triangle is 180. (100)
6. Theorem 4–2: The degree measure of each angle of an equiangular triangle is 60. (100)
7. Theorem 4–3: The acute angles of a right triangle are complementary. (101)
8. Definition of Congruent Triangles: Two triangles are congruent if and only if there is a correspondence such that their corresponding parts are congruent. (104)
9. Theorem 4–4: Congruence of triangles is reflexive, symmetric, and transitive. (105)
10. SSS: If each side of one triangle is congruent to the corresponding side of another triangle, then the triangles are congruent. (107)

11. SAS: If two sides and the included angle of one triangle are congruent to the corresponding sides and included angle of another triangle, then the triangles are congruent. (107)
12. ASA: If two angles and the included side of one triangle are congruent to the corresponding angles and included side of another triangle, then the triangles are congruent. (108)
13. AAS: If two angles and a nonincluded side of one triangle are congruent to the corresponding angles and nonincluded side of another triangle, then the triangles are congruent. (111)
14. Definition of Median of a Triangle: A segment is a median of a triangle if and only if it connects a vertex to the midpoint of the opposite side. (115)
15. Definition of Altitude of a Triangle: A segment is an altitude of a triangle if and only if it connects a vertex to the line containing the opposite side so that it is perpendicular to that line. (115)
16. Definition of Bisector of an Angle: A ray, $\overrightarrow{QS}$, is a bisector of $\angle PQR$ if and only if S is in the interior of the angle and $\angle PQS \cong \angle RQS$. (116)
17. An angle bisector of a triangle is a segment which is contained in a bisector of one of the angles of the triangle, and whose endpoints are on the triangle. (116)
18. Isosceles Triangle Theorem: If two sides of a triangle are congruent, then the angles opposite those sides are congruent. (118)
19. Theorem 4–7: If a triangle is equilateral, then the triangle is equiangular. (118)
20. Theorem 4–8: Each angle of an equilateral triangle has degree measure 60. (118)
21. Theorem 4–9: If two angles of a triangle are congruent, then the sides opposite those angles are congruent. (119)
22. Theorem 4–10: If a triangle is equiangular, then the triangle is equilateral. (119)
23. HA: If the hypotenuse and an acute angle of one right triangle are congruent to the corresponding hypotenuse and angle of another right triangle, then the triangles are congruent. (123)
24. LL: If two legs of one right triangle are congruent to the corresponding legs of another right triangle, then the triangles are congruent. (123)
25. LA: If one leg and an acute angle of one right triangle are congruent to the corresponding leg and angle of another right triangle, then the triangles are congruent. (124)
26. HL: If the hypotenuse and a leg of one right triangle are congruent to the corresponding hypotenuse and leg of another right triangle, then the triangles are congruent. (125)

Chapter Review

4–1 Use the figure at the right, to solve each of the following.

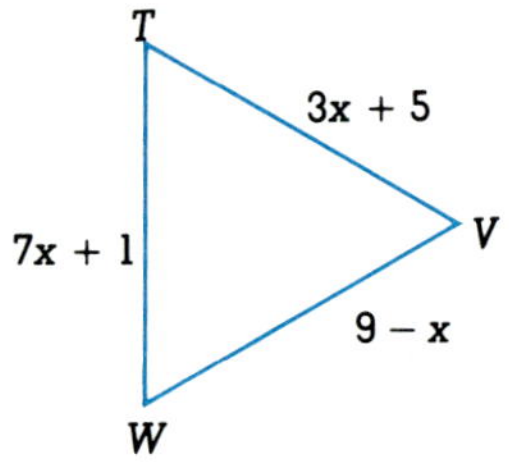

1. Name the triangle.
2. Name the sides of the triangle.
3. Name the angles of the triangle.
4. Suppose the triangle is equilateral. Solve for x and find the measure of each side.

4–2 In the figure at the right, $\overline{AB} \perp \overline{BC}$ and $\overline{BD} \perp \overline{AC}$. Also, $m\angle A = 62$. Find the degree measure for each of the following.

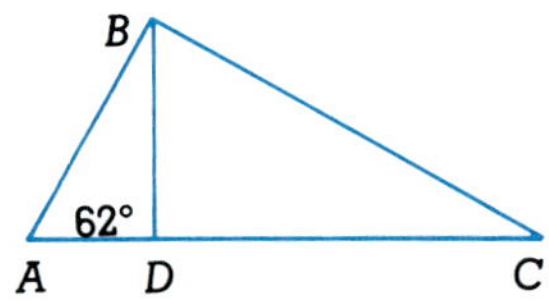

5. $\angle ABD$
6. $\angle ABC$
7. $\angle DBC$
8. $\angle BCD$

4–3 In the figure at the right, suppose $\triangle PQS \cong \triangle RQS$. Name the part of $\triangle RQS$ that is congruent to the part of $\triangle PQS$ given below.

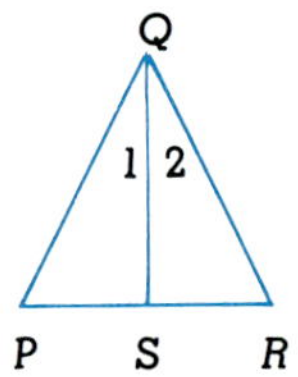

9. $\overline{PQ}$
10. $\angle P$
11. $\overline{QS}$
12. $\angle 1$
13. $\overline{SP}$
14. $\angle PSQ$

4–4 Use the figure at the right to solve each problem.

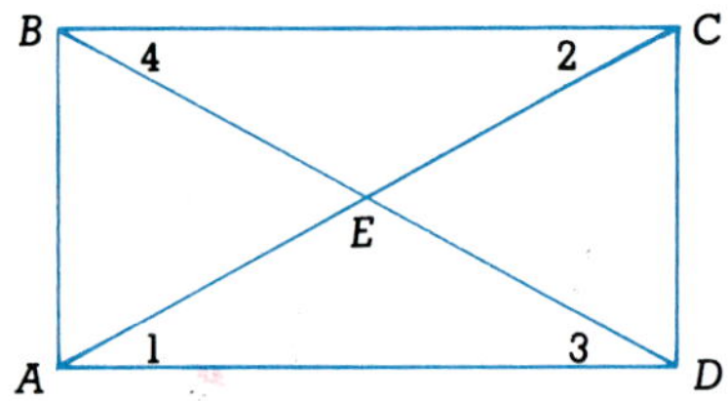

15. **Given:** $\overline{AC}$ and $\overline{BD}$ bisect each other
    Prove: $\overline{AB} \cong \overline{CD}$
16. **Given:** E is the midpoint of $\overline{AC}$
    $\angle 1 \cong \angle 2$
    Prove: $\angle 3 \cong \angle 4$

4–5 Use the figure at the right to solve each problem.

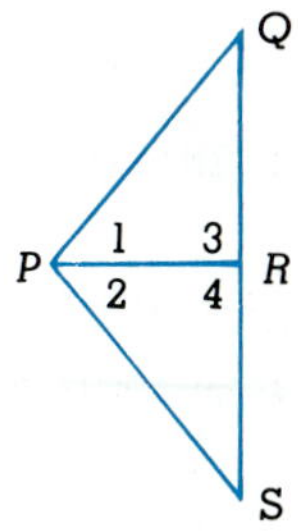

17. **Given:** $\overline{PR} \perp \overline{QS}$
$\angle Q \cong \angle S$
Prove: R is the midpoint of $\overline{QS}$

18. **Given:** $\overline{PQ} \cong \overline{PS}$
$\overline{PR}$ bisects $\overline{QS}$
Prove: $\angle 1 \cong \angle 2$

4–6 Use the figure at the right to solve each problem.

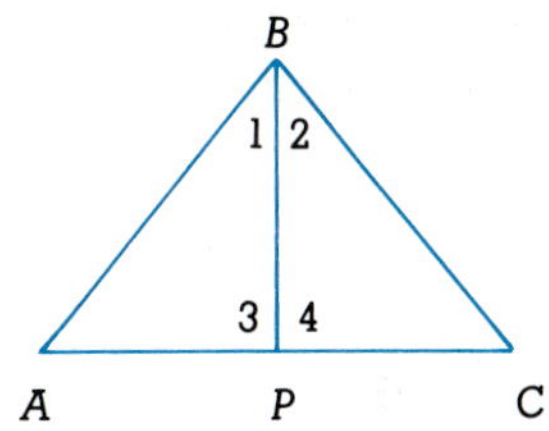

19. **Given:** $\overline{AB} \cong \overline{CB}$
$\overline{BP}$ bisects $\angle ABC$
Prove: $\overline{BP}$ is a median of $\triangle ABC$

20. **Given:** $\overline{BC} \cong \overline{BA}$
$\overline{PB}$ is a median of $\triangle ABC$
Prove: $\overline{PB}$ is an altitude of $\triangle ABC$

4–7 Complete a proof for the following.

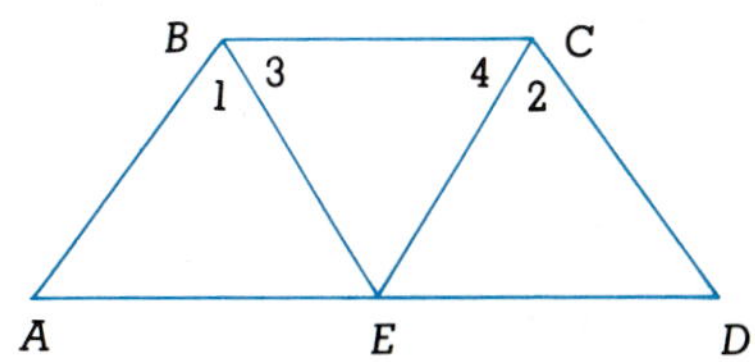

21. **Given:** $\angle A \cong \angle D$
$\overline{AB} \cong \overline{CD}$
E is the midpoint of $\overline{AD}$
Prove: $\angle 3 \cong \angle 4$

4–8 Complete a proof for the following.

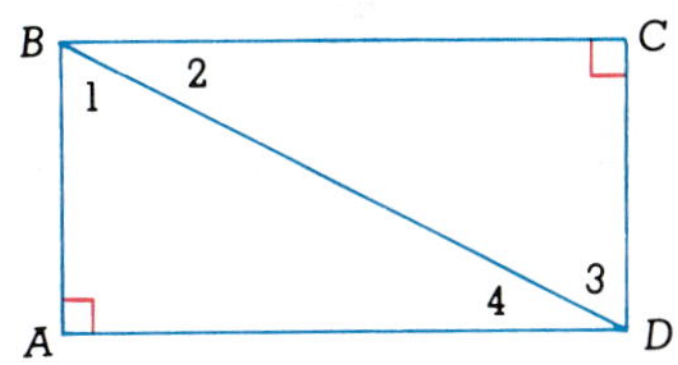

22. **Given:** $\angle A$ and $\angle C$ are right angles
$\angle 2 \cong \angle 4$
Prove: $\overline{AB} \cong \overline{CD}$

Chapter Test

Use the figure at the right to solve each of the following.

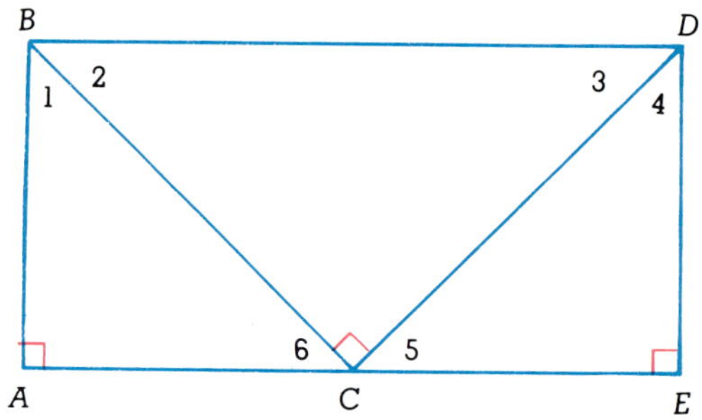

1. Name the triangle with A as one vertex.
2. Name the right angle in $\triangle DCE$.
3. Suppose $\triangle BAC \cong \triangle CED$. Name the part of $\triangle CED$ that is congruent to $\angle 1$.
4. Suppose $\triangle BCD$ is isosceles, with $\overline{BC} \cong \overline{CD}$. If $BC = 2x - 5$ and $CD = x + 3$, find x.
5. Suppose C is the midpoint of $\overline{AE}$, $\overline{BC}$ is congruent to $\overline{CD}$, and $\angle BDE$ is a right angle. If $m\angle 4 = 57$, find $m\angle 6$.

State whether or *not* each of the following represents a test for congruent triangles. Write *yes* or *no*.

6. HA
7. SSS
8. AA
9. ASA
10. HL
11. AAA
12. AAS
13. SSA
14. LA
15. LL

Complete a proof for the following.

16. **Given:** $\triangle ABC$ is equilateral
$\angle 1 \cong \angle 6$
$\angle 2 \cong \angle 3$
$\angle 4 \cong \angle 5$
Prove: $\overline{AD} \cong \overline{DC}$

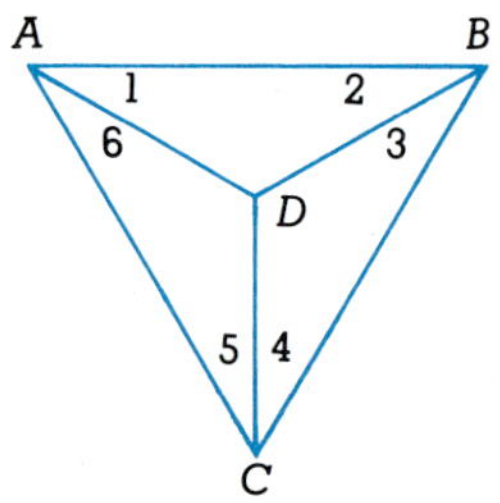

Chapter

5

Triangle Inequalities

Paths may cross to form triangular shapes. The properties of triangles can be used to decide which is the shortest path from one point to another, or how the distances compare.

5–1 Exterior Angles

In the figure below, $\angle 1$ is called an **exterior angle** of $\triangle PQR$.

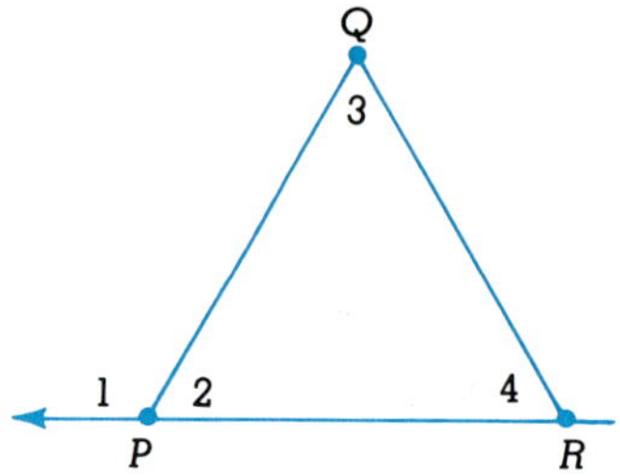

Notice that $\angle 1$ is adjacent to an angle of the triangle, namely $\angle 2$. The angles, $\angle 1$ and $\angle 2$, form a linear pair. The other angles of the triangle are *not* adjacent to $\angle 1$. These two angles, $\angle 3$ and $\angle 4$, are called **remote interior angles** with respect to $\angle 1$.

At each angle of a triangle is a pair of exterior angles. There are a total of six exterior angles for a given triangle.

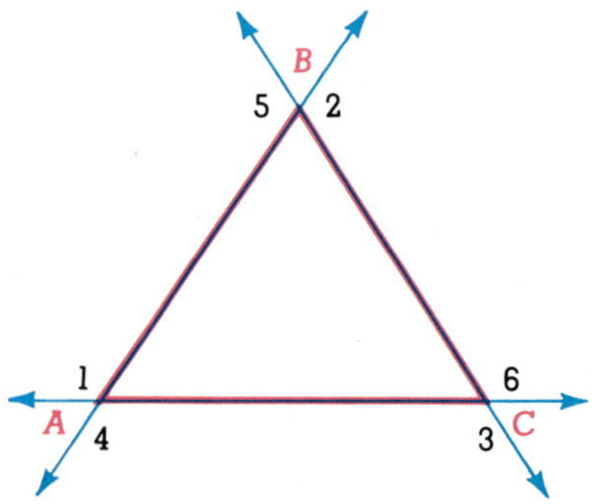

Exterior Angle	Remote Interior Angles
$\angle 1$	$\angle B$ and $\angle C$
$\angle 2$	$\angle A$ and $\angle C$
$\angle 3$	$\angle A$ and $\angle B$
$\angle 4$	$\angle B$ and $\angle C$
$\angle 5$	$\angle A$ and $\angle C$
$\angle 6$	$\angle A$ and $\angle B$

The angles in a linear pair are supplementary. Thus, the measures of exterior angles of a triangle are related to the measures of the remote interior angles.

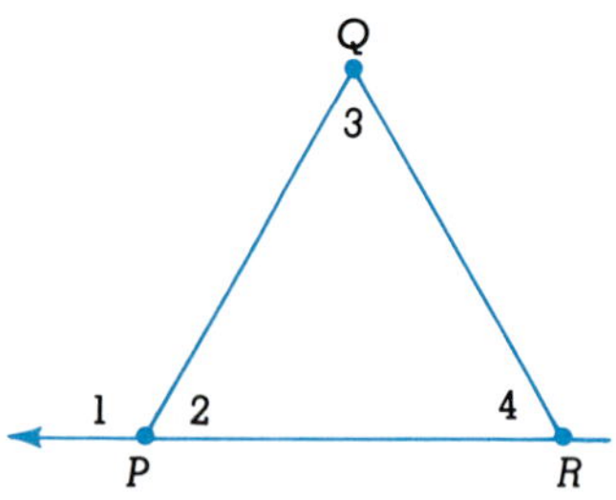

$m\angle 1 + m\angle 2 = 180$ *The angles in a linear pair are supplementary.*
$m\angle 2 + m\angle 3 + m\angle 4 = 180$ *Angle Sum Theorem*

$m\angle 1 + m\angle 2 = m\angle 2 + m\angle 3 + m\angle 4$ *Substitution*
$m\angle 1 = m\angle 3 + m\angle 4$ *Subtraction Property of Equality*

Theorem 5–1
Exterior Angle Theorem

The measure of an exterior angle of a triangle is equal to the sum of the measures of the two remote interior angles.

example 1

Use the information from the figure to find x, y, and z.

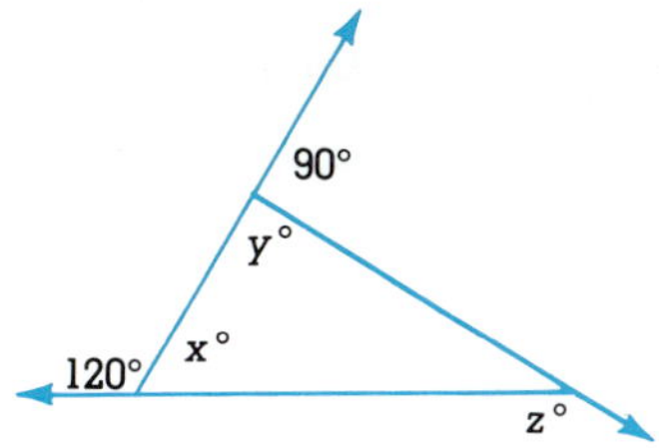

$120 + x = 180$
$x = 60$

$90 + y = 180$
$y = 90$

$z = x + y$
$= 60 + 90$
$= 150$

Inequality is defined by its properties. The following theorem is a direct result of the inequality properties. It is useful in proofs involving inequalities.

The Inequality Theorem could be used as a definition for inequality because it is a biconditional.

For any numbers a and b, $a > b$ if and only if there is a positive number c such that $a = b + c$.

Theorem 5–2
Inequality Theorem

example 2

Complete the following proof.

Given: $\angle 4$ is an exterior angle of $\triangle PQR$

Prove: $m\angle 4 > m\angle 1$
$m\angle 4 > m\angle 2$

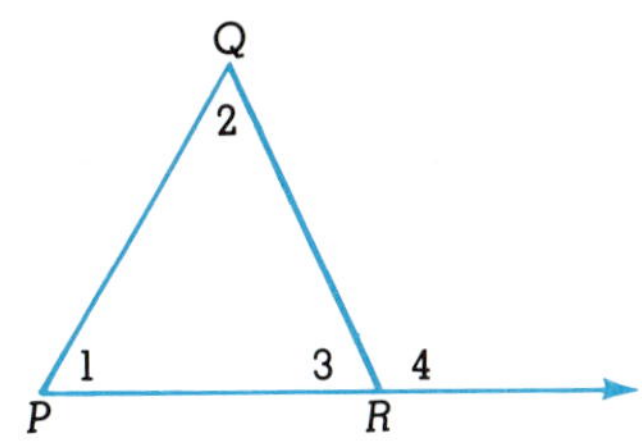

Proof:

STATEMENTS	REASONS
1. $\angle 4$ is an exterior angle of $\triangle PQR$	1. Given
2. $m\angle 4 = m\angle 1 + m\angle 2$	2. Theorem 5–1: Exterior Angle Theorem
3. $m\angle 4 > m\angle 1$ $m\angle 4 > m\angle 2$	3. Theorem 5–2: Inequality Theorem

The situation in Example **2** can be stated as the following theorem.

Theorem 5–3 **The measure of an exterior angle is greater than the measure of either remote interior angle.**

exercises

Exploratory For each of the following write an inequality that relates only x and y.

1. $x = 45 + y$
2. $x - y = 25$
3. $180 - x = 53 - y$
4. $75 - y = 180 - x$
5. $x + 30 = y$
6. $60 + 45 + x = 180 + y$
7. $46 + x + 89 = 91 + y$
8. $y + 85 = x$

For each of the following, write an inequality that relates only PQ and PR.

9. $PQ + QR = PR$
10. $PR - PQ = QR$
11. $RQ + PQ + RS = SR + PR$
12. $TV + PR + ST = PQ + TV$
13. $4 + PQ - PR = 10$
14. $PQ - PR = 25$

For each of the following, write an inequality that relates only $m\angle 1$ and $m\angle 2$.

15. $m\angle 1 = 180 + m\angle 2$
16. $m\angle 1 = 46 + m\angle 2$
17. $55 + m\angle 1 = m\angle 2$
18. $25 + 30 + m\angle 1 = m\angle 2$
19. $180 - m\angle 1 = 55 + 70 - m\angle 2$
20. $180 + m\angle 1 - 91 = m\angle 2 - 46$
21. $m\angle 1 = m\angle 2 + m\angle 3 + m\angle 4$
22. $m\angle 1 + m\angle 3 = m\angle 2 - m\angle 4$

Written For each of the following, find **x**.

1.
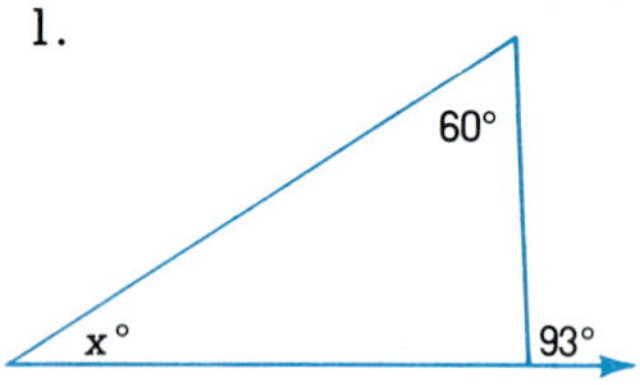

2.
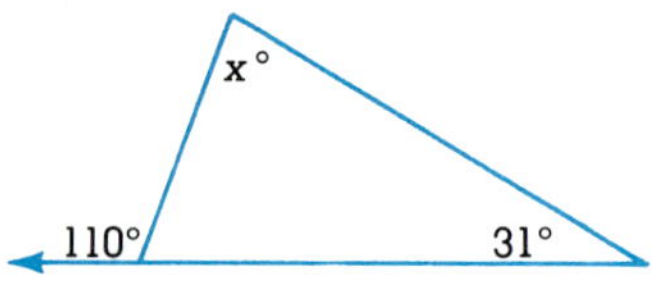

3.
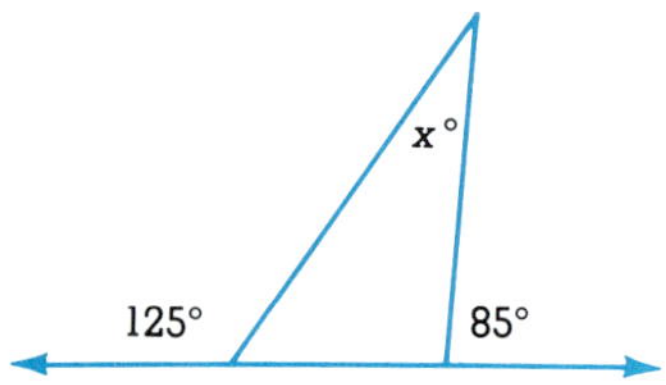

4.
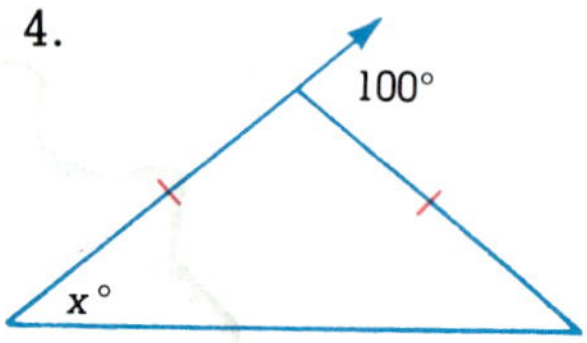

5.
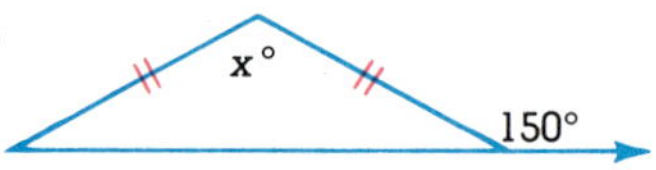

6.
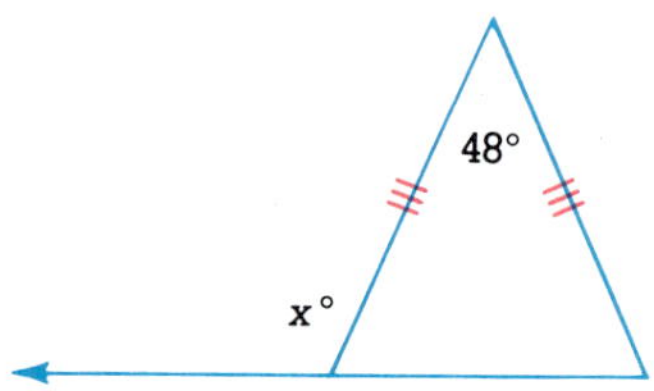

7. 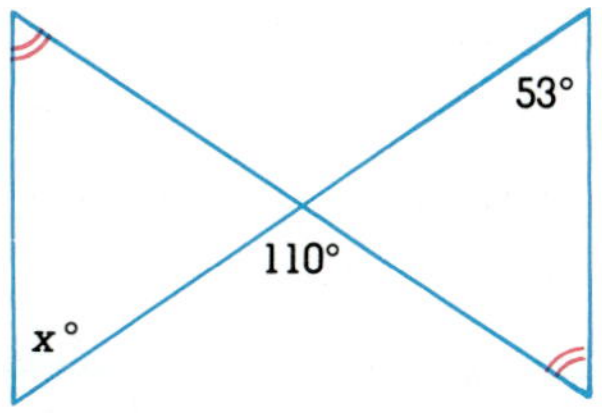

8. 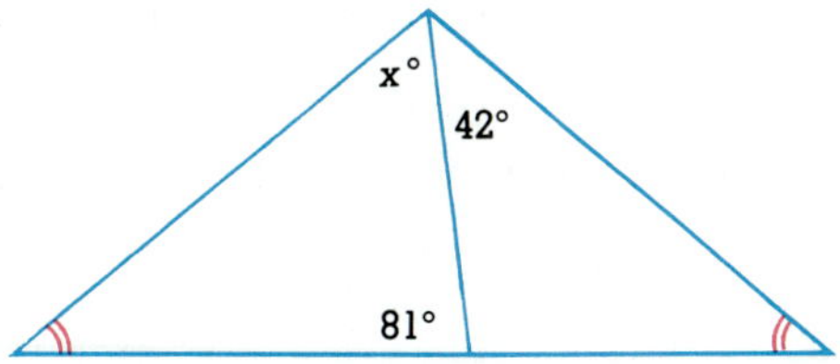

Use the information given in the figure to find each of the following.

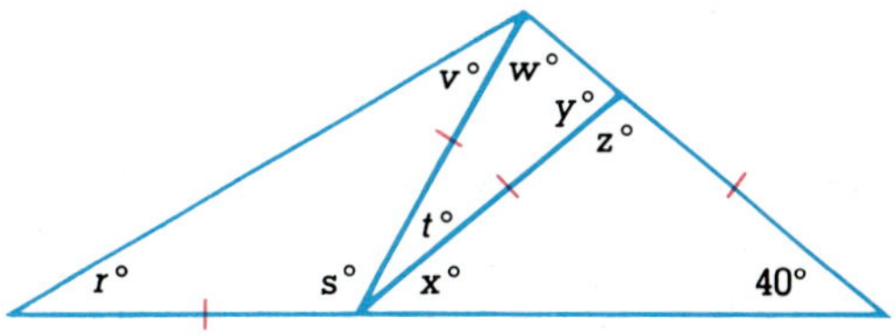

9. x 10. y
11. z 12. w
13. t 14. s
15. v 16. r

Use the information given in the figure to find each of the following.

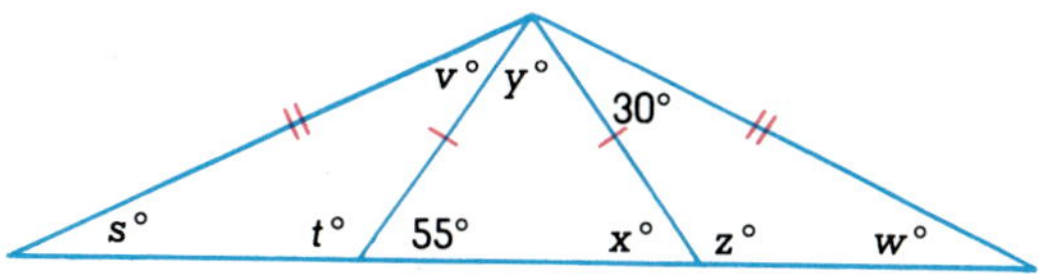

17. x 18. z
19. y 20. w
21. t 22. v

Use the figure at the right to prove each of the following.

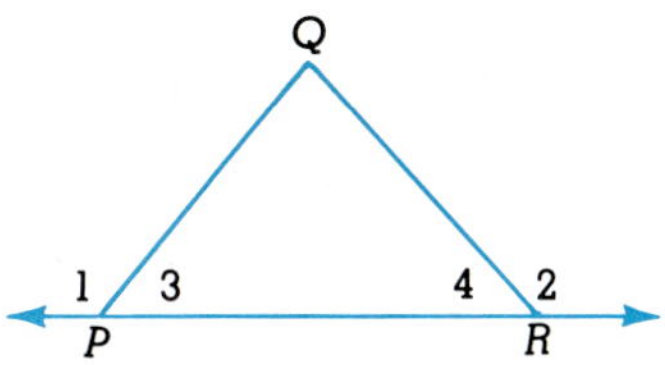

23. **Given:** $\angle 1 \cong \angle 2$
    Prove: $\triangle PQR$ is isosceles
24. **Given:** $\overline{PQ} \cong \overline{QR}$
    Prove: $\angle 1 \cong \angle 2$

25. Prove Theorem 5–1.
26. Complete the reasons for the following proof.

Theorem: For any numbers a and b, if there is a positive number c such that $a = b + c$, then $a > b$.

Given: $c > 0$ and $a = b + c$

Prove: $a > b$

Proof:

STATEMENTS	REASONS
1. $c > 0$ and $a = b + c$	1. ________
2. $a - b = c$	2. ________
3. $a - b > 0$	3. ________
4. $a > b$	4. ________

27. Complete the reasons for the following proof.

Theorem: For any numbers a and b, if $a > b$, then there is a positive number c such that $a = b + c$

Given: $a > b$

Prove: There is a c such that $c > 0$, and $a = b + c$, namely c is $a - b$.

Proof:

STATEMENTS	REASONS
1. $a > b$	1. ________
2. $a - b > 0$	2. ________
3. $a = a + 0$	3. ________
4. $a = a + (b - b)$	4. ________
5. $a = (a + b) - b$	5. ________
6. $a = (b + a) - b$	6. ________
7. $a = b + (a - b)$	7. ________

28. Given the figure at the right, plan a proof to show that $x > y$.

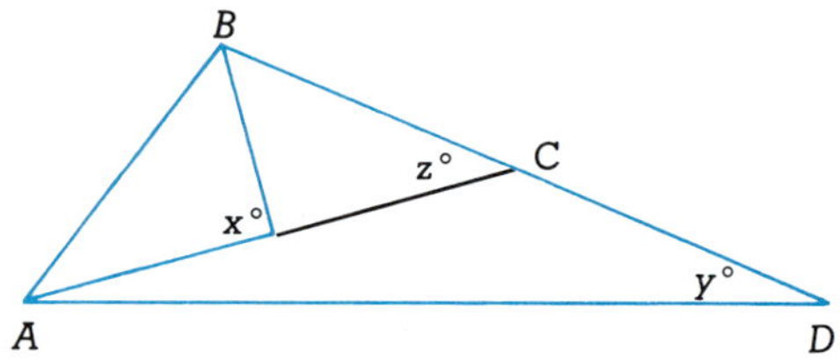

29. In the figure at the right, $b = x + c$. Plan a proof to show that $y > x$.

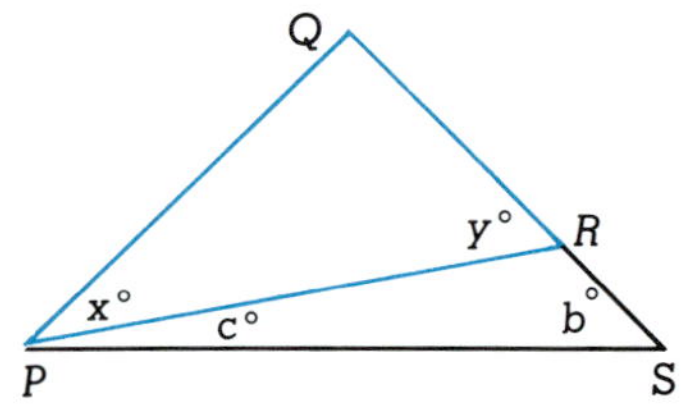

30. In the figure at the right, $\overline{PQ} \cong \overline{PS}$. Plan a proof to show that $m\angle SQR > m\angle 2$.

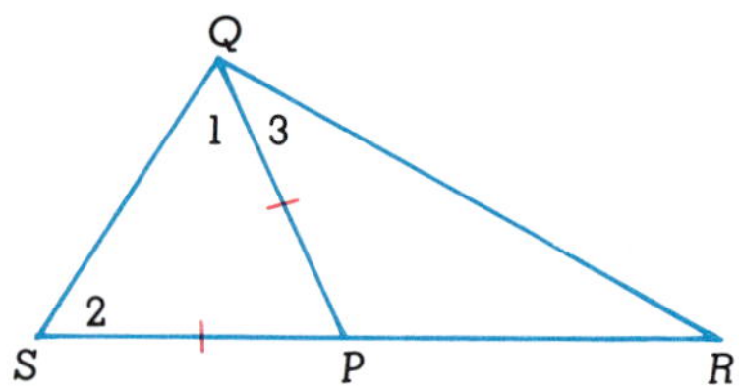

Excursions in Geometry

Memory

Study the following figures for two minutes. Then, close the book. On a piece of paper, draw as many of the figures as you can, in any order.

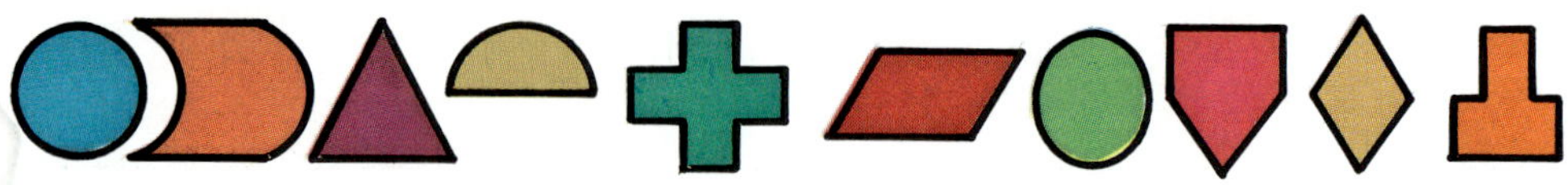

5–2 Using Diagrams

Diagrams help state given information. They may be used to show any of the following.

Existence *When points, lines, rays, angles, triangles, and so on appear on a figure, you may assume they exist.*

Relative Position *When points appear on a line or part of a line, you may assume they are collinear. Also, you may assume betweenness with respect to points on a line, and location with respect to half-planes, interiors, and exteriors. Adjacent angles and linear pairs are shown in figures, too.*

Intersection *When lines, rays or segments appear to intersect, you may assume that they do intersect.*

Diagrams are marked to give further information about congruence and measure. Such information normally is written in the *given* statements, if the diagram is with a proof.

DO *NOT* ASSUME ANY OF THE FOLLOWING UNLESS A FIGURE IS MARKED.

- congruence
- equality of measure
- inequality of measure
- bisectors or midpoints
- perpendiculars or right angles
- specific measures

example 1

Can the following statements be assumed from the appearance of the figure? Write *yes* or *no* for each statement.

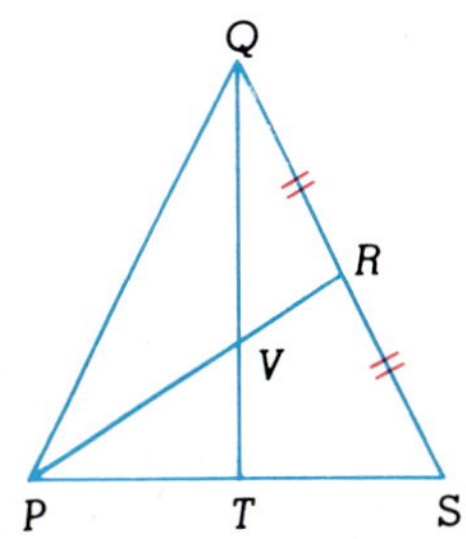

1. V is between Q and T
2. $\overline{PR}$ bisects $\angle QPS$
3. $\overline{PT} \cong \overline{TS}$
4. $\overline{PR}$ bisects $\overline{QS}$
5. $\overline{QR} \cong \overline{RS}$
6. $\overline{QT} \perp \overline{PS}$
7. $\angle QRV$ is a right angle
8. $\angle QTP$ and $\angle QTS$ form a linear pair

The answer to statements **1**, **4**, **5**, and **8** is *yes*.
The answer to statements **2**, **3**, **6**, and **7** is *no*.

Often, a geometric proof shows that parts of a figure are congruent. One method of proving parts congruent is to show they are corresponding parts of congruent triangles. Thus, the problem becomes one of locating two triangles that contain the parts, and, then, proving that the triangles are congruent. For example, consider the following.

Given: $\overline{RP} \cong \overline{RT}$
$\overline{RQ} \cong \overline{RS}$

Prove: $\overline{PS} \cong \overline{TQ}$

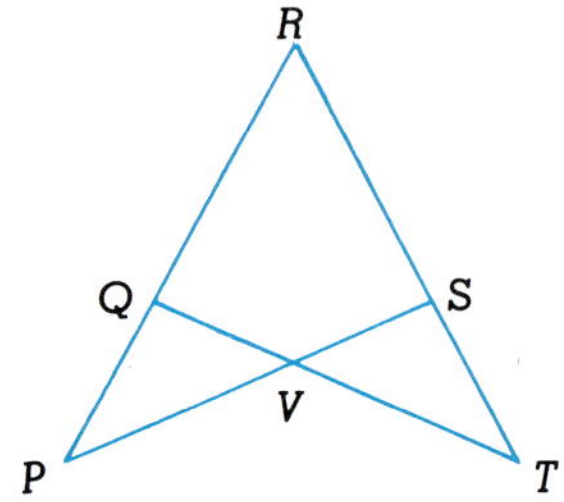

Altogether there are four triangles in the figure. They are indicated in different colors below.

To see overlapping triangles in a diagram, it is helpful to outline the separate triangles in different colors. Another way to see the triangles is to draw them separately.

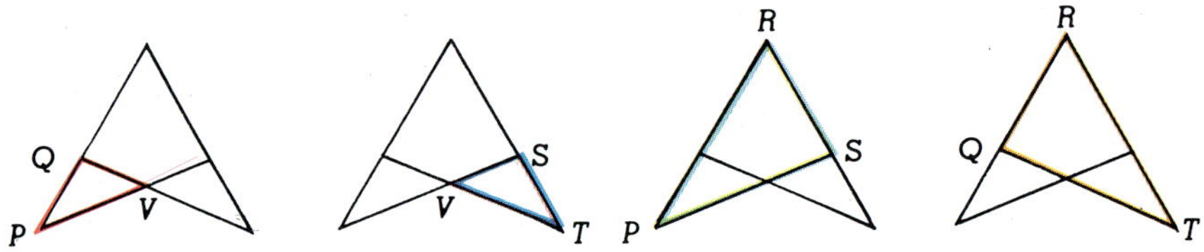

Two overlapping triangles, $\triangle PSR$ and $\triangle TQR$, contain the corresponding parts $\overline{PS}$ and $\overline{TQ}$.

To complete the proof, show that $\triangle PSR \cong \triangle TQR$ by using SAS. Then, conclude $\overline{PS} \cong \overline{TQ}$ since they are corresponding parts of congruent triangles. ***CPCTC***

example 2

One way to complete the following proof is first to show that $\angle 1$ and $\angle 2$ are corresponding parts of congruent triangles. Name the two triangles.

Given: $\overline{QP} \perp \overline{PT}$
$\overline{ST} \perp \overline{PT}$
$\angle Q \cong \angle S$

Prove: $\overline{PR} \cong \overline{TR}$

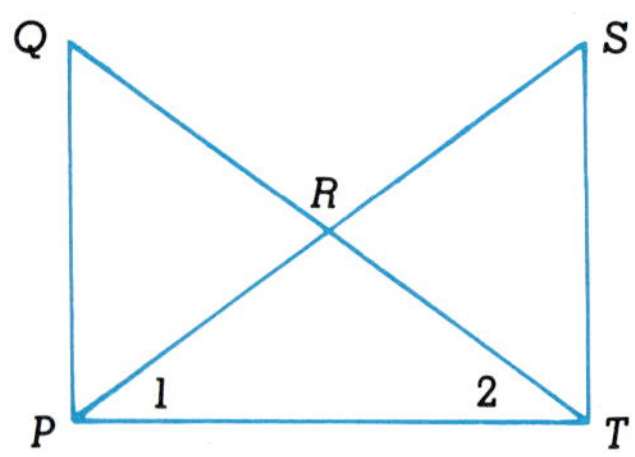

$\triangle TQP$ and $\triangle PST$ are right triangles that are congruent by LA.
$\triangle TQP$ contains $\angle 2$. $\triangle PST$ contains $\angle 1$.

Triangles are not easily seen when they overlap.

example 3 **Identify the triangles used in Reasons 2 and 7 of the following proof.**

Given: $\overline{PQ} \cong \overline{SQ}$

Prove: $m\angle 2 > m\angle 1$

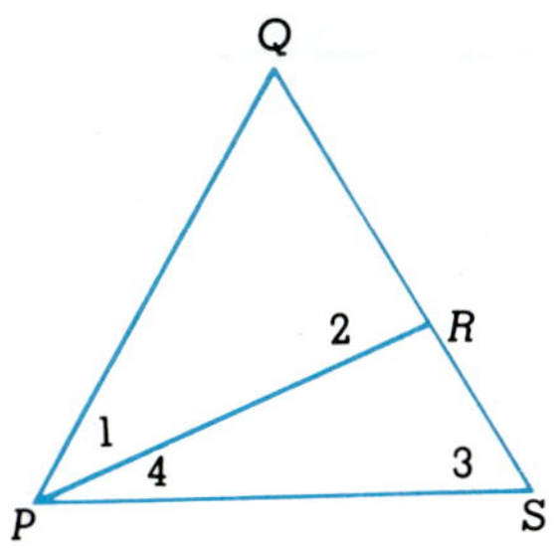

Proof:

STATEMENTS	REASONS
1. $\overline{PQ} \cong \overline{SQ}$	1. Given
2. $\angle QPS \cong \angle 3$	2. Theorem 4–6: Isosceles Triangle Theorem
3. $m\angle QPS = m\angle 3$	3. Definition of congruent angles
4. $m\angle QPS = m\angle 1 + m\angle 4$	4. Postulate 3–3: Angle Addition Postulate
5. $m\angle 3 = m\angle 1 + m\angle 4$	5. Postulate 2–9: Substitution
6. $m\angle 2 = m\angle 3 + m\angle 4$	6. Theorem 5–1: Exterior Angle Theorem
7. $m\angle 2 = m\angle 1 + m\angle 4 + m\angle 4$	7. Postulate 2–9: Substitution
8. $m\angle 2 > m\angle 1$	8. Theorem 5–2: Inequality Theorem

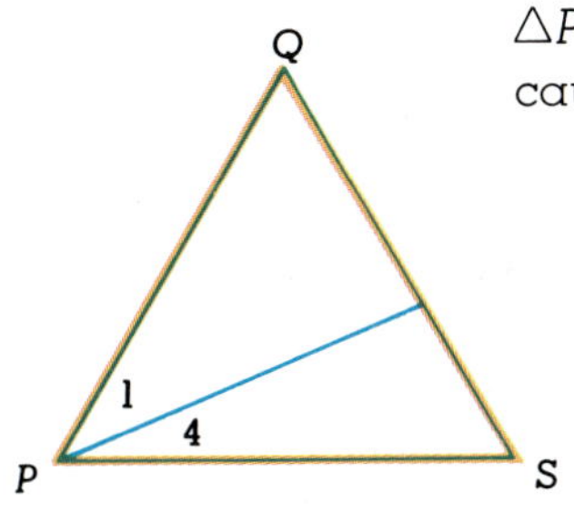

$\triangle PQS$ is the isosceles triangle used for Reason **2**, because $\overline{PQ} \cong \overline{QS}$.

$\triangle PRS$ is the triangle used for Reason **7**. The exterior angle, $\angle 2$, has remote interior angle, $\angle 3$.

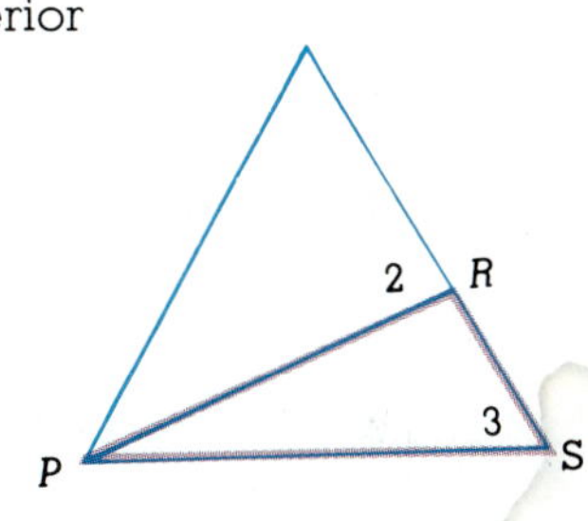

exercises

Exploratory Tell whether or *not* each of the following may be assumed from a diagram.

1. adjacent angles
2. existence of a point
3. congruent line segments
4. perpendicular lines
5. equality of distance measure
6. intersection of rays
7. existence of a ray
8. perpendicular rays
9. existence of a triangle
10. collinearity of segments
11. linear pair
12. vertical angles
13. intersection of lines
14. perpendicular segments
15. right angles
16. angle bisectors
17. existence of a line
18. betweenness of points
19. equality of angle measure
20. intersection of a line and a ray
21. inequality of distance measure
22. collinearity of points
23. congruent triangles
24. congruent angles
25. existence of segments
26. bisector of line segments
27. existence of an angle

Written Can the following statements be assumed from the appearance of the figure? Write *yes* or *no* for each statement.

1. $\angle PST$ and $\angle RST$ form a linear pair
2. $\overline{QR} \perp \overline{PV}$
3. $RV + VQ = RQ$
4. T is between P and V
5. $\angle SPT$ and $\angle VQT$ are congruent
6. $\overline{PS} \cong \overline{QV}$
7. $\overline{PT} \cong \overline{QT}$
8. $PV = TV + PT$
9. $\angle QVT$ is a right angle
10. $\angle PTS$ is adjacent to $\angle STV$
11. $\overline{PV}$ bisects $\overline{QS}$
12. T is the midpoint of $\overline{PV}$

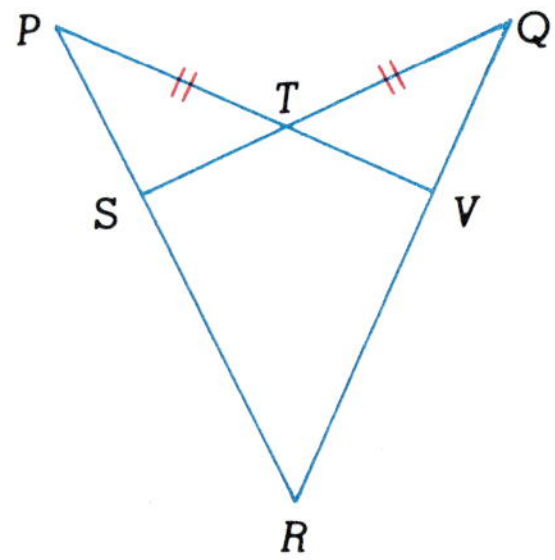

Locate and name all the triangles contained in each of the following figures.

13.
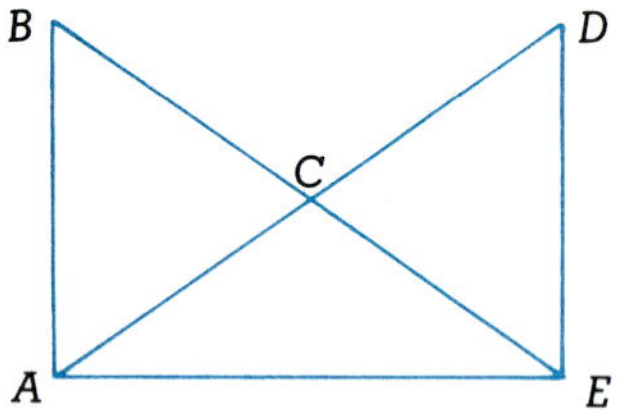

14.
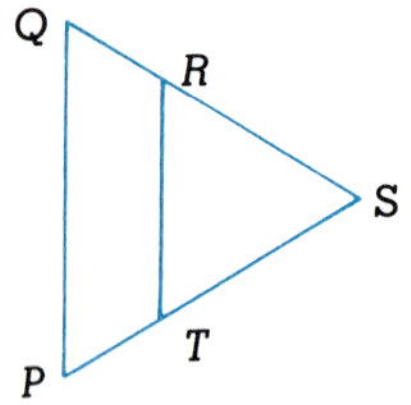

15.
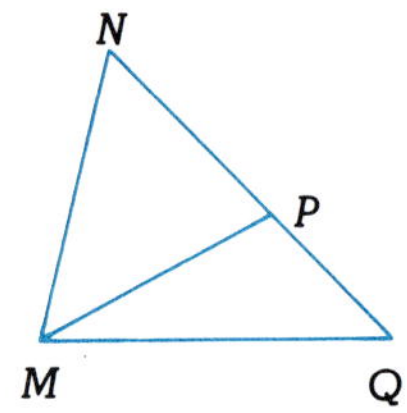

16.
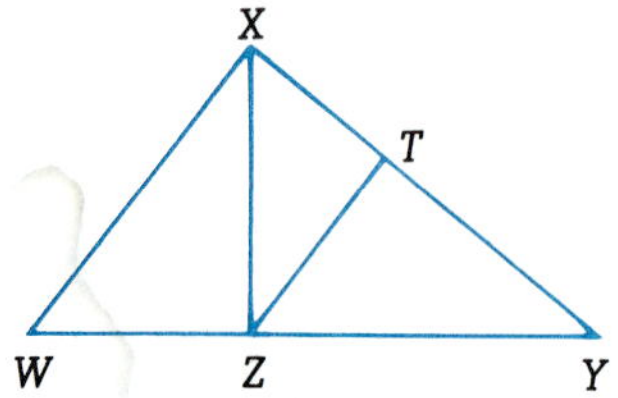

17.
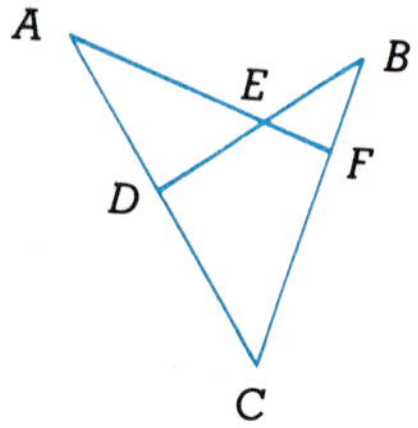

18.
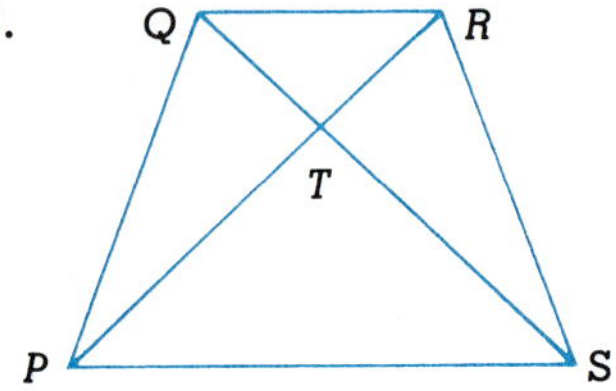

19. Identify the triangles used in Reasons **2** and **4** of the following proof.

Given: $\overline{QS} \cong \overline{QR}$

Prove: $m\angle 3 > m\angle 1$

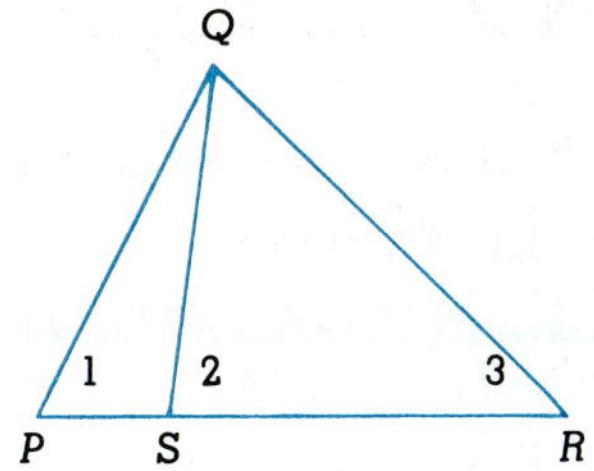

Proof:

STATEMENTS	REASONS
1. $\overline{QS} \cong \overline{QR}$	1. Given
2. $\angle 3 \cong \angle 2$	2. Theorem 4–6: Isosceles Triangle Theorem
3. $m\angle 3 = m\angle 2$	3. Definition of congruent angles
4. $m\angle 2 > m\angle 1$	4. Theorem 5–3: The measure of an exterior angle of a triangle is greater than the measure of either remote interior angle.
5. $m\angle 3 > m\angle 1$	5. Postulate 2–9: Substitution

20. Identify the triangles used in Reasons **4, 5, 7, 8,** and **9** of the following proof. List the triangles according to Reason number.

Given: $\overline{AB} \cong \overline{CD}$

$\angle BAD \cong \angle CDA$

Prove: $\overline{BE} \cong \overline{CE}$

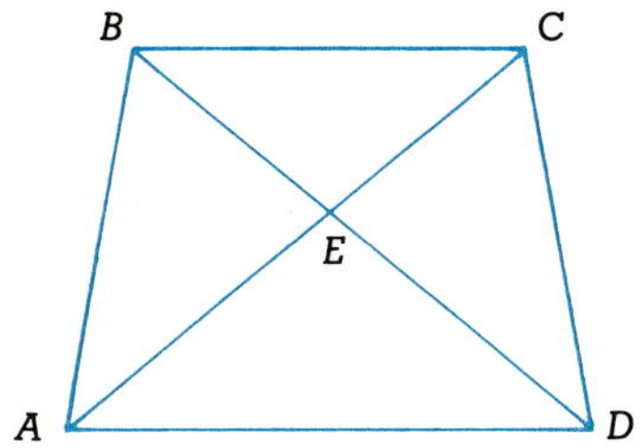

Proof:

STATEMENTS	REASONS
1. $\overline{AB} \cong \overline{CD}$	1. Given
2. $\angle BAD \cong \angle CDA$	2. Given
3. $\overline{AD} \cong \overline{AD}$	3. Theorem 2–2: Congruence of line segments is reflexive.
4. $\triangle BAD \cong \triangle CDA$	4. Postulate 4–2: SAS
5. $\overline{AC} \cong \overline{DB}$	5. Definition of congruent triangles *CPCTC*
6. $\overline{BC} \cong \overline{BC}$	6. Theorem 2–2: Congruence of line segments is reflexive.
7. $\triangle BCA \cong \triangle CBD$	7. Postulate 4–1: SSS
8. $\angle BCA \cong \angle CBD$	8. Definition of congruent triangles *CPCTC*
9. $\overline{CE} \cong \overline{BE}$	9. Theorem 4–9: If two angles of a triangle are congruent, then the sides opposite those angles are congruent.

Using Geometry

Solar Collectors

Solar collectors are devices for collecting energy from the sun. Collectors can be used for the heating and cooling of homes. Other uses include generating power to operate irrigation systems.

To catch the sun's rays, collectors must be tilted. Some collectors are tilted at a fixed angle to the sun. Other collectors move to follow the sun.

The fixed angle for a collector is determined from local latitude.

Albany, New York has a latitude of 42° 40′. The collector heating angle, and the heating and cooling angle are determined in the following way.

heating angle = *latitude* + 15° 00′
= 42° 40′ + 15° 00′
= 57° 40′

heating and cooling angle = *latitude* + 5° 00′
= 42° 40′ + 5° 00′
= 47° 40′

Exercises **Find the heating angle for the following locations with the given latitude.**

1. Amarillo, Texas 35° 11′
2. Atlantic City, N.J. 39° 22′
3. Anchorage, Alaska 61° 13′
4. Austin, Nevada 39° 29′
5. Atlanta, Georgia 33° 45′
6. Baker, Oregon 44° 47′

7–12. Find the heating and cooling angle for the locations in problems **1–6**.

5–3 Sides and Angles

If two sides of a triangle are *not* congruent, then the angles opposite those sides are *not* congruent. The following diagram suggests how the sides and angles are related.

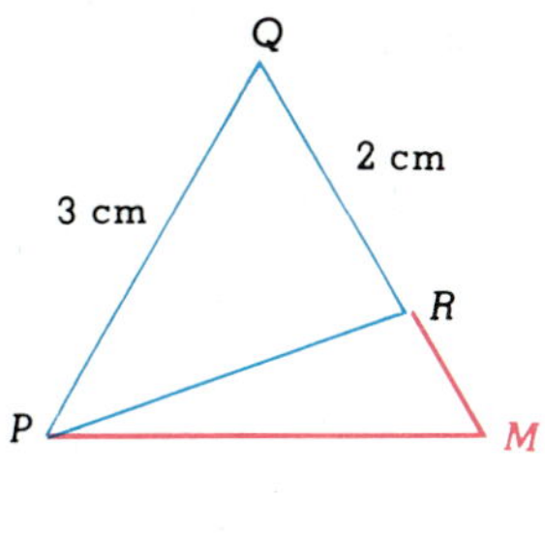

$PQ > QR$
$m\angle QRP > m\angle QPR$

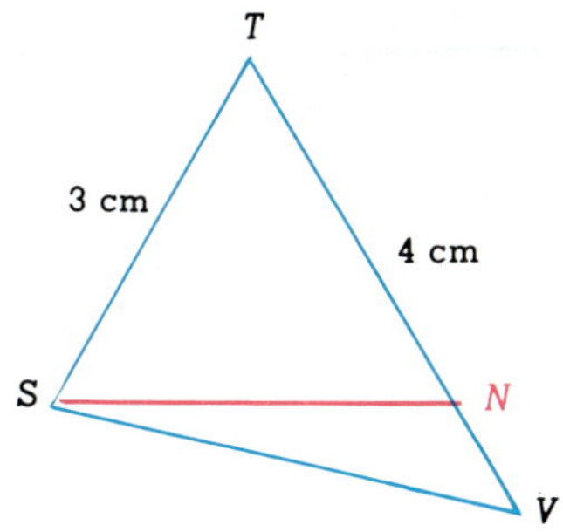

$ST < TV$
$m\angle TVS < m\angle TSV$

If the measures of two sides of a triangle are unequal, then the measure of the angles opposite those sides are unequal in the same order.

Theorem 5–4

example 1

Plan a proof for Theorem 5–4.

Given: $\triangle PQR$
$PQ > QR$

Prove: $m\angle 2 > m\angle 1$

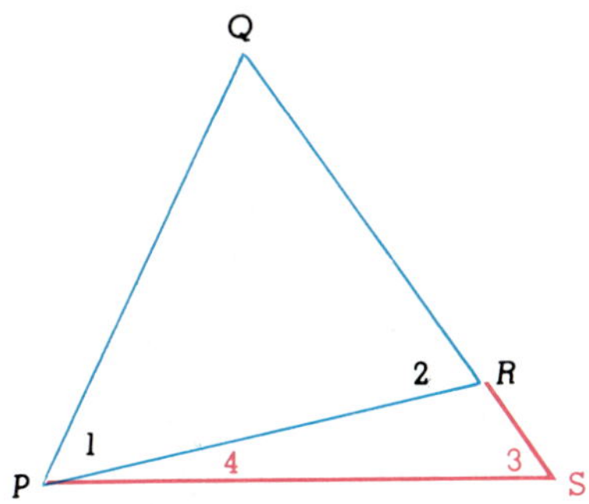

Extend $\overline{QR}$ to S so that $\overline{QS} \cong \overline{PQ}$. The new triangle formed, $\triangle PQS$, is isosceles. Thus, the base angles are congruent. Using this information, show that $m\angle 1 + m\angle 4 = m\angle 3$.

Also, $m\angle 2 = m\angle 3 + m\angle 4$, since $\angle 2$ is an exterior angle of $\triangle PRS$. From substitution, conclude that $m\angle 2 = m\angle 1 + m\angle 4 + m\angle 4$. As a result, $m\angle 2 > m\angle 1$ by the Inequality Theorem.

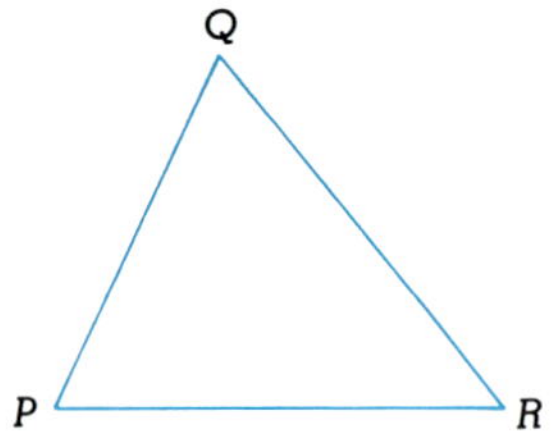

$m\angle P > m\angle R$
QR ? PQ

Suppose two angles of a triangle are *not* congruent. The figure at the left shows such a situation. The sides opposite those angles can be related in exactly one of the following ways.

$QR = PQ$	$QR < PQ$	$QR > PQ$
If the two sides have equal measures, then they are congruent. Thus, the triangle is isosceles and the two angles must be congruent, which is impossible.	*According to Theorem 5–4, if two sides have unequal measures, then the measures of the angles opposite those sides are unequal in the same order. Thus, $m\angle P < m\angle R$, which is impossible.*	*The measures of the two sides must be related in this way, since the other two ways are impossible.*

Theorem 5–5

If the measures of two angles of a triangle are unequal, then the measures of the sides opposite those angles are unequal in the same order.

example 2

Prove that the segment from a point to a line perpendicular to the line is the shortest segment joining the point and the line.

Given: $\overline{PQ} \perp t$
$\overline{PR}$ is any other segment joining P to t

Prove: $PR > PQ$

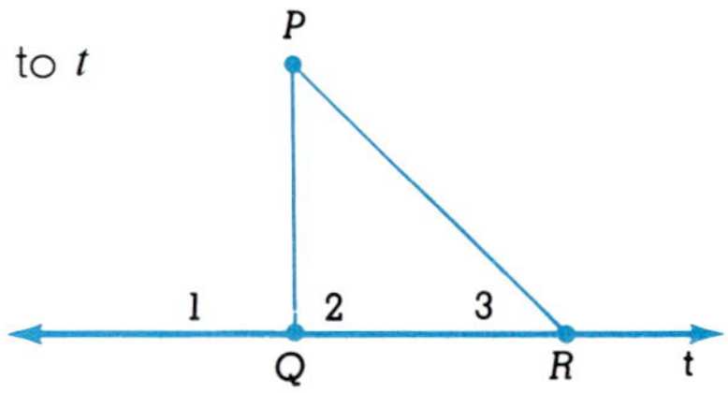

Proof:

STATEMENTS	REASONS
1. $\overline{PQ} \perp t$	1. Given
2. $\angle 1 \cong \angle 2$	2. Theorem 3–13: If two lines are perpendicular, then they form congruent adjacent angles.
3. $m\angle 1 = m\angle 2$	3. Definition of congruent angles
4. $m\angle 1 > m\angle 3$	4. Theorem 5–3: The measure of an exterior angle of a triangle is greater than the measure of either remote interior angle.
5. $m\angle 2 > m\angle 3$	5. Theorem 2–9: Substitution *Steps 3 & 4*
6. $PR > PQ$	6. Theorem 5–5: If the measures of two angles of a triangle are unequal, then the measures of the sides opposite those angles are unequal in the same order.

The proofs of the following theorems are almost identical. The first theorem is proved in Example **2**.

The shortest segment from a point to a line is the perpendicular segment. — Theorem 5–6

The shortest segment from a point to a plane is the perpendicular segment. — Theorem 5–7

exercises

Exploratory For each of the following triangles, list the angles in order from smallest to largest.

1.

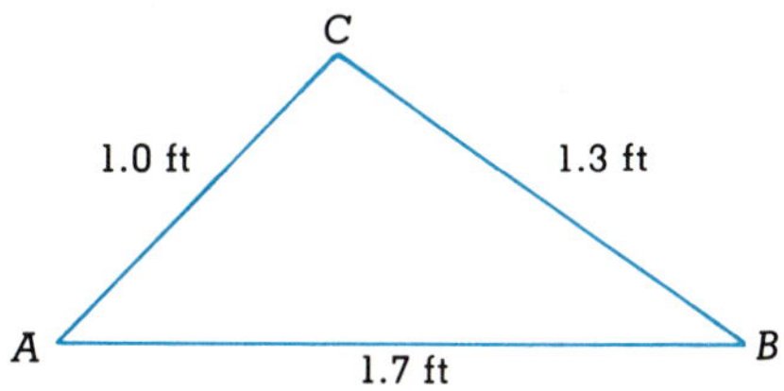

2.

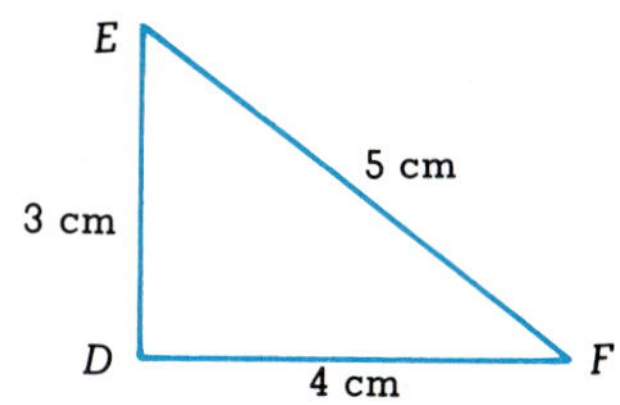

3.

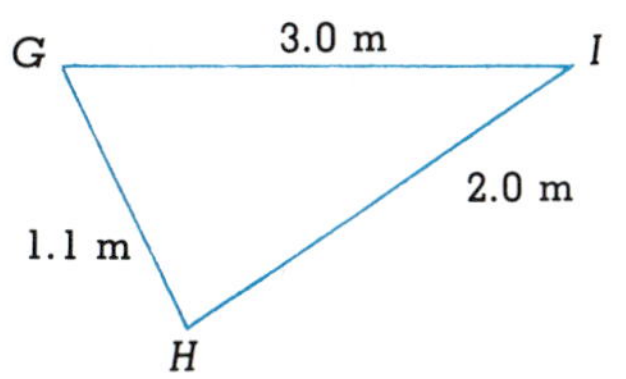

4.

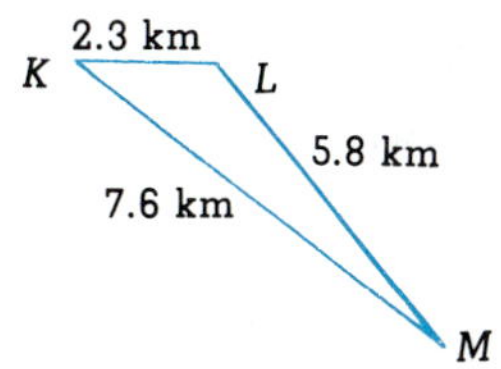

For each of the following triangles, list the sides in order from shortest to longest.

5.

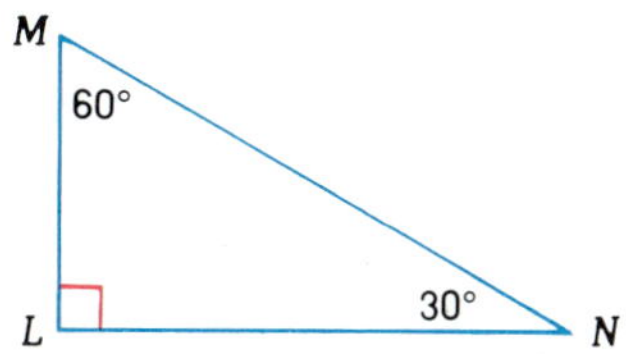

6.

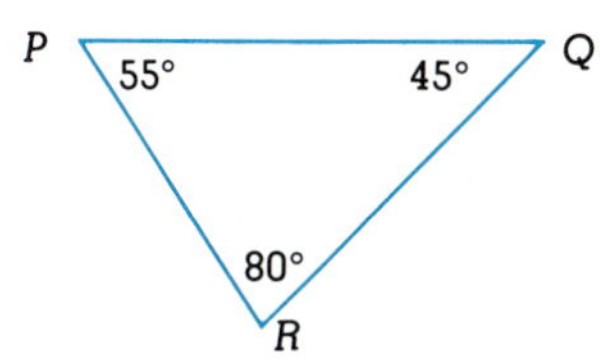

7.

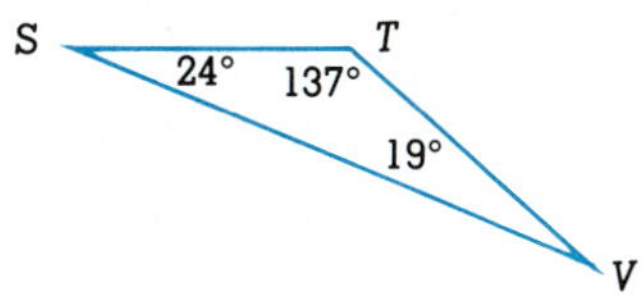

8.

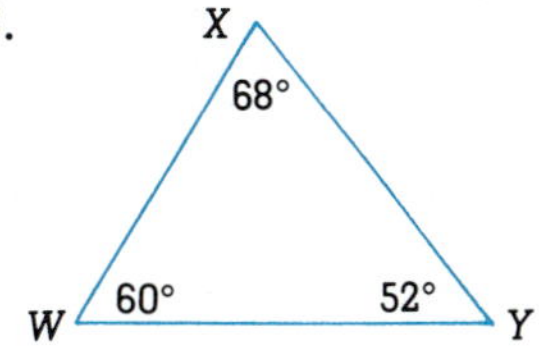

Written Plan a proof for each of the following.

1. **Given:** $\triangle ABC$
   $\angle A$ is a right angle
   Prove: $BC > AB$

2. **Given:** $\triangle ABC$
   $\angle A$ is a right angle
   Prove: $BC > AC$

3. **Given:** $\triangle PQR$
   $\overline{QR} > \overline{PQ}$
   $\overline{PR} \cong \overline{PQ}$
   Prove: $QR > PR$

4. **Given:** $\triangle PQR$
   $\overline{QR} > \overline{PQ}$
   $\overline{PR} \cong \overline{PQ}$
   Prove: $m\angle P > m\angle Q$

5–8. Write a formal proof for each of problems **1–4**.

9. Use the results of Theorems 5–4 and 5–5 to write a new theorem in *if and only if* form that relates noncongruent sides and angles of a triangle.

10. Write a formal proof for Theorem 5–4.

11. Write a formal proof for Theorem 5–7.

Excursions in Geometry

Inside and Outside

Each of the following figures can be traced so that, starting at one point, every other point is touched exactly once before reaching the starting point again. These figures are called **simple closed curves.**

Simple closed curves separate a plane into the inside, the outside, and the curve. One way to test whether a point is inside or outside is as follows.

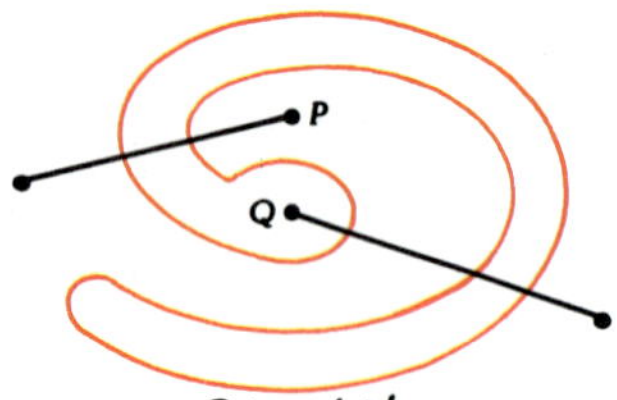

P is outside
Q is inside

Draw a line segment from the point to another point you know is outside the curve.

Count the number of times the line segment crosses the curve.

If the number is *odd*, the point is *inside*.
If the number is *even*, the point is *outside*.

Exercises For each of the following, tell whether P is *inside* or *outside*.

1. 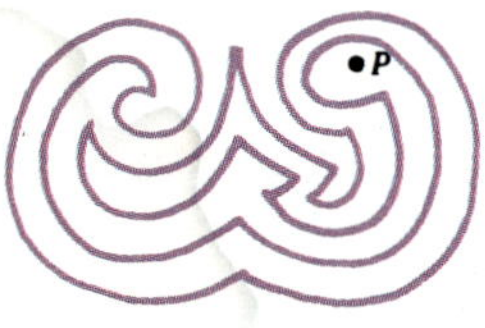

2. 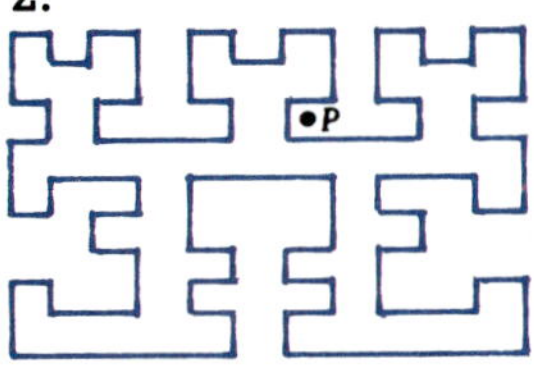

3. 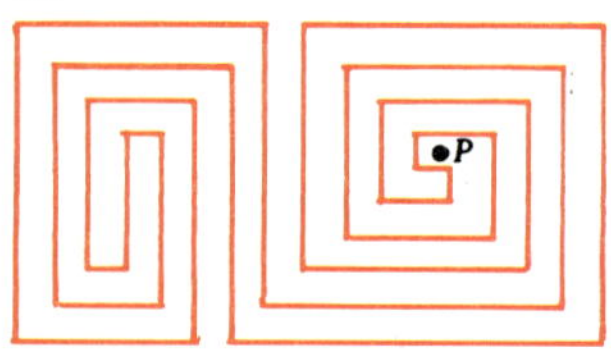

4. 

Using Money

Plane Fares

George Kim is planning a trip to Cincinnati to visit his grandmother. He called the airlines and asked about the cost of a flight from Los Angeles to Cincinnati and back. He was given the following information.

Days and Times \ Fares		First Class	Regular Coach	Excursion Fare (Coach)	Super Saver Fare (Coach)
Mon.–Thurs.	Day	$456 round trip	$380 round trip	save 20%	save 40%
	Night			save 30%	save 50%
Fri.–Sun.	Day			save 15%	save 30%
	Night			save 25%	save 40%

If George flies first class he will stay in a special section of the airplane which has extra large seats and special service. The greater cost pays for the seats, the service, and more expensive food.

All other fares are for coach class. Both excursion fare and super saver fare are discounted from the regular coach fare.

Example Find the cost of George's ticket if he flies excursion fare and leaves at noon on a Thursday.

$ 380	*regular coach*	$ 380	*regular coach*
× 0.20	*save 20%*	− $ 76	*discount*
$ 76.00	*discount*	$ 304	*cost of ticket*

An excursion fare ticket must be round trip, and reserved and bought at least 7 days before the flight. The return flight must be at least a week later than the arrival. A super saver fare ticket must be round trip with at least a week between arrival and return, too. Also, the super saver fare ticket must be reserved and bought at least one month before the flight.

A regular fare ticket, has no restriction on the days or time of day to fly, nor on when to reserve and buy the ticket.

Exercises **Find the cost for each of the following Cincinnati—Los Angeles round trip flights.**

1. excursion fare, weekday, daytime
2. super saver fare, weekday, daytime
3. excursion fare, weekday, nighttime
4. super saver fare, weekday, nighttime
5. excursion fare, weekend, daytime
6. super saver fare, weekend, daytime
7. excursion fare, weekend, nighttime
8. super saver fare, weekend, nighttime

5–4 The Triangle Inequality

You probably have heard that a straight line is the shortest distance between two points. According to this principle, which path should the woman in the picture take?

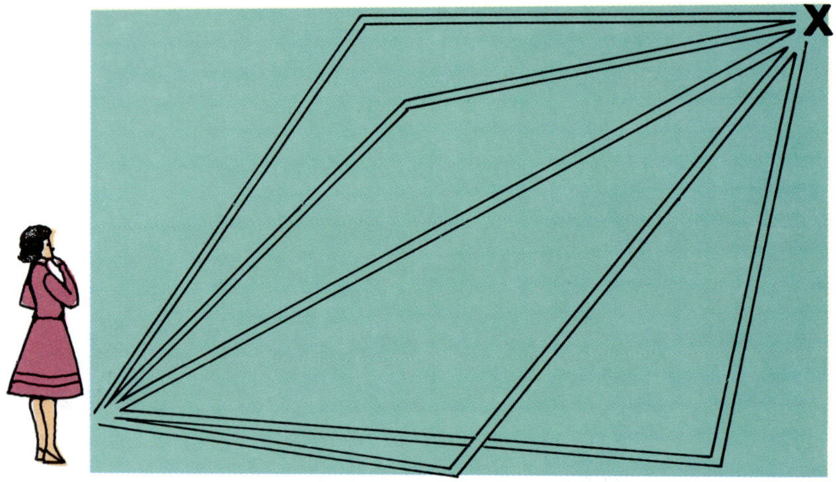

The following theorem applies the principle to triangles.

Theorem 5–8
Triangle Inequality

The sum of the measures of any two sides of a triangle is greater than the measure of the third side.

According to The Triangle Inequality, certain sets of measures *cannot* be used to form a triangle.

example 1

Tell whether or *not* it is possible to draw a triangle with sides of 2 units, 3 units, and 6 units.

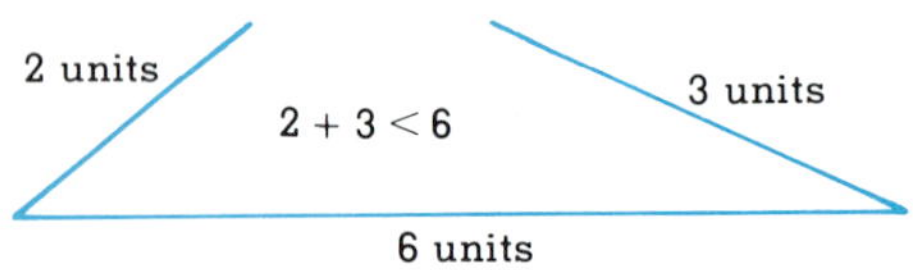

The sum of the measures of two sides *must* be greater than the measure of the third side.

No, it is *not* possible to draw a triangle with the given sides.

The proof of the Triangle Inequality uses an auxiliary figure.

example 2

Plan a proof for the Triangle Inequality.

Given: $\triangle PQR$

Prove: $RP + PQ > RQ$

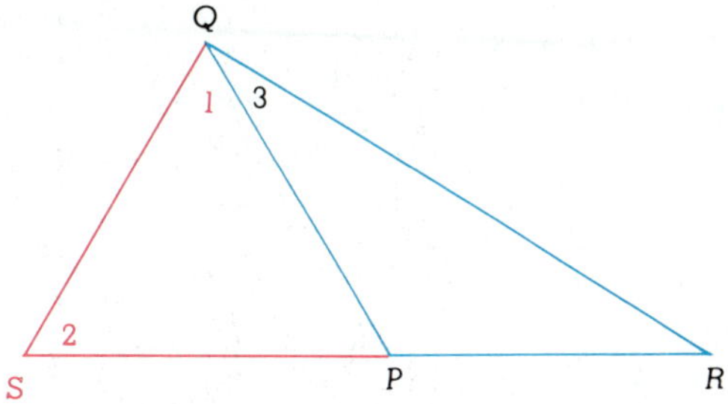

Draw $\overline{PS}$ so that P is between S and R, and $PQ = PS$. By betweenness, $RP + PS = RS$, and, then, by substitution, $RP + PQ = RS$. Thus, the proof must show that $RS > RQ$ to conclude that $RP + PQ > RQ$.

Triangle PQS is isosceles. Thus, the base angles are congruent and $m\angle 1 = m\angle 2$. According to the Angle Addition Postulate, $m\angle SQR = m\angle 1 + m\angle 3$. From substitution, conclude that $m\angle SQR = m\angle 2 + m\angle 3$. As a result, $m\angle SQR > m\angle 2$ by the Inequality Theorem.

Since $m\angle SQR > m\angle 2$, the measures of the two sides are unequal in the same order. That is, $RS > RQ$. Thus, by substitution, $RP + PQ > RQ$.

exercises

Exploratory Two sides of a triangle measure 14 centimeters and 15 centimeters. Tell whether or *not* each of the following can be a measurement for the third side.

1. 5 centimeters
2. 17 centimeters
3. 9 centimeters
4. 16 centimeters
5. 29 centimeters
6. 34 centimeters
7. 10 centimeters
8. 11 centimeters
9. 37 centimeters
10. 23 centimeters
11. 1 centimeter
12. 14 centimeters
13. 30 centimeters
14. 27 centimeters
15. 28 centimeters

Two sides of a triangle measure 23 feet and 7 feet. Tell whether or *not* each of the following can be a measurement for the third side.

16. 45 feet
17. 29 feet
18. 9 feet
19. 37 feet
20. 27 feet
21. 24 feet
22. 30 feet
23. 19 feet
24. 33 feet
25. 17 feet
26. 16 feet
27. 25 feet
28. 13 feet
29. 23 feet
30. 3 feet

Written For each of the following, tell whether or *not* it is possible to draw a triangle with sides of the given measures. Write *yes* or *no*.

1. 1, 2, 3
2. 11, 12, 17
3. 2.5, 6, 6.5
4. 9, 40, 41
5. 16, 12, 17
6. 4.7, 9, 4.1
7. 2.2, 12, 14.3
8. 20, 48, 52
9. 14, 15, 30
10. 2.3, 12, 12.2
11. 215, 204, 7
12. 100, 100, 5

Use the Triangle Inequality to complete each of the following.

13. If a triangle has two sides measuring 3 units and 7 units, then the third side is longer than ____ units and shorter than ____ units.
14. If a triangle has two sides measuring 10 units and 6 units, then the third side is longer than ____ units and shorter than ____ units.
15. If a triangle has two sides measuring 12 units and 15 units, then the third side is longer than ____ units and shorter than ____ units.
16. If a triangle has two sides measuring 11 units and 23 units, then the third side is longer than ____ units and shorter than ____ units.

Plan a proof for each of the following.

17. **Given:** $\angle ABC \cong \angle ACB$
Prove: $AD + AB > CD$

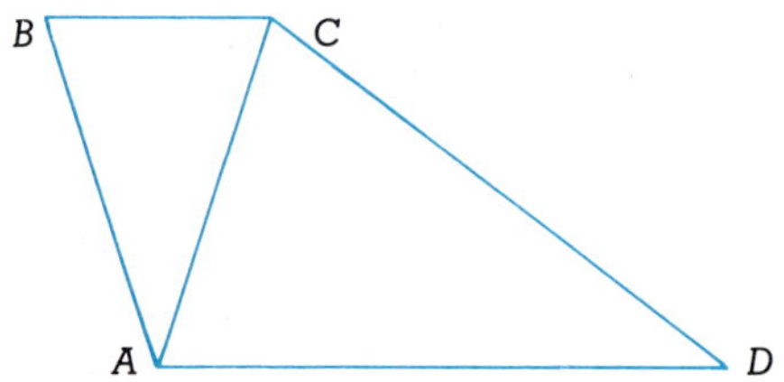

18. **Given:** $\triangle PTS \cong \triangle QRS$
Prove: $PR > QR$

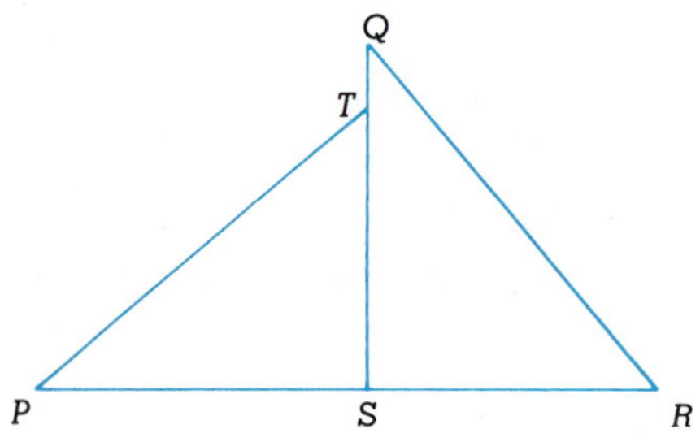

19. **Given:** $\overline{ED}$ bisects $\overline{AC}$
$\overline{ED} \perp \overline{AC}$
Prove: $BC > BA$

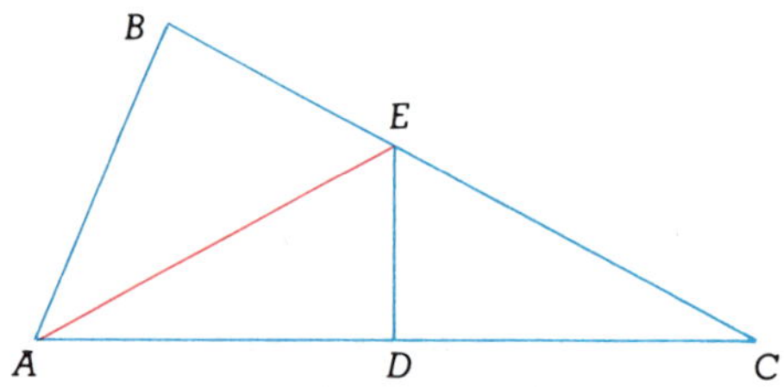

20–22. Write a formal proof for each of problems **17–19**.

23. Write a formal proof for Theorem 5–8.

Many steps are involved in transforming written words into a book. One of these steps is called **typesetting.** Below is an example of original material before and after typesetting.

```
   When material is typed on a
typewriter, each line has a differ-
ent length.  The typist must decide
how to hyphenate words.
   When material is typeset, the
spacing between words is adjusted
to make each line the same length.
Often, computers are used to deter-
mine the amount of spacing between
words.
   Also, computers may be used to
hyphenate words by using a diction-
ary and a list of hyphenation
rules.
```

When material is typed on a typewriter, each line has a different length. The typist must decide how to hyphenate words.

When material is typeset, the spacing between words is adjusted to make each line the same length. Often, computers are used to determine the amount of spacing between words.

Also, computers may be used to hyphenate words by using a dictionary and a list of hyphenation rules.

One reason computers are used for typesetting is that they are very fast. A computer can typeset an entire newspaper page in a few minutes or less. Without a computer it would take several hours. Some computers can typeset a single line of material in micro-seconds. *Micro-seconds are millionths of a second.*

Exercises Suppose a particular computer can typeset 5 three-inch lines of a newspaper article in 4 seconds. Calculate how much time the computer takes to typeset a newspaper article for each of the following. Round your answer to one decimal place.

1. 70 three-inch lines
2. 50 three-inch lines
3. 45 three-inch lines
4. 35 three-inch lines
5. 68 three-inch lines
6. 42 three-inch lines

5–5 The Hinge Theorem

Drafting artists sometimes use compasses to compare lengths. If the angle of the compass is changed, the distance between the compass points changes.

The wider the angle of the compass, the longer the distance between its points. The narrower the angle of the compass, the shorter the distance between its points.

Theorem 5–9
Hinge Theorem

If two sides of one triangle are congruent to two sides of another triangle, but the measures of the included angles are unequal, then the measures of the third sides are unequal in the same order.

In the following diagram, notice that $AB < CD$. If the compass is *not* changed after matching the points to A and B, then it will *not* match the points C and D. How do you change the angle of the compass so its points match C and D? The converse of the Hinge Theorem describes this situation.

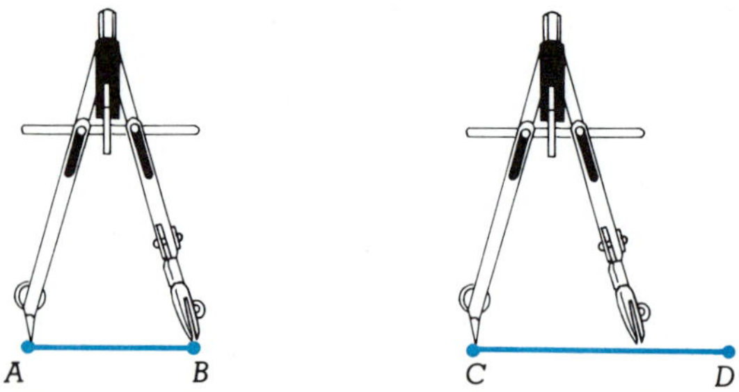

Theorem 5–10
Converse of the
Hinge Theorem

If two sides of one triangle are congruent to two sides of another triangle, but the measures of the third sides are unequal, then the measures of the angles included between the pairs of congruent sides are unequal in the same order.

examples

1 **Complete the proof given below.**

Given: $\overline{AB} \cong \overline{BC}$

$m\angle 1 < m\angle 2$

Prove: $AD < DC$

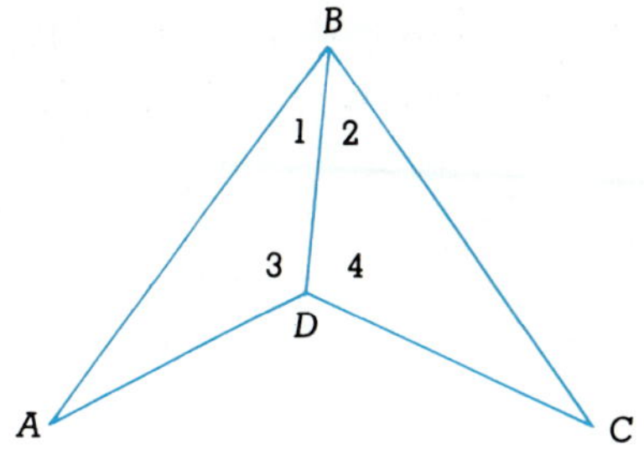

Proof:

STATEMENTS	REASONS
1. $\overline{AB} \cong \overline{BC}$	1. Given
2. $\overline{BD} \cong \overline{BD}$	2. Theorem 2–2: Congruence of line segments is reflexive.
3. $m\angle 1 < m\angle 2$	3. Given
4. $AD < DC$	4. Theorem 5–9: The Hinge Theorem

2 **In the figure below, $\triangle QTS$ has $\angle 1 \cong \angle 2$. Also, $\overline{PQ} \cong \overline{SR}$ and $PT < QR$. Write an inequality to show how the measures of $\angle PQT$ and $\angle QSR$ are related.**

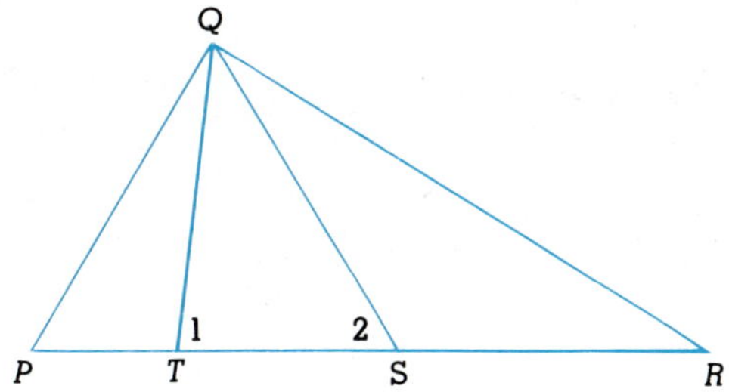

$\angle 1 \cong \angle 2$	*Given*
$\overline{QT} \cong \overline{QS}$	*Theorem 4–9 Converse of Isosceles Triangle Theorem*
$\overline{PQ} \cong \overline{SR}$	*Given*
$PT < QR$	*Given*

By the statements indicated in color, $\triangle PQS$ and $\triangle TQR$ satisfy the hypothesis of Theorem 5–10. Thus, the measures of the angles are related in the following way.

$$m\angle PQT < m\angle QSR$$

exercises

Exploratory For each of the following, name the theorem that describes the situation.

1. When a door opens, the measure of the distance between the edge of the door and the frame increases.
2. A pendulum swings in a triangular frame. The distances of the weight from the endpoints of the frame depend on the angles formed at the vertex of the frame.
3. To mark off a segment shorter than a given segment, match the points of a compass with the endpoints, then narrow the angle of the compass.
4. To mark off a segment longer than a given segment, match the points of a compass with the endpoints, then widen the angle of the compass.
5. If two noncongruent triangles have two sides of one congruent to two sides of the other, then the triangles with the smaller included angle has the shorter third side.
6. If two noncongruent triangles have two sides of one congruent to two sides of the other, then the triangle with the shorter third side has the smaller included angle.

Written Write a formal proof for each of the following.

1. **Given:** $\overline{PQ} \cong \overline{SQ}$
   Prove: $PR > SR$

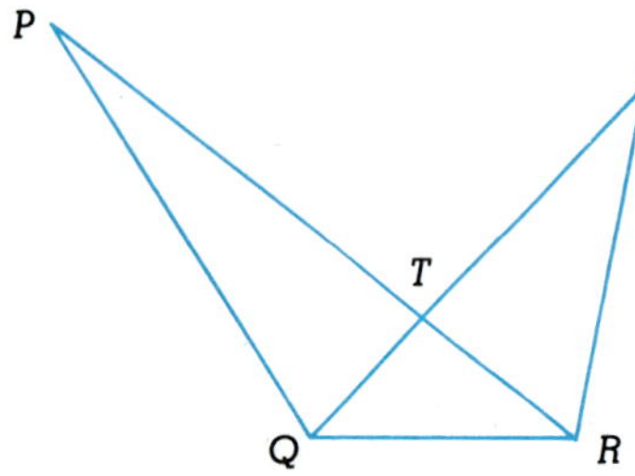

2. **Given:** $\overline{PQ} \cong \overline{RS}$
   $QR < PS$
   Prove: $m\angle 3 < m\angle 1$

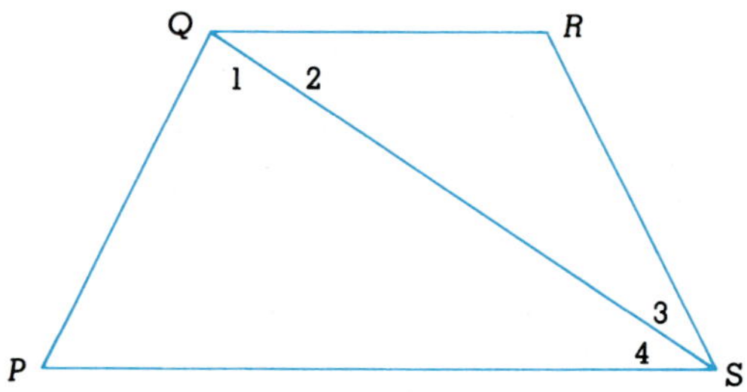

Use the figure at the right for each of the following.

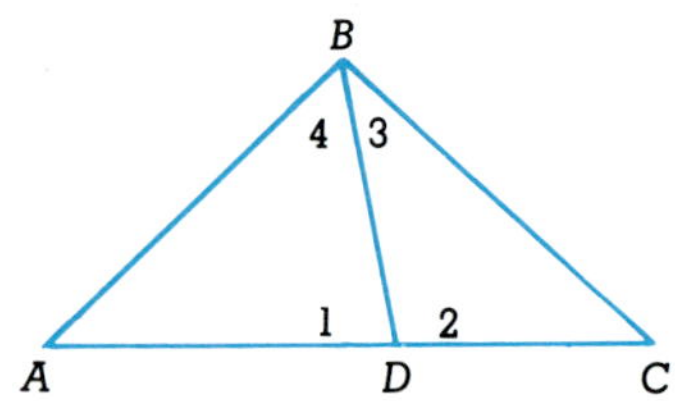

3. **Given:** $\overline{AB} \cong \overline{BC}$
   $m\angle 4 > m\angle 3$
   Prove: $AD > DC$
4. **Given:** $\overline{AB} \cong \overline{BC}$
   $AD > DC$
   Prove: $m\angle 4 > m\angle 3$
5. **Given:** D is the midpoint of $\overline{AC}$
   $AB > BC$
   Prove: $m\angle 1 > m\angle 2$
6. **Given:** $\overline{BD}$ bisects $\overline{AC}$
   $AB > BC$
   Prove: $m\angle 1 > m\angle 2$
7. **Given:** D is the midpoint of $\overline{AC}$
   $AB > BC$
   Prove: $m\angle 1 > m\angle 4$
8. **Given:** $\overline{BD}$ bisects $\overline{AC}$
   $AB > BC$
   Prove: $m\angle 1 > m\angle 4$
9. **Given:** D is the midpoint of $\overline{AC}$
   $AB > BC$
   Prove: $m\angle 1 > m\angle A$
10. **Given:** $\overline{BD}$ bisects $\overline{AC}$
    $AB > BC$
    Prove: $m\angle 1 > m\angle A$

Use the diagram and given statement below for each of the following.

Given: $\overline{PQ} \cong \overline{AB}$
$\overline{PR} \cong \overline{AC}$
$QR > BC$

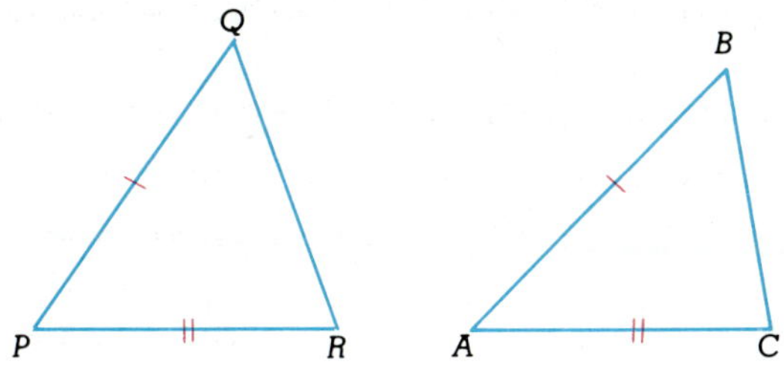

11. It is impossible for $\angle P$ and $\angle A$ to have the same measures. Show that assuming $m\angle P = m\angle A$, implies $QR = BC$, which contradicts $QR > BC$.

12. It is impossible for $\angle P$ to be smaller than $\angle A$. Show that assuming $m\angle P < m\angle A$, implies $QR < BC$, which contradicts $QR > BC$.

Challenge

The following is a plan for a proof of the Hinge Theorem.

Given: $\triangle PQR$ and $\triangle ABC$
$\overline{PQ} \cong \overline{AB}$
$\overline{PR} \cong \overline{AC}$
$m\angle P > m\angle A$

Prove: $QR > BC$

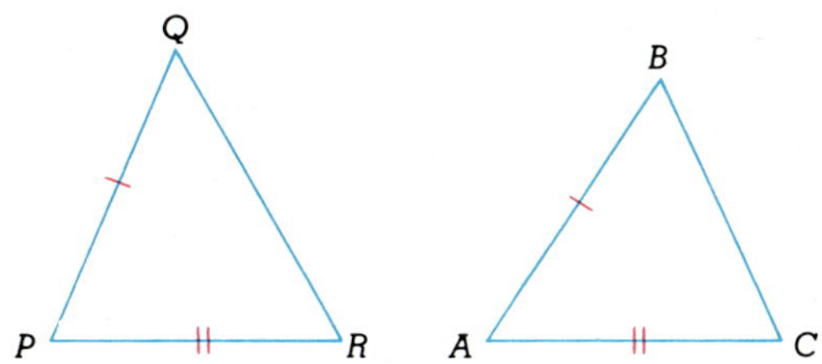

Plan for proof: On $\triangle PQR$ draw $\overrightarrow{PS}$ so that $\angle SPR \cong \angle BAC$ and $\overline{PS} \cong \overline{AB}$.

Case 1: S lies on $\overline{QR}$.
By SAS, show that $\triangle SPR \cong \triangle BAC$. Conclude $\overline{SR} \cong \overline{BC}$. $QR = QS + SR$, since S is between Q and R. Thus, $QR > SR$ and, by substitution, $QR > BC$.

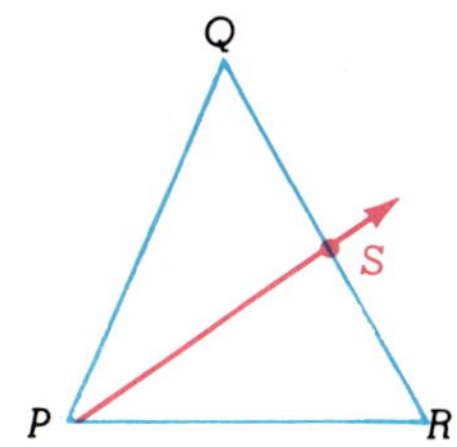

Case 2: S does *not* lie on $\overline{QR}$.
Suppose $\overrightarrow{PS}$ intersects $\overline{QR}$ at T and draw $\overrightarrow{PV}$ so that $\angle QPT$ is bisected and V lies on $\overline{QR}$. By SAS, show that $\triangle QPV \cong \triangle SPV$ and conclude $\overline{QV} \cong \overline{VS}$. The Triangle Inequality implies $VR + VS > SR$. Thus, by substitution, $VR + QV > SR$, or $QR > SR$. Next, use SAS to show $\triangle SPR \cong \triangle BAC$ and conclude $\overline{SR} \cong \overline{BC}$. Thus, using substitution, $QR > BC$.

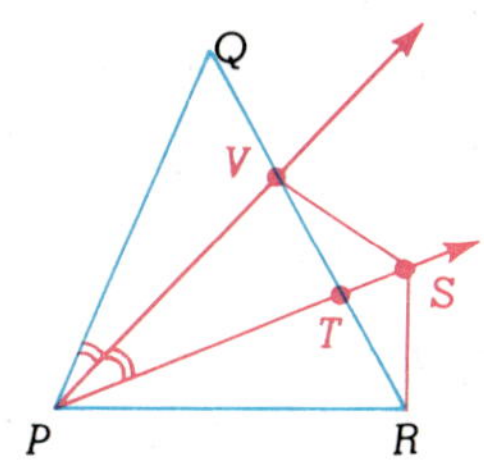

1. Write a formal proof for **Case 1** of the Hinge Theorem. The first statement should be the following.

   On $\triangle PQR$, draw $\overrightarrow{PS}$ so that $\angle SPR \cong \angle BAC$, $\overline{PS} \cong \overline{AB}$, and S lies on $\overline{QR}$.

2. Write a formal proof for **Case 2** of the Hinge Theorem. The first statement should be the following.

   On $\triangle PQR$, draw $\overrightarrow{PS}$ so that $\angle SPR \cong \angle BAC$ and $\overline{PS} \cong \overline{AB}$. Let T be the intersection of $\overrightarrow{PS}$ and $\overline{QR}$.

Vocabulary

exterior angle (132)
remote interior angles (132)
Triangle Inequality (148)
Hinge Theorem (152)

Chapter Summary

1. Exterior Angle Theorem: The measure of an exterior angle of a triangle is equal to the sum of the measures of the remote interior angles. (132)
2. Inequality Theorem: For any numbers a and b, $a > b$ if and only if there is a positive number c such that $a = b + c$. (133)
3. Theorem 5–3: The measure of an exterior angle of a triangle is greater than the measure of either remote interior angle. (134)
4. Diagrams may be used to show the existence of figures, the relative position of figures, or intersections. (137)
5. To see overlapping triangles in a diagram, it is helpful to outline each triangle in a color. (138)
6. Theorem 5–4: If the measures of two sides of a triangle are unequal, then the measures of the angles opposite those sides are unequal in the same order. (143)
7. Theorem 5–5: If the measures of two angles of a triangle are unequal, then the measures of the sides opposite those angles are unequal in the same order. (144)
8. Theorem 5–6: The shortest segment from a point to a line is the perpendicular segment. (145)
9. Theorem 5–7: The shortest segment from a point to a plane is the perpendicular segment. (145)
10. Triangle Inequality: The sum of the measures of any two sides of a triangle is greater than the measure of the third side. (148)
11. Hinge Theorem: If two sides of one triangle are congruent to two sides of another triangle, but the measures of the included angles are unequal, then the measures of the third sides are unequal in the same order. (152)
12. Converse of the Hinge Theorem: If two sides of one triangle are congruent to two sides of another triangle, but the measures of the third sides are unequal, then the measures of the angles included between the pairs of congruent sides are unequal in the same order. (152)

Chapter Review

5–1 Use the information given in the figure to find each of the following.

1. x
2. y
3. z
4. v
5. w
6. r
7. s
8. t

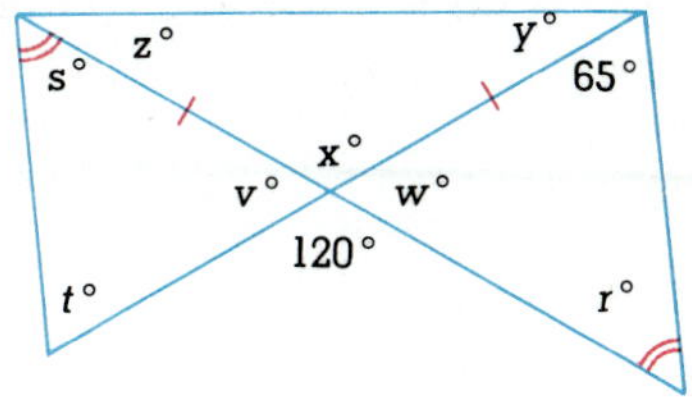

5–2 Can the following statements be assumed from the appearance of the figure? Write *yes* or *no*.

9. E is between A and D
10. $\overline{BD}$ bisects $\overline{EC}$
11. $\overline{BF} \cong \overline{CF}$
12. $\overline{EA} \cong \overline{DA}$
13. $\overline{EA} \cong \overline{ED}$
14. $\angle BED$ is a right angle

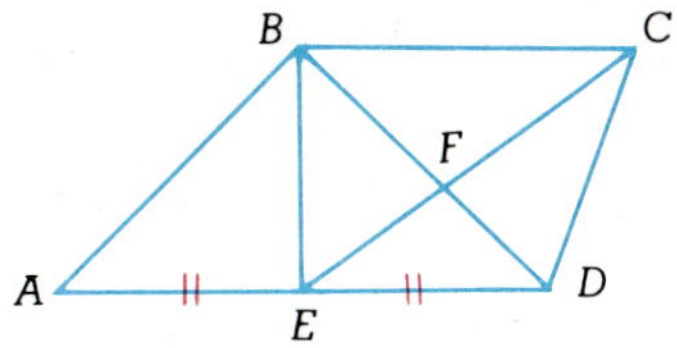

15. For the figure above, state how many triangles appear. Name each triangle.

5–3 The figure at the right below is *not* correctly drawn. Use the information given on the figure to solve each problem.

16. Name the sides of $\triangle WXY$ in order from longest to shortest.
17. Name the sides of $\triangle WYZ$ in order from longest to shortest.
18. Do the longest sides of $\triangle WXY$ and $\triangle WYZ$ have the same length? Write *yes*, *no*, or *not enough information*.
19. Are $\triangle WXY$ and $\triangle WYZ$ congruent? Write *yes* or *no*.

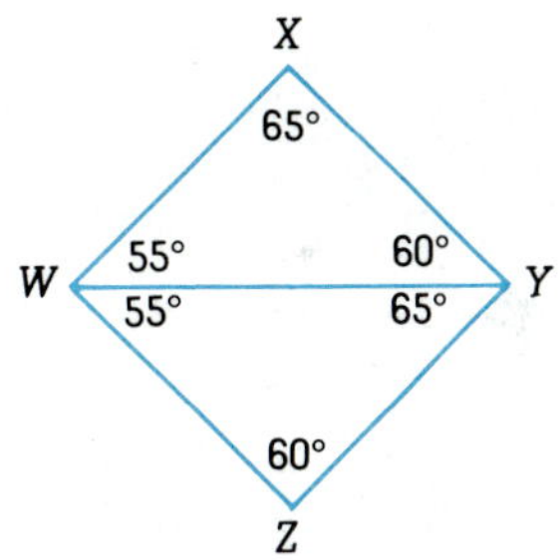

5–4 For each of the following, tell whether or *not* it is possible to draw a triangle with sides of the given measures.

20. 2.1, 2.8, 3.5
21. 41, 42, 43
22. 10, 15, 25
23. 106, 94, 10

5–5 Write a formal proof for each of the following.

24. **Given:** $\overline{AB} \cong \overline{BC}$
    $m\angle ABE > m\angle CBE$
    Prove: $AE > EC$
25. **Given:** $\overline{AB} \cong \overline{CD}$
    $AD > BC$
    Prove: $m\angle ACD > m\angle BAC$

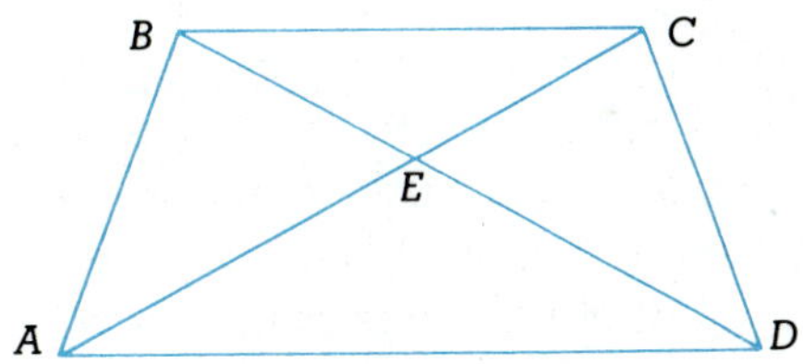

Chapter Test

Use the figure below to solve each problem of this test.

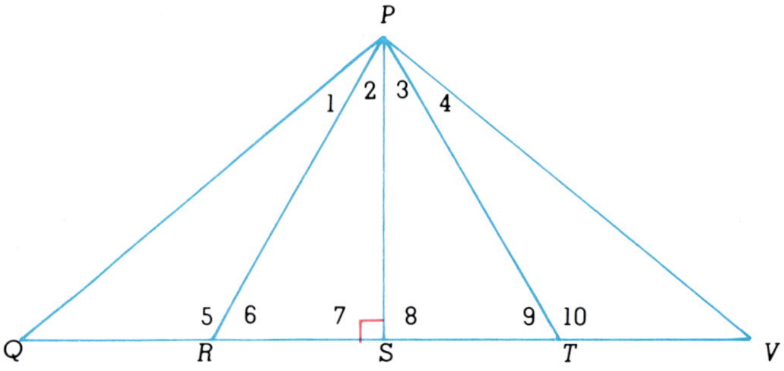

For each of the following, suppose $\overline{PQ} \cong \overline{PV}$, $\overline{PR} \cong \overline{PT}$, $m\angle 9 = 60$, and $m\angle 4 = 20$. Find each measure.

1. $m\angle 7$
2. $m\angle 8$
3. $m\angle 3$
4. $m\angle 6$
5. $m\angle 2$
6. $m\angle 5$
7. $m\angle 10$
8. $m\angle V$
9. $m\angle Q$
10. $m\angle 1$
11. State how many triangles appear. Name each triangle.

Name the postulates or theorems that justify each of the following statements. Use the figure at the top of the page.

12. $QS > QR$ since $QR + RS = QS$.
13. If $m\angle V > m\angle 4$, then $PT > TV$.
14. $RP + PT > RT$
15. $PS < PV$
16. If $\overline{PR} \cong \overline{PT}$ and $m\angle 3 > m\angle 2$, then $ST > RS$.
17. If $\overline{PR} \cong \overline{PT}$, $\overline{PQ} \cong \overline{PV}$, and $RV > QT$, then $m\angle RPV > m\angle QPT$.
18. $m\angle 5 = m\angle 2 + m\angle 7$
19. $m\angle RPT = m\angle 2 + m\angle 3$
20. $m\angle 8 > m\angle 6$
21. $m\angle 10 > 90$

Prove each of the following. Use the figure at the top of the page.

22. **Given:** $m\angle 7 = 90$
    Prove: $m\angle 5 > m\angle 4$

23. **Given:** $\overline{PR} \cong \overline{PT}$
    $m\angle 2 > m\angle 3$
    Prove: $RS > ST$

24. **Given:** $m\angle 8 = 90$
    Prove: $PV > PT$

Chapter 6

Parallels

Many real-life phenomena may be described using parallel lines. Fences are stretched on parallel posts. The sun's rays are all parallel. What other parallels can you see in this photograph?

6–1 Parallels and Transversals

The rails on a railroad track never meet. What would happen if they did meet?

In a plane, lines that never meet are called **parallel lines.**

Definition of Parallel Lines

Two lines are parallel if and only if they lie in the same plane and they do *not* intersect.

The symbol $\parallel$ means *is parallel to*. In the figure below, the two lines are parallel.

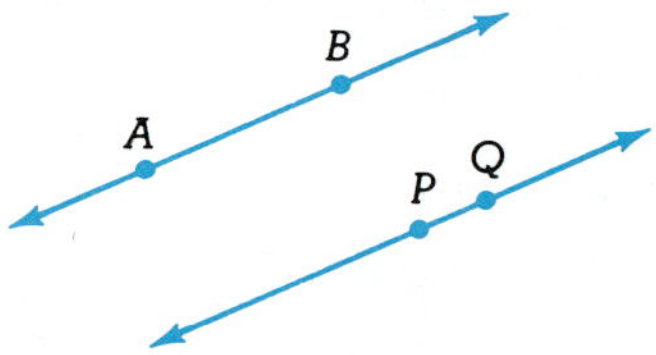

$\overleftrightarrow{AB} \parallel \overleftrightarrow{PQ}$

$\overleftrightarrow{AB}$ is parallel to $\overleftrightarrow{PQ}$

Parts of lines are parallel to each other if the lines containing them are parallel. For example, a ray can be parallel to a line segment. In the figure above, $\overrightarrow{AB} \parallel \overline{PQ}$.

The same term, parallel, and the same notation, $\parallel$, is used for lines, segments, and rays.

Some lines do *not* intersect and yet are *not* parallel. They are called **skew lines.**

Definition of Skew Lines

Two lines are skew if and only if they do *not* intersect and are *not* in the same plane.

Thus, there are three different ways that two lines can be positioned in space, intersecting, parallel, or skew.

example 1

In the figure below, name all pairs of lines which appear to be parallel. Then, name all pairs of lines which appear to be skew.

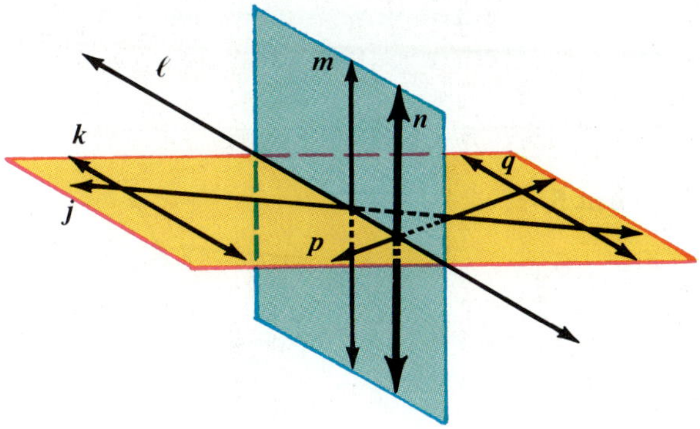

The pairs of parallel lines are as follows.		The pairs of skew lines are as follows.	
$k \parallel \ell$	$m \parallel n$	k and m	p and m
$k \parallel q$	$\ell \parallel q$	k and n	q and m
		j and n	q and n

In a plane, a transversal is a line that intersects two other lines in two different points.

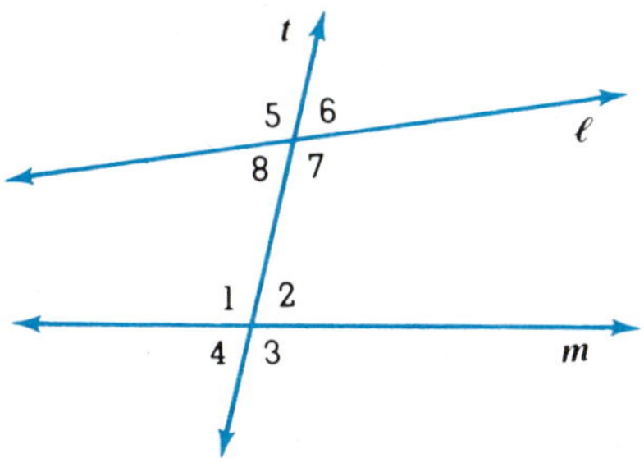

The transversal t *intersects*, or *cuts*, lines ℓ and m.

Eight angles are formed by the intersection of a transversal and two lines. Different names are used for various sets of these angles.

Interior Angles	$\angle 1$, $\angle 2$, $\angle 7$, $\angle 8$
Alternate Interior Angles	$\angle 1$ and $\angle 7$, $\angle 2$ and $\angle 8$
Exterior Angles	$\angle 3$, $\angle 4$, $\angle 5$, $\angle 6$
Corresponding Angles	$\angle 1$ and $\angle 5$, $\angle 2$ and $\angle 6$
	$\angle 3$ and $\angle 7$, $\angle 4$ and $\angle 8$

The theorems on the next page state how some of the pairs of angles are related.

Theorem 6–1 **If two lines are cut by a transversal and one pair of alternate interior angles are congruent, then the other pair of alternate interior angles also are congruent.**

Theorem 6–2 **If two lines are cut by a transversal and one pair of corresponding angles are congruent, then all pairs of corresponding angles are congruent.**

example 2

Prove Theorem 6–1.

Given: $\angle 4 \cong \angle 6$

Prove: $\angle 3 \cong \angle 5$

Proof:

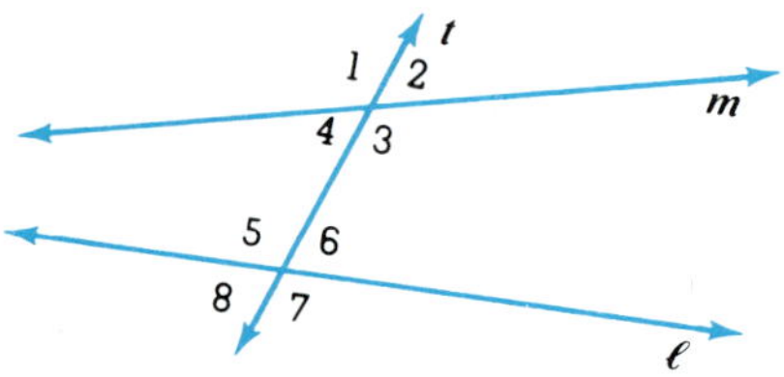

STATEMENTS	REASONS
1. $\angle 4$ and $\angle 3$ form a linear pair $\angle 5$ and $\angle 6$ form a linear pair	1. Given
2. $\angle 4$ and $\angle 3$ are supplementary $\angle 5$ and $\angle 6$ are supplementary	2. Postulate 3–4: Supplement Postulate
3. $\angle 4 \cong \angle 6$	3. Given
4. $\angle 3 \cong \angle 5$	4. Theorem 3–3: If two angles are supplementary to two congruent angles, then the two angles are congruent to each other.

exercises

Exploratory State whether each of the following suggest *intersecting*, *parallel*, or *skew*.

1. rungs on a ladder
2. railroad crossing sign
3. airline flight paths
4. lines on a football field
5. airport runways
6. electric power lines
7. lines on writing paper
8. guitar strings
9. spokes on a wheel
10. lines in a parking lot
11. siderails on highway
12. joints in a sidewalk
13. artists T-square
14. rows of corn
15. marks on a ruler
16. lines on a baseball diamond
17. slats on blinds
18. bowling alleys
19. skis on a skier
20. columns on the front of a building

Written Using the diagram at the right, find an example for each of the following.

1. Two coplanar lines that intersect
2. Two coplanar lines that do *not* intersect
3. Two parallel lines
4. Two skew lines
5. Two noncoplanar lines
6. Two parallel line segments
7. Two parallel lines and a transversal
8. Two parallel rays
9. A ray and a segment that are *not* parallel
10. Two coplanar lines

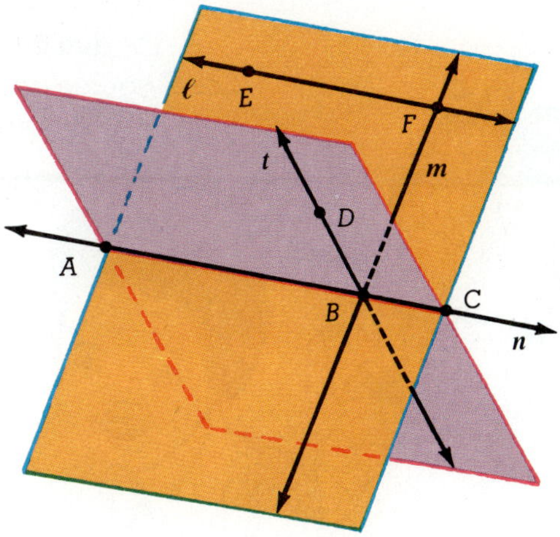

Draw a diagram to illustrate each of the following.

11. $\overleftrightarrow{AB} \parallel \overleftrightarrow{CD}$
12. ℓ and m are skew
13. t is a transversal to ℓ and m
14. $\overleftrightarrow{RS} \parallel \overleftrightarrow{TV}$, both lines are cut by $\overleftrightarrow{AB}$

Tell whether or *not* each of the following is true. Write *yes* or *no*.

15. Two lines are parallel if they do *not* intersect.
16. A line that intersects two skew lines is a transversal.
17. Skew lines are parallel.
18. If a line intersects two parallel lines, it is a transversal.
19. Alternate interior angles are on the same side of a transversal.
20. Corresponding angles are on the same side of a transversal.
21. Alternate interior angles have a line segment in common.
22. Corresponding angles have a ray in common.
23. It is possible for a transversal to be perpendicular to two lines.
24. It is possible for two parallel lines both to be parallel to a third line.

Write an *if and only if* definition for each of the following.

25. parallel line segments
26. parallel rays
27. alternate interior angles
28. corresponding angles

29. Prove Theorem 6–2.
30. Prove that if two lines are cut by a transversal and one pair of consecutive interior angles is supplementary, then the other pair of consecutive interior angles is supplementary.

Challenge For each of the following write *yes* or *no*. Then, explain.

1. Parallelism is reflexive.
2. Parallelism is symmetric.
3. Parallelism is transitive.

6–2 Indirect Proof

A TV detective, Sheerluck Ames, discovers that a suspect in a Los Angeles crime was in San Francisco when the crime was committed. The detective makes the following statement.

If the suspect was in San Francisco, then she did *not* commit the crime.

Then, the detective presents an argument to support the statement.

Suppose the suspect did commit the crime.
The crime was committed in Los Angeles.
The suspect was in San Francisco.
She could *not* be in Los Angeles and San Francisco at the same time.
Therefore, the suspect did *not* commit the crime.

The detective's argument is an example of **indirect reasoning.** The detective assumed the alternative of what he wanted to prove. Then, he showed that his assumption led to a contradiction. He concluded that his assumption must be false. Thus, the statement to be proved must be true. The following is an example of indirect reasoning in geometry.

example 1

Prove that a right triangle has no more than one right angle.

Suppose a right triangle has more than one right angle.

If all three angles are right, then the sum of the degree measures of the angles is 90 + 90 + 90 or 270. This is a contradiction, since the sum of the degree measures of the angles of a triangle must be 180.

If two of the angles are right, then the third angle must measure 0, because the sum of the degree measures is 180. This is a contradiction, since the degree measure of an angle is greater than 0.

Therefore, a right triangle has no more than one right angle.

A right triangle has no more than one right angle. — Theorem 6–3

You can use indirect reasoning in geometry proofs. Use the following steps to write an indirect proof.

How to Write an Indirect Proof

1. Assume the alternatives of the statement to be proved.
2. Show that the assumptions lead to a contradiction.
3. State that the assumptions are false.
4. Conclude that the statement to be proved is true.

examples

2

Prove that, in a plane, if two lines are cut by a transversal so that a pair of alternate interior angles are congruent, then the two lines are parallel.

Given: $\angle 1 \cong \angle 2$

Prove: $\ell \parallel m$

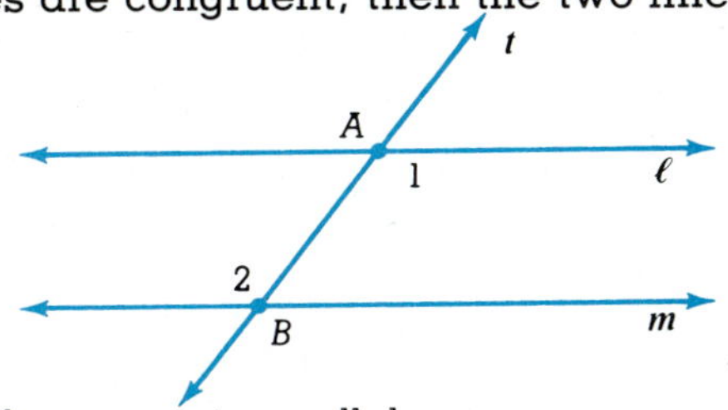

Proof:

Step 1 Suppose ℓ and m are *not* parallel.

Step 2 Then, ℓ and m intersect, say at point P. Now, $\angle 2$ is exterior to $\triangle BAP$. Thus, $m\angle 2 > m\angle 1$ because the measure of an exterior angle of a triangle is greater than the measure of either remote interior angle. But, $\angle 1 \cong \angle 2$ implies that $m\angle 1 = m\angle 2$.
The comparison property has been contradicted. Two measures *cannot* be both equal and unequal.

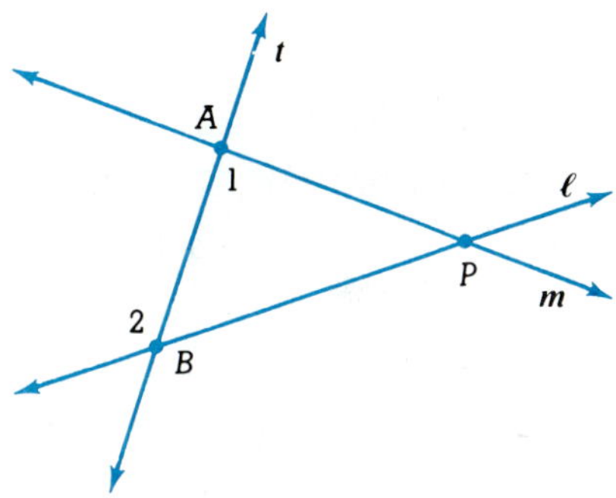

Step 3 The original assumption, ℓ and m are *not* parallel, is false.

Step 4 Thus, $\ell \parallel m$.

3

Prove that $\sqrt{10} \neq 2\sqrt{5}$.

Step 1 Suppose $\sqrt{10} = 2\sqrt{5}$.

Step 2 Then, $(\sqrt{10})^2 = (2\sqrt{5})^2$.

So, $10 = 4 \cdot 5$, or $10 = 20$. *A false statement*

Steps 3 & 4 This statement contradicts the definition of equality. Thus, $\sqrt{10} \neq 2\sqrt{5}$.

The following theorem uses the phrase, *exactly one*.

If two lines intersect, then
they intersect in exactly one point.

Any statement containing the phrase *exactly one* actually has two parts. The two parts for the theorem stated above are as follows.

Exactly one means at least one and no more than one.

If two lines intersect, then
they intersect in *at least one* point.

If two lines intersect, then
they intersect in *no more than one* point.

To prove a theorem that uses *exactly one*, you prove both parts. Often, the proof of the *no more than one* part is indirect.

example 4

Prove that if two lines intersect, then they intersect in exactly one point.

Given: ℓ and m intersect

Prove: ℓ and m intersect in exactly one point

Proof:

Part 1 Show that ℓ and m intersect in *at least one* point.

By the definition of *intersection*, ℓ and m have points in common. Thus, they have at least one point in common.

Part 2 Show that ℓ and m intersect in *no more than one* point.

Suppose ℓ and m have more than one point in common. Call two of these common points P and Q. Both ℓ and m go through P and Q. But Postulate 1–1 states that through any two points there is exactly one line. This is a contradiction since ℓ and m are *not* the same line. Therefore, ℓ and m intersect in no more than one point.

exercises

Exploratory State the alternatives you would use in indirect proofs for each of the following.

1. If the leaves of a plant are in groups of three, then it is poison ivy.
2. If a mushroom is red, then it is poisonous.

3. If the radio does *not* play well, then it is defective.
4. If the lamp won't turn on, then the light bulb is burned out.
5. If two lines intersect, then they intersect in no more than one point.
6. If a line *not* in a plane intersects the plane, then they intersect in no more than one point.
7. A right triangle has at least one acute angle.
8. A right triangle has at most one obtuse angle.
9. In a plane, if two lines are parallel to the same line, then they are parallel to each other.
10. If two lines do *not* intersect, then they are skew lines.
11. If the measures of two angles of a triangle are unequal, then the measures of the sides opposite those angles are unequal in the same order.
12. If two sides of one triangle are congruent to two sides of another triangle, but the lengths of the third sides are unequal, then the measures of the angles included between the pairs of congruent sides are unequal in the same order.

Written **Many tests include multiple-choice questions. One way to choose the correct answer is to use indirect reasoning and, eliminate answers you know to be incorrect. Choose the correct answer for each of the following.**

1. An exterior angle at the base of an isosceles triangle is always ______.
   a. an acute angle
   b. an obtuse angle
   c. a right angle
   d. a 60° angle
2. If two parallel lines are cut by a transversal, the corresponding angles are always ______.
   a. supplementary
   b. congruent
   c. acute
   d. complementary
3. If the measure of an acute angle increases, the measure of its supplement ______.
   a. increases
   b. decreases
   c. remains the same
   d. changes irregularly
4. If $AB > CD$ and $EF > GH$, then which statement is *not* always true? ______
   a. $\frac{1}{2}AB > \frac{1}{2}CD$
   b. $2EF > 2GH$
   c. $AB + EF > CD + GH$
   d. $AB - EF > CD - GH$
5. If two angles of a triangle are complementary, then the triangle must be ______.
   a. scalene
   b. isosceles
   c. right
   d. equiangular
6. In a plane, if line a is perpendicular to line b, and line c is parallel to line a, then b and c are ______.
   a. parallel
   b. perpendicular
   c. the same line
   d. oblique

7. If S is the midpoint of $\overline{RT}$, which relationship is *incorrect*? ______
   a. $RS < RT$ b. $RS = ST$
   c. $RS > RT$ d. $RS + ST = RT$

8. Distinct points which lie on the same line are said to be ______.
   a. coincident b. intersecting
   c. collinear d. vertical

9. Which of the following is the converse of the statement, *Vertical angles are congruent*? ______
   a. Vertical angles have the same measure.
   b. Congruent angles are vertical.
   c. Angles with equal measures are congruent.
   d. Congruent angles have the same measure.

10. *If p, then q* is a conditional statement. Then ______.
   a. *if not q,* then *p* is the inverse of the converse
   b. *if p, then not q* is the inverse
   c. *if q, then p* is the converse
   d. *if not p, then q* is the converse of the inverse

Prove each of the following.

11. A right triangle has at least one acute angle.
12. A right triangle has at most one obtuse angle.
13. In a plane, if two lines are parallel to the same line, then they are parallel to each other.
14. If two lines are cut by a transversal so that a pair of consecutive interior angles are supplementary, then the two lines are parallel.
15. In a plane, through a given point on a line, there is exactly one line perpendicular to the given line.
16. If a plane and a line *not* in the plane intersect, then they intersect in exactly one point.
17. If there is a line and a point *not* on a line, then there is exactly one plane that contains them.
18. If two lines intersect, then exactly one plane contains both lines.

Use indirect reasoning to solve each problem.

19. Jonah Johnson brought home a radio that played well when he bought it. At home, he plugged in the radio and found it did *not* play. Argue that the outlet is defective.
20. A crime suspect claims to be deaf. Detective Ann Lorca stood behind the suspect and fired a shot in the air. The suspect was startled and jumped. Argue that the suspect is *not* deaf.

Excursions in Geometry

Emeralds

An object is *grue* if and only if it is green until the year 2000 A.D., and blue after 2000 A.D. Below are two different assumptions about grue and green emeralds. Which do you think is the best assumption, and why?

All emeralds are green. All emeralds are grue.

Classifying Ideas

Reading Geometry

A good definition identifies the set to which the term being defined belongs, and states the properties which distinguish the term from others in its set. Satisfying these two requirements may involve the grouping or classifying of ideas. Consider the following definition of *angle*.

An angle is the union of two noncollinear rays which have the same endpoint. The rays are called sides and their common endpoint is the vertex of the angle.

This definition of angle includes rays, noncollinear, endpoint, sides and vertex.

The ideas included in the definition can be classified as follows.

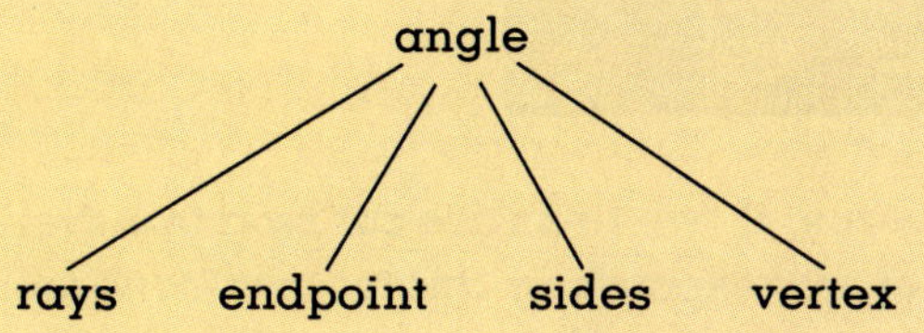

The word angle includes all the other ideas.

Example Find the term from the following group which includes all others.

snow winter
Christmas cold

Winter includes all the other ideas.

Exercises Select the word in each group which includes all the others.

1. points, lines, geometry, planes

2. acute, triangle, obtuse, right

3. algebra, arithmetic, geometry, mathematics

4. parallelogram, rectangle, rhombus, square

5. vertex, side, angle, triangle

6. hexagon, octagon, polygon, pentagon

Write the definition for each of the following words. List the ideas included in each.

7. diameter **8.** rectangle **9.** line segment **10.** transversal

6–3 Testing for Parallels

Many copy machines have a series of sorting trays that holds the copies. Viewed from the side, they look like a row of parallel lines.

Notice that each tray is set at an angle. The angles all have the same degree measure. This situation suggests the following test for parallel lines.

Theorem 6–4 **In a plane, if two lines are cut by a transversal so that a pair of corresponding angles are congruent, then the two lines are parallel.**

example 1

Prove Theorem 6–4.

Given: $\angle 1 \cong \angle 2$

Prove: $\ell \parallel m$

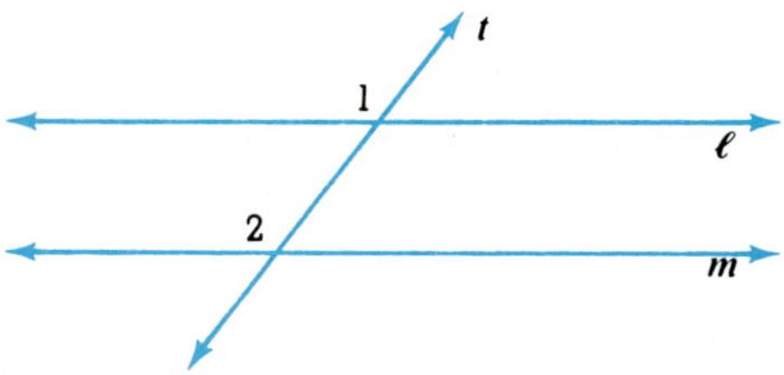

Proof:

Suppose ℓ and m are *not* parallel. Then ℓ and m intersect, say at point C. Now, $m\angle 2 > m\angle 3$ since $\angle 2$ is exterior to $\triangle ABC$. Also, $\angle 1 \cong \angle 3$ because they are vertical angles, and $m\angle 1 = m\angle 3$. Thus, by substitution, $m\angle 2 > m\angle 1$. But, $\angle 1 \cong \angle 2$ implies that $m\angle 2 = m\angle 1$. The Comparison Property has been contradicted.

The original assumption is false.

Thus, $\ell \parallel m$.

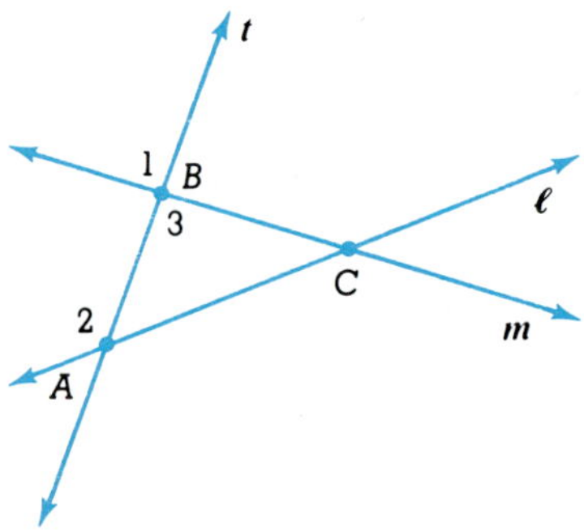

Using Theorem 6–4, the following tests can be derived.

In a plane, if two lines are cut by a transversal so that a pair of alternate interior angles are congruent, then the two lines are parallel. — Theorem 6–5

If two lines are cut by a transversal so that a pair of consecutive interior angles are supplementary, then the lines are parallel. — Theorem 6–6

If two lines are perpendicular to the same line, then the two lines are parallel. — Theorem 6–7

Example **2** shows how the tests are applied to formal proofs.

example 2

Complete the following proof.

Given: $\overline{BD}$ and $\overline{EA}$ bisect each other

Prove: $\overline{AB} \parallel \overline{ED}$

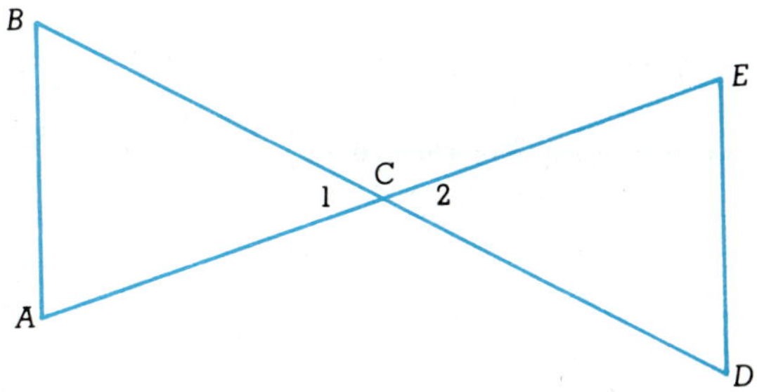

Proof:

STATEMENTS	REASONS
1. $\overline{BD}$ and $\overline{EA}$ bisect each other	1. Given
2. $\overline{BC} \cong \overline{CD}$ $\overline{AC} \cong \overline{CE}$	2. Theorem 2–6: If $\overline{PQ}$ is bisected at point M, then $\overline{PM} \cong \overline{MQ}$.
3. $\angle 1 \cong \angle 2$	3. Theorem 3–10: Vertical angles are congruent.
4. $\triangle BCA \cong \triangle DCE$	4. Postulate 4–2: SAS
5. $\angle A \cong \angle E$	5. Definition of congruent triangles.
6. $\overline{AB} \parallel \overline{ED}$	6. Theorem 6–5: In a plane, if two lines are cut by a transversal so that a pair of alternate interior angles are congruent, then the two lines are parallel.

example 3

Prove Theorem 6–7.

Given: $\ell \perp t$
$m \perp t$

Prove: $\ell \parallel m$

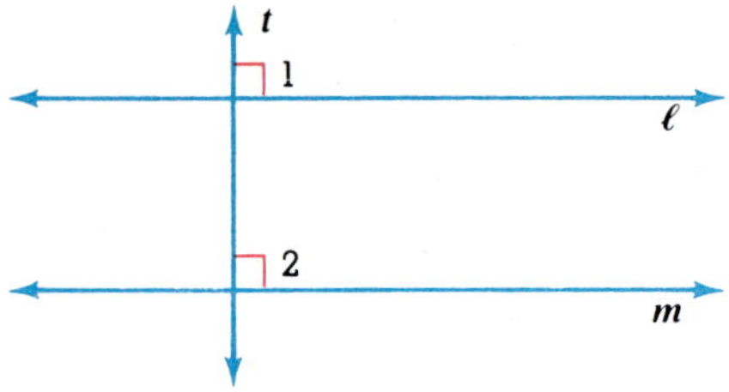

Proof:

STATEMENTS	REASONS
1. $\ell \perp t$ $m \perp t$	1. Given
2. $\angle 1$ is a right angle $\angle 2$ is a right angle	2. Theorem 3–11: If two lines are perpendicular, then they form four right angles.
3. $\angle 1 \cong \angle 2$	3. Theorem 3–6: If two angles are right angles, then the angles are congruent.
4. $\ell \parallel m$	4. Theorem 6–4: In a plane, if two lines are cut by a transversal so that a pair of corresponding angles are congruent, then the two lines are parallel.

exercises

Exploratory For each of the following, tell which theorem or theorems you would use to prove $\ell \parallel m$.

1. 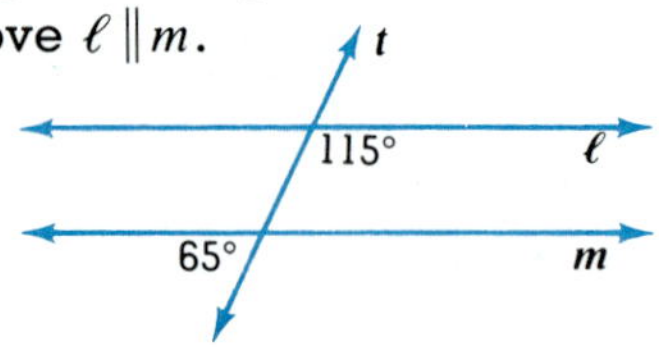

2. 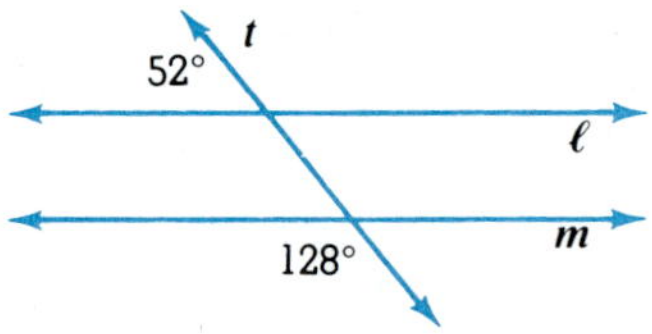

3. 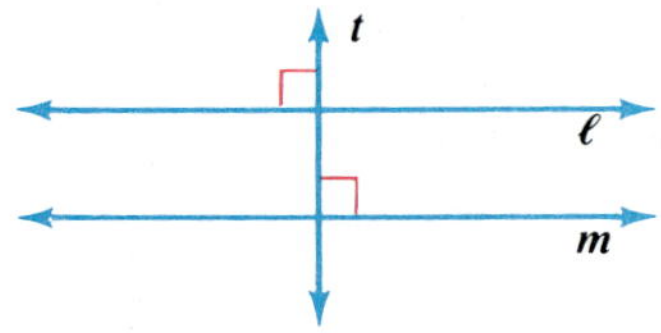

4. 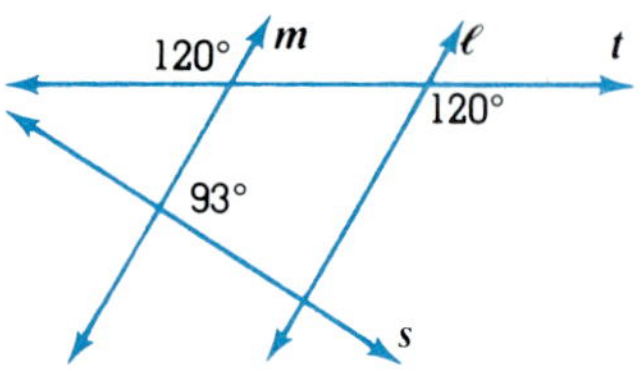

5. 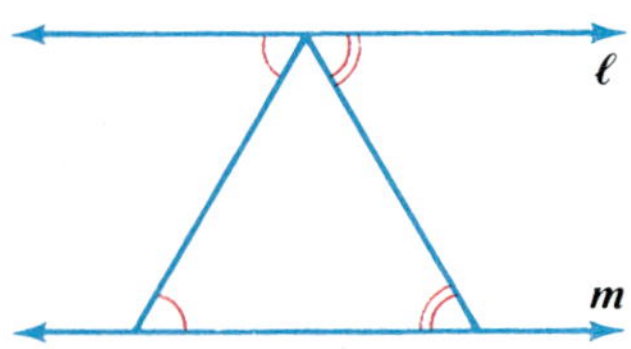

6. 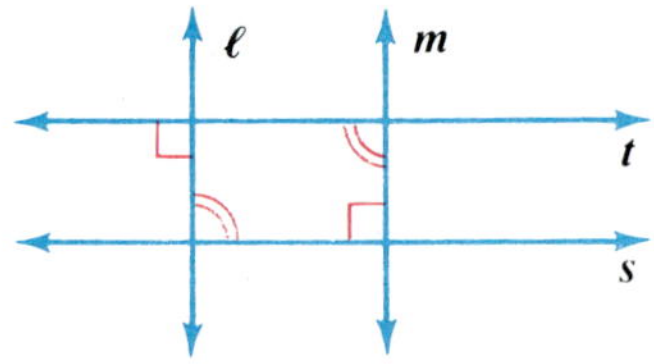

7. 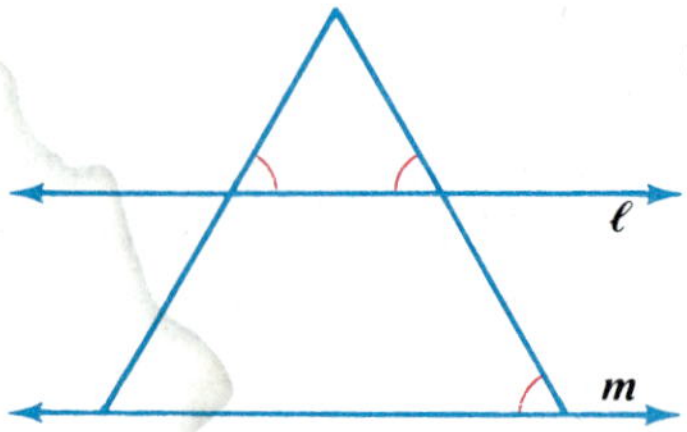

8. 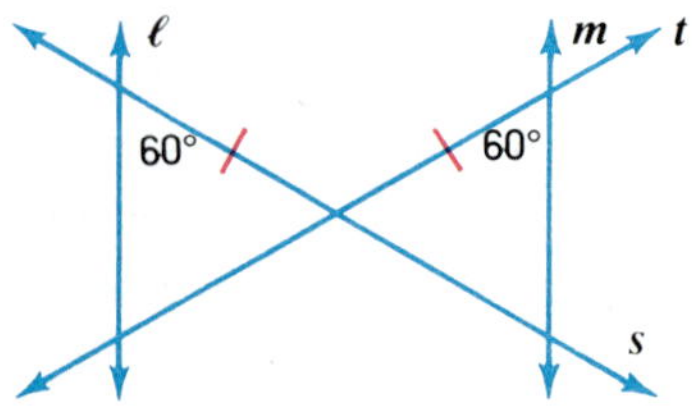

9. 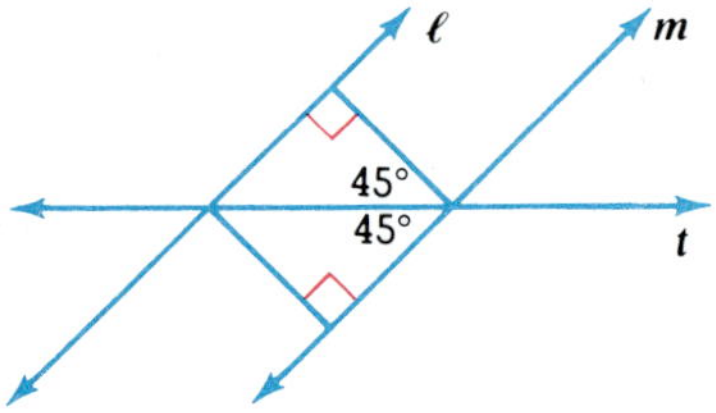

Written Use the figure at the right to solve each problem.

1. **Given:** $\angle R$ and $\angle S$ are supplementary
   $\angle Q \cong \angle S$
   Prove: $\overline{QP} \parallel \overline{RS}$
2. **Given:** $\angle Q$ and $\angle R$ are supplementary
   $\angle S \cong \angle Q$
   Prove: $\overline{QR} \parallel \overline{PS}$

Use the figure at the right to solve each problem.

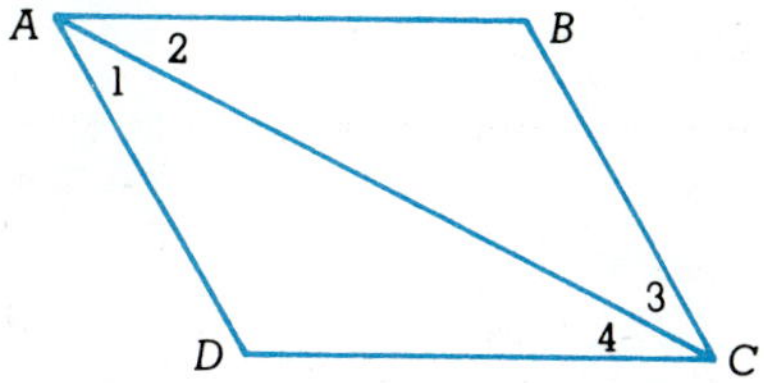

3. **Given:** $\angle 1 \cong \angle 3$
   $\overline{AD} \cong \overline{BC}$
   Prove: $\overline{DC} \parallel \overline{AB}$
4. **Given:** $\angle 2 \cong \angle 4$
   $\overline{AB} \cong \overline{DC}$
   Prove: $\overline{AD} \parallel \overline{BC}$
5. **Given:** $\overline{AD} \cong \overline{BC}$
   $\overline{DC} \cong \overline{AB}$
   Prove: $\overline{AD} \parallel \overline{BC}$
6. **Given:** $\overline{AD} \cong \overline{BC}$
   $\overline{DC} \cong \overline{AB}$
   Prove: $\overline{DC} \parallel \overline{AB}$

Use the figure at the right to solve each problem.

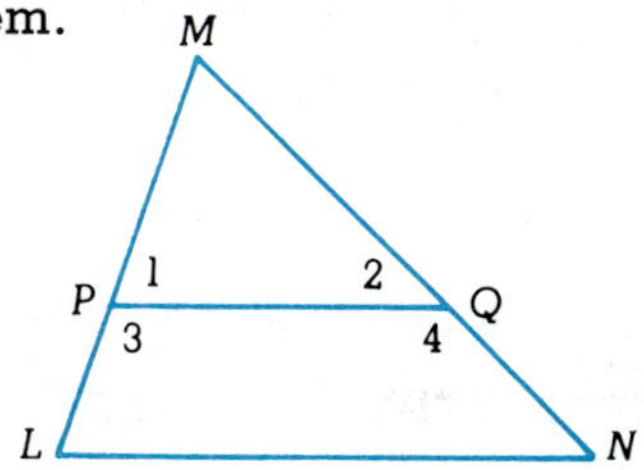

7. **Given:** $\overline{MP} \cong \overline{MQ}$
   $\angle 1 \cong \angle N$
   Prove: $\overline{PQ} \parallel \overline{LN}$
8. **Given:** $\overline{MP} \cong \overline{MQ}$
   $\angle L$ and $\angle 4$ are supplementary
   Prove: $\overline{PQ} \parallel \overline{LN}$

Use the figure at the right to solve each problem.

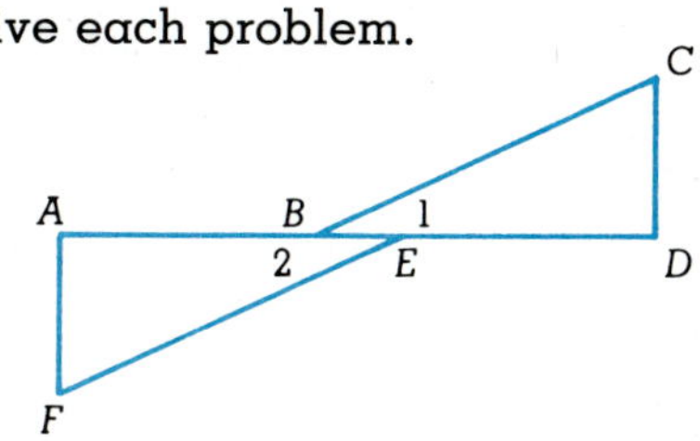

9. **Given:** $\angle 1 \cong \angle 2$
   Prove: $\overline{BC} \parallel \overline{FE}$
10. **Given:** $\overline{AF} \perp \overline{AD}$
    $\overline{CD} \perp \overline{AD}$
    Prove: $\overline{AF} \parallel \overline{CD}$
11. **Given:** $\angle F$ and $\angle 2$ are complementary
    $\angle C$ and $\angle 1$ are complementary
    Prove: $\overline{AF} \parallel \overline{CD}$
12. **Given:** $\overline{AB} \cong \overline{ED}$
    $\overline{AF} \cong \overline{CD}$
    $\angle A$ and $\angle D$ are right angles
    Prove: $\overline{BC} \parallel \overline{FE}$
13. Prove Theorem 6–5.
14. Prove Theorem 6–6.
15. Prove that, in a plane, if two lines are parallel to a third line, then they are parallel to each other.
16. Prove that if two lines are cut by a transversal so a pair of corresponding angles are congruent, then the bisectors of those angles are parallel.

CAREERS

Soil Scientist

Claudine Williams is a soil scientist for Anderson Brothers' Farms. She helps plan land use, including preventing soil erosion and providing irrigation. Claudine plans the location of parallel rows of pipelines to irrigate the fields such as the one pictured at the right.

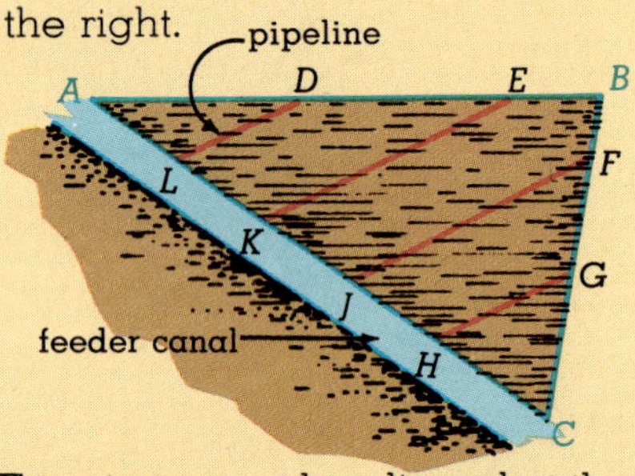

The pipes must be aligned at the correct angles to be parallel. For example, in the field pictured, suppose $m\angle C = 65$, $m\angle B = 80$, and pipeline $\overline{GH}$ is laid so that $m\angle GHC = 60$. Find the degree measure of $\angle AEK$.

$$\begin{aligned} m\angle A + m\angle B + m\angle C &= 180 \\ m\angle A + 80 + 65 &= 180 \\ m\angle A &= 35 \end{aligned} \qquad \begin{aligned} \angle CKE &\cong \angle CHG \\ m\angle CKE &= m\angle CHG \\ m\angle CKE &= 60 \end{aligned}$$

$$\begin{aligned} m\angle CKE + m\angle AKE &= 180 \\ 60 + m\angle AKE &= 180 \\ m\angle AKE &= 120 \end{aligned}$$

$$\begin{aligned} m\angle A + m\angle AKE + m\angle AEK &= 180 \\ 35 + 120 + m\angle AEK &= 180 \\ m\angle AEK &= 25 \end{aligned}$$

Exercises For each of the following, find $m\angle AEK$.

1. 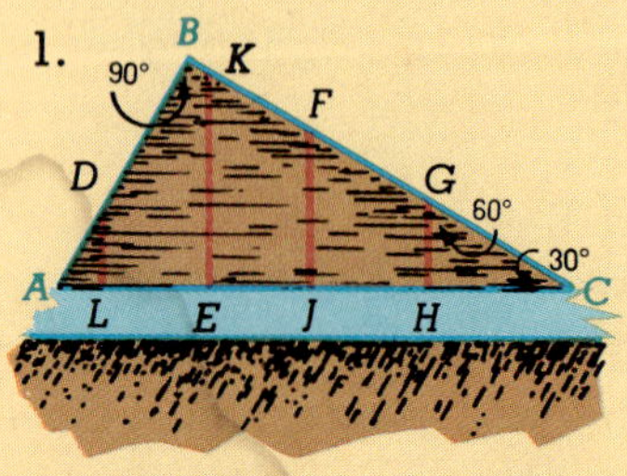

2. 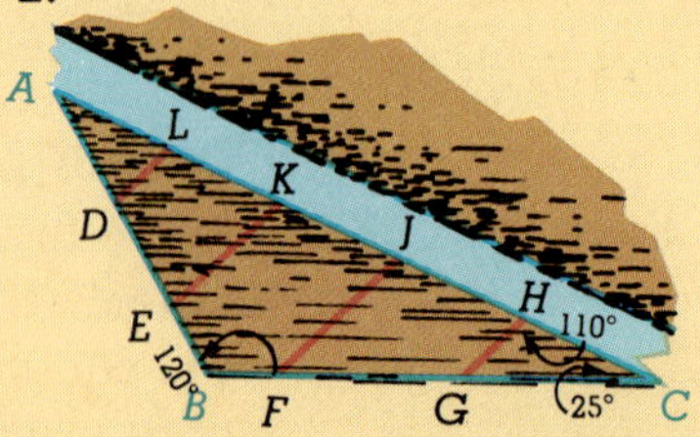

3. 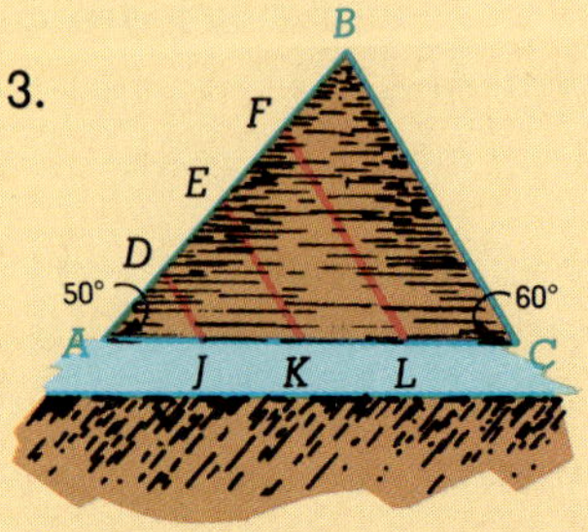

6–4 The Parallel Postulate

Around 300 B.C., Euclid (YOO klid) developed a system for geometry similar to what is used today. He wrote five postulates. His first four postulates were very basic and simple. In contrast, the fifth postulate was complex.

Many mathematicians throughout history thought Euclid's fifth postulate should be *proved*. They tried to prove it, without success.

The Parallel Postulate is one version of Euclid's fifth postulate.

Postulate 6–1
Parallel Postulate

If there is a line and a point *not* on the line, then there is exactly one line through the point and parallel to the given line.

Notice that *exactly one* is used in the Parallel Postulate. Thus, the Parallel Postulate actually has two parts. They are as follows.

> If there is a line and a point *not* on the line, then there is *at least* one line through the point and parallel to the given line.
>
> If there is a line and a point *not* on the line, then there is *at most* one line through the point and parallel to the given line.

Using the Parallel Postulate, it is possible to prove the following properties of parallel lines.

Theorem 6–8

If two parallel lines are cut by a transversal, then each pair of corresponding angles are congruent.

Theorem 6–9

If two parallel lines are cut by a transversal, then each pair of alternate interior angles are congruent.

Theorem 6–10

If two parallel lines are cut by a transversal, then each pair of consecutive interior angles are supplementary.

example 1

Prove Theorem 6–8.

Given: $\ell \parallel m$

Prove: $\angle 1 \cong \angle 2$
$\angle 3 \cong \angle 4$
$\angle 5 \cong \angle 6$
$\angle 7 \cong \angle 8$

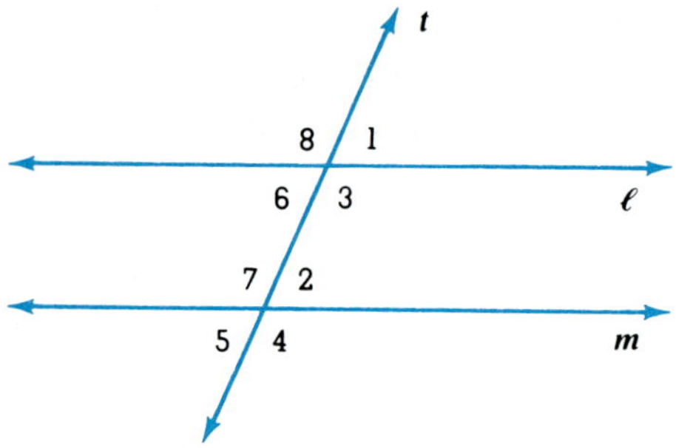

Proof:

Suppose $\angle 1$ is *not* congruent to $\angle 2$.

Draw $\overleftrightarrow{QP}$ so that $m\angle RQP = m\angle 2$. The Protractor Postulate justifies such a construction, and guarantees that $\overleftrightarrow{QP}$ and ℓ are different lines. By the definition of congruent angles, $\angle RQP \cong \angle 2$. Thus, since they are corresponding angles, $\overleftrightarrow{QP} \parallel m$. But, $\ell \parallel m$. Since $\overleftrightarrow{QP}$ and ℓ are different lines, and both go through Q, the Parallel Postulate is contradicted.

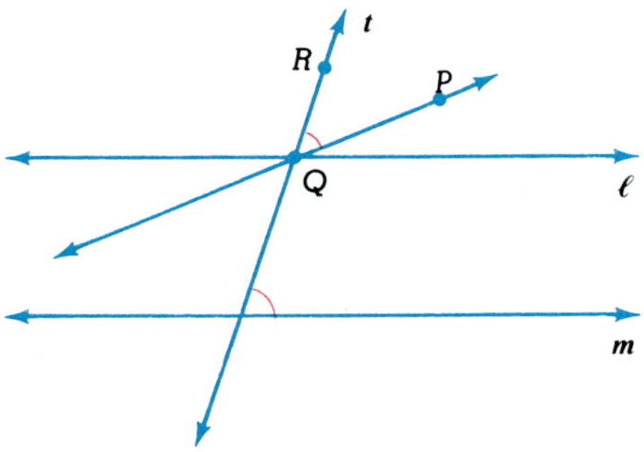

The original assumption is false.

Thus, $\angle 1 \cong \angle 2$.

If one pair of corresponding angles is congruent, then all pairs of corresponding angles are congruent. Since $\angle 1 \cong \angle 2$, all other pairs are congruent. Thus, $\angle 3 \cong \angle 4$, and $\angle 5 \cong \angle 6$, and $\angle 7 \cong \angle 8$.

Many results of Euclidean geometry depend on the Parallel Postulate.

example 2

Prove that the sum of the degree measures of the angles of a triangle is 180.

Given: $\triangle PQR$

Prove: $m\angle 5 + m\angle 4 + m\angle 3 = 180$

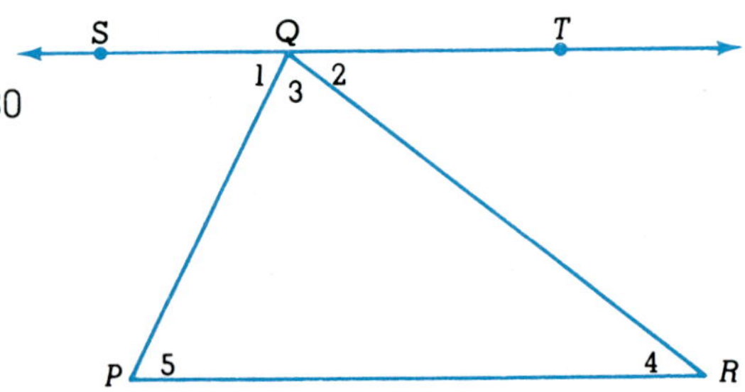

Proof:

STATEMENTS	REASONS
1. Draw $\overleftrightarrow{ST}$ through Q and parallel to $\overline{PR}$	1. Postulate 6–1: Parallel Postulate
2. $m\angle 1 + m\angle PQT = 180$	2. Postulate 3–4: The angles in a linear pair are supplementary.
3. $m\angle 2 + m\angle 3 = m\angle PQT$	3. Postulate 3–3: The Angle Addition Postulate
4. $m\angle 1 + m\angle 2 + m\angle 3 = 180$	4. Postulate 2–9: Substitution *Steps 2 & 3*
5. $\angle 2 \cong \angle 4$ $\angle 1 \cong \angle 5$	5. Theorem 6–9: If two parallel lines are cut by a transversal, then each pair of alternate interior angles is congruent. *Step 1*
6. $m\angle 2 = m\angle 4$ $m\angle 1 = m\angle 5$	6. Definition of congruent angles.
7. $m\angle 5 + m\angle 4 + m\angle 3 = 180$	7. Postulate 2–9: Substitution *Steps 4 & 6*

exercises

Exploratory Use the figure at the right to solve each problem.

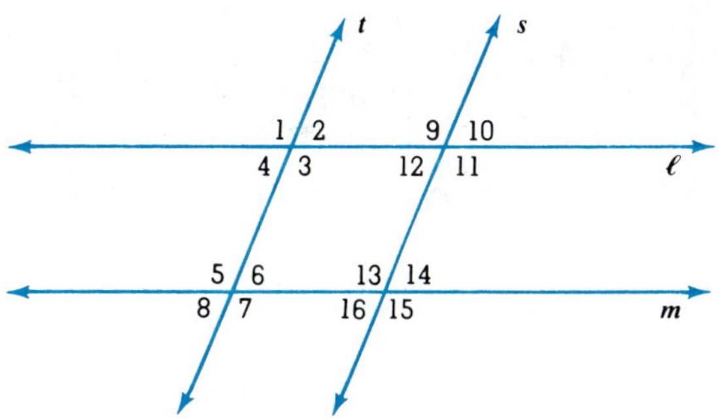

1. Suppose $t \parallel s$. Name four pairs of congruent alternate interior angles.
2. Suppose $t \parallel s$. Name four pairs of congruent corresponding angles.
3. Suppose $t \parallel s$. Name four pairs of supplementary consecutive interior angles.
4. Suppose $\ell \parallel m$. Name four pairs of congruent alternate interior angles.
5. Suppose $\ell \parallel m$. Name four pairs of congruent corresponding angles.
6. Suppose $\ell \parallel m$. Name four pairs of supplementary consecutive interior angles.

In the figure at the right, $\overline{ST} \parallel \overline{PR}$.

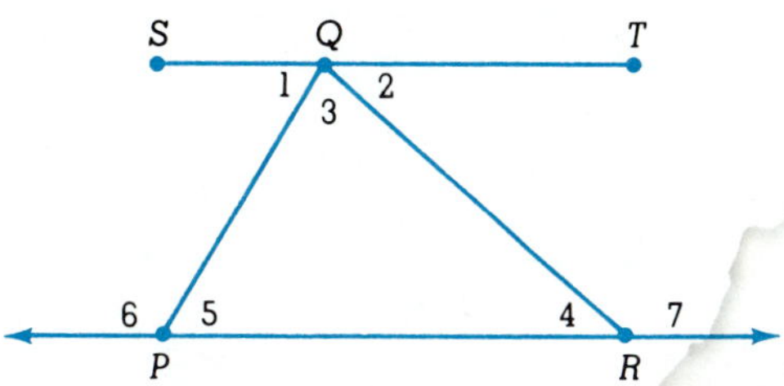

7. Name four pairs of congruent alternate interior angles.
8. Name four pairs of supplementary consecutive interior angles.

Written Each of the following might be a statement in a formal proof. For each statement, name a postulate or theorem that justifies the statement.

1. Draw $\overleftrightarrow{QT}$ so that $\overleftrightarrow{QT} \parallel \overline{PR}$.
2. Draw $\overline{SV}$ so that $\overline{PR} \parallel \overline{SV}$.
3. Draw $\overline{SV}$ so that $\angle QSV \cong \angle P$.
4. Call W the midpoint of $\overline{PR}$.
5. On $\overline{RQ}$, call A the point such that $\overline{RA} \cong \overline{AQ}$.
6. Draw $\overline{RM}$ such that $\overline{RM} \parallel \overline{PQ}$.

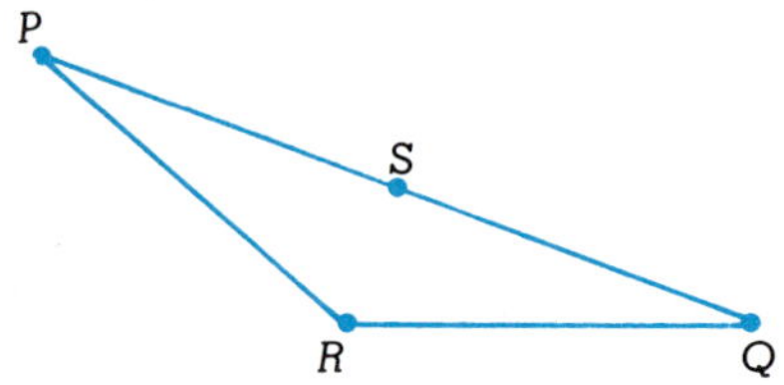

7. On $\overline{RQ}$, call B the point such that $\overline{SB}$ bisects $\overline{RQ}$.
8. On $\overline{RQ}$, call C the point such that $\overline{SC} \perp \overline{RQ}$.
9. On $\overline{PR}$, call D the point such that $\overline{SD} \parallel \overline{RQ}$.
10. Draw $\overline{RS}$.
11. On $\overrightarrow{QR}$, call E the point such that $\overline{PE} \perp \overrightarrow{QR}$.
12. Draw $\overline{PF}$ so that $\angle SPF \cong \angle SQR$.

In the diagram at the right, $\overline{QR} \parallel \overline{TV}$. Find the degree measures for each of the following.

13. $m\angle 1$
14. $m\angle 4$
15. $m\angle 2$
16. $m\angle 6$
17. $m\angle 3$
18. $m\angle 5$
19. $m\angle 7$
20. $m\angle Q$

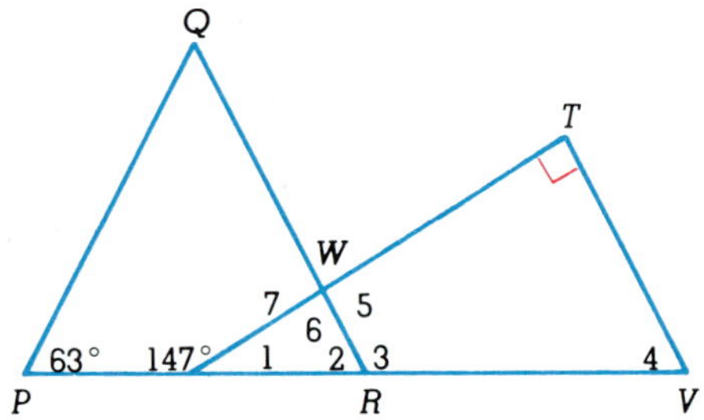

Solve each problem.

21. Prove Theorem 6–9 by contradicting the Parallel Postulate.
22. Prove Theorem 6–10 by contradicting the Parallel Postulate.

Challenge Solve each problem.

1. In the diagram below, $\overrightarrow{QT} \parallel \overrightarrow{RS}$. Find $m\angle P + m\angle Q + m\angle R$.

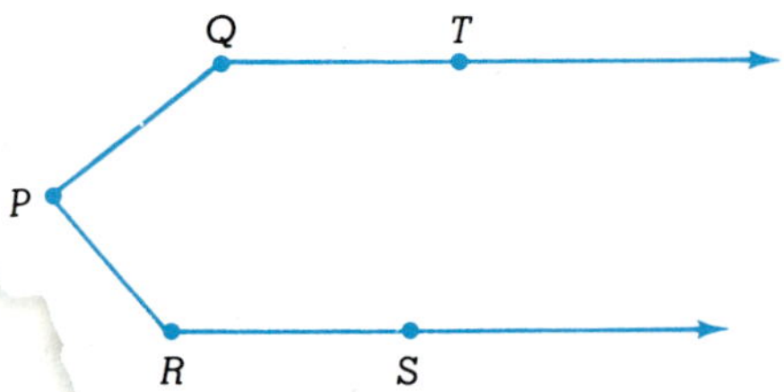

2. Use the diagram below to find $m\angle D$.

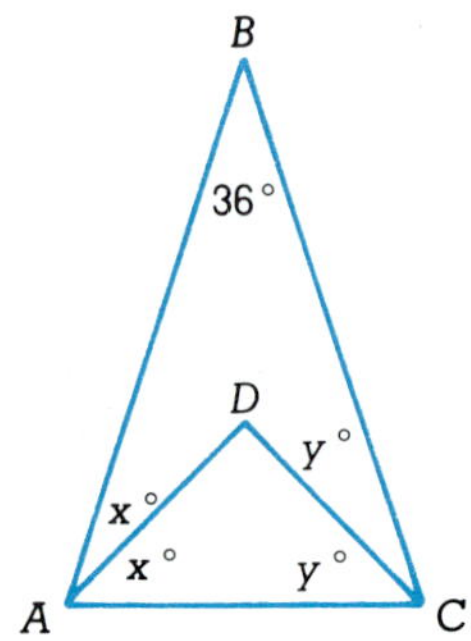

Non-Euclidean Geometry

Using Geometry

There are two possible alternatives to the Parallel Postulate. Neither contradicts any of Euclid's postulates, *except* the Parallel Postulate. In fact, each alternative can be used in *place* of the Parallel Postulate to form new geometries.

Hyperbolic Geometry

If there is a line and a point *not* on the line, then there is **more than one line** through the point and parallel to the given line.

Elliptic Geometry

If there is a line and a point *not* on the line, then there is **no line** through the point and parallel to the given line.

Geometries using alternatives to the Parallel Postulate are called **non-Euclidean** geometries. Because non-Euclidean geometries use alternatives to the Parallel Postulate, many of their theorems contradict theorems of Euclidean geometry. In hyperbolic geometry, the sum of the degree measures of the angles of a triangle is always *less than 180*. In elliptic geometry, the sum of the degree measures of the angles of a triangle is always *greater than 180*.

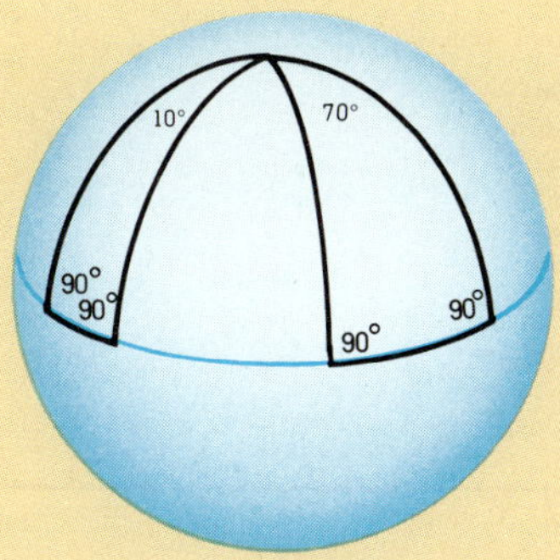

The great circles on a sphere can represent lines in elliptic geometry. The figure at the left shows models for two different triangles. The sum of the degree measures of one triangle is 190. The sum for the other triangle is 250.

Exercises Do the following hold for elliptic geometry? Write *yes* or *no*.

1. Through any two points there is exactly one line.
2. A line contains at least two points.
3. Two lines intersect in exactly one point.

4–6. Do the statements in problems 1–3 hold for hyperbolic geometry? Write *yes* or *no*.

6–5 Using Parallels

Viewed from the sides, the floors of a building represent parallel line segments. In most buildings, the outside walls are vertical. They are perpendicular to the ground floor. What angle do they form with each of the other floors?

The sides of the John Hancock Building in Chicago are not perpendicular to the ground floor.

Theorem 6–11

In a plane, if a line is perpendicular to one of two parallel lines, then it is perpendicular to the other.

example 1

Prove Theorem 6–11.

Given: $t \perp m$
$\ell \parallel m$

Prove: $t \perp \ell$

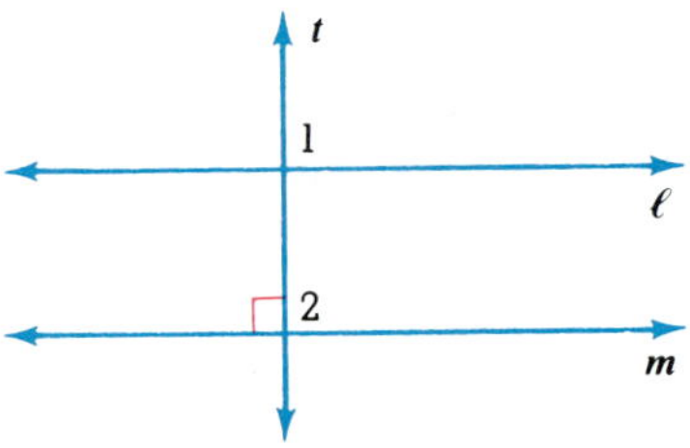

Proof:

STATEMENTS	REASONS
1. $t \perp m$	1. Given
2. $\angle 2$ is a right angle	2. Theorem 3–11: If two lines are perpendicular, then they form four right angles.
3. $m\angle 2 = 90$	3. Definition of right angle
4. $\ell \parallel m$	4. Given
5. $\angle 1 \cong \angle 2$	5. Theorem 6–8: If two parallel lines are cut by a transversal, then each pair of corresponding angles is congruent.
6. $m\angle 1 = m\angle 2$	6. Definition of congruent angles
7. $m\angle 1 = 90$	7. Postulate 2–9: Substitution *Steps 3 & 6*
8. $\angle 1$ is a right angle	8. Definition of right angle
9. $t \perp \ell$	9. Definition of perpendicular lines

Many geometric figures have parallel parts. By using the properties of parallel lines, properties of the figures themselves can be discovered.

examples

2 **In the figure at the right, opposite sides are parallel. How are the measures of the opposite sides related?**

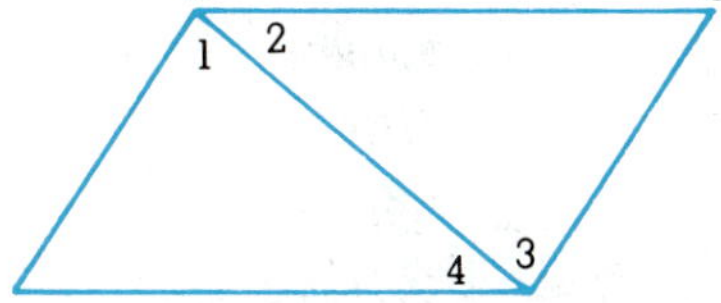

Draw $\overline{BD}$.
Since $\overline{BC} \parallel \overline{AD}$, you can show that $\angle 2 \cong \angle 4$. Similarly, since $\overline{AB} \parallel \overline{CD}$, you can show that $\angle 1 \cong \angle 3$. Finally, $\overline{BD} \cong \overline{BD}$. Thus, the two triangles formed are congruent. That is, $\triangle ABD \cong \triangle CDB$.

The opposite sides of the figure are corresponding parts of congruent triangles, and are congruent. Since congruent segments have the same measures, you can conclude the following.

$$AB = CD \text{ and } BC = AD$$

3 **Complete the following proof.**

Given: $\overline{QS} \cong \overline{QT}$
$\overline{ST} \parallel \overline{PR}$

Prove: $\angle P \cong \angle R$

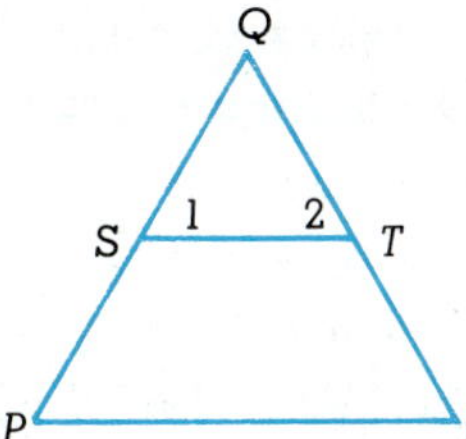

Proof:

STATEMENTS	REASONS
1. $\overline{QS} \cong \overline{QT}$	1. Given
2. $\angle 1 \cong \angle 2$	2. Theorem 4–6: Isosceles Triangle Theorem
3. $\overline{ST} \parallel \overline{PR}$	3. Given
4. $\angle P \cong \angle 1$ $\angle R \cong \angle 2$	4. Theorem 6–8: If two parallel lines are cut by a transversal, then each pair of corresponding angles is congruent.
5. $\angle P \cong \angle R$	5. Theorem 3–1: Congruence of angles is reflexive, symmetric, and transitive. *Steps 2 & 4*

exercises

Exploratory In the diagram at the right, $m \parallel n$. For each of the following, state *yes* or *no*.

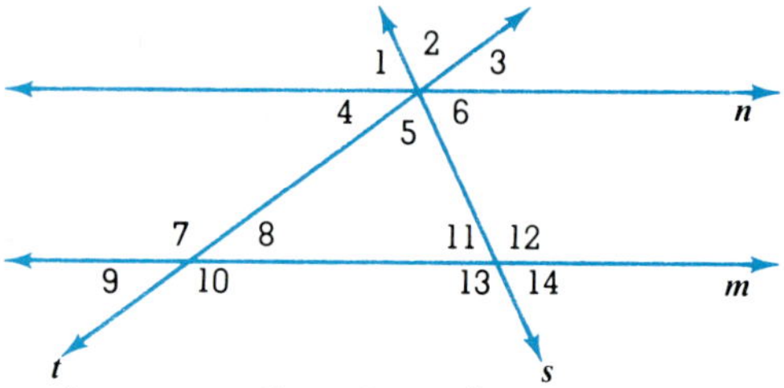

1. $\angle 5 \cong \angle 2$
2. $\angle 3 \cong \angle 8$
3. $\angle 3 \cong \angle 4$
4. $\angle 2 \cong \angle 6$
5. $\angle 4 \cong \angle 8$
6. $m\angle 5 + m\angle 6 = m\angle 7$
7. $m\angle 3 + m\angle 6 = m\angle 12$
8. $\angle 4$ and $\angle 12$ are alternate interior angles.
9. $\angle 2$ and $\angle 4$ are corresponding angles.
10. $\angle 4$ and $\angle 7$ are supplementary angles.
11. $\angle 5$ and $\angle 11$ are supplementary angles.

In the diagram at the right, $\ell \parallel m$ and $r \parallel s$. For each of the following, state *yes* or *no* or *sometimes*.

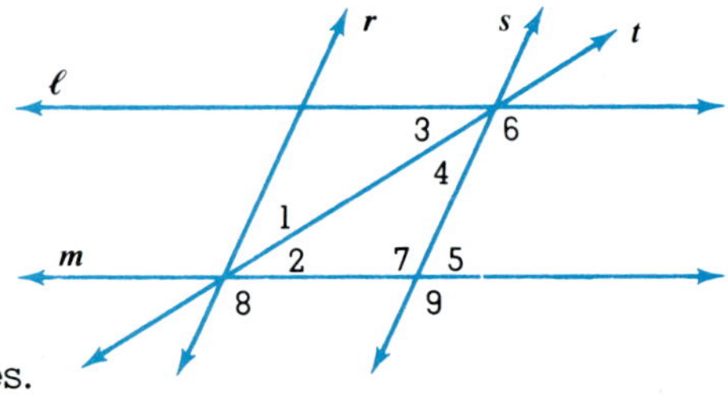

12. $\angle 2 \cong \angle 3$
13. $\angle 1 \cong \angle 4$
14. $\angle 7 \cong \angle 8$
15. $\angle 6 \cong \angle 9$
16. $m\angle 3 + m\angle 4 + m\angle 7 = 180$
17. $m\angle 1 + m\angle 2 + m\angle 7 = 180$
18. $\angle 5$ and $\angle 2$ are corresponding angles.
19. $\angle 7$ and $\angle 8$ are alternate interior angles.
20. $\angle 4$ and $\angle 7$ are supplementary angles.
21. $\angle 6$ and $\angle 5$ are supplementary angles.
22. $\angle 3$ and $\angle 4$ are complementary angles.

Written Use the figure at the right to solve each problem.

1. **Given:** $\overline{AB} \parallel \overline{CD}$
   $\overline{BC} \perp \overline{CD}$, $\overline{AB} \perp \overline{AD}$
   Prove: $\overline{BC} \parallel \overline{AD}$
2. **Given:** $\overline{BC} \parallel \overline{AD}$
   $\overline{CD} \perp \overline{BC}$
   $\overline{AB} \perp \overline{AD}$
   Prove: $\overline{AB} \parallel \overline{CD}$

Use the figure at the right to solve each problem.

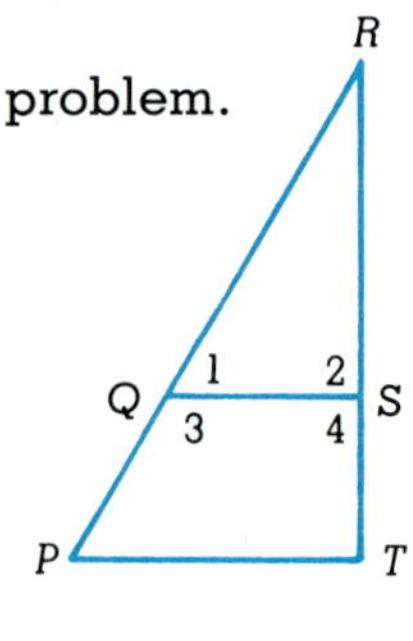

3. **Given:** $\overline{QS} \parallel \overline{PT}$
   $\angle T$ is a right angle
   Prove: $\triangle RQS$ is a right triangle
4. **Given:** $\overline{QS} \parallel \overline{PT}$
   $\angle T$ is a right angle
   Prove: $\overline{RS} \perp \overline{QS}$

Use the figure at the right to solve each problem.

5. Given: $\overline{BA} \parallel \overline{ED}$
   $\overline{BA} \cong \overline{ED}$
   Prove: $\overline{EC} \cong \overline{CA}$

6. Given: $\overline{BA} \parallel \overline{ED}$
   C is the midpoint of $\overline{BD}$
   Prove: $\overline{EC} \cong \overline{CA}$

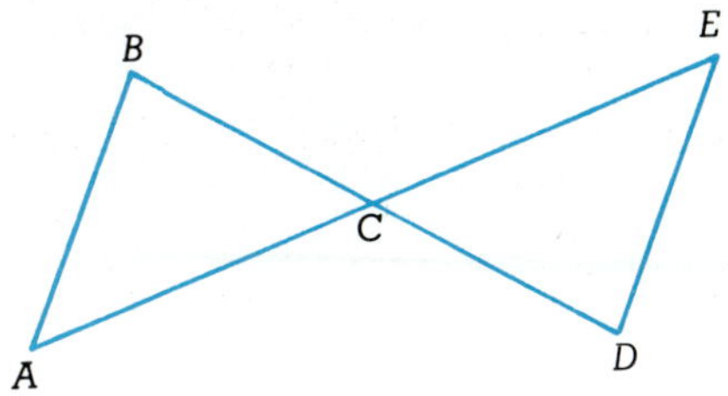

Use the figure at the right to solve each problem.

7. Given: $\overline{QS} \parallel \overline{PT}$
   $\triangle QRS$ is equilateral
   R is the midpoint of $\overline{PT}$
   Prove: $\overline{PQ} \cong \overline{TS}$

8. Given: $\overline{QS} \parallel \overline{PT}$
   $\triangle QRS$ is equilateral
   R is the midpoint of $\overline{PT}$
   Prove: $\angle P \cong \angle T$

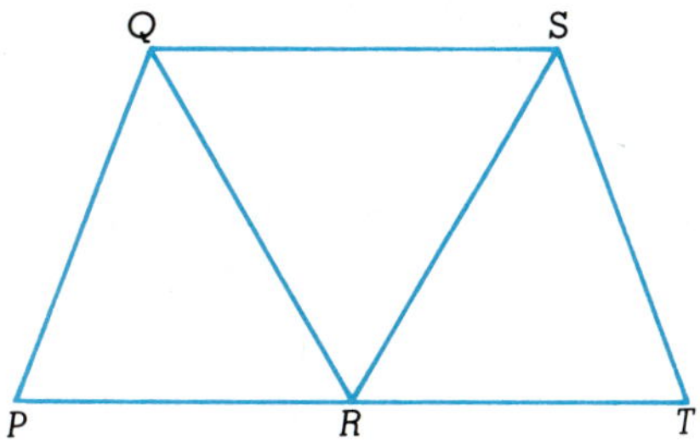

Use the figure at the right to solve each problem.

9. Given: $\overline{BC} \parallel \overline{AE}$
   $\angle 1 \cong \angle 2$
   $\angle 3 \cong \angle 4$
   Prove: $\overline{BD} \perp \overline{AD}$

10. Given: $\overline{AB} \parallel \overline{CE}$
    $\overline{BD} \cong \overline{AD}$
    Prove: $\angle BDC \cong \angle ADE$

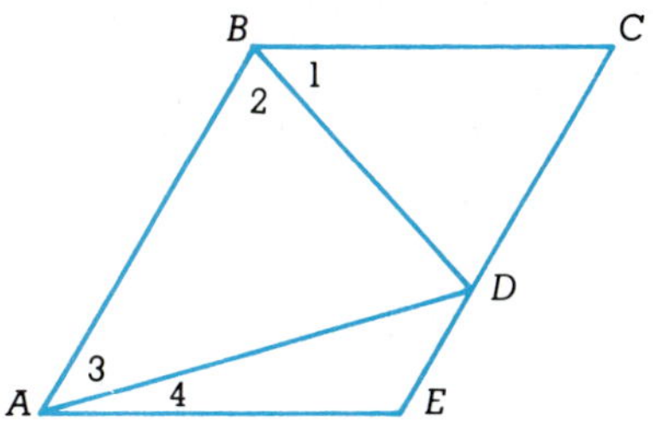

Drafting artists normally have two different triangles. For each of the following, name the degree measure of the angle the parallel lines form with the T-square.

11.

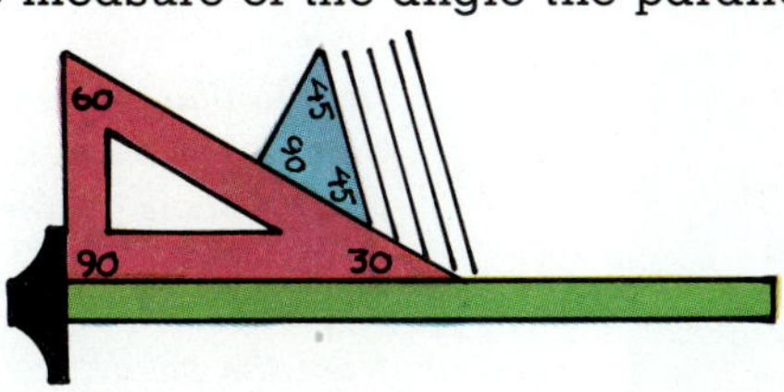

12.

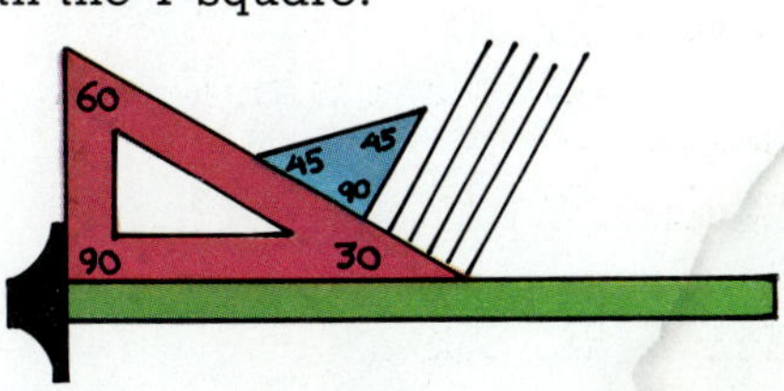

6–6 Distance

Given a line and a point *not* on the line, it is possible to construct another line going through the point and perpendicular to the given line. In fact, there is only one such line.

Theorem 6–12

Given a line and a point *not* on the line, there is exactly one line through the point and perpendicular to the given line.

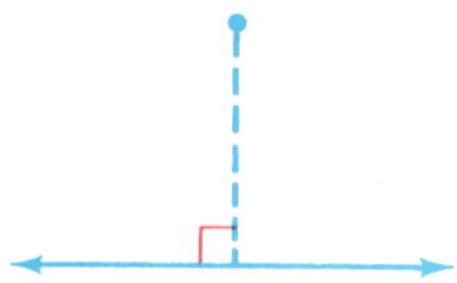

Because there is only one perpendicular segment between a point and a line, it can be used to define the distance from a point to a line.

Definition of the Distance Between a Point and a Line

The distance between a point and a line is the length of the segment perpendicular to the line from the point. The measure of the distance between a line and a point on the line is zero.

Can you talk about the distance between two lines? If the two lines are parallel, the answer is *yes*.

Definition of the Distance Between Parallel Lines

The distance between two parallel lines is the distance between any point of one line and the other line.

When weavers stretch yarn on a frame, they adjust the distance between strips of yarn both at the top and at the bottom of the frame. Then they know the strips of yarn are the same distance apart from top to bottom.

In a similar manner, the distance between a point and a plane is the length of the segment perpendicular to the plane from the point.

Two parallel lines are everywhere equidistant. Theorem 6–13

Equidistant means at the same distance.

example 1

Prove Theorem 6–13.

Given: $\ell \parallel m$
$\overline{AC} \perp m$
$\overline{BD} \perp m$

Prove: $AC = BD$

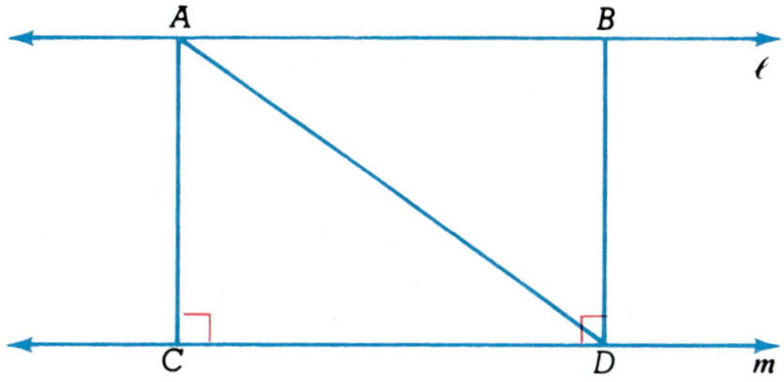

Proof:

STATEMENTS	REASONS
1. Draw $\overline{AD}$	1. Postulate 1–1: Through any two points there is exactly one line.
2. $\overline{AC} \perp m$	2. Given
3. $\angle ACD$ is a right angle	3. Theorem 3–11: If two lines are perpendicular, then they form four right angles.
4. $\triangle ACD$ is a right triangle	4. Definition of right triangle
5. $\ell \parallel m$	5. Given
6. $\overline{BD} \perp m$	6. Given
7. $\overline{BD} \perp \ell$	7. Theorem 6–11: If a line is perpendicular to one of two parallel lines, then it is perpendicular to the other.
8. $\angle DBA$ is a right angle	8. Theorem 3–11: If two lines are perpendicular, then they form four right angles.
9. $\triangle DBA$ is a right triangle	9. Definition of right triangle
10. $\angle BAD \cong \angle ADC$	10. Theorem 6–9: If two parallel lines are cut by a transversal, then each pair of alternate interior angles is congruent. *Step 5*
11. $\overline{AD} \cong \overline{AD}$	11. Theorem 2–2: Congruence of line segments is reflexive.
12. $\triangle ACD \cong \triangle DBA$	12. Theorem 4–11: HA *Steps 4, 9, 10, 11*
13. $\overline{AC} \cong \overline{BD}$	13. Definition of congruent triangles *CPCTC*
14. $AC = BD$	14. Definition of congruent segments

exercises

Exploratory Use the theorem to solve each of the following.

Given a line and a point *not* on the line, there is exactly one line through the point and perpendicular to the given line.

1. State the hypothesis of the theorem.
2. Draw and label a diagram to illustrate the hypothesis of the theorem.
3. State the *given* for a proof of the theorem.
4. State the conclusion of the theorem.
5. State the *prove* for a proof of the theorem.
6. The conclusion of the theorem contains *exactly*. Thus, the conclusion actually has two parts. State the two parts.

Two parallel lines are everywhere equidistant.

7. State the theorem in *if-then* form.
8. State the hypothesis of the theorem.
9. State the conclusion of the theorem.
10. State the converse of the theorem.

Written Use the following proof of Theorem 6–12 to solve each problem.

Theorem: Given a line and a point *not* on the line, there is exactly one line through the point and perpendicular to the given line.

Given: ℓ and P *not* on ℓ

Prove: There is exactly one line through P and perpendicular to ℓ.

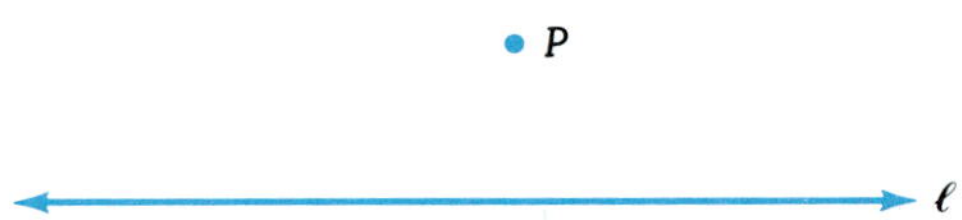

Proof:

Part 1 proves that there is *at least one* perpendicular.

1. Through P draw m so that $m \parallel \ell$.
2. Through P draw $\overleftrightarrow{PQ}$ so that $\overleftrightarrow{PQ} \perp m$.
3. $\overleftrightarrow{PQ} \perp \ell$.

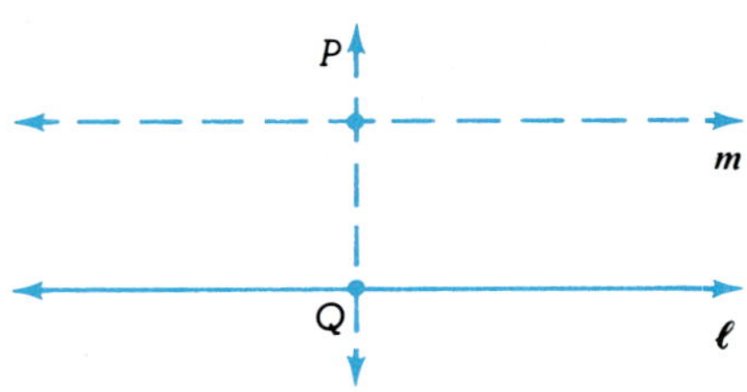

Part 2 proves that there is *no more than one* perpendicular.

Suppose there is more than one perpendicular. Call $\overleftrightarrow{PQ}$ and $\overleftrightarrow{PR}$ two different lines through P and perpendicular to ℓ.
Then, $\angle 1$ and $\angle 2$ are right angles.
Thus, $\triangle PQR$ has two right angles.

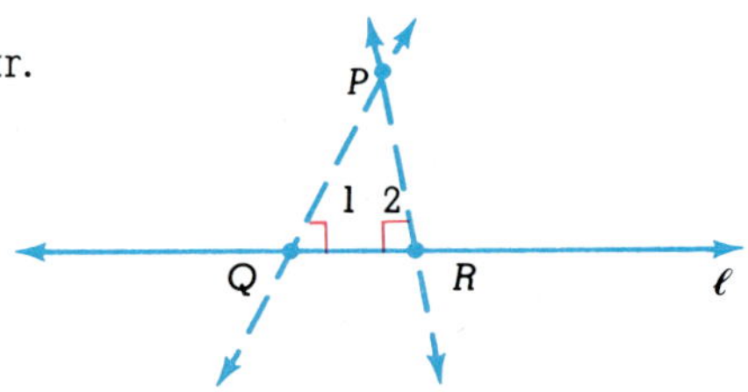

This is a contradiction, since Theorem 6–3 states that a right triangle has only one right angle. Therefore, there is no more than one line through P perpendicular to ℓ.

1. State whether *Part 1* is proved by *direct* or *indirect* reasoning.
2. State the postulate or theorem that justifies the *Part 1* statement, *Through P draw m so that* $m \parallel \ell$.
3. State the postulate or theorem that justifies the *Part 1* statement, *Through P draw* $\overleftrightarrow{PQ}$ *so that* $\overleftrightarrow{PQ} \perp m$.
4. State the postulate or theorem that justifies the *Part 1* statement, $\overleftrightarrow{PQ} \perp \ell$.
5. State whether *Part 2* is proved by *direct* or *indirect* reasoning.
6. State the assumption made in *Part 2* of the proof.

Use a ruler and Theorem 6–13 to tell whether or *not* the lines in each of the following are parallel. Write *yes* or *no*, and explain your answer.

7. 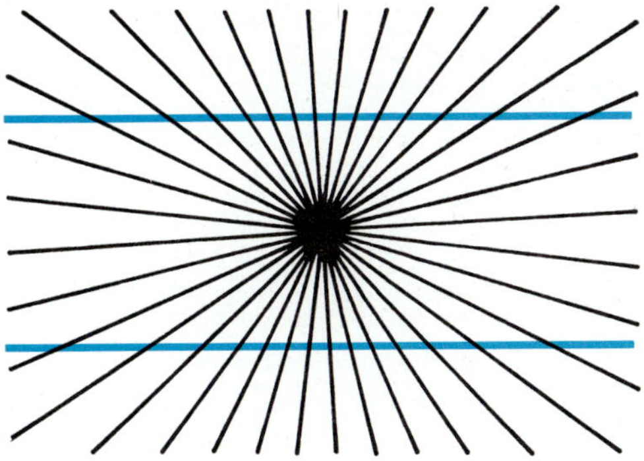

8. 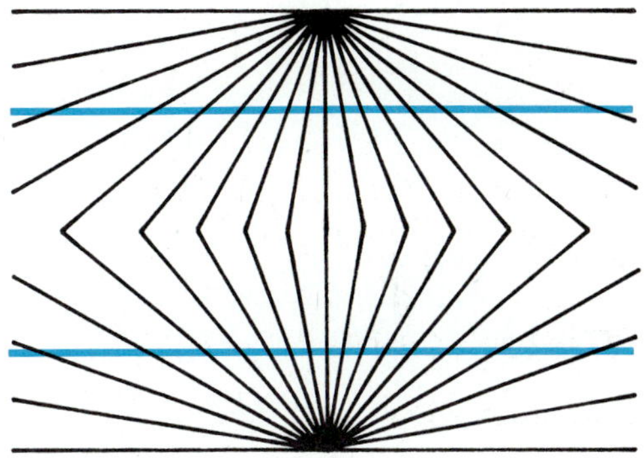

Use the theorem to solve each of the following.

If a point is on the bisector of an angle, then it is equidistant from the sides of the angle.

9. State the *given* for a proof of this theorem.
10. Draw and label a diagram to illustrate the hypothesis of the theorem.
11. Plan a proof of the theorem.
12. Write a formal proof for the theorem.
13. State the converse of the theorem.
14. Plan a proof for the converse of the theorem.
15. Write a formal proof for the converse of the theorem.
16. Write an *if and only if* definition for the bisector of an angle, using equidistance.

Challenge **Two planes, or a plane and a line, are parallel if and only if they do *not* intersect. Solve each problem about parallel planes.**

1. If two lines lie in parallel planes, are the lines parallel? Write *yes* or *no*, and explain.
2. If a line is perpendicular to one of two parallel planes, is it perpendicular to the other? Write *yes* or *no*, and explain.
3. Explain how you would find the distance between two parallel planes.
4. Are parallel planes everywhere equidistant? Write *yes* or *no*, and explain.

Using Money

Renting Apartments

WEST
Near bus. 2 bedrm, range, refrig., 1 yr. lease, deposit. $250 per mo. 1852 W. Third Ave. Call 451-1306

WEST
1773 W. Fifth Ave. Available May 1. 2 bedrm., gas, air cond., carpet. $265 plus dep. 889-6196.

Sunny Fran is looking for an apartment on the West Side. She called and made appointments to see each of the apartments described in these ads. To help her decide whether she is interested in either apartment, she asked for the following information.

	1852 W. Third	1773 W. Fifth
security deposit	$ 250	$ 250
rent	$ 250	$ 265
gas	$ 50	$ 35
electric	$ 13	$ 20
water	$ 12	paid
range	yes	yes
refrigerator	yes	no
air conditioner	no	yes
dishwasher	yes	no
washer-dryer	coin operated no hookups	hookups laundry near
shower	yes	yes

To figure her monthly payments, Ms. Fran *adds* rent, gas, electric, and water. The first month she will pay a security deposit *in addition* to the regular monthly rent. If she rents the apartment on 1773 W. Fifth Ave., she *also* must buy a refrigerator.

Exercises Solve each problem.

1. Find the monthly payment for the apartment on 1852 W. Third Ave.
2. Find the monthly payment for the apartment on 1773 W. Fifth Avenue.
3. How can Ms. Fran check to find out the correctness of the gas payments information?
4. How can Ms. Fran check to find out the correctness of the electric payments information?
5. How can Ms. Fran check to find out the correctness of the water payments information?
6. How can Ms. Fran find out the monthly cost of a telephone?
7. What is the cost of a new refrigerator?
8. What is the cost of a new air conditioner?

Vocabulary

parallel lines (160)
skew lines (160)
transversal (161)
interior angles (161)
alternate interior angles (161)
exterior angles (161)
corresponding angles (161)
indirect reasoning (164)
exactly one (166)
equidistant (185)

Chapter Summary

1. There are three different ways that lines can be positioned in space, intersecting, parallel, or skew. (160)
2. Theorem 6–1: If two lines are cut by a transversal and one pair of alternate interior angles is congruent, then the other pair of angles also is congruent. (162)
3. Theorem 6–2: If two lines are cut by a transversal and one pair of corresponding angles is congruent, then the other pair of angles also is congruent. (162)
4. Theorem 6–3: A right triangle has no more than one right angle. (165)
5. How to Write an Indirect Proof: (165)
    1. Assume the alternatives of the statements to be proved.
    2. Show that the assumptions lead to a contradiction.
    3. State that the assumptions are false.
    4. Conclude that the statement to be proved is true.
6. Any statement containing the phrase *exactly one* actually has two parts. To prove a theorem that uses *exactly one*, you prove both parts. (166)
7. Theorem 6–4: In a plane, if two lines are cut by a transversal so that a pair of corresponding angles is congruent, then the two lines are parallel. (170)
8. Theorem 6–5: In a plane, if two lines are cut by a transversal so that a pair of alternate interior angles is congruent, then the two lines are parallel. (171)
9. Theorem 6–6: If two lines are cut by a transversal so that a pair of consecutive interior angles is supplementary, then the lines are parallel. (171)
10. Theorem 6–7: If two lines are perpendicular to the same line, then the two lines are parallel. (171)
11. The Parallel Postulate: If there is a line and a point *not* on the line, then there is exactly one line through the point and parallel to the given line. (175)

12. Theorem 6–8: If two parallel lines are cut by a transversal, then each pair of corresponding angles is congruent. (175)
13. Theorem 6–9: If two parallel lines are cut by a transversal, then each pair of alternate interior angles is congruent. (175)
14. Theorem 6–10: If two parallel lines are cut by a transversal, then each pair of consecutive interior angles is supplementary. (175)
15. Theorem 6–11: In a plane, if a line is perpendicular to one of two parallel lines, then it is perpendicular to the other. (180)
16. Theorem 6–12: Given a line and a point *not* on the line, there is exactly one line through the point and perpendicular to the given line. (184)
17. Definition of the Distance Between a Point and a Line: The distance between a point and a line is the length of the segment perpendicular to the line from the point. The measure of the distance between a line and a point *on* the line is zero. (184)
18. Definition of the Distance Between Parallel Lines: The distance between two parallel lines is the distance between any point of one line and the other line. (184)
19. Theorem 6–13: Two parallel lines are everywhere equidistant. (185)

Chapter Review

6–1 Using the diagram at the right, find an example for each of the following.

1. Two parallel lines
2. Two parallel rays
3. Two skew lines
4. Two noncoplanar lines
5. Two parallel lines and a transversal
6. Two coplanar lines that intersect

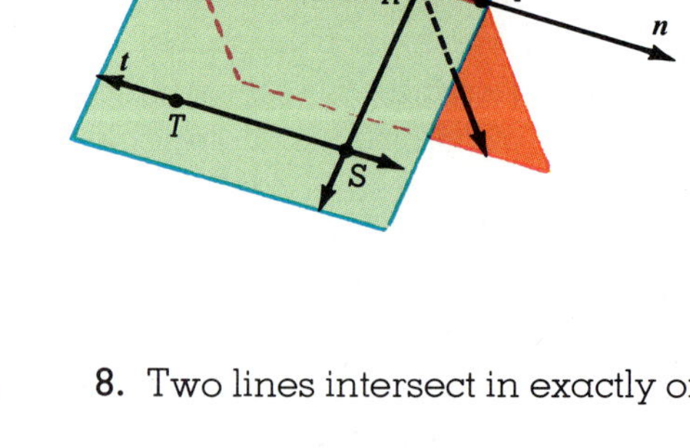

6–2 Prove each of the following.

7. An isosceles triangle has *no* more than one obtuse angle.
8. Two lines intersect in exactly one point.

6–3 Use the figure at the right to solve each problem.

9. Given: T is the midpoint of $\overline{QS}$ and $\overline{PR}$
Prove: $\overline{QP} \parallel \overline{RS}$

10. Given: $\triangle QPT \cong \triangle SRT$
Prove: $\overline{QR} \parallel \overline{PS}$

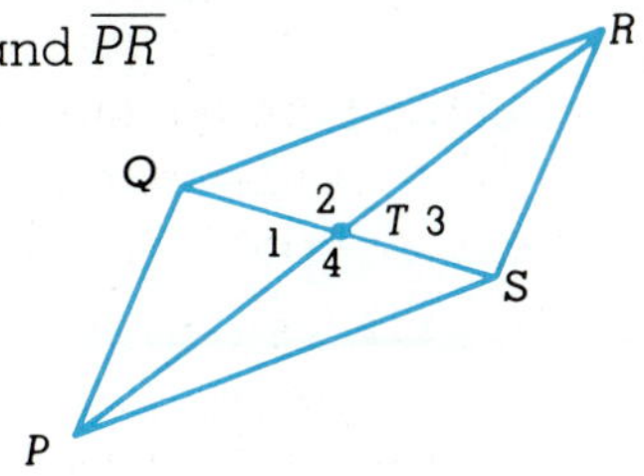

6–4 11. Complete Reasons 1 and 5 for the following proof.

Given: $\triangle PQR$

Prove: $m\angle 5 + m\angle 4 + m\angle 3 = 180$

Proof:

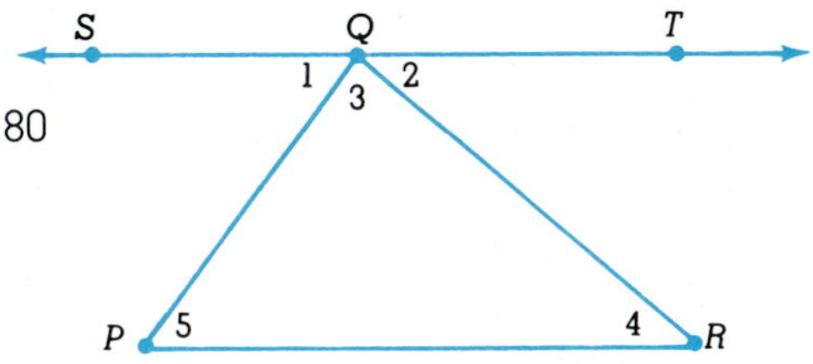

STATEMENTS	REASONS
1. Draw $\overleftrightarrow{ST}$ through Q and parallel to $\overleftrightarrow{PR}$.	1. ________________
2. $m\angle 1 + m\angle PQT = 180$	2. Postulate 3–4: The angles in a linear pair are supplementary.
3. $m\angle 2 + m\angle 3 = m\angle PQT$	3. Postulate 3–3: The Angle Addition Postulate
4. $m\angle 1 + m\angle 2 + m\angle 3 = 180$	4. Postulate 2–9: Substitution
5. $\angle 2 \cong \angle 4$ $\angle 1 \cong \angle 5$	5. ________________
6. $m\angle 2 = m\angle 4$ $m\angle 1 = m\angle 5$	6. Definition of Congruent Angles
7. $m\angle 5 + m\angle 4 + m\angle 3 = 180$	7. Postulate 2–9: Substitution

6–5 Use the figure at the right to solve the following problem.

12. Given: $\overline{DE} \parallel \overline{AC}$
E is the midpoint of $\overline{BC}$
$\overline{EF} \cong \overline{BE}$
Prove: D is the midpoint of $\overline{AB}$

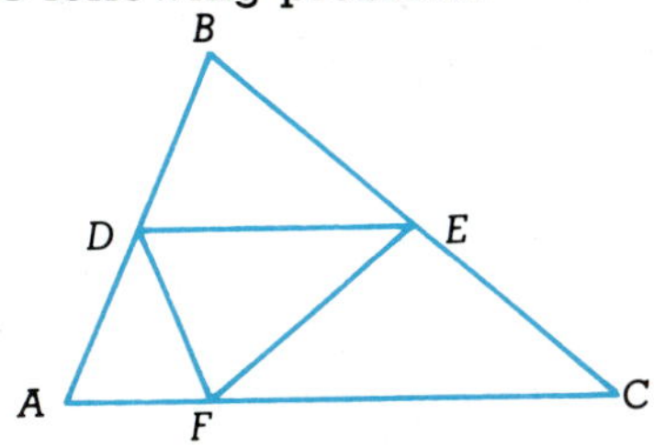

6–6 Prove each of the following.

13. If two lines are everywhere equidistant, then the lines are parallel.

14. If two lines are parallel, then they are everywhere equidistant.

Chapter Test

Use the diagram to find an example for each of the following.

1. Two parallel segments
2. Two parallel lines and a transversal
3. Two skew lines
4. Two coplanar lines that are *not* parallel
5. A ray and a segment that are *not* parallel
6. Two noncoplanar lines

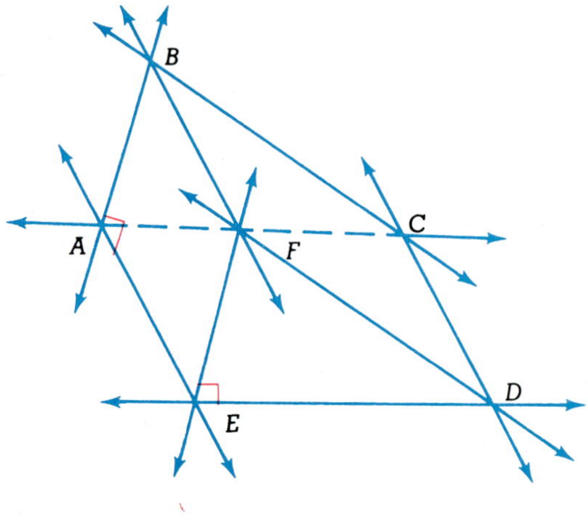

Prove each of the following.

7. If the measures of two angles of a triangle are unequal, then the measures of the sides opposite those angles are unequal in the same order.

8. In a plane, through a given point on a line, there is exactly one line perpendicular to the given line.

Use the figure at the right to solve each problem.

9. **Given:** $\triangle ABC \cong \triangle DEF$
   Prove: $\overline{AB} \parallel \overline{DE}$

10. **Given:** $\overline{AB} \parallel \overline{DE}$
    $\triangle ABC$ is a right triangle
    $\angle E$ is a right angle
    Prove: $\overline{BC} \parallel \overline{EF}$

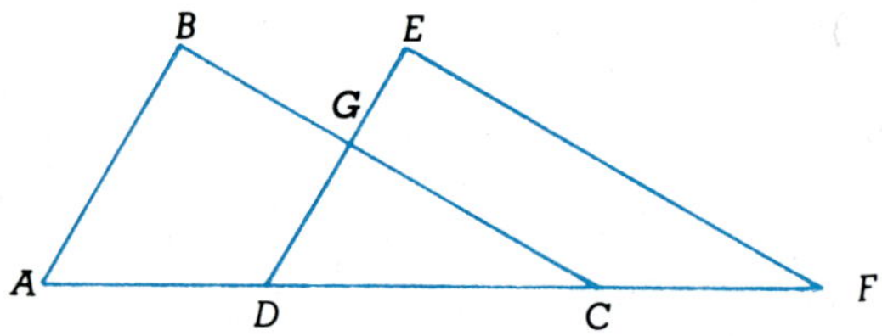

Chapter

7

Polygons

Designs in Native American weavings often contain polygon shapes. The rugs in this photograph are designed by Navajos.

7–1 Classifying Polygons

The following figures are formed by connecting a series of points. Each figure is called a **polygon.**

Definition of Polygon

A figure is a polygon if and only if it meets each of the following conditions.

1. **It is formed by three or more coplanar segments called sides.**
2. **Sides which have a common endpoint are noncollinear.**
3. **Each side intersects exactly two other sides, but only at their endpoints.**

If a figure fails to meet one of the conditions, then it is *not* a polygon.

example 1

Explain why the figure at the right is *not* a polygon.

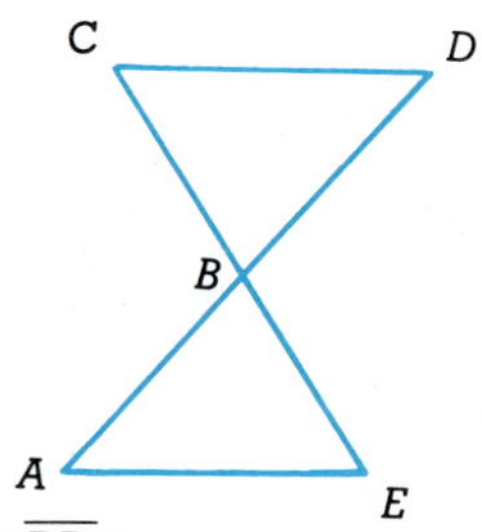

The figure fails to meet condition 2 because $\overline{AB}$ and $\overline{BD}$ have a common endpoint and are collinear.

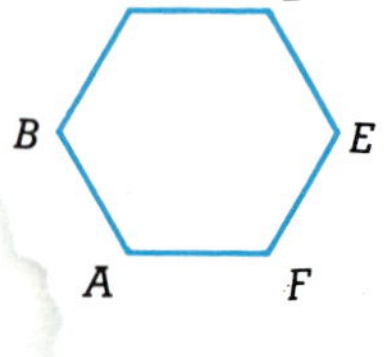

The endpoints of the sides of a polygon are the *vertices*. The angles formed at each vertex are called *angles of the polygon*.

Consecutive vertices are used when referring to a polygon. For example, the polygon at the left is called *polygon ABCDEF*.

Polygons can be classified by the number of sides. The chart at the right gives some common names.

In general, a polygon with n sides is called an *n-gon*. Thus, an octagon also is called an 8-gon. A polygon with 13 sides is called a 13-gon.

Number of Sides	Polygon
3	triangle
4	quadrilateral
5	pentagon
6	hexagon
7	heptagon
8	octagon
9	nonagon
10	decagon
12	dodecagon

Another way to classify polygons is **convex** or **nonconvex.** A polygon is convex if and only if any line containing a side of the polygon does *not* contain a point interior to the polygon. A polygon is nonconvex if and only if it is *not* a convex polygon.

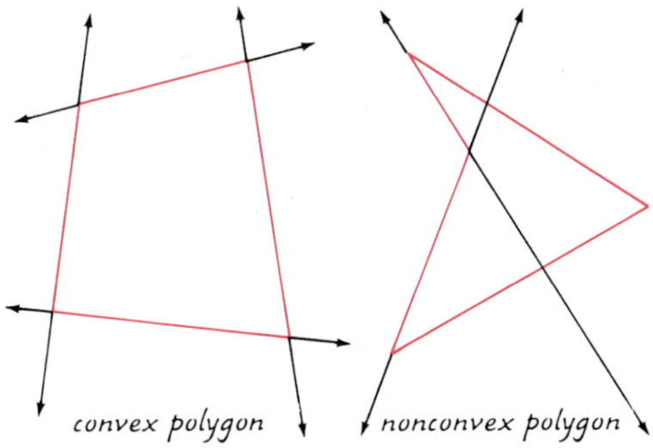

A convex polygon is classified as a **regular polygon** when all sides are congruent and all the angles are congruent.

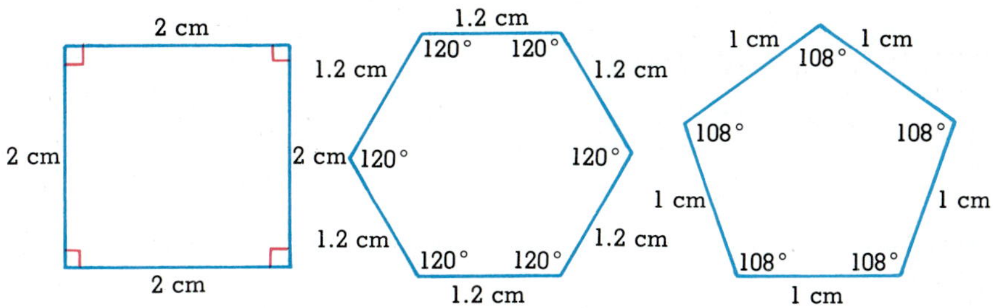

For a convex polygon to be a regular polygon, it must be *both* equilateral *and* equiangular. A polygon meeting one of these conditions but *not* the other is *not* a regular polygon.

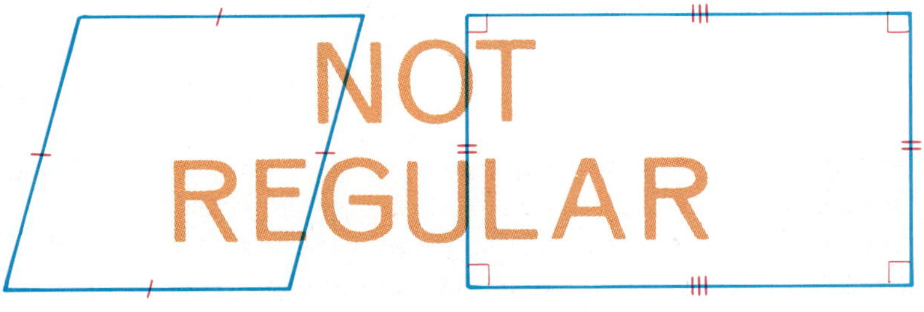

example 2

Classify the following polygons by the number of sides, as convex or nonconvex, and as regular or *not* regular.

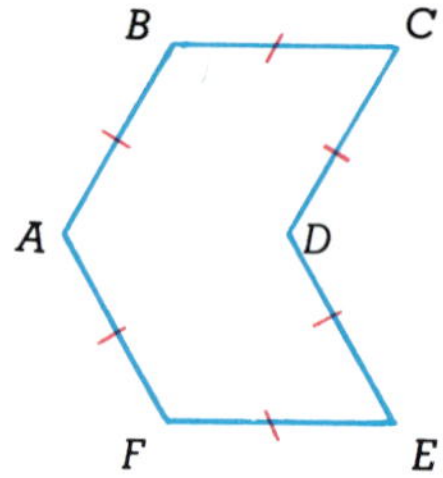

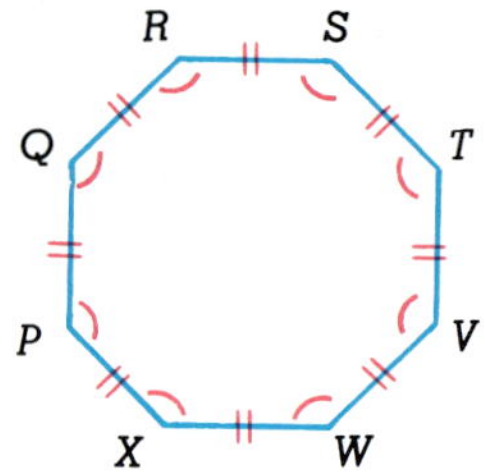

Polygon *ABCDEF* has 6 sides. It is a hexagon.

If $\overline{CD}$ is extended through *D*, it passes through the interior of the hexagon. Thus, it is nonconvex.

It is nonconvex. Thus, it is *not* regular.

Polygon *PQRSTVWX* has 8 sides. It is an octagon.

No side of the octagon can be extended to pass through the interior. Thus, it is convex.

All sides and all angles of the convex octagon are congruent. Thus, it is regular.

exercises

Exploratory Tell whether or *not* each of the following are polygons. State *yes* or *no*.

1. 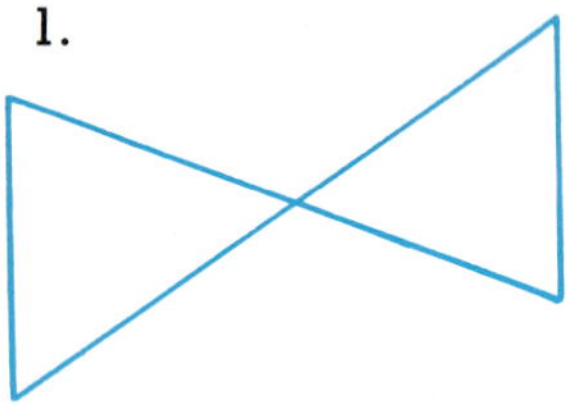

2. 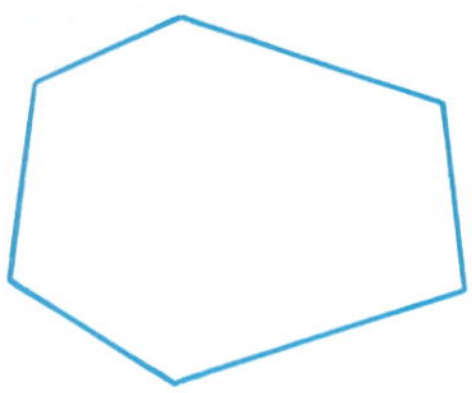

3. 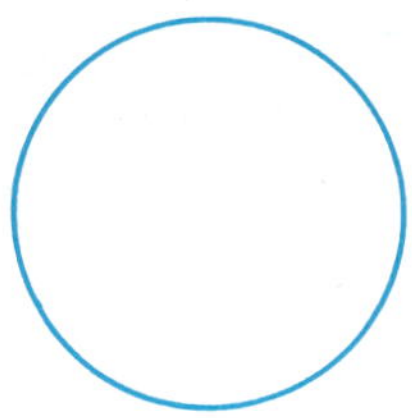

4. 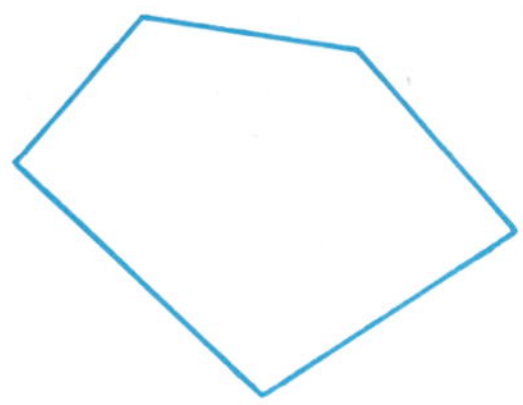

5. 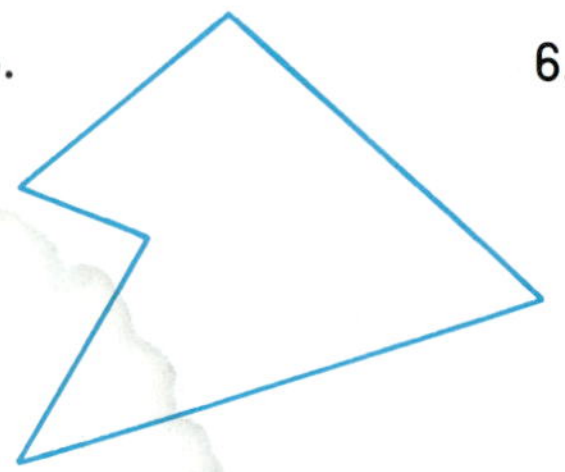

6. 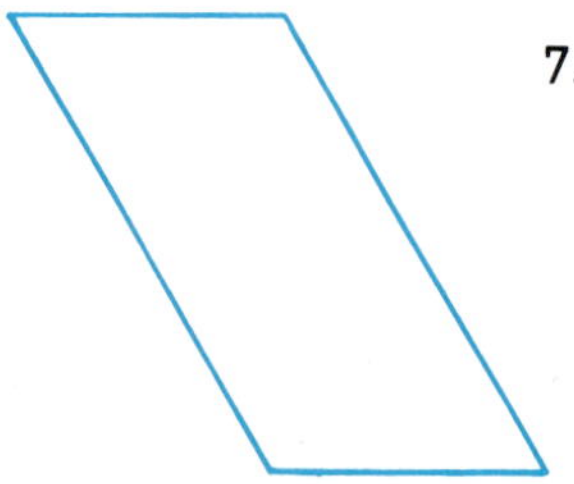

7. 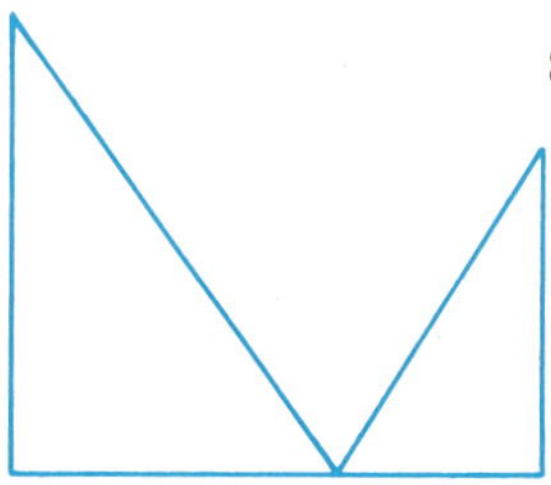

8. 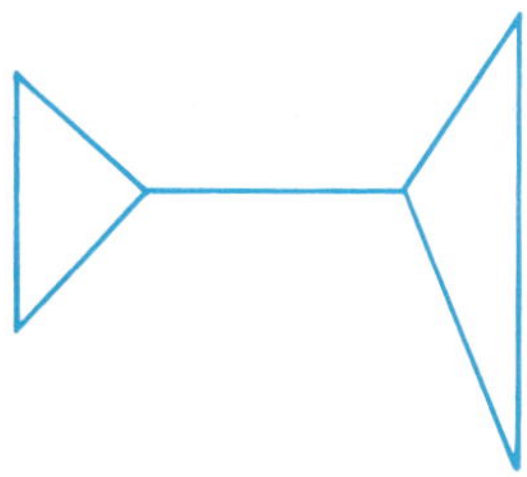

Written For each of the following, classify the polygon with the given number of sides.

1. 3	**2.** 8	**3.** 10	**4.** 4
5. 5	**6.** 18	**7.** 7	**8.** 9
9. 24	**10.** 6	**11.** n	**12.** x

Classify each of the following polygons by the number of sides.

13.

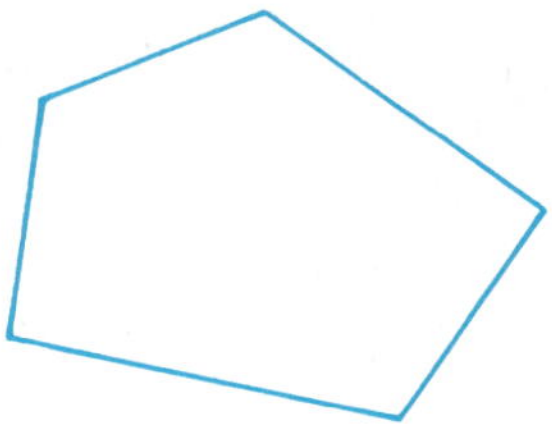

14.

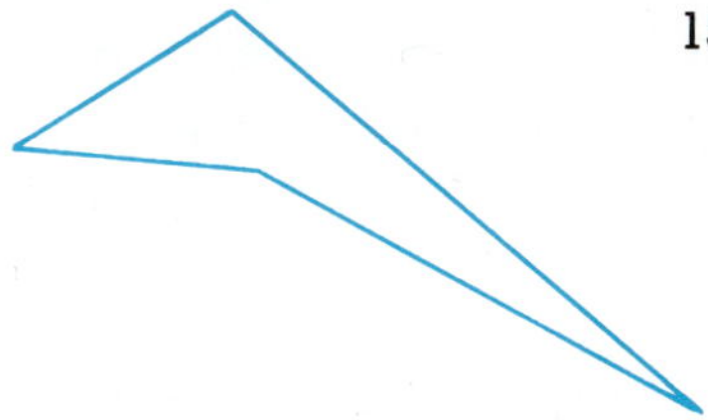

15.

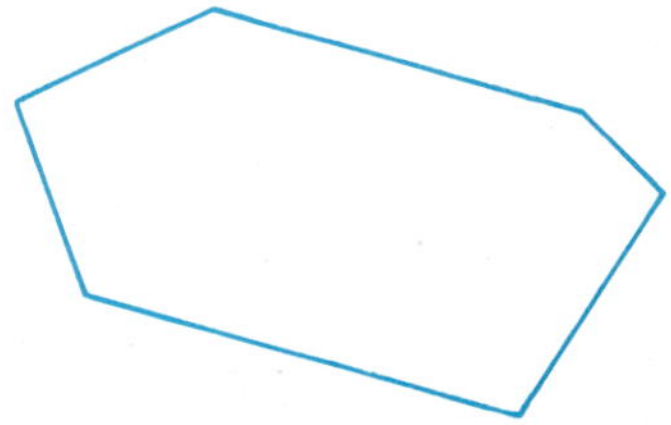

16.

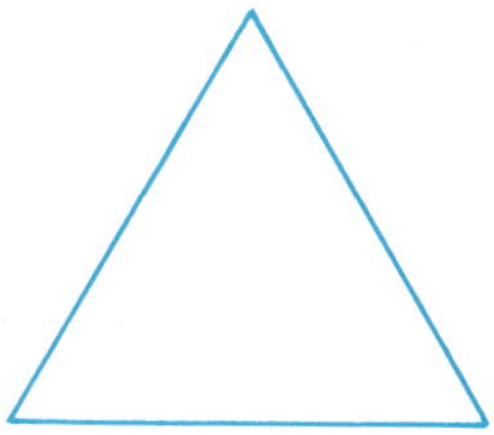

17.

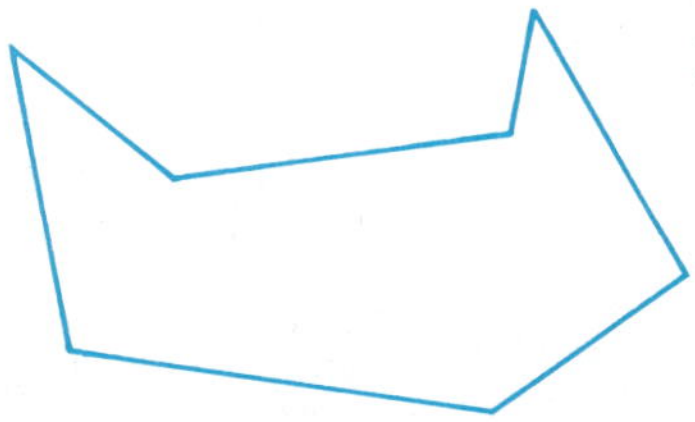

18.

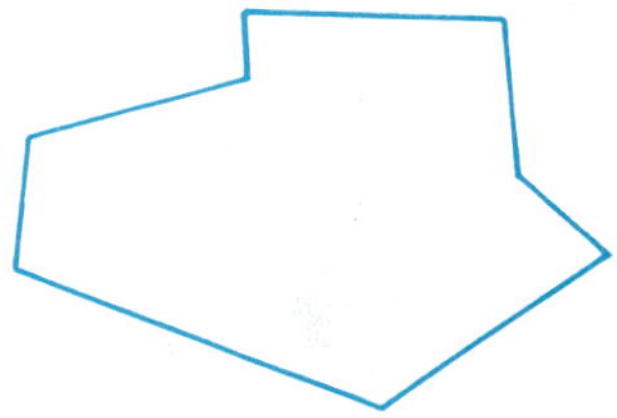

19–24. Classify each polygon in problems **13–18** as convex or nonconvex.

Use the polygon at the right to solve each problem.

25. Name the vertices.
26. Name the angles.
27. Name the sides.
28. Name the polygon.
29. Classify the polygon by number of sides.
30. Classify the polygon as convex or nonconvex.

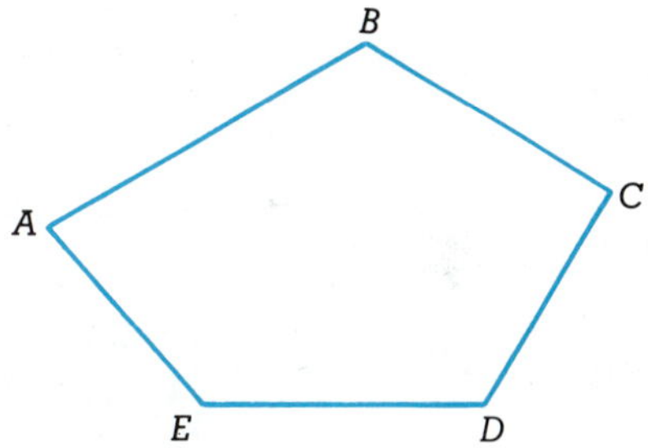

Consecutive **means** ***following one after the other in order, without gaps*****.**

31. Name a pair of consecutive angles for the polygon in problems **25–30.**

32. Name a pair of consecutive sides for the polygon in problems **25–30.**

33. Give an *if and only if* definition for a pair of consecutive angles of a polygon.

34. Give an *if and only if* definition for a pair of consecutive sides of a polygon.

Classify each of the following as ***regular*** **or** ***not regular*****. Then explain your answer.**

35.

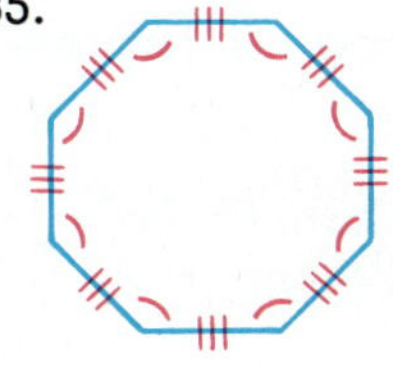

36.

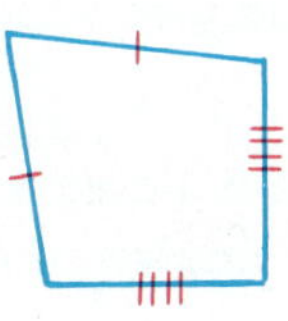

37.

38.

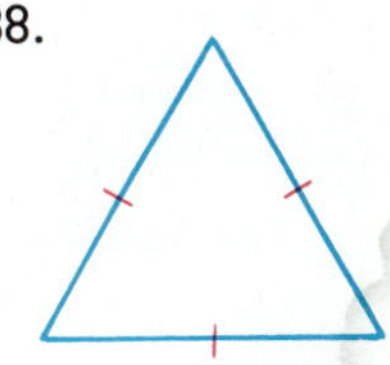

7–2 Angles of Polygons

According to the Angle Sum Theorem, the degree measures of the angles of a triangle total 180. What about the angles of a quadrilateral, or the angles of a pentagon? Is there a fixed total for these polygons?

A line segment joining two nonconsecutive vertices of a convex polygon is called a **diagonal** of the polygon. In each of the following polygons, all possible diagonals from one vertex are shown.

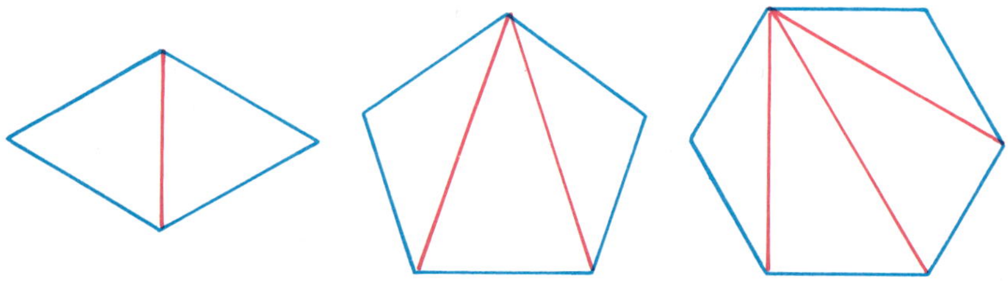

Notice that in each case, the polygon is separated into triangles. The sum of the measures of the angles of each polygon can be found by adding the measures of the angles of the triangles.

Polygon	Number of Sides	Number of Triangles	Sum of Degree Measures of Angles
triangle	3	1	$1 \cdot 180$ or 180
quadrilateral	4	2	$2 \cdot 180$ or 360
pentagon	5	3	$3 \cdot 180$ or 540
hexagon	6	4	$4 \cdot 180$ or 720
heptagon	7	5	$5 \cdot 180$ or 900
octagon	8	6	$6 \cdot 180$ or 1080
nonagon	9	7	$7 \cdot 180$ or 1260
decagon	10	8	$8 \cdot 180$ or 1440
.	.	.	.
.	.	.	.
.	.	.	.
n-gon	n	$n - 2$	$(n - 2)180$

These and many other examples suggest the following theorem.

Theorem 7–1

If a convex polygon has n sides, and S is the sum of the degree measures of its angles, then $S = (n - 2)180$.

examples

1 Find the sum of the measures of the angles of a hexagon.

$$\begin{aligned} S &= (n-2)180 \\ &= (6-2)180 && \textit{A hexagon has 6 sides.} \\ &= 4 \cdot 180 \\ &= 720 \end{aligned}$$

2 The degree measure of one angle of a regular polygon is 140. Find the number of sides.

Define a variable. Let n stand for the number of sides.

Write an equation. A polygon has the same number of angles as it has sides. Also, since a regular polygon is equiangular, each of its angles has the same measure.

$$\text{measure of angle} = \frac{\text{sum of measures of angles}}{\text{number of angles}}$$

$$140 = \frac{S}{n}$$

Solve the equation.

$$\begin{aligned} 140 &= \frac{(n-2)180}{n} && \textit{Substitution} \\ 140n &= (n-2)180 && \textit{Multiply by } n. \\ 140n &= 180n - 360 && \textit{Distributive Property} \\ {}^{-}40n &= {}^{-}360 && \textit{Subtract } 180n. \\ n &= 9 && \textit{Divide by } {}^{-}40. \end{aligned}$$

Check the solution.

$$\begin{aligned} 140 &= \frac{(n-2)180}{n} \\ 140 &\stackrel{?}{=} \frac{(9-2)180}{9} \\ 140 &\stackrel{?}{=} \frac{(7)180}{9} \\ 140 &\stackrel{?}{=} \frac{1260}{9} \\ 140 &= 140 \end{aligned}$$

Answer the problem. The polygon has 9 sides

1. Define a variable.
2. Write an equation.
3. Solve the equation.
4. Check the solution.
5. Answer the problem.

Problem Solving Procedure

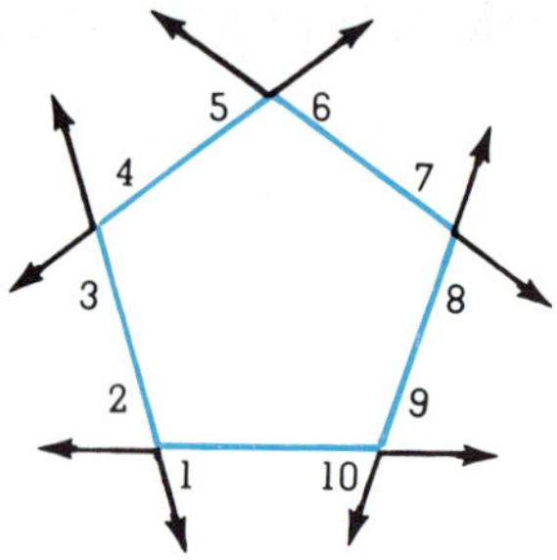

A pair of exterior angles is formed at each vertex of any convex polygon. A pentagon, for example, has a total of 10 exterior angles.

In the hexagon at the right, the sides are extended to form one exterior angle at each vertex. At each vertex, the angle of the hexagon and the exterior angle form a linear pair. Altogether, there are six linear pairs.

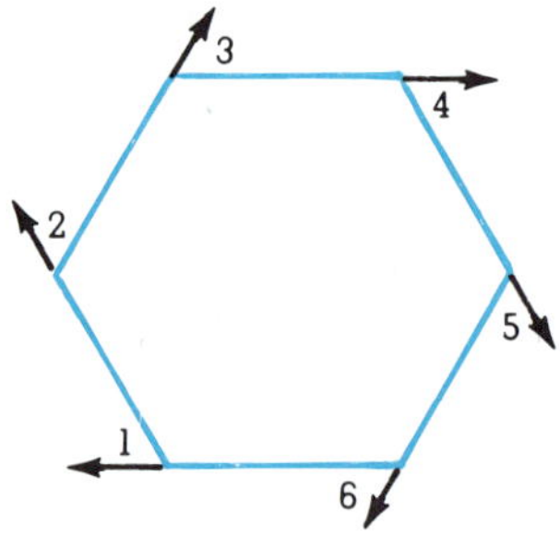

The total degree measures of these exterior angles can be found as follows.

sum of measures of exterior angles = *sum of measures of linear pairs* − *sum of measures of angles of hexagon*

$$= 6 \cdot 180 - (6 - 2)180$$
$$= 1080 - 720$$
$$= 360$$

What sum would you find for an n-gon?

Theorem 7–2 **The sum of the degree measures of the exterior angles of a convex polygon, one at each vertex, is 360.**

exercises

Exploratory **For each of the following convex polygons, find the number of diagonals that can be drawn from one vertex.**

1. triangle
2. quadrilateral
3. pentagon
4. hexagon
5. heptagon
6. octagon
7. decagon
8. 15-gon
9. n-gon

10–18. For each of the convex polygons in problems **1–9,** find the number of triangles that can be formed by drawing all possible diagonals from one vertex.

19–27. For each of the convex polygons in problems **1–9,** find the total number of diagonals that can be formed.

Written Find the sum of the degree measures of the angles of a convex polygon for each number of sides.

1. 17
2. 20
3. 12
4. 13
5. 59
6. 15
7. x
8. t

For each of the following, the degree measure of one angle of a regular polygon is given. Find the number of sides.

9. 150
10. 160
11. 120
12. 60
13. 165
14. 156
15. 144
16. 179

For each of the following, the number of sides of a regular polygon is given. Find the degree measure of each angle.

17. 4
18. 5
19. 8
20. 10
21. 13
22. 17
23. d
24. x

25–32. Find the degree measure of one exterior angle for each regular polygon in problems **17–24**.

Solve each problem.

33. The sum of the degree measures of the angles of a convex polygon is 3240. Find the number of sides.

34. The sum of the degree measures of the angles of a convex polygon is 1260. Find the number of sides.

35. The degree measure of one exterior angle of a regular polygon is 30. Find the number of sides.

36. The degree measure of one exterior angle of a regular polygon is 45. Find the number of sides.

Each of the following figures was formed by extending the sides of a regular polygon. Find the degree measure of the angles formed at each vertex.

37.

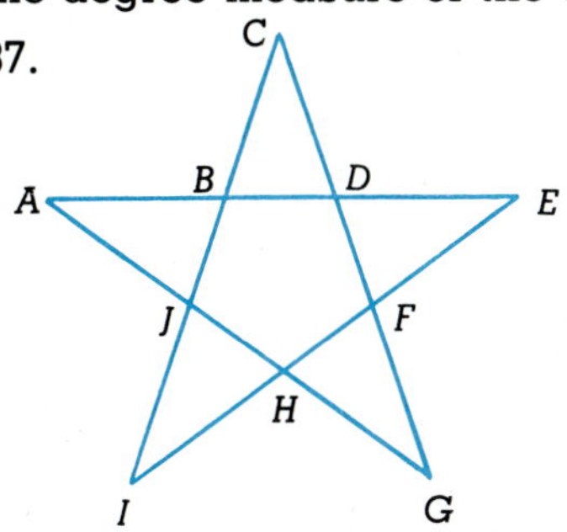

38.

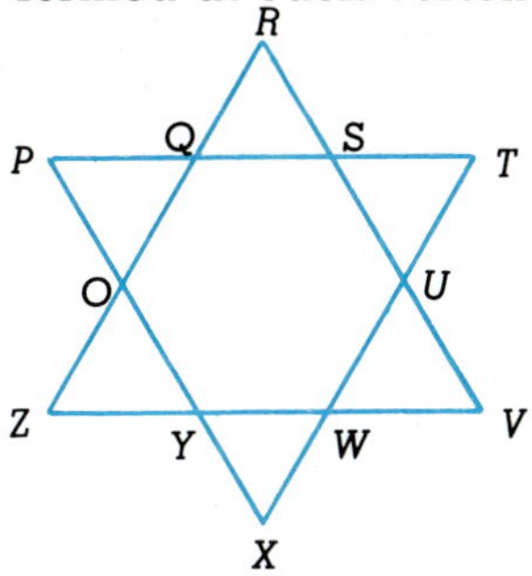

39.

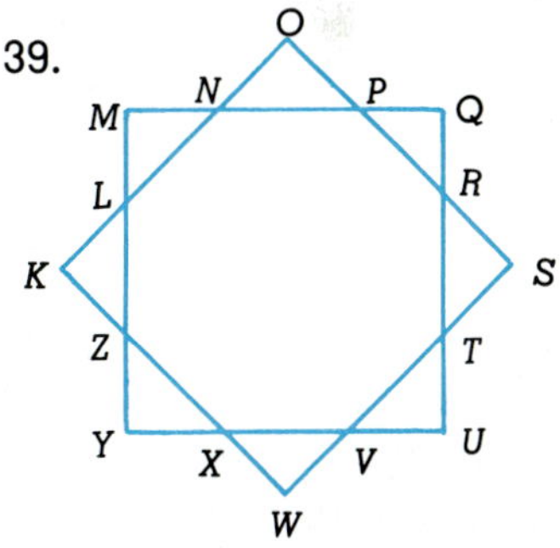

Write a proof using algebra for each of the following.

40. The sum of the degree measures of a regular pentagon is 540.

41. The sum of the degree measures of the exterior angles, one at each vertex, of a convex quadrilateral is 360.

42. The sum of the degree measures of the exterior angles, one at each vertex, of a convex n-gon is 360.

43. The sum of the degree measures of the exterior angles, two at each vertex, of a convex n-gon is 720.

7–3 Parallelograms

The pieces in the stained glass window represent a special kind of quadrilateral. Notice that in each piece both pairs of opposite sides are parallel. Thus, the quadrilateral represented by the pieces is called a **parallelogram.**

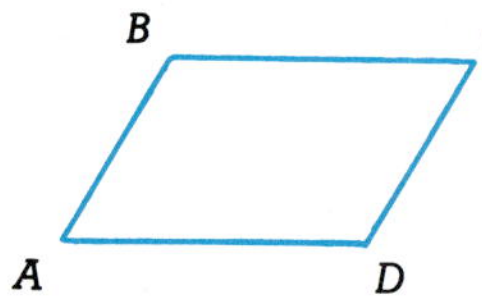

This figure is called *parallelogram ABCD* and written $\square ABCD$.

Definition of Parallelogram

A quadrilateral is a parallelogram if and only if both pairs of opposite sides are parallel.

The following theorem states an important characteristic of parallelograms.

Theorem 7–3

A diagonal of a parallelogram separates the parallelogram into two congruent triangles.

example 1

Prove Theorem 7–3.

Given: $\square ABCD$

Prove: $\triangle ABD \cong \triangle CDB$

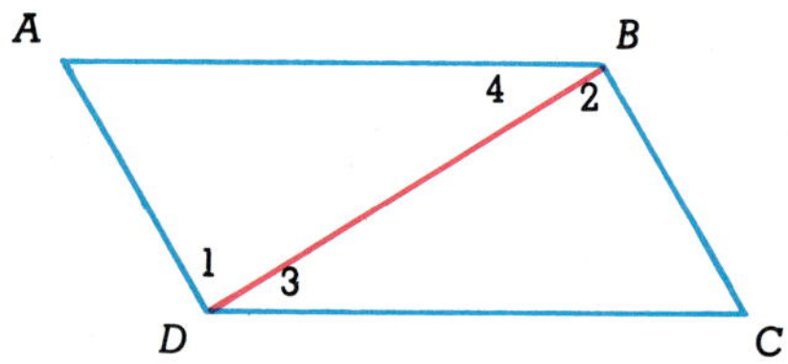

Proof:

STATEMENTS	REASONS
1. $\square ABCD$	1. Given
2. $\overline{AB} \parallel \overline{DC}$ $\overline{AD} \parallel \overline{BC}$	2. Definition of parallelogram
3. $\angle 1 \cong \angle 2$ $\angle 3 \cong \angle 4$	3. Theorem 6–9: If two parallel lines are cut by a transversal, then each pair of alternate interior angles is congruent.
4. $\overline{BD} \cong \overline{BD}$	4. Theorem 2–2: Congruence of line segments is reflexive.
5. $\triangle ABD \cong \triangle CDB$	5. Postulate 4–3: ASA *Steps 3 and 4*

In the parallelogram at the right, $\angle R$ and $\angle T$ are called *opposite angles*. Also, $\angle Q$ and $\angle S$ are called *opposite angles*. The following theorems are about opposite angles and opposite sides. Their proofs use congruent triangles formed by a diagonal of a parallelogram.

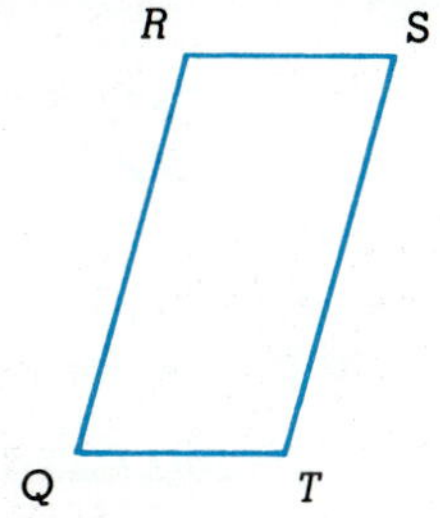

Opposite angles of a parallelogram are congruent. — Theorem 7–4

Opposite sides of a parallelogram are congruent. — Theorem 7–5

Notice that the diagonals of a parallelogram intersect to form four triangles. The triangles outlined in color are congruent. The other pair of triangles is congruent, too. The proof of the following theorem about parallelograms is based on these pairs of triangles.

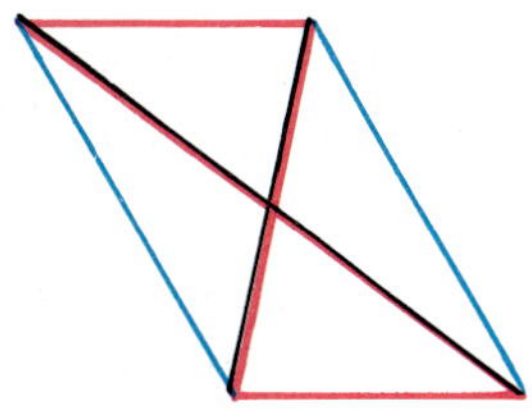

The diagonals of a parallelogram bisect each other. — Theorem 7–6

example 2

Complete the following proof.

Given: $\square EFGH$

Prove: $\overline{ED} \cong \overline{DG}$
$\overline{DH} \cong \overline{FD}$

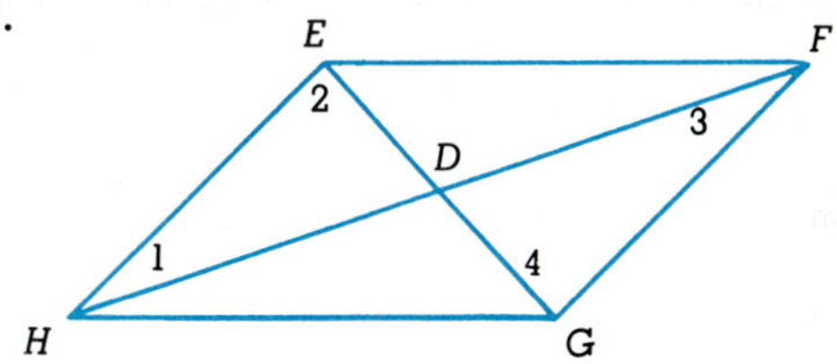

Proof:

STATEMENTS	REASONS
1. $\square EFGH$	1. Given
2. $\overline{FG} \cong \overline{EH}$	2. Theorem 7–5: Opposite sides of a parallelogram are congruent.
3. $\overline{FG} \parallel \overline{EH}$	3. Definition of parallelogram
4. $\angle 1 \cong \angle 3$ $\angle 2 \cong \angle 4$	4. Theorem 6–9: If two parallel lines are cut by a transversal, then each pair of alternate interior angles is congruent.
5. $\triangle EDH \cong \triangle GDF$	5. Postulate 4–3: ASA
6. $\overline{ED} \cong \overline{DG}$ $\overline{DH} \cong \overline{FD}$	6. Definition of congruent triangles ***CPCTC***

exercises

Exploratory Tell whether or *not* each of the following statements must be true about the parallelogram at the right. State *yes* or *no*.

1. $\overline{FE} \parallel \overline{GH}$
2. $\angle FGH \cong \angle FEH$
3. $\angle FDE \cong \angle GDH$
4. $\overline{EH} \parallel \overline{FG}$
5. $\overline{GH} \cong \overline{FE}$
6. $\overline{FD} \cong \overline{DG}$
7. $\overline{FD} \cong \overline{DH}$
8. $\overline{FD} \cong \overline{DE}$
9. $\overline{FG} \cong \overline{FD}$
10. $\overline{FG} \cong \overline{EH}$
11. $\triangle GDH \cong \triangle EDF$
12. $\triangle FHE \cong \triangle HFG$
13. $\triangle GDH \cong \triangle FDE$
14. $\triangle FHE \cong \triangle GEH$
15. D is the midpoint of $\overline{EG}$
16. D is the midpoint of $\overline{FH}$

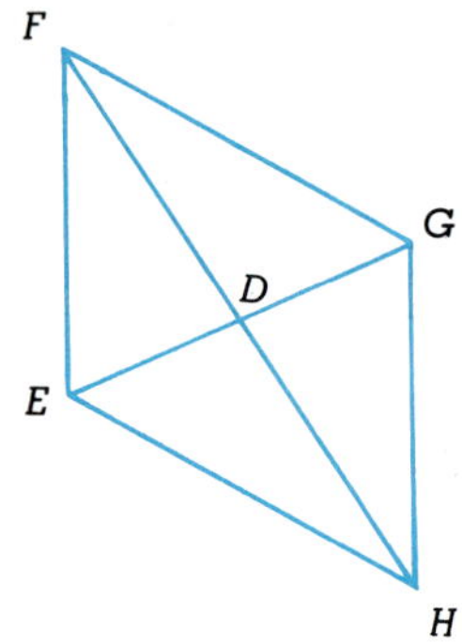

Written For each of the following, use the parallelogram at the right and the given information to find x.

1. $QR = 16$, $PS = x$
2. $PT = 7$, $PR = x$
3. $QS = 2.4$, $TQ = x$
4. $RS = 3.9$, $PQ = x$
5. $RQ = 4x + 9$, $PS = 7x - 6$
6. $QP = 2x + 5$, $SR = 3x$
7. $TR = 3x - 10$, $PR = 28$
8. $QS = 2x + 7$, $TS = 5x - 8$

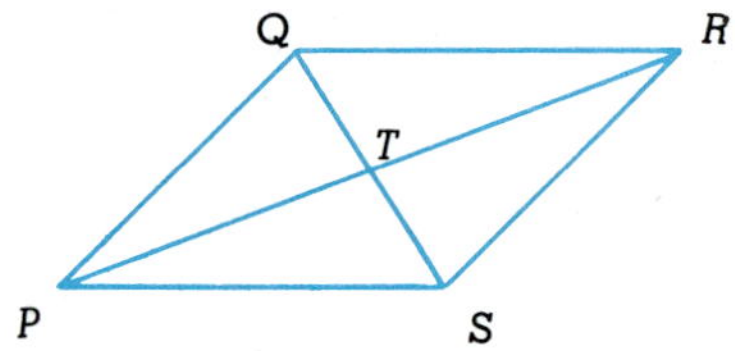

Complete a proof for each of the following.

9. Theorem 7–4.
10. Theorem 7–5.
11. **Given:** $\square PQST$
    $\overline{RP}$ bisects $\angle QPT$
    $\overline{VS}$ bisects $\angle QST$
    Prove: $\overline{RP} \parallel \overline{VS}$
12. **Given:** $\square PQST$
    $\overline{RP}$ bisects $\angle QPT$
    $\overline{VS}$ bisects $\angle QST$
    Prove: $\overline{RP} \cong \overline{VS}$

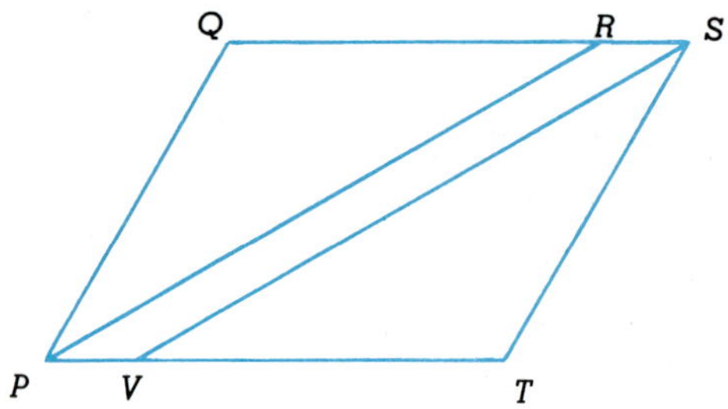

13. **Given:** $\square ABCD$
    $\overline{DE} \perp \overline{AC}$
    $\overline{BF} \perp \overline{AC}$
    Prove: $\overline{DE} \cong \overline{BF}$
14. **Given:** $\square ABCD$
    $\overline{DE} \perp \overline{AC}$
    $\overline{BF} \perp \overline{AC}$
    Prove: $\overline{DE} \parallel \overline{BF}$

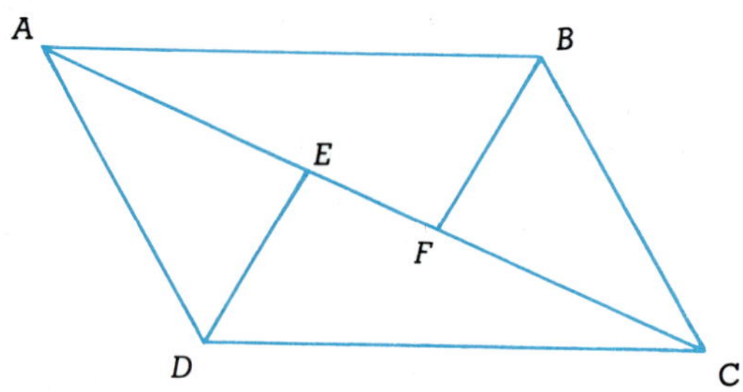

Using Money

Taxes

Social security taxes and local taxes are withheld from most paychecks. Social security taxes, called FICA, are 6.05% of the first $17,700 (1978) of gross pay. Local tax rates vary. Paycheck statements show both these deductions.

Brian Wagoner is single, claims one exemption, and earns $12,000 per year. He receives 26 paychecks each year. He checks his statement to be sure the tax deductions are correct.

CAPITAL WAREHOUSE NO. 08683

NUMBER	PERIOD ENDING	REGULAR HOURS	OVERTIME	TOTAL GROSS
856	Feb. 27	40.0		461.54
GROSS PAY	FED. INCOME TAX	FICA	LOCAL TAX	TOTAL TAX
461.54	74.35	27.92	6.92	109.19
DISAB.	CREDIT UNION	INSURANCE	OTHER	PERSONAL DEDUCTIONS
				NET PAY
				352.35

This statement covers your gross earnings less authorized deductions and those required by law. It is your receipt for deductions and should be kept as your record of payments.

The FICA rate is 6.05% of yearly earnings.

$$12{,}000 \times 6.05\% = 12{,}000 \times 0.0605$$
$$= 726$$

Since there are 26 pay periods, divide by 26.

$$726 \div 26 = 27.923$$

The FICA tax is $27.92.

The local tax rate is 1.5% of yearly earnings.

$$12{,}000 \times 1.5\% = 12{,}000 \times 0.015$$
$$= 180$$

Since there are 26 pay periods, divide by 26.

$$180 \div 26 = 6.923$$

The local tax is $6.92.

Exercises For each annual income, find the FICA tax at 6.05% and the local tax at 1.5% for one pay period. Assume there are 26 paychecks per year. Round each answer to the nearest cent.

1. $8000 **2.** $9500 **3.** $6200

4. $12,000 **5.** $14,600 **6.** $17,000

7–4 Tests for Parallelograms

Hazel Thompson is cutting glass pieces in the shape of parallelograms. She must know how to be sure her cutting pattern is a parallelogram.

There are several tests for a parallelogram. Hazel is most likely to use the following.

Theorem 7–7

If both pairs of opposite sides of a quadrilateral are congruent, then the quadrilateral is a parallelogram.

The proofs of this test and the next are very similar. They both use the triangles formed by a diagonal and sides of a given quadrilateral.

Theorem 7–8

If two sides of a quadrilateral are parallel and congruent, then the quadrilateral is a parallelogram.

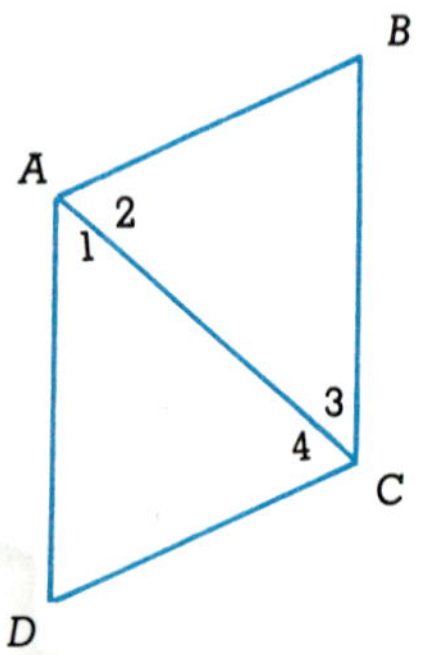

Suppose that $\overline{BC} \cong \overline{AD}$ and $\overline{AB} \cong \overline{DC}$, then show that the triangles are congruent by SSS, since $\overline{AC} \cong \overline{AC}$. Then, show that alternate interior angles are congruent, and conclude that opposite sides are parallel.

Suppose that $\overline{BC} \parallel \overline{AD}$ and $\overline{BC} \cong \overline{AD}$, then show $\angle 3 \cong \angle 1$ and, thus, the triangles are congruent by SAS, since $\overline{AC} \cong \overline{AC}$. Thus, $\angle 2 \cong \angle 4$ and $\overline{AB} \cong \overline{CD}$.

example 1

Prove that if the diagonals of a quadrilateral bisect each other, then the quadrilateral is a parallelogram.

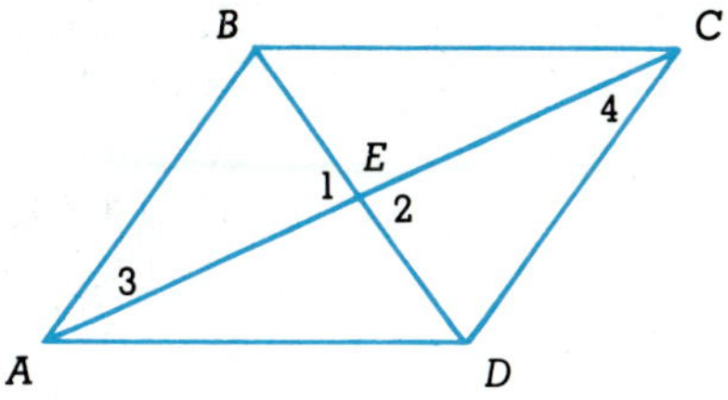

Given: $\overline{BD}$ bisects $\overline{AC}$
$\overline{AC}$ bisects $\overline{BD}$

Prove: quadrilateral $ABCD$ is a parallelogram

Proof:

STATEMENTS	REASONS
1. $\overline{BD}$ bisects $\overline{AC}$ $\overline{AC}$ bisects $\overline{BD}$	1. Given
2. $\overline{AE} \cong \overline{EC}$ $\overline{BE} \cong \overline{ED}$	2. Theorem 2–6: Bisector Theorem
3. $\angle 1 \cong \angle 2$	3. Theorem 3–10: Vertical angles are congruent.
4. $\triangle BEA \cong \triangle DEC$	4. Postulate 4–2: SAS *Steps 2 and 3*
5. $\angle 3 \cong \angle 4$ $\overline{AB} \cong \overline{DC}$	5. Definition of congruent triangles *CPCTC*
6. $\overline{AB} \parallel \overline{DC}$	6. Theorem 6–5: In a plane, if two lines are cut by a transversal so that a pair of alternate interior angles are congruent, then the lines are parallel.
7. quadrilateral $ABCD$ is a parallelogram	7. Theorem 7–8: If two sides of a quadrilateral are parallel and congruent, then the quadrilateral is a parallelogram. *Steps 5 and 6*

If the diagonals of a quadrilateral bisect each other, then the quadrilateral is a parallelogram.

Theorem 7–9

exercises

Exploratory **Tell whether or *not* each of the following tests could be used to prove that a quadrilateral is a parallelogram. State *yes* or *no*.**

1. Both pairs of opposite sides are parallel.
2. Both pairs of opposite sides are congruent.
3. The quadrilateral is regular.
4. Both pairs of consecutive sides are congruent.
5. Both pairs of consecutive angles are congruent.
6. Both pairs of opposite angles are congruent.

For each of the following conditions, tell whether or *not* quadrilateral *PQRS* must be a parallelogram. State *yes* or *no*. Then, explain.

7. $\overline{PQ} \cong \overline{QR}$
   $\overline{RS} \cong \overline{PS}$
8. $\overline{RQ} \cong \overline{SP}$
   $\angle 2 \cong \angle 6$
9. T is the midpoint of $\overline{QS}$.
   T is the midpoint of $\overline{PR}$.
10. $\angle 2 \cong \angle 3$
    $\overline{QR} \cong \overline{PS}$
11. $QT = 5$
    $RT = 9$
    $PT = 5$
    $ST = 5$
12. $m\angle 1 = 60$
    $m\angle 2 = 70$
    $m\angle 5 = 60$
    $m\angle 6 = 70$
13. $m\angle PQR = 71$
    $m\angle QPS = 109$
    $\angle PQR \cong \angle RSP$
14. $QR = 8$
    $PQ = 8$
    $RS = 4$
    $SP = 4$
15. $\triangle PQR \cong \triangle RSP$
16. $\triangle PQT \cong \triangle RST$

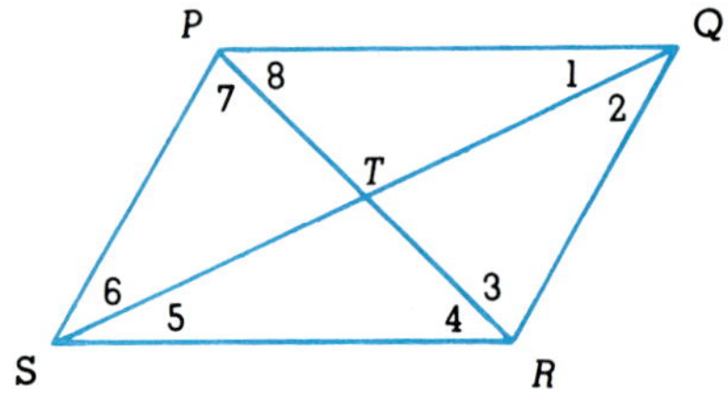

Written Prove each of the following.

1. Theorem 7–7.
2. Theorem 7–8.
3. If both pairs of opposite angles of a quadrilateral are congruent, then the quadrilateral is a parallelogram.
4. If one pair of opposite angles of a quadrilateral is congruent and one pair of opposite sides is parallel, then the quadrilateral is a parallelogram.

5. **Given:** $\overline{AE} \cong \overline{EC}$
   $\angle ECD \cong \angle EAB$
   Prove: quadrilateral $ABCD$ is a parallelogram
6. **Given:** $\overline{BA} \parallel \overline{CD}$
   $\overline{BD} \cong \overline{AC}$
   $\angle DBC \cong \angle ADB$
   Prove: quadrilateral $ABCD$ is a parallelogram

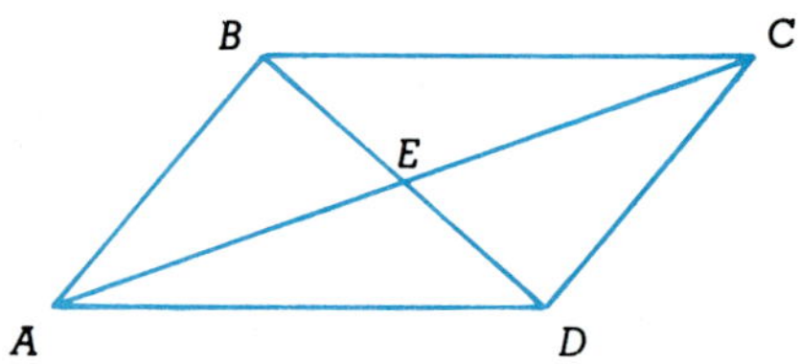

7. **Given:** $\triangle PQR \cong \triangle STV$
   $\overline{PR} \parallel \overline{VS}$
   Prove: quadrilateral $PRSV$ is a parallelogram
8. **Given:** $\square PQST$
   $\overline{QR} \cong \overline{VT}$
   Prove: quadrilateral $PRSV$ is a parallelogram

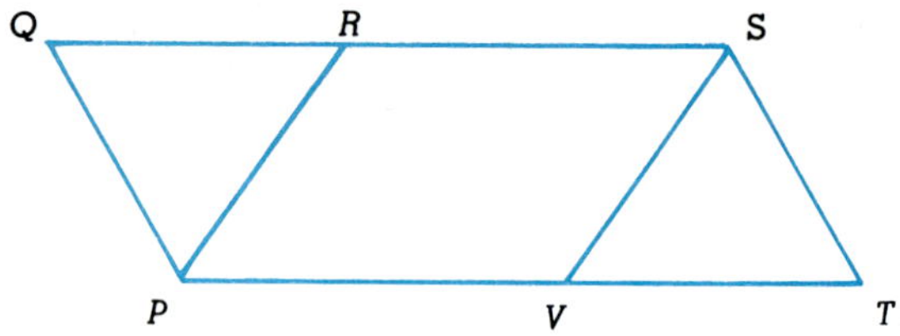

9. **Given:** $\square ABCD$
   E is the midpoint of $\overline{AB}$
   F is the midpoint of $\overline{BC}$
   G is the midpoint of $\overline{CD}$
   H is the midpoint of $\overline{DA}$
   Prove: quadrilateral $EFGH$ is a parallelogram

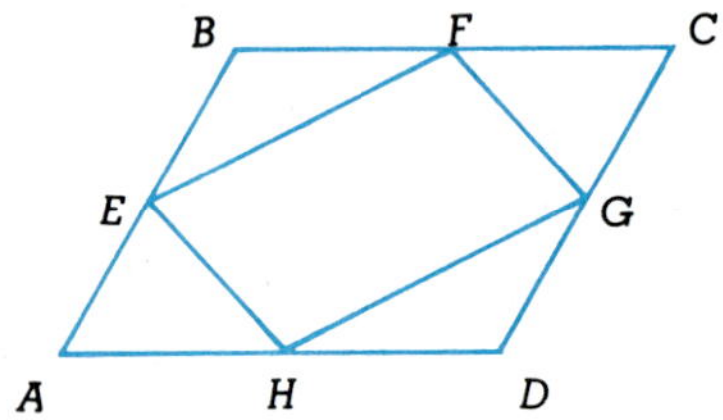

10. **Given:** $\overline{AF} \cong \overline{FC}$
$\overline{BE} \cong \overline{GD}$
F is the midpoint of $\overline{EG}$
Prove: quadrilateral $ABCD$ is a parallelogram

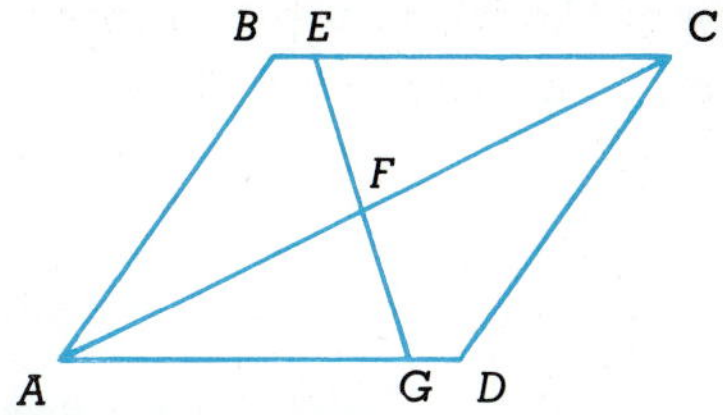

11. **Given:** $\overline{AB} \cong \overline{BC}$
D is the midpoint of $\overline{CF}$
D is the midpoint of $\overline{BE}$
Prove: quadrilateral $ABEF$ is a parallelogram

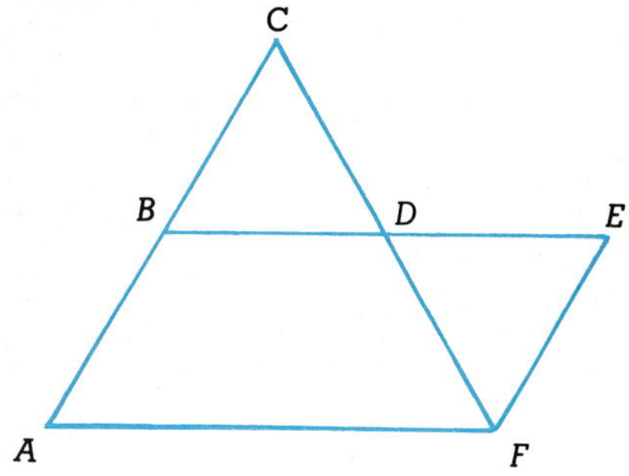

12. **Given:** $\overline{VQ} \perp \overline{PR}$
$\overline{VS} \perp \overline{RT}$
$\overline{VQ} \perp \overline{VS}$
Prove: quadrilateral $VQRS$ is a parallelogram

13. **Given:** $\angle PQV \cong \angle TSV$
$\overline{VS} \parallel \overline{RP}$
Prove: quadrilateral $VQRS$ is a parallelogram

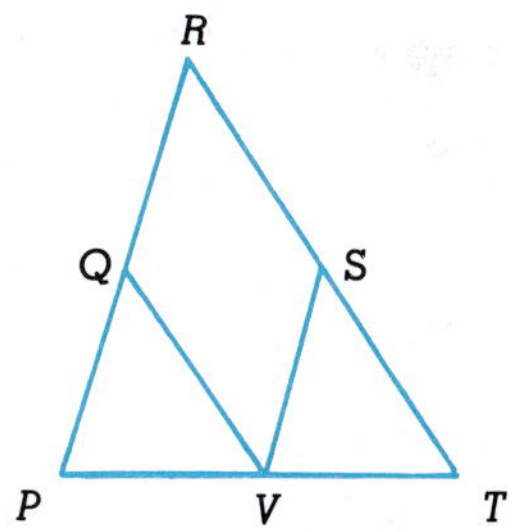

14. **Given:** $\square ACEF$
$\angle G \cong \angle F$
$\overline{BG} \parallel \overline{CD}$
Prove: quadrilateral $BCDG$ is a parallelogram

15. **Given:** $\square BCDG$
$\angle E \cong \angle A$
$\overline{CE} \parallel \overline{AF}$
Prove: quadrilateral $ACEF$ is a parallelogram

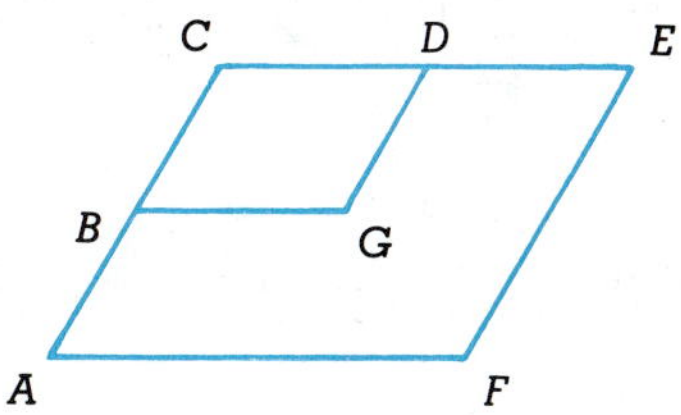

Loops

Excursions in Geometry

Which of the following loops of string will make a knot when the ends are pulled tight?

7–5 Special Parallelograms

The shapes of the figures below are examples of three special types of parallelograms. These parallelograms are classified by their sides and their angles.

A rectangle can be classified by its angles. *Can you think of other ways?*

Definition of Rectangle

A quadrilateral is a rectangle if and only if it is a parallelogram with four right angles.

Because rectangle $ABCD$ at the right is a parallelogram, opposite sides are congruent, so $\overline{AB} \cong \overline{CD}$. Also, $\angle BAD$ and $\angle CDA$ are right angles, and $\overline{AD} \cong \overline{AD}$. Thus, $\triangle BAD \cong \triangle CDA$, and the following conclusion can be made.

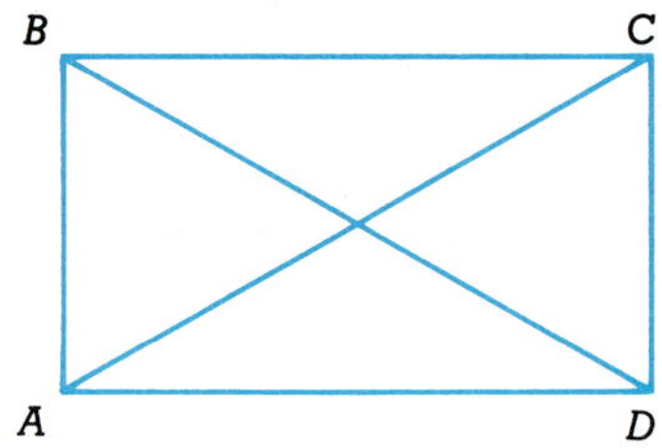

Theorem 7–10

The diagonals of a rectangle are congruent.

A rhombus can be classified by its sides.

Definition of Rhombus

A quadrilateral is a rhombus if and only if it is a parallelogram with all four sides congruent.

The diagonals of a rhombus *not* only bisect each other, but also bisect opposite angles of the rhombus. This property can be proved by showing pairs of triangles congruent.

Also, the diagonals of a rhombus intersect to form four right angles.

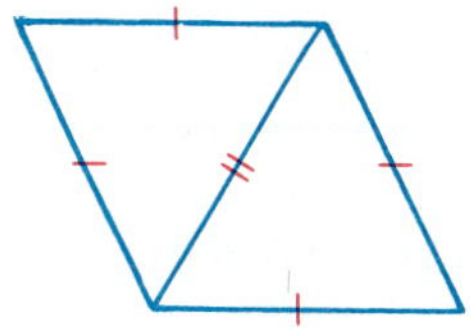

The diagonals of a rhombus bisect the angles of the rhombus. Theorem 7–11

The diagonals of a rhombus are perpendicular. Theorem 7–12

example 1

Prove Theorem 7–12.

Given: rhombus $ABCD$

Prove: $\overline{AC} \perp \overline{BD}$

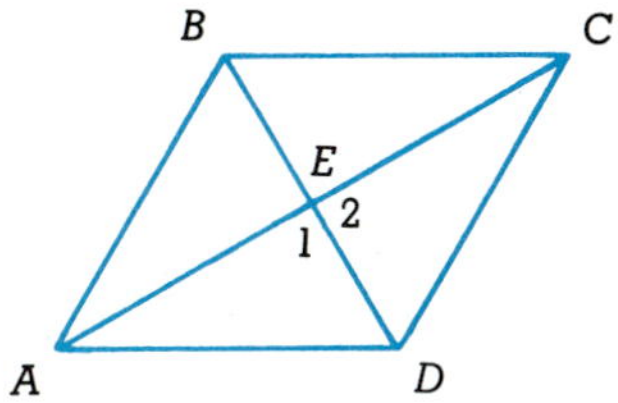

Proof:

STATEMENTS	REASONS
1. rhombus $ABCD$	1. Given
2. $\overline{AD} \cong \overline{DC}$	2. Definition of rhombus
3. $\overline{BD}$ bisects $\overline{AC}$	3. Theorem 7–6: Diagonals of a parallelogram bisect each other.
4. $\overline{AE} \cong \overline{EC}$	4. Theorem 2–6: Bisector Theorem
5. $\overline{DE} \cong \overline{DE}$	5. Theorem 2–2: Congruence of line segments is reflexive.
6. $\triangle ADE \cong \triangle CDE$	6. Postulate 4–1: SSS *Steps 2, 4, and 5*
7. $\angle 1 \cong \angle 2$	7. Definition of congruent triangles *CPCTC*
8. $\overline{AC} \perp \overline{BD}$	8. Theorem 3–14: If two intersecting lines form congruent adjacent angles, then they are perpendicular.

A rhombus may or may *not* be a rectangle. Also, a rectangle may or may *not* be a rhombus. A quadrilateral that is *both* a rectangle and a rhombus is called a **square**.

A quadrilateral is a square if and only if it is a rectangle and all four sides are congruent. Definition of Square

exercises

Exploratory Name which quadrilaterals have each of the following properties.

1. All angles are right angles.
2. The diagonals bisect each other.
3. The opposite sides are congruent.
4. The opposite sides are parallel.
5. The diagonals bisect the opposite angles.
6. The diagonals are perpendicular.
7. The opposite angles are congruent.
8. All sides are congruent.
9. It is equiangular.
10. It is equilateral.
11. It is equiangular and equilateral.
12. It is regular.

Written Use the rhombus at the right and the information given to solve each problem.

1. $ST = 10\frac{7}{8}$. Find TV.
2. $ST = 12.3$. Find TV.
3. $SP = 52.1$. Find PV.
4. $SP = 22\frac{5}{8}$. Find PV.
5. $PT = 6\frac{3}{4}$. Find RT.
6. $PT = 7.6$. Find RT.
7. $m\angle RST = 112.6$. Find $m\angle RSV$.
8. $m\angle STV = 75.0$. Find $m\angle STR$.
9. $m\angle SVT = 25.9$. Find $m\angle RVT$.
10. $m\angle TVR = 123.4$. Find $m\angle VST$.
11. Find $m\angle SPR$.
12. Find $m\angle VPT$.

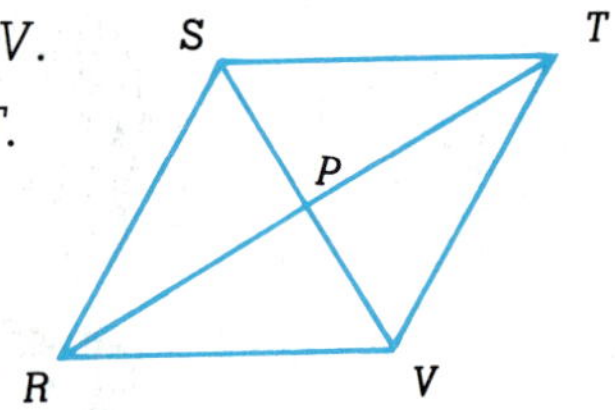

Prove each of the following.

13. If a polygon is a square, then its diagonals bisect each other.
14. If a polygon is a square, then its diagonals are perpendicular.
15. **Given:** $\square PRTW$ is a square
    Q is the midpoint of $\overline{PR}$
    S is the midpoint of $\overline{RT}$
    V is the midpoint of $\overline{TW}$
    X is the midpoint of $\overline{PW}$
    Prove: $\overline{QV} \perp \overline{SX}$

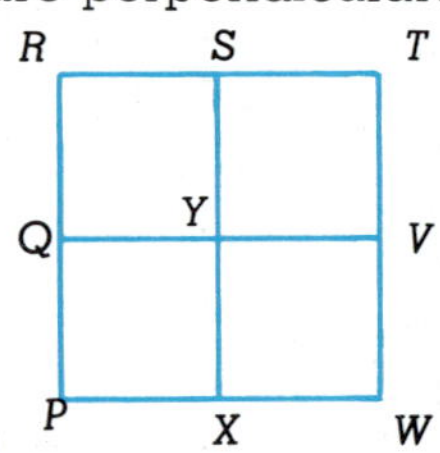

16. **Given:** $\square ABCD$ is a rhombus
    $\overline{AF} \cong \overline{BG}$
    $\overline{BG} \cong \overline{CH}$
    $\overline{CH} \cong \overline{DE}$
    $\overline{DE} \cong \overline{AF}$
    Prove: quadrilateral $EFGH$ is a parallelogram

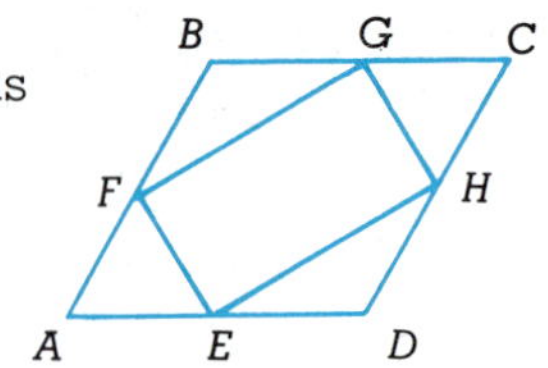

17. If the diagonals of a parallelogram are congruent, then it is a rectangle.
18. If the diagonals of a parallelogram are perpendicular, then it is a rhombus.
19. If the diagonals of a parallelogram bisect its angles, then it is a rhombus.
20. If the diagonals of a parallelogram are congruent, then it is a rectangle.
21. Theorem 7–10.
22. Theorem 7–11.

Mineralogist

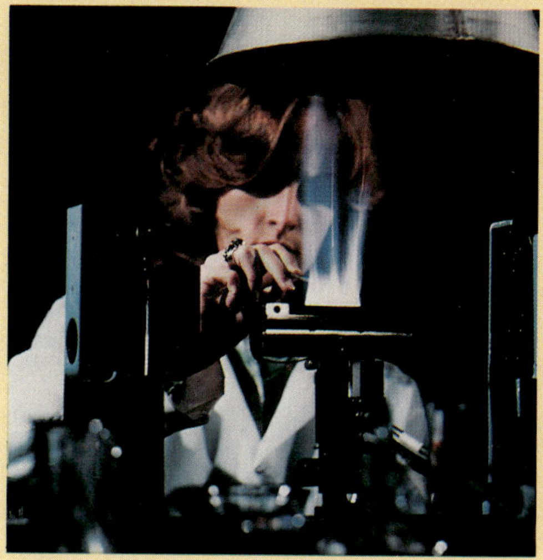

Kathy Seall is a mineralogist for B. T. Duvall Power Co. As a mineralogist she studies the various crystals found in nature. Crystals can be used to change sunshine into electric current.

In their original state, crystals are solids with flat surfaces. The flat surfaces, called faces, have different geometric shapes, depending on the type of crystal.

One way to classify crystals is by the number of faces, the shape of the faces, and the congruent faces. For example, alum crystals have eight faces, each of which are equilateral triangles.

The picture at the right shows wulfenite crystals. How many faces do you see? What are the shapes of each of the faces? Are any faces congruent?

Exercises For each of the following crystals, name the number of faces, the shape of the faces, and identify any congruent faces.

1. borax

2. calcite

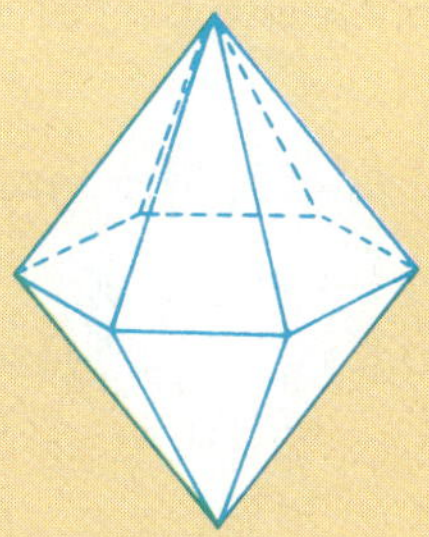

3. quartz

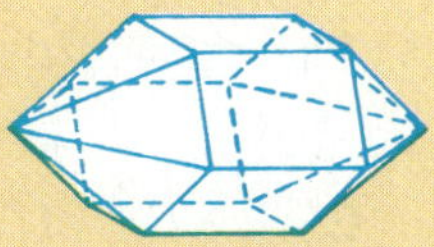

4. gypsum

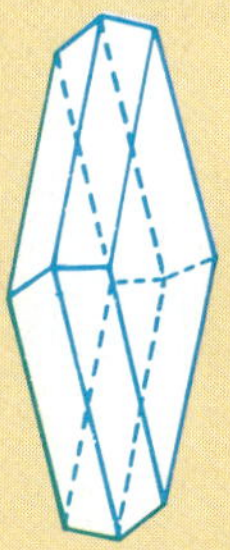

7–6 Trapezoids

The design of the College Park Pyramids in Indianapolis is based on a quadrilateral called a **trapezoid.**

Definition of Trapezoid

A quadrilateral is a trapezoid if and only if it has exactly one pair of parallel sides.

The parallel sides of a trapezoid are called *bases*. The nonparallel sides of a trapezoid are called *legs*. If the legs are congruent, then the trapezoid is called an *isosceles trapezoid*.

Together ∠B and ∠C are called base angles. Also, ∠A and ∠D together are called base angles.

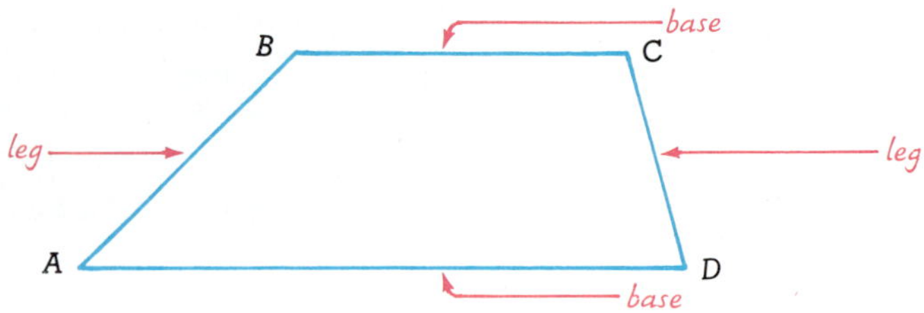

Isosceles trapezoids have special properties.

Theorem 7–13

The diagonals of an isosceles trapezoid are congruent.

Theorem 7–14

Each pair of base angles of an isosceles trapezoid is congruent.

example 1

Complete the following proof.

Given: $\overline{PS} \parallel \overline{QR}$
$\overline{PQ} \cong \overline{SR}$

Prove: $\angle P \cong \angle S$

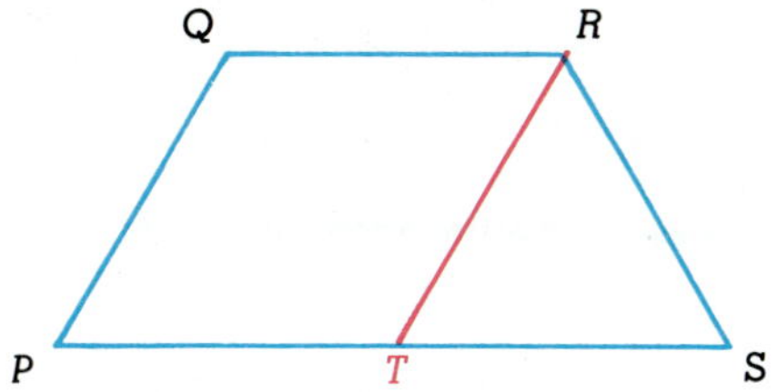

Proof:

STATEMENTS	REASONS
1. Draw $\overline{RT}$ so that T is between P and S, and $\overline{PT} \cong \overline{QR}$	1. Postulate 2–2: Ruler Postulate
2. $\overline{QR} \parallel \overline{PS}$	2. Given
3. Quadrilateral $PQRT$ is a parallelogram	3. Theorem 7–8: If two sides of a quadrilateral are parallel and congruent, then the quadrilateral is a parallelogram.
4. $\overline{TR} \cong \overline{PQ}$	4. Theorem 7–5: Opposite sides of a parallelogram are congruent.
5. $\overline{PQ} \cong \overline{SR}$	5. Given
6. $\overline{TR} \cong \overline{SR}$	6. Theorem 2–4: Congruence of line segments is transitive.
7. $\angle RTS \cong \angle S$	7. Theorem 4–6: Isosceles Triangle Theorem
8. $\overline{QP} \parallel \overline{RT}$	8. Definition of parallelogram
9. $\angle P \cong \angle RTS$	9. Theorem 6–8: If two parallel lines are cut by a transversal, then each pair of corresponding angles is congruent.
10. $\angle P \cong \angle S$	10. Theorem 3–1: Congruence of angles is transitive. *Steps 7 and 9*

The line segment joining the midpoints of the legs of a trapezoid is called the **median**. In the trapezoid at the right, the median is $\overline{BE}$.

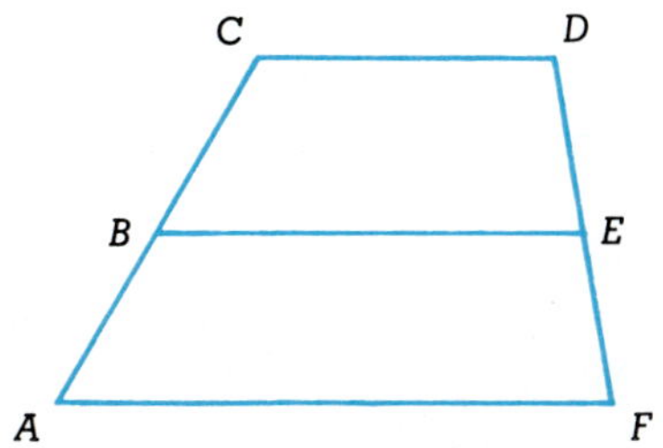

The median of a trapezoid is parallel to the bases, and its measure is one-half the sum of the measures of the bases.

Theorem 7–15

example 2 The bases of a trapezoid have lengths of 22 inches and 14 inches. Find the length of the median.

measure of median $= \frac{1}{2}$ *(measure of short base + measure of long base)*

$= \frac{1}{2}(14 + 22)$

$= \frac{1}{2}(36)$

$= 18$ The median is 18 inches long.

exercises

Exploratory Tell whether or *not* it is possible for a trapezoid to have the following. State *yes* or *no*.

1. a leg longer than either base
2. two congruent sides, but *not* be isosceles
3. three congruent sides
4. congruent bases
5. congruent diagonals
6. bisecting diagonals
7. one pair of opposite angles congruent
8. both pairs of opposite angles congruent
9. two right angles
10. two obtuse angles
11. three obtuse angles
12. four acute angles

Written For each of the following, the lengths of the bases of a trapezoid are given. Find the length of the median.

1. 6 cm, 18 cm
2. 4 in., 18 in.
3. $3\frac{1}{4}$ ft, $2\frac{3}{4}$ ft
4. $4\frac{1}{3}$ yd, $5\frac{1}{3}$ yd
5. 9.6 mm, 4.4 mm
6. 7.3 m, 2.5 m
7. 87.6 cm, 91.4 cm
8. 14.3 mm, 32.1 mm
9. $23\frac{1}{4}$ mi, $18\frac{1}{4}$ mi
10. 6.0 km, 3.5 km
11. 54.1 m, 42.7 m
12. $33\frac{1}{3}$ ft, $12\frac{2}{3}$ ft

The figure at the right is an isosceles trapezoid with bases $\overline{AD}$ and $\overline{BC}$.

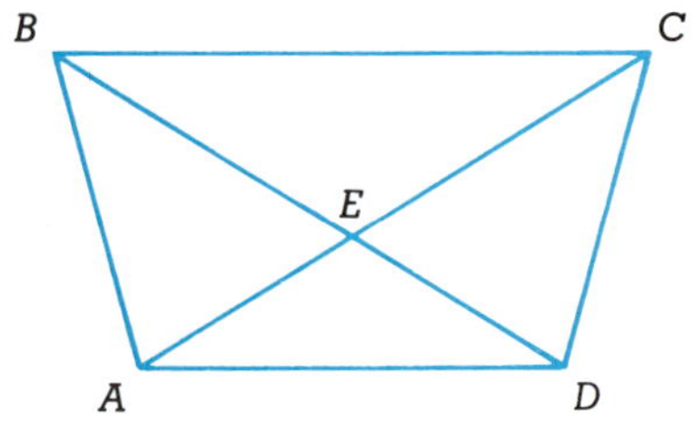

13. $BD = 8.2$. Find AC.
14. $AB = 14.6$. Find CD.
15. $CD = 13.5$. Find AB.
16. $CA = 16.7$. Find BD.
17. $m\angle BAD = 115$. Find $m\angle CDA$.
18. $m\angle ABC = 60$. Find $m\angle DCB$.
19. $m\angle BAD = 132$. Find $m\angle DAB$.
20. $m\angle BCD = 53$. Find $m\angle ADC$.
21. $m\angle EAD = 25$. Find $m\angle ECB$.
22. $m\angle EAD = 32$. Find $m\angle EDA$.
23. $m\angle BDA = 27$. Find $m\angle ACB$.
24. $m\angle BDA = 35$. Find $m\angle AED$.
25. $m\angle ABD = 23$. Find $m\angle DCA$.
26. $m\angle ACD = 21$. Find $m\angle DBA$.
27. $m\angle BEC = 126$. Find $m\angle ECB$.
28. $m\angle AED = 131$. Find $m\angle DBC$.

Use the figure at the right to complete each proof.

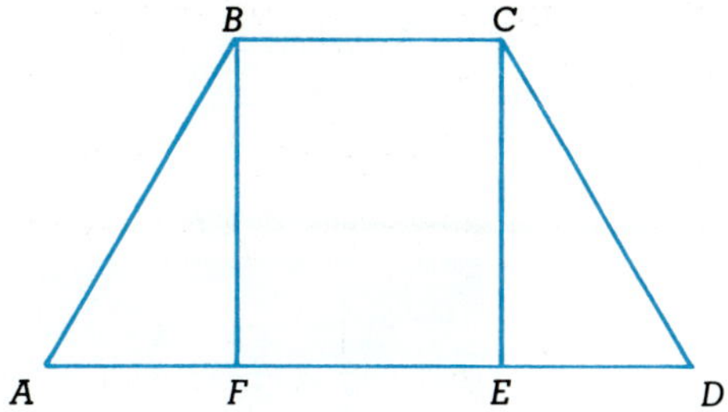

29. **Given:** $\overline{BC} \parallel \overline{AD}$
$\overline{AB} \cong \overline{DC}$
$\overline{BF} \perp \overline{AD}$
$\overline{CE} \perp \overline{AD}$
Prove: $\overline{BF} \cong \overline{CE}$

30. **Given:** $\overline{BC} \parallel \overline{AD}$
$\overline{AB} \cong \overline{DC}$
$\overline{BF} \perp \overline{AD}$
$\overline{CE} \perp \overline{AD}$
Prove: quadrilateral $FBCE$ is a parallelogram

31. **Given:** $\overline{BC} \parallel \overline{AD}$
$\overline{AB} \cong \overline{DC}$
$\overline{BF} \perp \overline{AD}$
$\overline{CE} \perp \overline{AD}$
Prove: $\triangle ABF \cong \triangle DCE$

32. **Given:** $\overline{BC} \parallel \overline{AD}$
$\overline{AB} \cong \overline{DC}$
$\overline{BF} \perp \overline{AD}$
$\overline{CE} \perp \overline{AD}$
Prove: $\angle BAF \cong \angle CDE$

Use the figure at the right to complete each proof.

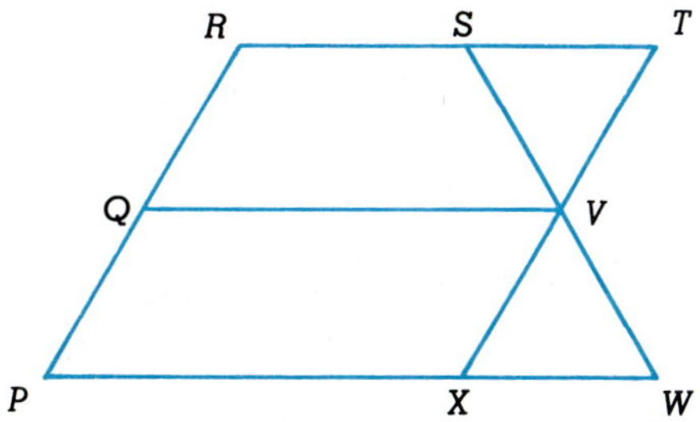

33. **Given:** $\overline{RT} \parallel \overline{PW}$
$\overline{TX} \parallel \overline{RP}$
Q is the midpoint of $\overline{PR}$
V is the midpoint of $\overline{SW}$
Prove: V is the midpoint of $\overline{TX}$

34. **Given:** $\overline{RT} \parallel \overline{PW}$
$\overline{TX} \parallel \overline{RP}$
Q is the midpoint of $\overline{PR}$
V is the midpoint of $\overline{SW}$
Prove: $\triangle SVT \cong \triangle WVX$

35. **Given:** $\overline{RT} \parallel \overline{PW}$
$\overline{TX} \parallel \overline{RP}$
Q is the midpoint of $\overline{PR}$
V is the midpoint of $\overline{SW}$
Prove: quadrilateral $QRTV$ is a parallelogram

36. **Given:** $\overline{RT} \parallel \overline{PW}$
$\overline{TX} \parallel \overline{RP}$
Q is the midpoint of $\overline{PR}$
V is the midpoint of $\overline{SW}$
Prove: quadrilateral $PQVX$ is a parallelogram

37. **Given:** $\overline{RT} \parallel \overline{PW}$
$\overline{TX} \parallel \overline{RP}$
Q is the midpoint of $\overline{PR}$
V is the midpoint of $\overline{SW}$
Prove: $\overline{QV} \parallel \overline{RT}$

38. **Given:** $\overline{RT} \parallel \overline{PW}$
$\overline{TX} \parallel \overline{RP}$
Q is the midpoint of $\overline{PR}$
V is the midpoint of $\overline{SW}$
Prove: $\overline{QV} \parallel \overline{PW}$

Solve each problem.

39. Plan a proof for Theorem 7–13.
40. Prove Theorem 7–13.

7–7 Perimeter

Julie Newton is a carpenter. The floor plan below shows a room she will panel on her next job.

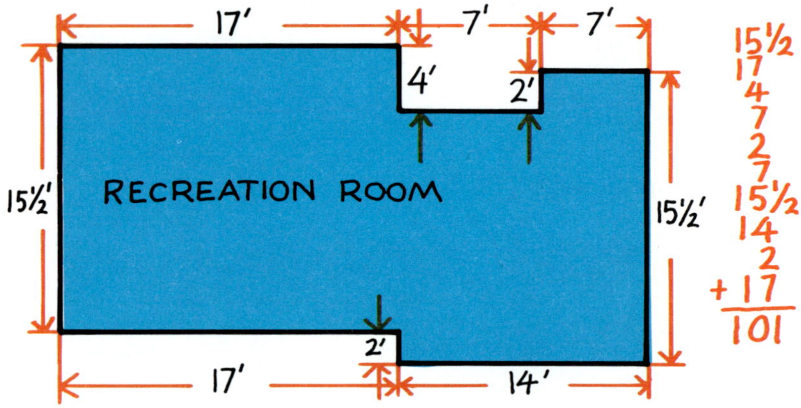

The perimeter of a polygon is the total length of its sides.

To determine the amount of molding she needs for ceiling trim, Ms. Newton first finds the **perimeter** of the room. She adds the measures for each side of the room. The perimeter is 101 feet.

Definition of Perimeter

The sum of the measures of the sides of a polygon is the measure of the perimeter of the polygon.

Each pair of opposite sides of a rectangle is congruent. Thus, only two measurements, the length and the width, are needed to find the perimeter of a rectangle.

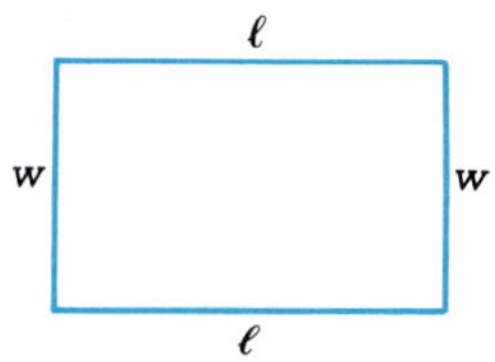

Let l be the measure of the length.
Let w be the measure of the width.
Let P be the measure of the perimeter.

$P = l + w + l + w$	*Definition of perimeter*
$= l + l + w + w$	*Commutative property of addition*
$= 2l + 2w$	*Substitution*
$= 2(l + w)$	*Distributive property*

Theorem 7–16
Perimeter of Rectangle

If a rectangle has perimeter P units, length l units, and width w units, then $P = 2(l + w)$.

example 1

Find the amount of metal edging needed for a rectangular table measuring 12.4 meters by 4.7 meters.

$P = 2(l + w)$
$= 2(12.4 + 4.7)$ *Substitute 12.4 for l and 4.7 for w.*
$= 2(17.1)$
$= 34.2$

The amount of metal edging needed is 34.2 meters.

All sides of a regular polygon are congruent. Thus, only the measure of one side and the number of sides of the polygon are needed to find the perimeter. For example, consider a regular pentagon.

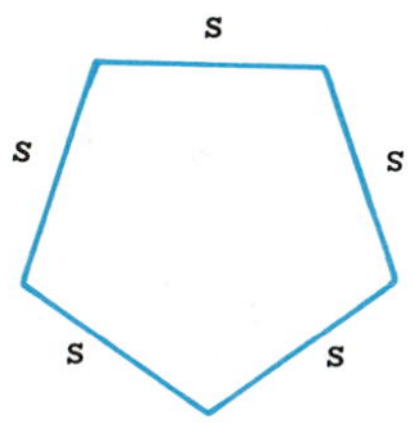

Let s stand for the measure of a side.
Let P stand for the measure of the perimeter.

$P = s + s + s + s + s$	Definition of perimeter
$= 5s$	Substitution

This, and many other examples, suggest the following theorem.

If a regular n-gon has perimeter P units, and a side measures s units, then $P = ns$.

Theorem 7–17
Perimeter of Regular Polygon

example 2

The perimeter of a regular octagon is 32 feet. Find the length of a side.

$P = ns$
$32 = 8s$ *Substitute 32 for P and 8 for n.*
$4 = s$ *Solve for s.*

The length of one side is 4 feet.

exercises

Exploratory Find the perimeter for each of the following.

1.
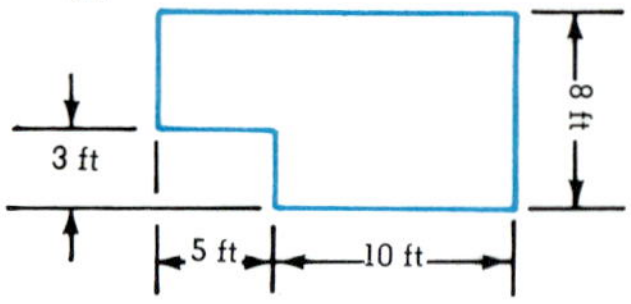

2.
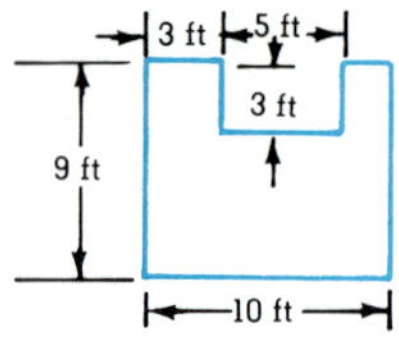

3.
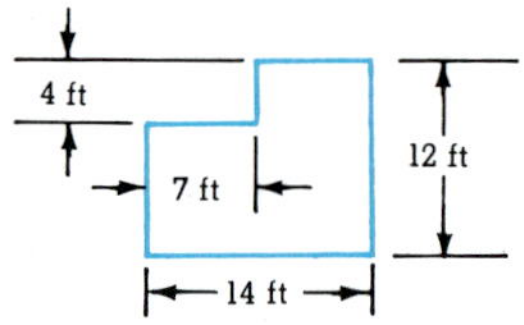

4.
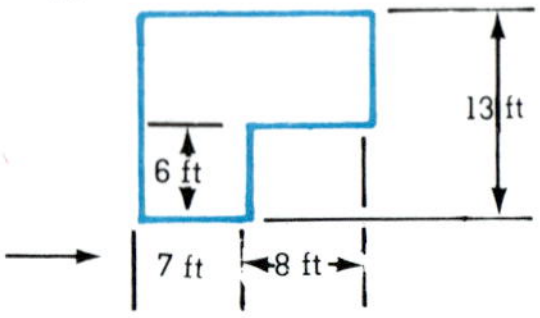

5.
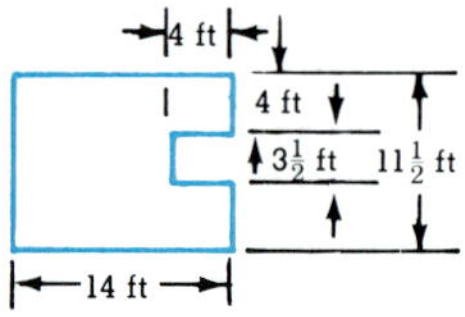

6.
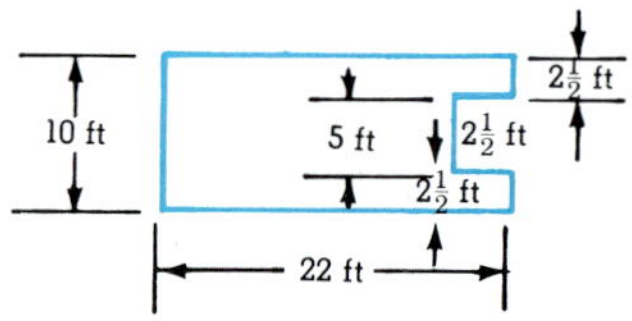

7.
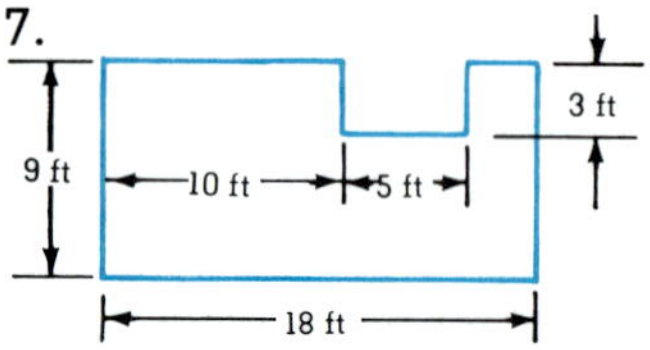

8.
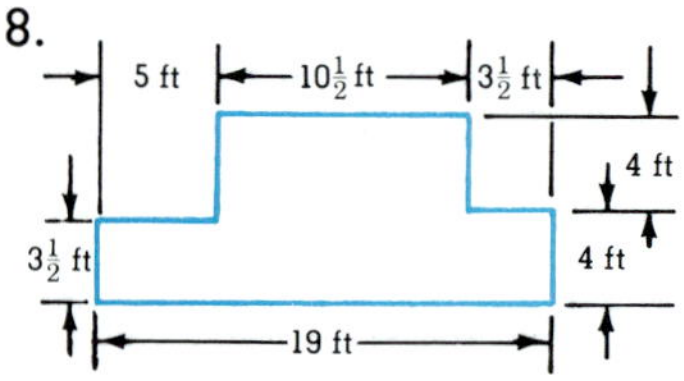

9.
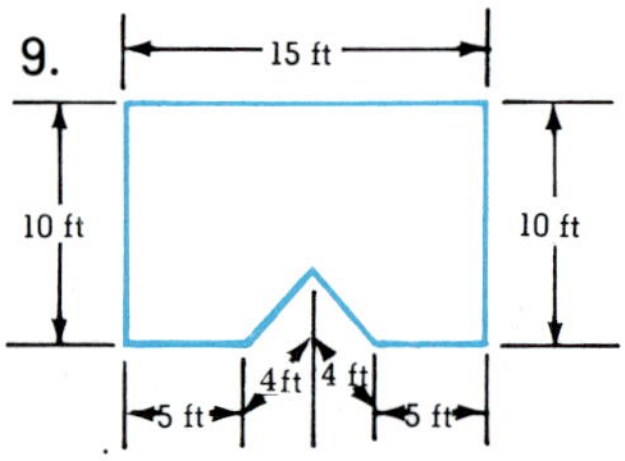

Written The length and width of rectangles are given below. Find the perimeter.

1. 3.4 m, 1.6 m
2. 3.2 m, 1.4 m
3. $3\frac{1}{8}$ in., $1\frac{1}{4}$ in.
4. $17\frac{1}{2}$ ft, $16\frac{2}{3}$ ft
5. 43.8 mi, 38 mi
6. 34.5 mm, 27.9 mm
7. 25.7 cm, 2.1 cm
8. $22\frac{1}{2}$ yd, 15 yd
9. $5\frac{1}{4}$ in., $4\frac{3}{4}$ in.

For each of the following, the perimeter and width of a rectangle are given. Find the length.

10. 141.4 km, 33.4 km
11. 21.6 mi, 6.9 mi
12. 19.4 cm, 9.3 cm
13. 12 in., $3\frac{1}{2}$ in.
14. $18\frac{1}{4}$ ft, $7\frac{3}{4}$ ft
15. $10\frac{2}{3}$ yd, $3\frac{1}{4}$ yd

For each of the following regular polygons, the length of a side is given. Find the perimeter.

16. decagon, 15.4 cm
17. hexagon, 12 ft
18. pentagon, $4\frac{1}{4}$ in.
19. square, 24.9 mm
20. octagon, 32.3 mi
21. nonagon, $3\frac{1}{3}$ yd

For each of the following regular polygons, the perimeter is given. Find the length of a side.

22. square, 15.40 mm
23. pentagon, 12.0 m
24. hexagon, 69 yd
25. heptagon, 21.7 cm
26. octagon, $55\frac{1}{2}$ ft
27. nonagon, 6.3 km

28. How many inches of lead are needed to put around 24 hexagons and 15 rhombuses of stained glass with sides measuring $1\frac{1}{2}$ inches?

29. How many feet of chain link fencing are needed for an L-shaped yard measuring 35 feet by 20 feet by 32 feet by 20 feet by 67 feet by 40 feet?

7–8 Solids

The soup can, the basketball, and the pyramid are examples of geometric solids. Solids enclose a part of space.

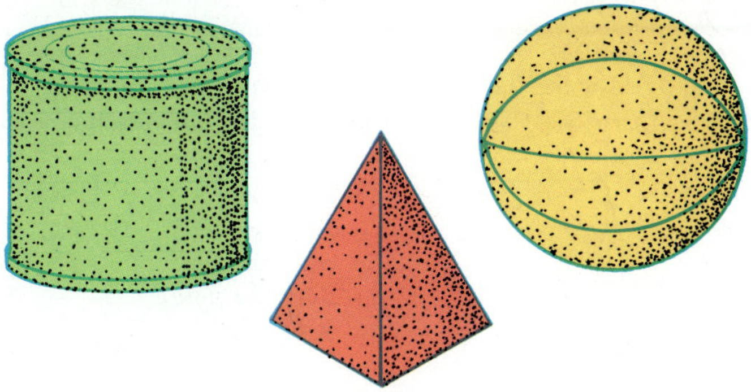

The surfaces of the soup can and the basketball are curved. The surfaces of the pyramid are flat. These flat surfaces form polygons. Solids with flat surfaces that form polygons are called **polyhedra.**

A polygon separates a plane into three parts, the interior, the exterior, and the polygon itself.

The flat surfaces formed by polygons and their interiors are called *faces*. The four triangles, $\triangle ABC$, $\triangle CBD$, $\triangle CDA$, $\triangle ABD$, and their interiors are the faces of the polyhedron at the right.

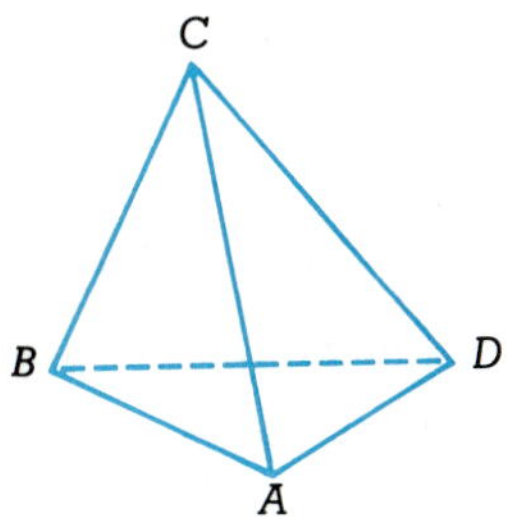

Pairs of faces intersect at *edges*. The six line segments, $\overline{AB}$, $\overline{BD}$, $\overline{AD}$, $\overline{DC}$, $\overline{AC}$, and $\overline{BC}$ are the edges of the polyhedron at the right.

Three or more edges intersect at a *vertex*. The four points, A, B, C, and D, are the vertices of the polyhedron at the right.

example 1

Name the faces, edges, and vertices of the polyhedron at the right.

The 7 faces are quadrilaterals $FAEJ$, $JEDI$, $IDCH$, $HCBG$, $GBAF$ and pentagons $ABCDE$ and $FGHIJ$.

The 15 edges are $\overline{AB}$, $\overline{BC}$, $\overline{CD}$, $\overline{DE}$, $\overline{EA}$, $\overline{FG}$, $\overline{GH}$, $\overline{HI}$, $\overline{IJ}$, $\overline{JF}$, $\overline{FA}$, $\overline{GB}$, $\overline{HC}$, $\overline{ID}$, and $\overline{JE}$.

The 10 vertices are A, B, C, D, E, F, G, H, I, and J.

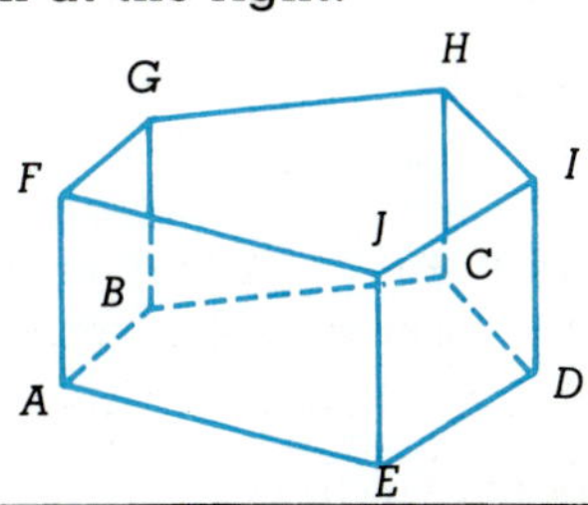

Each face of the polyhedron at the right forms a regular polygon. Also, each edge has the same measure. Polyhedra with such properties are called **regular polyhedra.**

A polyhedron must have *at least* three polygons intersecting at each vertex. Also, the total degree measures of the angles formed must be *less than* 360. Thus, there are only five regular polyhedra.

Two polygons do not form a polyhedron.

If the angle measures at a vertex total 360, the figure is flat.

The different regular polyhedra are determined by the regular polygons that form the faces.

regular polygon	measure of one angle	number of angles at vertex	total measures of angles at vertex
triangle	60	3	180
triangle	60	4	240
triangle	60	5	300
square	90	3	270
pentagon	108	3	324

The five regular polyhedra often are called *Platonic solids*. Each are named according to the number of faces.

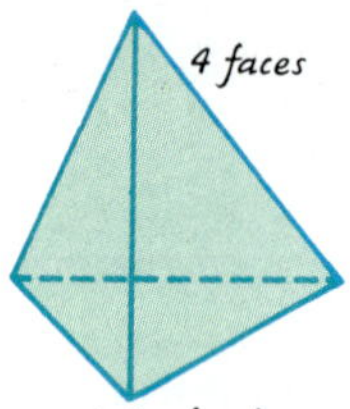

tetrahedron

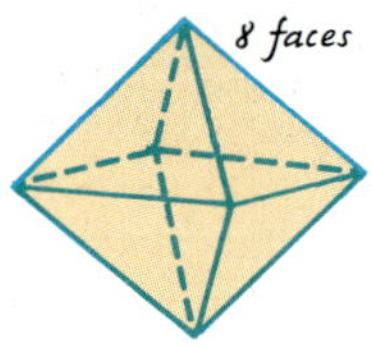

octahedron

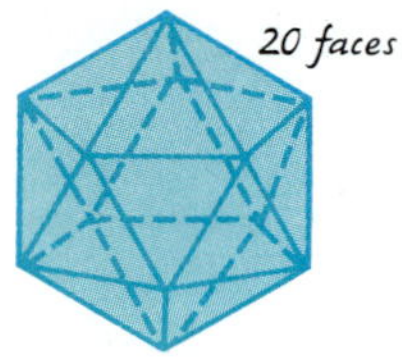

icosahedron

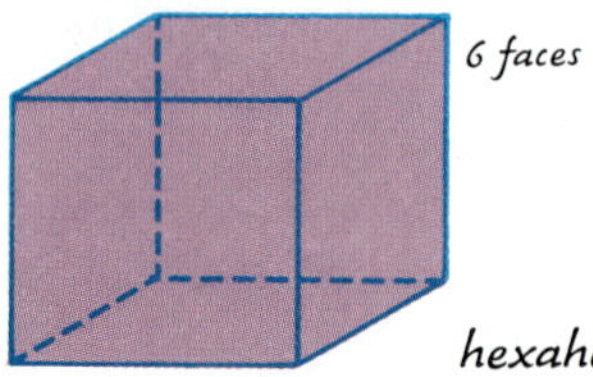

hexahedron

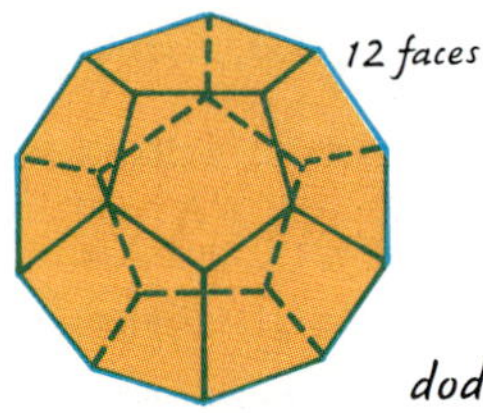

dodecahedron

exercises

Exploratory Name the faces, edges, and vertices for each of the following polyhedra.

1. 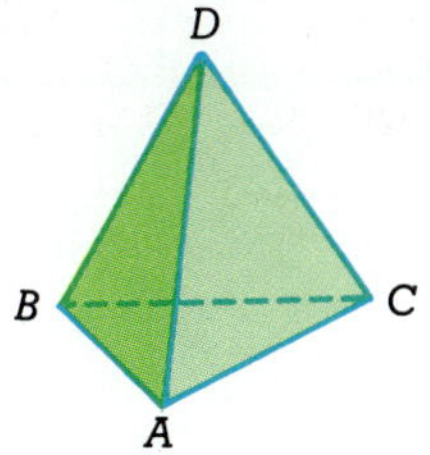

2. 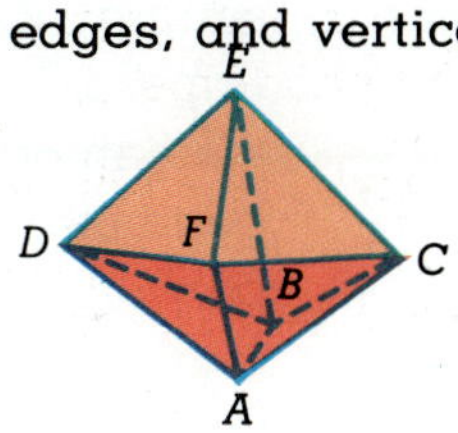

3. 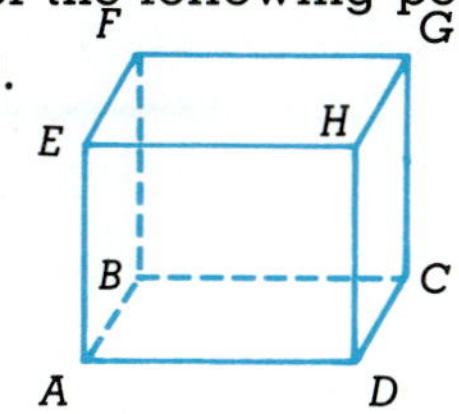

4. The sum of the measures of the angles formed at one vertex of a polyhedron must be less than what number?

5–7. Find the number of polygons intersecting at vertex D for each polyhedron in problems 1–3.

8–10. Find the sum of the angle measures at vertex D for each regular polyhedron in problems 1–3.

Written For each of the following, name the type of polygons which form the faces.

1. tetrahedron
2. hexahedron
3. octahedron
4. dodecahedron

5–8. Find the degree measure of one angle for a face of each polyhedron in problems 1–4.

For each of the following, given are a possible number of polygons intersecting at one vertex of a polyhedron. Then, find the sum of the measures of the angles formed at that vertex.

9. 4 equilateral triangles
10. 6 equilateral triangles
11. 3 squares
12. 4 squares
13. 5 squares
14. 3 regular pentagons
15. 3 regular hexagons
16. 2 regular octagons

17–24. Given the polygons in problems 9–16, can a polyhedron be formed? Write *yes* or *no*.

25–27. Complete the chart for tetrahedron.

Polyhedron	Number of Faces	Number of Vertices	Number of Edges
Tetrahedron			
Hexahedron			
Octahedron			

28–30. Complete the chart for hexahedron.

31–33. Complete the chart for octahedron.

34–36. The number of faces, F, the number of vertices, V, and the number of edges, E, of a regular polyhedron are related by the formula of Swiss mathematician Leonhard Euler, $F + V = E + 2$. Verify this formula by substituting the values from the table for tetrahedron, hexahedron, and octahedron in the formula.

Vocabulary

polygon (194)
n-gon (195)
convex polygon (195)
nonconvex polygon (195)
regular polygon (195)
diagonal (198)
parallelogram (202)
rectangle (210)
rhombus (210)
square (211)
trapezoid (214)
median of trapezoid (215)
perimeter (218)
polyhedron (221)
regular polyhedron (222)
Platonic solids (222)

Chapter Summary

1. Definition of Polygon: A figure is a polygon if and only if it meets all of the following conditions.
   1. It is formed by three or more coplanar segments called sides.
   2. Sides which have a common endpoint are noncollinear.
   3. Each side intersects exactly two of the other sides, but only at their endpoints. (194)
2. Theorem 7–1: If a convex polygon has n sides and S is the sum of the degree measures of its angles, then $S = (n - 2)180$. (198)
3. Problem Solving Procedure: 1. Define a variable. 2. Write an equation. 3. Solve the equation. 4. Check the solution. 5. Answer the problem. (199)
4. Theorem 7–2: The sum of the degree measures of the exterior angles of a convex polygon, one at each vertex, is 360. (200)
5. Definition of Parallelogram: A quadrilateral is a parallelogram if and only if each pair of opposite sides are parallel. (202)
6. Theorem 7–3: A diagonal of a parallelogram separates the parallelogram into two congruent triangles. (202)

7. Theorem 7–4: Opposite angles of a parallelogram are congruent. (203)
8. Theorem 7–5: Opposite sides of a parallelogram are congruent. (203)
9. Theorem 7–6: The diagonals of a parallelogram bisect each other. (203)
10. Theorem 7–7: If both pairs of opposite sides of a quadrilateral are congruent, then the quadrilateral is a parallelogram. (206)
11. Theorem 7–8: If two sides of a quadrilateral are parallel and congruent, then the quadrilateral is a parallelogram. (206)
12. Theorem 7–9: If the diagonals of a quadrilateral bisect each other, then the quadrilateral is a parallelogram. (207)
13. Definition of Rectangle: A quadrilateral is a rectangle if and only if it is a parallelogram with four right angles. (210)
14. Theorem 7–10: The diagonals of a rectangle are congruent. (210)
15. Definition of Rhombus: A quadrilateral is a rhombus if and only if it is a parallelogram with all four sides congruent. (210)
16. Theorem 7–11: The diagonals of a rhombus bisect the angles of the rhombus. (211)
17. Theorem 7–12: The diagonals of a rhombus are perpendicular. (211)
18. Definition of Square: A quadrilateral is a square if and only if it is a rectangle and all four sides are congruent. (211)
19. Definition of Trapezoid: A quadrilateral is a trapezoid if and only if it has exactly one pair of parallel sides. (214)
20. Theorem 7–13: The diagonals of an isosceles trapezoid are congruent. (214)
21. Theorem 7–14: Each pair of base angles of an isosceles trapezoid are congruent. (214)
22. Theorem 7–15: The median of a trapezoid is parallel to the bases, and its measure is one-half the sum of the measures of the bases. (215)
23. Definition of Perimeter: The sum of the measures of the sides of a polygon is the measure of the perimeter of the polygon. (218)
24. Theorem 7–16: If a rectangle has perimeter P units, length l units, and width w units, then $P = 2(l + w)$. (218)
25. Theorem 7–17: If a regular n-gon has perimeter P units, and a side measures s units, then $P = ns$. (219)
26. Polyhedrons are geometric solids with flat surfaces that form polygons. (221)
27. Regular polyhedra have faces that are regular polygons. Each edge has the same measure. There are only five regular polyhedra. (222)

Chapter Review

7–1 Use the polygon at the right to solve each problem.

1. Name the vertices.
2. Name the sides.
3. Name the angles.
4. Classify the polygon by the number of sides.
5. Classify the polygon as *convex* or *non-convex*.
6. Classify the polygon as *regular* or *not regular*.

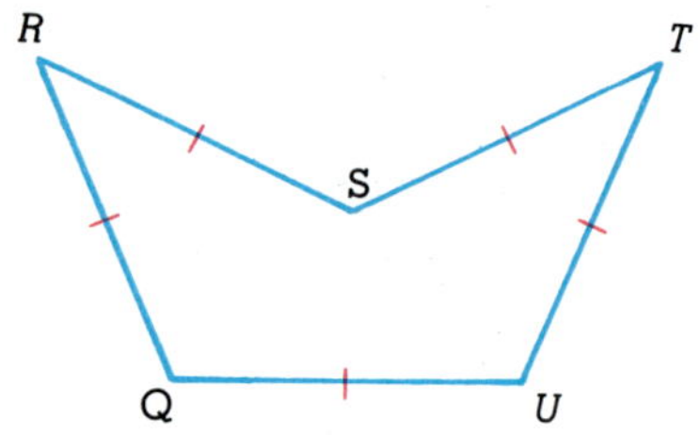

7–2 For each of the following convex polygons, find the sum of the degree measures of the angles.

7. hexagon
8. pentagon
9. 12-gon

10–12. Suppose each polygon in problems **7–9** are regular polygons. Find the measure of one angle.

13. Find the measure of one exterior angle of a regular nonagon.

7–3 Use the parallelogram at the right to solve each of the following.

14. If $BE = 45$, find BD.
15. If $CD = 7.2$, find BA.
16. If $m\angle 1 = 2a + 5$ and $m\angle 2 = 3a - 7$, find a.
17. If $m\angle BCD = 6x + 3$ and $m\angle DAB = 3x + 12$, find x.
18. If $m\angle ABC = 4y + 9$ and $m\angle BCD = 3y + 24$, find y.
19. If $AC = 16.4$, find EC.

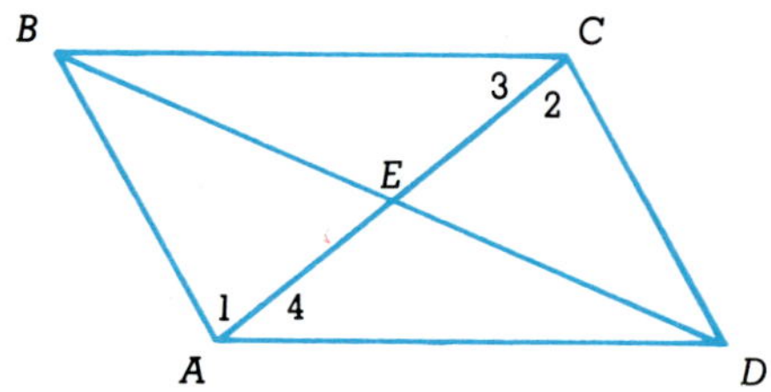

7–4 Prove each of the following.

20. **Given:** $\angle 1$ and $\angle 2$ are supplementary
$\angle 2 \cong \angle 3$
Prove: quadrilateral *PQRS* is a parallelogram

21. **Given:** $\square ABCD$
$\overline{AE} \cong \overline{CF}$
Prove: quadrilateral *EBFD* is a parallelogram

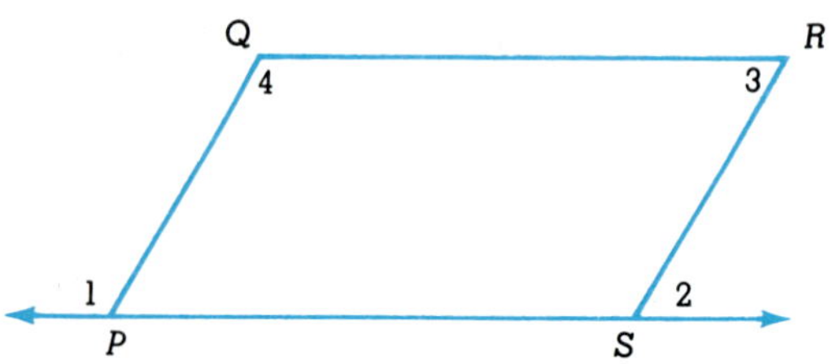

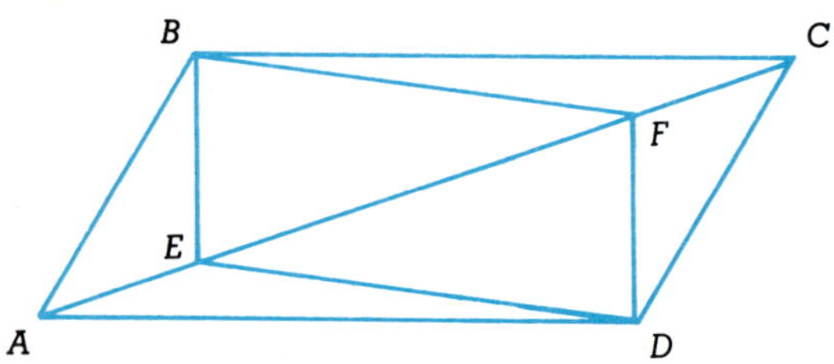

7–5 Tell whether or *not* each of the following statements must be true. Write *yes* or *no*.

22. The diagonals of a rhombus bisect each other.

23. The diagonals of a rectangle bisect the opposite angles.

24. All sides of parallelograms are congruent.

25. Every square is a rhombus.

Use the rhombus at the right to solve each of the following.

26. Suppose $AE = 6$. Find AC.

27. Suppose $m\angle ABE = 23\frac{3}{4}$. Find $m\angle ABC$.

28. Find $m\angle CED$.

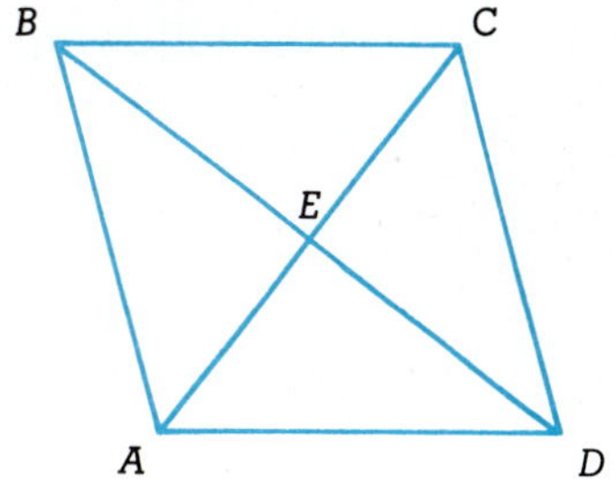

7–6 Prove each of the following.

29. Given: $\overline{BC} \parallel \overline{AD}$, $\overline{AB} \nparallel \overline{CD}$
$\overline{AB} \cong \overline{DC}$

Prove: $\triangle AED$ is isosceles

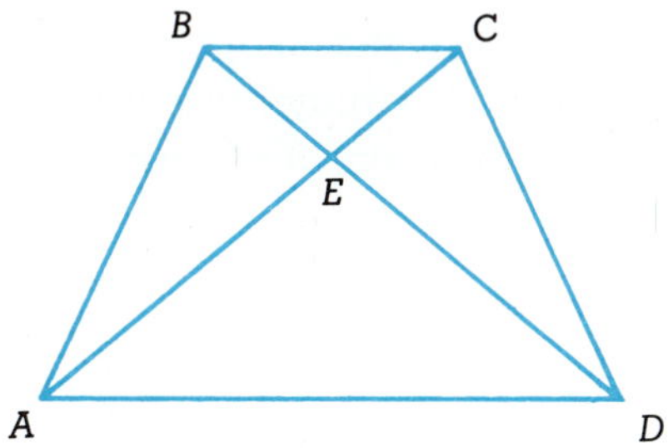

30. Given: $\overline{RS} \parallel \overline{PV}$
$\overline{QT}$ is a median of trapezoid $PRSV$

Prove: W is the midpoint of $\overline{RV}$

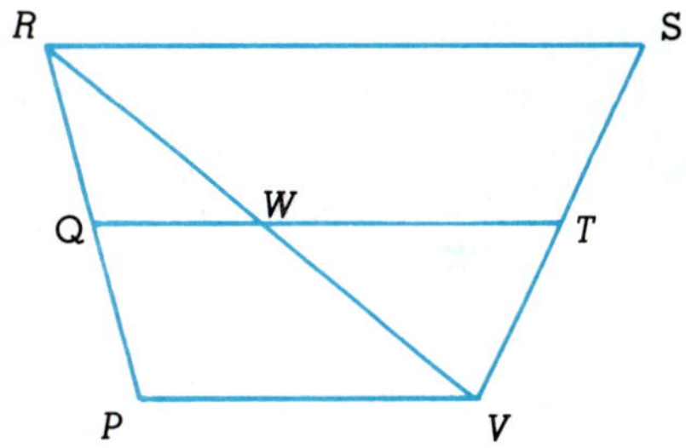

7–7 Find the perimeter for each of the following.

31. rectangle with length 4.2 m and width 3.8 m

32. regular octagon with side measuring $4\frac{1}{4}$ ft

33. How many centimeters of lead are needed to put around 25 hexagons of stained glass each with sides measuring 5.7 centimeters?

7–8 34. Find the sum of the measures of the angles formed at each vertex of an icosahedron.

Chapter Test

Solve each of the following.

1. Find the number of diagonals that can be drawn for a decagon.
2. Find the sum of the degree measures of the angles of a 15-gon.
3. Find the degree measure of one exterior angle of a regular pentagon.
4. Find the degree measure of one angle of a regular octagon.

Use $\square QRST$ for each of the following.

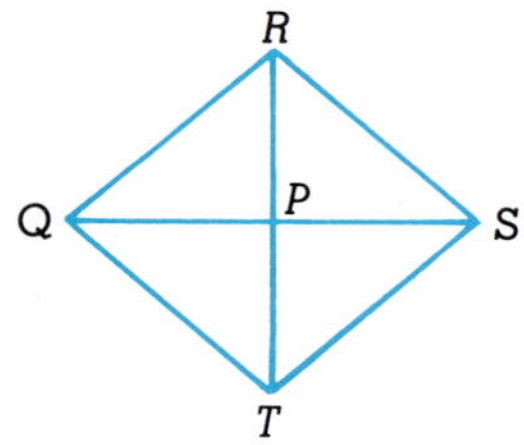

5. If $m\angle RST = 45$, find $m\angle QRS$.
6. If $m\angle QRS = 6x + 4$, find $m\angle QTS$.
7. If $RT = 18$, find PR.
8. If $QP = 12$, find SQ.
9. If $m\angle QTS = 4x - 2$ and $m\angle TSR = 5x - 7$, find x.
10. If $RP = 8x - 3$ and $PT = 6x + 5$, find x.
11. **Given:** $\overline{RT} \cong \overline{QS}$
    Prove: $\square QRST$ is a rectangle
12. **Given:** $\overline{RT} \perp \overline{QS}$
    Prove: $\square QRST$ is a rhombus

Prove each of the following.

13. **Given:** $\overline{BC} \parallel \overline{AD}$, $\overline{AB} \nparallel \overline{CD}$
    $\overline{AB} \cong \overline{CD}$
    Prove: $\triangle ABC \cong \triangle DCB$

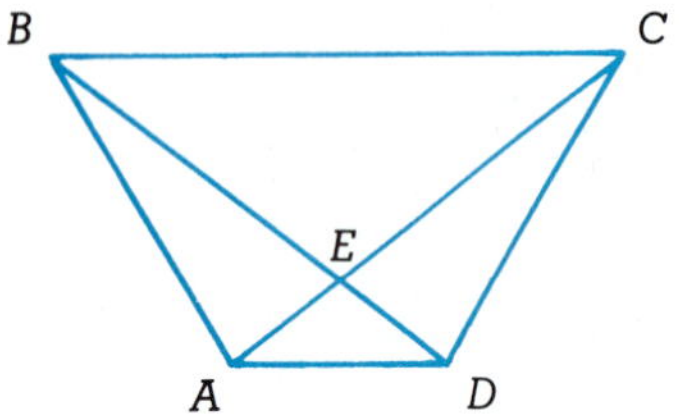

14. **Given:** $\overline{PQ} \parallel \overline{RS}$, $\overline{PQ} \cong \overline{RS}$
    T is the midpoint of $\overline{QP}$
    V is the midpoint of $\overline{RS}$
    Prove: quadrilateral $TQRV$ is a parallelogram

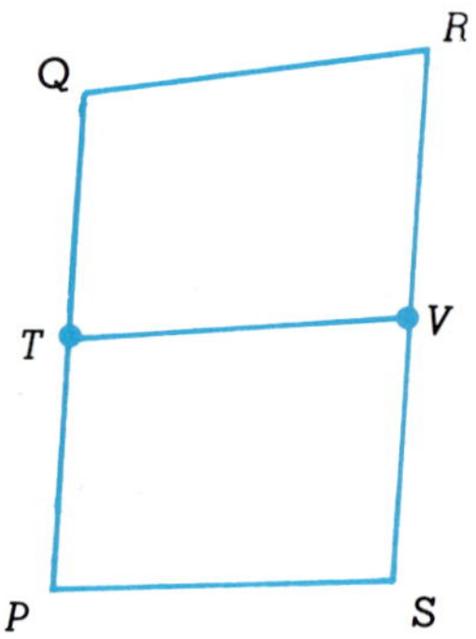

Solve each problem.

15. Suppose the bases of a trapezoid measure 6.8 mm and 15.7 mm. Find the measurement of the median.
16. Suppose a rectangle measures $2\frac{1}{2}$ inches by $1\frac{3}{4}$ inches. Find its perimeter.
17. Find the perimeter of a regular decagon if one side measures 84.65 centimeters.
18. Suppose four squares intersect at a common vertex. Can a polyhedron be formed that includes these squares? Write *yes* or *no*.

Chapter 8

Similarity

Plastic parts for models of the latest-style planes, cars, and ships are built to scale. Model makers plan models using the concepts of similarity.

8–1 Ratios and Proportions

Team records often are compared using **ratios.** For example, the newspaper records show that the National Basketball Association team from Kansas City won 34 of the 54 games it played. The ratio of games won to games played is $\frac{34}{54}$ or about 0.630. The ratio of games won to games played for the Denver team is $\frac{30}{54}$ or about 0.56. Kansas City won a greater percentage of its games than Denver.

WESTERN CONFERENCE Midwest Division	W	L
Milwaukee	39	15
Kansas City	34	20
Denver	30	24
Chicago	18	36
Utah	13	41

example 1

The pitch of a roof is the ratio of the rise to the run. If a roof has a span measuring 6.75 m and a rise measuring 2.7 m, what is the pitch?

$$pitch = \frac{rise}{run} = \frac{2.7}{3.375} = 0.8 \qquad run = \frac{span}{2}$$

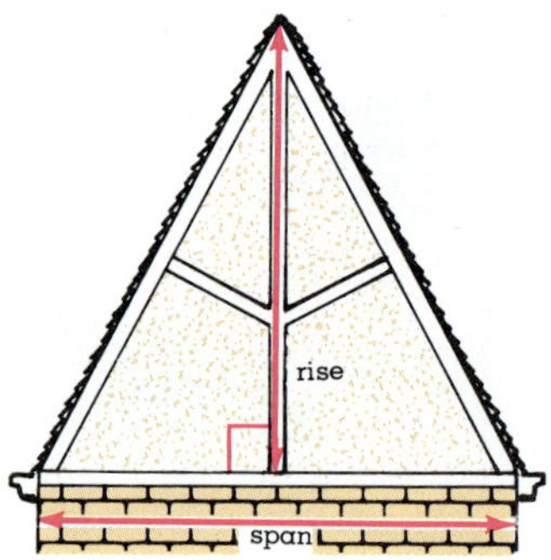

The equation that shows two ratios equivalent is a proportion.

A **proportion** shows that two ratios are equivalent. For example, suppose Boston and New York are tied for first place in the Atlantic Division. The two teams have the same percentage. The ratios of games won to games played are the same.

$$\frac{24}{48} = \frac{27}{54}$$

Every proportion has two **cross products.** In the proportion $\frac{24}{48} = \frac{27}{54}$, the cross products are 24 times 54 and 48 times 27. The cross products of a proportion are equal.

$$\frac{24}{48} \times \frac{27}{54}$$
$$24 \times 54 = 48 \times 27$$
$$1296 = 1296$$

For any numbers a and c, and any nonzero numbers b and d, $\frac{a}{b} = \frac{c}{d}$ if and only if $ad = bc$.

Theorem 8–1
Equality of
Cross Products

examples

2 **A racing car uses alcohol for fuel. The alcohol weighs 4.8 pounds per gallon and is burned with a fuel to air ratio of 1 to 15. Use a proportion to find how many pounds of air are used in burning 1 gallon of alcohol.**

Let x stand for the amount of air used in pounds.

$$\frac{4.8}{x} = \frac{1}{15}$$
$$x = 4.8 \times 15$$
$$= 72$$

The car will use 72 pounds of air.

3 **Prove that for any numbers a and c, and any nonzero numbers b and d, if $\frac{a}{b} = \frac{c}{d}$, then $ad = bc$.**

Given: $\frac{a}{b} = \frac{c}{d}$, b and d are nonzero.

Prove: $ad = bc$

Proof:

STATEMENTS	REASONS
1. $\frac{a}{b} = \frac{c}{d}$, b and d are nonzero	1. Given
2. $bd\frac{a}{b} = bd\frac{c}{d}$	2. Postulate 2–8: Multiplication Property of Equality
3. $da = bc$	3. Postulate 2–9: Substitution
4. $ad = bc$	4. Postulate 2–10: Commutative Property for Multiplication

exercises

Exploratory Use the standings listed at the right to find the ratio of games won to games played for each team.

1. Kansas City
2. Denver
3. Milwaukee
4. Indiana
5. Chicago
6. Utah
7. New York
8. Atlanta
9. Boston
10. Detroit
11. Houston
12. Washington

EASTERN CONFERENCE		
Atlantic Division	W	L
Boston	20	6
Philadelphia	20	7
New York	14	13
Washington	10	13
New Jersey	11	16
Central Division		
Atlanta	17	12
Houston	14	12
San Antonio	14	13
Cleveland	13	17
Indiana	12	16
Detroit	8	19
WESTERN CONFERENCE		
Midwest Division		
Milwaukee	19	10
Kansas City	15	14
Denver	10	18
Chicago	8	20
Utah	5	21

Gears on bicycles are called sprockets. To find bicycle gear ratios, you must first find the ratio of the number of rear sprocket teeth to the number of front sprocket teeth. Find the ratio for each of the following to two decimal places.

13. 12 rear sprocket teeth
24 front sprocket teeth

14. 12 rear sprocket teeth
38 front sprocket teeth

15. 13 rear sprocket teeth
26 front sprocket teeth

16. 13 rear sprocket teeth
52 front sprocket teeth

17. 14 rear sprocket teeth
42 front sprocket teeth

18. 14 rear sprocket teeth
28 front sprocket teeth

19. 15 rear sprocket teeth
45 front sprocket teeth

20. 15 rear sprocket teeth
55 front sprocket teeth

21. 20 rear sprocket teeth
50 front sprocket teeth

22. 20 rear sprocket teeth
30 front sprocket teeth

23. 24 rear sprocket teeth
44 front sprocket teeth

24. 24 rear sprocket teeth
54 front sprocket teeth

25. 26 rear sprocket teeth
46 front sprocket teeth

26. 28 rear sprocket teeth
56 front sprocket teeth

Written Write the cross products for each of the following proportions.

1. $\frac{11}{12} = \frac{x}{24}$
2. $\frac{5}{8} = \frac{20}{x}$
3. $\frac{t}{18} = \frac{5}{6}$
4. $\frac{14}{b} = \frac{7}{8}$
5. $\frac{1.2}{1.6} = \frac{k}{4}$
6. $\frac{7}{12} = \frac{9.8}{m}$
7. $\frac{b}{3.24} = \frac{1}{8}$
8. $\frac{7.29}{a} = \frac{27}{9}$
9. $\frac{x}{30 - x} = \frac{2}{3}$
10. $\frac{5 + x}{8 + x} = \frac{3}{4}$
11. $\frac{3 - n}{n + 1} = \frac{2}{1}$
12. $\frac{n + 3}{10} = \frac{3n - 2}{8}$

13–24. Solve each proportion in problems 1–12.

Proportions can be used to change a fraction to a percent. For example, to change $\frac{5}{6}$ to a percent, you solve the proportion $\frac{5}{6} = \frac{n}{100}$. For each of the following, write the proportion you would use to change the fraction to a percent.

25. $\frac{5}{8}$
26. $\frac{1}{12}$
27. $\frac{1}{3}$
28. $\frac{3}{16}$
29. $\frac{3}{8}$
30. $\frac{7}{8}$
31. $\frac{5}{12}$
32. $\frac{3}{4}$
33. $\frac{4}{5}$
34. $\frac{3}{5}$
35. $\frac{1}{10}$
36. $\frac{6}{10}$

37–48. Change each fraction in problems **25–36** to a percent with one decimal place.

Use proportions to solve each of the following.

49. Ann Towns plays basketball for a university team. Last season she attempted 156 field goals and made 117 field goals. What percent of the field goals attempted did she make?

50. The mixture for a finish coat of concrete is 1 part cement to 2 parts sand. How much cement should be mixed with 15 pounds of sand?

51. The sales tax on \$140 is \$8.40. What is the rate of sales tax? (Hint: Find a percent.)

52. A designated hitter made 8 hits in 9 games. If he continues hitting at that rate, how many hits will he make in 108 games?

53. A recipe for preparing material to dye calls for 4 parts alum to 1 part washing soda. How much washing soda should be used for 150 g of alum?

54. An old Shaker cake recipe calls for 2 parts butter to 3 parts sugar. How much butter should be used with $4\frac{1}{2}$ cups of sugar?

55. Prove that for any numbers a and b, and any nonzero numbers b and d,

if $ad = bc$, then $\frac{a}{b} = \frac{c}{d}$.

Challenge Solve each proportion.

1. $\frac{x}{3} = \frac{2}{5 - x}$
2. $\frac{x}{-3} = \frac{2}{x - 5}$
3. $\frac{x}{4} = \frac{^-4}{x - 10}$
4. $\frac{x}{6} = \frac{1}{x + 5}$
5. $\frac{x}{2} = \frac{2}{6x - 5}$
6. $\frac{2x}{1} = \frac{^-8}{x - 4}$
7. $\frac{^-1}{x - 1} = \frac{2}{6x - 5}$
8. $\frac{x - 4}{x - 2} = \frac{x - 2}{x + 2}$
9. $\frac{x - 3}{2x} = \frac{x - 2}{2x + 1}$
10. $\frac{x + 3}{x - 4} = \frac{x - 5}{x + 2}$

Using Money

Deductions

The monies withheld from a paycheck are called **deductions.** Deductions from paychecks include federal, state, local, and FICA taxes. Other deductions are made for medical and accident insurance, credit unions, and so on.

WELLS WALLS

This statement covers your gross earnings less authorized deductions and those required by law. It is your receipt for deductions and should be kept as your record of payments.

NUMBER	FED. INCOME TAX	DISABILITY
15477	60.10	
PERIOD ENDING	**FICA**	**INSURANCE**
Nov. 25	23.27	6.00
REGULAR HOURS	**STATE TAX**	**CREDIT UNION**
37.5	2.89	
OVERTIME HOURS	**LOCAL TAX**	**OTHER**
	5.77	

TOTAL GROSS	TOTAL TAX	TOTAL PERSONAL	NET PAY
384.68	92.03	6.00	286.65

TOTAL GROSS −	TOTAL TAX		−	TOTAL PERSONAL		= NET PAY
	$60.10	*Federal*		*$0.00*	*Disability*	
	23.27	*FICA*		*6.00*	*Insurance*	
	2.89	*State*		*0.00*	*Credit Union*	
	+ 5.77	*Local*		*+ 0.00*	*Other*	
$384.68 −	*$92.03*		−	*$6.00*		= *$286.65*

Exercises Find the total tax, the total personal, and the net pay for each statement.

1.

NUMBER	FED. INCOME TAX	DISABILITY
12470	12.40	
PERIOD ENDING	**FICA**	**INSURANCE**
Mar. 14	10.51	2.19
REGULAR HOURS	**STATE TAX**	**CREDIT UNION**
40.0		
OVERTIME HOURS	**LOCAL TAX**	**OTHER**
2.5		

TOTAL GROSS	TOTAL TAX	TOTAL PERSONAL	NET PAY
179.60			

2.

NUMBER	FED. INCOME TAX	DISABILITY
66272	72.73	
PERIOD ENDING	**FICA**	**INSURANCE**
Oct. 15	27.92	
REGULAR HOURS	**STATE TAX**	**CREDIT UNION**
75.0	4.42	5.00
OVERTIME HOURS	**LOCAL TAX**	**OTHER**
	6.92	

TOTAL GROSS	TOTAL TAX	TOTAL PERSONAL	NET PAY
461.54			

3.

NUMBER	FED. INCOME TAX	DISABILITY
17373	1.50	
PERIOD ENDING	**FICA**	**INSURANCE**
Jan. 23	2.82	
REGULAR HOURS	**STATE TAX**	**CREDIT UNION**
12.0		
OVERTIME HOURS	**LOCAL TAX**	**OTHER**
	0.07	

TOTAL GROSS	TOTAL TAX	TOTAL PERSONAL	NET PAY
46.73			

4.

NUMBER	FED. INCOME TAX	DISABILITY
4119	137.12	
PERIOD ENDING	**FICA**	**INSURANCE**
May 2	41.19	17.20
REGULAR HOURS	**STATE TAX**	**CREDIT UNION**
160.0	9.62	25.00
OVERTIME HOURS	**LOCAL TAX**	**OTHER**

TOTAL GROSS	TOTAL TAX	TOTAL PERSONAL	NET PAY
692.31			

8–2 Properties of Proportions

Proportions have many interesting properties. They can be derived using cross products and the properties of equality. The following chart summarizes these properties. *Notice that the terms are defined only when denominators have values that are not zero.*

For any numbers a, b, c, and d, the following properties hold whenever all terms are defined.	
If $\frac{a}{b} = \frac{c}{d}$, then $\frac{a}{c} = \frac{b}{d}$.	Theorem 8–2
If $\frac{a}{b} = \frac{c}{d}$, then $\frac{b}{a} = \frac{d}{c}$.	Theorem 8–3
$\frac{a}{b} = \frac{c}{d}$ if and only if $\frac{a+b}{b} = \frac{c+d}{d}$.	Theorem 8–4 Addition Property of Proportions
$\frac{a}{b} = \frac{c}{d}$ if and only if $\frac{a-b}{b} = \frac{c-d}{d}$.	Theorem 8–5 Subtraction Property of Proportions
$\frac{a}{b} = \frac{c}{d}$ if and only if $\frac{a}{b} = \frac{a+c}{b+d}$ or $\frac{c}{d} = \frac{a+c}{b+d}$.	Theorem 8–6 Summation Property of Proportions

example 1

Prove that for any numbers a, b, c, and d, if $\frac{a}{b} = \frac{c}{d}$, then $\frac{a+b}{b} = \frac{c+d}{d}$, whenever all terms are defined.

Given: $\frac{a}{b} = \frac{c}{d}$

Prove: $\frac{a+b}{b} = \frac{c+d}{d}$

Proof:

STATEMENTS	REASONS
1. $\frac{a}{b} = \frac{c}{d}$	1. Given
2. $ad = bc$	2. Theorem 8–1: Equality of Cross Products
3. $ad + bd = bc + bd$	3. Postulate 2–7: Addition Property of Equality
4. $(a + b)d = b(c + d)$	4. Postulate 2–14: Distributive Property
5. $\frac{a+b}{b} = \frac{c+d}{d}$	5. Theorem 8–1: Equality of Cross Products

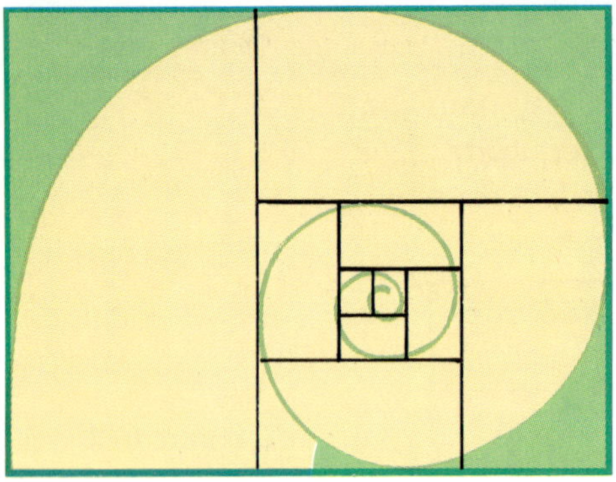

The figures above show how special rectangles, called *golden rectangles*, can be used to approximate the spiral of a chambered nautilus shell. Golden rectangles have interesting mathematical properties. For example, the measures of their sides are always in the ratio of about 1 to 1.618. *This ratio is called the golden ratio.*

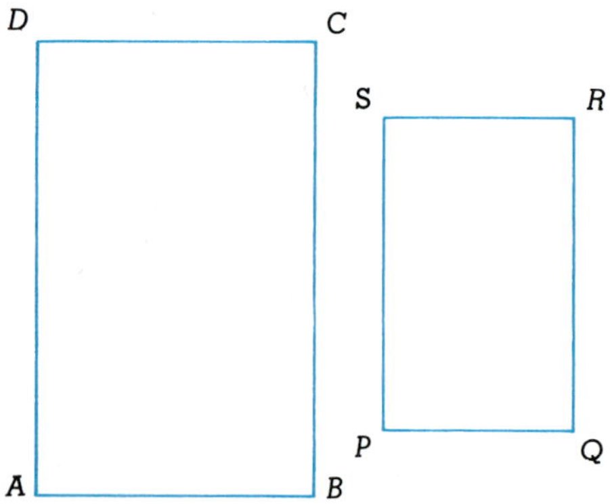

The two rectangles at the left are golden rectangles.

$$\frac{AB}{BC} = \frac{AB}{DA} = \frac{CD}{BC} = \frac{CD}{DA} \approx \frac{1}{1.618}$$

$$\frac{PQ}{QR} = \frac{PQ}{SP} = \frac{RS}{QR} = \frac{RS}{SP} \approx \frac{1}{1.618}$$

Using properties of proportions, you can show that there is a correspondence between the rectangles so that the measures of their corresponding sides are *proportional*.

$$\frac{AB}{PQ} = \frac{BC}{QR} = \frac{CD}{RS} = \frac{DA}{SP}$$

example 2 **Given that $\frac{AB}{BC} \approx \frac{1}{1.618}$ and $\frac{PQ}{QR} \approx \frac{1}{1.618}$, show $\frac{AB}{PQ} = \frac{BC}{QR}$.**

$\frac{AB}{BC} \approx \frac{1}{1.618}$	$\frac{PQ}{QR} \approx \frac{1}{1.618}$	*Given*
	$\frac{AB}{BC} = \frac{PQ}{QR}$	*Substitution*
	$\frac{AB}{PQ} = \frac{BC}{QR}$	*Theorem 8–2*

exercises

Exploratory Name the property shown in each of the following.

1. If $\frac{3}{1} = \frac{6}{2}$, then $\frac{3-1}{1} = \frac{6-2}{2}$.
2. If $\frac{3}{1} = \frac{6}{2}$, then $\frac{2}{1} = \frac{4}{2}$.
3. If $\frac{3}{1} = \frac{6}{2}$, then $\frac{4}{1} = \frac{8}{2}$.
4. If $\frac{3+1}{1} = \frac{6+2}{2}$, then $\frac{3}{1} = \frac{6}{2}$
5. If $\frac{6}{8} = \frac{7}{x}$, then $\frac{8}{6} = \frac{x}{7}$
6. If $\frac{2}{10} = \frac{1}{y}$, then $\frac{y}{10} = \frac{1}{2}$.
7. If $\frac{3}{4} = \frac{3+y}{8}$, then $\frac{3}{4} = \frac{y}{4}$.
8. If $\frac{2}{3} = \frac{x}{6}$, then $\frac{2}{3} = \frac{x+2}{9}$.
9. If $\frac{12}{b} = \frac{11}{a}$, then $\frac{a}{b} = \frac{11}{12}$.
10. If $\frac{x}{27} = \frac{y}{13}$, then $\frac{x}{y} = \frac{27}{13}$.
11. If $\frac{25}{14} = \frac{7}{AB}$, then $\frac{14}{25} = \frac{AB}{7}$.
12. If $\frac{25}{14} = \frac{7}{AB}$, then $\frac{AB}{14} = \frac{7}{25}$.
13. If $\frac{PQ}{QR} = \frac{PQ}{SP}$, then $\frac{QR}{SP} = \frac{PQ}{PQ}$.
14. If $\frac{CD}{BC} = \frac{RS}{QR}$, then $\frac{CD}{RS} = \frac{BC}{QR}$.
15. If $\frac{CD}{RS} = \frac{BC}{QR}$, then $\frac{RS}{CD} = \frac{QR}{BC}$.
16. If $\frac{DF+BD}{BD} = \frac{DG+FH}{FH}$, then $\frac{DF}{BD} = \frac{DG}{FH}$.

Written Prove each of the following.

1. Theorem 8–2.
2. Theorem 8–3.
3. For any numbers a, b, c, and d, if $\frac{a}{b} = \frac{c}{d}$, then $\frac{a-b}{b} = \frac{c-d}{d}$ whenever all terms are defined.
4. For any numbers a, b, c, and d, if $\frac{a-b}{b} = \frac{c-d}{d}$, then $\frac{a}{b} = \frac{c}{d}$ whenever all terms are defined.
5. For any numbers a, b, c, and d, if $\frac{a}{b} = \frac{a+c}{b+d}$, then $\frac{a}{b} = \frac{c}{d}$.
6. For any numbers a, b, c, and d, if $\frac{a}{b} = \frac{c}{d}$, then $\frac{c}{d} = \frac{a+c}{b+d}$.

Suppose the measures of corresponding sides of the polygons at the right are proportional. For each of the following, find PS.

7. $AB = 5$, $AD = 2$, $PQ = 3$
8. $BC = 7$, $DA = 14$, $RQ = 3$
9. $CD = 6.3$, $RS = 4.5$, $AD = 7.0$
10. $QP = 2.1$, $BA = 2.8$, $DA = 3.2$
11. $AD = 91.0$, $AB = 84.7$, $PQ = 12.1$
12. $CD = 43.6$, $DA = 21.8$, $SR = 33.0$

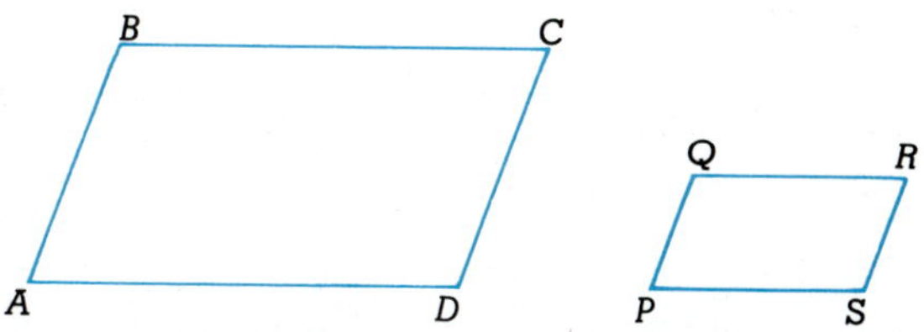

A corresponds to P
B corresponds to Q
C corresponds to R
D corresponds to S

8–3 Similar Polygons

The clothes in the photograph all have the same shape, but they differ in size. In geometry, figures that have the same shape, but may differ in size are called **similar figures.**

Corresponding parts of polygons are used to tell if they are similar. For example, the parallelograms *ABCD* and *PQRS* below are similar.

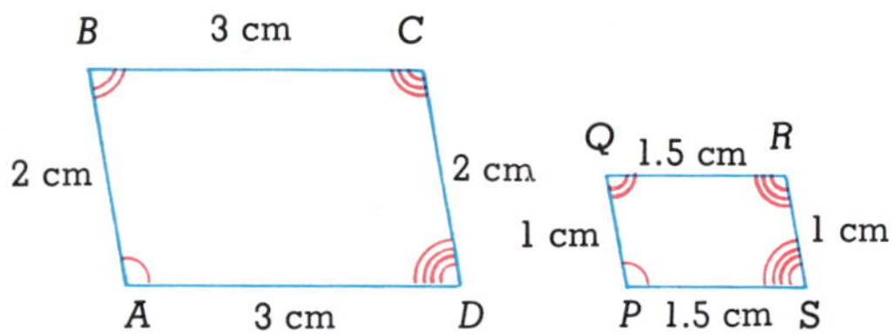

$\angle A \cong \angle P$
$\angle B \cong \angle Q$
$\angle C \cong \angle R$
$\angle D \cong \angle S$

$$\frac{AB}{PQ} = \frac{BC}{QR} = \frac{CD}{RS} = \frac{DA}{SP} = \frac{2}{1}$$

The symbol $\sim$ means *similar* or *is similar to*. We write, $\square ABCD \sim \square PQRS$, and we say *parallelogram ABCD is similar to parallelogram PQRS*. The order of letters indicates the vertices that correspond.

Definition of Similar Polygons

Two polygons are similar if and only if there is a correspondence such that their corresponding angles are congruent and the measures of their corresponding sides are proportional.

For two polygons to be similar, there are two conditions to be met. If *only one* of the conditions is met, then the polygons may *not* be similar.

The corresponding angles for these two figures are congruent. The measures of the corresponding sides are *not* proportional.

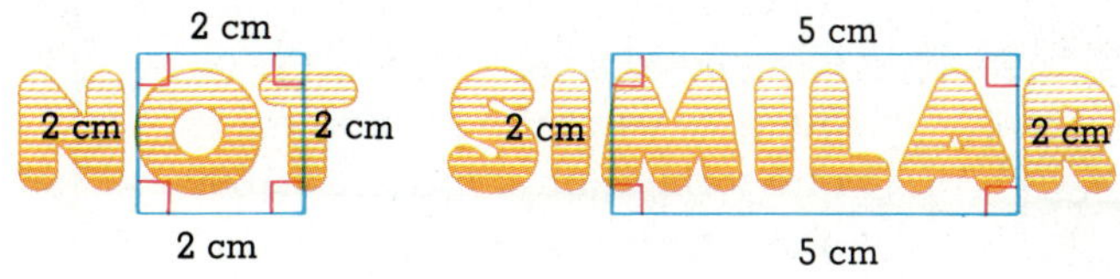

The measures of corresponding sides are proportional for these two figures. The corresponding angles are *not* congruent.

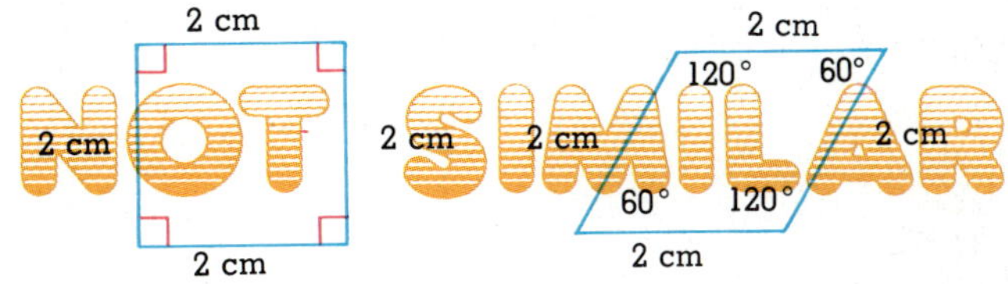

example 1

Tell whether or *not* all squares are similar.

All squares have four right angles, and all right angles are congruent. Thus, any correspondence between squares will have corresponding angles congruent.

All four sides of a square are congruent. Thus, the measures of each set of corresponding sides form equal ratios. That is, the measures of the corresponding sides are proportional.

Therefore, all squares are similar.

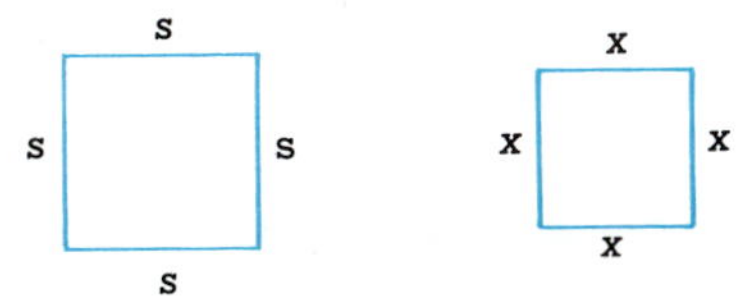

$$\frac{s}{x} = \frac{s}{x} = \frac{s}{x} = \frac{s}{x}$$

If the measures of some parts of two similar figures are known, then it may be possible to find the measures of other parts.

example 2

Suppose quadrilaterals *ABCD* and *PQRS* are similar. Find x.

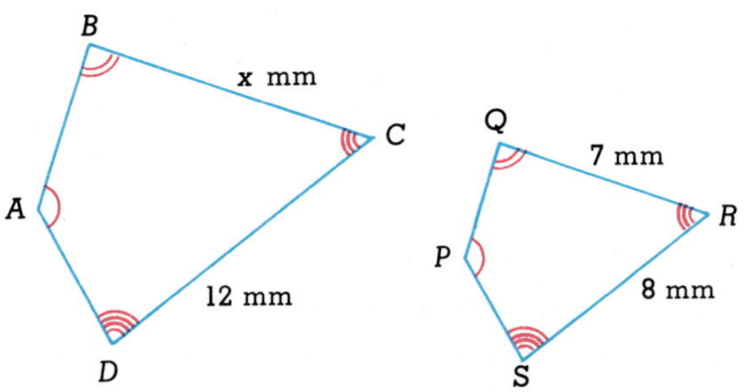

$$\frac{BC}{QR} = \frac{DC}{SR}$$

$$\frac{x}{7} = \frac{12}{8}$$

$$8x = 84$$

$$x = 10.5$$

example 3

Imagine that a triangle is formed by connecting Atlanta, Cleveland, and New York. Such a triangle is similar to one formed on a map. It is 600 miles by air from Cleveland to New York. How far is it from New York to Atlanta?

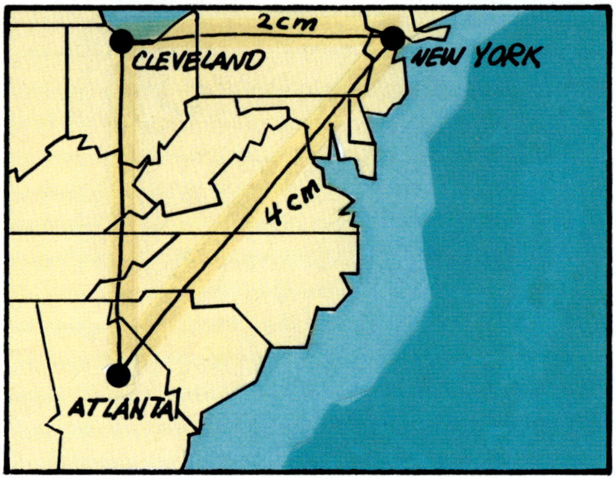

$$\frac{\text{map distance Cle.–N. Y.}}{\text{actual distance Cle.–N. Y.}} = \frac{\text{map distance N. Y.–Atl.}}{\text{actual distance N. Y.–Atl.}}$$

$$\frac{2}{600} = \frac{4}{x}$$

$$2x = 2400$$

$$x = 1200$$

exercises

Exploratory For each of the following list the information needed to show that the figures in each pair are similar.

1. $\triangle ABC \sim \triangle DEF$

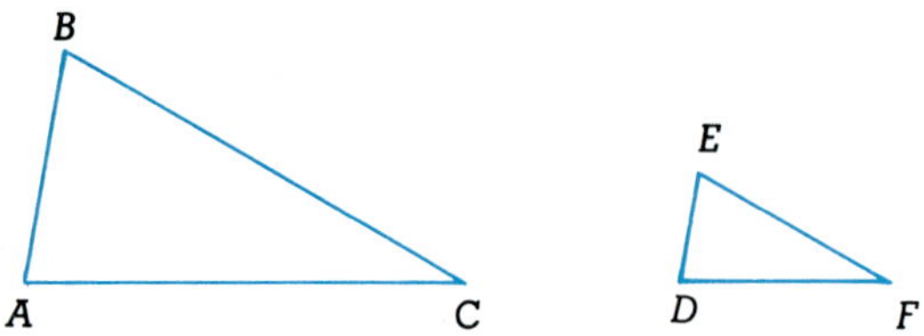

2. polygon $PQRS$ ~ polygon $ABCD$

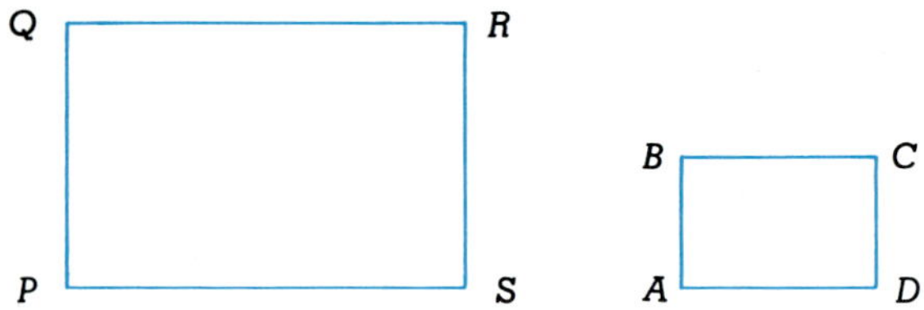

3. polygon $RSTVWX$ ~ polygon $LMNOPQ$

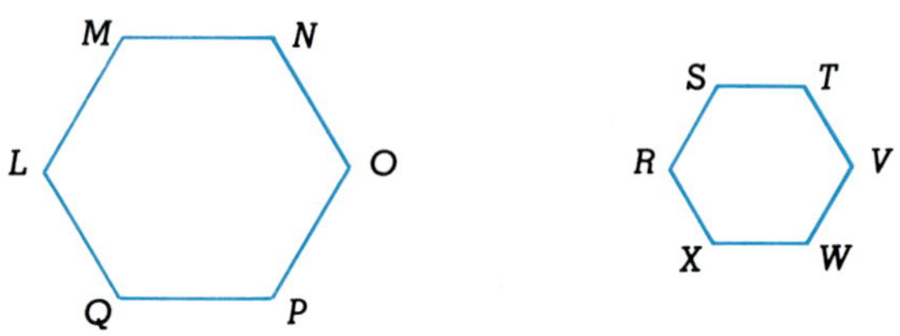

4. polygon $ABCD$ ~ polygon $NMLP$

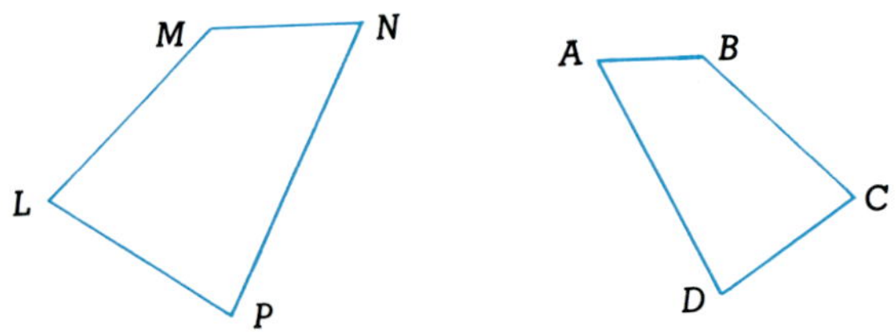

Tell whether or *not* the figures in each pair are similar.

5.

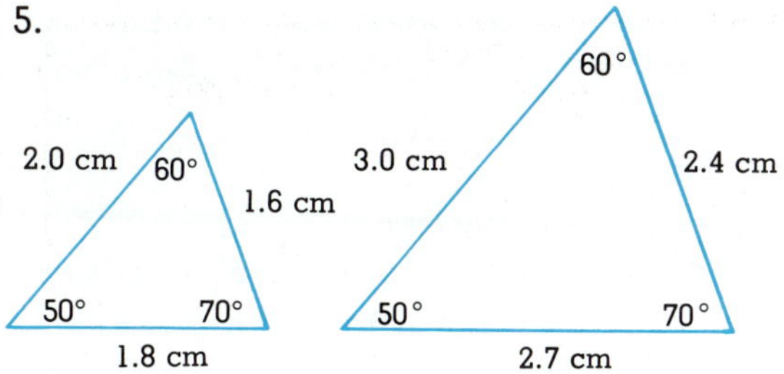

6.

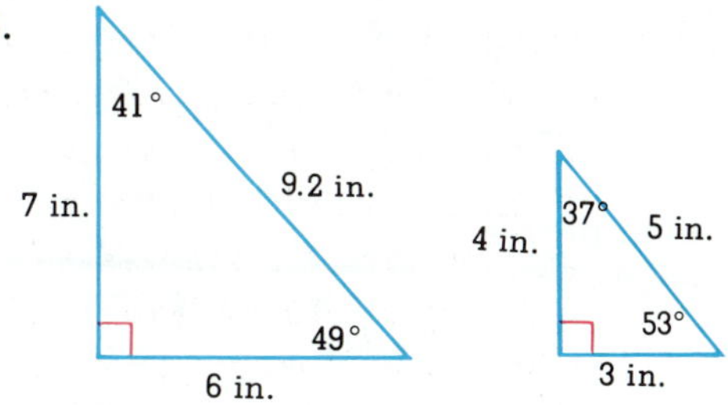

7.

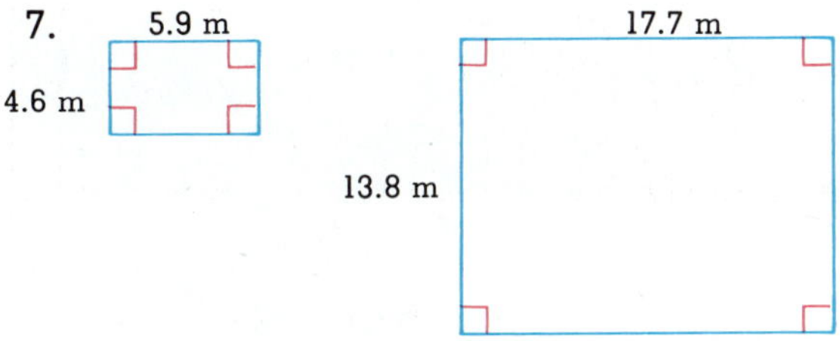

8.

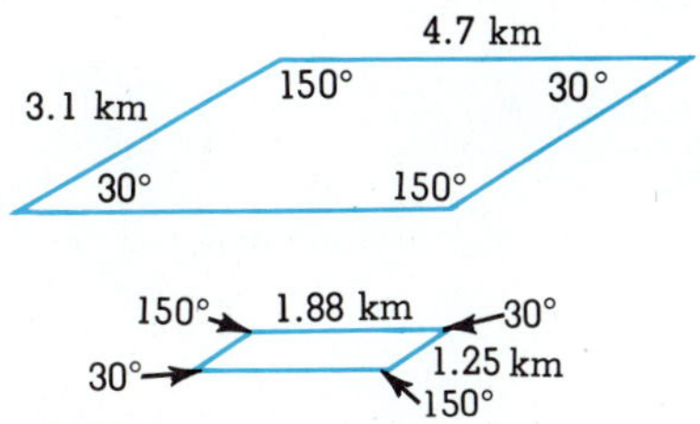

Written For each of the following, write *yes* or *no*. Then, explain your answer.

1. All equilateral triangles are similar.
2. All isosceles triangles are similar.
3. All rectangles are similar.
4. All regular hexagons are similar.
5. All rhombuses are similar.
6. Congruent triangles are similar.
7. All trapezoids are similar.
8. All parallelograms are similar.
9. Similar quadrilaterals are congruent.
10. Congruent quadrilaterals are similar.

For each of the following, find **x**.

11. $\triangle ABC \sim \triangle FED$

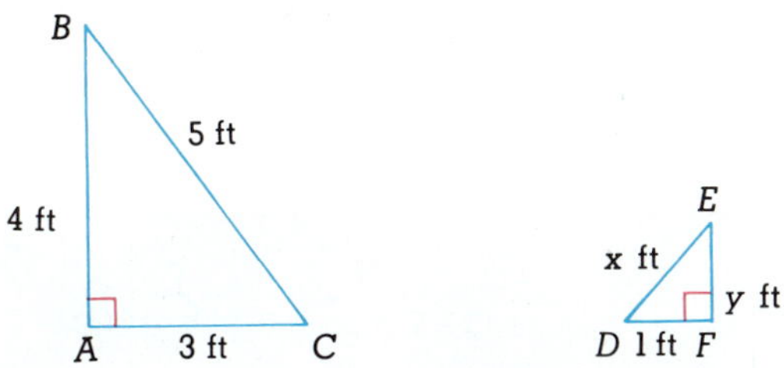

12. $\square ABCD \sim \square FGHE$

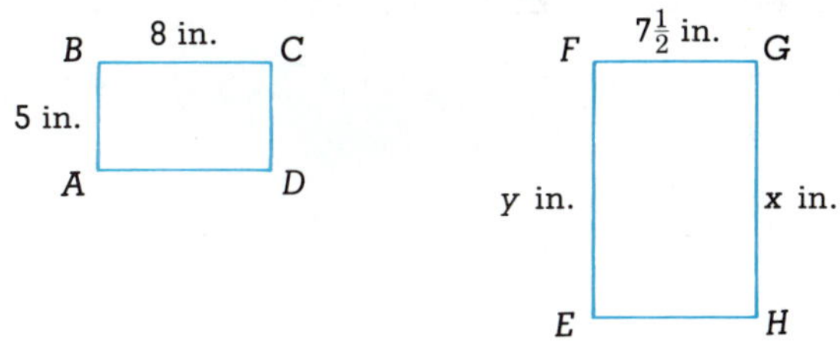

13. trapezoid $PQRS$ ~ trapezoid $VWTU$

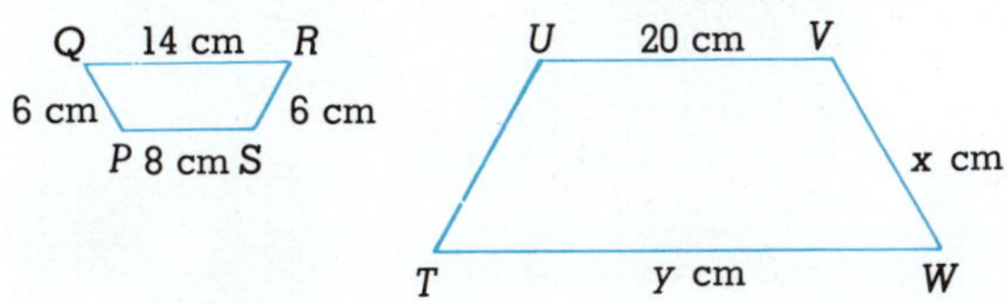

14. pentagon $ABCDE$ ~ pentagon $PQRST$

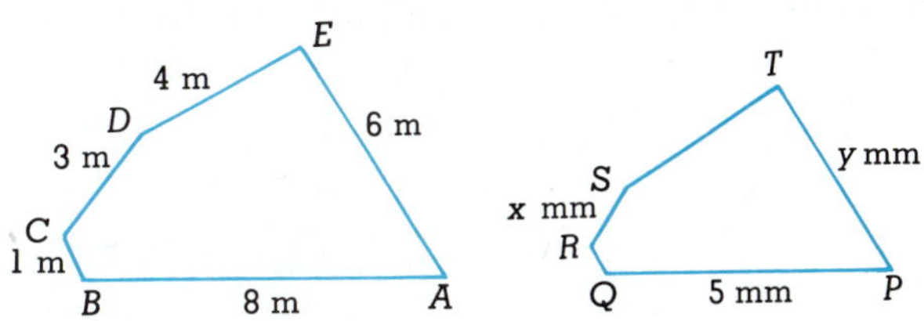

15–18. In problems 11–14, find *y*.

Prove each of the following.

19. Similarity of triangles is reflexive.

20. Similarity of triangles is symmetric.

21. Similarity of triangles is transitive.

22. If two triangles are congruent, then the triangles are similar.

Solve each problem.

23. Figures in a scale drawing are similar to the actual figures. Suppose the side of a pentagon that is 12 m long corresponds to a side on a scale drawing 3 cm long. Find the length of a side that measures 2 cm on the scale drawing.

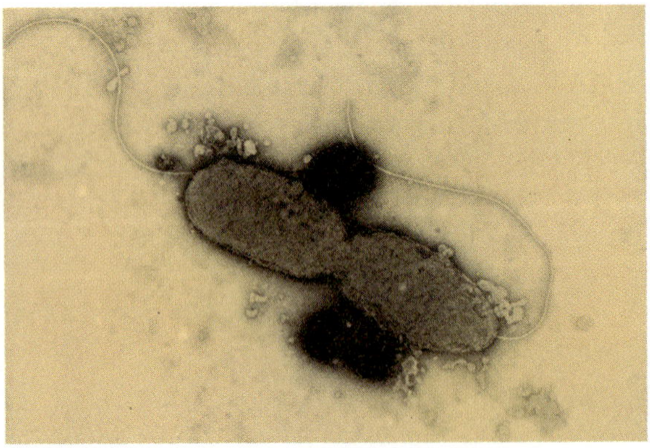

24. The figures seen in electron microscope photographs are similar to the actual figures. Suppose the length of a cell that is 2.4 cm in a photograph corresponds to an actual cell 3.1 microns long. Find the length of a cell that is 0.2 cm in the photograph.

25. Figures seen in photographs are similar to the actual figures. Suppose the width of a pyramid is actually 208 m and its photograph is 2.6 cm. If one side of the pyramid measures 3.0 cm, find its actual length.

8–4 Similar Triangles

To show that two triangles are similar, it is *not* necessary to meet all the conditions of the definition. The following postulate tells you that the shape of a triangle is completely determined by the measures of its angles.

If two angles of one triangle are congruent to two angles of another triangle, then the triangles are similar.

Postulate 8–1
AA Similarity

example 1

Complete the following proof.

Given: $\overline{ST} \parallel \overline{PR}$
$QS = 3$
$SP = 1$
$TR = 1.2$

Prove: $QT = 3.6$

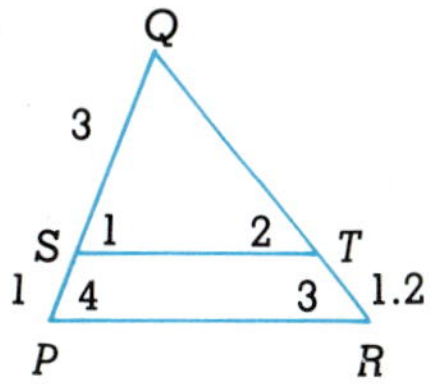

Proof:

STATEMENTS	REASONS
1. $\overline{ST} \parallel \overline{PR}$	1. Given
2. $\angle 1 \cong \angle 4$ $\angle 2 \cong \angle 3$	2. Theorem 6–8: If two parallel lines are cut by a transversal, then each pair of corresponding angles is congruent.
3. $\triangle SQT \sim \triangle PQR$	3. Postulate 8–1: AA Similarity
4. $\dfrac{QT}{QR} = \dfrac{QS}{QP}$	4. Definition of Similar Polygons
5. $QS = 3$ $SP = 1$ $TR = 1.2$	5. Given
6. $QP = QS + SP$ $= 3 + 1$ $= 4$ $QR = QT + TR$ $= QT + 1.2$	6. Definition of Between and Postulate 2–9: Substitution
7. $\dfrac{QT}{QT + 1.2} = \dfrac{3}{4}$	7. Postulate 2–9: Substitution *Steps 4, 5, and 6*
8. $4QT = 3(QT + 1.2)$ $= 3QT + 3.6$	8. Theorem 8–1: Equality of Cross Products and Postulate 2–14: Distributive Property
9. $QT = 3.6$	9. Postulate 2–7: Subtraction Property of Equality

To check triangles for similarity, it is only necessary to test angle measures. Could similarity of triangles be checked by testing the measures of corresponding sides for proportionality?

Theorem 8–7
SSS Similarity

If there is a correspondence between two triangles so that the measures of their corresponding sides are proportional, then the two triangles are similar.

example 2

Prove Theorem 8–7.

Given: $\frac{AB}{PQ} = \frac{BC}{QR} = \frac{CA}{RP}$

Prove: $\triangle BAC \sim \triangle QPR$

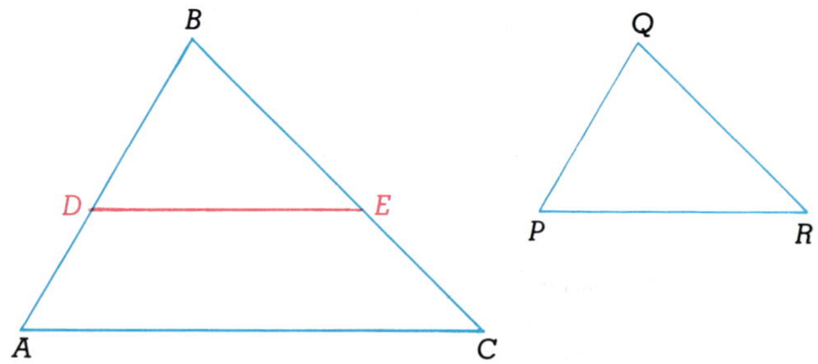

Proof:

On $\overline{AB}$ locate D so that $\overline{DB} \cong \overline{PQ}$. Draw $\overline{DE}$ so that $\overline{DE} \parallel \overline{AC}$.

Since parallel lines cut by a transversal form congruent corresponding angles, $\angle BDE \cong \angle A$ and $\angle BED \cong \angle C$. Thus, $\triangle BDE \sim \triangle BAC$ by AA Similarity.

By the definition of similar polygons, the following proportions hold.

$$\frac{AB}{BD} = \frac{BC}{BE} = \frac{CA}{ED}$$

Using these proportions and the multiplication property of equality, you can show the following.

$$AB = DB \cdot \frac{BC}{BE} \quad \text{and} \quad AB = DB \cdot \frac{CA}{ED}$$

Using the proportions in the given statement and the multiplication property of equality, you can show the following.

$$AB = PQ \cdot \frac{BC}{QR} \quad \text{and} \quad AB = PQ \cdot \frac{CA}{RP}$$

Next, use substitution.

$$DB \cdot \frac{BC}{BE} = PQ \cdot \frac{BC}{QR} \quad \text{and} \quad DB \cdot \frac{CA}{ED} = PQ \cdot \frac{CA}{RP}$$

But, $\overline{DB} \cong \overline{PQ}$, so $DB = PQ$.

$$\frac{BC}{BE} = \frac{BC}{QR} \qquad \frac{CA}{ED} = \frac{CA}{RP}$$

$$BC \cdot QR = BE \cdot BC \qquad CA \cdot RP = ED \cdot CA$$

$$QR = BE \qquad RP = ED$$

$$\overline{QR} \cong \overline{BE} \qquad \overline{RP} \cong \overline{ED}$$

By SSS congruence, $\triangle BDE \cong \triangle QPR$. Recall that $\triangle BDE \sim \triangle BAC$. Thus, by using AA similarity, you can show that $\triangle BAC \sim \triangle QPR$.

If the measures of two sides of a triangle are proportional to the measures of two corresponding sides of another triangle, and the included angles are congruent, then the triangles are similar.

Theorem 8–8
SAS Similarity

exercises

Exploratory Use each figure and the information given to tell whether or *not* each pair of triangles are similar.

1.

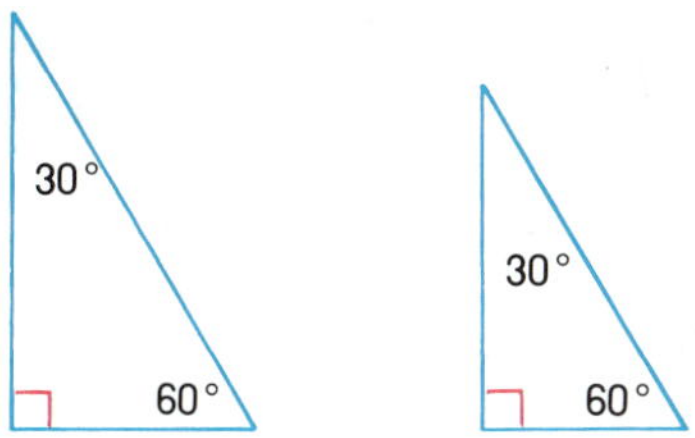

2.

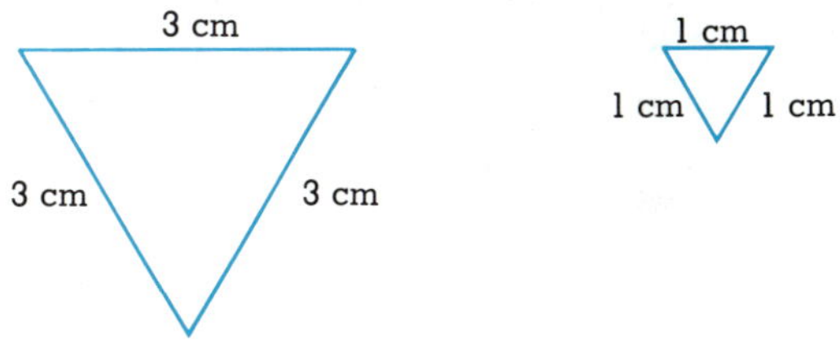

3.

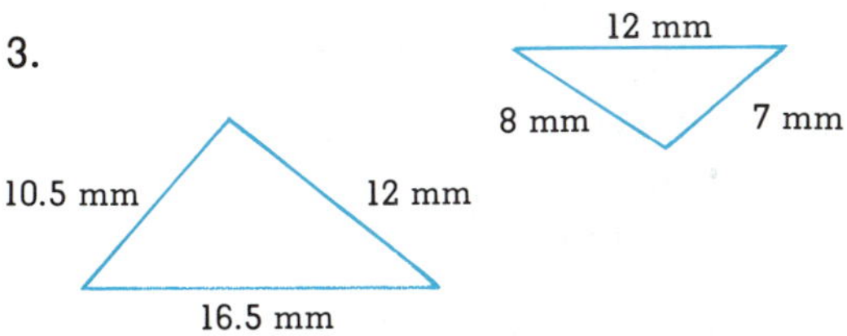

4.

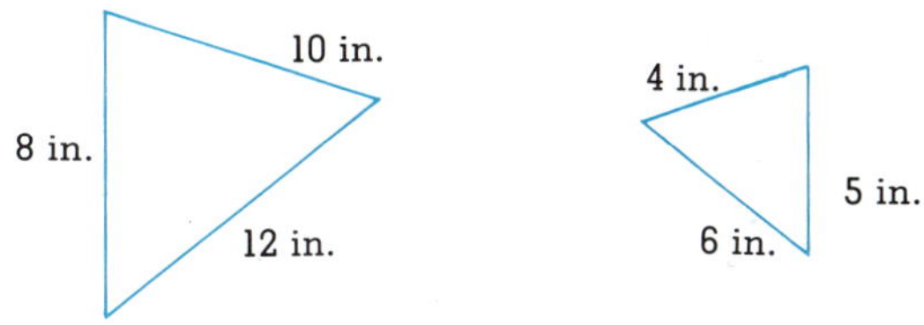

5.

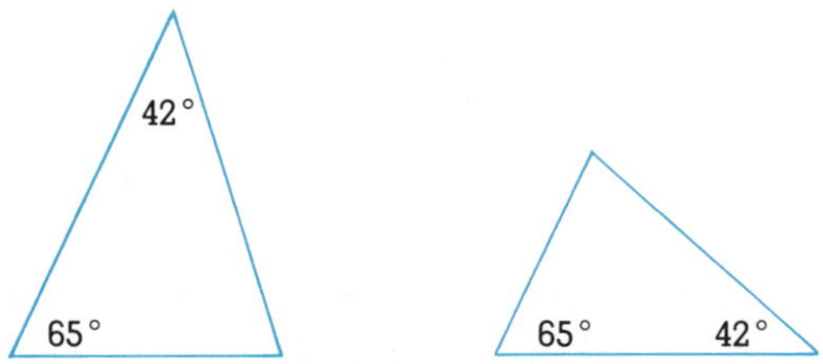

6.

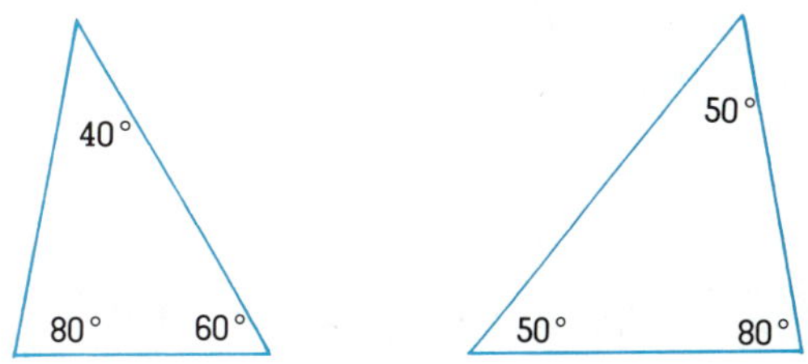

Written For each of the following, use the figure at the right and the given information to find **x**.

1. $QR = x + 4$
   $RS = 2x + 3$
   $QP = 3$
   $TS = 5$

2. $TS = 6$
   $QP = 4$
   $RS = \frac{1}{4}x + 1$
   $QR = 3x - 5$

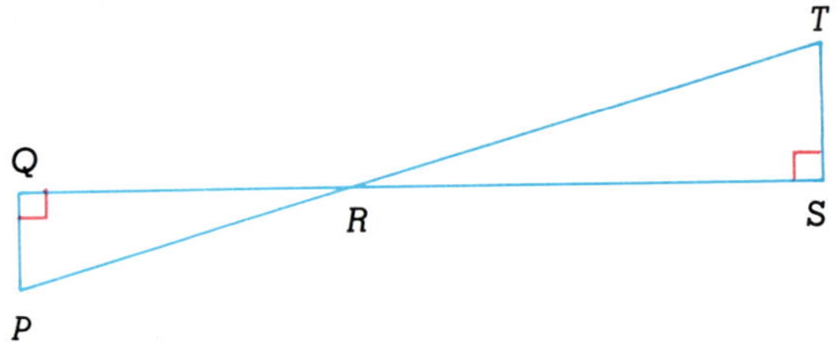

Prove each of the following.

3. **Given:** $\angle D$ is a right angle.
   $\overline{BE} \perp \overline{AC}$
   Prove: $\triangle ADC \sim \triangle ABE$

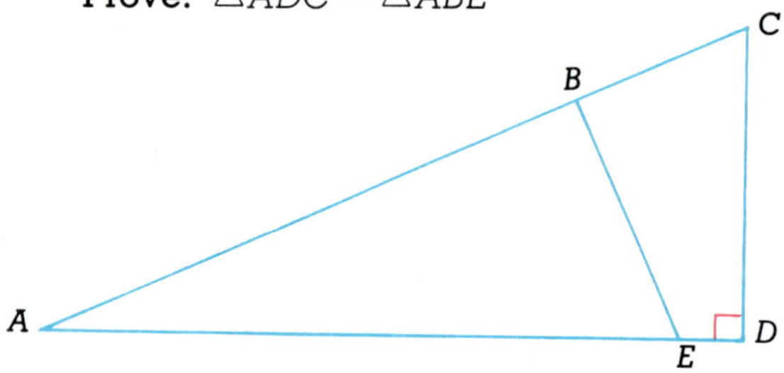

4. **Given:** $\overline{QS} \parallel \overline{PT}$
   Prove: $\triangle QRS \sim \triangle TRP$

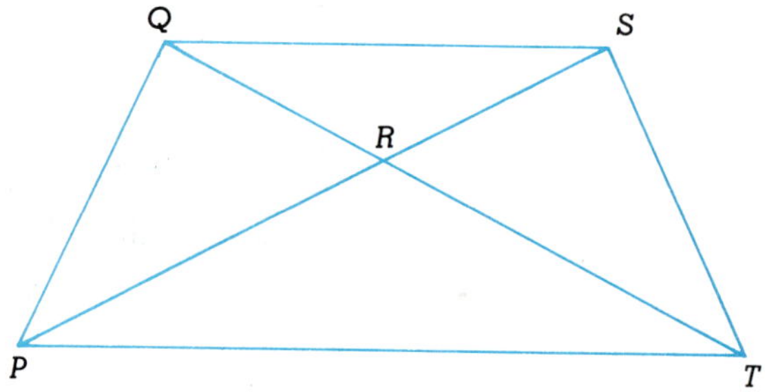

5. **Given:** $\angle Q$ is a right angle.
   $\square WSTV$ is a square.
   Prove: $\triangle PWS \sim \triangle TVR$

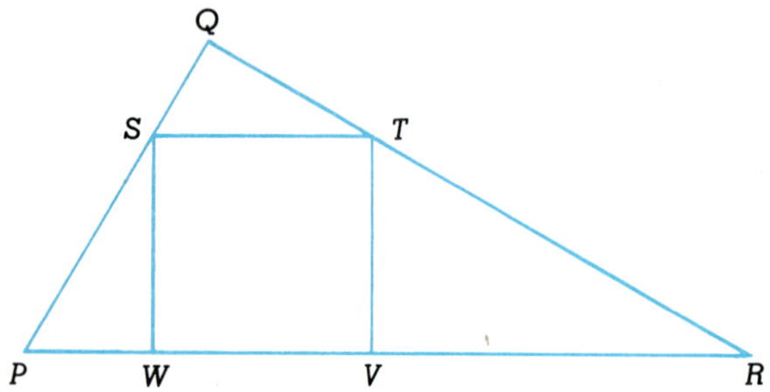

6. **Given:** $\overline{PR} \cong \overline{TR}$
   $\overline{VQ} \perp \overline{PR}$
   $\overline{VS} \perp \overline{RT}$
   Prove: $\frac{PQ}{TS} = \frac{VQ}{VS}$

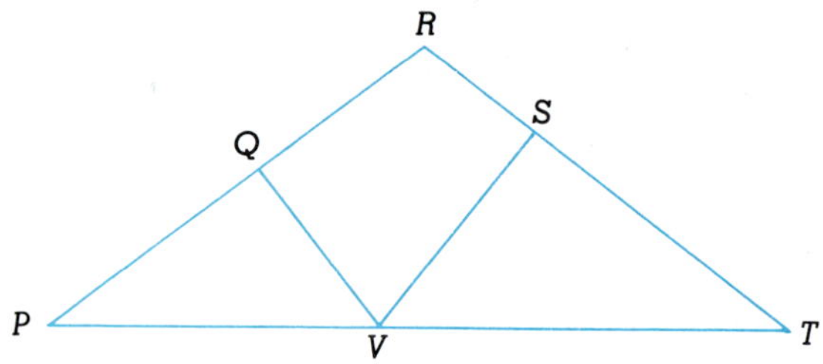

7. Prove Theorem 8–8.

8. If the measures of the sides adjacent to the right angles of two right triangles are proportional, then the triangles are similar.

CAREERS

Remote Sensing Specialist

Jim Gibbons is a remote sensing specialist. He uses the camera shown below to take aerial photographs for land use maps. With such maps, local agencies can study residential growth, water pollution, soil erosion, and so on.

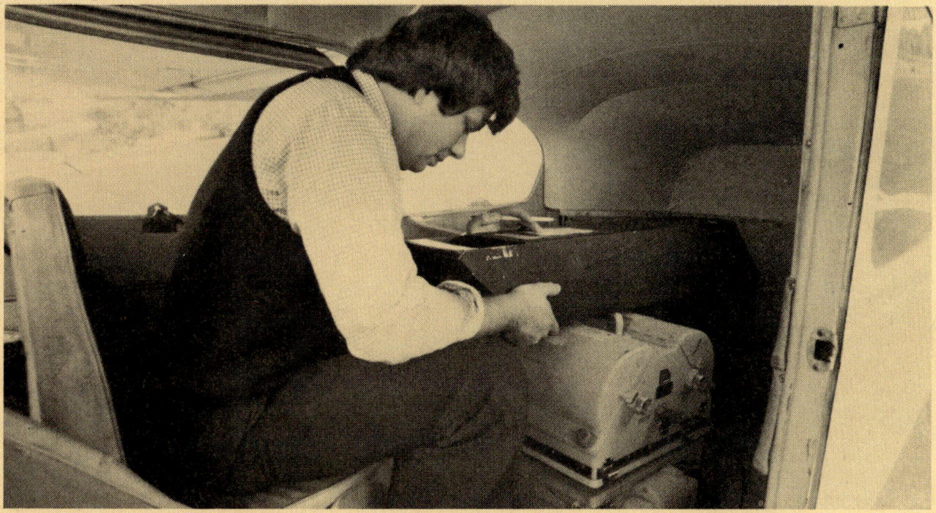

Proportions are used to determine the actual dimensions of regions shown in the photographs. Suppose a photograph is taken from a height of 914 meters. Then, 1 centimeter in the photograph is the same as 240 meters on the land.

Example Find the actual dimensions of a region that is 0.25 centimeters long and 0.14 centimeters wide in an aerial photograph.

Let ℓ stand for the actual length and w stand for the actual width.

$$\frac{\ell}{0.25} = \frac{240}{1} \qquad \frac{w}{0.14} = \frac{240}{1}$$

$$\ell = 240 \times 0.25 \qquad w = 240 \times 0.14$$

$$= 60 \qquad = 33.6$$

The region is 60 meters long and 33.6 meters wide.

Exercises Find the actual dimensions of regions for each of the following dimensions in an aerial photograph.

1. 0.85 centimeters by 0.7 centimeters
2. 0.68 centimeters by 0.91 centimeters
3. 3.5 centimeters by 8.2 centimeters
4. 5.2 centimeters by 1.5 centimeters

8–5 Proportional Parts

Proportional parts can be used to prove that lines are parallel.

Theorem 8–9

If a line intersects two sides of a triangle, cutting off segments with measures proportional to the measures of these two sides, then it is parallel to the third side.

example 1

Prove Theorem 8–9.

Given: $\frac{BA}{CB} = \frac{DE}{CD}$

Prove: $\overline{BD} \parallel \overline{AE}$

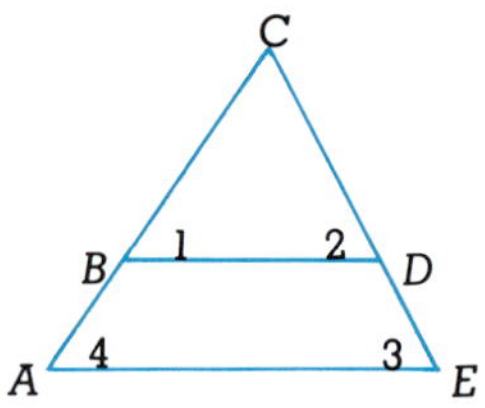

Proof:

STATEMENTS	REASONS
1. $\frac{BA}{CB} = \frac{DE}{CD}$	1. Given
2. B is between C and A. D is between C and E.	2. Given
3. $CA = CB + BA$ $CE = CD + DE$	3. Definition of between
4. $BA = CA - CB$ $DE = CE - CD$	4. Postulate 2–7: Subtraction Property of Equality
5. $\frac{CA - CB}{CB} = \frac{CE - CD}{CD}$	5. Postulate 2–9: Substitution *Step 1*
6. $\frac{CA}{CB} = \frac{CE}{CD}$	6. Theorem 8–5: Subtraction Property of Proportions
7. $\angle C \cong \angle C$	7. Theorem 3–1: Congruence of angles is reflexive.
8. $\triangle CBD \sim \triangle CAE$	8. Theorem 8–8: SAS Similarity
9. $\angle 1 \cong \angle 4$	9. Definition of similar polygons
10. $\overline{BD} \parallel \overline{AE}$	10. Theorem 6–4: In a plane, if two lines are cut by a transversal so that a pair of corresponding angles are congruent, then the two lines are parallel.

Each floor of an A-frame is the base of a triangle with part of the roof as sides. Also, each floor is parallel to the ground floor. The following theorem tells you that, because the floors are parallel, the triangles formed are similar.

If a line is parallel to one side of a triangle and intersects the other two sides, then it divides them proportionally.

Theorem 8–10

example 2

Prove Theorem 8–10.

Given: $\overline{BD} \parallel \overline{AE}$

Prove: $\frac{BA}{CB} = \frac{DE}{CD}$

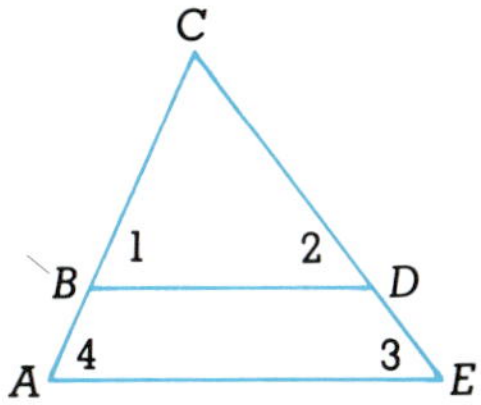

Proof:

STATEMENTS	REASONS
1. $\overline{BD} \parallel \overline{AE}$	1. Given
2. $\angle 1 \cong \angle 4$ $\angle 2 \cong \angle 3$	2. Theorem 6–8: If two parallel lines are cut by a transversal, then each pair of corresponding angles is congruent.
3. $\triangle ACE \sim \triangle BCD$	3. Postulate 8–1: AA Similarity
4. $\frac{CA}{CB} = \frac{CE}{CD}$	4. Definition of Similar Polygons
5. B is between A and C. D is between C and E.	5. Given
6. $CA = BA + CB$ $CE = DE + CD$	6. Definition of between
7. $\frac{BA + CB}{CB} = \frac{DE + CD}{CD}$	7. Postulate 2–9: Substitution
8. $\frac{BA}{CB} = \frac{DE}{CD}$	8. Theorem 8–4: Addition Property of Proportions

The proofs of the following theorems are based on Theorems 8–9 and 8–10.

Theorem 8–11

If a segment joins the midpoints of two sides of a triangle, then it is parallel to the third side and its length is one-half the length of the third side.

Theorem 8–12

If three parallel lines intersect two transversals, then they divide the transversals proportionally.

Theorem 8–13

If three parallel lines cut off congruent segments on one transversal, then they cut off congruent segments on any transversal.

exercises

Exploratory In the figure at the right, $\overline{BD} \parallel \overline{AE}$. Use this figure to tell whether or *not* each of the following must be true.

1. $\frac{CB}{BA} = \frac{CD}{DE}$
2. $\frac{CD}{CB} = \frac{DE}{BA}$
3. $\frac{AC}{BC} = \frac{ED}{DC}$
4. $\frac{CB}{BA} = \frac{DE}{DC}$
5. $\frac{BC}{ED} = \frac{AB}{CD}$
6. $\frac{CB}{CD} = \frac{CA}{CE}$
7. $\frac{CB}{CD} = \frac{BA}{DE}$
8. $\frac{AB}{BC} = \frac{DE}{CD}$
9. $\frac{AC}{DC} = \frac{EC}{BC}$
10. $\frac{AC}{EC} = \frac{DC}{BC}$
11. $\frac{CA}{BA} = \frac{CE}{DE}$
12. $\frac{BA}{DE} = \frac{CA}{CE}$
13. $\frac{ED}{BC} = \frac{CD}{AB}$
14. $\frac{EC}{AC} = \frac{BC}{DC}$
15. $\frac{AB}{AE} = \frac{DE}{DB}$

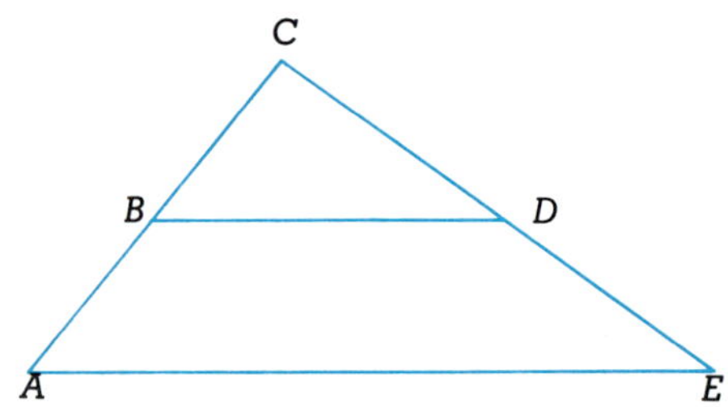

Written For each of the following, tell whether or *not* $\overline{QT} \parallel \overline{RS}$. Write *yes* or *no*.

1. $PR = 30$, $PQ = 9$, $PT = 12$, $PS = 18$
2. $QR = 22$, $TS = 9$, $RP = 65$, $SP = 27$
3. $RP = 13.5$, $RQ = 7.2$, $TP = 4.2$, $SP = 9.0$
4. $PQ = 34.88$, $RQ = 18.32$, $PS = 33.25$, $TS = 11.45$

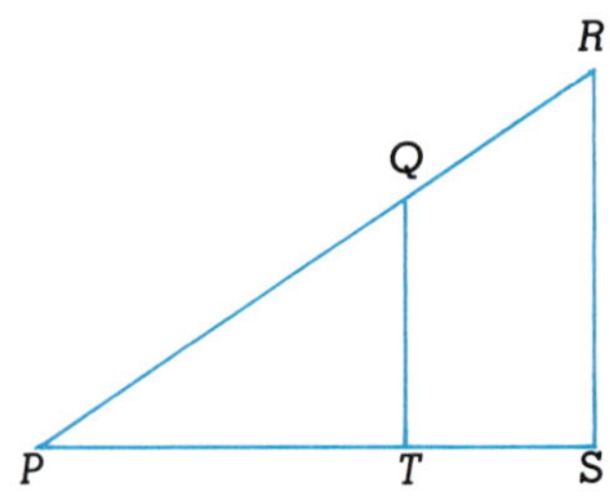

For each of the following, use the figure at the right and the given information to find x.

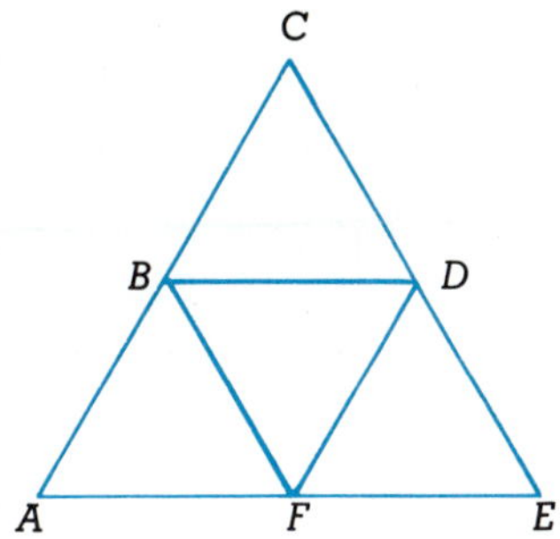

5. $\overline{BD} \parallel \overline{AE}$
$AB = 6$
$DE = 8$
$DC = 4$
$BC = x$

6. $\overline{AC} \parallel \overline{DF}$
$DC = 7$
$DE = 5$
$FA = 8$
$FE = x$

7. $\overline{DF} \parallel \overline{CA}$
$AF = 3.42$
$FE = 3.35$
$DE = 6.7$
$CD = x$

8. $\overline{FB} \parallel \overline{CE}$
$AB = 4.90$
$AF = 5.62$
$FE = 8.43$
$BC = x$

9. $\overline{EC} \parallel \overline{FB}$
$AF = x + 2$
$FE = 5$
$AB = x$
$BC = 3$

10. $\overline{AE} \parallel \overline{DB}$
$CB = x$
$CD = x - 2$
$BA = 5$
$DE = 1$

11. Prove Theorem 8–12.

12. Prove Theorem 8–13.

Use the following to solve problems 13–18.

Given: Q is the midpoint of $\overline{RP}$.
S is the midpoint of $\overline{RT}$.

Prove: $\overline{QS} \parallel \overline{PT}$
$QS = \frac{1}{2}PT$

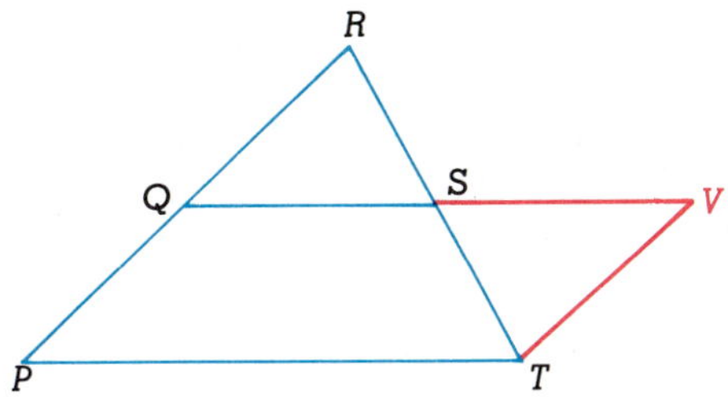

13. The first part of a proof shows that $\overline{QS} \parallel \overline{PT}$. Plan a proof for this part based on Theorem 8–9.

14. The second part of a proof shows that $QS = \frac{1}{2}PT$. A statement in this part of the proof says, *Extend* $\overline{QS}$ *to* V *so that* $\overline{SV} \cong \overline{QS}$. State the reason.

15. Assume the results of problems **13** and **14**, and explain why $\triangle RSQ \cong \triangle TSV$.

16. Assume the results of problems **13–15**, and explain why $\overline{QP} \parallel \overline{TV}$.

17. Assume the results from problems **13–16**, and explain why $\overline{QV} \cong \overline{PT}$.

18. Assume the results from problems **13–17**, and explain why $QS = \frac{1}{2}PT$.

Visual Thinking

Excursions in Geometry

Ask another person to read the following instructions, as you follow.

Study the figure at the right for one minute. Then, close your eyes.
In your mind, switch the black square and the blue circle.
Next, in your mind, switch the yellow triangle and the red diamond.
Open your eyes and draw the new arrangement. Then check your drawing.

8-6 Parts of Similar Polygons

When two triangles are similar, the measures of their corresponding sides are proportional. What about their corresponding altitudes, medians, and so on? Do these measures form the same ratio as do the measures of the corresponding sides?

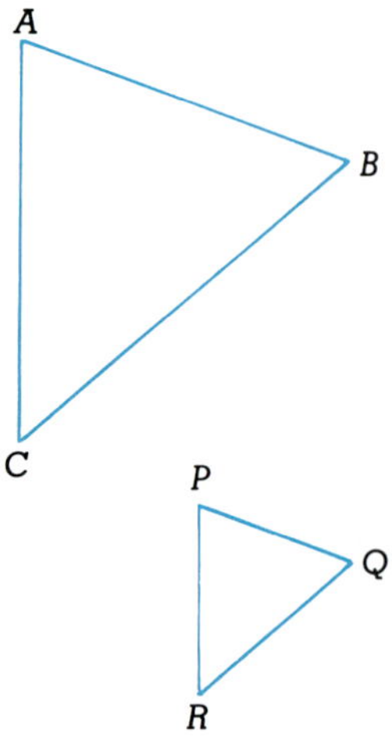

By using the Summation Property of Proportions, you can show that the measures of corresponding perimeters are proportional to the measures of corresponding sides of similar triangles.

In the figure at the left, suppose $\triangle ABC \sim \triangle PQR$.

$\frac{AB}{PQ} = \frac{BC}{QR} = \frac{CA}{RP}$	*Definition of similar triangles*
$\frac{AB}{PQ} = \frac{AB + BC}{PQ + QR}$	*Summation Property of Proportions*
$\frac{AB + BC}{PQ + QR} = \frac{CA}{RP}$	*Substitution*
$\frac{AB + BC + CA}{PQ + QR + RP} = \frac{CA}{RP}$	*Summation Property of Proportions*

Theorem 8–14

If two triangles are similar, then the measures of corresponding perimeters are proportional to the measures of corresponding sides.

example 1

In the figure at the right, suppose $\triangle ABD \sim \triangle CDA$. If the perimeter of $\triangle ABD$ measures 42.2, the perimeter of $\triangle CDA$ measures 63.3, and $CD = 21.0$, find AB.

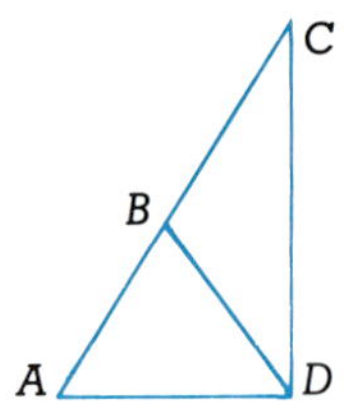

$$\frac{AB}{CD} = \frac{AB + BD + DA}{CD + DA + AC}$$

$$\frac{AB}{21.0} = \frac{42.2}{63.3}$$

$$63.3AB = 21.0 \times 42.2$$

$$AB = \frac{21.0 \times 42.2}{63.3}$$

$$= 14.0$$

In the figure below, suppose $\triangle ABC \sim \triangle PQR$. Then $\angle A \cong \angle P$ because they are corresponding angles. Since $\overline{BD}$ is an altitude for $\triangle ABC$, it forms a right angle, $\angle BDA$. Similarly, $\angle QSP$ is a right angle. By AA Similarity, $\triangle ABD \sim \triangle PQS$. Thus, $\frac{BD}{QS} = \frac{BA}{QP}$.

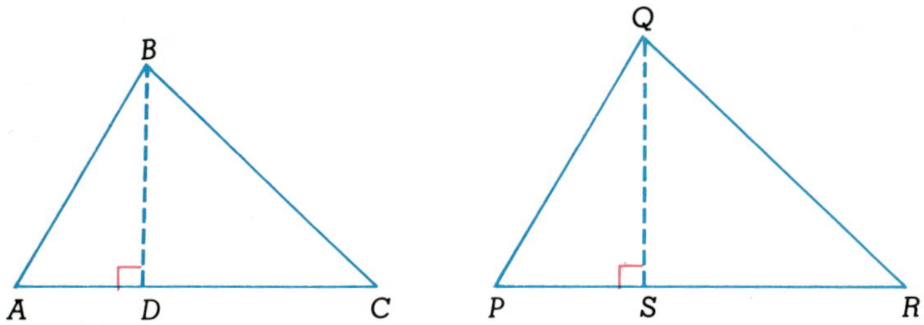

Theorem 8–15

If two triangles are similar, then the measures of corresponding altitudes are proportional to the measures of corresponding sides.

Theorem 8–16

If two triangles are similar, then the measures of corresponding angle bisectors of the triangles are proportional to the measures of corresponding sides.

Theorem 8–17

If two triangles are similar, then the measures of corresponding medians are proportional to the measures of corresponding sides.

example 2

Plan a proof for Theorem 8–17.

Given: $\triangle ABC \sim \triangle PQR$
D is the midpoint of $\overline{AC}$.
S is the midpoint of $\overline{PR}$.

Prove: $\frac{BD}{QS} = \frac{BA}{QP}$

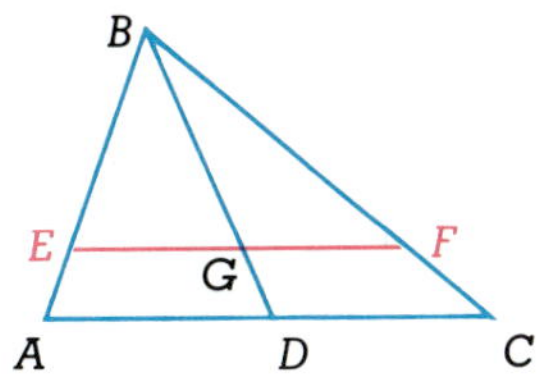

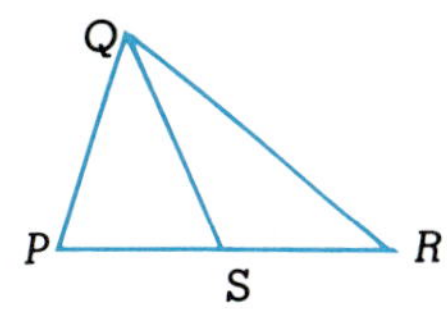

Proof:

Draw $\overline{EF}$ so that $\overline{BE} \cong \overline{QP}$ and $\angle BEF \cong \angle QPR$.

Since $\triangle ABC \sim \triangle PQR$, conclude that $\angle QPR \cong \angle BAC$. Thus, by transitivity of angle congruence, $\angle BEF \cong \angle BAC$. Since $\angle ABD$ is congruent to itself, AA Similarity implies that $\triangle BEG \sim \triangle BAD$.

Using proportions and substitution, show that $\triangle BEG \cong \triangle QPS$ by SAS. Then, substitute in proportions to show that $\frac{BD}{QS} = \frac{BA}{QP}$.

exercises

Exploratory In the figure at the right, $\triangle ABC \sim \triangle PQR$. Also, $\overline{BD}$ is an altitude of $\triangle ABC$, and $\overline{QS}$ is an altitude of $\triangle PQR$. Tell whether or *not* each of the following *must be* true.

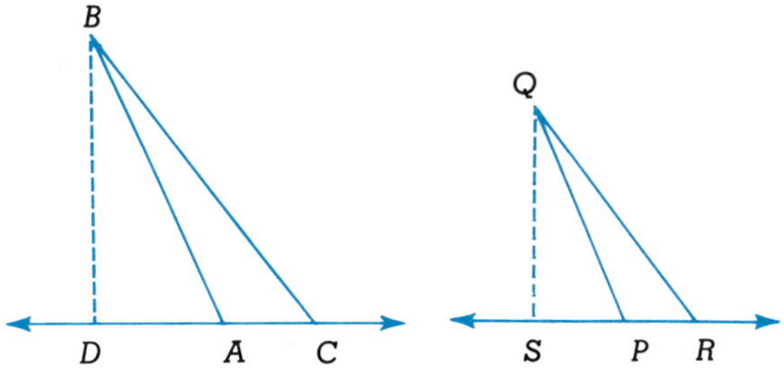

1. $\frac{BD}{QS} = \frac{AB}{PQ}$
2. $\frac{AD}{PS} = \frac{QR}{BC}$
3. $\frac{QP}{AB} = \frac{BD}{QS}$
4. $\frac{QR}{BC} = \frac{QS}{BD}$
5. $\frac{BD}{QS} = \frac{AC}{PR}$
6. $\frac{AB}{BD} = \frac{PQ}{QS}$
7. $\frac{PS}{PR} = \frac{QS}{PR}$
8. $\frac{BC}{QR} = \frac{BD}{QS}$
9. $\frac{PR}{AC} = \frac{QS}{BD}$
10. $\frac{PR}{AC} = \frac{BD}{QS}$
11. $\frac{PQ}{BD} = \frac{AB}{QS}$
12. $\frac{PQ}{AB} = \frac{QS}{BD}$

Written In the figure at the right, $\triangle TVX \sim \triangle LMO$.

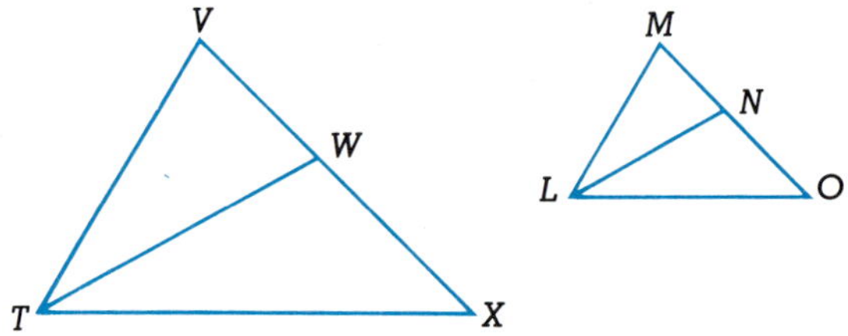

1–8. Suppose $\overline{WT}$ bisects $\angle VTX$, and $\overline{NL}$ bisects $\angle MLO$. State eight different proportions that follow from this information.

9–16. Suppose $\overline{TW}$ is a median of $\triangle TVX$, and $\overline{LN}$ is a median of $\triangle LMO$. State eight different proportions that follow from this information.

In the figure at the right, $\triangle ABE \sim \triangle DEC$. Also, E is the midpoint of $\overline{AD}$, and C is the midpoint of $\overline{BD}$. Use the figure and the given information to find x.

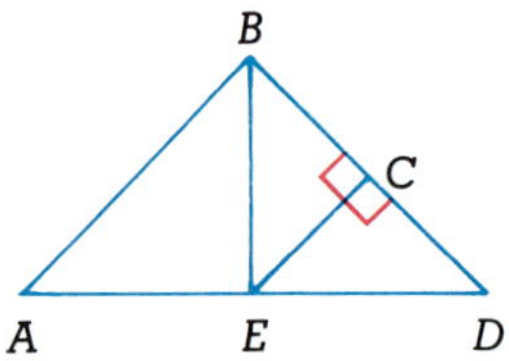

17. $CE = 2$, $BE = 2 + x$, $AB = 12$, $DE = 3$

18. $BD = 2.2$, $BE = 1.5$, $AD = 3.3$, $CE = x - 1$

19. $AB = 13.2$, $BE = 8.4$, $EC = 7.0$, $AD = 0.5x$

20. $EC = 4.5$, $DC = 4.3$, $ED = 10.32$, $BE = 5x - 17$

Solve each problem.

21. Suppose $\triangle PQR \sim \triangle ABC$. If $PR = 1.8$, $AC = 1.2$, and the perimeter of $\triangle ABC$ measures 3.4, find the measure of the perimeter of $\triangle PQR$.

22. Suppose $\triangle LMN \sim \triangle XYZ$. If the perimeter of $\triangle LMN$ measures 7.6, $LM = 3.0$, $MN = 2.8$ and $XY = 2.7$, find XZ.

23. The perimeters of two equilateral triangles measure 18.0 and 41.4. Suppose an altitude of the first triangle measures 5.2. Find the measure of the altitude of the second triangle.

24. The perimeters of two similar triangles measure 51.3 and 27.0. Suppose an angle bisector of the first triangle measures 17.1. Find the measure of the corresponding angle bisector in the second triangle.

25. Plan a proof for Theorem 8–15.

26. Prove Theorem 8–15.

27. Plan a proof for Theorem 8–16.

28. Prove Theorem 8–17.

8–7 Using Similarity

Properties of similar triangles can be used to solve practical problems such as the following.

Jeanetta Johnston measured the height of an oak tree by placing a mirror a short distance from the tree, on a level surface, and facing up. She stood where she could see the top of the tree in the mirror. The mirror was 6 feet from the tree. Jeanetta was $\frac{1}{2}$ foot from the mirror and her eyes were 5 feet from the ground. A beam of light forms the same angle with a mirror as its reflection. Find the height of the tree.

Define a variable. The problem asks for the height of the tree. Let x stand for the measure of the height.

Make a drawing.

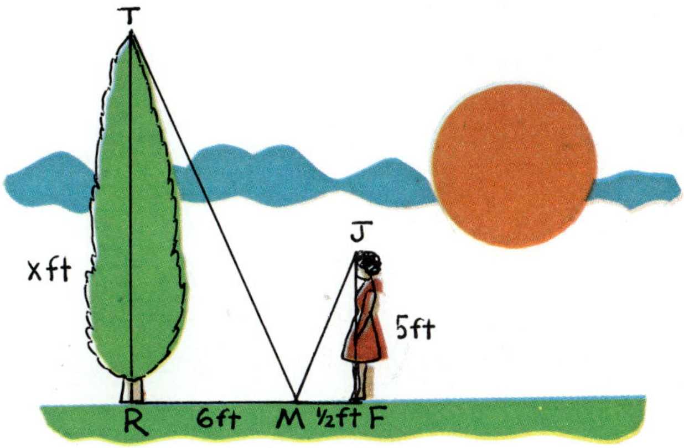

Notice the two triangles in the drawing.

Write a proportion. Assume that Jeanetta and the tree are both perpendicular to the ground. Also, $\angle TMR \cong \angle JMF$ since they represent the angles formed by a beam of light and its reflection. Thus, $\triangle TMR \sim \triangle JMF$ by AA Similarity and the measures of corresponding sides are proportional.

$$\frac{x}{5} = \frac{6}{\frac{1}{2}}$$

Solve the proportion.

$$\frac{x}{5} = \frac{6}{\frac{1}{2}}$$

$$\frac{1}{2}x = 30$$

$$x = 2 \cdot 30 \text{ or } 60$$

Answer the problem. The tree is 60 feet tall.

Problem Solving Procedure for Proportions

1. **Define a variable.**
2. **Make a drawing.**
3. **Write a proportion.**
4. **Solve the proportion.**
5. **Answer the problem.**

example 1

In 12 inches length, a piece of metal tapers from 8 inches wide to a point. Find the width of the tapered metal for a length of 15 inches.

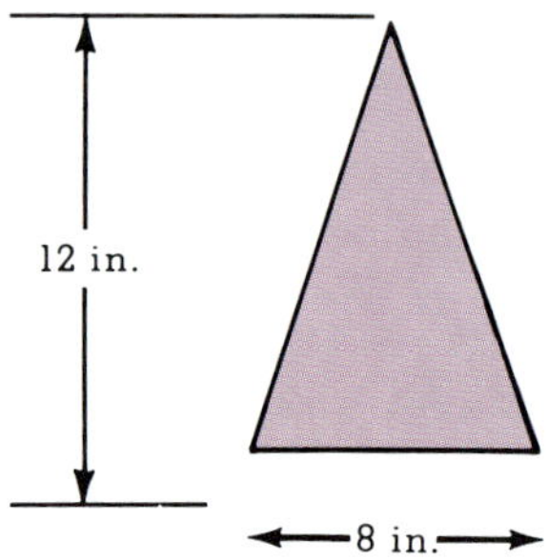

Define a variable. Let x stand for the width of the metal for a length of 15 inches.

Make a drawing.

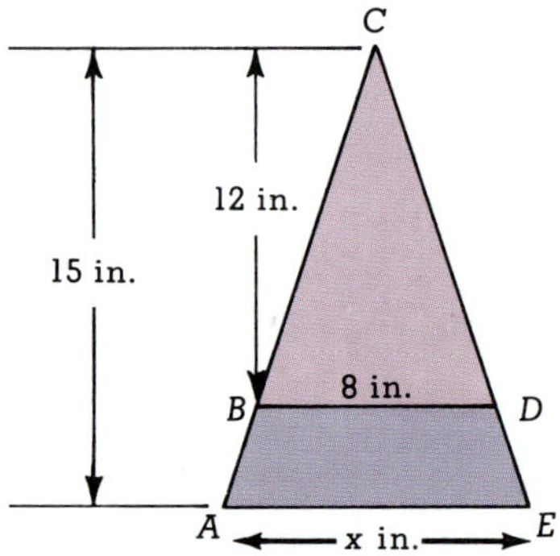

Write a proportion. Assume $\overline{BD} \parallel \overline{AE}$. Thus, $\triangle BCD \sim \triangle ACE$. By Theorem 8–15, the measures of corresponding altitudes of similar triangles are proportional to the measures of corresponding sides.

$$\frac{x}{8} = \frac{15}{12}$$

Solve the proportion.

$$\frac{x}{8} = \frac{15}{12}$$

$$12x = 120$$

$$x = 10$$

Answer the problem. The width of the tapered metal is 10 inches.

example 2

A hypsometer shown at the right can be used to measure the height of a tree. Sight through the straw to the top of the tree. Note where the free-hanging string crosses the scale. Suppose Gene Pulski had the readings shown. His eye was 167 cm from the ground, and he was 15 m, or 1500 cm, from the tree. Find the height of the tree.

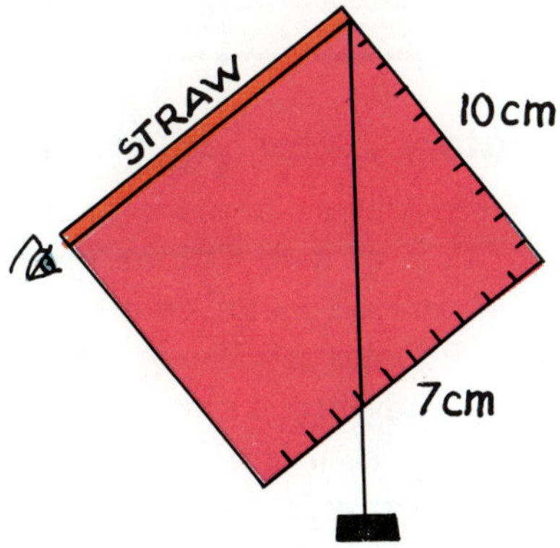

Define a variable. Let x stand for the measure of the distance from eye level to the tip of the tree.

Make a drawing.

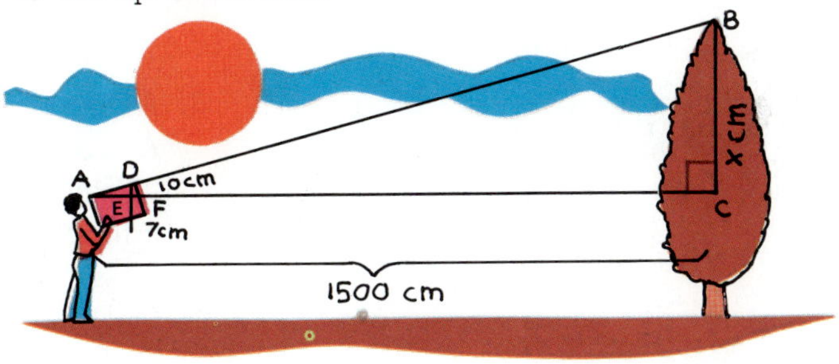

Write a proportion. First, assume that $\angle DFE$ and $\angle BCA$ are right angles, $\overline{DE} \parallel \overline{BC}$, and $\overline{AD} \parallel \overline{EF}$.

$\angle FED \cong \angle ADE$	*They are alternate interior angles of $\overline{AD}$ and $\overline{EF}$.*
$\angle ADE \cong \angle ABC$	*They are corresponding angles of $\overline{DE}$ and $\overline{BC}$.*
$\angle FED \cong \angle ABC$	*Angle congruence is transitive.*
$\angle DFE \cong \angle BCA$	*All right angles are congruent.*
$\triangle DEF \sim \triangle ABC$	*AA Similarity*

The measures of corresponding sides of similar triangles are proportional.

$$\frac{BC}{EF} = \frac{AC}{DF}$$

Solve the proportion.

$$\frac{x}{7} = \frac{1500}{10}$$

$$10x = 10{,}500$$

$$x = 1050$$

Answer the problem. The tree is 1050 cm + 167 cm or 1217 cm tall.

exercises

Exploratory Solve each problem.

1. In 8 cm length, a piece of metal tapers from 4 cm wide to a point. Find the width of the tapered metal for a length of 9 cm.
2. In 5 cm length, a piece of metal tapers from 3 cm wide to a point. Find the width of the tapered metal for a length of 6 cm.
3. Elena Rogelio used a hypsometer to measure the height of a chestnut tree. She had a reading of 10 cm on the vertical scale and 5.5 cm on the horizontal scale. She was 20 m from the tree. Find the height of the tree in meters. Elena's eyes are 1.5 m from the ground.
4. Lamar Presley measured the height of a pecan tree by sighting the top of the tree in a mirror that was a short distance from the tree and facing up. The mirror was 6.0 m from the tree. Lamar was 0.9 m from the mirror and his eyes were 1.8 m from the ground. Find the height of the tree.

Written Solve each problem.

1. A tower casts a shadow 64 feet long. A 6 foot tall pole near the tower casts a shadow 8 feet long. How tall is the tower?
2. A flag pole casts a shadow 3 meters long. A woman near the pole casts a shadow 0.75 meters long. The woman is 1.5 meters tall. How tall is the flag pole?
3. Use similar triangles to find the length of the lake.

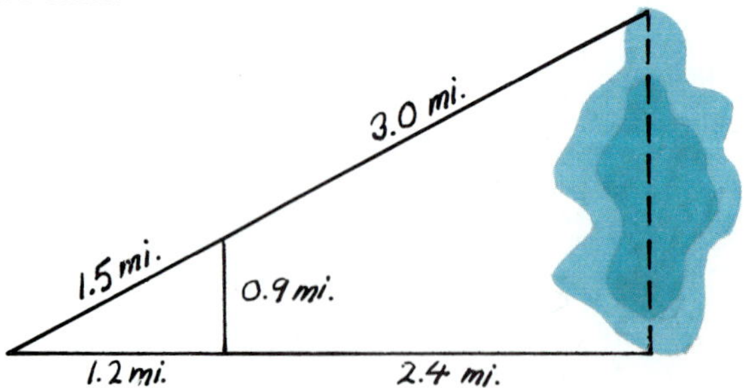

4. Most T. V. screens have similar shapes. The measure of the diagonal is used to give screen size. Suppose the dimensions of a 9 inch screen are $5\frac{1}{2}$ inches by $7\frac{1}{2}$ inches. Find the dimensions of an 18 inch screen.
5. Suppose a person is 300 cm from a camera lens, and the film is 1.3 cm from the lens. If the person is 180 cm tall, how tall is her image on the film?
6. Suppose film is 1.3 cm from a camera lens and can have an image no more than 4.5 cm tall. If the person in front of the lens is 180 cm tall, how far from the lens should she be?

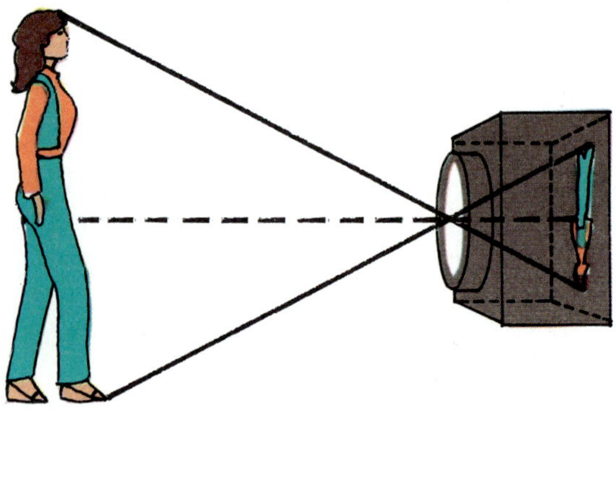

Vocabulary

ratios (230)
proportion (230)
cross products (231)
similar figures (238)

Chapter Summary

1. A ratio is a comparison of two numbers by division. (230)
2. An equation of the form $\frac{a}{b} = \frac{c}{d}$ which states that two ratios are equal is called a proportion. (230)
3. Equality of Cross Products: For any numbers a and c, and any nonzero numbers b and d, $\frac{a}{b} = \frac{c}{d}$ if and only if $ad = bc$. (231)
4. Properties of proportion: (235)

For any numbers a, b, c, and d, the following properties hold whenever all terms are defined.	
Theorem 8–2	If $\frac{a}{b} = \frac{c}{d}$, then $\frac{a}{c} = \frac{b}{d}$.
Theorem 8–3	If $\frac{a}{b} = \frac{c}{d}$, then $\frac{b}{a} = \frac{d}{c}$.
Theorem 8–4: Addition Property of Proportions	$\frac{a}{b} = \frac{c}{d}$ if and only if $\frac{a+b}{b} = \frac{c+d}{d}$
Theorem 8–5: Subtraction Property of Proportions	$\frac{a}{b} = \frac{c}{d}$ if and only if $\frac{a-b}{b} = \frac{c-d}{d}$.
Theorem 8–6: Summation Property of Proportions	$\frac{a}{b} = \frac{c}{d}$ if and only if $\frac{a}{b} = \frac{a+c}{b+d}$ or $\frac{c}{d} = \frac{a+c}{b+d}$.

5. Definition of Similar Polygons: Two polygons are similar if and only if there is a correspondence such that their corresponding angles are congruent and the measures of their corresponding sides are proportional. (238)
6. AA Similarity: If two angles of one triangle are congruent to the corresponding angles of another triangle, then the triangles are similar. (243)
7. SSS Similarity: If there is a correspondence between two triangles so that the measures of their corresponding sides are proportional, then the two triangles are similar. (244)
8. SAS Similarity: If the measures of two sides of a triangle are proportional to the measures of two corresponding sides of another triangle, and the included angles are congruent, then the triangles are similar. (245)
9. Theorem 8–9: If a line intersects two sides of a triangle, cutting off segments with measures proportional to the measures of these two sides, then it is parallel to the third side. (248)
10. Theorem 8–10: If a line is parallel to one side of a triangle and intersects the other two sides, then it divides them proportionally. (249)
11. Theorem 8–11: If a segment joins the midpoints of two sides of a triangle, then it is parallel to the third side and its length is one-half the length of the third side. (250)
12. Theorem 8–12: If three parallel lines intersect two transversals, then they divide the transversals proportionally. (250)
13. Theorem 8–13: If three parallel lines cut off congruent segments on one transversal, then they cut off congruent segments on any transversal. (250)
14. Theorem 8–14: If two triangles are similar, then the measures of corresponding perimeters are proportional to the measures of corresponding sides. (252)
15. Theorem 8–15: If two triangles are similar, then the measures of corresponding altitudes are proportional to the measures of corresponding sides. (253)
16. Theorem 8–16: If two triangles are similar, then the measures of corresponding angle bisectors of the triangles are proportional to the measures of corresponding sides. (253)
17. Theorem 8–17: If two triangles are similar, then the measures of corresponding medians are proportional to the measures of corresponding sides. (253)
18. Problem Solving Procedure for Proportions: (255)
    1. Define a variable. **2.** Make a drawing. **3.** Write a proportion. **4.** Solve the proportion. **5.** Answer the problem.

Chapter Review

8–1 Write the cross products for each of the following.

1. $\frac{5}{x} = \frac{2}{3}$
2. $\frac{x}{9} = \frac{7}{15}$
3. $\frac{1}{x} = \frac{5}{x + 5}$
4. $\frac{n + 4}{3} = \frac{5n - 3}{8}$

5–8. Solve each proportion in problems **1–4.**

8–2 Solve each problem.

9. Prove that for any numbers, a, b, c and d, if $\frac{a}{b} = \frac{c}{d}$, then $\frac{a - b}{c - d} = \frac{b}{d}$ whenever all terms are defined.
10. Prove that for any numbers a, b, c, and d, if $\frac{a + c}{b + d} = \frac{a}{b}$, then $\frac{a}{b} = \frac{c}{d}$.

8–3 For each of the following, write *yes* or *no*. Then explain your answer.

11. All congruent triangles are similar.
12. All similar triangles are congruent.

8–4 Prove each of the following.

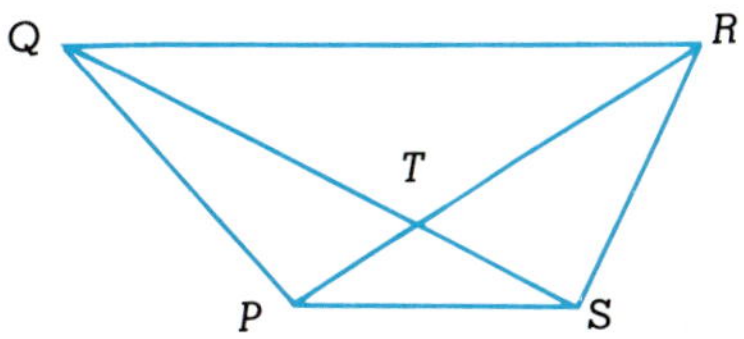

13. **Given:** $\frac{PR}{QS} = \frac{RS}{QP}$
    trapezoid $PQRS$
    Prove: $\triangle PQR \sim \triangle SRQ$
14. **Given:** $\overline{QR} \parallel \overline{PS}$
    Prove: $\frac{QT}{TS} = \frac{TR}{PT}$

8–5 Use the figure at the right and the given information to find x.

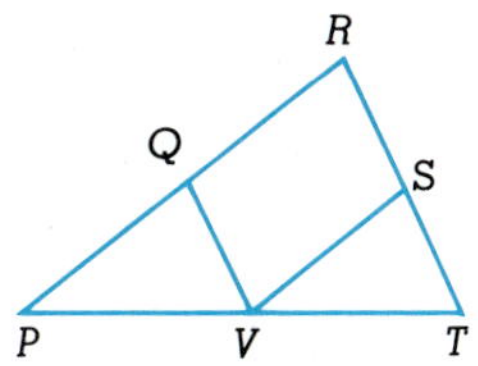

15. $\overline{SV} \parallel \overline{PR}$
    $TS = 5 + x$
    $TV = 8 + x$
    $VP = 4$
    $SR = 3$
16. $\overline{RT} \parallel \overline{QV}$
    $TV = 7.29$
    $PV = x$
    $PQ = 9$
    $QR = 27$

8–6 In the figure at the right, $\triangle ABE \sim \triangle AEC$ and $\triangle CEA \sim \triangle CBE$. Use the figure at the right and the given information to find x.

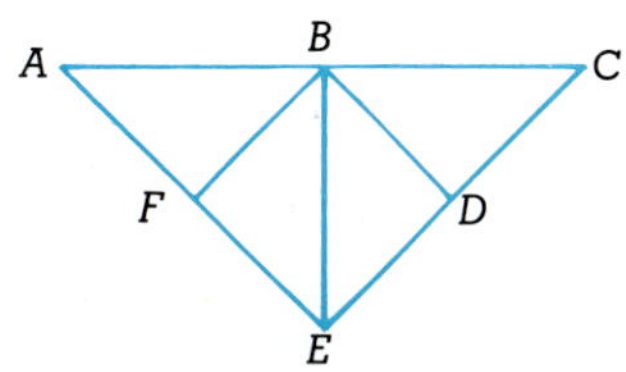

17. $\overline{BE}$ bisects $\angle AEC$.
    $\overline{BF}$ bisects $\angle ABE$.
    $AB = 4.5$
    $FB = 1.4$
    $AE = x$
    $EB = 1$
18. B is the midpoint of $\overline{AC}$.
    $\overline{BD}$ is a median of $\triangle EBC$.
    $BE = x + 3$
    $BD = 4$
    $CE = 6$
    $CB = 5$

8–7 Solve each problem.

19. A light pole casts a shadow 42 feet long. A man standing next to the pole casts a shadow 12 feet long. If the man is $5\frac{3}{4}$ feet tall, how tall is the light pole?
20. Before photographers make enlarged prints of negatives, they make small contact prints. Suppose a contact print is 50.8 mm wide and 58.4 mm high. If the enlargement is to be 190.5 mm wide, how high will it be?

Chapter Test

Solve each proportion.

1. $\frac{x}{28} = \frac{60}{15}$
2. $\frac{15}{33} = \frac{50}{x}$
3. $\frac{1}{15} = \frac{x}{7}$
4. $\frac{5}{5x + 4} = \frac{2}{3}$
5. $\frac{21}{1 - x} = \frac{7}{x}$
6. $\frac{x - 1}{5 - x} = \frac{3}{5}$

Use the figure at the right and the given information to find **x**.

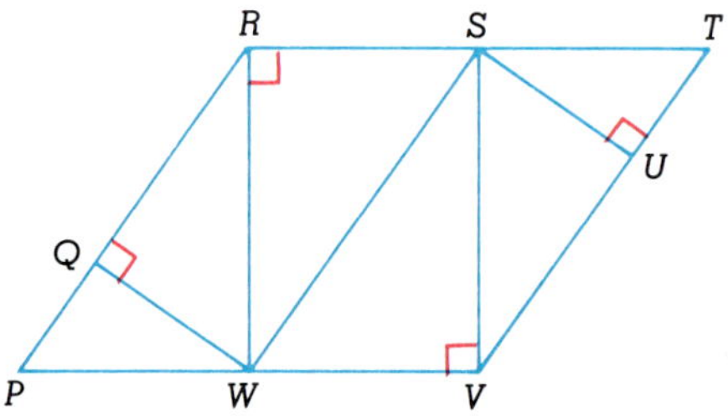

7. $\triangle RWP \sim \triangle VST$
   $PW = x$
   $QW = 1$
   $SU = 5$
   $ST = x + 5$
8. $\triangle WRS \sim \triangle VUS$
   $SW = 3$
   $RS = 1\frac{1}{3}$
   $SU = x + \frac{2}{3}$
   $SV = 2$

Prove each of the following.

9. **Given:** $\triangle ABC \sim \triangle RSP$
   D is the midpoint of $\overline{AC}$.
   Q is the midpoint of $\overline{PR}$.
   $\overline{TP}$ bisects $\angle SPQ$.
   $\overline{EC}$ bisects $\angle BCD$.

   Prove: $\frac{QT}{DE} = \frac{SP}{BC}$

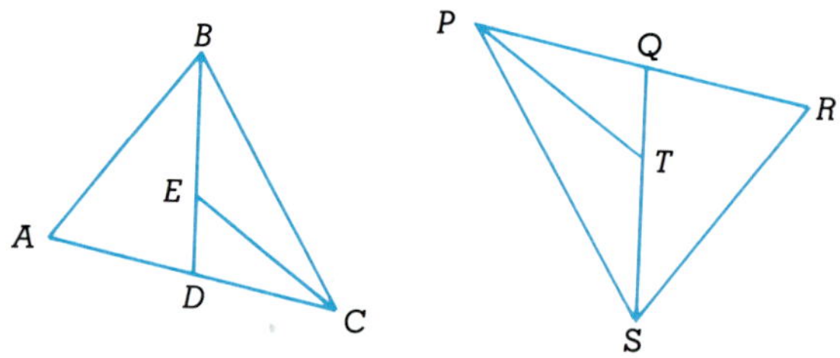

10. If two triangles are similar, then the measures of corresponding altitudes are proportional to the measures of corresponding perimeters.

Solve each of the following.

11. Use similar triangles to find the distance across the river.

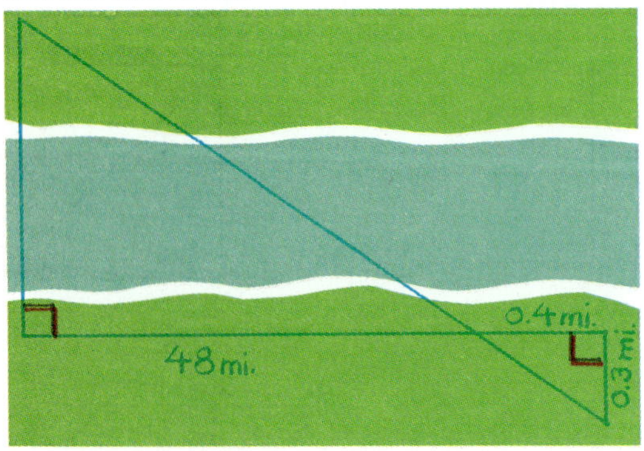

12. Fritz Gundlach measured the height of a building by sighting the top of the building in a mirror that was on the ground and facing up. The mirror was $17\frac{1}{2}$ ft from the building and Fritz was $2\frac{5}{8}$ ft from the mirror. If Fritz's eyes were 6 ft from the ground, how tall is the building?

Chapter

9

Right Triangles

The part of a sundial that casts the shadow has a triangular shape. Its construction is based on several right triangles that represent its position with respect to the sun.

9–1 Square Roots

Suppose $\triangle ABD \sim \triangle BCD$ in the figure at the right. Then, the following proportion can be written.

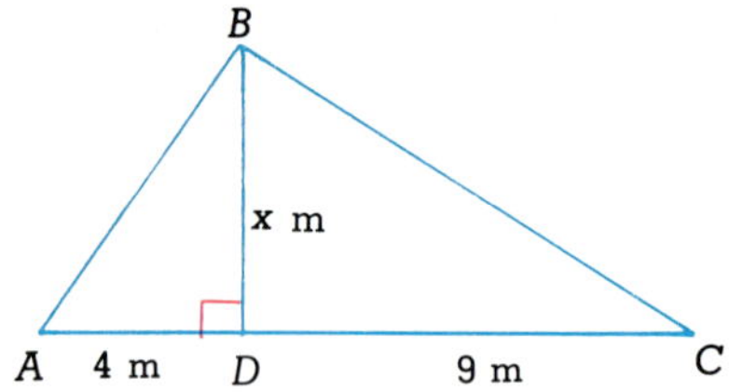

$$\frac{DC}{BD} = \frac{BD}{DA}$$

To find the value of x, make the following substitutions in the proportion.

$$\frac{9}{x} = \frac{x}{4}$$

Then, cross multiply.

$x \cdot x = 9 \cdot 4$ or $x^2 = 36$

Squaring a number means using that number as a factor two times. The inverse of squaring is finding the **square root.** To find the square root of 36, you must find two equal factors whose product is 36.

What is the value of x in the problem above?

Since 6 times 6 is 36, one square root of 36 is 6. Since $^-6$ times $^-6$ is 36, another square root of 36 is $^-6$.

Definition of Square Root

For any numbers a and b, if $a^2 = b$, then a is a square root of b.

radical sign → $\sqrt{36}$ ← *radicand*

The symbol $\sqrt{}$, called a **radical sign,** indicates a nonnegative square root. The expression under the radical sign is called the radicand (rad-i-*KAND*).

$\sqrt{36} = 6$ *$\sqrt{36}$ indicates the principal, or nonnegative, square root of 36.*

$^-\sqrt{36} = {^-6}$ *$^-\sqrt{36}$ indicates the negative square root of 36.*

$^\pm\sqrt{36} = \pm 6$ *$^\pm\sqrt{36}$ indicates both square roots of 36.*

examples

1 Find $\sqrt{49}$.

$\sqrt{49} = 7$ since $7^2 = 49$.

2 Find $^-\sqrt{100}$.

$^-\sqrt{100} = {^-10}$ since the square root of 100 is 10.

To *simplify* a square root such as $\sqrt{196}$, find the square root of any perfect square under the radical sign. Use the prime factorization of 196 to do this.

$$\begin{aligned}\sqrt{196} &= \sqrt{2 \cdot 2 \cdot 7 \cdot 7} \\ &= \sqrt{2^2 \cdot 7^2} \\ &= \sqrt{2^2} \cdot \sqrt{7^2} \\ &= 2 \cdot 7 \text{ or } 14 \quad \textbf{Check: } 14^2 = 196\end{aligned}$$

An important property of square roots was used to simplify $\sqrt{196}$.

For any nonnegative numbers a and b, $\sqrt{ab} = \sqrt{a} \cdot \sqrt{b}$.

Postulate 9–1 Product Property of Square Roots

A similar property for quotients of square roots also can be used to simplify square roots.

For any nonnegative numbers a and b $\sqrt{\frac{a}{b}} = \frac{\sqrt{a}}{\sqrt{b}}$ as long as b is *not* zero.

Postulate 9–2 Quotient Property of Square Roots

examples

3 Simplify $\sqrt{300}$.

$$\begin{aligned}\sqrt{300} &= \sqrt{2 \cdot 2 \cdot 3 \cdot 5 \cdot 5} && \textit{Find the prime factorization of 300.} \\ &= \sqrt{2^2 \cdot 3 \cdot 5^2} \\ &= \sqrt{2^2} \cdot \sqrt{3} \cdot \sqrt{5^2} && \textit{Use the product property of square roots.} \\ &= 2 \cdot \sqrt{3} \cdot 5 \\ &= 10\sqrt{3}\end{aligned}$$

Check: $(10\sqrt{3})^2 = 100 \cdot 3$ or 300

4 Simplify $\sqrt{\frac{16}{9}}$.

$$\begin{aligned}\sqrt{\frac{16}{9}} &= \frac{\sqrt{16}}{\sqrt{9}} && \textit{Use the quotient property of square roots.} \\ &= \frac{4}{3}\end{aligned}$$

Check: $\left(\frac{4}{3}\right)^2 = \frac{16}{9}$

The method used to simplify $\frac{\sqrt{5}}{\sqrt{2}}$ is called **rationalizing the denominator.** Notice that the denominator becomes a rational number.

$$\frac{\sqrt{5}}{\sqrt{2}} = \frac{\sqrt{5}}{\sqrt{2}} \cdot \frac{\sqrt{2}}{\sqrt{2}}$$

Notice that $\frac{\sqrt{2}}{\sqrt{2}}$ is equal to 1.

$$= \frac{\sqrt{10}}{\sqrt{4}}$$

$$= \frac{\sqrt{10}}{2}$$

A square root is said to be simplified when the following conditions are met.

1. The radicand has no factor other than 1 that is a perfect square.
2. The radicand does *not* contain a fraction.
3. No radical appears in the denominator of a fraction.

Some numbers such as 5 and 13 have square roots that are *not* integers or rationals. The square roots are **irrational numbers.**

Such a table is given at the back of this book.

To compute with irrational numbers, you often use decimal approximations. Tables of square roots give approximations.

example 5

Use the table to find a decimal approximation for $\sqrt{5}$.

number → n *square of the number* → n^2 *square root of the number* → $\sqrt{n}$

n	n^2	$\sqrt{n}$
1	1	1.000
2	4	1.414
3	9	1.732
4	16	2.000
5	25	2.236
6	36	2.449
7	49	2.646
8	64	2.828

Locate 5 in the column marked n.

Locate the value for $\sqrt{5}$ in the column marked $\sqrt{n}$.

exercises

Exploratory State the square of each number.

1. 8
2. 12
3. $^-7$
4. $^-11$
5. 0.2
6. 0.1
7. $\frac{2}{5}$
8. $\frac{4}{9}$
9. 1.3
10. $^-2.5$

Simplify.

11. $\sqrt{64}$
12. $^-\sqrt{144}$
13. $^-\sqrt{49}$
14. $\sqrt{49}$
15. $\sqrt{0.04}$
16. $\sqrt{0.01}$
17. $^\pm\sqrt{1.69}$
18. $\sqrt{6.25}$
19. $\sqrt{\frac{4}{25}}$
20. $^\pm\sqrt{\frac{16}{81}}$

Written State the principal square root of each expression.

1. 1
2. 100
3. 0.25
4. 1.21
5. 0.09
6. $\frac{16}{121}$
7. $\frac{49}{4}$
8. 0.0025
9. 169
10. 16,900

Simplify.

11. $\sqrt{256}$
12. $\sqrt{361}$
13. $\sqrt{900}$
14. $\sqrt{484}$
15. $\sqrt{80}$
16. $\sqrt{72}$
17. $\sqrt{75}$
18. $\sqrt{54}$
19. $\sqrt{5}\sqrt{15}$
20. $\sqrt{8}\sqrt{2}$
21. $\sqrt{\frac{5}{4}}$
22. $\sqrt{\frac{2}{9}}$
23. $\sqrt{\frac{1}{8}}$
24. $\frac{1}{\sqrt{2}}$
25. $\frac{5}{\sqrt{3}}$
26. $\sqrt{\frac{1}{3}}$
27. $\sqrt{\frac{5}{3}}$
28. $\sqrt{\frac{7}{2}}$
29. $\frac{\sqrt{21}}{\sqrt{14}}$
30. $\sqrt{\frac{2}{3}}\sqrt{\frac{5}{2}}$

31–50. Tell whether each expression in problems **11–30** is a *rational number* or an *irrational number*.

Find the approximate value of each expression. Use the table of Squares and Approximate Square Roots at the back of this book.

51. $\sqrt{76}$
52. $\sqrt{21}$
53. $\sqrt{17}$
54. $\sqrt{61}$
55. $\sqrt{6} - \sqrt{12}$
56. $\sqrt{15} + \sqrt{19}$
57. $\sqrt{\frac{7}{2}}$
58. $\sqrt{\frac{5}{3}}$
59. $\sqrt{200}$
60. $\sqrt{175}$

Solve for x. Approximate the value of x to three decimal places. Use the table of Squares and Approximate Square Roots.

61. $x^2 - 40 = 0$
62. $x^2 = \frac{3}{2}$
63. $x^2 + 16 = 64$
64. $100 + x^2 = 121$
65. $x^2 + (\sqrt{5})^2 = (\sqrt{30})^2$
66. $(\sqrt{11})^2 + (4)^2 = x^2$
67. $\frac{3}{x} = \frac{x}{12}$
68. $\frac{3}{x} = \frac{x}{7}$
69. $\frac{5}{x} = \frac{x}{9}$

9–2 The Geometric Mean

When the altitude to the hypotenuse of a right triangle is drawn, two smaller triangles are formed. These two triangles are similar to each other, and both are similar to the original triangle.

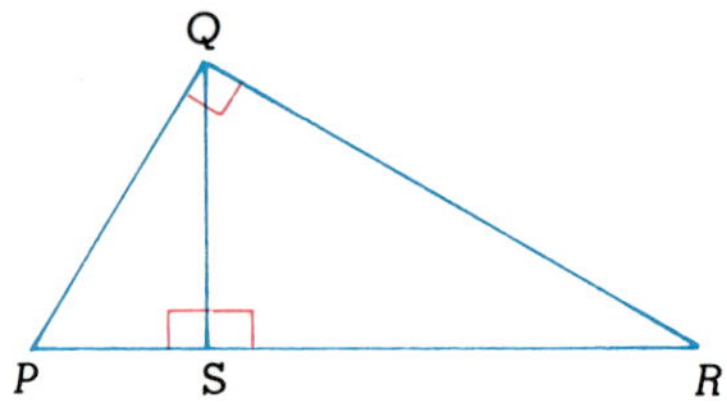

Theorem 9–1

If the altitude is drawn from the right angle to the hypotenuse of a right triangle, then the two triangles formed are similar to the given triangle and to each other.

example 1

Prove Theorem 9–1.

Given: $\angle PQR$ is a right angle
$\overline{QS}$ is an altitude of $\triangle PQR$

Prove: $\triangle PQR \sim \triangle PSQ$
$\triangle PQR \sim \triangle QSR$
$\triangle PSQ \sim \triangle QSR$

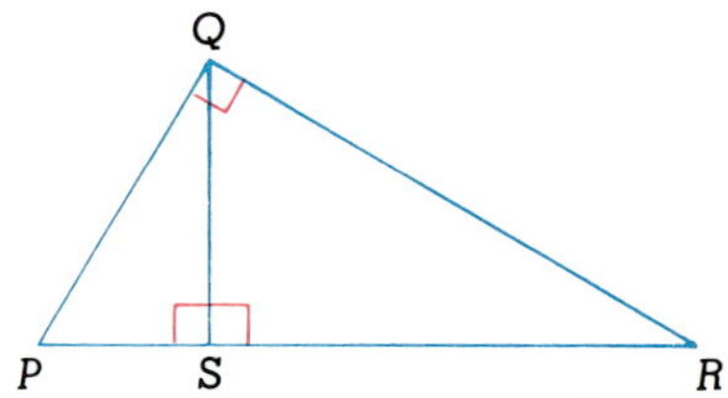

Proof:

STATEMENTS	REASONS
1. $\angle PQR$ is a right angle	1. Given
2. $\overline{QS}$ is an altitude of $\triangle PQR$	2. Given
3. $\overline{QS} \perp \overline{PR}$	3. Definition of altitude
4. $\angle PSQ$ is a right angle $\angle QSR$ is a right angle	4. Definition of perpendicular
5. $\angle PQR \cong \angle PSQ$ $\angle PQR \cong \angle QSR$ $\angle PSQ \cong \angle QSR$	5. Theorem 3–6: If two angles are right angles, then the angles are congruent.
6. $\angle P \cong \angle P$ $\angle R \cong \angle R$	6. Theorem 3–1: Congruence of angles is reflexive.
7. $\triangle PQR \sim \triangle PSQ$ $\triangle PQR \sim \triangle QSR$	7. Postulate 8–1: AA Similarity
8. $\angle PQS \cong \angle QRP$	8. Definition of similar triangles
9. $\triangle PSQ \sim \triangle QSR$	9. Postulate 8–1: AA Similarity

By Theorem 9–1, the two smaller triangles in the figure at the right are similar. Thus, corresponding sides are proportional, and the following proportion can be written.

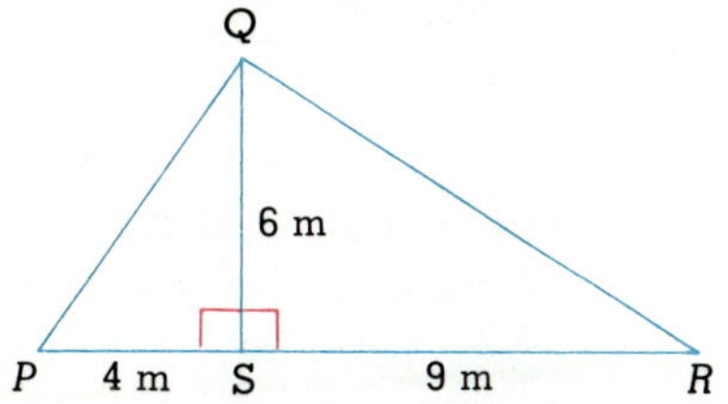

$$\frac{4}{6} = \frac{6}{9}$$

$\triangle PSQ \sim \triangle QSR$

Notice that the denominator of the first fraction is 6 and the numerator of the second fraction also is 6. Six is called the **geometric mean** between 4 and 9.

For any positive numbers a and b, x is the geometric mean between a and b if and only if $\frac{a}{x} = \frac{x}{b}$ and x is positive.

Definition of Geometric Mean

examples

2 **Find the geometric mean between 4 and 10.**

Let x stand for the geometric mean.

$\frac{4}{x} = \frac{x}{10}$ *Definition of geometric mean*

$x^2 = 40$ *Equality of cross products*

$x = \sqrt{40}$ *Definition of square root and geometric mean is positive*

$x = 2\sqrt{10}$

The geometric mean is $2\sqrt{10}$.

3 **Show that the value of the geometric mean between a and b is $\sqrt{ab}$.**

Let x stand for the geometric mean.

$\frac{a}{x} = \frac{x}{b}$ *Definition of geometric mean*

$x^2 = ab$ *Equality of cross products*

$x = \sqrt{ab}$ *Definition of square root and geometric mean is positive*

The proofs of the following theorems are based on the fact that the triangles formed are similar.

Theorem 9–2

The measure of the altitude drawn from the right angle to the hypotenuse of a right triangle is the geometric mean between the measure of the two segments of the hypotenuse.

Theorem 9–3

If the altitude is drawn to the hypotenuse of a right triangle, then the measure of a leg of the triangle is the geometric mean between the measure of the hypotenuse and the measure of the segment of the hypotenuse adjacent to that leg.

exercises

Exploratory For each of the following pairs of numbers, find the geometric mean.

1. 2 and 8 **2.** 3 and 12 **3.** 2 and 18 **4.** 5 and 20
5. 9 and 16 **6.** 2 and 50 **7.** 1 and 36 **8.** 16 and 1
9. 4 and 4 **10.** 2 and 32 **11.** 72 and 2 **12.** 8 and 8
13. 27 and 3 **14.** 5 and 45 **15.** 16 and 4 **16.** 18 and 8

Written For each of the following pairs of numbers, find the geometric mean.

1. 3 and 5 **2.** 4 and 6 **3.** 5 and 8 **4.** 2 and 10
5. 24 and 6 **6.** 5 and 4 **7.** 12 and 4 **8.** 25 and 4
9. 15 and 5 **10.** 8 and 10 **11.** 18 and 3 **12.** 4 and 18
13. $\frac{1}{4}$ and 9 **14.** 4 and $\frac{1}{9}$ **15.** $\frac{2}{5}$ and $\frac{5}{2}$ **16.** $\frac{2}{3}$ and $\frac{1}{3}$
17. $\frac{1}{4}$ and $\frac{3}{4}$ **18.** $\frac{2}{3}$ and $\frac{1}{2}$ **19.** $\frac{1}{2}$ and $\frac{1}{4}$ **20.** $\frac{1}{3}$ and $\frac{1}{4}$

Use the figure at the right to solve each of the following. Approximate each answer to three decimal places.

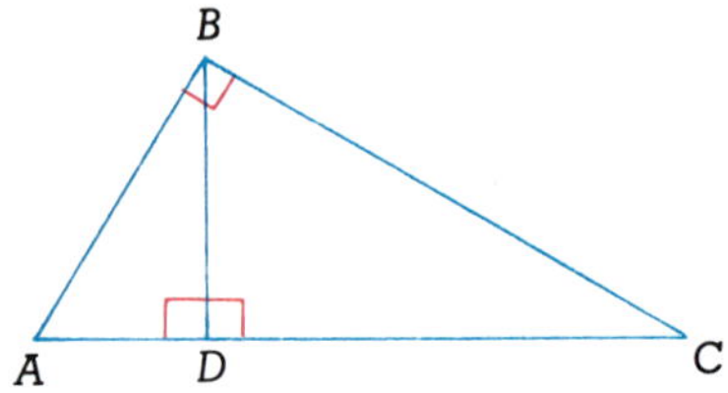

21. Find BD if $AD = 3$ and $DC = 10$.
22. Find BD if $AD = 5$ and $DC = 9$.
23. Find AB if $AC = 8$ and $AD = 3$.
24. Find BD if $DC = 12$ and $AD = 3$.
25. Find BC if $AD = 3$ and $DC = 4$.
26. Find BA if $DA = 4$ and $DC = 4$.

27. Prove Theorem 9–2. **28.** Prove Theorem 9–3.

Reading Geometry

Prefixes

Many words used in mathematics provide information through prefixes. Compare the meaning of the prefixes with its use in each definition below.

The prefix semi means half

A semicircle is the union of the endpoints of a diameter and all points of the circle lying on one side of the diameter. $\overset{\frown}{CZL}$ and $\overset{\frown}{CWL}$ are semicircles of $\odot P$.
Semicircles are half circles.

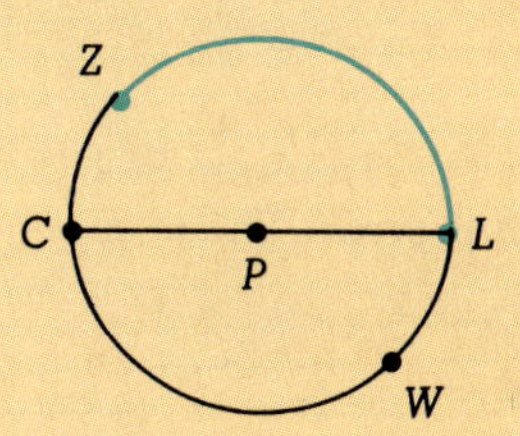

The prefix tri means three

A triangle is the union of three segments determined by three noncollinear points.
Triangles have three angles.

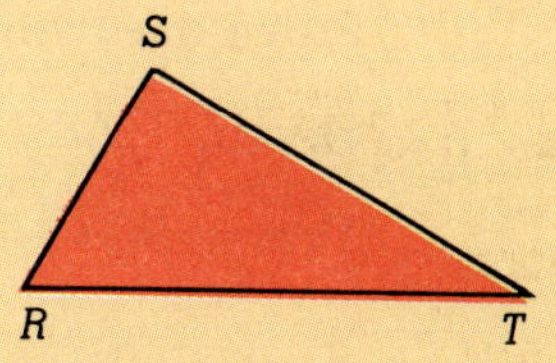

Exercises For each pair of definitions, choose the word that most closely fits the meaning of the underlined prefix.

1. Bisector of an arc: A line that contains the midpoint of the arc.
   Bisector of a segment: A line, segment, or plane that intersects a segment $\overline{XY}$ in the midpoint of $\overline{XY}$.
   a. beside b. two c. around
2. Circumscribed polygon: A polygon is said to be circumscribed about a circle if all of its sides are tangent to the circle.
   Circumference: The distance around a circle is the circumference, C.
   a. around b. across c. above
3. Equiangular triangle: A triangle with three congruent angles.
   Equilateral triangle: A triangle with all sides congruent.
   a. equal b. half c. between
4. Polygon: A closed plane figure formed by three or more line segments, called sides. No two sides with a common endpoint are collinear.
   Polygonal region: The union of a polygon and its interior.
   a. two b. four c. many
5. Quadrants: The four regions into which the coordinate plane is separated by the x-axis and y-axis.
   Quadrilateral: A four-sided polygon.
   a. half b. four c. many

9–3 The Pythagorean Theorem

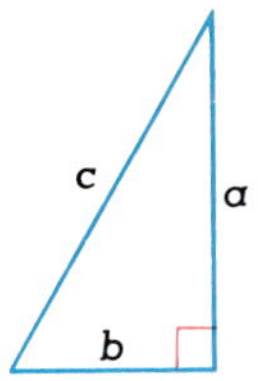

$c^2 = a^2 + b^2$

The following result has been used and studied for thousands of years. A Chinese manuscript from about 1000 B.C. illustrates the theorem. It is named after a Greek mathematician from the sixth century. A Hindu mathematician of the twelfth century presented an interesting proof. President Garfield presented his own special proof in 1876.

Theorem 9–4
Pythagorean Theorem

For any right triangle, the square of the measure of the hypotenuse equals the sum of the squares of the measures of the legs.

example

1 Prove the Pythagorean Theorem.

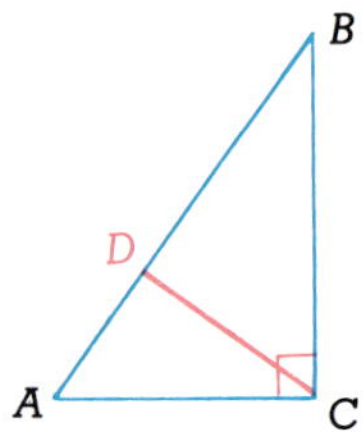

Given: $\triangle ABC$ is a right triangle

Prove: $BC^2 + CA^2 = AB^2$

Proof:

STATEMENTS	REASONS
1. Draw altitude $\overline{CD}$.	1. Theorem 6–12: Given a line and a point *not* on the line, there is exactly one line through the point and perpendicular to the given line.
2. $\triangle ABC$ is a right triangle.	2. Given
3. $\frac{AB}{BC} = \frac{BC}{BD}$ $\frac{AB}{CA} = \frac{CA}{DA}$	3. Theorem 9–3: If the altitude is drawn to the hypotenuse of a right triangle, then the measure of a leg of the triangle is the geometric mean between the measure of the hypotenuse and the measure of the segments of the hypotenuse adjacent to that leg.
4. $BC^2 = AB \cdot BD$ $CA^2 = AB \cdot DA$	4. Theorem 8–1: Equality of Cross Products
5. $BC^2 + CA^2 = AB \cdot BD + AB \cdot DA$	5. Postulate 2–7: Addition
6. $BC^2 + CA^2 = AB(BD + DA)$	6. Postulate 2–14: Distributive Property
7. $AB = BD + DA$	7. Definition of between
8. $BC^2 + CA^2 = AB^2$	8. Postulate 2–9: Substitution

If the square of the measure of one side of a triangle does *not* equal the sum of the squares of the measures of the other two sides, then the triangle is *not* a right triangle. Suppose such an equation *is* true for a given triangle. Can you conclude that the triangle is a right triangle?

The answer is *yes*.

Recall that the contrapositive of a true statement also is a true statement. This sentence is the contrapositive of the Pythagorean Theorem.

If the square of the measure of one side of a triangle equals the sum of the squares of the measures of the other two sides, then the triangle is a right triangle.

Theorem 9–5 Converse of Pythagorean Theorem

Together, the Pythagorean Theorem and its converse provide a useful test for right triangles.

examples

2 **Tell whether or *not* a triangle with sides measuring 11.5 centimeters, 16.1 centimeters, and 20.7 centimeters is a right triangle.**

Recall that if the measures of two angles of a triangle are unequal, then the measures of the sides opposite those angles are unequal in the same order. As a result, the hypotenuse of a right triangle is its longest side. Thus, it is only necessary to check one equation.

$$(20.7)^2 \stackrel{?}{=} (16.1)^2 + (11.5)^2$$
$$428.49 \stackrel{?}{=} 259.21 + 132.25$$
$$428.49 \neq 391.46$$

Since the equation is *not* true, the triangle is *not* a right triangle.

3 **Tell whether or *not* a triangle with sides measuring 4 inches, $5\frac{1}{3}$ inches, and $6\frac{2}{3}$ inches is a right triangle.**

The hypotenuse of a right triangle is its longest side. Thus, it is only necessary to check one equation.

$$\left(6\tfrac{2}{3}\right)^2 \stackrel{?}{=} (4)^2 + \left(5\tfrac{1}{3}\right)^2$$
$$\left(\tfrac{20}{3}\right)^2 \stackrel{?}{=} (4)^2 + \left(\tfrac{16}{3}\right)^2$$
$$\frac{400}{9} \stackrel{?}{=} 16 + \frac{256}{9}$$
$$\frac{400}{9} = \frac{400}{9}$$

Since the equation is true, the triangle is a right triangle.

Many practical problems can be solved by using the Pythagorean Theorem or its converse.

example

4 The diagram below shows a roof truss. The height of the roof is 9 feet. It spans 18 feet on one part and 12 feet on the other. Find the lengths of both the long rafters and the short rafters, *not* counting the overhang. Round the answer to the nearest foot.

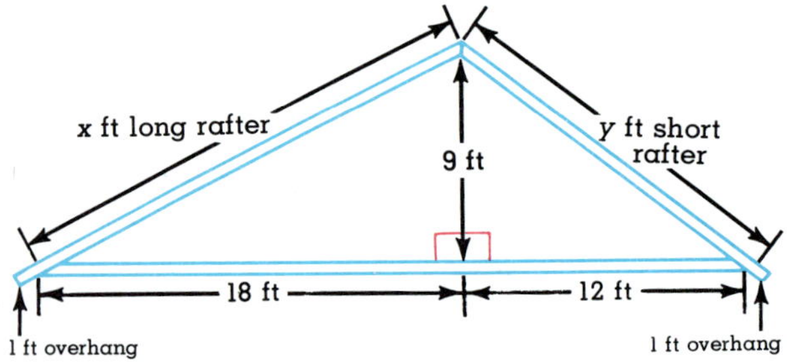

Define variables. Let x be the measure of the long rafter. Let y be the measure of the short rafter.

Write equations. Two right triangles are formed. One measures 18 feet by 9 feet by x feet. The other measures 12 feet by 9 feet by y feet.

$$x^2 = 18^2 + 9^2 \qquad y^2 = 12^2 + 9^2$$

Solve the equations.

$$\begin{aligned} x^2 &= 18^2 + 9^2 \\ x^2 &= 324 + 81 \\ x^2 &= 405 \\ x &= 9\sqrt{5} \\ x &\approx 9 \cdot 2.236 \\ &\approx 20.124 \\ &\approx 20 \end{aligned} \qquad \begin{aligned} y^2 &= 12^2 + 9^2 \\ y^2 &= 144 + 81 \\ y^2 &= 225 \\ y &= 15 \end{aligned}$$

Answer the problem. The long rafter is about 20 feet and the short rafter is 15 feet.

exercises

Exploratory Three integers that can be measures for the three sides of a right triangle are called a Pythagorean triple. Tell whether or *not* each of the following is a Pythagorean triple.

1. 5, 10, 12
2. 12, 16, 20
3. 15, 20, 25
4. 24, 32, 40
5. 9, 12, 15
6. 8, 15, 17
7. 3, 4, 5
8. 9, 12, 14
9. 30, 40, 50
10. 3, 6, 8
11. 20, 21, 29
12. 9, 40, 41
13. 16, 30, 34
14. 5, 12, 13
15. 14, 17, 19
16. 20, 21, 25
17. 5, 7, 9
18. 10, 24, 26
19. 15, 36, 39
20. 12, 35, 37
21. 6, 8, 10
22. 20, 48, 52
23. 7, 24, 25
24. 8, 8, 11
25. 18, 24, 30
26. 1, 3, 3
27. 6, 8, 12
28. 21, 28, 35
29. 11, 60, 61
30. 27, 36, 45

Written For each of the following, tell whether or *not* it is possible to draw a right triangle with sides of the given measures. Write *yes* or *no*.

1. 1.6, 3.0, 3.4
2. 2.2, 2.4, 3.3
3. 3.87, 4.47, 5.91
4. 0.27, 0.36, 0.45
5. $1, \sqrt{2}, \sqrt{3}$
6. $\sqrt{3}, \sqrt{4}, \sqrt{5}$
7. $\sqrt{5}, \sqrt{6}, \sqrt{7}$
8. $2, 2\frac{1}{3}, 3\frac{2}{3}$
9. $2, 2\frac{2}{3}, 3\frac{1}{3}$
10. $6\frac{1}{2}, 7\frac{1}{2}, 8\frac{1}{2}$
11. $1\frac{1}{2}, 2, 2\frac{1}{2}$
12. $2, 2, 2\sqrt{2}$

For each of the following, find the measure of the hypotenuse for a right triangle with legs of the given measures. Round each answer to one decimal place.

13. 5, 12
14. 12, 16
15. 3, 6
16. 5, 8
17. 11.2, 10
18. 6, 10.3
19. 1.7, 1.1
20. 3.4, 5.7
21. 7.8, 4.2

For each of the following find x. Round each answer to one decimal place.

22.
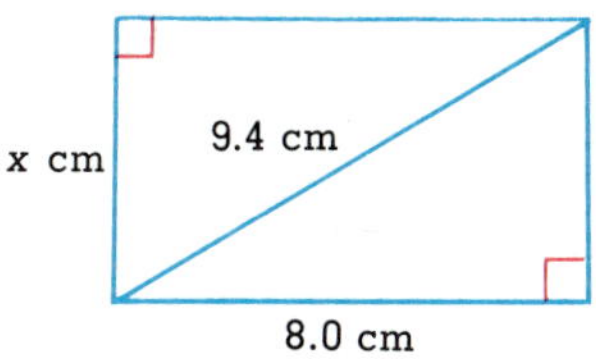

23.
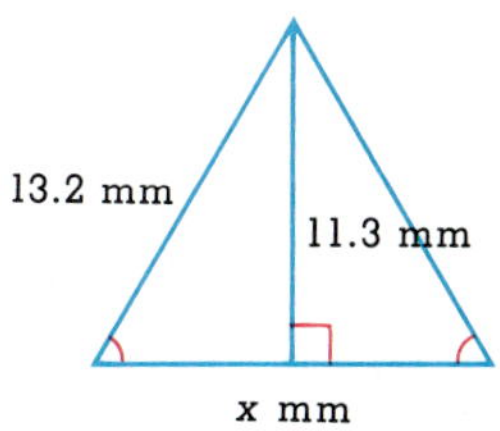

24.
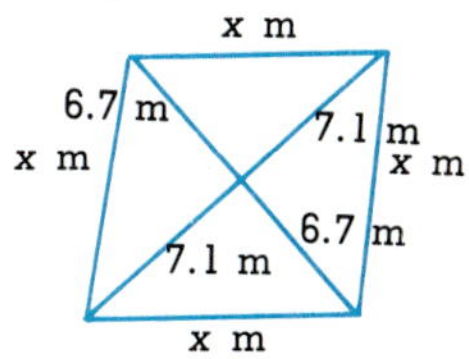

25. A stair stringer is a board which supports stairs. Suppose a set of stairs is to rise 8 feet over a length of 15 feet. Find the length of the stair stringer to the nearest tenth of a foot.

26. A picket fence is to have a gate 42 inches wide. The gate is 54 inches high. Find the length of a diagonal brace for the gate to the nearest inch.

27. Prove Theorem 9–5.

9–4 Special Right Triangles

The diagonal of a square separates the square into two isosceles right triangles. By using the Pythagorean Theorem, it is possible to develop a formula relating the length of the diagonals to the length of a side.

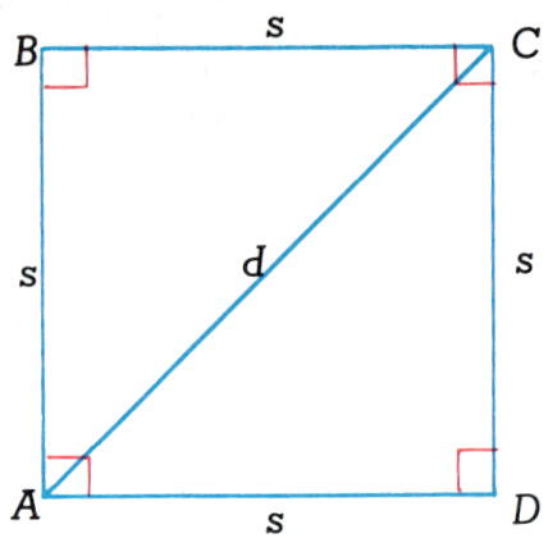

Let s be the measure of each side of a square.
Let d be the measure of a diagonal of the square.

$$d^2 = s^2 + s^2$$

$$d^2 = 2s^2$$

$$d = s\sqrt{2}$$

Theorem 9–6

If each acute angle of a right triangle has degree measure 45, then the hypotenuse measures $\sqrt{2}$ times the measure of a leg.

The altitude of an equilateral triangle separates the triangle into two congruent right triangles. Using the Pythagorean Theorem, it is possible to develop a formula relating the length of the altitude to the length of a side.

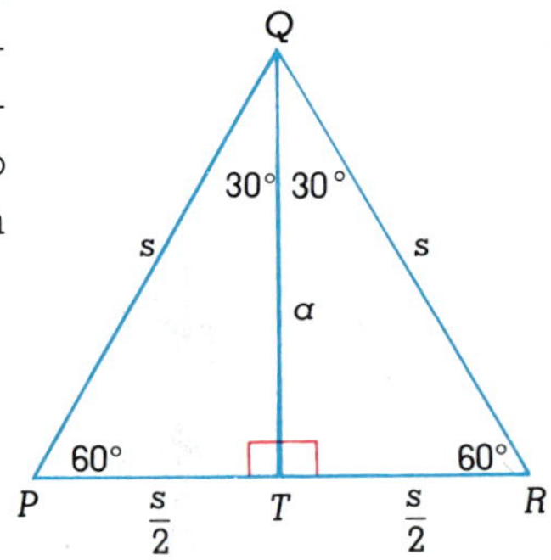

Let s be the measure of each side of an equilateral triangle.
Let a be the measure of an altitude of the triangle.

$$s^2 = a^2 + \left(\frac{s}{2}\right)^2$$

$$a^2 = s^2 - \left(\frac{s}{2}\right)^2$$

$$a^2 = \frac{3s^2}{4}$$

$$a = \frac{s\sqrt{3}}{2}$$

If the acute angles of a right triangle have degree measures 30 and 60, then the measure of the hypotenuse is 2 times the measure of the shorter leg and the measure of the longer leg is $\sqrt{3}$ times the measure of the shorter leg.	Theorem 9–7

The properties of the special right triangles can be used to solve problems.

example 1

Consecutive bases of a square-shaped baseball diamond are 90 feet apart. Find the distance from home plate to second base. Round your answer to the nearest foot.

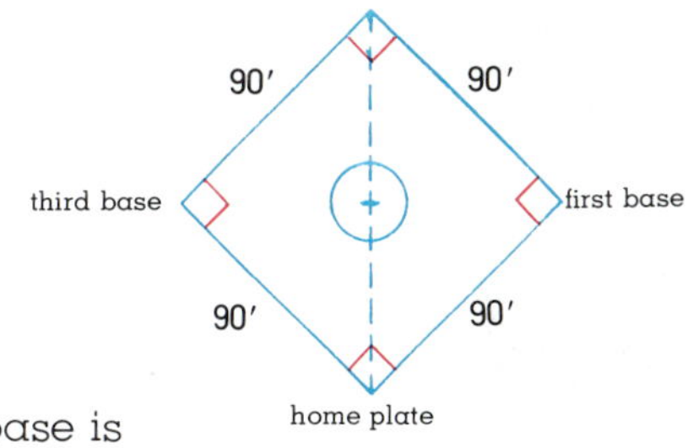

A line segment from home plate to second base is the diagonal of a square with side measuring 90 feet.

$$d = s\sqrt{2}$$
$$= 90\sqrt{2}$$
$$\approx 90 \cdot 1.414 \text{ or } 127.26$$

The distance from home plate to second base is about 127 feet.

exercises

Exploratory For each of the following, lengths of the sides of squares are given. Find the lengths of the diagonals.

1. 1 ft	**2.** 3 cm	**3.** 2 in.	**4.** 12 in.
5. 45 mm	**6.** 10 yd	**7.** 1.8 m	**8.** 2.5 cm
9. 31.2 m	**10.** $\frac{3}{4}$ ft	**11.** $4\frac{2}{3}$ yd	**12.** $1\frac{1}{2}$ ft

For each of the following, lengths of the sides of equilateral triangles are given. Find the lengths of the altitudes.

13. 1 in.	**14.** 1 m	**15.** 4 cm	**16.** 6 yd
17. 25 cm	**18.** 14 m	**19.** 2.5 cm	**20.** 1.4 m
21. $\frac{2}{3}$ yd	**22.** $\frac{1}{4}$ ft	**23.** $2\frac{3}{4}$ ft	**24.** $2\frac{1}{3}$ yd

Written For each of the following, find x. Round each answer to one decimal place.

1\. 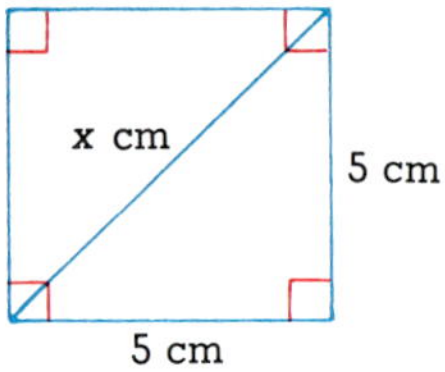

2\. 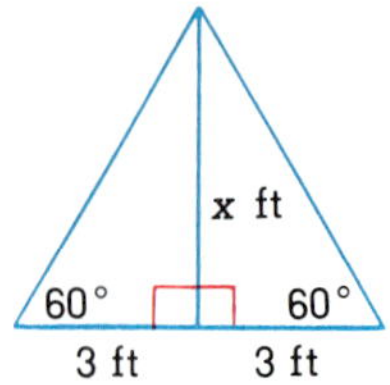

3\. 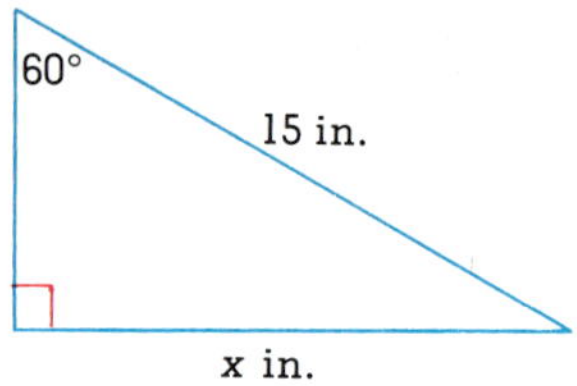

4\. 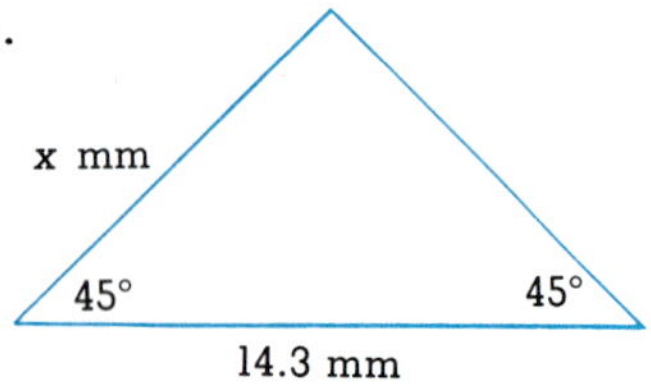

5\. 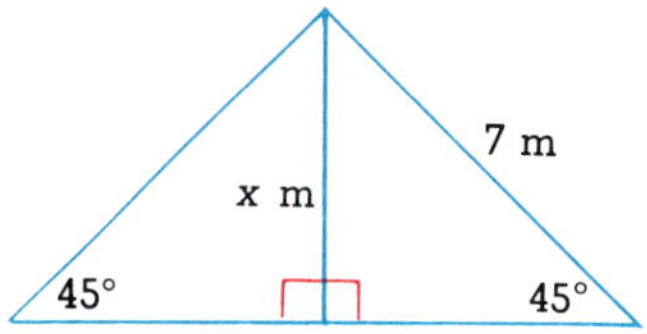

6\. 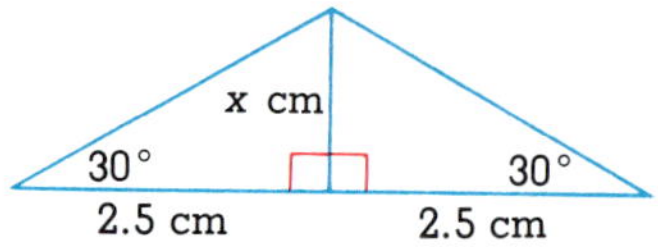

7\. 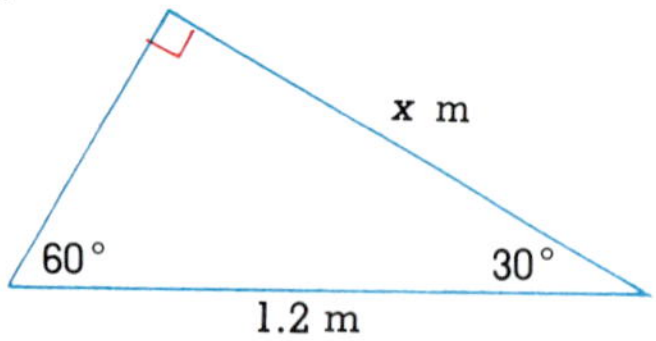

8\. 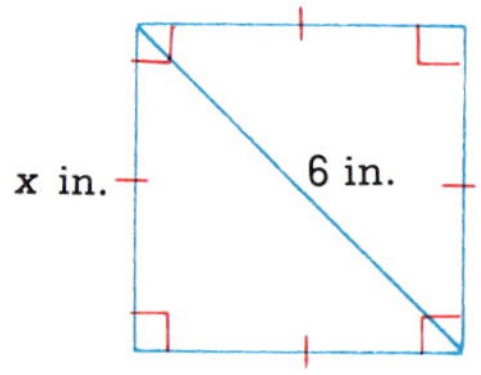

9\. 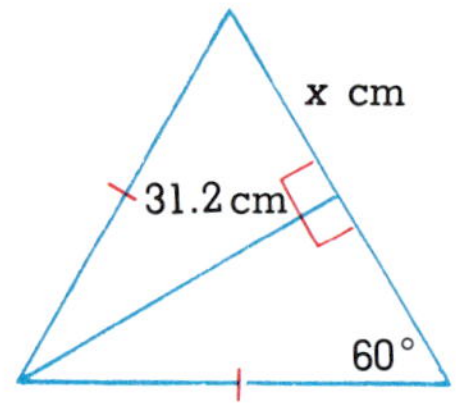

Solve each problem. Round each answer to one decimal place.

10. Consecutive bases of a square-shaped baseball diamond are 90 feet apart. Find the distance from first base to third base.
11. Find the measure of the diagonal of a square nut if each side measures 3 centimeters.
12. The perimeter of an equilateral triangle is 2.4. Find the measure of an altitude of the triangle.
13. The measure of an altitude of an equilateral triangle is 5.2. Find the perimeter of the triangle.

Use the figure at the right to find the value for each of the following.

14. u
15. v
16. w
17. x
18. y
19. z

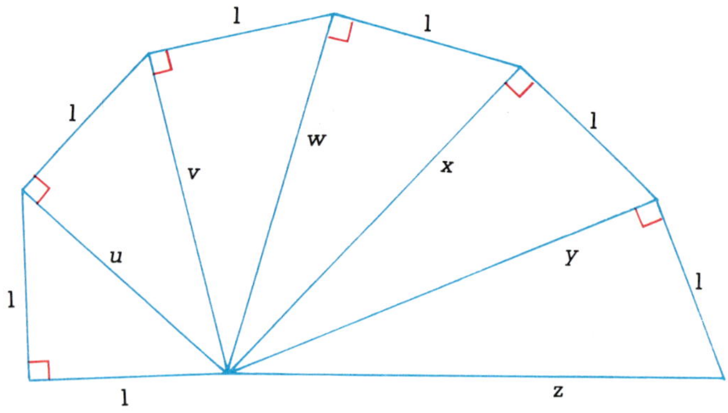

Challenge Suppose each side of the cube at the right measures s units. Develop a formula for the measure of the distance from A to F in terms of s.

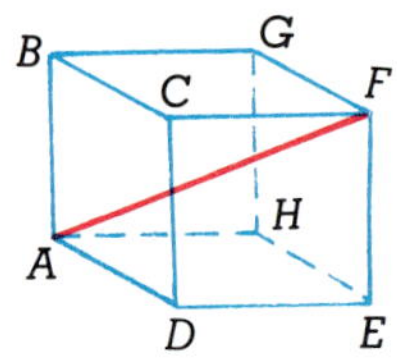

Using Geometry

Building Swings

David Wagoner builds swings for a city Parks and Recreation Department. Plans for building the swings use properties of triangles. Mr. Wagoner must construct a set of swings that is 2 meters tall for Schiller Park. He needs to determine the lengths of the legs. The *legs* and *supporting base* of the swings form *equilateral triangles*.

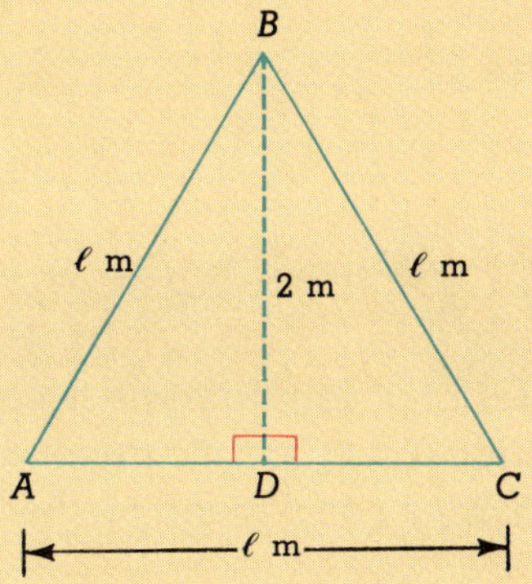

The equilateral triangle at the left represents the swings. Since $\overline{BD}$ is an altitude of the triangle the following equation can be written.

$$BD = \frac{\ell\sqrt{3}}{2}$$

To find ℓ, the measure of a leg, substitute 2 for BD and solve the equation for ℓ.

$$2 = \frac{\ell\sqrt{3}}{2}$$
$$4 = \ell\sqrt{3}$$
$$\ell = \frac{4}{\sqrt{3}}$$
$$\approx \frac{4}{1.7}$$
$$\approx 2.4$$

Each leg is approximately 2.4 meters long.

Exercises Given each of the following heights for swings, find the lengths of the legs. Round each answer to one decimal place.

1. 3 meters
2. 4 meters
3. 8 feet
4. 6 feet

9–5 Trigonometry

Notice that for the 30°–60° right triangles below, the ratios of the side opposite the 30° angle to the measure of the hypotenuse are equal.

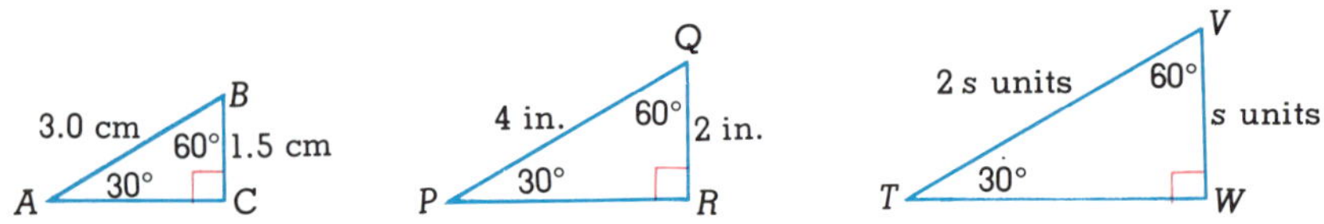

$\frac{BC}{BA} = \frac{1.5}{3.0}$ or $\frac{1}{2}$ $\quad$ $\frac{QR}{QP} = \frac{2}{4}$ or $\frac{1}{2}$ $\quad$ $\frac{VW}{VT} = \frac{s}{2s}$ or $\frac{1}{2}$

Consider the two triangles at the right. They are right triangles and $\angle A \cong \angle P$. Thus, the triangles are similar. How do the ratios $\frac{BC}{AB}$ and $\frac{QR}{PQ}$ compare?

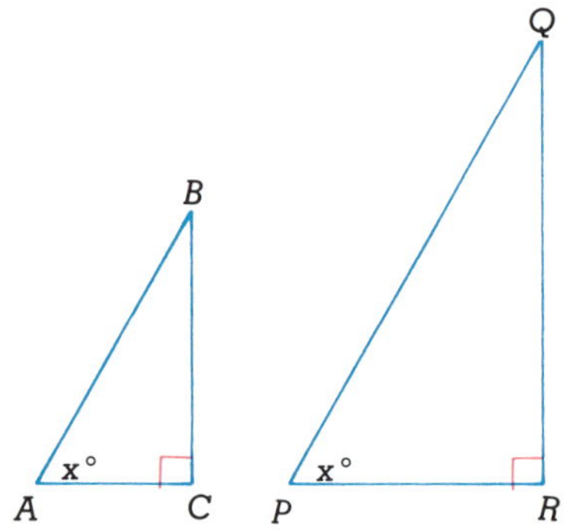

Because the triangles are similar, the ratios are equal.

A ratio of the measures of two sides of a right triangle is called a **trigonometric ratio.** The value of a trigonometric ratio depends only on the measure of an acute angle. It does *not* depend on the size of the triangle.

The three most common trigonometric ratios are defined in the following way.

Definition of Sine

A ratio is the sine of an acute angle of a right triangle if and only if it is the ratio of the measure of the leg opposite the acute angle to the measure of the hypotenuse.

Definition of Cosine

A ratio is the cosine of an acute angle of a right triangle if and only if it is the ratio of the measure of the leg adjacent to the acute angle to the measure of the hypotenuse.

A ratio is the tangent of an acute angle of a right triangle if and only if it is the ratio of the measure of the leg opposite the acute angle to the measure of the leg adjacent to the acute angle.

Definition of Tangent

Usually, we write sin A to mean *the sine of angle A*, and write the following equation.

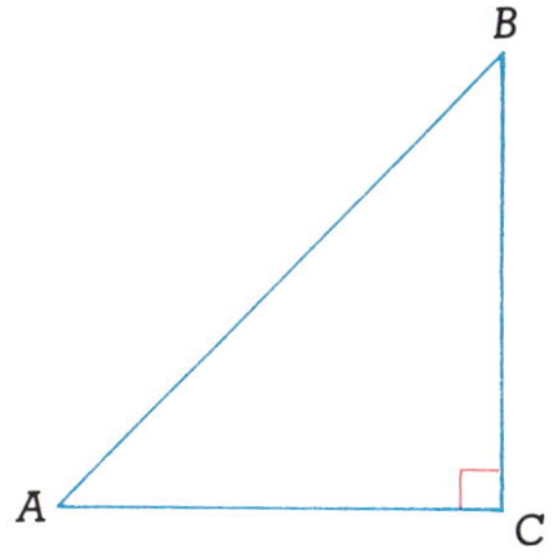

$$\sin A = \frac{BC}{AB}$$

Similarly, we use cos for *cosine*, and *tan* for *tangent*.

$$\cos A = \frac{CA}{AB}$$

$$\tan A = \frac{BC}{CA}$$

If the degree measure of $\angle A$ is 45°, then we write *sin 45°*, *cos 45°*, and *tan 45°*.

Given the measures of the sides of a triangle, the values of sine, cosine, and tangent can be calculated.

example 1

Find sin 45°, cos 45°, and tan 45°.

An isosceles right triangle has a 45° angle. According to Theorem 9–6, the hypotenuse measures $\sqrt{2}$ times the measure of a side.

$$\sin 45° = \sin A = \frac{BC}{AB} = \frac{s}{s\sqrt{2}} = \frac{s}{s\sqrt{2}} \cdot \frac{\sqrt{2}}{\sqrt{2}} = \frac{\sqrt{2}}{2}$$

$$\cos 45° = \cos A = \frac{CA}{BA} = \frac{s}{s\sqrt{2}} = \frac{\sqrt{2}}{2}$$

$$\tan 45° = \tan A = \frac{BC}{CA} = \frac{s}{s} = 1$$

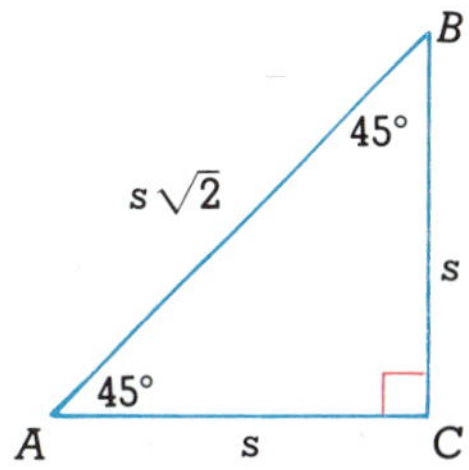

Thus, $\sin 45° = \frac{\sqrt{2}}{2}$, $\cos 45° = \frac{\sqrt{2}}{2}$, and $\tan 45° = 1$.

exercises

Exploratory For the three triangles at the right, express each of the following as a fraction in simplest form.

1. $\sin A$	2. $\cos A$	3. $\tan A$
4. $\sin B$	5. $\cos B$	6. $\tan B$
7. $\sin E$	8. $\cos E$	9. $\tan E$
10. $\sin F$	11. $\cos F$	12. $\tan F$
13. $\sin G$	14. $\cos G$	15. $\tan G$
16. $\sin I$	17. $\cos I$	18. $\tan I$

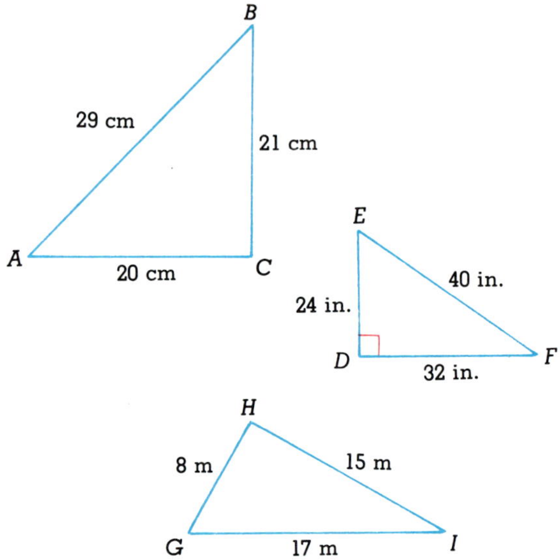

Written For each of the following triangles, express the sine, cosine, and tangent of each acute angle to three decimal places.

1.
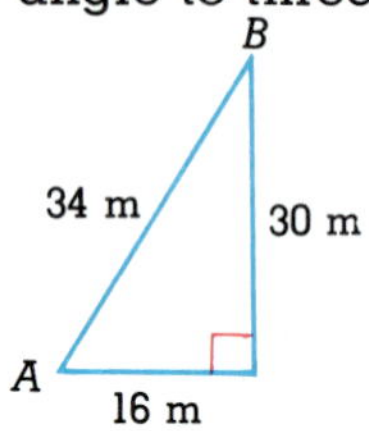

2.
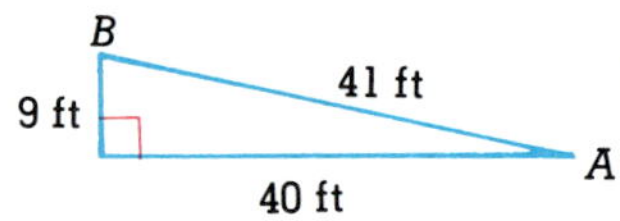

3.
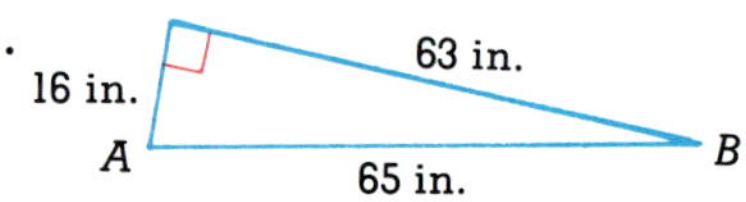

4.
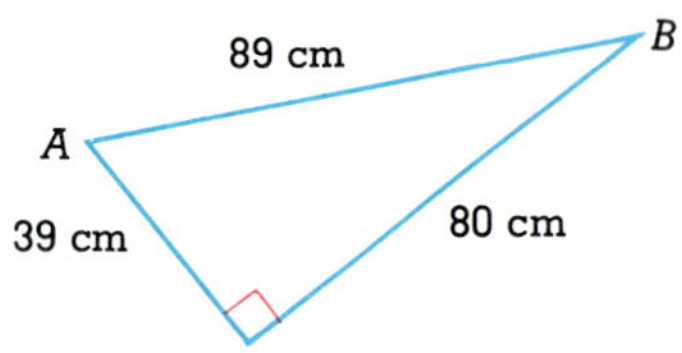

5.
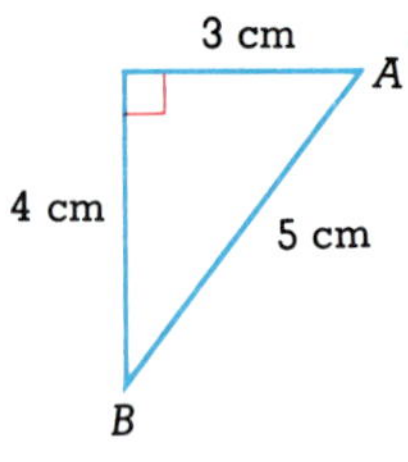

6.
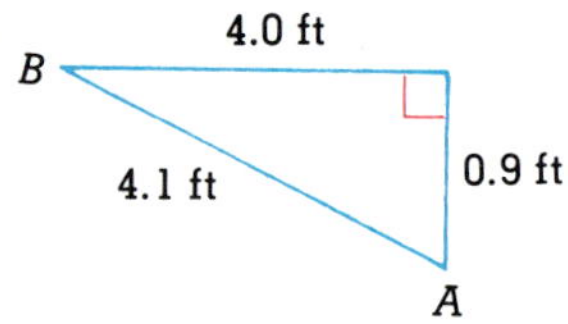

Solve each problem.

7. Find sin 30°.
8. Find cos 30°.
9. Find tan 30°.
10. Find sin 60°.
11. Find cos 60°.
12. Find tan 60°.
13. Suppose $\angle A$ and $\angle B$ are complementary angles of a triangle. How are $\sin A$ and $\cos B$ related?

9–6 Trigonometric Tables

A table can be used to find decimal approximations for values of trigonometric ratios. Such a table is given at the back of this book.

example 1

Find cos 48°.

Locate 48° in the column marked Angle.

Angle	sin	cos	tan
45°	0.7071	0.7071	1.0000
46°	0.7193	0.6947	1.0355
47°	0.7314	0.6820	1.0724
48°	0.7431	0.6691	1.1106
49°	0.7547	0.6561	1.1504
50°	0.7660	0.6428	1.1918
51°	0.7771	0.6293	1.2349
52°	0.7880	0.6157	1.2799
53°	0.7986	0.6018	1.3270

Locate the value for cos 48° in the column marked cos.

An approximate value for cos 48° is 0.6691.

A table also can be used to find the measure of an angle, given a ratio.

example 2

Given cos A = 0.5446, find $m\angle A$.

Locate 0.5446 in the column marked cos.

Locate the measure of the angle in the column marked Angle.

Angle	sin	cos	tan
54°	0.8090	0.5878	1.3764
55°	0.8192	0.5736	1.4281
56°	0.8290	0.5592	1.4826
57°	0.8387	0.5446	1.5399
58°	0.8480	0.5299	1.6003
59°	0.8572	0.5150	1.6643
60°	0.8660	0.5000	1.7321
61°	0.8746	0.4848	1.8040
62°	0.8829	0.4695	1.8807
63°	0.8910	0.4540	1.9626
64°	0.8988	0.4384	2.0503
65°	0.9063	0.4226	2.1445

The measure of $\angle A$ is 57°.

If a given value does *not* match any entry in a table, choose the closest value from the table.

example 3

Given $\tan B = 0.1234$, find $m\angle B$.

Look in the column marked *tan*.

Angle	sin	cos	tan
0°	0.0000	1.0000	0.0000
1°	0.0175	0.9998	0.0175
2°	0.0349	0.9994	0.0349
3°	0.0523	0.9986	0.0524
4°	0.0698	0.9976	0.0699
5°	0.0872	0.9962	0.0875
6°	0.1045	0.9945	0.1051
7°	0.1219	0.9925	0.1228
8°	0.1392	0.9903	0.1405
9°	0.1564	0.9877	0.1584
10°	0.1736	0.9848	0.1763
11°	0.1908	0.9816	0.1944
12°	0.2079	0.9781	0.2126

Locate the measure of the angle in the column marked Angle.

The value of 0.1234 is between 0.1228 and 0.1405. It is closer to 0.1228.

The measure of $\angle B$ is about 7°.

exercises

Exploratory **Use the table at the back of this book to answer each of the following.**

1. Name the *least* possible value for sine.
2. Name the measurement of an angle for which the sine is at its *minimum*.
3. Name the *greatest* possible value for sine.
4. Name the measurement of an angle for which the sine is at its *maximum*.
5. Name the *least* possible value for cosine.
6. Name the measurement of an angle for which the cosine is at its *minimum*.
7. Name the *greatest* possible value for cosine.
8. Name the measurement of an angle for which the cosine is at its *maximum*.
9. Name the *least* possible value for tangent.
10. Name the *greatest* possible value for tangent.
11. Name the measurement of the angle for which sine and cosine are *equal*.
12. Name the tangent of the angle for which sine and cosine are *equal*.

Written Use the table at the back of this book to find the value for each of the following ratios.

1. $\sin 0°$
2. $\sin 32°$
3. $\cos 0°$
4. $\tan 0°$
5. $\cos 69°$
6. $\tan 29°$
7. $\sin 58°$
8. $\cos 10°$
9. $\tan 38°$
10. $\sin 44°$
11. $\tan 45°$
12. $\tan 89°$
13. $\cos 22°$
14. $\tan 79°$
15. $\cos 36°$
16. $\cos 45°$
17. $\sin 13°$
18. $\cos 90°$
19. $\tan 50°$
20. $\sin 63°$
21. $\tan 84°$
22. $\sin 18°$
23. $\cos 16°$
24. $\tan 90°$
25. $\tan 51°$
26. $\cos 58°$
27. $\sin 89°$
28. $\sin 90°$

Use the table to find the measurement of each angle to the nearest degree.

29. $\sin A = 0.1045$
30. $\sin B = 0.5000$
31. $\cos C = 0.8988$
32. $\tan D = 0.3839$
33. $\tan E = 2.0503$
34. $\sin F = 0.7431$
35. $\cos P = 0.2588$
36. $\tan Q = 3.0777$
37. $\sin R = 0.9994$
38. $\tan S = 0.7002$
39. $\cos T = 0.8988$
40. $\sin V = 0.9205$
41. $\cos M = 0.3000$
42. $\tan N = 0.1500$
43. $\cos A = 0.7777$
44. $\cos P = 0.6$
45. $\tan L = 4.00$
46. $\tan T = 25$
47. $\sin X = 0.3$
48. $\tan Z = 0.15$
49. $\cos W = 0.65$
50. $\tan F = 1.75$
51. $\sin Q = 0.25$
52. $\cos E = 0.25$
53. $\tan R = 1.500$
54. $\cos G = 0.999$
55. $\sin H = 0.999$
56. $\sin A = \frac{1}{2}$
57. $\cos T = \frac{4}{5}$
58. $\tan Z = \frac{3}{4}$

Find values for each of the following to the nearest four decimal places.

59. $\cos 30° - \sin 30°$
60. $\cos 60° - \sin 30°$
61. $\sin 90° + \cos 30°$
62. $\cos 30° + \cos 60°$
63. $\tan 45° + \tan 0°$
64. $\tan 0° + \cos 90°$
65. $3 \sin 30° - \sin 90°$
66. $\cos 60° - 2 \cos 30°$
67. $2 \tan 30° - \tan 60°$
68. $\tan 60° - 2 \tan 30°$
69. $2 \sin 45° - \sin 90°$
70. $2 \cos 45° - \cos 90°$
71. $\frac{1}{2} \sin 60° - \sin 30°$
72. $\frac{1}{2} \cos 60° + \sin 30°$
73. $\tan 45° + \tan 60°$
74. $\tan 45° - \frac{1}{2} \tan 60°$
75. $\frac{1}{2}(\sin 60° + \sin 3C°)$
76. $\frac{1}{2}(\tan 45° - \tan 30°)$

Excursions in Geometry

Barbers

The sign on a barber's window says, "I shave all those people in town, and *only* those people, who do *not* shave themselves."

Who shaves the barber?

9–7 Using Trigonometry

Trigonometric ratios can be used to find missing measures of right triangles.

examples

1 Find $m\angle A$ to the nearest degree for the triangle at the right.

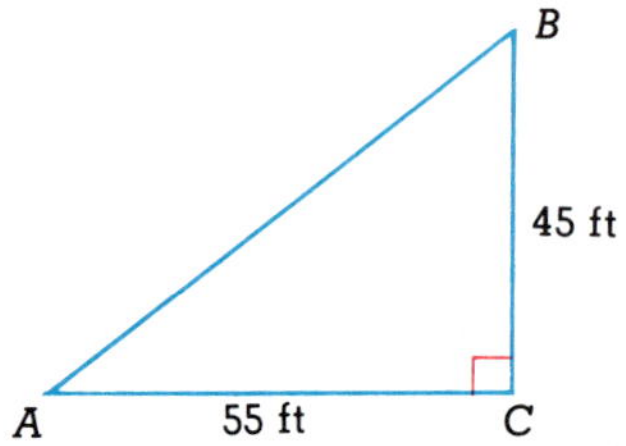

$$\tan A = \frac{BC}{AC}$$

$$\tan A = \frac{45}{55}$$

$$\tan A \approx 0.8182$$

$$A \approx 39°$$

2 Find QR to the nearest tenth for the triangle at the right.

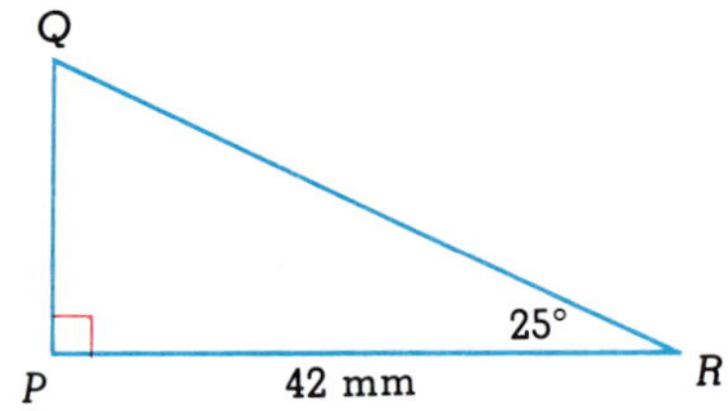

$$\cos R = \frac{PR}{QR}$$ *Could sin R or tan R be used?*

$$\cos 25° = \frac{42}{QR}$$

$$QR = \frac{42}{\cos 25°}$$

$$\approx \frac{42}{0.9063}$$

$$QR \approx 46.3$$

3 Find TV and VW to the nearest tenth for the triangle at the right.

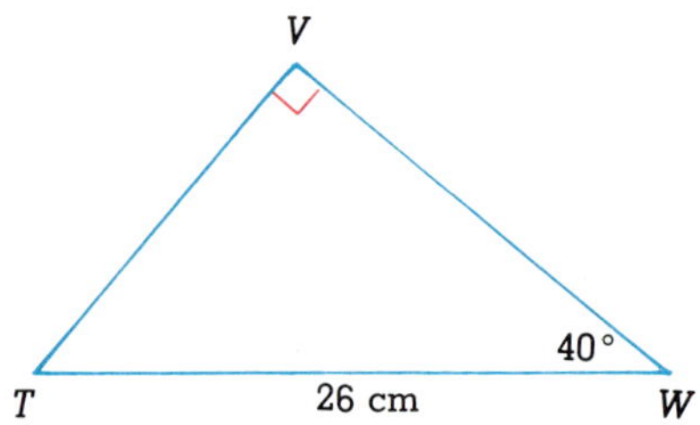

$$\sin W = \frac{TV}{TW} \qquad \cos W = \frac{VW}{TW}$$

$$\sin 40° = \frac{TV}{26} \qquad \cos 40° = \frac{VW}{26}$$

$$TV = 26 \sin 40° \qquad VW = 26 \cos 40°$$

$$\approx 26\ (0.6428) \qquad \approx 26\ (0.7660)$$

$$TV \approx 16.7 \qquad VW \approx 19.9$$

To see the top of a peak or an object in the sky, a person must look up rather than looking straight ahead. An angle, called the *angle of elevation*, is formed by the line of sight and a horizontal line.

A person in a tower or on a cliff must look down to see an object below. This person's line of sight forms an *angle of depression* with a horizontal line.

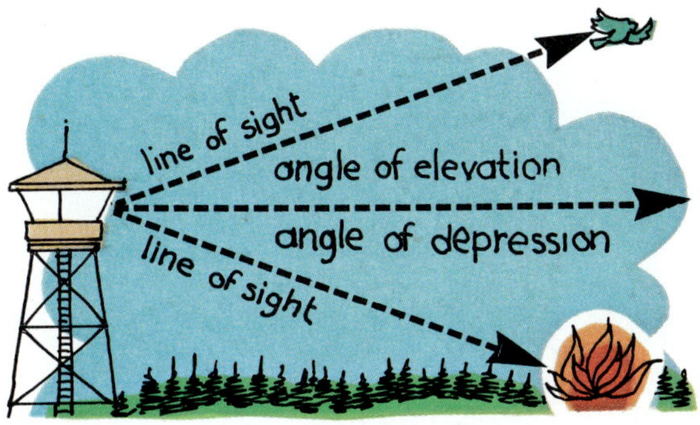

example 4

A surveyor is 100 meters from a dam. The angle of elevation to the top of the dam is 26°. The surveyor's instrument is 1.73 meters above the ground. Find the height of the dam to the nearest two decimal places.

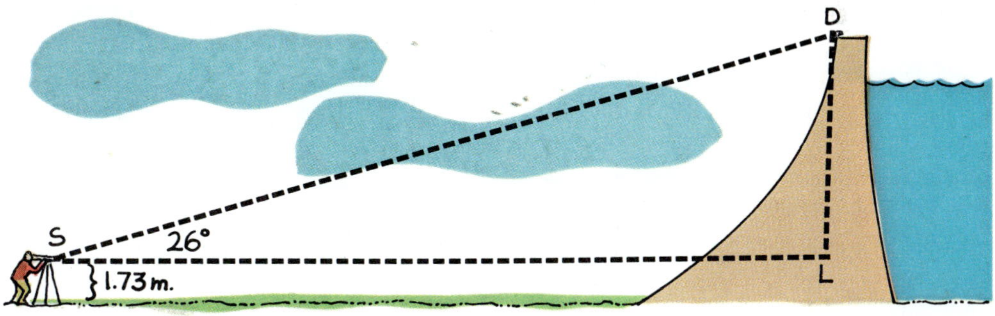

$$\tan 26° = \frac{DL}{100}$$

$$0.4877 \approx \frac{DL}{100}$$

$$48.77 \approx DL$$

The height of the dam is about 48.77 + 1.73 or 50.50 meters.

exercises

Exploratory For each of the following, state an equation that would enable you to solve each problem. Use the triangle at the right.

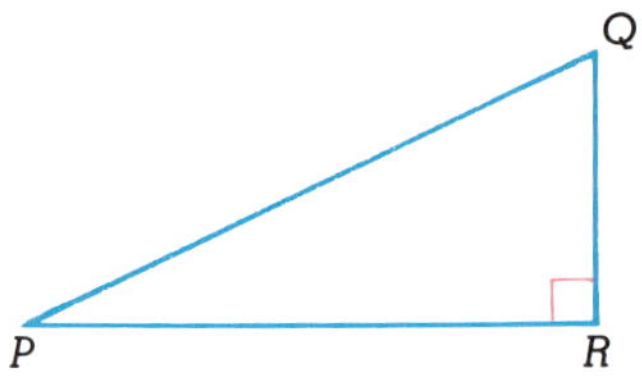

1. Given $m \angle P = 15$ and $PQ = 37$, find QR.
2. Given $m \angle Q = 72$ and $PR = 13$, find QR.
3. Given $m \angle P = 47$ and $QR = 10$, find PQ.
4. Given $QR = 21.5$ and $m \angle Q = 87$, find PR.
5. Given $PR = 13.4$ and $m \angle P = 16$, find PQ.
6. Given $PR = 31.8$ and $m \angle P = 19$, find PQ.

Written Use $\triangle ABC$ at the right to help solve each problem.

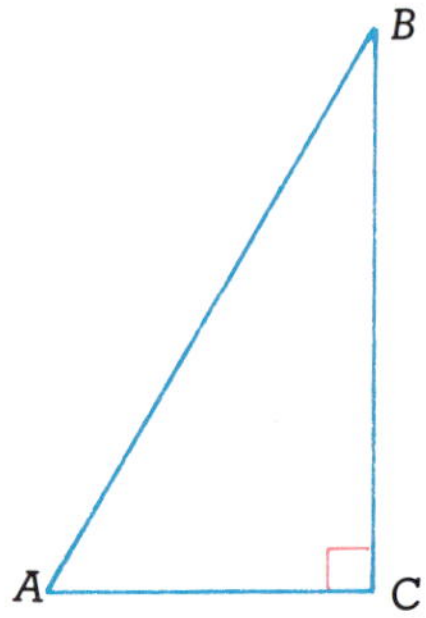

1. Given $m \angle B = 32$ and $AB = 74$, find AC.
2. Given $m \angle B = 43$ and $AB = 90$, find BC.
3. Given $AB = 14$ and $AC = 6$, find $m \angle A$.
4. Given $AB = 60$ and $AC = 20$, find $m \angle A$.
5. Given $AC = 5$ and $BC = 12$, find $m \angle B$.
6. Given $BC = 4$ and $AC = 3$, find $m \angle B$.
7. Given $m \angle B = 34$ and $BC = 15$, find AC.
8. Given $m \angle A = 62$ and $AC = 20$, find BC.
9. Given $m \angle A = 68$ and $BC = 98$, find AB.
10. Given $m \angle B = 32$ and $AC = 10$, find AB.

Solve each problem. Round all answers to two decimal places.

11. A surveyor is 100 meters from a bridge. The angle of elevation to the top of the bridge is 35°. The surveyor's instrument is 1.45 meters above the ground. Find the height of the bridge.
12. A surveyor is 100 meters from a building. The angle of elevation to the top of the building is 23°. The surveyor's instrument is 1.55 meters above the ground. Find the height of the building.
13. To secure a 500 meter radio tower against high winds, guy wires are attached to a ring on the tower. The ring is 5 meters from the top. The wires form a 15° angle with the tower. Find the distance from the tower to the guy wire anchor in the ground.
14. In a parking garage, each level is 20 feet apart. Each ramp to a level is 130 feet long. Find the measure of the angle of elevation for each ramp.
15. Before Apollo 11 descended to the surface of the moon, it made one orbit at a distance of 3 miles from the surface. At one point in its orbit, the on board guidance system measured the angles of depression to the near and far edges of a huge crater. The angles measured 25° and 18°. Find the distance across the crater.

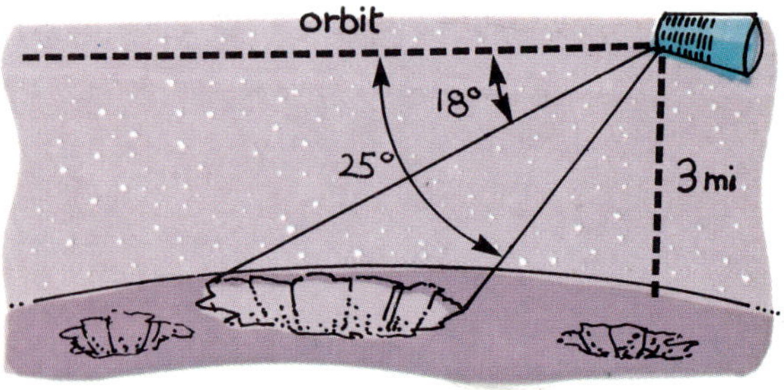

Vocabulary

square root (264)
radical sign (264)
principal square root (264)
rationalizing the denominator (266)
irrational numbers (266)
geometric mean (268)
trigonometric ratio (280)
sine (280)
cosine (280)
tangent (281)

Chapter Summary

1. Definition of square root: For any numbers a and b, if $a^2 = b$, then a is a square root of b. (264)
2. The radical sign indicates a nonnegative square root. (264)
3. Product Property of Square Roots: For any nonnegative numbers a and b, $\sqrt{ab} = \sqrt{a} \cdot \sqrt{b}$. (265)
4. Quotient Property of Square Roots: For any nonnegative numbers a and b, $\sqrt{\frac{a}{b}} = \frac{\sqrt{a}}{\sqrt{b}}$ as long as b is *not* zero. (265)
5. A square root is said to be simplified when the following conditions are met. (266)
   1. The radicand has no factor other than 1 that is a perfect square.
   2. The radicand does *not* contain a fraction.
   3. No radical appears in the denominator of a fraction.
6. Theorem 9–1: If the altitude is drawn from the right angle to the hypotenuse of a right triangle, then the two triangles formed are similar to the given triangle and to each other. (268)
7. Definition of Geometric Mean: For any positive numbers a and b, x is the geometric mean between a and b if and only if $\frac{a}{x} = \frac{x}{b}$ and x is positive. (269)
8. Theorem 9–2: The measure of the altitude drawn from the right angle to the hypotenuse of a right triangle is the geometric mean between the measure of the two segments of the hypotenuse. (270)
9. Theorem 9–3: If the altitude is drawn to the hypotenuse of a right triangle, then the measure of a leg of the triangle is the geometric mean between the measure of the hypotenuse and the measure of the segment of the hypotenuse adjacent to that leg. (270)

10. The Pythagorean Theorem: For any right triangle, the square of the measure of the hypotenuse equals the sum of the squares of the measures of the legs. (272)
11. Converse of Pythagorean Theorem: If the square of the measure of one side of a triangle equals the sum of the squares of the measures of the other two sides, then the triangle is a right triangle. (273)
12. Theorem 9–6: If each acute angle of a right triangle has degree measure 45, then the hypotenuse measures $\sqrt{2}$ times the measure of a leg. (276)
13. Theorem 9–7: If the acute angles of a right triangle have degree measures 30 and 60, then the hypotenuse measures 2 times the measure of the shorter leg and the longer leg measures $\sqrt{3}$ times the measure of the shorter leg. (277)
14. Definition of Sine: A ratio is the sine of an acute angle of a right triangle if and only if it is the ratio of the measure of the leg opposite the acute angle to the measure of the hypotenuse. (280)
15. Definition of cosine: A ratio is the cosine of an acute angle of a right triangle if and only if it is the ratio of the measure of the leg adjacent to the acute angle to the measure of the hypotenuse. (280)
16. Definition of Tangent: A ratio is the tangent of an acute angle of a right triangle if and only if it is the ratio of the measure of the leg opposite the acute angle to the measure of the leg adjacent to the acute angle. (281)
17. A table can be used to find decimal approximations for values of trigonometric ratios. (283)

Chapter Review

9–1 **State the principal square root of each expression.**

1. 25 2. 0.04 3. $\frac{81}{16}$ 4. 75 5. $\frac{5}{3}$ 6. $\frac{1}{2}$

Find the approximate value of each expression. Use the table of Squares and Approximate Square Roots.

7. $\sqrt{175}$ 8. $\sqrt{6} - \sqrt{15}$ 9. $\sqrt{\frac{7}{3}}$

9–2 **Use the figure at the right to solve each of the following. Approximate each answer to three decimal places.**

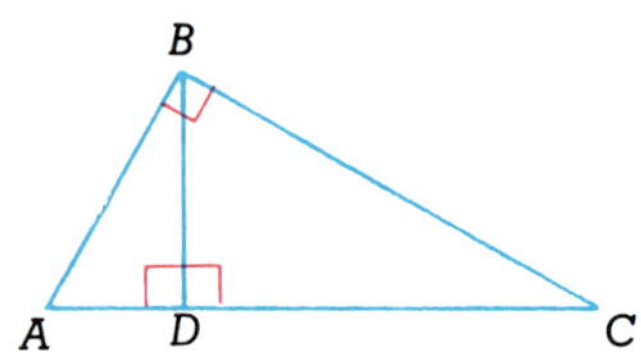

10. Find BD if $AD = 4$ and $DC = 6$.
11. Find AB if $AC = 18$ and $AD = 4$.
12. Find BC if $AD = 4\frac{8}{9}$ and $DC = \frac{1}{9}$.
13. Find BA if $AD = \frac{1}{3}$ and $DC = \frac{1}{4}$.

9–3 For each of the following, tell whether or *not* it is possible to draw a right triangle with sides of the given measures. Write *yes* or *no*.

14. 0.9, 2.1, 2.3 **15.** 0.9, 4.0, 4.1 **16.** 4, 4, $4\sqrt{2}$

17. 4, $7\frac{1}{2}$, $8\frac{1}{2}$ **18.** $\sqrt{5}$, $\sqrt{7}$, $\sqrt{12}$ **19.** $\sqrt{5}$, $\sqrt{6}$, $\sqrt{11}$

For each of the following, find the measure of the hypotenuse for a right triangle with legs of the given measure. Round each answer to one decimal place.

20. 9.4, 8.0 **21.** 7.1, 6.7 **22.** 8, 15

9–4 For each of the following, find x. Round each answer to one decimal place.

23.

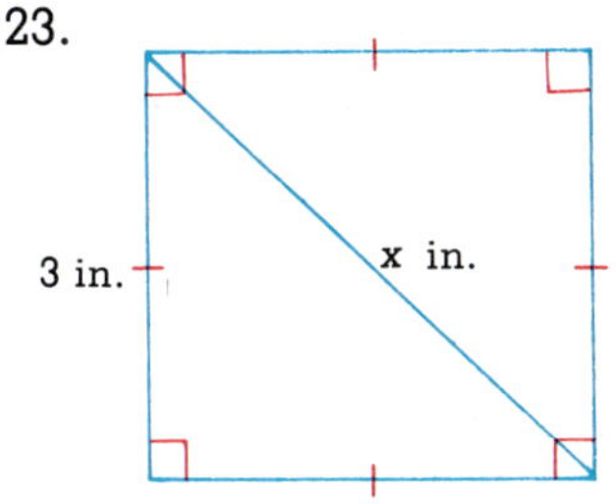

24.

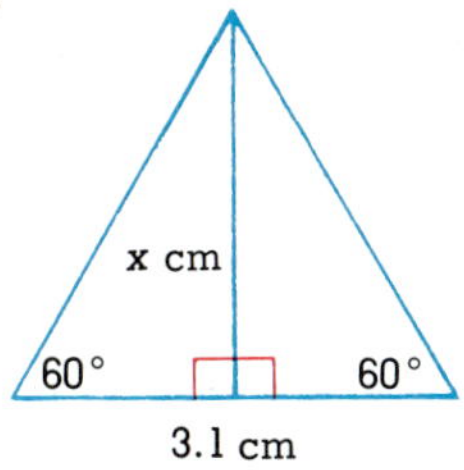

25.

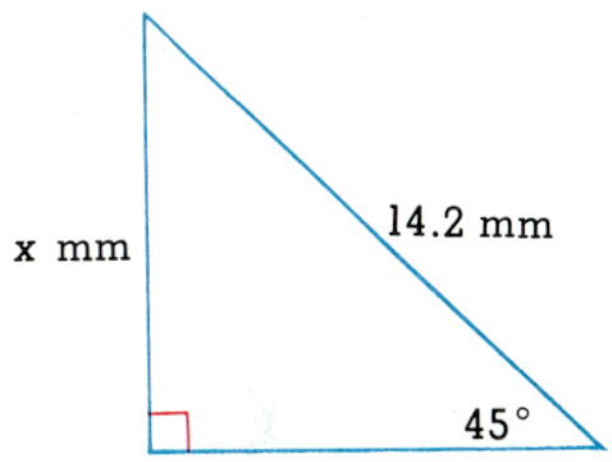

9–5 For each of the following triangles, express the sine, cosine, and tangent of each acute angle to three decimal places.

26.

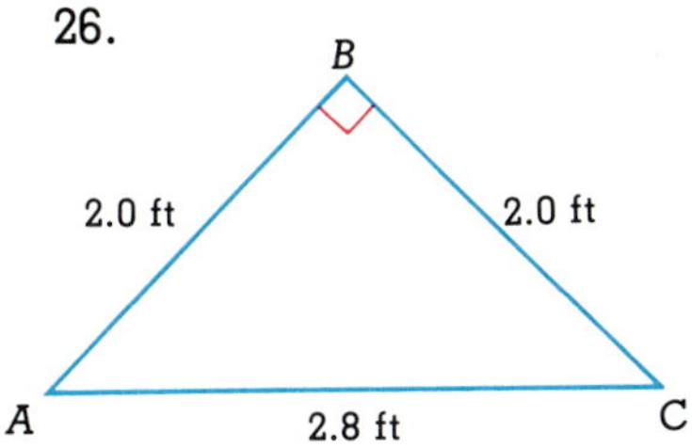

27.

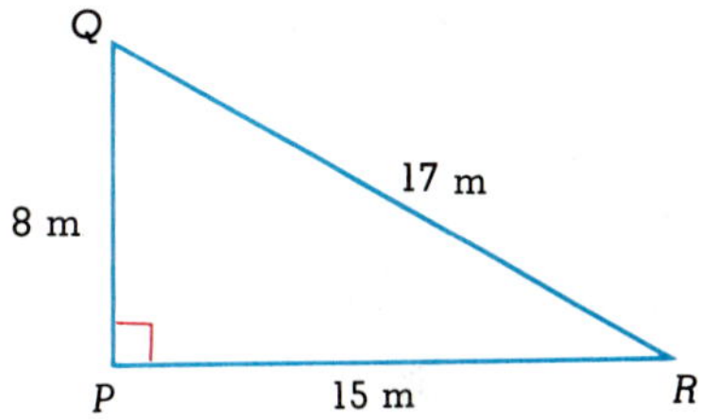

9–6 Find the approximate value of each expression. Use the table of Trigonometric Ratios.

28. sin 14° **29.** cos 33° **30.** tan 51°

Find the measure of each angle to the nearest degree.

31. $\sin A = 0.5000$ **32.** $\cos B = 0.3200$ **33.** $\tan C = 0.95$

9–7 Use $\triangle ABC$ at the right to help solve each problem.

34. given $m\angle B = 25$ and $AB = 32$, find AC.

35. given $m\angle A = 52$ and $AB = 7.5$, find BC.

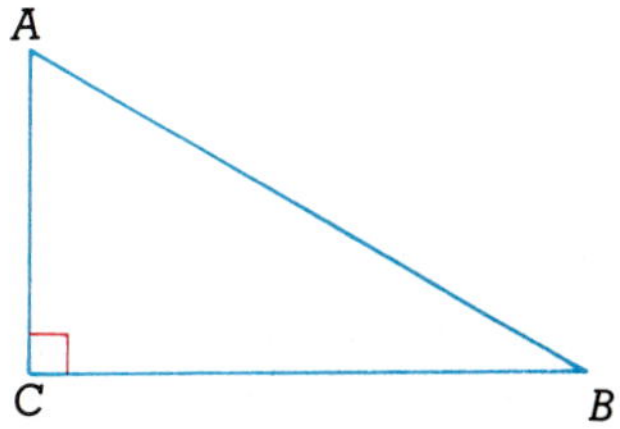

36. A surveyor is 350 m from a mountain peak. The angle of elevation to the top of the peak is 85°. The surveyor's instrument is 1.7 m above the ground. Find the height of the mountain peak to the nearest decimal place.

Chapter Test

For each of the following pairs of numbers, find the geometric mean. Approximate each answer to three decimal places.

1. 3 and 12
2. 11 and 11
3. 5 and 4
4. $\frac{1}{3}$ and $\frac{1}{2}$
5. $\sqrt{2}$ and $2\sqrt{2}$
6. $\frac{3}{4}$ and $\frac{4}{3}$

For each of the following, tell whether or *not* it is possible to draw a right triangle with sides of the given measures. Write *yes* or *no*.

7. 16, 30, 34
8. 0.3, 0.4, 0.5
9. $\sqrt{7}$, $2\sqrt{6}$, 5
10. $\frac{1}{3}$, $\frac{1}{4}$, $\frac{1}{5}$

For each of the following, find x. Round each answer to one decimal place.

11.

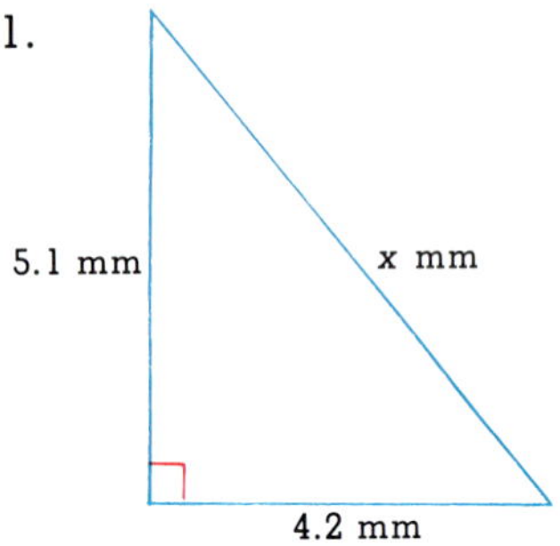

12.

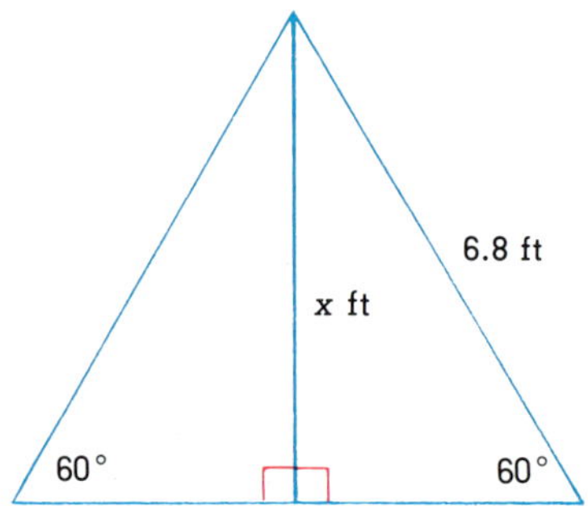

13.

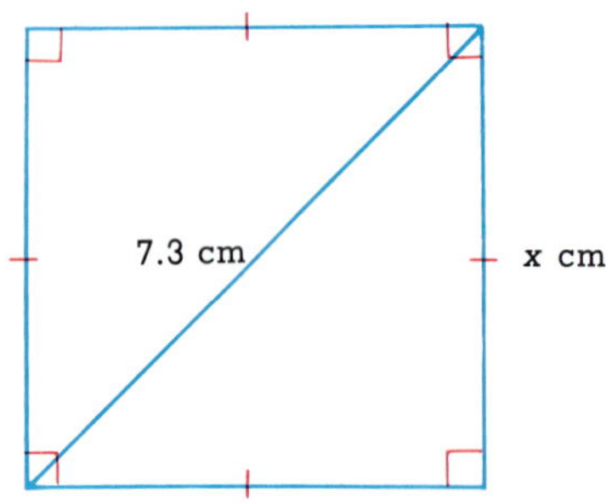

14.

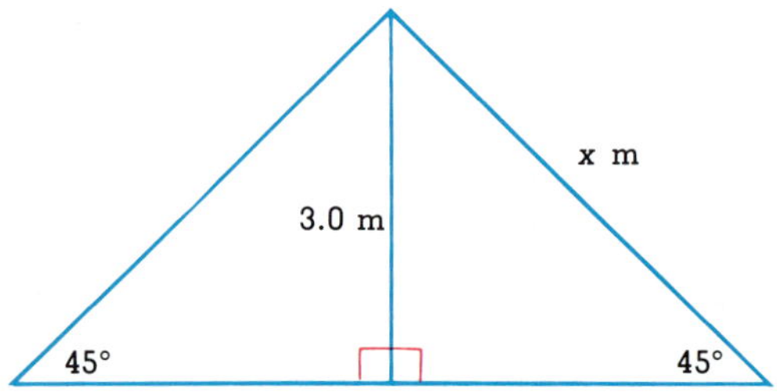

Use the figure at the right to solve each problem.

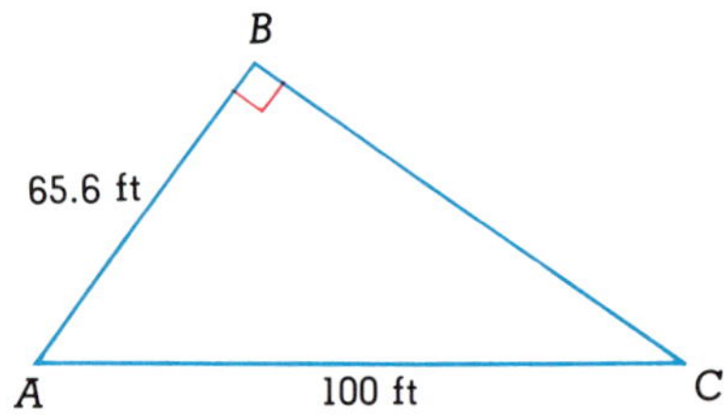

15. Find $\sin A$ to four decimal places.
16. Find $\sin C$ to four decimal places.
17. Find $\tan A$ to four decimal places.
18. Find $\cos C$ to four decimal places.
19. Find $m\angle A$ to the nearest degree.
20. Find $m\angle C$ to the nearest degree.
21. A flag pole casts a shadow 20 feet long when the angle of elevation of the sun measures 43°. Find the height of the pole to the nearest foot.

Chapter

10

Circles and Spheres

The photograph above provides a graphic illustration of circles. A circle is the set of points in a plane that are a given distance from a given point in the plane.

10–1 Parts of Circles

To irrigate the cornfields shown in the photograph, water is sprayed from pipes that rotate about central points. Circular patterns result. Notice that all the points along the edge of a field are the same distance from the center. Each edge forms a **circle.**

Definition of Circle

In a plane, a figure is a circle if and only if it is the set of all points that are a given distance from a given point, called the center.

Usually, circles are named by their centers. For example, the circle at the right is called *circle P*. The symbol $\odot$ means *circle*. Thus, $\odot P$ means *circle P*.

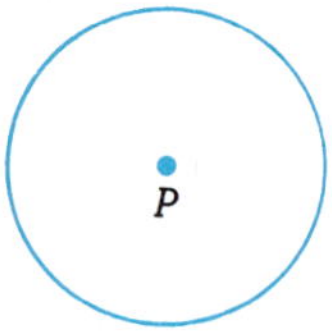

The following figures show line segments that are related to circles.

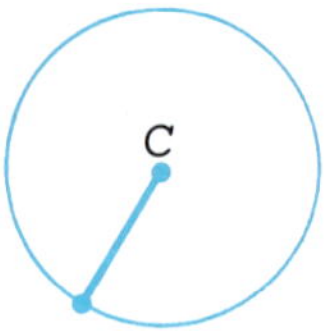

A **radius** of a circle is a line segment whose endpoints are the center of the circle and a point on the circle.

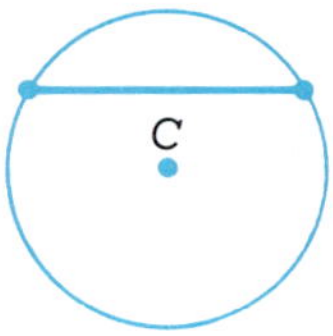

A **chord** of a circle is a line segment whose endpoints are points on the circle.

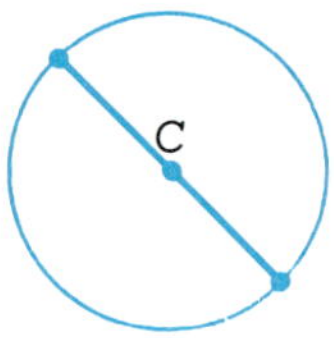

A **diameter** of a circle is a line segment whose endpoints are points on the circle and which contains the center of the circle.

It follows from the definition of a circle that all the radii of a circle are congruent. Also, all the diameters of a circle are congruent.

How are the measure of a radius and the measure of a diameter of a circle related?

A circle separates a plane into three parts. The parts are the *interior*, the *exterior*, and the circle itself.

Suppose a point is in the *interior* of a circle. The measure of the segment joining the point to the center is *less than* the measure of the radius.

Suppose a point is in the *exterior* of a circle. The measure of the segment joining the point to the center is *greater than* the measure of the radius.

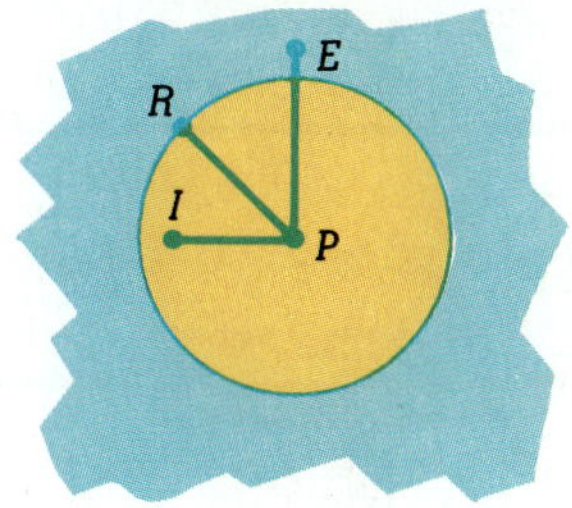

$PI < PR$

$PE > PR$

A line can intersect a circle in one of two ways.

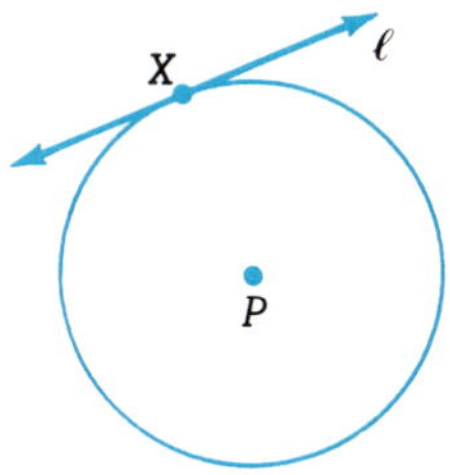

A line can intersect a circle in exactly one point. Such a line is called a **tangent** to a circle. The point of intersection is called the *point of tangency*. The line, ℓ, is tangent to $\odot P$ at X.

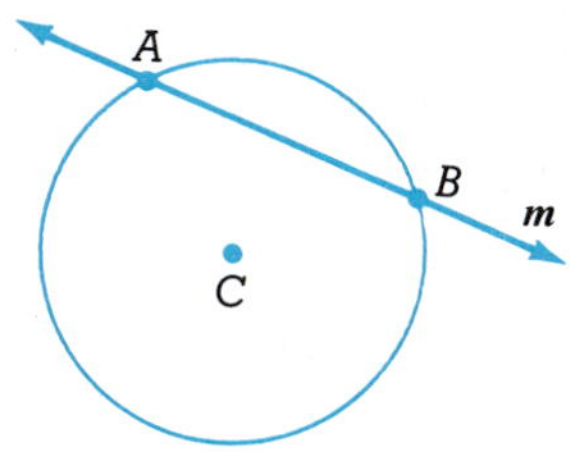

A line can intersect a circle in exactly two points. Such a line is called a **secant** of a circle. A secant of a circle contains a chord of the circle. The line, m, contains the chord $\overline{AB}$.

The following theorem guarantees that these are the only ways lines can intersect a circle.

Theorem 10–1

In a plane, if a line contains a point in the interior of a circle, then the line intersects the circle in exactly two points.

example 1

All points of $\odot P$ are 4 units from P. Suppose $PA = 5$ and $PB = 2$. Tell whether $\overleftrightarrow{AB}$ intersects $\odot P$ in *zero points*, *one point*, or *two points*.

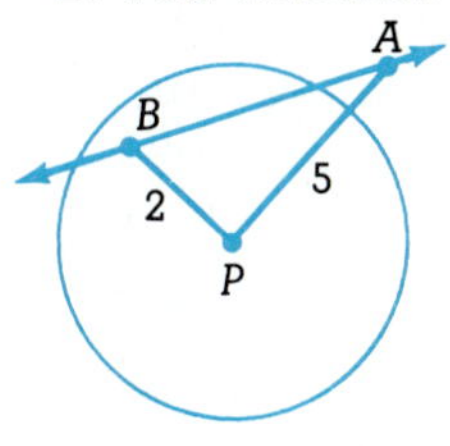

B must be in the interior of $\odot P$ because the measure of the distance from P to B is less than the measure of the radius of the circle. Thus, by Theorem 10–1, $\overleftrightarrow{AB}$ intersects the circle in exactly two points.

exercises

Exploratory Use the figure at the right to solve each problem.

1. Name the center of the circle.
2. Name the circle.
3. Name three radii of the circle.
4. Name a diameter of the circle.
5. Name two chords of the circle.
6. Name a tangent of the circle.
7. Name a secant of the circle.
8. Name two points in the interior of the circle.
9. Name two points in the exterior of the circle.
10. State how many of the points named in the figure lie on the circle.

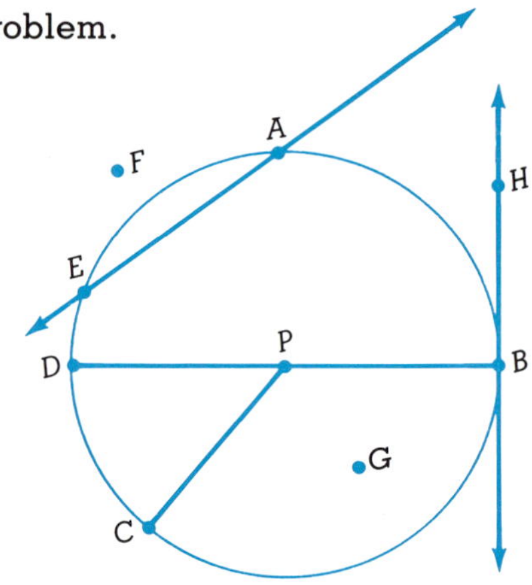

Written For each of the following, write *yes* or *no*.

1. A diameter of a circle is a chord of the circle.
2. A radius of a circle is a chord of the circle.
3. A chord of a circle is a secant of the circle.
4. A chord of a circle is a diameter of the circle.
5. Two radii of a circle always form a diameter of the circle.
6. A radius of a circle is tangent to the circle.
7. Any point interior to a circle is between two points on the circle.
8. A secant of a circle contains a chord of the circle.
9. The line connecting a point interior to a circle to a point exterior to a circle intersects the circle in exactly one point.
10. A line connecting two points exterior to a circle *cannot* intersect the circle.

Suppose all points of $\odot R$ are 6 units from R. For each of the following, tell whether $\overleftrightarrow{AB}$ intersects $\odot R$ in *zero points*, *one point*, or *two points*. There may be more than one answer.

11. $RA = 5.2$, $RB = 7.9$
12. $RA = 9.0$, $RB = 4$
13. $RA = 6.0$, $RB = 3.3$
14. $RA = 6.0$, $RB = 6.0$
15. $RA = 2.1$, $RB = 4.5$
16. $RA = 6.0$, $RB = 11.3$
17. $RA = 8.7$, $RB = 6.4$
18. $RA = 7.9$, $RB = 9.7$

Solve each problem.

19. Write an if and only if definition for a radius of a circle.
20. Write an if and only if definition for a diameter of a circle.
21. Show that if the diameter of a circle measures d units and the radius of the circle measures r units, then $d = 2r$.
22. Show that if X is a point in the interior of $\odot P$, then there are points A and B on $\odot P$ so that X is between A and B.

10–2 Angles and Arcs

Two rays can be drawn from the center of a circle to form an angle. Such an angle is called a **central angle.**

A central angle separates a circle into **arcs.** For example, the figure at the right shows $\odot P$ with a central angle, $\angle APB$. The points A and B, and all points of the circle interior to the angle form a *minor arc* called $\widehat{AB}$ or $\widehat{ADB}$. The points A and B, and all points of the circle exterior to $\angle APB$ form a *major arc* called $\widehat{ACB}$.
Three letters are needed to name a major arc.

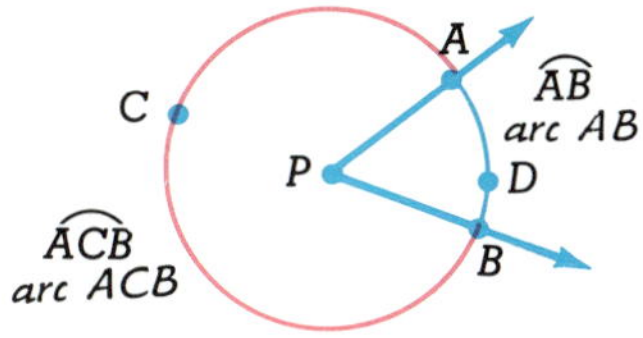

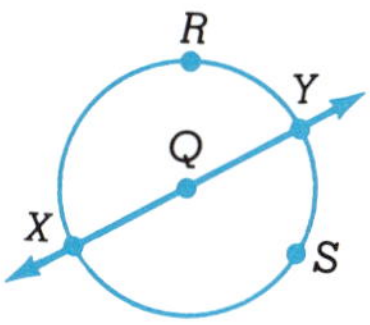

A straight angle with its vertex at the center of a circle forms two **semicircles** with the circle. In the figure at the left, $\widehat{XRY}$ and $\widehat{XSY}$ are semicircles.
A diameter forms two semicircles too.

Arcs are measured by their corresponding central angles. In the figure at the right, we say the degree measure of $\widehat{RS}$ is 90 and we write $m\widehat{RS} = 90$.

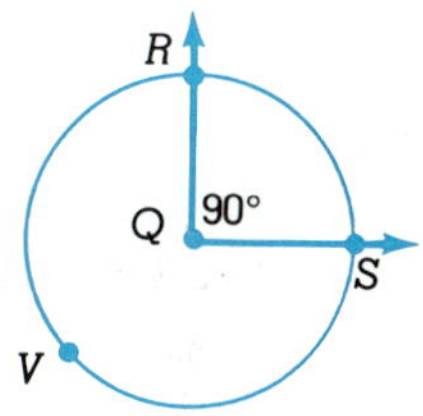

The degree measure of a minor arc is the degree measure of its central angle. The degree measure of a major arc is 360 minus the degree measure of its central angle. The degree measure of a semicircle is 180.

Definition of Arc Measure

As with angles, measures of arcs can be added to find measures of other arcs. For example, $m\widehat{PQ} = 110$ and $m\widehat{QR} = 120$. Thus, $m\widehat{PQR} = 230$.

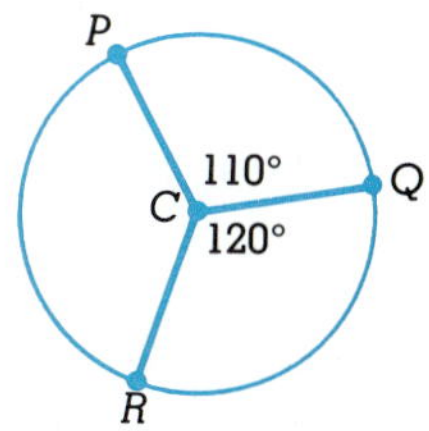

If Q is a point on $\widehat{PQR}$, then $m\widehat{PQ} + m\widehat{QR} = m\widehat{PQR}$.
The sum of the degree measures of the arcs of a circle is 360. Why?

Postulate 10–1
Arc Addition Postulate

example 1

In $\odot P$ at the right, $m\angle APB = 30$. Find $m\widehat{AB}$, $m\widehat{ACB}$, $m\widehat{BC}$ and $m\widehat{BAC}$.

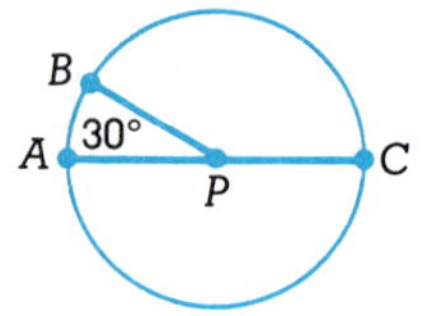

The central angle of $\widehat{AB}$ and $\widehat{ACB}$ is $\angle APB$. Thus, $m\widehat{AB} = 30$ and $m\widehat{ACB} = 360 - 30$ or 330.

By the Arc Addition Postulate, $m\widehat{AB} + m\widehat{BC} = m\widehat{ABC}$. Since $m\widehat{AB} = 30$ and $\widehat{ABC}$ is a semicircle, the following holds.

$$30 + m\widehat{BC} = 180$$
$$m\widehat{BC} = 180 - 30 \text{ or } 150$$

Finally, $\widehat{BAC}$ is a major arc for $\angle BPC$ which measures 150. Thus, $m\widehat{BAC} = 360 - 150$ or 210.

All circles are similar.

All circles have the same shape. But *not* all circles have the same size. Notice that the tree ring circles share a common center but have radii of different lengths.

Circles that have the same center are called <u>concentric circles</u>.

As the tree rings show, circles with radii having different lengths are *not* congruent. For two circles to have the same size, their radii must have the same length.

Definition of Congruent Circles

Two circles are congruent if and only if their radii are congruent.

Two arcs of a circle are congruent if and only if they have the same measure.

Theorem 10–2

In a circle or in congruent circles, two central angles are congruent if and only if their minor arcs are congruent.

exercises

Exploratory For each of the following, tell whether it is a *minor arc*, a *major arc*, or a *semicircle*. Use $\odot P$ at the right.

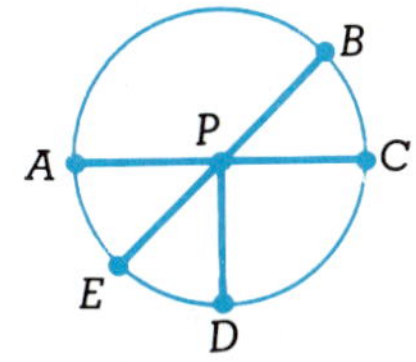

1. $\widehat{AB}$
2. $\widehat{ECA}$
3. $\widehat{BAE}$
4. $\widehat{BDE}$
5. $\widehat{DCE}$
6. $\widehat{CBD}$
7. $\widehat{DAB}$
8. $\widehat{AE}$
9. $\widehat{BC}$
10. $\widehat{BCD}$
11. $\widehat{BDC}$
12. $\widehat{AD}$
13. $\widehat{CBA}$
14. $\widehat{DC}$
15. $\widehat{ACD}$
16. $\widehat{CAE}$
17. $\widehat{CDE}$
18. $\widehat{ACB}$
19. $\widehat{DE}$
20. $\widehat{ADC}$
21. $\widehat{DAE}$
22. $\widehat{CE}$
23. $\widehat{BCE}$
24. $\widehat{ABE}$

Written Use the figure at the right to solve each of the following. Assume P is the center of the circle.

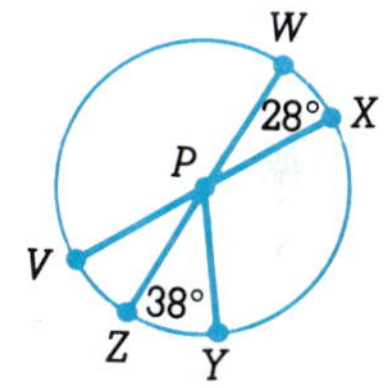

1. Find $m\widehat{YZ}$.
2. Find $m\widehat{WX}$.
3. Find $m\angle VPZ$.
4. Find $m\widehat{VZ}$.
5. Find $m\widehat{VWX}$.
6. Find $m\widehat{ZVW}$.
7. Find $m\widehat{WYZ}$.
8. Find $m\widehat{ZXW}$.
9. Find $m\angle XPY$.
10. Find $m\widehat{XY}$.
11. Find $m\widehat{XWY}$.
12. Find $m\widehat{WZX}$.
13. Find $m\widehat{VW}$.
14. Find $m\angle VPW$.

Use the figure at the right to solve each of the following. Assume that $\overline{XY}$ and $\overline{AB}$ are diameters of $\odot P$.

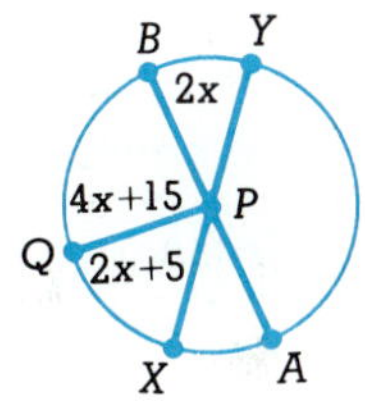

15. Find $m\widehat{BY}$.
16. Find $m\widehat{XQ}$.
17. Find $m\widehat{BX}$.
18. Find $m\widehat{QYA}$.
19. Find $m\widehat{AY}$.
20. Find $m\widehat{XAY}$.
21. Find $m\widehat{AQ}$.
22. Find $m\widehat{YQ}$.
23. Find $m\angle APX$.
24. Find $m\angle BPY$.
25. Find $m\angle YPQ$.
26. Find $m\angle BPX$.

For each of the following, write *yes* or *no*.

27. If $m\widehat{AB} = 32$ and $m\widehat{XY} = 32$, then $\widehat{AB} \cong \widehat{XY}$.
28. If $\widehat{AB} \cong \widehat{XY}$, and $m\widehat{AB} = 32$, then $m\widehat{XY} = 32$.
29. Two congruent circles have congruent radii.
30. All radii of a circle are congruent.
31. Two concentric circles have congruent radii.
32. If two circles have the same center, they are congruent.
33. If two central angles are congruent, then their corresponding minor arcs are congruent.
34. If two minor arcs are congruent, then their corresponding central angles are congruent.

Solve each of the following.

35. Prove that in two congruent circles, if two arcs are congruent then their corresponding central angles are congruent.
36. Prove that in two congruent circles, if two central angles are congruent, then their corresponding arcs are congruent.
37. Suppose a wheel has 30 spokes evenly spaced and numbered consecutively from 1 through 30. Find the measure of the central angle formed by spokes 1 and 14.

10–3 Arcs and Chords

When a minor arc and a chord have the same endpoints, we call the arc, the *arc of the chord*. For example, in the figure at the right, $\widehat{PQ}$ is the arc of $\overline{PQ}$.

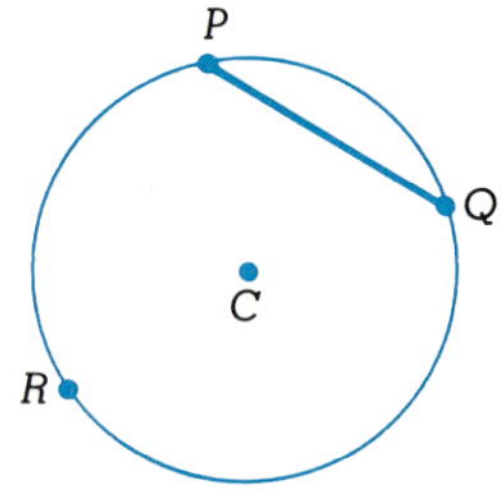

Chords and their arcs are related in the following way.

Theorem 10–3

In a circle or in congruent circles, two minor arcs are congruent if and only if their corresponding chords are congruent.

The proof of this theorem is based on congruent triangles. The following example shows how they are used to prove one part of the theorem.

example 1

Prove that if two arcs of a circle are congruent, then their corresponding chords are congruent.

Given: $\widehat{AB} \cong \widehat{PQ}$
C is center of circle

Prove: $\overline{AB} \cong \overline{PQ}$

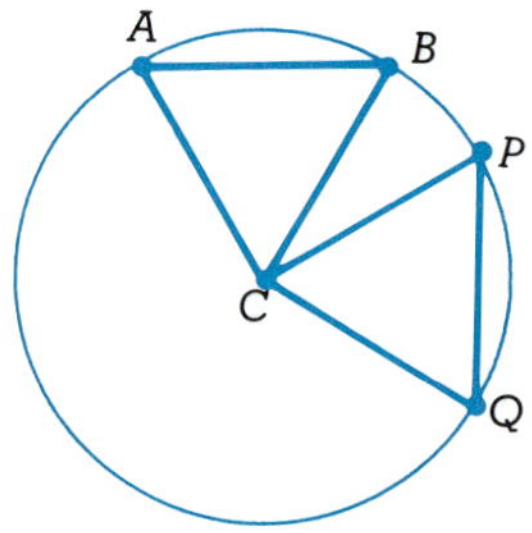

Proof:

STATEMENTS	REASONS
1. Draw radii $\overline{AC}$, $\overline{BC}$, $\overline{PC}$, and $\overline{QC}$ and chords $\overline{AB}$ and $\overline{PQ}$	1. Postulate 1–1: Through any two points there is exactly one line.
2. $\overline{AC} \cong \overline{PC}$ $\overline{BC} \cong \overline{QC}$	2. Definition of circle
3. $\widehat{AB} \cong \widehat{PQ}$, $\odot C$	3. Given
4. $\angle ACB \cong \angle PCQ$	4. Theorem 10–2: In a circle or in congruent circles, if two minor arcs are congruent, then their central angles are congruent.
5. $\triangle ACB \cong \triangle PCQ$	5. Postulate 4–3: SAS *steps 2 and 4*
6. $\overline{AB} \cong \overline{PQ}$	6. Definition of congruent triangles *CPCTC*

The midpoint of an arc separates the arc into two congruent arcs. In $\odot C$ at the right, M is the midpoint of $\overset{\frown}{AB}$ because $\overset{\frown}{AM} \cong \overset{\frown}{MB}$, and we say that $\overline{CM}$ bisects $\overset{\frown}{AB}$. By using congruent triangles, it can be shown that $\overline{CM}$ also bisects $\overline{AB}$. In addition, $\overline{CM} \perp \overline{AB}$.

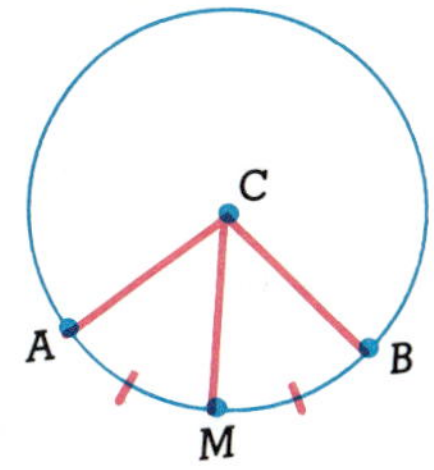

Suppose a diameter or radius of a circle is perpendicular to a chord. Does it bisect the chord or its arc?

In a circle, if a diameter is perpendicular to a chord, then it bisects the chord and its arcs. Theorem 10–4

example 2

Suppose a chord of a circle is 24 centimeters long and is 9 centimeters from the center of the circle. Find the length of the radius.

Let $\overline{AD}$ represent the chord, and draw $\overline{BC}$ perpendicular to $\overline{AD}$. Then, $BC = 9$.

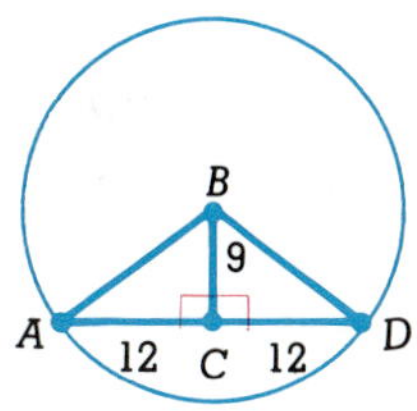

By Theorem 10–4, $\overline{BC}$ bisects $\overline{AD}$. Thus, AC and CD are $\frac{24}{2}$ or 12.

$\overline{AB}$ is a radius of the circle and the hypotenuse of a right triangle with sides measuring 12 and 9.

$$\begin{aligned} AB^2 &= 12^2 + 9^2 \\ &= 225 \\ AB &= 15 \end{aligned}$$

Thus, the radius is 15 centimeters long.

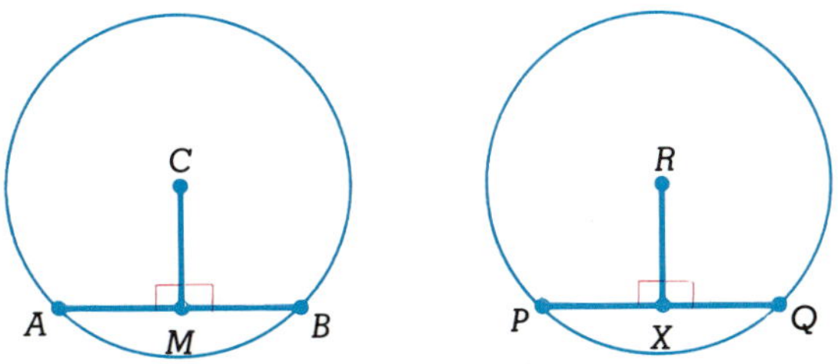

Suppose $\odot C$ and $\odot R$ are congruent, and $\overline{AB} \cong \overline{PQ}$. Are the chords the same distance from the centers of the circles?

Suppose, instead, that you knew the chords to be equidistant from the centers of the circles. Could you conclude that the chords are congruent?

In a circle or in congruent circles, two chords are congruent if and only if they are equidistant from the center. Theorem 10–5

exercises

Exploratory For each of the following, name the theorem or theorems that justify the statement. Use $\odot P$ at the right.

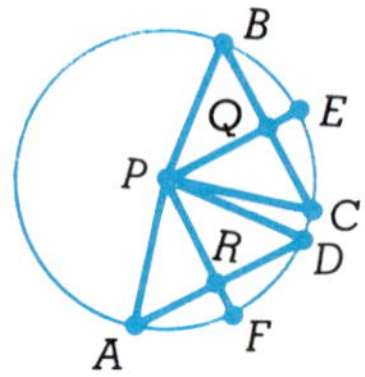

1. If $\overline{BC} \cong \overline{AD}$, then $\overset{\frown}{BC} \cong \overset{\frown}{AD}$.
2. If $\overset{\frown}{BC} \cong \overset{\frown}{AD}$, then $\overline{BC} \cong \overline{AD}$.
3. If $\overline{BC} \cong \overline{AD}$, then $PQ = PR$.
4. If $\overline{PF} \perp \overline{AD}$, then $\overset{\frown}{AF} \cong \overset{\frown}{FD}$.
5. If $\overline{PQ} \perp \overline{BC}$, then $\overline{BQ} \cong \overline{QC}$.
6. If $PQ = PR$, then $\overline{CB} \cong \overline{DA}$.

Written In the figure at the right, P is the center of the circle and $\overline{SY} \perp \overline{QT}$. Use the figure to solve each problem.

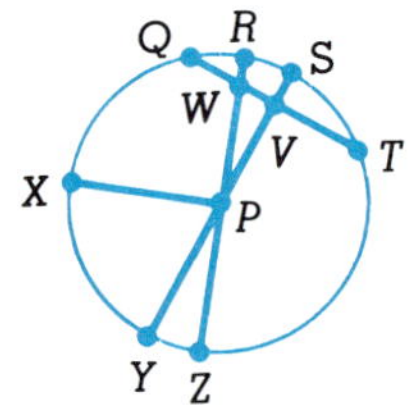

1. Name a segment congruent to $\overline{VT}$.
2. Name the midpoint of $\overset{\frown}{QT}$.
3. Name the midpoint of $\overline{TQ}$.
4. Name an arc congruent to $\overset{\frown}{QY}$.
5. Name an arc congruent to $\overset{\frown}{ST}$.
6. Name a segment congruent to $\overline{RZ}$.
7. Which segment is longer, $\overline{WP}$ or $\overline{VP}$?
8. Which segment is longer, $\overline{QT}$ or $\overline{YS}$?
9. If W is the midpoint of $\overline{QV}$, then R the midpoint of $\overset{\frown}{QS}$? Write *yes* or *no*.
10. If X is the midpoint of $\overset{\frown}{ZXR}$, how is $\overline{XP}$ related to $\overline{ZR}$?

Solve each problem.

11. Suppose a chord of a circle is 10 inches long and is 12 inches from the center of the circle. Find the length of the radius.
12. Suppose a chord of a circle is 18 centimeters long and is 12 centimeters from the center of the circle. Find the length of the radius.
13. Suppose the diameter of a circle is 20 centimeters long and a chord is 16 centimeters long. Find the distance between the chord and the center of the circle.
14. Suppose the diameter of a circle is 10 inches long and a chord is 6 inches long. Find the distance between the chord and the center of the circle.

Prove each of the following.

15. If two chords of a circle are congruent, then their corresponding arcs are congruent.
16. If two congruent circles each contain a chord that is congruent to the other chord, then the corresponding arcs are congruent.
17. If a radius of a circle bisects an arc of a circle, then it bisects the corresponding chord.
18. Theorem 10–4.

19. If two chords of a circle are congruent, then they are equidistant from the center of the circle.

20. If two chords of a circle are equidistant from the center of the circle, then the two chords are congruent.

21. In a circle, if a chord is a perpendicular bisector of another chord, then the first chord is a diameter of the circle.

22. In a circle, if one chord is longer than another chord, then the longer chord is closer to the center of the circle.

23. **Given:** $\overline{AC}$ is a diameter
$\overline{AC} \perp \overline{BD}$
Prove: $\widehat{BC} \cong \widehat{DC}$

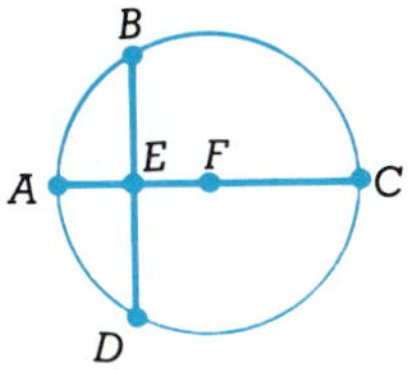

24. **Given:** $\widehat{PTS} \cong \widehat{QST}$
$\overline{RS} \cong \overline{RT}$
Prove: $\triangle PQR$ is isosceles

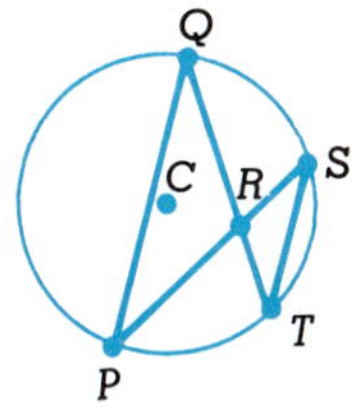

Centers of Circles

Excursions in Geometry

The following steps can be used to find the center of a circle.

Place a sheet of paper over the circle so that one corner touches the circle.

The paper intersects the circle at two other points, A and B. Draw the chord, $\overline{AB}$.

Place the paper so that it touches the circle at a different point. Repeat step two, and draw $\overline{EF}$.

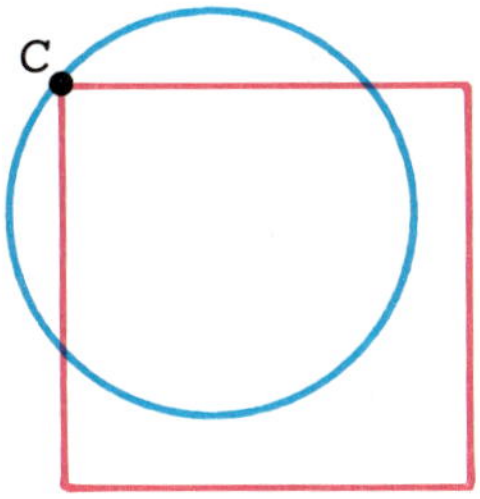

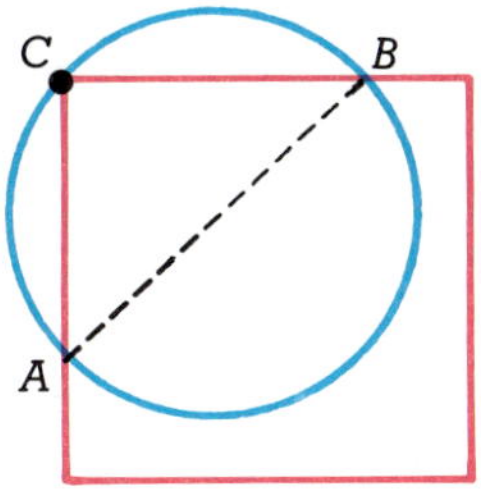

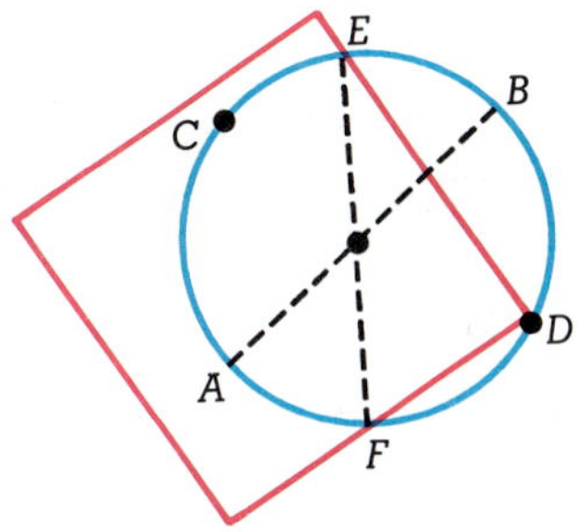

The intersection of $\overline{AB}$ and $\overline{EF}$ is the center of the circle.

Exercises Trace each of the following circles. Find the center and the length of the radius to the nearest millimeter.

1.

2.

3.

4.

5.

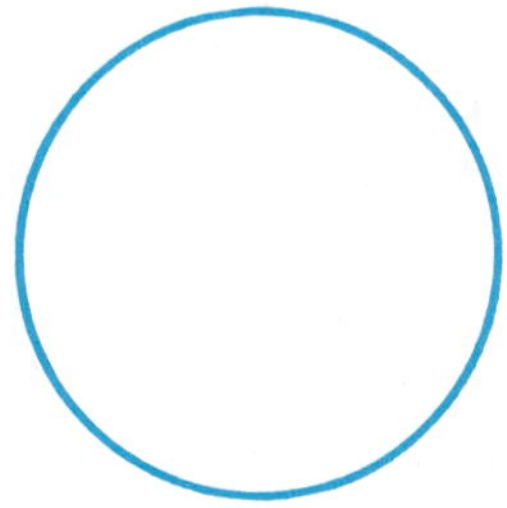

10–4 Inscribed Angles

When a football pass is intercepted it is usually caught between the passer and the receiver.

Intercept has a different meaning in geometry. In each of the figures below, the angles intercept arcs of the circles. The **intercepted arcs** are shown in color.

Note that some angles intercept two arcs.

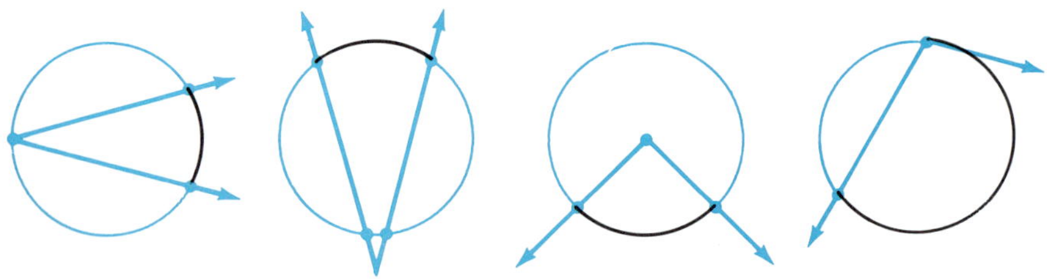

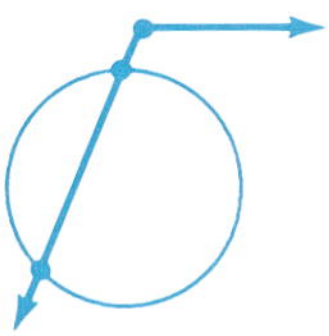

An angle intercepts an arc if and only if each of the following characteristics hold.

1. The endpoints of the arc lie on the angle.
2. All points of the arc, except the endpoints, are in the interior of the angle.
3. Each side of the angle contains an endpoint of the arc.

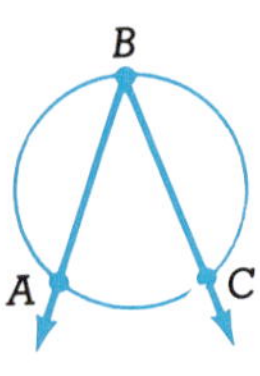

In the figure at the left, $\angle ABC$ intercepts an arc of a circle, and the vertex of the angle lies on the circle. Such an angle is called an inscribed angle. We say that $\angle ABC$ intercepts $\widehat{AC}$ and is inscribed in $\widehat{ABC}$.

The measures of inscribed angles are related to the measures of their intercepted arcs.

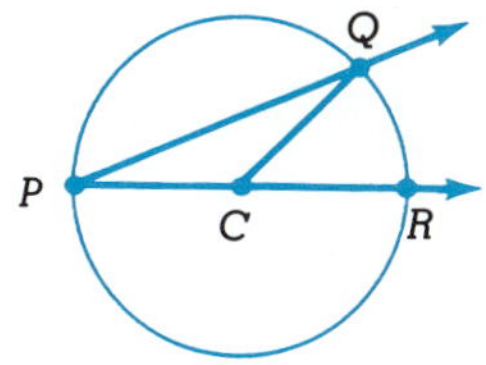

In the figure at the right, $\angle QPR$ is an inscribed angle which intercepts $\widehat{QR}$. By using the Isosceles Triangle Theorem and the Exterior Angle Theorem, it is possible to show that $m\angle QPR = \frac{1}{2}m\angle QCR$.

Since, $\angle QCR$ is a central angle, $m\angle QCR = m\widehat{QR}$. Thus, $m\angle QPR = \frac{1}{2}m\widehat{QR}$.

The following deductions show how to find the measures of inscribed angles that do *not* contain diameters of a circle.

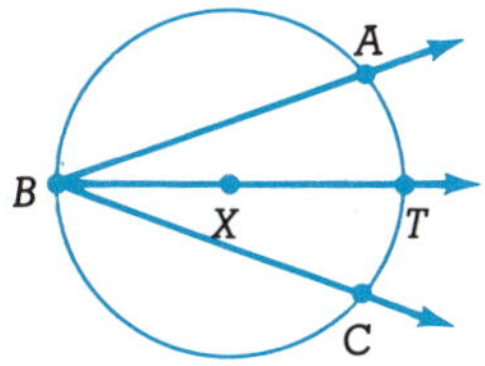

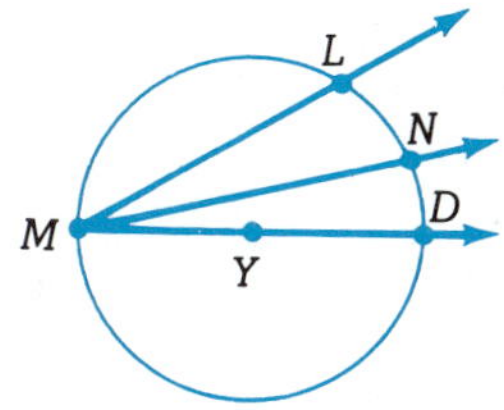

$$\begin{aligned} m\angle ABC &= m\angle ABT + m\angle TBC \\ &= \tfrac{1}{2}m\widehat{AT} + \tfrac{1}{2}m\widehat{TC} \\ &= \tfrac{1}{2}(m\widehat{AT} + m\widehat{TC}) \\ &= \tfrac{1}{2}m\widehat{AC} \end{aligned}$$

$$\begin{aligned} m\angle LMN &= m\angle LMD - m\angle NMD \\ &= \tfrac{1}{2}m\widehat{LD} - \tfrac{1}{2}m\widehat{ND} \\ &= \tfrac{1}{2}(m\widehat{LD} - m\widehat{ND}) \\ &= \tfrac{1}{2}m\widehat{LN} \end{aligned}$$

The measure of an inscribed angle of a circle equals one-half the measure of its intercepted arc.

Theorem 10–6

example 1

In the figure at the right, $m\widehat{PQ} = 112$, $m\widehat{QS} = 54$, and $m\widehat{ST} = 88$. Find $m\angle 1$, $m\angle 2$, and $m\angle 3$.

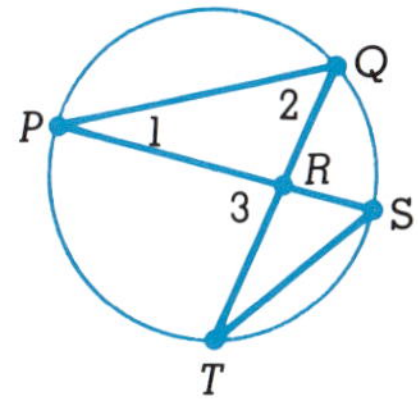

$$\begin{aligned} m\widehat{PQ} + mQS + m\widehat{ST} + m\widehat{TP} &= 360 \\ 112 + 54 + 88 + m\widehat{TP} &= 360 \\ m\widehat{TP} &= 106 \end{aligned}$$

$$\begin{aligned} m\angle 1 &= \tfrac{1}{2}\widehat{QS} \\ &= \tfrac{1}{2}(54) \\ &= 27 \end{aligned} \qquad \begin{aligned} m\angle 2 &= \tfrac{1}{2}\widehat{TP} \\ &= \tfrac{1}{2}(106) \\ &= 53 \end{aligned}$$

$$\begin{aligned} m\angle 3 &= m\angle 1 + m\angle 2 \\ &= 27 + 53 \\ &= 80 \end{aligned}$$

The proofs of the following theorems are based on Theorem 10–6.

Theorem 10–7

If two inscribed angles of a circle or congruent circles intercept congruent arcs, then the angles are congruent.

Theorem 10–8

If an angle is inscribed in a semicircle, then the angle is a right angle.

example 2

Prove Theorem 10–8.

Given: $\widehat{PQR}$ is a semicircle of $\odot C$

Prove: $\angle PQR$ is a right angle

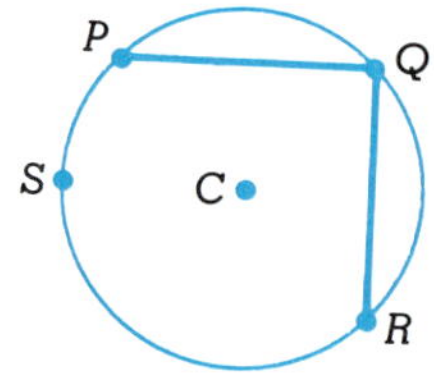

Proof:

STATEMENTS	REASONS
1. $\widehat{PQR}$ is a semicircle of $\odot C$	1. Given
2. $\widehat{PSR}$ is a semicircle of $\odot C$	2. Definition of semicircle
3. $m\widehat{PSR} = 180$	3. Definition of arc measure
4. $m \angle PQR = \frac{1}{2}m\widehat{PSR}$	4. Theorem 10–6: The measure of an inscribed angle of a circle equals one-half the measure of its intercepted arc.
5. $m \angle PQR = \frac{1}{2}(180)$ or 90	5. Postulate 2–9: Substitution
6. $\angle PQR$ is a right angle	6. Definition of right angle

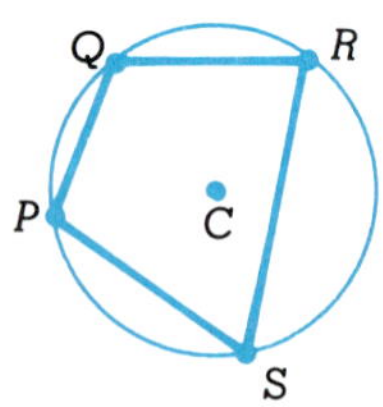

A polygon is inscribed in a circle if each of its vertices lies on the circle. For example, in the figure at the left, quadrilateral $PQRS$ is inscribed in $\odot C$.

Since the measures of the arcs of a circle total 360, the opposite angles of an inscribed quadrilateral are related in a special way.

Theorem 10–9

If the angles of a quadrilateral are inscribed in a circle, then each pair of opposite angles are supplementary.

exercises

Exploratory For each of the following, name the arc or arcs intercepted by the given angle.

1.
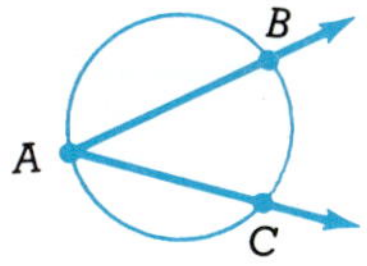

2.
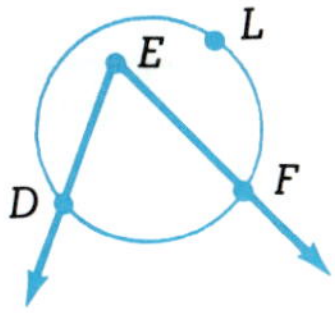

3.
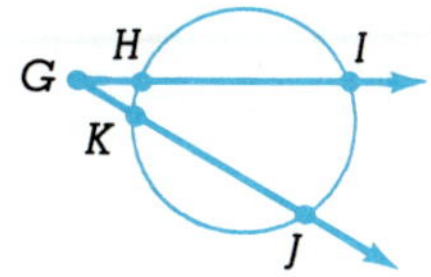

4.
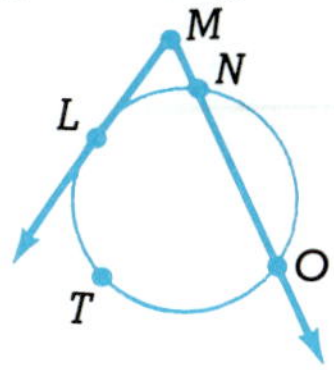

5.
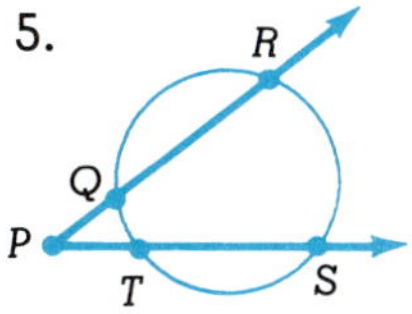

6.
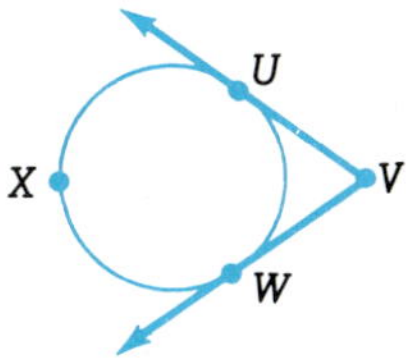

7.
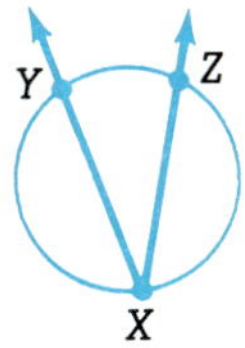

8.
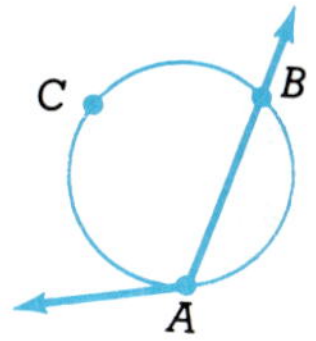

9–16. For each of the figures in problems **1–8**, tell whether or *not* the angle is an inscribed angle. State *yes* or *no*.

Written In the figure at the right, P is the center of the circle, $\overline{AB} \parallel \overline{DC}$, $m\widehat{BC} = 94$, and $m\angle APB = 104$. Find the measures for each of the following.

1. $\widehat{AB}$	2. $\angle BAC$	3. $\angle BDC$
4. $\angle BCA$	5. $\angle ADB$	6. $\angle ADC$
7. $\angle APB$	8. $\angle PAB$	9. $\angle ABP$
10. $\angle ACD$	11. $\angle BCD$	12. $\angle DEC$
13. $\angle AED$	14. $\angle EAD$	15. $\angle BAD$
16. $\angle DBC$	17. $\widehat{AD}$	18. $\widehat{DC}$

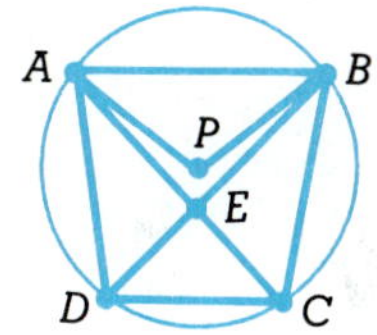

The figure at the right shows a regular hexagon inscribed in $\odot P$ with radius measuring 12 units. Solve each of the following problems about the figure.

19. Find the length of each side.
20. Find the distance of each side from the center of the circle.
21–22. Solve problems **19** and **20** about an equilateral triangle inscribed in the same circle.

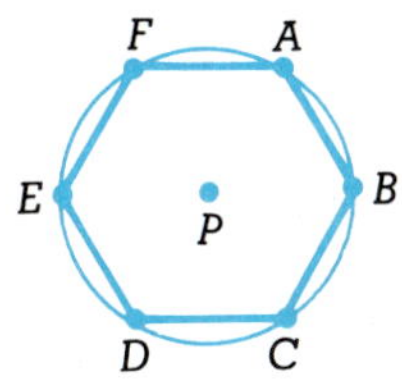

Prove each of the following.

23. If an angle is inscribed in a circle so that one side contains a diameter, then the measure of the angle is one-half the measure of its intercepted arc.
24. The bisector of an inscribed angle separates the intercepted arc into two congruent arcs.
25. Theorem 10–7.
26. Theorem 10–9.

10–5 Tangents

In a plane, a line that intersects a circle in exactly one point is called a *tangent*. Also, segments and rays that are contained in the tangent and intersect the circle are said to be tangent to the circle.

In the figure at the right, S is the center of the circle, and T is the point of tangency. Since X is in the exterior of the circle and $\overline{ST}$ is a radius, $SX > ST$. A similar inequality holds for any point in the exterior of the circle.

The shortest segment from a point to a line is the perpendicular segment. Thus, $\overline{ST} \perp \overleftrightarrow{TX}$.

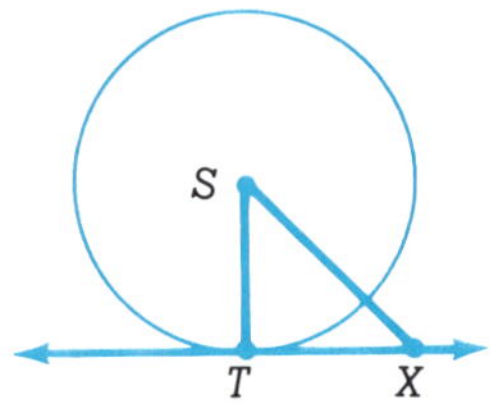

Theorem 10–10

A tangent to a circle is perpendicular to the radius drawn to the point of tangency.

example 1

In the figure at the right, $\overline{QR}$ and $\overline{QP}$ are tangents and $m\angle RPC = 15$. Find $m\angle Q$.

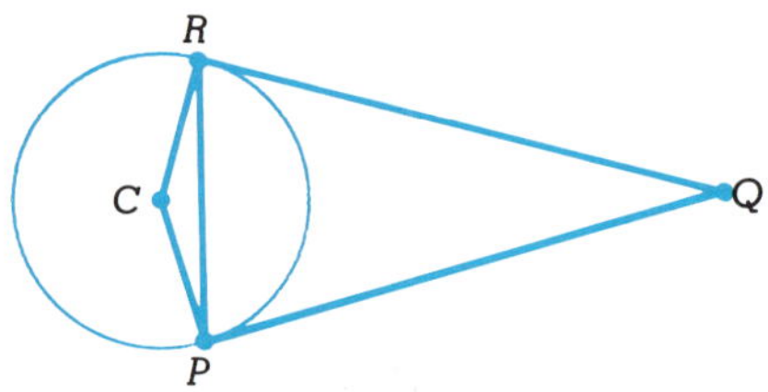

Since $\overline{CR}$ and $\overline{CP}$ are radii, they are congruent. Thus, by the Isosceles Triangle Theorem, $\angle RPC \cong \angle PRC$, and $m\angle PRC = 15$.

Since $\overline{QR}$ and $\overline{QP}$ are tangents, they are perpendicular to the radii, and $m\angle CPQ = 90$ and $m\angle CRQ = 90$. Thus, $m\angle PRQ = 90 - 15$ or 75 and $m\angle QPR = 90 - 15$ or 75.

The sum of the angle measures of a triangle is 180.

$$m\angle Q + m\angle PRQ + m\angle QPR = 180$$
$$m\angle Q + 75 + 75 = 180$$
$$m\angle Q = 30$$

The converse of Theorem 10–10 also holds, and provides one method for identifying tangents to a circle.

Theorem 10–11

In a plane, if a line is perpendicular to a radius of a circle at its endpoint on the circle, then the line is a tangent.

From a point exterior to a circle it is possible to draw two tangents. For example, $\overrightarrow{PQ}$ and $\overrightarrow{PR}$ are both tangent to $\odot C$ at the right. *$\overline{PQ}$ and $\overline{PR}$ are tangent segments.*

By drawing $\overline{PC}$, two right triangles are formed. These triangles are congruent by HL, and the following conclusion can be made.

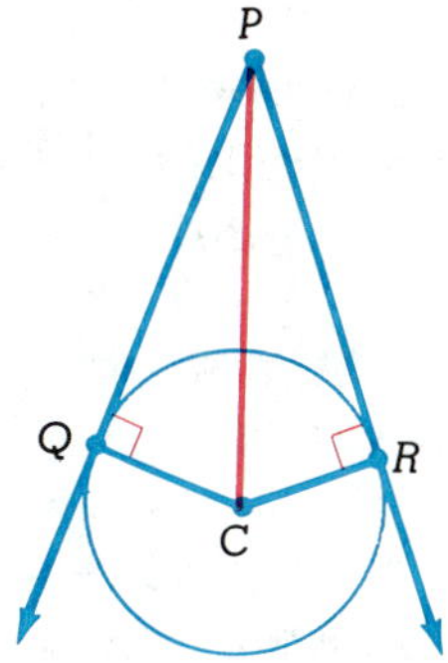

Theorem 10–12

The two tangent segments from the same exterior point of a circle are congruent.

example 2

In the figure at the right, both $\overline{AB}$ and $\overline{AC}$ are tangent to $\odot P$. Suppose $PB = 10$ and $AP = 26$. Find AC.

By Theorem 10–10, a tangent to a circle is perpendicular to the radius drawn to the point of tangency. Thus, $\overline{BP} \perp \overline{AB}$ and $\overline{CP} \perp \overline{AC}$, and $\triangle ABP$ and $\triangle ACP$ are right triangles.

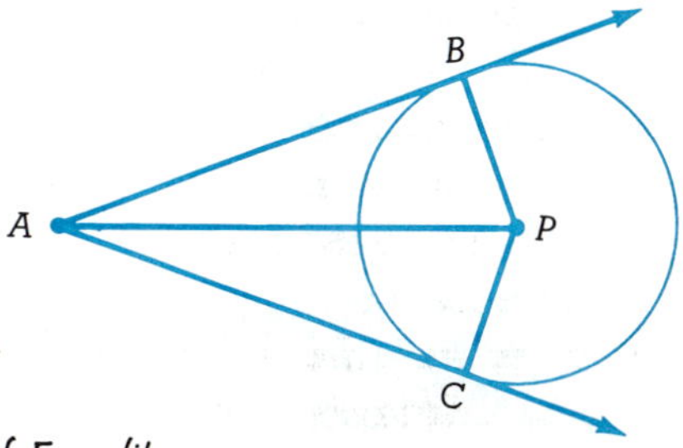

$AB^2 + PB^2 = AP^2$	*Pythagorean Theorem*
$AB^2 + 10^2 = 26^2$	*Substitution*
$AB^2 = 26^2 - 10^2$	*Subtraction Property of Equality*
$AB = \sqrt{576}$ or 24	

By Theorem 10–12, the two tangent segments, $\overline{AB}$ and $\overline{AC}$, are congruent. Thus, $AC = AB = 24$.

A polygon is circumscribed about a circle if and only if each side of the polygon is tangent to the circle.

The circle is inscribed in the polygon.

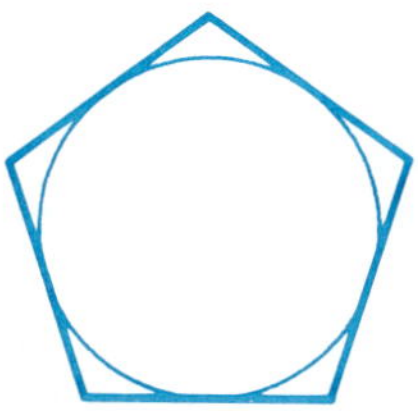

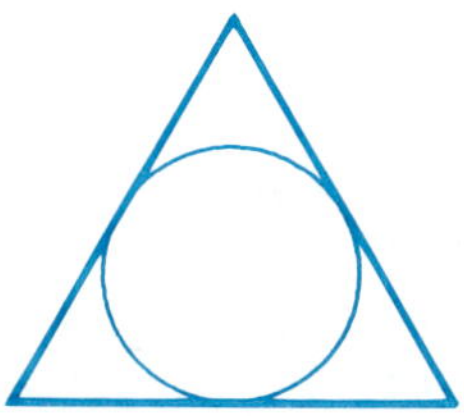

Must a circumscribed polygon be regular? The answer is *no*. But, it is possible to show that a polygon is regular if and only if it can be inscribed in and circumscribed about two circles with a common center.

exercises

Exploratory Solve each problem.

1. Tell how many tangents can be drawn to a circle from a point outside the circle.
2. Tell how many tangents can be drawn to a circle from a point inside the circle.
3. Tell how many tangents can be drawn to a circle from a point on the circle.
4. Can a radius of a circle be a tangent of the circle? State *yes* or *no*.
5. Tell whether the sides of a circumscribed polygon are *chords* or *tangents* of the circle.
6. Tell whether the sides of an inscribed polygon are *chords* or *tangents* of the circle.

Written For each of the following, find x. Assume C is the center of the circle.

1.

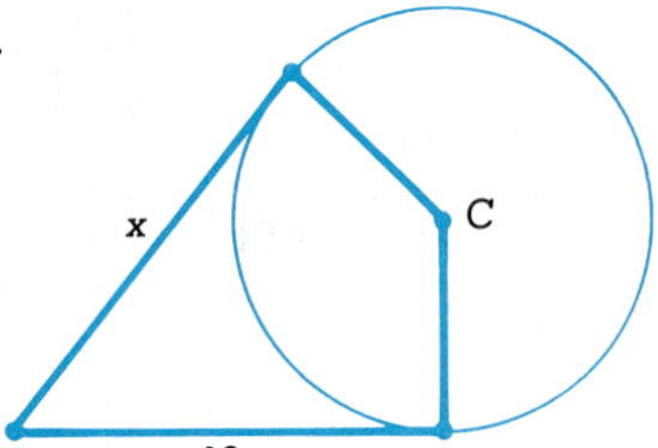

2.

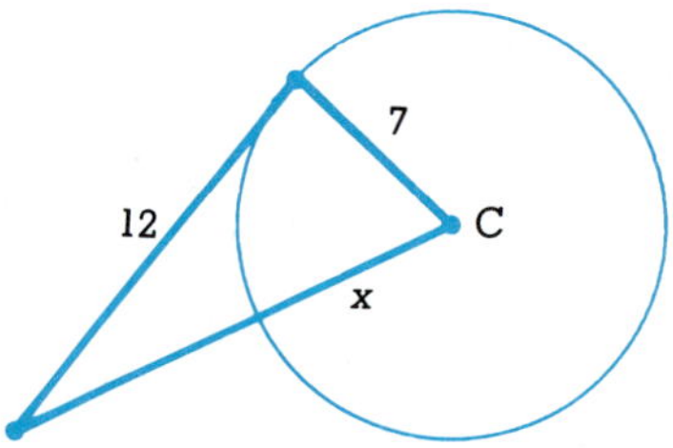

3.

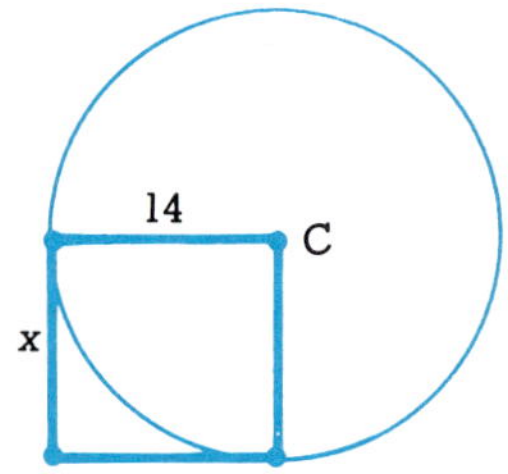

4.

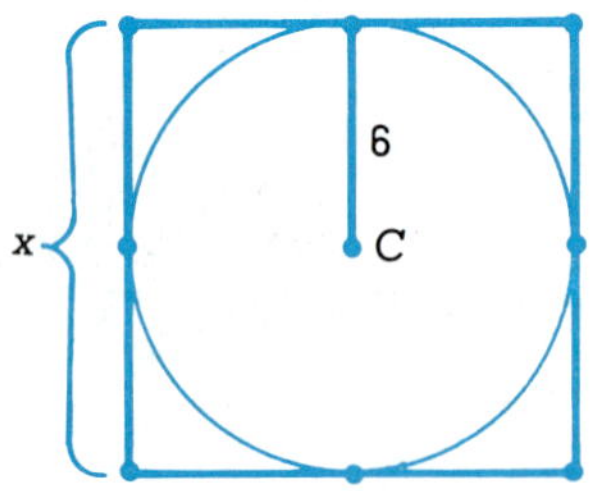

5.

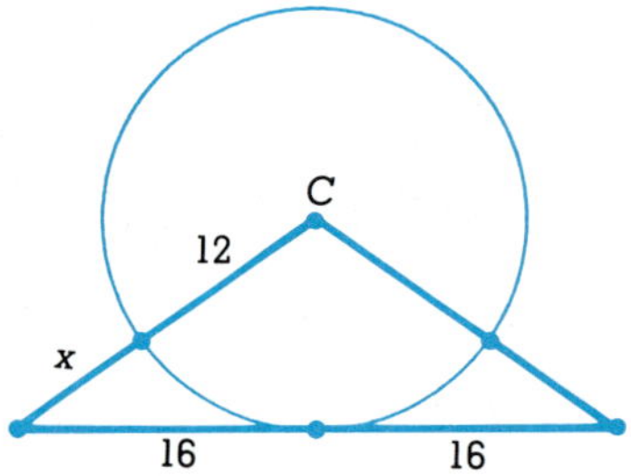

6.

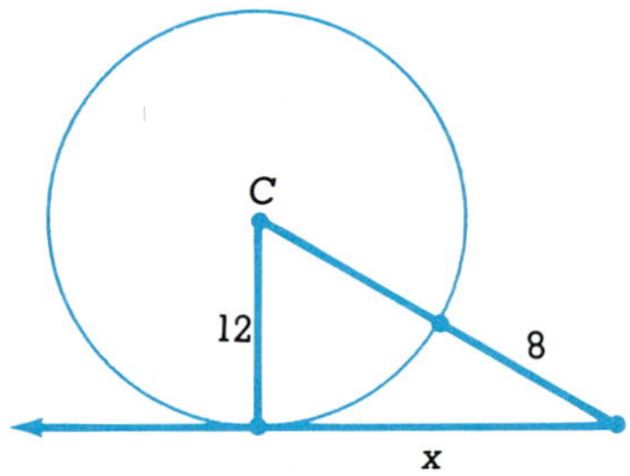

In the figure at the right, $\overline{AB}$ and $\overline{CD}$ both are tangent to $\odot P$ and $\odot Q$. Also, $AP = 8$, $BQ = 5$, and $m\angle CPE = 45$. Find the measures for each of the following.

7. $\widehat{CE}$
8. $\angle PCG$
9. $\angle CGP$
10. $\overline{CG}$
11. $\angle FGD$
12. $\angle QDG$
13. $\angle GQD$
14. $\widehat{DF}$
15. $\overline{DQ}$
16. $\overline{DG}$
17. $\overline{DC}$
18. $\overline{PG}$
19. $\overline{GQ}$
20. $\overline{PQ}$
21. $\overline{AB}$

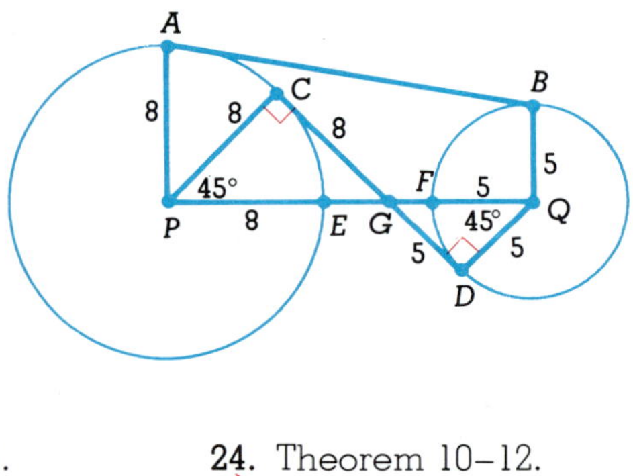

Prove each of the following.

22. Theorem 10–10.
23. Theorem 10–11.
24. Theorem 10–12.
25. If a hexagon can be inscribed in and circumscribed about two concentric circles, then the polygon is regular.

CAREERS

Wall Contractor

Jack Stanley worked for Monticello Builders on the construction of the curved wall shown at the right. He had to order the correct number of bricks for the wall. To complete this task, he determined the length of each curve.

The plan below shows some curves of the wall.

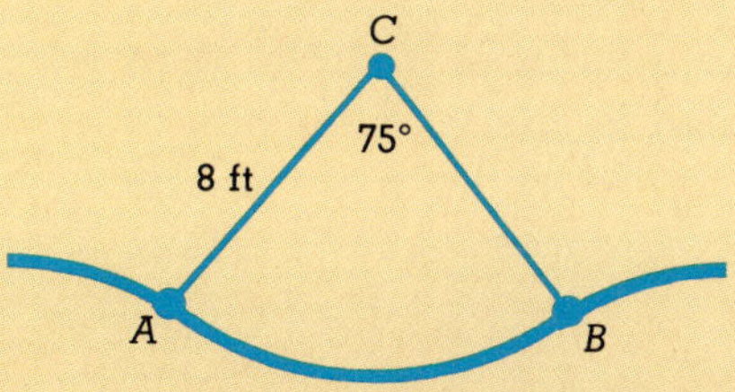

Suppose $\widehat{AB}$ is an arc of a circle with center C. Then the length of $\widehat{AB}$ is found in the following way using 3.14 for π.

$$\frac{\textit{measure of } \widehat{AB}}{\textit{measure of circumference}} = \frac{\textit{degree measure of } \angle ACB}{360}$$

$$\frac{\textit{measure of } \widehat{AB}}{2(3.14)(8)} \approx \frac{75}{360} \qquad C = 2\pi r$$

$$\textit{measure of } \widehat{AB} \approx 2(3.14)(8)\frac{75}{360}$$

$$\approx 10.5$$

The arc is about 10.5 feet long.

Exercises Find the length of an arc, given each of the following measurements. Use 3.14 for π and round to the nearest tenth.

1. radius: 9 feet
   central angle: 80°
2. radius: 11 feet
   central angle: 75°
3. radius: 8 meters
   central angle: 50°
4. radius: 4 meters
   central angle: 70°

10–6 Measuring Angles

Two secants intersect to form angles that intercept arcs in three different ways.

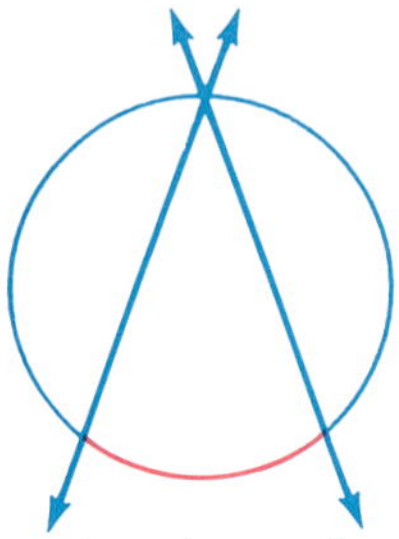

intersect on circle

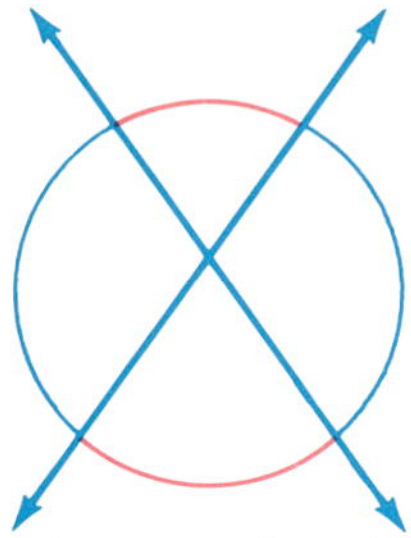

intersect inside circle

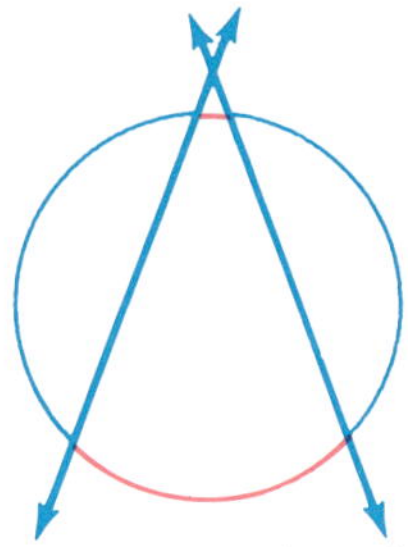

intersect outside circle

In each case, the measure of the angle formed is related to the measures of the intercepted arcs. If the secants intersect on the circle, an inscribed angle is formed. Its measure is one-half the measure of the intercepted arc. The following theorems describe the other two cases.

Theorem 10–13

If two secants intersect in the interior of a circle, then the measure of an angle formed is one-half the sum of the measures of the arcs intercepted by the angle and its vertical angle.

Theorem 10–14

If two secants intersect in the exterior of a circle, then the measure of an angle formed is one-half the positive difference of the measures of the intercepted arcs.

example 1

In $\odot P$ at the right, $m\angle CBD = 52$ and $m\widehat{CD} = 39$. Find $m\widehat{DE}$.

$$m\angle CBD = \tfrac{1}{2}(m\widehat{AE} + m\widehat{CD})$$
$$52 = \tfrac{1}{2}(m\widehat{AE} + 39)$$
$$104 = m\widehat{AE} + 39$$
$$65 = m\widehat{AE}$$

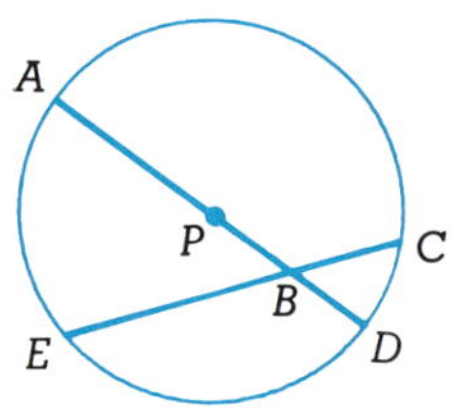

$$m\widehat{AED} = 180$$
$$m\widehat{DE} + m\widehat{AE} = 180$$
$$m\widehat{DE} + 65 = 180$$
$$m\widehat{DE} = 115$$

example 2

Prove Theorem 10–14.

Given: $\overline{AC}$ and $\overline{AD}$ intersect at A
A is exterior to a circle

Prove: $m\angle BAD = \frac{1}{2}(m\widehat{CD} - m\widehat{BE})$

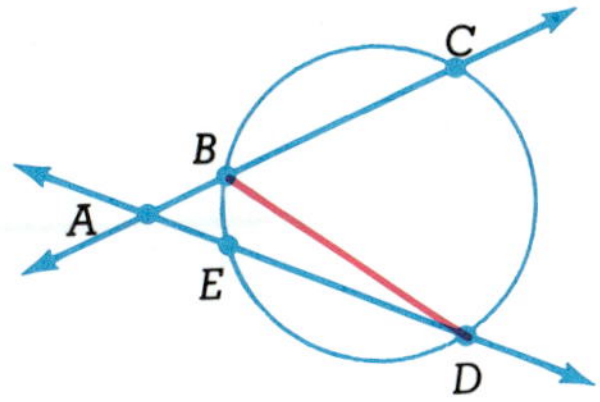

Proof:

Draw $\overline{BD}$. Then, use the Exterior Angle Theorem and the Subtraction Property of Equality to show that $m\angle BAD = m\angle CBD - m\angle BDA$.

Next, since $\angle CBD$ and $\angle BDA$ are inscribed angles, their measures are one-half the measures of their intercepted arcs. Finally, by using substitution and the Distributive Property, obtain the following results.

$$\begin{aligned} m\angle BAD &= m\angle CBD - m\angle BDA \\ &= \tfrac{1}{2}m\widehat{CD} - \tfrac{1}{2}m\widehat{BE} \\ &= \tfrac{1}{2}(m\widehat{CD} - m\widehat{BE}) \end{aligned}$$

A tangent and a secant, or two tangents intercept arcs in several ways. In each case, the measure of the angle formed is related to the intercepted arcs.

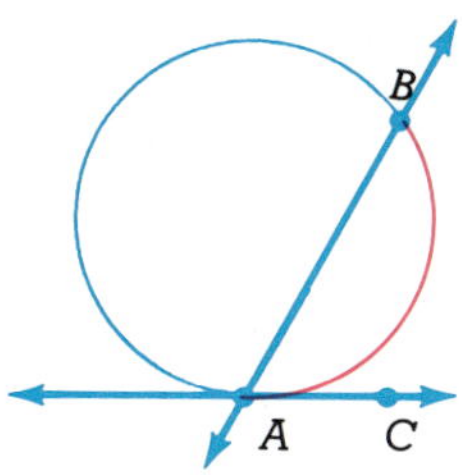

$m\angle BAC = \frac{1}{2}m(\widehat{BA})$

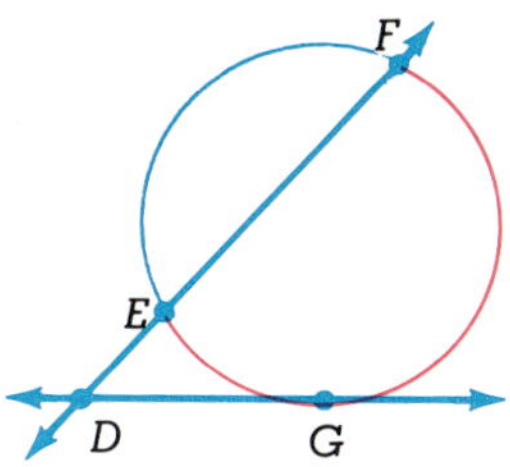

$m\angle FDG = \frac{1}{2}(m\widehat{FG} - m\widehat{GE})$

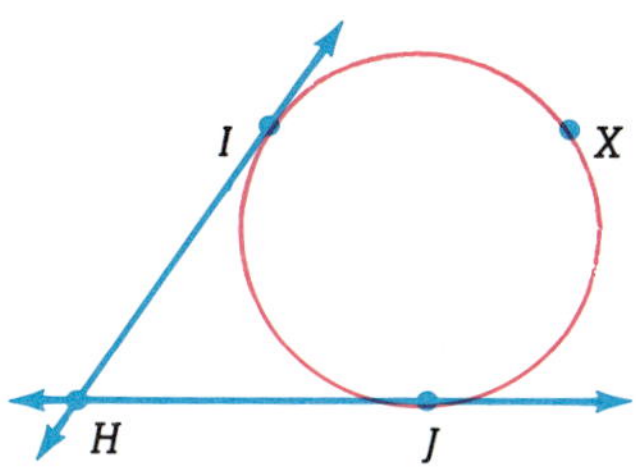

$m\angle IHJ = \frac{1}{2}(m\widehat{IXJ} - m\widehat{IJ})$

If a secant and a tangent intersect at the point of tangency, then the measure of the angles formed is one-half the measure of its intercepted arc.

Theorem 10–15

If a secant and a tangent, or two tangents, intersect in the exterior of a circle, then the measure of the angle formed is one-half the positive difference of the measures of the intercepted arcs.

Theorem 10–16

exercises

Exploratory For each of the following, measurements of certain arcs are given. Find the degree measure of each numbered angle.

1.

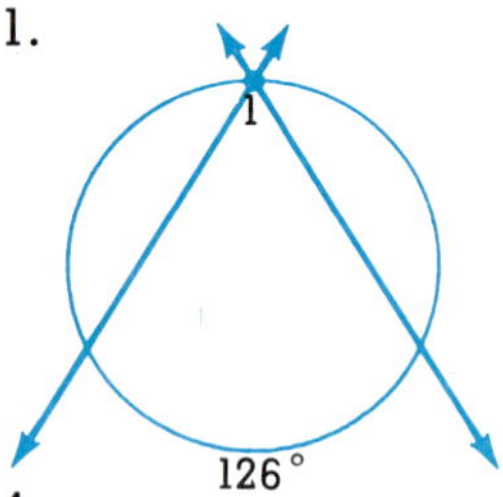

2.

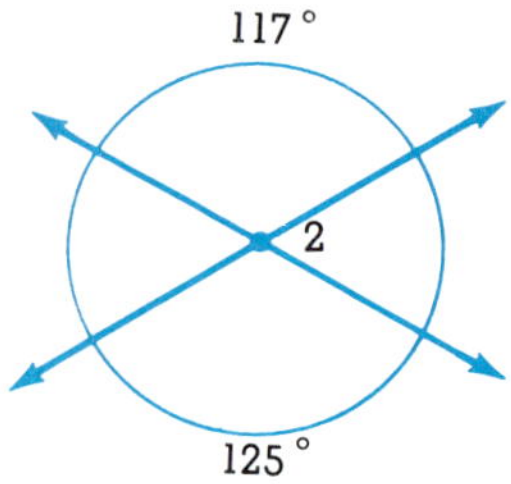

3.

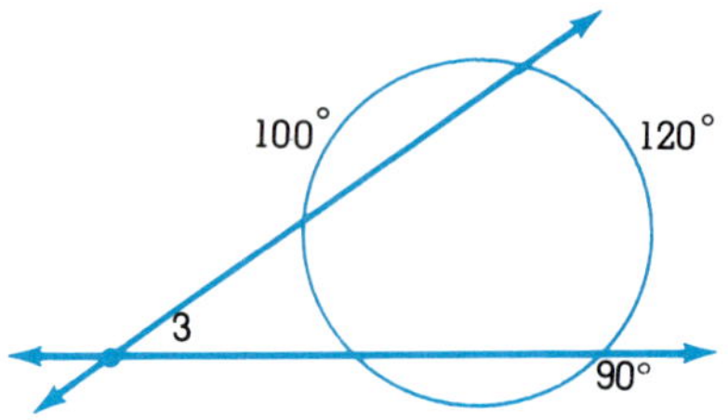

4.

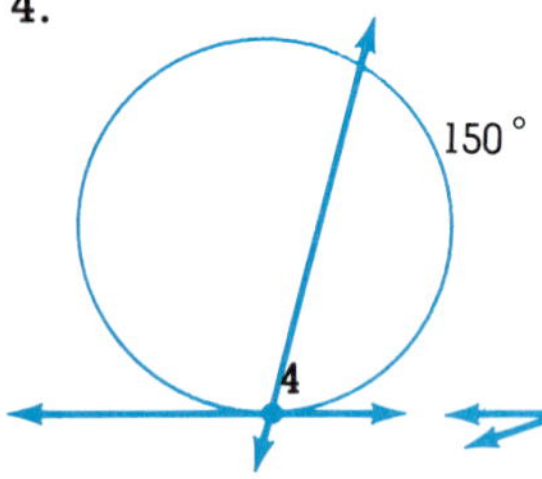

5.

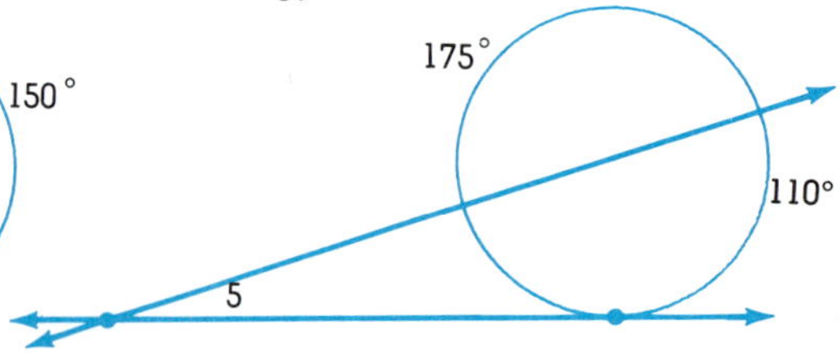

6.

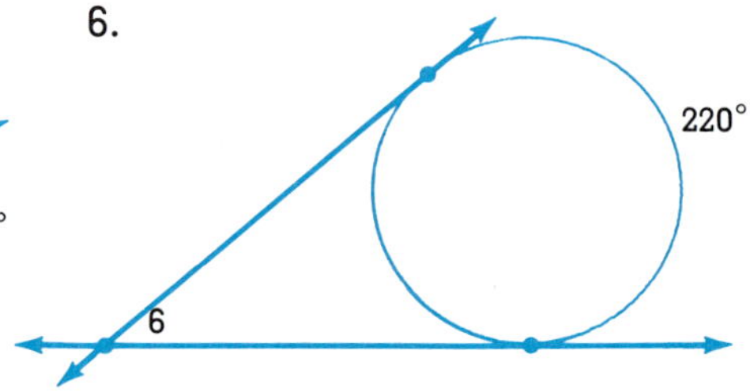

Written In the figure at the right, $m\widehat{BC} = 84$, $m\widehat{CD} = 38$, $m\widehat{DE} = 64$, and $m\widehat{EF} = 60$. Find each of the following.

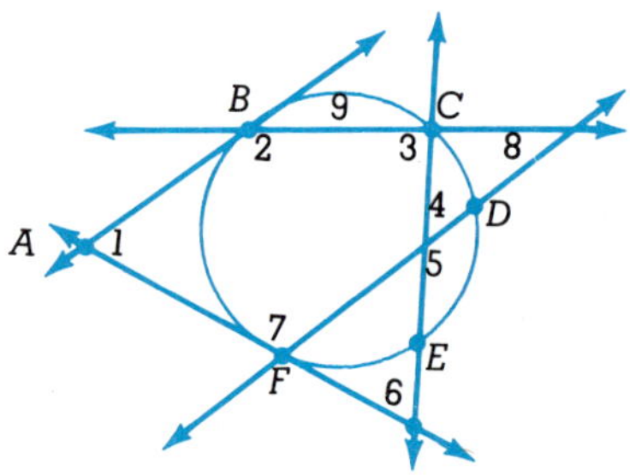

1. $m\widehat{BF}$	2. $m\widehat{BDF}$	3. $m\angle 1$
4. $m\widehat{BFC}$	5. $m\angle 2$	6. $m\angle 9$
7. $m\widehat{BFE}$	8. $m\angle 3$	9. $m\angle 4$
10. $m\angle 5$	11. $m\widehat{FBC}$	12. $m\angle 6$
13. $m\widehat{FBD}$	14. $m\angle 7$	15. $m\angle 8$

Solve each problem.

16. Prove Theorem 10–13.

17. Prove Theorem 10–15.

18. Prove that if a secant and a tangent intersect in the exterior of a circle, then the measure of the angle formed is one-half the positive difference of the measures of the intercepted arcs.

19. Prove that if two tangents intersect in the exterior of a circle, then the measure of the angle formed is one-half the positive difference of the measures of the intercepted arcs.

20. **Given:** $\odot P$ and $\odot E$ tangent at A.
$\overline{AE} \perp \overline{EB}$
Prove: $m\widehat{AD} = 90$

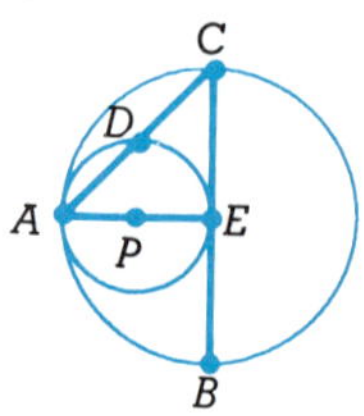

21. **Given:** $\overline{RP} \perp \overline{QT}$
$\overline{RT} \perp \overline{QS}$
Prove: $m\widehat{PT} = m\widehat{ST}$

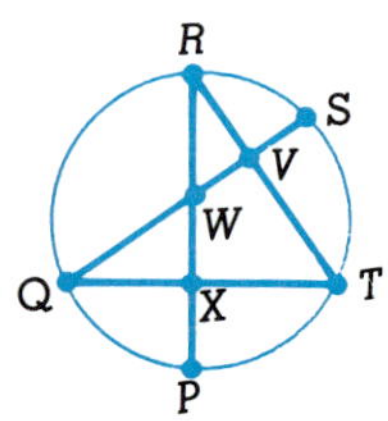

10–7 Segments

Angles formed by intersecting chords are related to arcs of the circle. In addition, the segments formed by the intersecting chords are related.

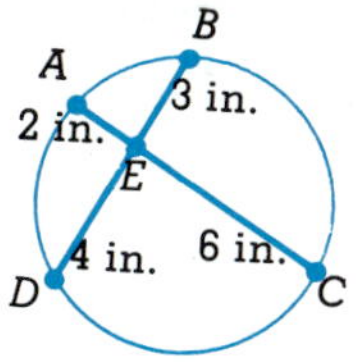

In the figure at the left, the segments of $\overline{AC}$ are $\overline{AE}$ and $\overline{EC}$. The segments of $\overline{BD}$ are $\overline{BE}$ and $\overline{ED}$.

$$AE \cdot EC = 2 \cdot 6 = 12 \qquad BE \cdot ED = 3 \cdot 4 = 12$$

If two chords intersect in a circle, then the product of the measures of the segments of one chord equals the product of the measures of the segments of the other chord.

Theorem 10–17

The proof of the theorem is based on properties of similar triangles.

example 1

Prove Theorem 10–17.

Given: $\overline{AC}$ and $\overline{BD}$ intersect at E

Prove: $AE \cdot EC = BE \cdot ED$

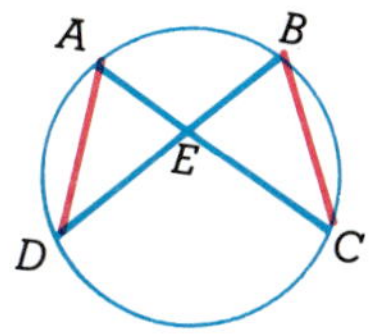

Proof:

STATEMENTS	REASONS
1. Draw $\overline{AD}$ and $\overline{BC}$.	1. Postulate 1–1: Through any two points there is exactly one line.
2. $\angle A \cong \angle B$ $\angle D \cong \angle C$	2. Theorem 10–7: If two inscribed angles intercept congruent arcs, then the angles are congruent.
3. $\triangle DAE \sim \triangle CBE$	3. Postulate 8–1: AA Similarity
4. $\dfrac{AE}{BE} = \dfrac{ED}{EC}$	4. Definition of similar polygons
5. $AE \cdot EC = BE \cdot ED$	5. Theorem 8–1: Equality of Cross Products

In the figure below, both $\overline{RP}$ and $\overline{RT}$ are called **secant segments** because they contain chords of the circle. The parts of these segments that are exterior to the circle are called **external secant segments.** In the figure, $\overline{RQ}$ and $\overline{RS}$ are external secant segments.

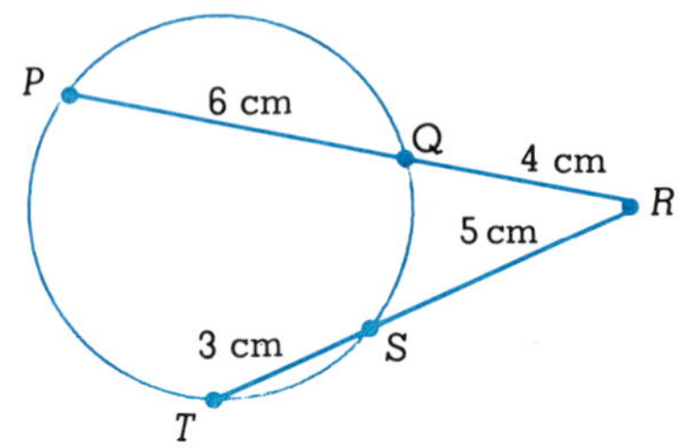

The measures of the segments of the figure are related in an interesting way.

$RQ \cdot RP$	$RS \cdot RT$
$4 \cdot 10$	$5 \cdot 8$
40	40

To prove the following theorem, draw a figure like the one shown. Connect P and S, and T and Q. Show that $\triangle PSR \sim \triangle TQR$. Then, write a proportion and find its cross product.

Theorem 10–18

If two secant segments are drawn to a circle from an exterior point, then the product of the measures of one secant segment and its external secant segment equals the product of the measures of the other secant segment and its external secant segment.

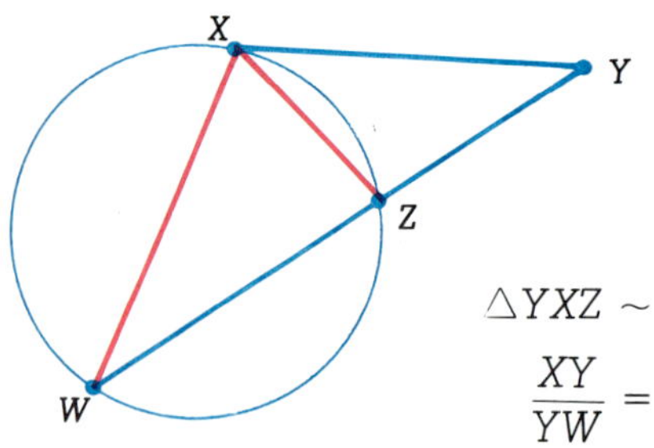

The figure at the left shows a tangent segment and a secant segment drawn to a circle from an exterior point. By drawing $\overline{XW}$ and $\overline{XZ}$ several triangles are formed.

$\triangle YXZ \sim \triangle YWX$	*AA Similarity*
$\frac{XY}{YW} = \frac{YZ}{XY}$	*Definition of Similarity*
$XY^2 = YW \cdot YZ$	*Equality of Cross Products*

Theorem 10–19

If a tangent segment and a secant segment are drawn to a circle from an exterior point, then the square of the measure of the tangent segment equals the product of the measures of the secant segment and its exterior secant segment.

example 2

In the figure at the right, B is a point of tangency. Find x.

$$AC \cdot DC = BC^2$$
$$(x + 6.6)(x) = (8)^2$$
$$x^2 + 6.6x = 64$$
$$x^2 + 6.6x - 64 = 0$$
$$x = \frac{^-(6.6) \pm \sqrt{(6.6)^2 - 4(1)(^-64)}}{2(1)}$$
$$= \frac{^-6.6 \pm \sqrt{43.56 + 256}}{2}$$
$$\approx \frac{^-6.6 \pm 17.3}{2}$$
$$\approx 5.35 \text{ or } ^-11.95$$

x is about 5.35.

a is 1, b is 6.6, and c is ⁻64.

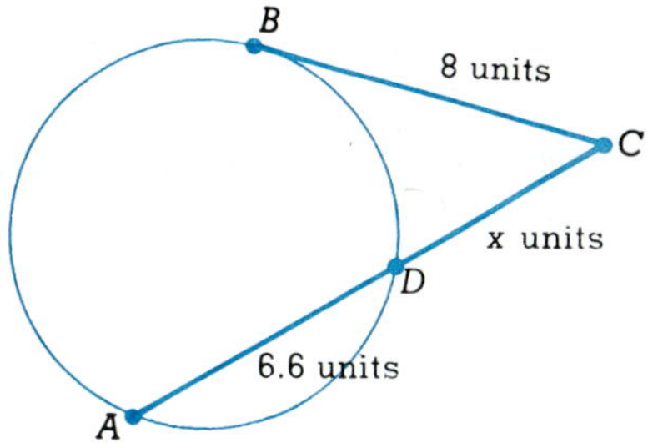

exercises

Exploratory For each of the following, state the equation you would use to find x.

1.
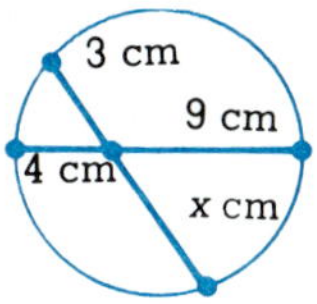

2.
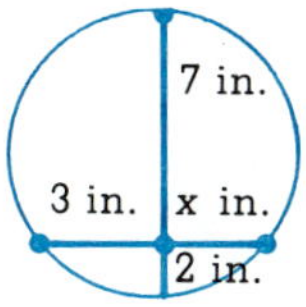

3.
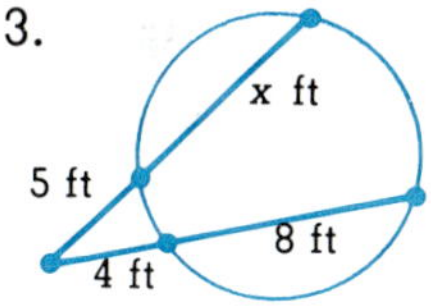

4.
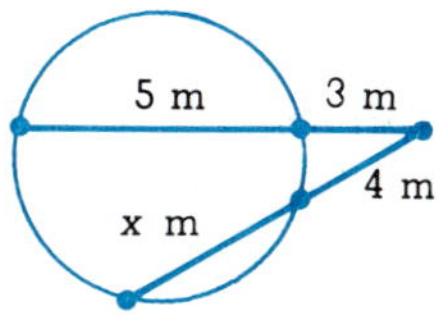

5.
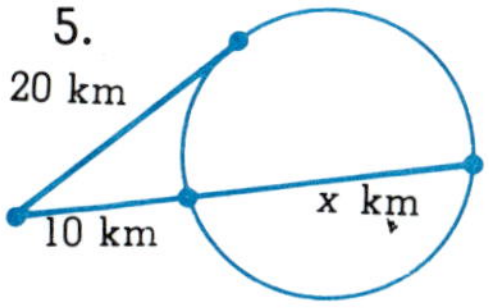

6.
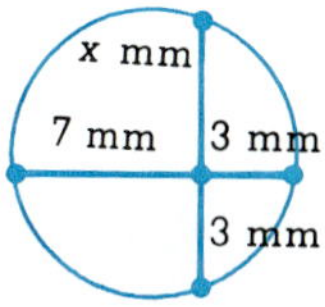

7.
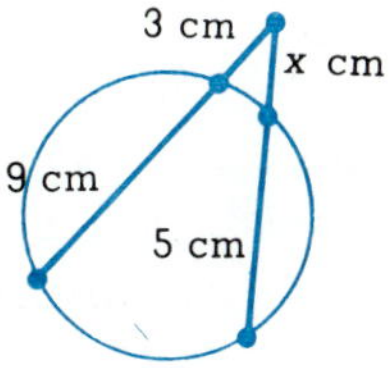

8.
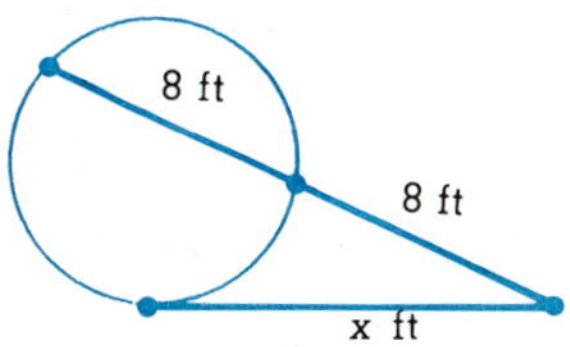

9.
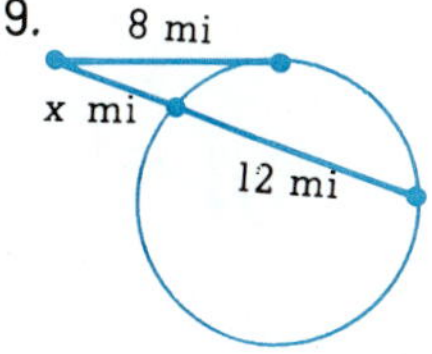

Written For each of the following, find x.

1.
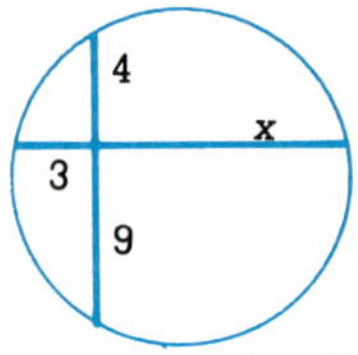

2.
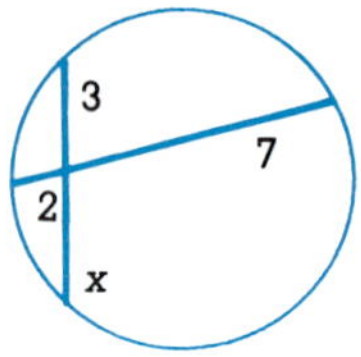

3.
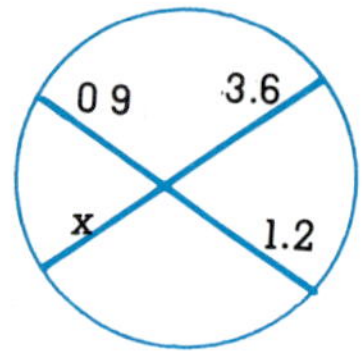

4.
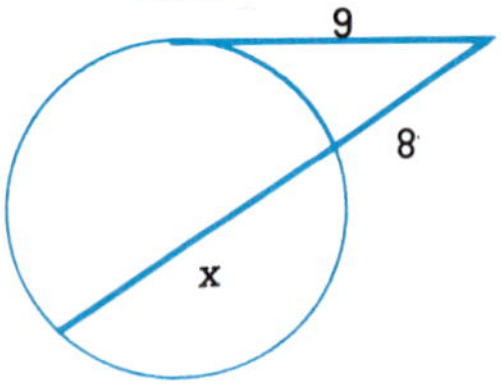

5.
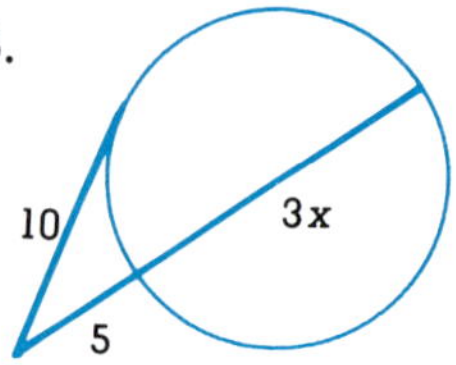

6.
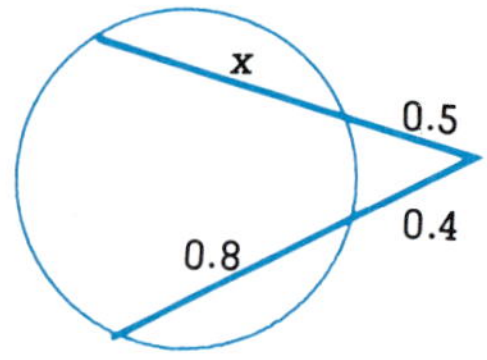

7.
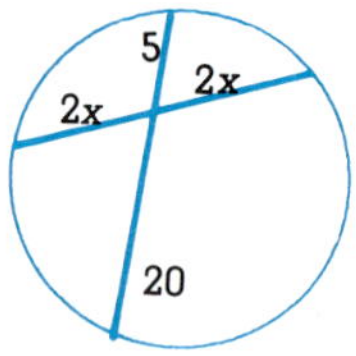

8.
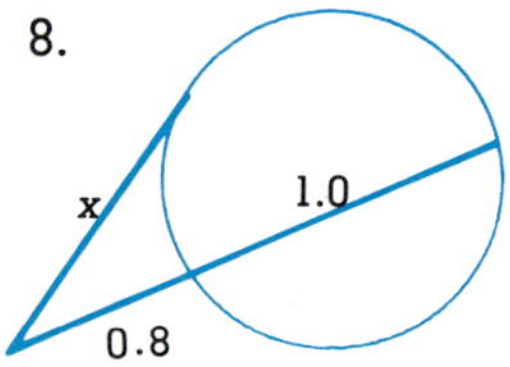

9.
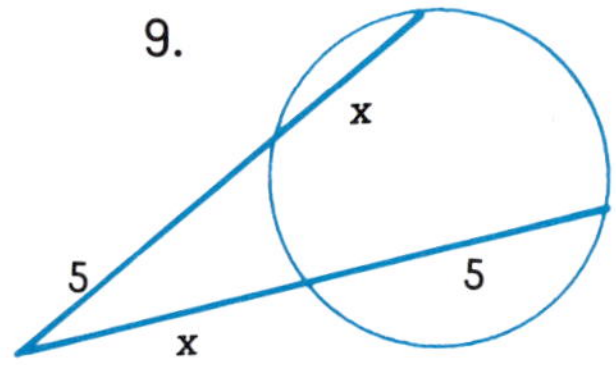

10.
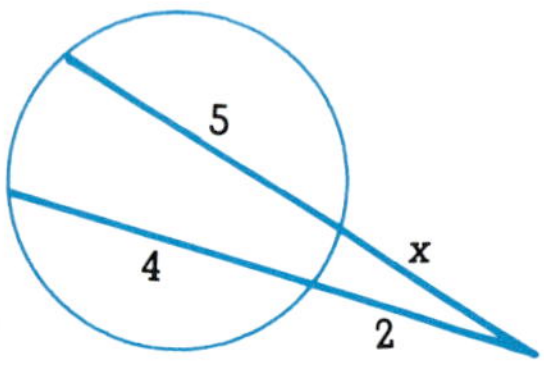

11.
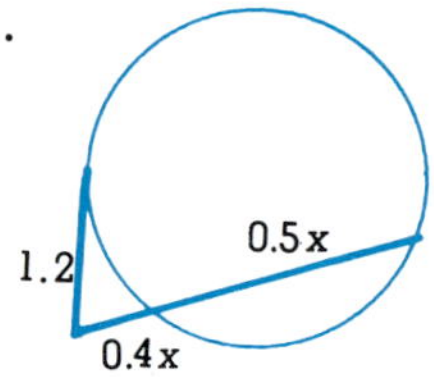

12.
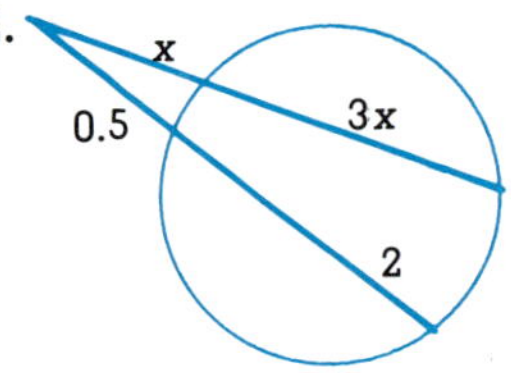

Use the figure at the right to solve each problem. Round each answer to one decimal place.

13. Suppose $FH = 6$, $HD = 2$, and $HE = 3$. Find CH.
14. Suppose $AD = 16$, $FD = 8$, and $AG = 6$. Find GC.
15. Suppose $AC = 21$, $AF = 8$ and $FD = 8$. Find AG.
16. Suppose $CH = 20$, $EH = 10.5$, and $DH = 8$. Find FH.
17. Suppose $AF = 16$, $FH = 6$, and $DH = 2$. Find BA.
18. Suppose $GC = 15.3$ and $GA = 6.0$. Find AB.

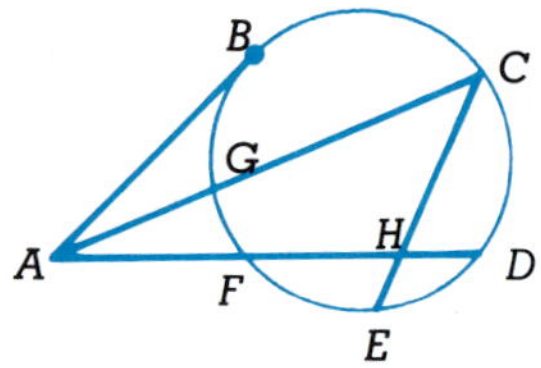

Solve each problem.

19. Prove Theorem 10–18.

20. Prove Theorem 10–19.

21. **Given:** $\overleftrightarrow{CF}$ is tangent to $\odot P$ at F
$\overleftrightarrow{CF}$ is tangent to $\odot Q$ at F
Prove: $CA \cdot CB = CE \cdot CD$

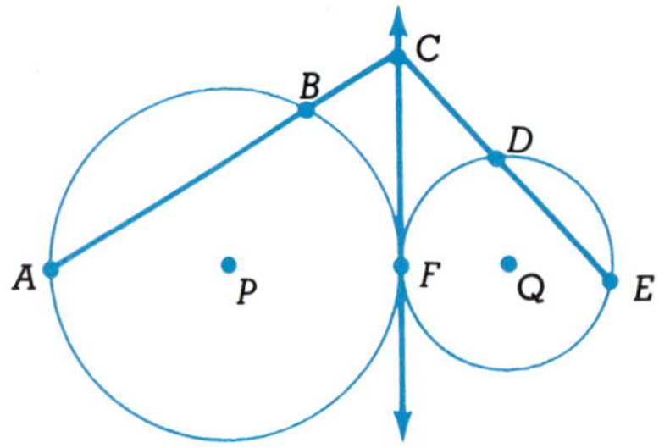

22. **Given:** $\overrightarrow{GM}$ is tangent to $\odot R$ at M
$\overrightarrow{GM}$ is tangent to $\odot S$ at M
Prove: $GH \cdot GL = GJ \cdot GK$

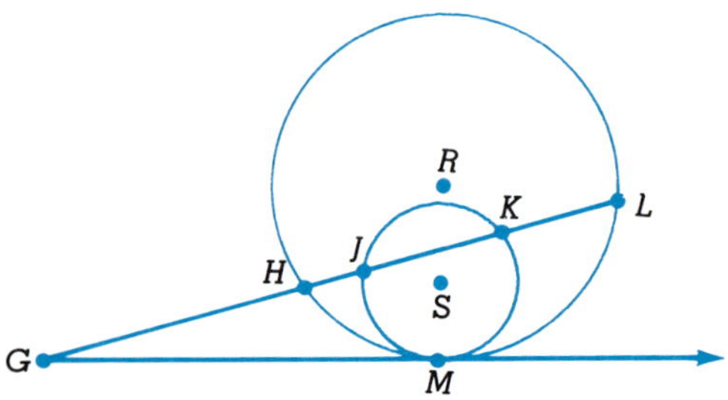

10–8 Spheres

In space, many congruent circles can be drawn using the same center. Taken together, all these circles form a **sphere.**

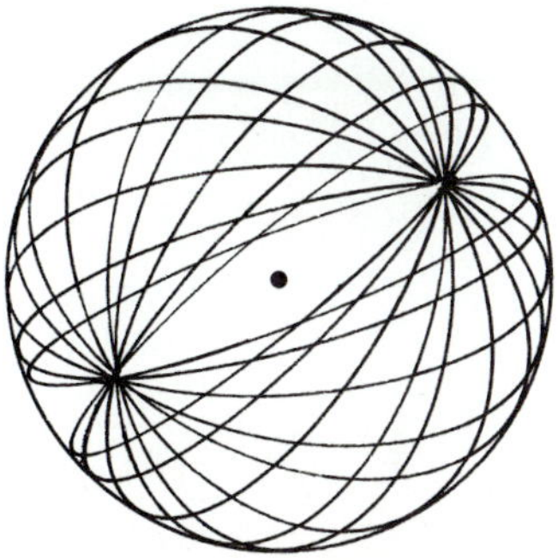

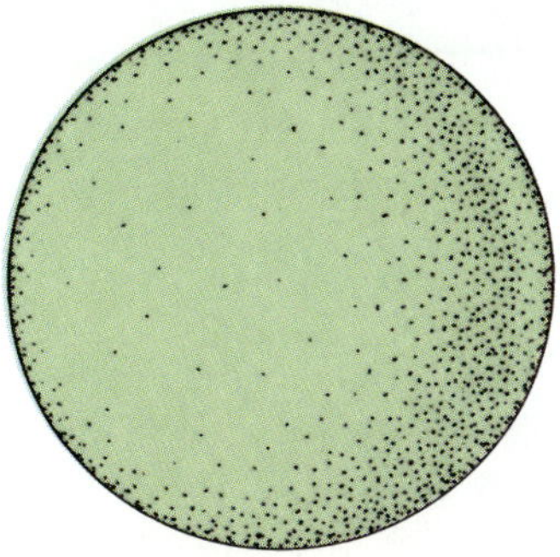

In space, a figure is a sphere if and only if it is the set of all points that are a given distance from a given point, called the center.

Definition of Sphere

A **radius** of a sphere is a line segment whose endpoints are the *center* and *a point on the sphere*. In the figure at the right, $\overline{CR}$ is a radius.

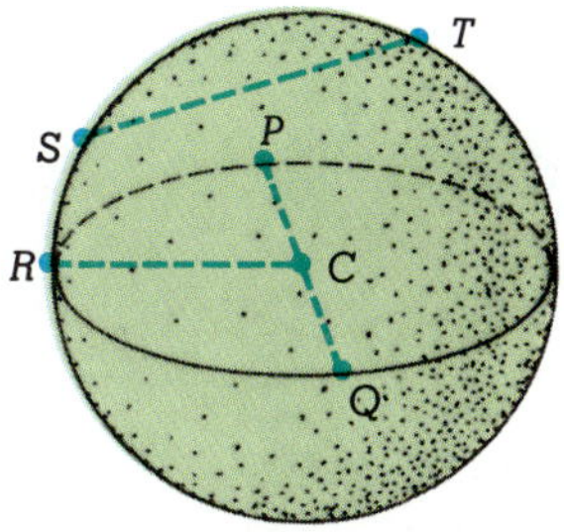

A **chord** of a sphere is a line segment whose endpoints are *points on the sphere*. In the figure at the right, $\overline{TS}$ is a chord.

A **diameter** of a sphere is a line segment which *contains the center*, and whose *endpoints are points on the sphere*. In the figure at the right, $\overline{PQ}$ is a diameter.

A plane can intersect a sphere in one of three ways.

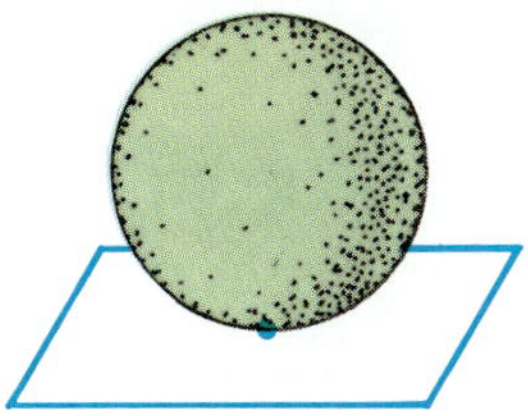

A plane may be tangent to a sphere.

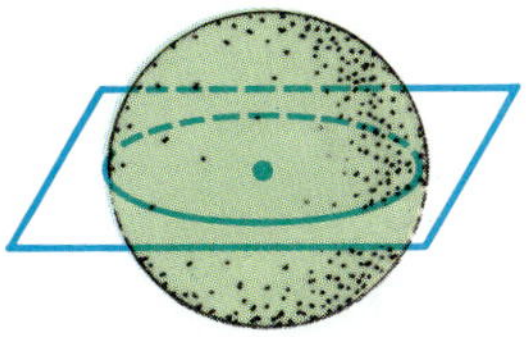

A plane may intersect the sphere and contain the center of the sphere.

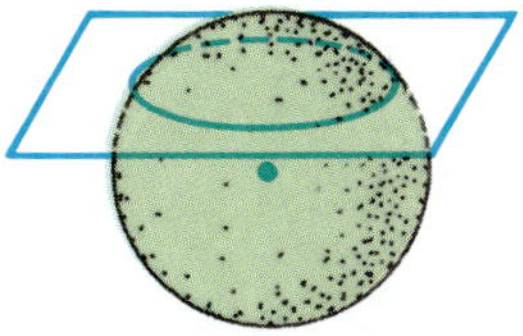

A plane may intersect the sphere but *not* contain the center of the sphere.

example 1

Show that the intersection of a sphere and a plane is a circle.

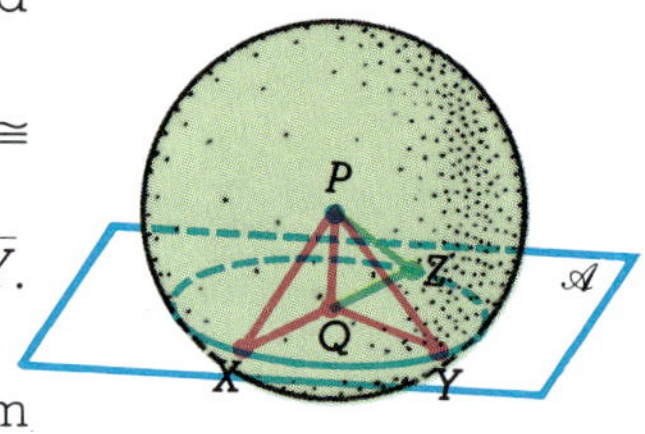

Suppose X and Y are two points in the intersection of the sphere and the plane. Draw $\overline{PQ}$ so that $\overline{PQ} \perp \mathscr{A}$ and Q lies in $\mathscr{A}$. Next, draw $\overline{QX}$, $\overline{QY}$, $\overline{PX}$ and $\overline{PY}$.

$\overline{PQ} \perp \overline{QX}$ and $\overline{PQ} \perp \overline{QY}$, since $\overline{PQ} \perp \mathscr{A}$. Thus, $\angle PQX \cong \angle PQY$ because they are right angles.

Since $\overline{PX}$ and $\overline{PY}$ are both radii of the sphere, $\overline{PX} \cong \overline{PY}$. Also, $\overline{PQ} \cong \overline{PQ}$. So, $\triangle PQX \cong \triangle PQY$ by HL.

As a result, $\overline{QX} \cong \overline{QY}$, and X and Y are equidistant from Q. They lie on the circle with Q as a center and radius measuring QY. In addition, any point on the circle is in $\mathscr{A}$.

Suppose Z is a point on $\odot$Q. Then, $\overline{QZ} \cong \overline{QY}$. Also, $\overline{PQ} \perp \overline{QZ}$ since $\overline{PQ} \perp \mathscr{A}$. Because, $\overline{PQ} \cong \overline{PQ}$, the two right triangles, $\triangle PQY$ and $\triangle PQZ$ are congruent. Thus, $PY = PZ$ and Z must be on the sphere.

Theorem 10–20

If a plane intersects a sphere in more than one point, then the intersection is a circle.

Each great circle separates a sphere into two hemispheres.

Suppose a plane intersects a sphere in more than one point *and* contains the center of the sphere. The intersection is called a **great circle.** A great circle has the same center as the sphere. Its radii also are radii of the sphere.

The surface of a globe is a model of a sphere. Each north-south meridian represents a great circle. Of the parallels of latitude, *only the equator* represents a great circle.

exercises

Exploratory Describe each of the following as a model of a *circle*, a *sphere*, or *neither*.

1. ball bearing
2. basketball hoop
3. Jupiter
4. basketball
5. car tire
6. corona of eclipse
7. orbit of electron
8. cross section of pipe
9. telephone dial
10. ping pong ball
11. helicopter pad
12. record

Written For each of the following, write *yes* or *no*.

1. A diameter of a sphere is a chord of the sphere.
2. A radius of a sphere is a chord of the sphere.
3. All radii of a sphere are congruent.
4. All diameters of a sphere are congruent.
5. All great circles of a sphere are congruent.
6. All circles of a sphere are congruent.
7. A diameter of a great circle is a diameter of the sphere.
8. A diameter of a sphere is a diameter of a great circle of the sphere.
9. A secant of a sphere intersects the sphere in exactly one point.
10. A radius of a sphere intersects the sphere in exactly one point.
11. Two spheres may intersect in exactly one point.
12. Two great circles of a sphere may intersect in exactly one point.
13. Two great circles of a sphere may intersect in exactly two points.
14. Two spheres may intersect in exactly two points.

Write *if and only if* definitions for each of the following.

15. radius of a sphere
16. chord of a sphere
17. great circle of a sphere
18. diameter of a sphere
19. interior of a sphere
20. exterior of a sphere
21. line tangent to a sphere
22. plane tangent to a sphere
23. secant of a sphere
24. hemisphere
25. congruent spheres
26. arc of a sphere

Suppose C is the center of a sphere. Suppose I is in the interior of the sphere and E is in the exterior of the sphere. Solve each problem about the sphere.

27. In how many points does $\overline{IE}$ intersect the sphere?
28. In how many points does $\overleftrightarrow{IE}$ intersect the sphere?
29. In how many points does $\overrightarrow{IE}$ intersect the sphere?
30. In how many points does $\overrightarrow{EI}$ intersect the sphere?
31. In how many points does $\overline{CE}$ intersect the sphere?
32. In how many points does $\overline{CI}$ intersect the sphere?
33. How many chords contain I?
34. How many radii contain I?

Challenge In the figure at the right, P is the center of the sphere. The plane, $\mathscr{B}$, intersects the sphere in $\odot R$.

1. Suppose $PS = 25$ and $PR = 7$. Find RS.
2. Suppose $PS = 15$ and $PR = 9$. Find RS.
3. Suppose $PS = 13$ and $RS = 12$. Find PR.
4. Suppose $PS = 26$ and $RS = 24$. Find PR.
5. Prove that $\overline{PR} \perp \overline{RS}$.
6. Prove that $PS > PR$.

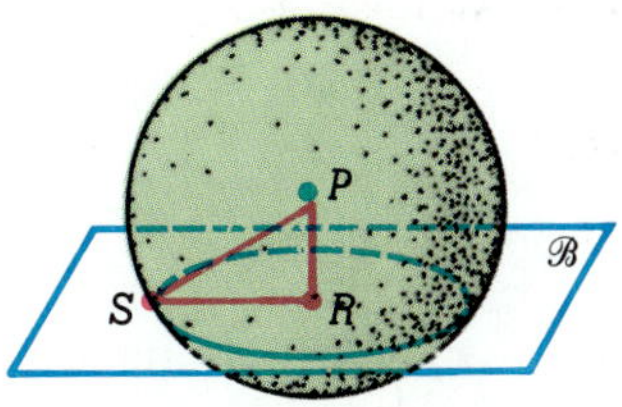

Vocabulary

circle (294)
radius (294)
chord (294)
diameter (294)
interior of a circle (295)
exterior of a circle (295)
tangent (295)
secant (295)
central angle (297)
minor arc (297)
major arc (297)
semicircle (297)
concentric circles (298)
intercepted arc (304)
inscribed angle (304)
secant segment (316)
external secant segment (316)
sphere (319)
great circle (320)
hemisphere (320)

Chapter Summary

1. In a plane, a figure is a circle if and only if it is the set of all points that are a given distance from a given point, called the center. (294)
2. Theorem 10–1: In a plane, if a line contains a point in the interior of a circle, then the line intersects the circle in exactly two points. (295)
3. The degree measure of a minor arc is the degree measure of its central angle. The degree measure of a major arc is 360 minus the degree measure of its central angle. The degree measure of a semicircle is 180. (297)
4. Postulate 10–1: Arc Addition Postulate: If Q is a point on $\widehat{PQR}$, then $m\widehat{PQ} + m\widehat{QR} = m\widehat{PQR}$. (297)
5. All circles are similar, but two circles are congruent if and only if their radii are congruent. (298)
6. Theorem 10–2: In a circle or in congruent circles, two central angles are congruent if and only if their minor arcs are congruent.
7. Theorem 10–3: In a circle or in congruent circles, two minor arcs are congruent if and only if their corresponding chords are congruent. (300)
8. Theorem 10–4: In a circle, if a diameter is perpendicular to a chord, then it bisects the chord and its arcs. (301)
9. Theorem 10–5: In a circle or in congruent circles, two chords are congruent if and only if they are equidistant from the center. (301)

10. Theorem 10–6: The measure of an inscribed angle of a circle equals one-half the measure of its intercepted arc. (305)
11. Theorem 10–7: If two inscribed angles intercept congruent arcs, then the angles are congruent. (306)
12. Theorem 10–8: If an angle is inscribed in a semicircle, then the angle is a right angle. (306)
13. Theorem 10–9: If the angles of a quadrilateral are inscribed in a circle, then each pair of opposite angles are supplementary. (306)
14. Theorem 10–10: A tangent to a circle is perpendicular to the point of tangency. (308)
15. Theorem 10–11: In a plane, if a line is perpendicular to a radius of a circle at its endpoint on the circle, then the line is a tangent. (308)
16. Theorem 10–12: The two tangent segments from the same exterior point of a circle are congruent. (309)
17. Theorem 10–13: If two secants intersect in the interior of a circle, then the measure of an angle formed is one-half the sum of the measures of the arcs intercepted by the angle and its vertical angle. (312)
18. Theorem 10–14: If two secants intersect in the exterior of a circle, then the measure of an angle formed is one-half the positive difference of the measures of the intercepted arcs. (312)
19. Theorem 10–15: If a secant and a tangent intersect at the point of tangency, then the measure of the angles formed is one-half the measure of its intercepted arc. (313)
20. Theorem 10–16: If a secant and a tangent, or two tangents, intersect in the exterior of a circle, then the measure of the angle formed is one-half the positive difference of the measures of the intercepted arcs. (313)
21. Theorem 10–17: If two chords intersect in a circle, then the product of the measures of the segments of one chord equals the product of the measures of the segments of the other chord. (315)
22. Theorem 10–18: If two secant segments are drawn to a circle from an exterior point, then the product of the measures of one secant segment and its external secant segment equals the product of the measures of the other secant segment and its external secant segment. (316)
23. Theorem 10–19: If a tangent segment and a secant segment are drawn to a circle from an exterior point, then the square of the measure of the tangent segment equals the product of the measures of the secant segment and its exterior secant segment. (316)

24. In space, a figure is a sphere if and only if it is the set of all points that are a given distance from a given point, called the center. (319)
25. Theorem 10–20: If a plane intersects a sphere in more than one point, then the intersection is a circle. (320)

Chapter Review

10–1 **Solve each of the following.**

1. A radius of a circle is a chord of a circle. Write *yes* or *no*.
2. A diameter of a circle is a chord of the circle. Write *yes* or *no*.
3. Write an if and only if definition for the diameter of a circle.

10–2 **Use the figure at the left to solve each of the following. Assume that $\overline{XY}$ and $\overline{AB}$ are diameters of $\odot P$.**

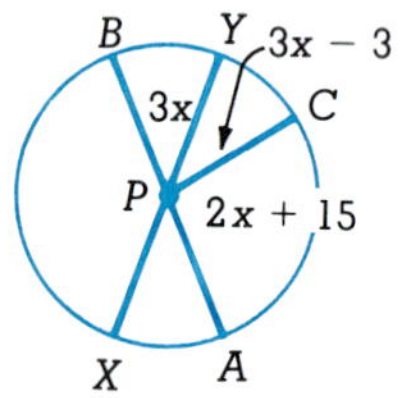

4. Find $m\angle BPY$.
5. Find $m\widehat{YAX}$.
6. Find $m\angle BPC$.
7. Find $m\widehat{BX}$.
8. Find $m\angle XPA$.
9. Find $m\widehat{CA}$.
10. Find $m\angle CPA$.
11. Find $m\widehat{BC}$.

10–3 **Solve each of the following.**

12. Given: $\overline{AB}$ is a diameter
$\overline{AB} \perp \overline{CD}$
Prove: $\widehat{CB} \cong \widehat{DB}$

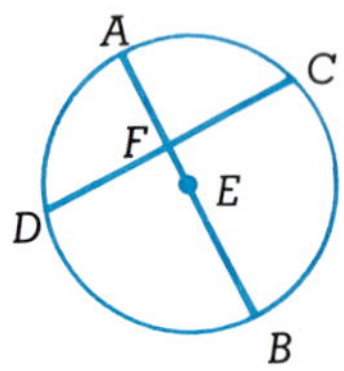

13. Suppose a chord of a circle is 16 centimeters long and is 6 centimeters from the center of the circle. Find the length of the radius.

10–4 **In the figure at the left, P is the center of the circle. $\overline{AB} \parallel \overline{CD}$, $m\widehat{BD} = 72$, and $m\angle CPD = 110$. Find the measure for each of the following.**

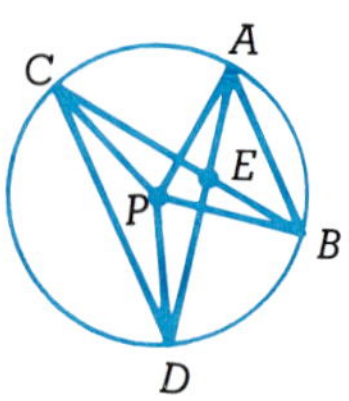

14. $\angle DPB$
15. $\widehat{AB}$
16. $\widehat{CD}$
17. $\angle APB$
18. $\angle CPA$
19. $\angle CDA$
20. $\widehat{CA}$
21. $\angle DAB$

10–5 For each of the following, find x. Assume C is the center of the circle.

22.

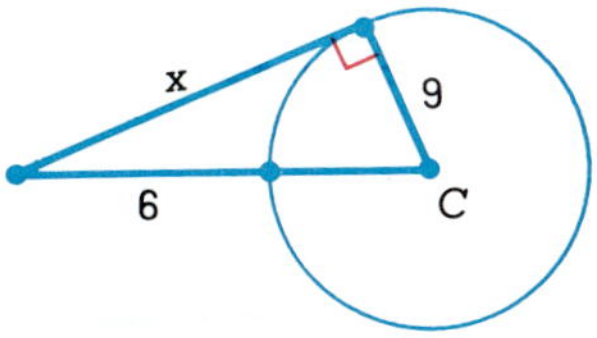

23.

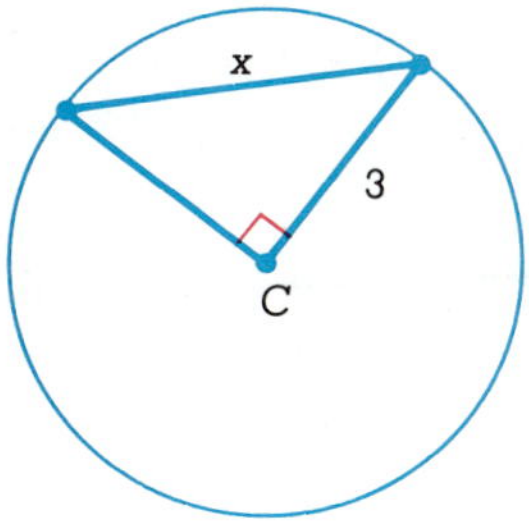

24.

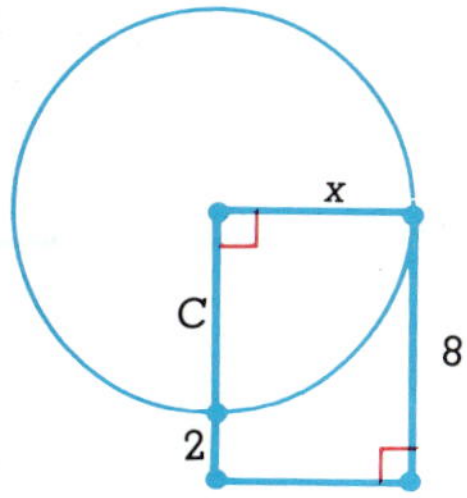

25. Tell whether the sides of a circumscribed polygon are *chords* or *tangents* of the circle.

10–6 In the figure at the right, P is the center of the circle and $m\widehat{AB} = 29$, $m\angle AEB = 42$, and $m\widehat{BG} = 18$. Find each of the following.

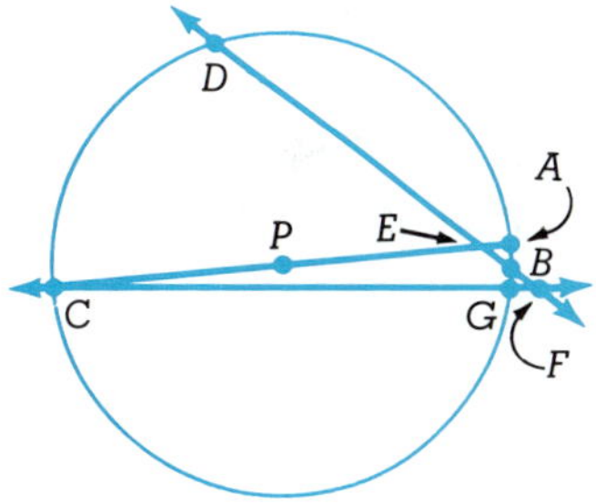

26. $m\angle DEC$
27. $m\widehat{CD}$
28. $m\widehat{AD}$
29. $m\angle GFD$
30. $m\angle AED$
31. $m\widehat{GC}$

10–7 For each of the following, find x.

32.

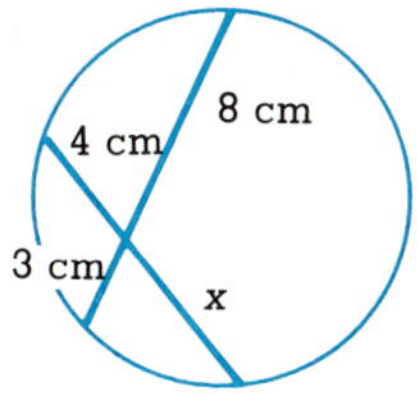

33.

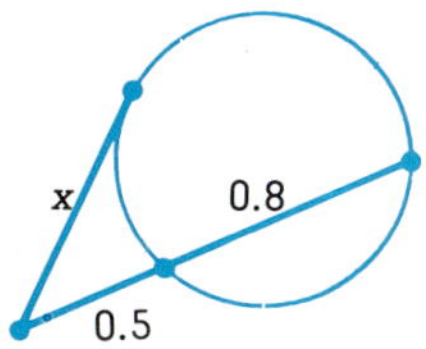

34.

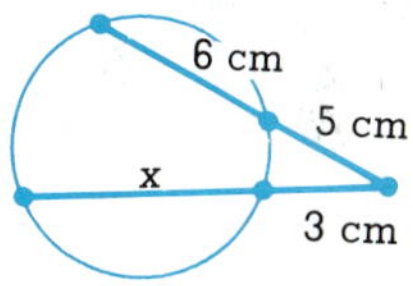

10–8 Solve each of the following.

35. A radius of a sphere is a chord of the sphere. Write *yes* or *no*.

36. All circles of a sphere are congruent. Write *yes* or *no*.

Suppose C is the center of a sphere. Suppose I is in the interior of the sphere and E is in the exterior of the sphere. Solve each problem about the sphere.

37. In how many points does $\overrightarrow{EI}$ intersect the sphere?

38. How many chords contain I?

39. In how many points does $\overline{CI}$ intersect the sphere?

40. In how many points does $\overline{CE}$ intersect the sphere?

Chapter Test

1. Write an if and only if definition for the radius of a circle.

For each of the following, find x. Assume C is the center of the circle.

2.

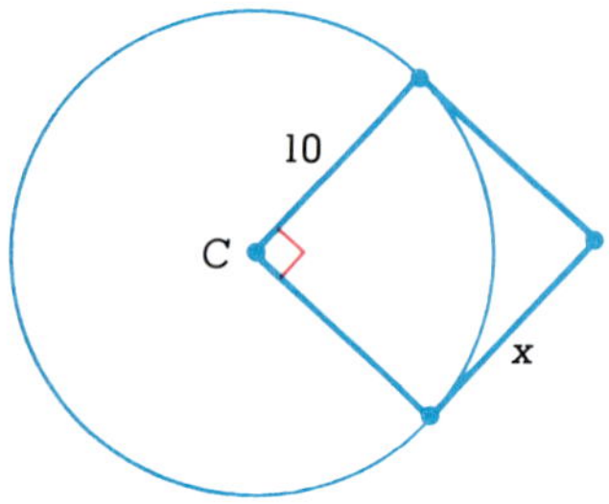

3.

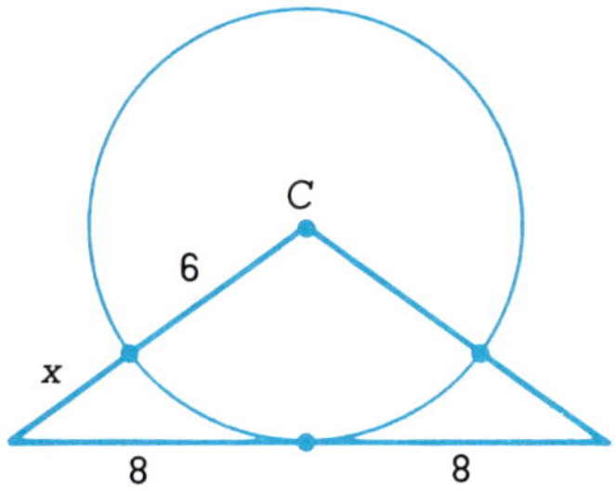

4.

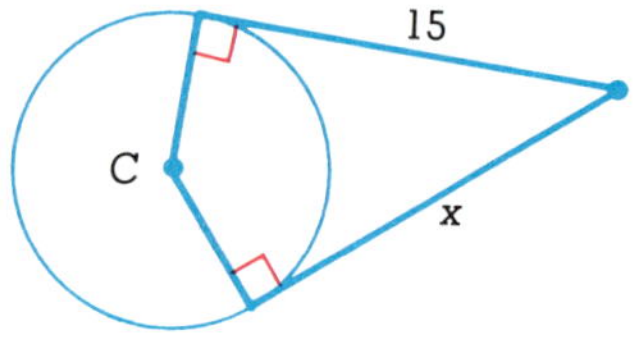

5.

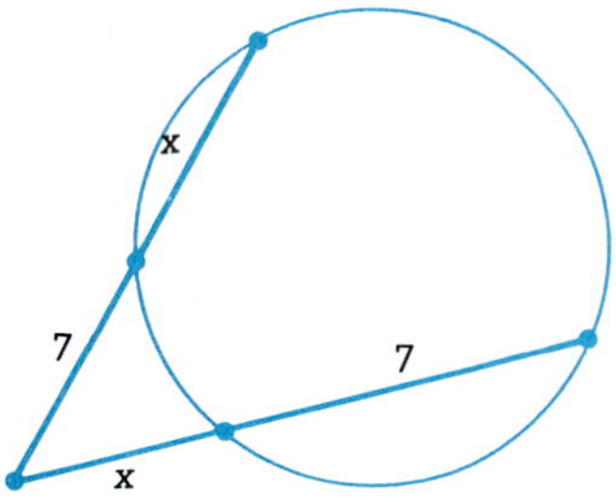

6.

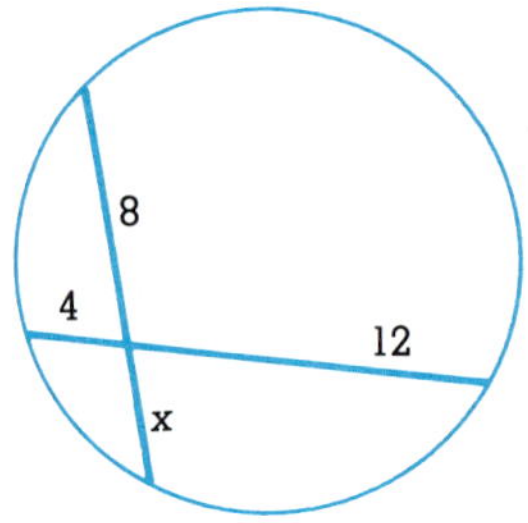

7.

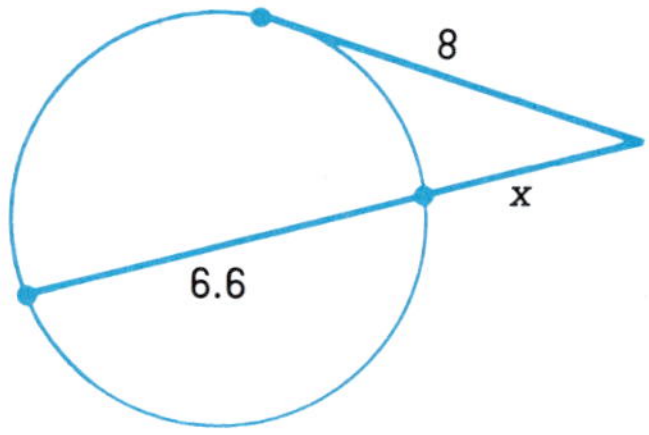

Solve each of the following.

8. Suppose the diameter of a circle is 10 inches long and a chord is 6 inches long. Find the distance between the chord and the center of the circle.

9. Tell whether the sides of an inscribed polygon are chords or tangents of the circle.

10. Two spheres may intersect in exactly two points. Write *yes* or *no*.

In the figure at the left, P is the center of the circle. $\overline{AB} \parallel \overline{CD}$, $m\widehat{BD} = 42$, $m\widehat{BE} = 12$. Find each of the following.

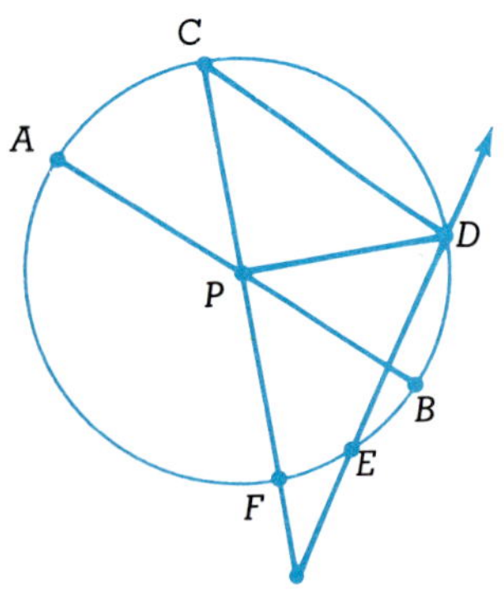

11. $m\angle BPD$
12. $m\angle APC$
13. $m\angle BPF$
14. $m\widehat{CD}$
15. $m\widehat{AC}$
16. $m\widehat{FB}$
17. $m\widehat{AF}$
18. $m\widehat{FE}$
19. $m\angle FCD$
20. $m\angle DCF$

Chapter 11

Area and Volume

Heavy snows increase the chances for spring floods. Experts must consider different types of snow when making their forecasts. An inch of dense, wet snow may contain 5300 gallons of water per acre. An inch of powdery, light snow may contain only 1300 gallons of water per acre.

11–1 Defining Area

The amount of surface a carpet covers is called its **area.** For example, a 9 foot (ft) by 12 foot carpet covers 108 square feet (ft^2) of area.

Normally carpets are sold by the square yard (yd^2).

Together, a polygon and its interior are called a **polygonal region.**

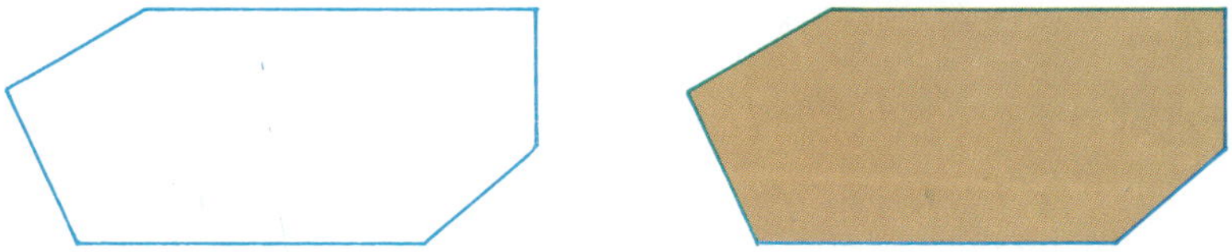

polygon   polygonal region

Area measurements are used to compare polygonal regions.

Postulate 11–1
Area Postulate

For any polygonal region and a given unit of measure, there is a unique positive number called the measure of the area of the region.

Area will mean the measure of the area of a polygonal region. Thus, for example, the area of a square is the region formed by the square and its interior.

If each side measures 1 unit, we say that the area is 1 square unit. Thus, a square one centimeter (cm) wide and one centimeter long has an area of 1 square centimeter (cm^2).

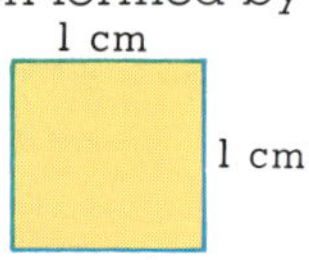

area is
1 square centimeter

The area of a region may differ depending on the unit of measure. For example, the area of the square region at the right is 1 square inch. The area of the region also is about 6.45 square centimeters.

In a given situation, we do *not* always state the unit of measure. If no units are stated, assume the same unit is used for all measurements.

Using the following postulates, it is possible to find the areas of various regions.

1 inch

1 inch

area is
1 square inch

If two figures are congruent, then they have exactly the same area.

Postulate 11–2

If a polygonal region is separated into nonoverlapping regions, then the sum of the areas of these regions equals the area of the entire region.

Postulate 11–3
Area Addition Postulate

The rectangular region at the right is separated into 12 congruent squares whose sides are 1 centimeter long. Thus, the area of each square is 1 square centimeter. By the Area Addition Postulate, the area of the rectangle is 12 square centimeters.

4 cm

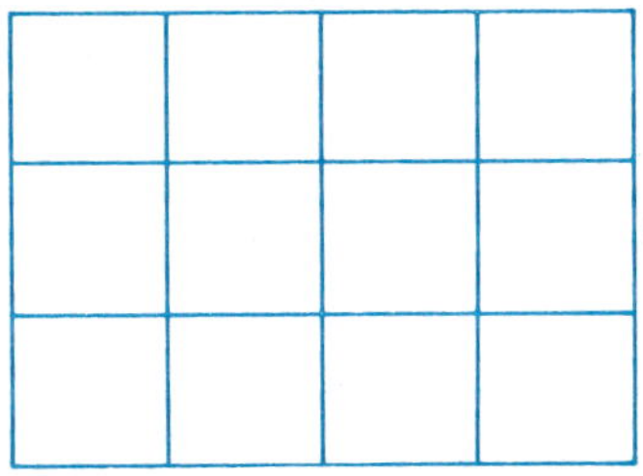

3 cm

This and many other examples suggest the following postulate.

If a rectangle has an area of A square units, its length is ℓ units, and its width is w units, then $A = \ell w$.

Postulate 11–4

example 1

Find the area of a rectangle that has a length of 6.1 meters (m) and a width of 3.7 meters.

$$\begin{aligned} A &= \ell w && \textit{Postulate 11–4} \\ &= 6.1 \times 3.7 \\ &= 22.57 \end{aligned}$$

The area of the rectangle is 22.57 square meters (m^2).

example 2 **Develop a formula for the area of a square, if each side is s units long.**

Let A stand for the area of the square.

$$\begin{aligned} A &= \ell w && \textit{A square is a rectangle.} \\ &= s \cdot s && \ell \textit{ is } s \textit{ and } w \textit{ is } s. \\ &= s^2 \end{aligned}$$

The formula is $A = s^2$.

Theorem 11–1 **If a square has an area of A square units, and each side is s units long, then $A = s^2$.**

exercises

Exploratory **Solve each of the following.**

1. Draw as many rectangles as you can that are separated into 12 one-centimeter squares. Name the dimensions for each rectangle.
2. Draw as many rectangles as you can that are separated into 16 one-centimeter squares. Name the dimensions for each rectangle.
3. Draw as many rectangles as you can that are separated into 24 one-centimeter squares. Name the dimensions for each rectangle.
4. Draw as many rectangles as you can that are separated into 36 one-centimeter squares. Name the dimensions for each rectangle.

Find the area and perimeter of each rectangle described below.

5. 6 units by 6 units
6. 9 units by 4 units
7. 12 units by 3 units
8. 18 units by 2 units
9. 5 units by 5 units
10. 6 units by 4 units
11. 7 units by 3 units
12. 8 units by 2 units

For each of the following state *yes* or *no*.

13. If two rectangles have the same area, then they have the same perimeter.
14. If two rectangles have the same perimeter, then they have the same area.
15. If two squares have the same area, then they have the same perimeter.
16. If two squares have the same perimeter, then they have the same area.

Written Find the area of each rectangle described below.

1. 5 centimeters by 4 centimeters
2. 7 inches (in.) by 4 inches
3. 10 feet by 12 feet
4. 3.5 inches by 10 inches
5. 42.5 centimeters by 53.4 centimeters
6. 16.3 meters by 21.2 meters
7. 29 miles by 23 miles
8. 48 yards (yd) by 38 yards
9. $5\frac{1}{2}$ miles by $3\frac{1}{3}$ miles
10. $4\frac{1}{3}$ yards by $2\frac{2}{3}$ yards
11. $5\frac{1}{2}$ inches by $4\frac{1}{4}$ inches
12. $\frac{3}{4}$ foot by $\frac{1}{4}$ foot
13. x feet by $(x + 3)$ feet
14. $2k$ centimeters by $(k + 1)$ centimeters
15. $(3a + 1)$ inches by $(4a + 2)$ inches
16. $(9w - 3)$ miles by $(4w - 1)$ miles

Find the width of each rectangle given the length and area.

17. 3.5 inches, 35 square inches
18. 48 yards, 1824 square yards
19. 16.3 meters, 345.56 square meters
20. 21 kilometers, 239.4 square kilometers
21. $2\frac{2}{3}$ yards, $11\frac{5}{9}$ square yards
22. $30\frac{1}{2}$ feet, 610 square feet
23. $3\frac{1}{2}$ feet, $7\frac{7}{8}$ square feet
24. $2\frac{1}{2}$ feet, $5\frac{5}{6}$ square feet

Find the area of each shaded region below.

25.
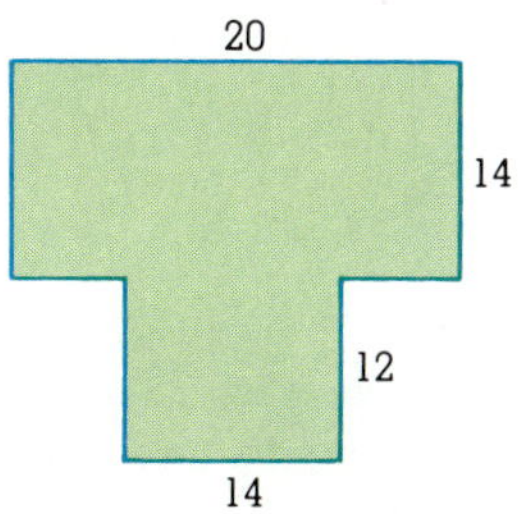

26.
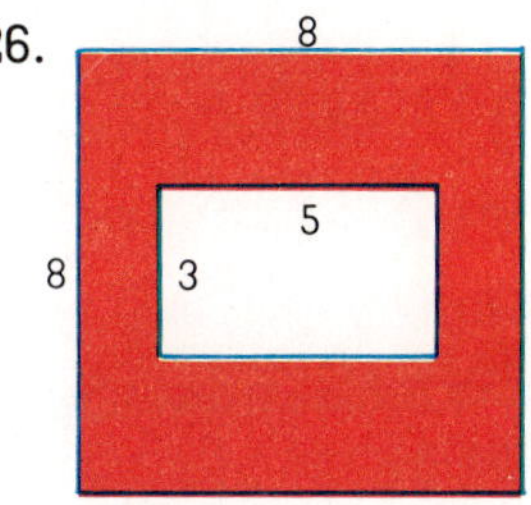

27.
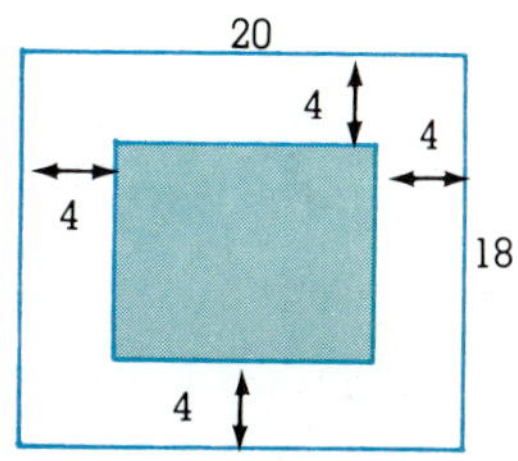

28.
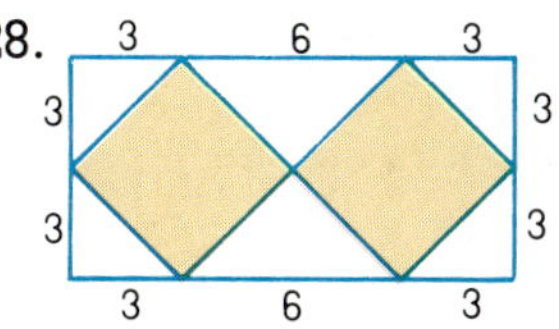

29.
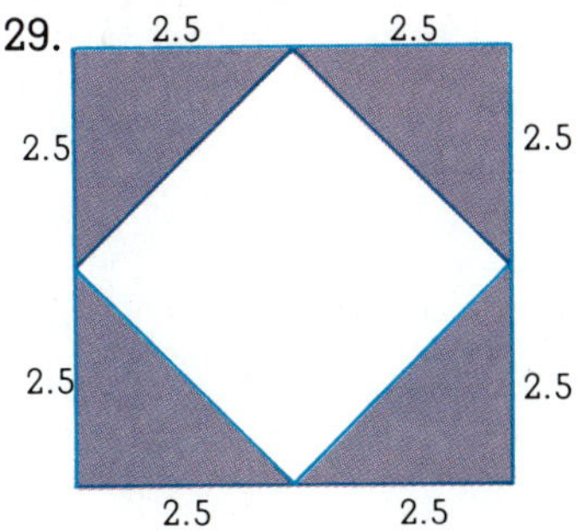

30.
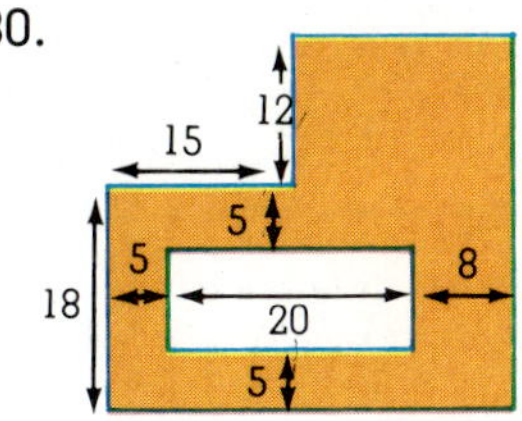

Solve each problem.

31. A rectangular field to be planted with grain is 150 meters long and 160 meters wide. One bushel bag of grain plants 100 square meters of land. How many bushel bags are needed to plant the field?

32. A rectangular window is $5\frac{1}{2}$ feet by 4 feet. The window is to be filled with stained glass. Most stained glass costs about $12 per square foot. Estimate the cost of stained glass for the window.

Using Geometry

Metric Units of Area

The following are the units of area most commonly used in the metric system.

Square kilometer is a unit of area used to measure regions such as surfaces of lakes, deserts, forests, and countries. One square kilometer is about the same as the area of 25 city blocks.

Square meter is the standard unit of area in the metric system. It is used to measure areas of carpets, floors, gardens, and fields. One square meter is about the area of the top of a card table.

Square centimeter is a unit of area used to measure smaller regions such as the area of pieces of paper, cards, footprints, and so on. One square centimeter is about the area of the hole in a telephone dial or a pushbutton on a telephone.

The common units are related.

1 square kilometer is the same as 1000 × 1000 or 1,000,000 square meters.

1 square centimeter is the same as 0.01 × 0.01 or 0.0001 square meters.

Exercises Tell which unit of area, *square kilometer*, *square meter*, or *square centimeter*, you would use to find the area for each of the following.

1. a thumbnail
2. a gymnasium floor
3. a blanket
4. a sheet of graph paper
5. a national park
6. a coin
7. a playing card
8. a parking lot
9. the surface of the moon
10. North Carolina
11. a side of a barn
12. a cow pasture

Tell which area is smaller.

13. 1 square centimeter or 1 square kilometer
14. 1 square meter or 1 square kilometer
15. 1 square centimeter or 1 square meter
16. 1 square meter or 10 square centimeters
17. 1000 square meters or 1 square kilometer
18. 1000 square centimeters or 10 square meters
19. 0.001 square kilometers or 10,000 square meters
20. 0.01 square meters or 10 square centimeters

11–2 Area Formulas

Any side of a parallelogram may be called the **base** of the parallelogram. For each base there is a corresponding **altitude.** The altitude of a parallelogram is a segment perpendicular to the base with endpoints on the base and its opposite side.

For each parallelogram shown below, one base is 10 units long, and its corresponding altitude is 7 units long.

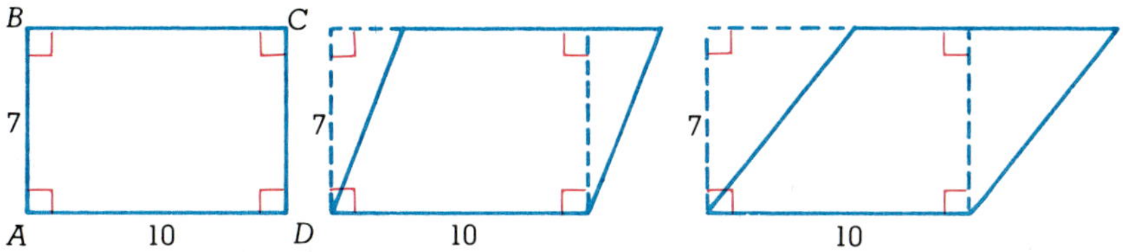

By using the Area Addition Postulate and congruence, it is possible to show that each region has the same area, 7×10 or 70 square units.

If a parallelogram has an area of A square units, a base of b units and a corresponding altitude of h units, then $A = bh$.

Theorem 11–2

example 1

Prove Theorem 11–2

Given: $\square PQRS$
$PS = b$
$TS = h$
area is A square units

Prove: $A = bh$

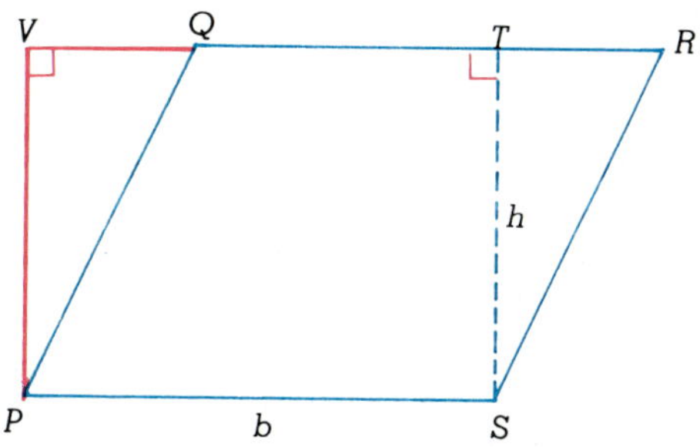

Proof:

Extend $\overline{RQ}$ and draw $\overline{PV}$ so that $\overline{PV} \perp \overline{RQ}$. Then $PV = h$ and quadrilateral $PVTS$ is a rectangle with a length of h units and a width of b units.

Since $\overline{PQ} \cong \overline{SR}$, show $\triangle PVQ \cong \triangle STR$ by HL. Thus, by the Area Addition Postulate, show that the area of the parallelogram is the same as the area of the rectangle. Thus, $A = bh$.

The area of a triangle is related to the area of a parallelogram. For example, consider $\triangle ABC$ below.

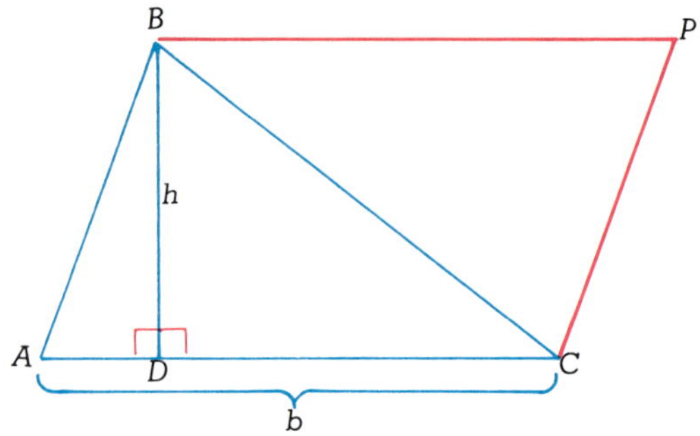

$\overline{BP}$ is drawn so that it is both parallel to and congruent to $\overline{AC}$. Similarly, $\overline{PC}$ is drawn parallel to $\overline{AB}$. Thus, the quadrilateral $ABPC$ is a parallelogram and its area is bh square units.

But, the two triangles, $\triangle ABC$ and $\triangle PCB$ are congruent and, thus, have the same areas. Together, their areas are the same as the parallelogram. This result suggests Theorem 11–3 which follows.

Theorem 11–3

If a triangle has an area of A square units, a base of b units, and a corresponding altitude of h units, then $A = \frac{1}{2}bh$.

A diagonal of a trapezoid separates the trapezoid into two triangles. The formula for the area of a trapezoid can be found by finding the areas of triangles.

An altitude of the trapezoid is a segment perpendicular to both bases of the trapezoid, and with endpoints on the bases. First, notice that $\overline{QT}$ is an altitude of both triangles.

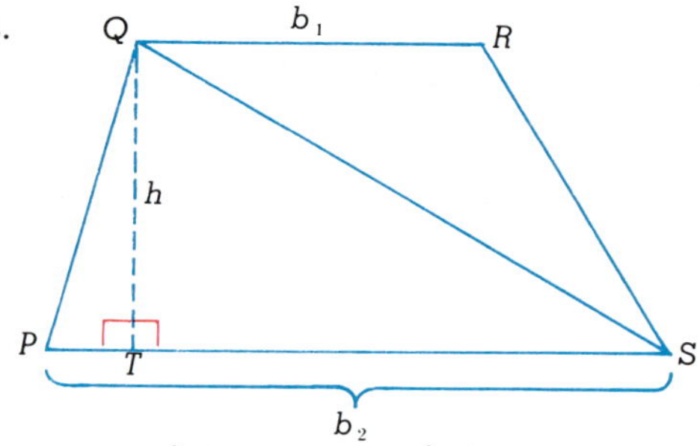

$$\text{area of trapezoid } PQRS = \text{area } \triangle PQS + \text{area } \triangle QRS$$
$$= \tfrac{1}{2}b_2h + \tfrac{1}{2}b_1h$$
$$= \tfrac{1}{2}h(b_2 + b_1)$$

Theorem 11–4

If a trapezoid has an area of A square units, and bases of b_1 units and b_2 units, and an altitude of h units, then $A = \frac{1}{2}h(b_2 + b_1)$.

example 2

Find the area of the isosceles trapezoid whose bases are 20 units and 32 units, and whose legs measure 10 units each.

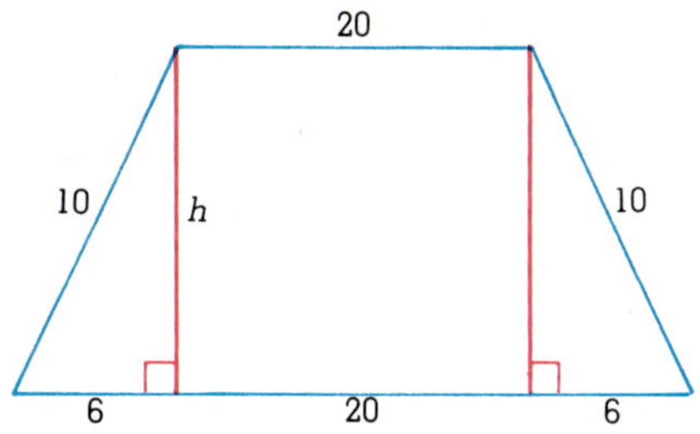

Draw the altitudes shown at the right. Two right triangles are formed with a leg of 6 units and hypotenuse of 10 units. Use the Pythagorean Theorem to find the length of the altitude.

$$h^2 + 6^2 = 10^2$$
$$h^2 = 10^2 - 6^2$$
$$= 100 - 36$$
$$= 64$$
$$h = 8$$

Then, use the formula to find the area of the trapezoid.

$$A = \frac{1}{2}h(b_1 + b_2)$$
$$= \frac{1}{2}(8)(20 + 32)$$
$$= \frac{1}{2}(8)(52)$$
$$= 208$$

The area is 208 square units.

exercises

Exploratory **Find the area of each parallelogram given the following measurements for base and altitude.**

1. 12 inches and 8 inches
2. 6 feet and 2 feet
3. 2 centimeters and 9 centimeters
4. 4 yards and 5 yards
5. 21 millimeters (mm) and 12 millimeters
6. 9 meters and 4 meters
7. 7 feet and 12 feet
8. 11 miles and 11 miles

9–16. Find the area of each triangle given the measurements for base and altitude in problems 1–8.

Find the area of each trapezoid described below.

17. bases 9 centimeters and 2 centimeters
altitude 7 centimeters
18. bases 16 meters and 11 meters
altitude 11 meters
19. bases 30 miles and 20 miles
altitude 34 miles
20. bases 55 centimeters and 40 centimeters
altitude 21 centimeters
21. bases 12 inches and 4 inches
altitude 8 inches
22. bases 7 feet and 5 feet
altitude 5 feet

Written Complete the chart for each parallelogram described below.

	base	altitude	area
1.	10 feet		40 square feet
2.		8 inches	20 square inches
3.	4.7 meters	2.8 meters	
4.	13.6 centimeters		68.68 square centimeters
5.		16 meters	20.8 square meters
6.		12 feet	15.6 square feet
7.	$4\frac{1}{2}$ inches	$4\frac{1}{2}$ inches	
8.	$1\frac{3}{4}$ feet		$3\frac{1}{3}$ feet^2
9.	$(2x + 8)$ meters	x meters	
10.	$3k$ inches	k inches	

11–20. Complete the chart in problems **1–10** for triangles.

Find the area of each of the following regions.

21.

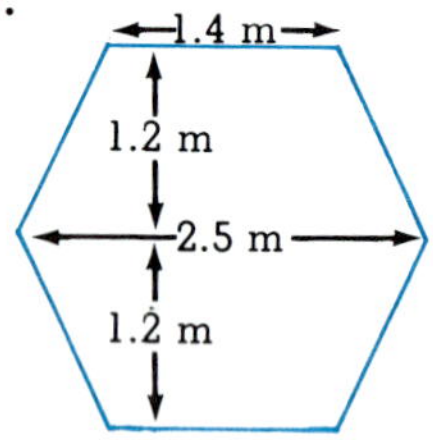

22.

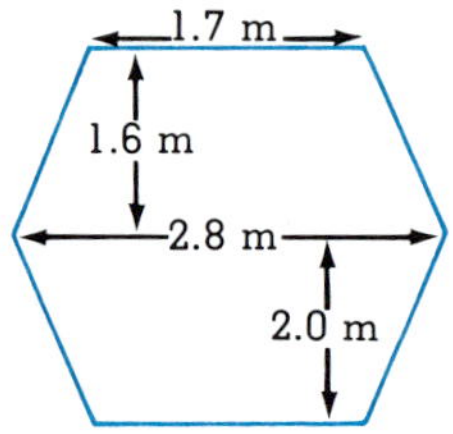

23.

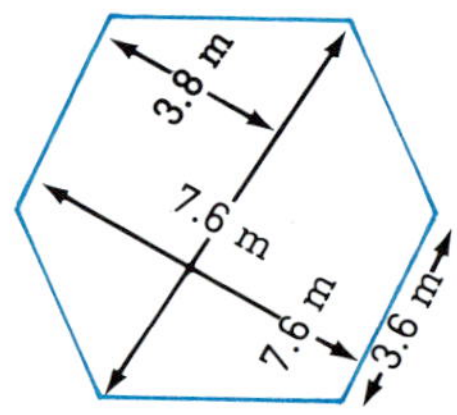

24.

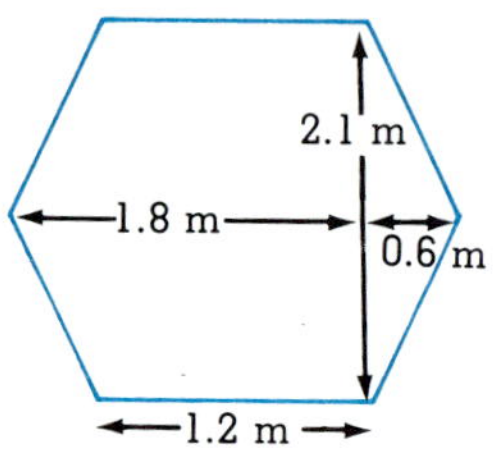

25.

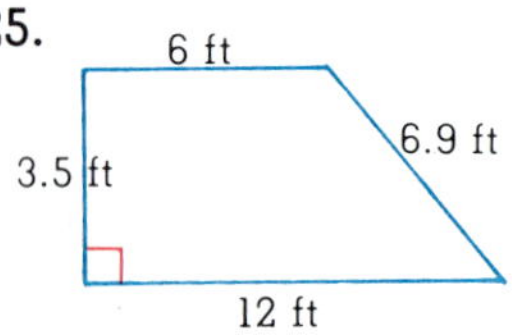

26.

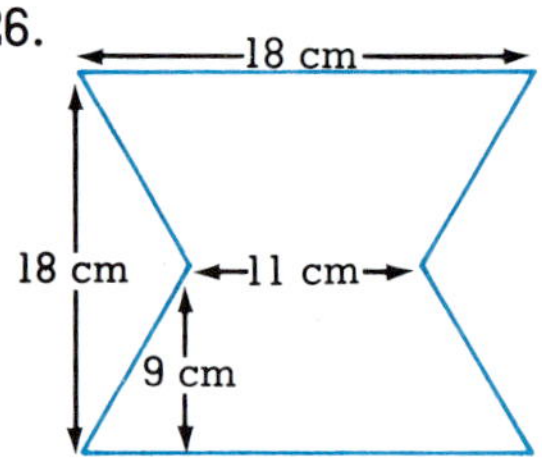

Solve each problem.

27. Prove Theorem 11–3.

28. Prove Theorem 11–4.

29. Find the area of an isosceles trapezoid whose bases are 8 units and 13 units and whose legs are 6.5 units.

30. Find the area of an isosceles trapezoid whose bases are 37 kilometers and 12 kilometers, and whose legs are 32.5 kilometers.

11–3 Regular Polygons

In the figure at the right, $\odot X$ is *circumscribed* about the regular hexagon. Notice that $\overline{XP}$, $\overline{XQ}$, $\overline{XR}$, $\overline{XS}$, $\overline{XT}$, and $\overline{XV}$ are radii of the circle. They also are called **radii of the regular hexagon.**

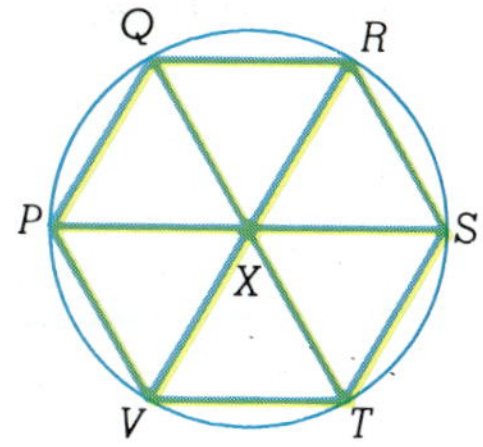

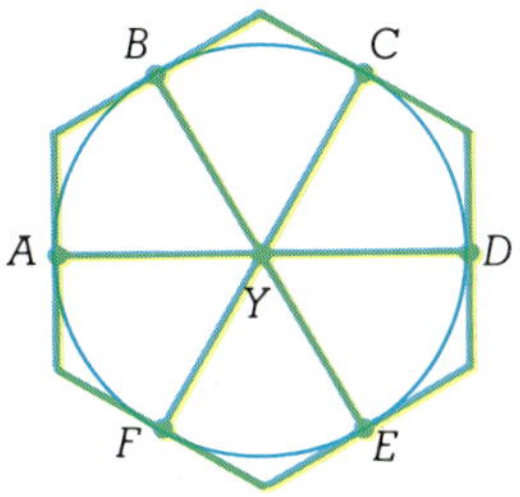

In the figure at the left, $\odot Y$ is *inscribed* in the regular hexagon. Notice that $\overline{YA}$, $\overline{YB}$, $\overline{YC}$, $\overline{YD}$, $\overline{YE}$, and $\overline{YF}$ are radii of the circle with endpoints at points of tangency. They are called **apothems of the regular hexagon.**

A segment is a radius of a regular polygon if and only if it is a radius of a circle circumscribed about the polygon.

Definition of Radius

A segment is an apothem of a regular polygon if and only if it is a radius of a circle inscribed in the polygon.

Definition of Apothem

For a regular polygon, a circle inscribed in the polygon and a circle circumscribed about the polygon have the same center. This point is also called the **center of the regular polygon.**

A point is a center of a regular polygon if and only if it is the common center of its inscribed and circumscribed circles.

Definition of Center

If a polygon is *not* regular, then the inscribed circle and the circumscribed circle do *not* have the same center.

A polygon is regular if and only if a circle inscribed in the polygon and a circle circumscribed about the polygon have the same center.

Theorem 11–5

Each side of a regular polygon is tangent to a circle inscribed in the polygon. A radius of a circle drawn to a point of tangency is perpendicular to the tangent. Thus, each apothem of a regular polygon is perpendicular to one side of the polygon.

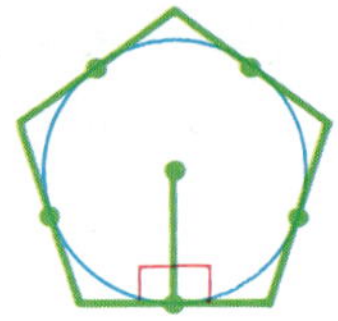

Theorem 11–6

If a segment is an apothem of a regular polygon, then it is perpendicular to a side of the polygon at the point of tangency.

Given the length of the apothem and of one side of a regular polygon, the area can be found.

example 1

Find the area of a regular hexagon if the apothem is a units long and one side is s units long.

By drawing all the radii of the hexagon, 6 triangles are formed. Each apothem of a regular polygon is perpendicular to a side of the polygon. Thus, the altitude of each triangle is an apothem of a units. The base of each triangle is a side of the hexagon, and is s units long.

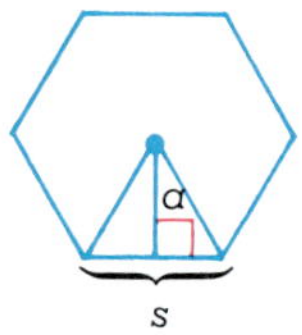

By the Area Addition Postulate, the area of the hexagon equals the sum of the areas of the triangles.

$$A = \tfrac{1}{2}sa + \tfrac{1}{2}sa + \tfrac{1}{2}sa + \tfrac{1}{2}sa + \tfrac{1}{2}sa + \tfrac{1}{2}sa$$
$$= 6\left(\tfrac{1}{2}sa\right)$$
$$= 3sa$$ The area is $3sa$ square units.

Suppose a regular polygon has an apothem of a units and a side of s units. Then, the following chart gives values for the area, based on the number of sides.

number of sides	3	4	5	6	7	8	9	10
area	$3\left(\frac{1}{2}sa\right)$	$4\left(\frac{1}{2}sa\right)$	$5\left(\frac{1}{2}sa\right)$	$6\left(\frac{1}{2}sa\right)$	$7\left(\frac{1}{2}sa\right)$	$8\left(\frac{1}{2}sa\right)$	$9\left(\frac{1}{2}sa\right)$	$10\left(\frac{1}{2}sa\right)$

The values in the chart suggest the following theorem.

Theorem 11–7

If a regular polygon with n sides has an area of A square units, an apothem of a units, and a side of s units, then $A = \frac{1}{2}san$.

A **central angle** of a regular polygon is formed by two radii drawn to consecutive vertices.

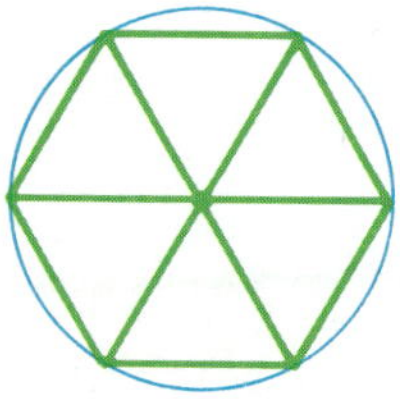

The central angles also are central angles of a circle. The sum of their degree measures must be 360. *Why?*

A regular hexagon has 6 central angles. Thus, the degree measure of each is $\frac{360}{6}$ or 60.

If a regular polygon has n sides, then the degree measure of each central angle is $\frac{360}{n}$.

Theorem 11–8

---

example 2

Find the area of a regular hexagon having sides 6 inches long.

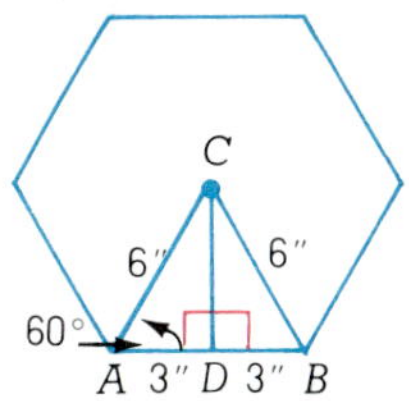

Since $\angle ACB$ is a central angle its degree measure is $\frac{360}{6}$ or 60. Also, $\overline{AC} \cong \overline{CB}$, so the triangle must be equilateral. Thus, AC and CB are 6.

Since the apothem $\overline{CD}$ also is an altitude, it bisects $\overline{AB}$. Thus AD and DB are 3.

The length of the apothem can be found using either trigonometry or the Pythagorean Theorem. For a 30°–60° right triangle, the length is known to be $AD\sqrt{3}$ or $3\sqrt{3}$. Thus, the area of the hexagon is found as follows.

$$\begin{aligned} A &= \tfrac{1}{2}san \\ &= \tfrac{1}{2}(3\sqrt{3})(6)(6) \\ &= 54\sqrt{3} \\ &\approx 93.5 \end{aligned}$$

The area is about 93.5 square inches.

exercises

Exploratory Find the degree measure of a central angle of each regular polygon with the given number of sides.

1. 3 **2.** 4 **3.** 5
4. 6 **5.** 8 **6.** 9
7. 10 **8.** 12 **9.** 14

Written In the figure at the right, $\triangle ABC$ is equilateral. Name each of the following.

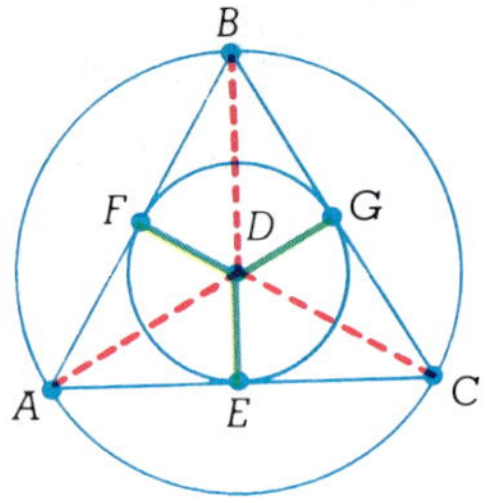

1. center of $\triangle ABC$
2. apothem of $\triangle ABC$
3. radius of $\triangle ABC$
4. central angle
5. base of $\triangle ADC$
6. base of $\triangle BDC$
7. altitude of $\triangle ADC$
8. altitude of $\triangle BDC$
9. base of $\triangle ADB$
10. altitude of $\triangle ADB$
11. circle inscribed in $\triangle ABC$
12. circle circumscribed about $\triangle ABC$

Find the area of each regular polygon with the following measurements. Round each answer to the nearest tenth.

13. equilateral triangle
apothem 5.8 centimeters
side 20 centimeters
14. equilateral triangle
apothem 9 centimeters
side 6.9 centimeters
15. square
side 4.2 feet
apothem 2.1 feet
16. square
apothem 8 inches
side 16 inches
17. pentagon
apothem 5.5 inches
side 8.0 inches
18. pentagon
apothem 8.9 miles
side 13.0 miles
19. hexagon
apothem 16.45 millimeters
side 19.0 millimeters
20. hexagon
apothem 8.7 meters
side 10 meters
21. octagon
apothem 7.5 feet
side 6.2 feet
22. octagon
apothem 20.3 centimeters
side 16.8 centimeters

Solve each problem.

23. Find the area of an equilateral triangle with a radius of $4\sqrt{3}$ units.
24. Find the area of an equilateral triangle with a radius of 6 units.
25. Find the area of a regular hexagon with an apothem of 12 centimeters.
26. Find the area of a regular hexagon with a side of 12 centimeters.
27. Find the area of a square inscribed in a circle with a radius of 10 feet.
28. Find the area of a square circumscribed about a circle with a radius of 10 feet.
29. Find the area of a regular pentagon with a perimeter of 16.5 inches and an apothem of 4 inches.
30. Find the area of a regular pentagon with a perimeter of 68.3 meters and an apothem of 5 meters.
31. Suppose an equilateral triangle has a side of 24 centimeters. Find the length of an apothem, a radius, the perimeter, and the area.
32. Suppose a regular hexagon has a side of 16 centimeters. Find the length of an apothem, a radius, the perimeter, and the area.
33. Prove Theorem 11–6.
34. Prove Theorem 11–7 for a regular pentagon.
35. Prove Theorem 11–8 for a regular hexagon.
36. Prove Theorem 11–8 for an equilateral triangle.

11–4 Circles

Suppose a wheel, such as a bicycle wheel, makes one complete revolution. The distance it travels is the same as the **circumference** of the wheel.

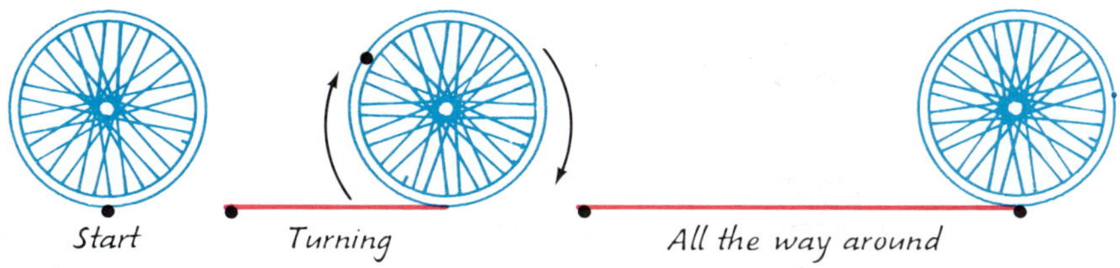

The circumference of a circle is related to the perimeter of regular polygons. The sequence of figures below suggest the relationship.

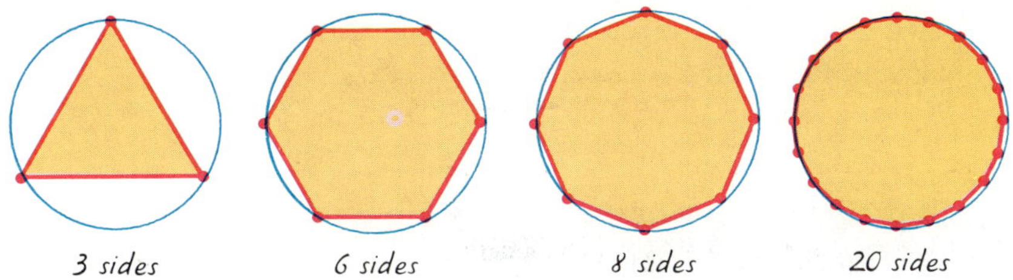

Notice that as the number of sides of the inscribed polygons increases, the polygons look more and more like a circle. Thus, as the number of sides increases, the perimeter becomes closer and closer to the circumference of the circle.

The circumference of a circle is the *limit* of the perimeters of the inscribed regular polygons as the number of sides increases.

Definition of Circumference

As the number of sides of the polygons increases, the area of the polygons becomes closer and closer to the area of the circle.

The area of a circle is the limit of the areas of the inscribed polygons as the number of sides increases.

Definition of Area of Circle

The Greek letter π stands for a number that *cannot* be named exactly by a decimal or fraction. One decimal approximation is given below.

π is *about* 3.14159265358979323846264338

Both the circumference and area of a circle are related to π.

Suppose the radius of a circle is r units. The following chart gives the approximate perimeter and area of each regular polygon inscribed in the circle.

Number of Sides	3	4	5	6	8	10	20	50	100
Perimeter	$5.20r$	$5.66r$	$5.88r$	$6.00r$	$6.12r$	$6.18r$	$6.27r$	$6.28r$	$6.28r$
Area	$1.30r^2$	$2.00r^2$	$2.38r^2$	$2.60r^2$	$2.83r^2$	$2.94r^2$	$3.09r^2$	$3.13r^2$	$3.14r^2$

The values in the chart suggest formulas for the circumference and area of a circle.

Theorem 11–9

If a circle has a circumference of C units and a radius of r units, then $C = 2\pi r$.

Theorem 11–10

If a circle has an area of A square units and a radius of r units, then $A = \pi r^2$.

example 1

Find the circumference and area of a circle with a radius of 5.6 centimeters. Use 3.14 for π and round your answer to one decimal place.

$$\begin{aligned} C &= 2\pi r \\ &\approx 2(3.14)(5.6) \\ &\approx 35.168 \\ &\approx 35.2 \end{aligned}$$

$$\begin{aligned} A &= \pi r^2 \\ &\approx (3.14)(5.6)^2 \\ &\approx 98.4704 \\ &\approx 98.5 \end{aligned}$$

The circumference is about 35.2 centimeters.
The area is about 98.5 square centimeters.

example 2

Find the area of a circle inscribed in an equilateral triangle whose sides are 5 centimeters. Use 3.14 for π and round your answer to the nearest hundredth.

By using a theorem about tangent segments and properties of congruence, it is possible to conclude that $\overline{AD} \cong \overline{DC}$. Thus, $AD = 2.5$. Also, $\overline{PA}$ bisects $\angle EAD$. Thus, $m\angle PAD = 30$. Finally $\overline{PD} \perp \overline{AD}$.

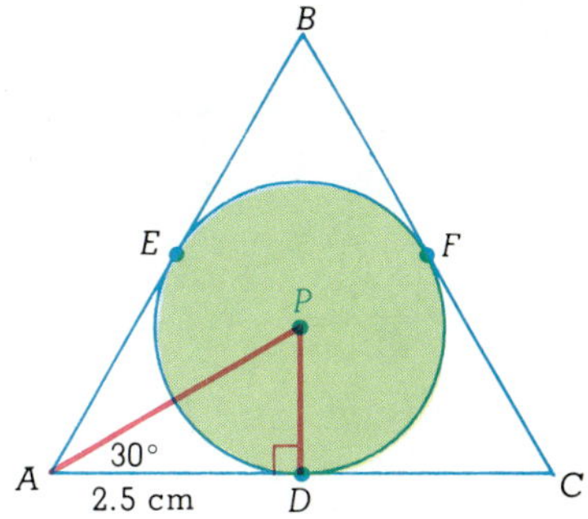

Using trigonometry, find PD.

$$\begin{aligned} \tan 30° &= \frac{PD}{2.5} \\ PD &= 2.5 \tan 30° \\ &\approx 2.5(0.5774) \\ &\approx 1.4435 \\ &\approx 1.4 \end{aligned}$$

Then, find the area of the circle.

$$\begin{aligned} A &= \pi r^2 \\ &\approx (3.14)(1.4)^2 \\ &\approx 6.1544 \\ &\approx 6.15 \end{aligned}$$

The area of the circle is about 6.15 square centimeters.

A **sector** of a circle is a region bounded by a central angle and the intercepted arc.

If a circle is separated by central angles, then the sum of their degree measures must be 360. Thus, a circle can be separated into exactly three sectors whose central angles measure 120°. Since the boundaries of the sectors are congruent, the sectors have the same area. As a result, the area of each sector is $\frac{1}{3}$ the area of the circle or $\frac{1}{3}\pi r^2$, if the radius is r units.

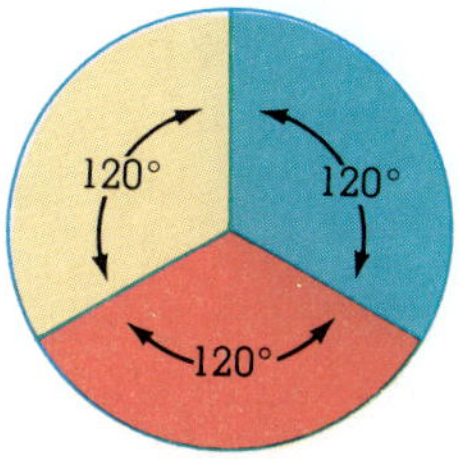

$$\frac{\textit{degree measure of central angle}}{\textit{degree measure of circle}} = \textit{fraction of area of circle}$$

In general, the area of a sector of a circle is the fraction of the area of the circle times the area of the circle.

exercises

Exploratory Find the circumference of each circle whose radius is given below. Give answers in terms of π.

1. 7 centimeters
2. 10 meters
3. 4 inches
4. 8 feet
5. 7.4 centimeters
6. 9.6 meters
7. 2.4 kilometers
8. 1.5 miles
9. $\frac{1}{2}$ foot
10. $3\frac{1}{3}$ yards
11. $4\frac{2}{3}$ yards
12. $5\frac{3}{4}$ feet
13. $5\sqrt{2}$ centimeters
14. $4\sqrt{3}$ inches
15. $3\sqrt{2}$ inches

16–30. Find the area of each circle described in problems 1–15. Give answers in terms of π.

Written Find the circumference of each circle whose radius is given below. Use 3.14 for π and round each answer to the nearest tenth.

1. 7 centimeters
2. 10 meters
3. 4 inches
4. 8 feet
5. 7.4 centimeters
6. 9.6 meters
7. 2.4 kilometers
8. 1.5 miles
9. $\frac{1}{2}$ foot
10. $3\frac{1}{3}$ yards
11. $4\frac{2}{3}$ yards
12. $5\frac{3}{4}$ feet
13. $5\sqrt{2}$ centimeters
14. $4\sqrt{3}$ inches
15. $3\sqrt{2}$ inches

16–30. Find the area of the circles described in problems 1–15. Use 3.14 for π and round each answer to the nearest hundredth.

Each of the following are measurements for central angles of a circle. The area of the corresponding sector is what fraction of the area of its circle?

31. 40°
32. 60°
33. 30°
34. 2°
35. 45°
36. 15°
37. 18°
38. 72°

39–46. Suppose a circle has a radius of 4.6 centimeters. Find the areas of sectors formed by central angles whose measurements are given in problems 31–38. Use 3.14 for π and round each answer to the nearest tenth.

Find each area. Use 3.14 for π and round each answer to the nearest tenth.

47. the area of a circle inscribed in a square whose sides are 7.2 meters
48. the area of a circle inscribed in a square whose diagonal is 6.5 feet
49. the shaded region shown below
50. the shaded region shown below

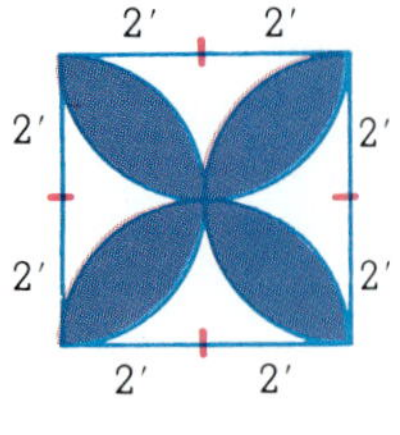

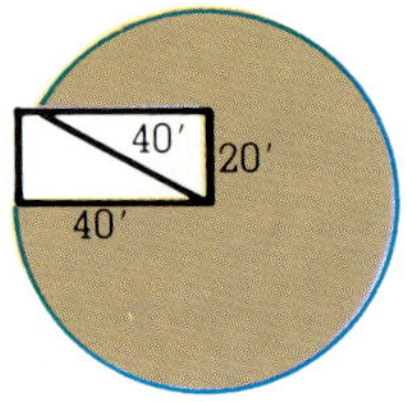

Three-dimensional Figures

Reading Geometry

In the real world, part of a figure cannot be both in front and in back at the same time. Yet it is possible to draw such figures.

When three-dimensional figures are shown on paper, they are drawn in perspective. The figure on the right shows the edges of a cube drawn in perspective. Although no edges of the cube cross each other, in the drawing they *appear* to cross in two places. Any parts of a figure that are directly in front or behind each other from a given point of view will *appear* to cross in a drawing. Artists use this property to create visual illusions.

Broken lines are used to show parts that are behind others. The drawings below show two different ways of viewing a cube.

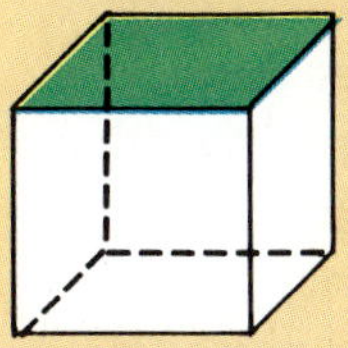

viewed from above

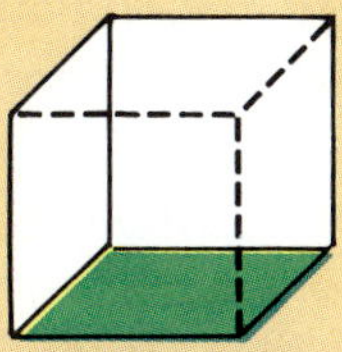

viewed from below

Exercises Use broken lines to show two views for each of the following.

1. 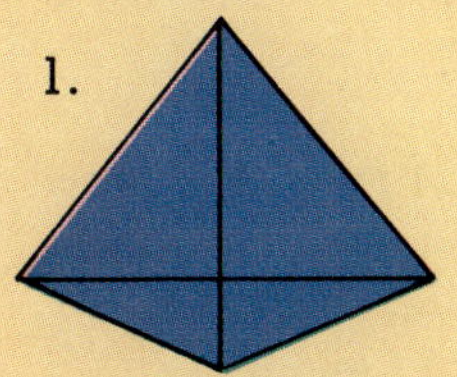

2. 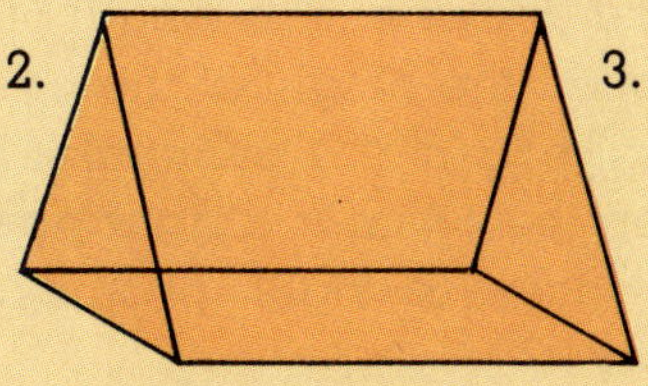

3. 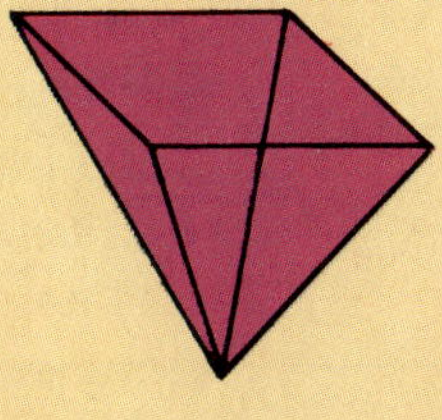

4. 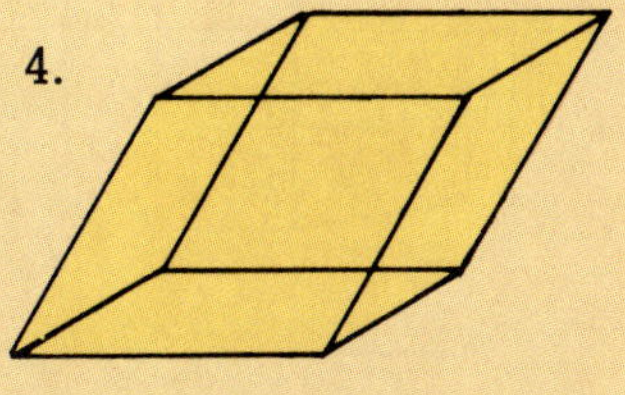

11–5 Surface Area

If the figure below is traced on a piece of paper, cut out, and folded along the broken lines, it will look like a cube. The area of the figure is called the *total* **surface area,** or total area, of the cube it forms.

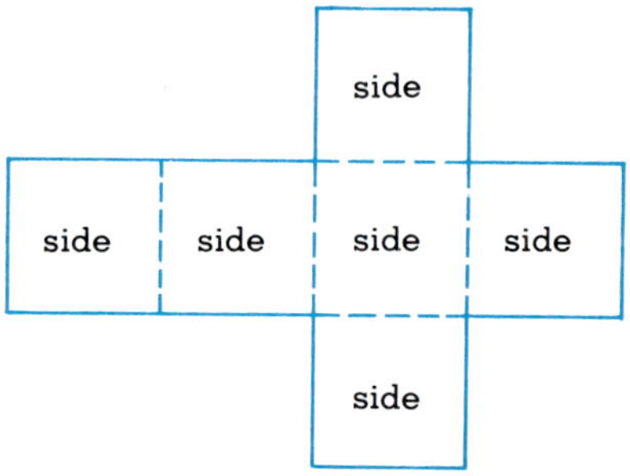

One way to find the surface area of a solid is to add the area of each face. For certain types of solids, another way is to use formulas.

One type of solid is called a **prism.** A cube and a rectangular box are both prisms. Some other prisms are shown below.

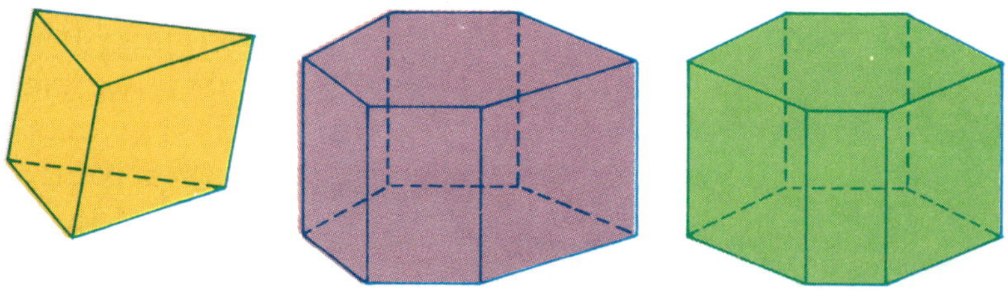

Prisms have the following characteristics.

1. Two faces, called **bases,** are formed by congruent polygons that lie in parallel planes.
2. The faces that are *not* bases, called **lateral faces,** are formed by parallelograms.
3. The intersections of two adjacent lateral faces, called **lateral edges,** are parallel segments.

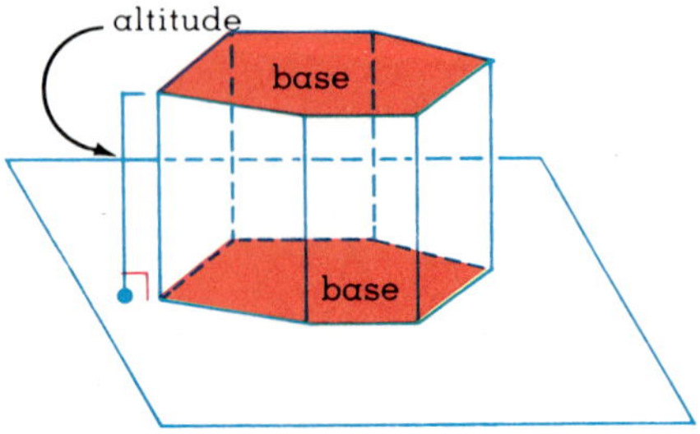

An **altitude** of a prism is a segment perpendicular to the base planes with an endpoint in each plane. The length of an altitude is called the height of the prism.

If the lateral edges of a prism also are altitudes, then the prism is a **right prism.** The lateral faces of right prisms are rectangles.

A prism that is not a right prism is called an oblique prism.

The **lateral area** of a prism is the area of *all* the lateral faces. As a result of the Distributive Property, the lateral area of a right prism can be found by multiplying the height by the perimeter of a base.

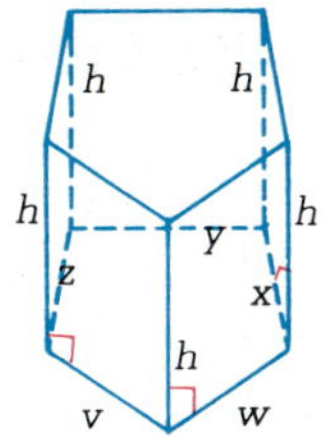

$$A = hv + hw + hx + hy + hz$$
$$= h(v + w + x + y + z)$$

Does this hold for oblique prisms?

Theorem 11–11

If a right prism has a lateral area of L square units, a height of h units, and its bases have a perimeter of p units, then $L = ph$.

The total surface area of a prism is found by adding the lateral area to the areas of both bases. Since the bases of a prism are congruent, they have the same area.

Theorem 11–12

If the total surface area of a right prism is T square units, the bases have an area of B square units and a perimeter of p units, and the height is h units, then $T = ph + 2B$.

example 1

Find the total surface area of a right prism whose base is a right triangle with legs of 5 inches and 12 inches, and whose height is 20 inches.

First, use the Pythagorean Theorem to find the measure of the hypotenuse.

$$a^2 = 5^2 + 12^2$$
$$= 169$$
$$a = 13$$

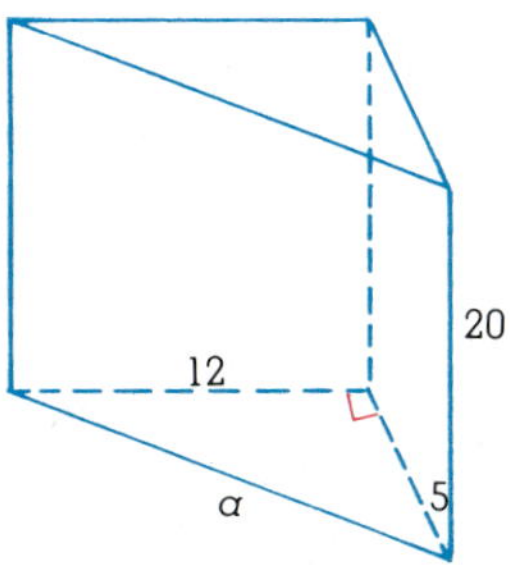

Then, find the total surface area.

$$T = ph + 2B$$
$$= (5 + 12 + 13)20 + 2\left(\frac{1}{2} \cdot 5 \cdot 12\right)$$
$$= (30) \cdot 20 + 2(30)$$
$$= 660$$

The total surface area is 660 square inches.

The bases of a cylinder are formed by two congruent circles that lie in parallel planes. The segment whose endpoints are centers of these circles is called the **axis** of the cylinder.

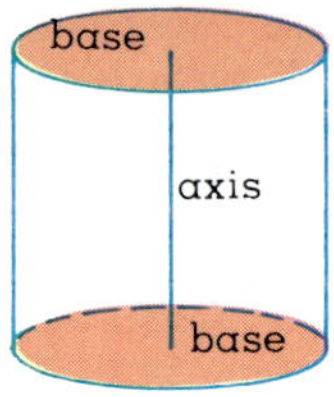

The **altitude** of a cylinder is a segment perpendicular to the base planes with an endpoint in each plane. If the axis of a cylinder also is an altitude of the cylinder, then the cylinder is called a **right cylinder.**

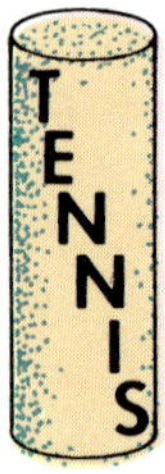

The lateral area of a cylinder is the area of the curved surface. Think of the label from a cylindrical can. Most have a rectangular shape. What is the height of the label on the can? What is the base?

Theorem 11–13

If a right cylinder has a lateral area of L square units, a height of h units, and the bases have radii of r units, then $L = 2\pi rh$.

Theorem 11–14

If a right cylinder has a total surface area of T square units, a height of h units, and the bases have radii of r units, then $T = 2\pi rh + 2\pi r^2$.

example 2

Find the number of square feet of cardboard needed to make the sides of 30,000 orange juice cans if they each are 6 inches tall and are 3 inches in diameter. Use 3.14 for π.

$$\begin{aligned} \textit{number square inches} &= \textit{number cans} \times \textit{lateral area one can} \\ &= 30{,}000 \times 2\pi rh \\ &\approx 30{,}000 \times 2(3.14)\left(\tfrac{3}{2}\right)(6) \\ &\approx 1{,}695{,}600 \end{aligned}$$

There are 12×12 or 144 square inches in 1 square foot.
Thus, about $1{,}695{,}600 \div 144$ or 11,775 square feet of cardboard is needed.

exercises

Exploratory **Use the prism at the right to help solve each problem.**

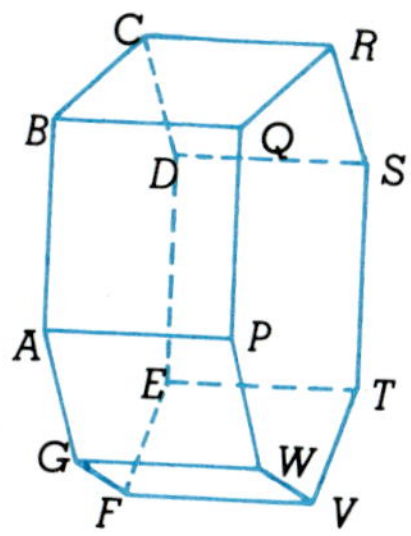

1. Name each base.
2. Tell how many sides to each base.
3. Tell how many lateral faces to the figure.
4. Name the lateral faces.
5. Tell whether or *not* the figure appears to be a right prism.
6. Suppose each polygon forming a base is regular with sides of s units. Suppose the height is h units and the lateral area is L square units. Write a formula for the lateral area of the prism.

Written **Find the lateral area of each right prism described below.**

1. rectangular base 2 centimeters by 3 centimeters; height 6 centimeters
2. rectangular base 9 meters by 6 meters; height 25 meters
3. right triangular base with legs 6 inches and 8 inches; height 11 inches
4. right triangular base with legs 18 millimeters and 12 millimeters; height 15 millimeters
5. equilateral triangular base with sides 2.7 meters; height 5 meters
6. equilateral triangular base with sides 8.1 centimeters; height 13 centimeters
7. regular hexagonal base with sides 7.3 units; height 4 units
8. regular hexagonal base with sides 11.1 units; height 6 units

9–16. Find the total surface area of each right prism described in problems **1–8.**

Find the lateral area of each right cylinder described below. Use 3.14 for π and round each answer to the nearest tenth.

17. radius: 7 centimeters
height: 10 centimeters

18. radius: 4 inches
height: 11 inches

19. radius: 6 feet
height: 10.2 feet

20. radius: 4.5 feet
height: 7 feet

21. radius: 11.8 meters
height: 23 meters

22. radius: 26.3 centimeters
height: 26.3 centimeters

23–28. Find the total surface area of each right cylinder described in problems **17–22.**

Solve each problem. Use 3.14 for π and round each answer to the nearest tenth.

29. Find the surface area of a cylindrical gas tank which is 25 feet tall and has a radius of 30 feet.

30. Find the surface area of a cylindrical water tower which is 8 meters tall and has a diameter of 8 meters.

31. Suppose two different crystals occur in the shape of right prisms. One crystal has rectangular bases of 2 units by 3 units with a height of 2.5 units. The other crystal has a regular hexagonal base with sides of 2.1 units and height 2.3 units. Which crystal has the greater surface area?

11–6 More about Surface Area

The figure at the right is called a pyramid. It has the following characteristics.

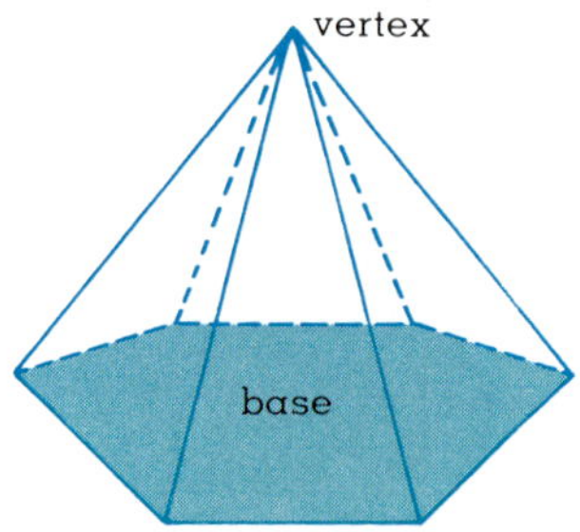

1. All the faces, except one face, intersect at a point called the **vertex.**
2. The face that does *not* intersect at the vertex is a polygon and is called the **base.**
3. The faces meeting at the vertex, called **lateral faces,** are triangles.

A pyramid is a **regular pyramid** if and only if its base is regular and the segment joining the base to the vertex is perpendicular to the base. This segment is called the **altitude.**

All the lateral faces of a regular pyramid are congruent isosceles triangles. The height of each lateral face is called the **slant height** of the pyramid.

The figure at the right is a regular hexagonal pyramid. Its lateral area can be found in the following way.

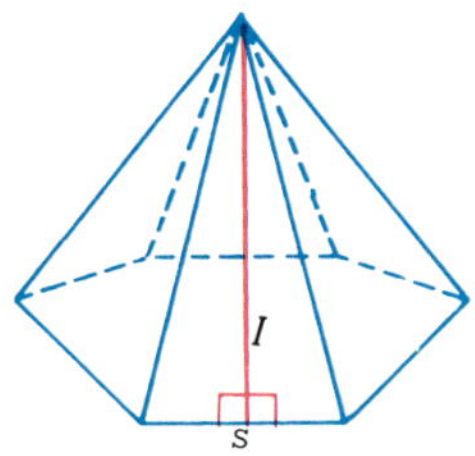

$$L = \frac{1}{2}ls + \frac{1}{2}ls + \frac{1}{2}ls + \frac{1}{2}ls + \frac{1}{2}ls + \frac{1}{2}ls$$
$$= \frac{1}{2}l(s + s + s + s + s + s) \text{ or } 3ls$$

Theorem 11–15

If a regular pyramid has a lateral area of L square units, a slant height of l units, and its base has perimeter p units, then $L = \frac{1}{2}pl$.

example 1

Find the total surface area of a regular square pyramid in which the slant height is 10 inches and the sides of the base are 7 inches.

total surface area = *lateral area* + *base area*

$$= \frac{1}{2}pl \quad + s^2$$
$$= \frac{1}{2}(4 \cdot 7)(10) + (7)^2$$
$$= 189$$

The total surface area is 189 square inches.

The figure shown below is a **right circular cone.** It has a circular **base,** and a **vertex** at T. Its **axis** is $\overline{TC}$ which is a segment joining the vertex to the center of the base.

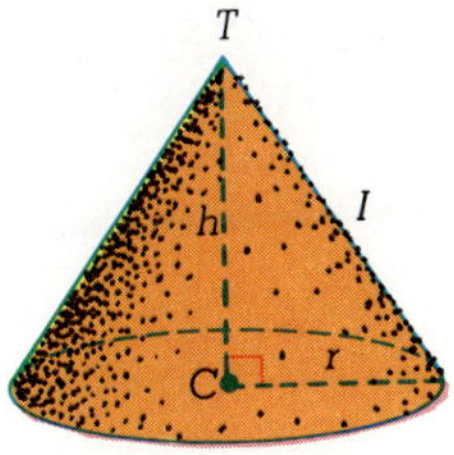

In a right circular cone, the axis is perpendicular to the base. Thus, it is the altitude of the cone. The length of any segment joining the vertex to the circle is called the slant height. In the cone at the left, the slant height is l units.

Finding the lateral area and total surface area of a right circular cone is similar to finding those measurements for a regular pyramid.

Theorem 11–16

If a right circular cone has a lateral area of L square units, a total surface area of T square units, a slant height of l units, and the radius of the base is r units, then $L = \pi r l$ and $T = \pi r l + \pi r^2$.

example 2

A right circular cone has a radius of 7.6 centimeters. Its altitude is 11.9 centimeters. Find the total surface area of the cone. Use 3.14 for π and round to the nearest tenth.

First, use the Pythagorean Theorem to find the slant height.

$$\begin{aligned} l^2 &= 11.9^2 + 7.6^2 \\ &= 141.61 + 57.76 \\ &= 199.37 \\ l &= \sqrt{199.37} \\ &\approx 14.1 \end{aligned}$$

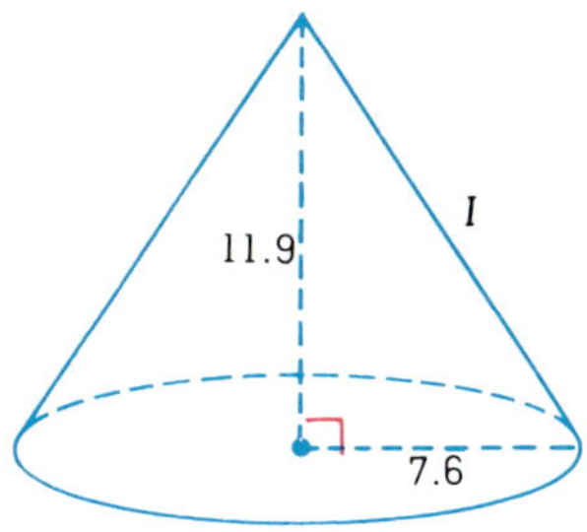

Then, find the total surface area.

$$\begin{aligned} T &= \pi r l + \pi r^2 \\ &\approx (3.14)(7.6)(14.1) + (3.14)(7.6)^2 \\ &\approx 336.5 + 181.4 \\ &\approx 517.9 \end{aligned}$$

The total surface area is about 517.9 square centimeters.

A formula for the surface area of a sphere is given in the following theorem.

Theorem 11–17

If a sphere has a surface area of A square units and a radius of r units, then $A = 4\pi r^2$.

exercises

Exploratory Find the lateral area of each regular pyramid described below.

1. triangular base with sides of 3 centimeters; slant height of 6 centimeters
2. triangular base with sides of 5 centimeters; slant height of 6 centimeters
3. square base with sides of 3 centimeters; slant height of 6 centimeters
4. square base with sides of 5 centimeters; slant height of 6 centimeters
5. pentagonal base with sides of 3 centimeters; slant height of 6 centimeters
6. pentagonal base with sides of 5 centimeters; slant height of 6 centimeters
7. hexagonal base with sides of 3 centimeters; slant height of 6 centimeters
8. hexagonal base with sides of 5 centimeters; slant height of 6 centimeters
9. octagonal base with sides of 3 centimeters; slant height of 6 centimeters
10. octagonal base with sides of 5 centimeters; slant height of 6 centimeters

Written Find the lateral area of each right circular cone described below. Use 3.14 for π and round to the nearest tenth.

1. base radius: 6 centimeters
   slant height: 10 centimeters
2. base radius: 4 feet
   slant height: 11 feet
3. base radius: 4.1 meters
   slant height: 10.3 meters
4. base radius: 3.2 millimeters
   slant height: 5.2 millimeters
5. base radius: 6.9 centimeters
   slant height: 6.9 centimeters
6. base radius: 4.4 meters
   slant height: 6.3 meters
7. base radius: $4\frac{1}{2}$ inches
   slant height: $5\frac{1}{2}$ inches
8. base radius: $3\frac{3}{4}$ yards
   slant height: $7\frac{1}{2}$ yards
9. base radius: 100 feet
   slant height: 1200 feet
10. base radius: 35.0 meters
    slant height: 67.2 meters

11–20. Find the total surface area for right circular cones given the information in problems **1–10**. Use 3.14 for π and round to the nearest tenth.

Find the surface area of each sphere given the following radii. Use 3.14 for π and round to the nearest tenth.

21. 10 centimeters
22. 2 inches
23. 2000 feet
24. 3.8 meters
25. 372 miles
26. 75.3 centimeters

Solve each problem.

27. The Great Pyramid covers about 570,000 square feet and is 480 feet tall. Find its lateral area.

28. The radius of the earth is about 4000 miles. Find its surface area to the nearest square mile (mi^2). Use 3.14 for π.

Excursions in Geometry

Pick's Theorem

A geoboard usually is a square piece of wood with an arrangement of nails on its surface. The nails are in rows and columns of 5 each for a total of 25 nails.

A variety of geometric figures can be formed by looping rubber bands around the nails on a geoboard. The figure at the right shows two such figures. The area of the region outlined in green is 1 square unit.

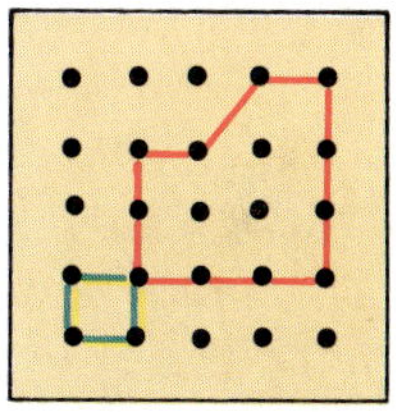

The following theorem (Pick's Theorem) gives a formula for the area of a region outlined on a geoboard.

If the area of a region on a geoboard is A square units, the borders of the region touch x nails, and there are y nails in the interior, then $A = \frac{x}{2} + y - 1$.

Example Find the area of the region outlined in red in the figure shown above.

$$A = \frac{x}{2} + y - 1$$
$$= \frac{11}{2} + 3 - 1 \quad \textit{x is 11 and y is 3.}$$
$$= 7\frac{1}{2}$$

Exercises Find the area for each of the following regions. Use Pick's Theorem.

1.
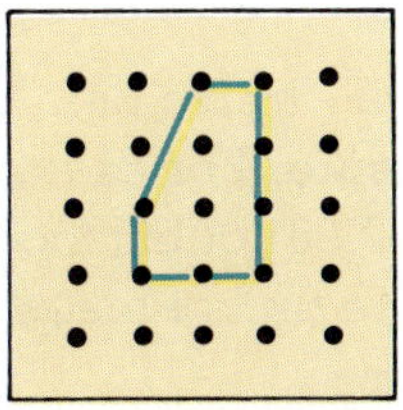

2.
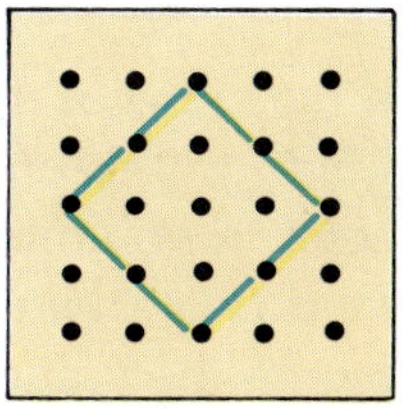

3.
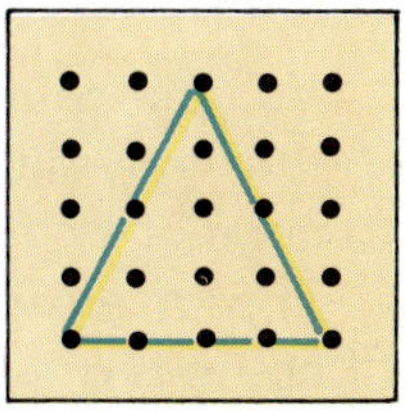

4.
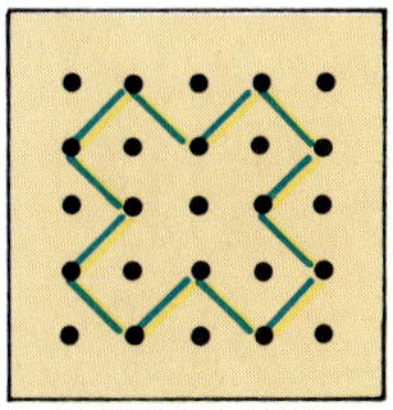

11–7 Defining Volume

The photograph below shows polluted air in a major city.

About 1,640,000,000 grams of sulfur dioxide are emitted each day into the air over the city. This means that each cubic meter (m^3) of air carries about 0.000082 grams of sulfur dioxide each day.

Volume is used to measure amounts of space. When each edge of a cube is 1 unit long, we say that the volume of the cube is 1 cubic unit. Thus, a cube one centimeter by one centimeter by one centimeter has a volume of one cubic centimeter (cm^3).

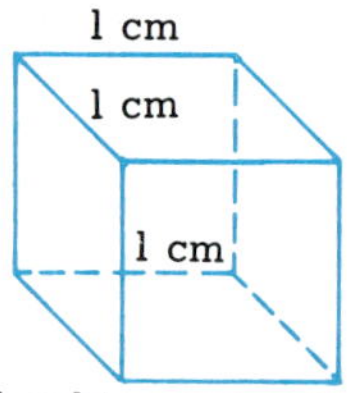

1 cubic centimeter

Postulate 11–5
Volume Postulate

For any solid region and a given unit of measure, there is a unique positive number called the measure of the volume of the region.

Volume will mean the measure of the volume. The volume of a figure may differ depending on the unit of measure. For example, a cube of 1 cubic centimeter in volume is about 0.061 cubic inches.

In a given situation, we do *not* always state the unit of measure. If no units are stated, assume the same unit is used.

Using the following postulates, it is possible to find the volumes of various regions.

If two solids are congruent, then they have exactly the same volume.

Postulate 11–6

If a solid region is separated into nonoverlapping regions, then the sum of the volumes of these regions equals the volume of the given region.

Postulate 11–7 Volume Addition Postulate

The rectangular box below is separated into 24 congruent cubes whose sides are 1 unit each. Thus, the area of each cube is 1 cubic unit. Notice that the box has three layers, each with 2 × 4 or 8 cubes. By the Volume Addition Postulate, the volume of the box is 24 cubic units.

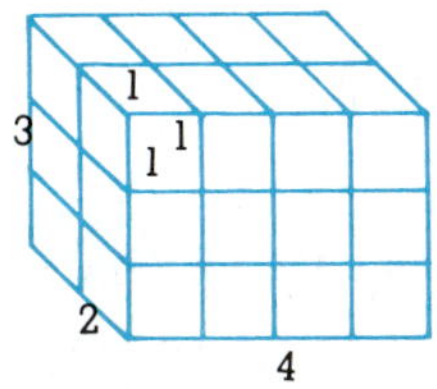

$$V = (2 \times 4) \times 3$$
$$= \textit{Base area} \times \textit{height}$$

In a similar manner, any right prism can be separated into layers to find its volume. The example shown and many others suggest the following.

If a right prism has a volume of V cubic units, a base with an area of B square units, and a height of h units, then $V = Bh$.

Postulate 11–8

example 1

Find the volume of a right hexagonal prism that has a base with sides of 8 centimeters and a height of 20 centimeters. Round your answer to the nearest unit.

$$V = Bh$$
$$= \left(\tfrac{1}{2}ans\right)h$$
$$= \tfrac{1}{2}(4\sqrt{3})(6)(8)20$$
$$\approx 3325.44$$

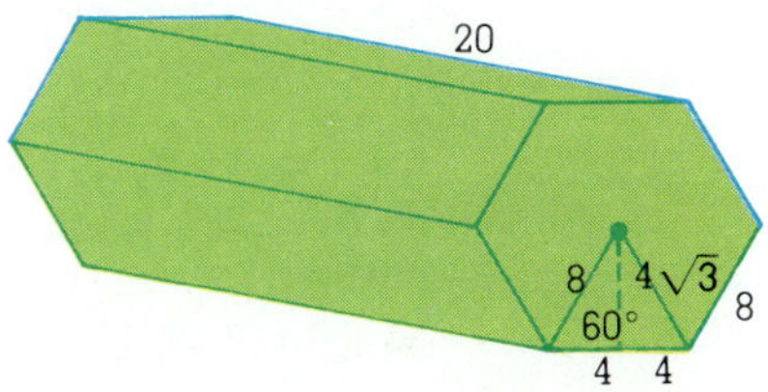

The volume is about 3325 cubic centimeters.

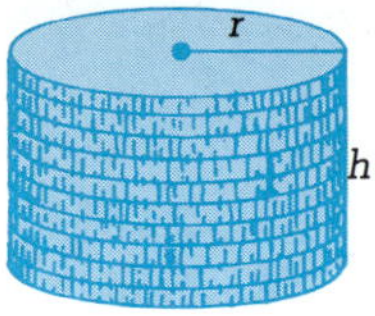

A right cylinder can be thought of as separated into circular layers similar to a stack of coins. The volume of the cylinder is the area of each layer times the height of the stack.

$$\text{Volume} = \text{Base area} \times \text{height}$$
$$= \pi r^2 \times h$$

Theorem 11–18

If a right cylinder has a volume of V cubic units, and its base has a radius of r units, then $V = \pi r^2 h$.

example 2

Hexagonal metal bars are made by cutting the sides from cylindrical bars. The radii of the hexagonal and cylindrical bars is 2.5 centimeters. Find the volume of waste metal from making a bar 60 centimeters long. Use 3.14 for π and round to the nearest hundredth.

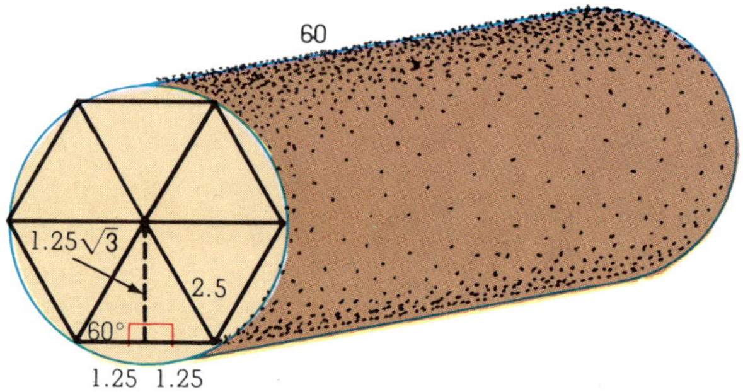

$$\begin{aligned}\text{volume waste metal} &= \text{volume cylindrical bar} - \text{volume hexagonal bar}\\ &= \pi r^2 h - Bh\\ &= \pi r^2 h - \tfrac{1}{2}ansh\\ &\approx (3.14)(2.5)^2(60) - \tfrac{1}{2}(1.25\sqrt{3})(6)(2.5)(60)\\ &\approx 203.25\end{aligned}$$

About 203.25 cubic centimeters of metal are wasted.

exercises

Exploratory Find the volume of each right prism described below.

1. area of base: 12 m^2
   height: 3.5 m
2. area of base: 17.5 cm^2
   height: 14 cm
3. area of base: 16 ft^2
   height: 4.2 ft
4. area of base: 48 m^2
   height: 3 m
5. area of base: 720 cm^2
   height: 20 cm
6. area of base: 25 m^2
   height: 7 m

Written Find the volumes for right prisms given the following information about them.

1. height of 3 meters; square bases with sides of 4 meters
2. height of 8 centimeters; rectangular bases 9 centimeters by 10 centimeters
3. height of 40 inches; hexagonal bases with sides 5 inches
4. height of 36 inches; hexagonal bases with radius of 0.5 inches
5. height of 20 centimeters; equilateral triangular bases with radius of 1.5 centimeters
6. height of 27 centimeters; equilateral triangular bases with sides of 2 centimeters

Find the volumes for right cylinders given the following information about them. Use 3.14 for π and round to the nearest tenth.

7. radius of 2 meters
   height of 8 meters
8. radius of 4 centimeters
   height of 7 centimeters
9. radius of 3.2 centimeters
   height of 10.5 centimeters
10. radius of 6.3 meters
    height of 12.1 meters
11. radius of $1\frac{1}{2}$ feet
    height of 4 feet
12. radius of $3\frac{1}{2}$ inches
    height of 2 inches

Solve each problem.

13. A cube is 1 centimeter by 1 centimeter by 1 centimeter. Suppose the length of each edge is doubled. How does the volume of the new cube compare to the original cube?
14. A cube is 1 centimeter by 1 centimeter by 1 centimeter. Suppose the length of each edge is tripled. How does the volume of the new cube compare to the original cube?
15. A right cylinder has a height of 1 centimeter and a radius of 1 centimeter. Suppose each is doubled. How does the volume of the new cylinder compare to the original cylinder?
16. A right cylinder has a height of 1 centimeter and a radius of 1 centimeter. Suppose each is tripled. How does the volume of the new cylinder compare to the original cylinder?
17. Suppose a classroom is 30 feet long, 24 feet wide, and 10 feet high. Each person in the room must have 200 cubic feet of air. Find the maximum number of people that should be allowed in the room.
18. A driveway is 80 feet long and 12 feet wide. Blacktop will be spread over the driveway at an average depth of 2 inches. Approximate the volume of blacktop needed. (Hint: Assume the driveway is a rectangular prism.)
19. A pipe is 100 feet long and has an inside diameter of $\frac{1}{2}$ foot. Find the number of cubic feet of oil that it can hold.
20. A cube of aluminum with each edge 15 inches long is melted. Then, it is rolled into a wire with a diameter of 0.2 inches. Find the length of the wire.

Using Geometry

Metric Units of Volume

In the metric system, volume is measured in cubic units such as cubic meters, or cubic centimeters.

Cubic meter is used to measure volumes such as the storage space in a grain elevator, amounts of gravel or soil, and the space in a refrigerator. A refrigerator packing crate has a volume of about 1 cubic meter.

Cubic centimeter is used to measure volumes of boxes, blocks, and so on. A common marble has a volume of about 1 cubic centimeter.

The volume of a cube with an edge 1 meter long is 1 cubic meter. Since 1 meter is 100 centimeters, the cube also has a volume of 100 × 100 × 100 or 1,000,000 cubic centimeters.

The volume of a cube with an edge 1 centimeter long is 1 cubic centimeter. Since 1 centimeter is 0.01 meters, the cube also has a volume of 0.01 × 0.01 × 0.01 of 0.000001 cubic meters.

Exercises Tell which unit, *cubic meter* or *cubic centimeter*, you would use to find the volume for each of the following.

1. a seam of coal
2. a railroad boxcar
3. a suitcase
4. a walnut shell
5. a baggage compartment on a bus
6. a pair of dice
7. building blocks
8. air in a small balloon

Cubic units are sometimes called "dry" units of volume. Capacity units are sometimes called "fluid" units of volume.

9. Name three common capacity units in the metric system.
10. Tell what kinds of volume measures are used with each capacity unit named in problem 9.
11. Explain how the capacity units named in problem 9 are related.
12. Explain how the capacity units named in problem 9 are related to cubic units.
13. Which units would you use to describe the volume of some milk, *cubic units* or *capacity units*?
14. Which units would you use to describe the volume of a brick, *cubic units* or *capacity units*?

11–8 More About Volume

The volume of a pyramid is less than the volume of a prism with the same base and height. Similarly, the volume of a cone is less than the volume of a cylinder with the same base and height.

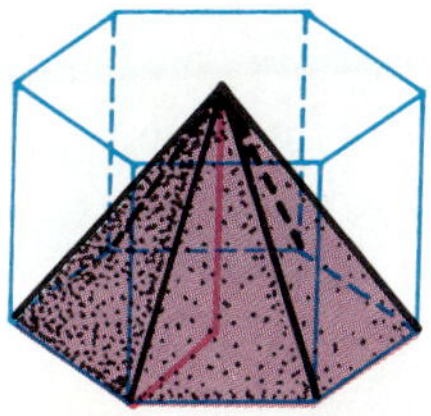

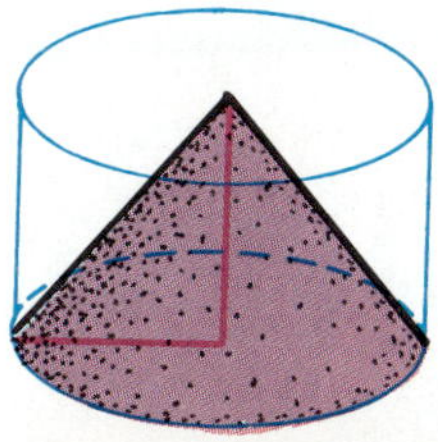

If a right pyramid has a volume of V cubic units, a height of h units, and the area of the base is B square units, then $V = \frac{1}{3}Bh$.

Theorem 11–19

If a right circular cone has a volume of V cubic units, a height of h units, and the area of the base is B square units, then $V = \frac{1}{3}Bh$.

Theorem 11–20

example 1

Find the total volume of the two cones inside the right circular cylinder shown at the right.

$$\text{Total Volume} = \text{Volume Top Cone} + \text{Volume Bottom Cone}$$
$$= \tfrac{1}{3}Bh \quad + \tfrac{1}{3}Bh$$
$$= \tfrac{1}{3}(\pi r^2)\left(\tfrac{1}{2}a\right) + \tfrac{1}{3}(\pi r^2)\left(\tfrac{1}{2}a\right)$$
$$= \tfrac{1}{3}\pi r^2 a$$

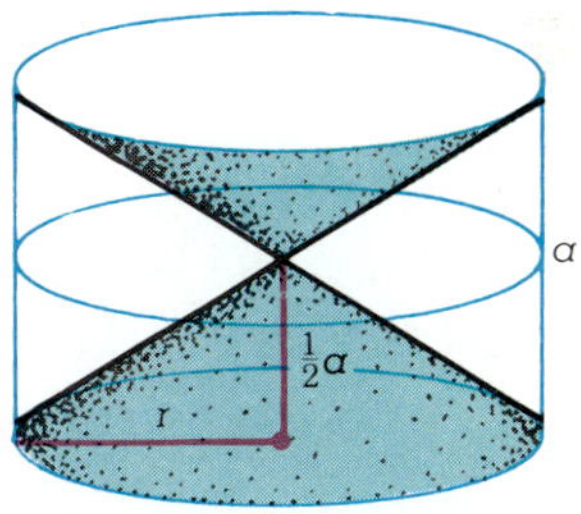

The volume of a sphere is less than the volume of a right circular cylinder with the same radius and a height which is the same length as the diameter of the sphere.

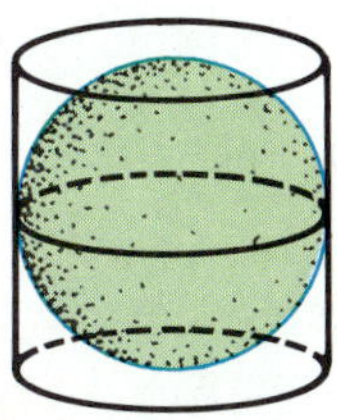

If a sphere has a volume of V cubic units and a radius of r units, then $V = \frac{4}{3}\pi r^3$.

Theorem 11–21

The photograph shows two matching decks of cards. One deck resembles a right rectangular prism. The other deck resembles an oblique rectangular prism. The photograph suggests that the two prisms have the same volume.

A **cross section** of a solid is the intersection of the solid with a plane that is parallel to the base of the solid. In the photograph, each card represents a plane parallel to the base of the deck. Thus, the area of each card represents the area of a cross section of the prism. Notice that *at each level*, the decks of cards have the same cross-sectional area.

The number of cards in a stack represents the measure of the height of that stack. Thus, the two decks have the same height.

Postulate 11–9
Cavalieri's Principle

If two solids have the same cross-sectional area at every level, and the same height, then they have the same volume.

As a result of Cavalieri's Principle, if a prism has a base with an area of B square units and a height of h units, then its volume is Bh cubic units, whether it is right or oblique. Similarly, the volume formulas for cylinders, cones, and pyramids hold whether they are right or oblique.

example 2

Find the volume of the cone. Use 3.14 for π and round to the nearest cubic meter.

$$\begin{aligned} V &= \tfrac{1}{3}Bh \\ &= \tfrac{1}{3}\left[\pi \cdot \left(\tfrac{12}{2}\right)^2\right](13) \\ &= 156\pi \\ &\approx 489.84 \text{ or } 490 \end{aligned}$$

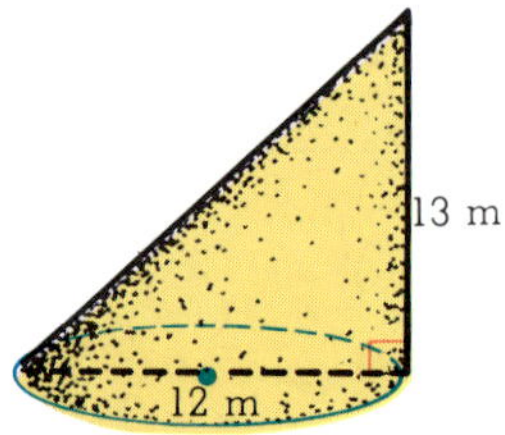

The volume is about 490 cubic meters.

exercises

Exploratory Find the volumes of pyramids, given the following information about them. Round your answer to the nearest tenth.

1. area of base: 15 ft^2
   height: 7 ft
2. area of base: 24 cm^2
   height: 5 cm
3. area of base: 10 ft^2
   height: 3 ft
4. area of base: 17 m^2
   height: 9 m
5. area of base: 31.2 cm^2
   height: 11.6 cm
6. area of base: 27.9 mm^2
   height: 18.5 mm
7. area of base: 63.5 cm^2
   height: 21.1 cm
8. area of base: 3.49 m^2
   height: 1.3 m
9. area of base: $9\frac{1}{2}$ in^2
   height: 2 in.
10. area of base: 7 ft^2
    height: $1\frac{3}{4}$ ft

Written Find the volumes for right circular cones, given the following information about them. Use 3.14 for π and round to the nearest tenth.

1. base radius: 5 ft
   height: 16 ft
2. base radius: 4 m
   height: 11 m
3. base radius: 5 m
   height: 13 m
4. base radius: 9 yd
   height: 16 yd
5. base radius: 8.6 cm
   height: 12.0 cm
6. base radius: 4.2 m
   height: 10.3 m
7. base radius: 100 ft
   height: 130 ft
8. base radius: 350 mm
   height: 670 mm

Find the volumes for spheres, given the following information about them. Use 3.14 for π and round to the nearest tenth.

9. radius: 6400 kilometers
10. diameter: 8000 miles
11. area of great circle:
    50.24 square centimeters
12. area of great circle:
    28.26 square meters

Find the volume of each shaded figure. Use 3.14 for π and round to the nearest tenth.

13.

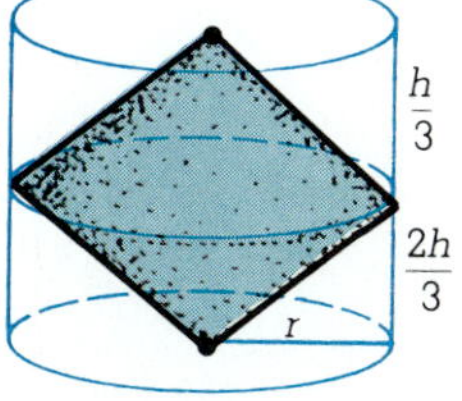

14.

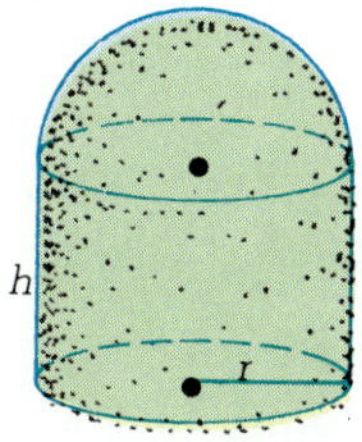

15.

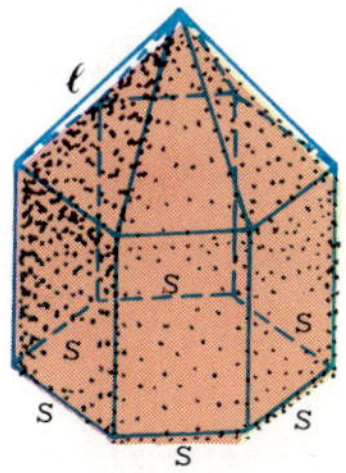

16.

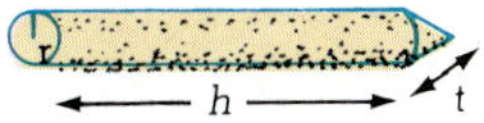

Using Computers

Approximating Area

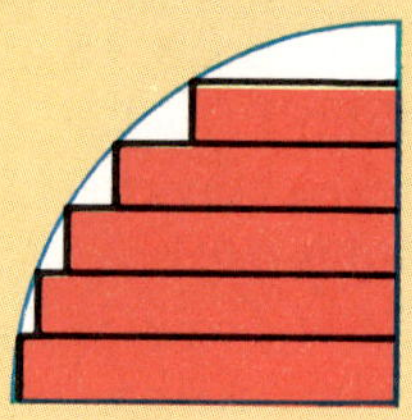

A radius of the sector at the right has been separated into 6 intervals of the same length. By drawing vertical lines from the base to the circle, 5 rectangles are formed. One way to approximate the area of the sector is to add the areas of the rectangles. The greater the number of rectangles, the better the approximation.

The following program approximates the area of a circle with a radius of 1 unit. It separates a sector, like the one shown, into n intervals, adds the areas of the rectangles formed, and multiplies by 4.

```
10 INPUT N
20 LET S = 0
30 FOR I = 1 TO N − 1
40 LET A = (1/N)*SQR(1−(I/N) ↑ 2)
50 LET S = S + A
60 NEXT I
70 PRINT "NUMBER OF INTERVALS IS"; N
80 PRINT "AREA OF CIRCLE IS"; 4*S; "SQUARE UNITS"
90 END
```

The formula for the area of a circle is $A = \pi r^2$. Using this formula, the area of a circle with a radius of 1 unit can be found.

$$\begin{aligned} A &= \pi r^2 \\ &= \pi(1)^2 \\ &= \pi \end{aligned}$$

Thus, the program also approximates values for π.

Exercises Find approximate values of π for each of the following numbers of intervals. Use a program similar to the one above.

1. 10 **2.** 20 **3.** 50 **4.** 100
5. 300 **6.** 500 **7.** 1000 **8.** 5000

Statement 40 in the program tells the computer to calculate a value for the area of the ith rectangle. Explain each of the following statements in the program.

9. Statement 10 **10.** Statement 30
11. Statement 50 **12.** Statement 80

Vocabulary

area (328)
polygonal region (328)
apothem (337)
circumference (341)
total surface area (346)
prism (346)
bases (346)
lateral faces (346)
lateral edges (346)
right prism (346)
oblique prism (346)
lateral area (347)
cylinder (348)
axis (348)
right cylinder (348)
pyramid (350)
regular pyramid (350)
slant height (350)
right circular cone (351)
volume (354)

Chapter Summary

1. Area Postulate: For any polygonal region and a given unit of measure, there is a unique positive number called the measure of the area of the region. (328)
2. Postulate 11–2: If two regions are congruent, then they have exactly the same area. (329)
3. Area Addition Postulate: If a region is separated into nonoverlapping regions, then the sum of the areas of these regions equals the area of the given region. (329)
4. If a circle has a circumference of C units and a radius of r units, then $C = 2\pi r$. (342)
5. Formulas for areas of polygons and circles:

   Rectangle $A = \ell w$ (329)
   Square $A = s^2$ (330)
   Parallelogram $A = bh$ (333)
   Triangle $A = \frac{1}{2}bh$ (334)
   Trapezoid $A = \frac{1}{2}h(b_1 + b_2)$ (334)
   Regular Polygon $A = \frac{1}{2}san$ (338)
   Circle $A = \pi r^2$ (342)

6. Sector: A sector of a circle is a region bounded by a central angle and the intercepted arc. (343)

7. Volume Postulate: For any solid region and a given unit of measure, there is a unique positive number called the measure of the volume of the region. (354)
8. Postulate 11–6: If two solids are congruent, then they have exactly the same volume. (355)
9. Volume Addition Postulate: If a solid region is separated into nonoverlapping regions, then the sum of the volumes of these regions equals the volume of the given region. (355)
10. Cavalieri's Principle: If two solids have the same cross-sectional area at every level, and the same height, then they have the same volume. (360)
11. Formulas for areas and volumes of solids:

Right Prism	Right Cylinder	Right Circular Cone
$L = ph$	$L = 2\pi rh$	$L = \pi rl$
$T = ph + 2B$	$T = 2\pi rh + 2\pi r^2$	$T = \pi rl + \pi r^2$
$V = Bh$	$V = \pi r^2 h$	$V = \frac{1}{3}Bh$
Regular Pyramid	**Sphere**	
$L = \frac{1}{2}pl$	$A = 4\pi r^2$	
$V = \frac{1}{3}Bh$	$V = \frac{4}{3}\pi r^3$	

Chapter Review

11–1 Find the area for each of the following.

1. a rectangle with length 6 feet and width 4 feet
2. a parallelogram with height 4 feet and base 10 feet

11–2 Find the area for each of the following. Round each answer to the nearest tenth.

3. a right triangle with a base of 6 meters and a hypotenuse of 10 meters
4. an equilateral triangle with each side 16 centimeters long
5. a trapezoid with bases of 14 feet and 12 feet, and a height of 10 feet
6. a rhombus with diagonals of 12 inches and 20 inches

11–3 Find the area for each of the following. Round each answer to the nearest tenth.

7. a square inscribed in a circle with radius of 25 centimeters
8. a regular hexagon with an apothem of 20 centimeters
9. a regular hexagon with a side of 32 centimeters

11–4 For each of the following, use 3.14 for π and round to the nearest tenth.

10. Find the circumference of a circle with a radius of 9.6 meters.

11. Find the area of a circle with a radius of 9.6 meters.

12. Find the area of a sector of a circle, if its central angle is 72° and the radius of the circle is 5 inches.

11–5 For each of the following, use 3.14 for π and round to the nearest tenth.

13. Find the lateral area of a regular octagonal prism if its base has sides of 2 centimeters and its height is 3 centimeters.

14. Find the surface area of a right cylindrical gas tank if it is 7 meters tall and has a diameter of 11 meters.

11–6 For each of the following, use 3.14 for π and round to the nearest tenth.

15. Find the total surface area of a square pyramid, if each side is 3 centimeters long and its slant height is 6 centimeters.

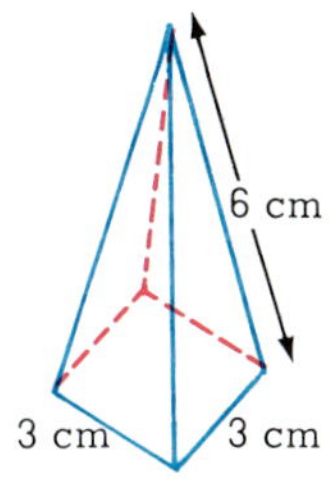

16. Find the total surface area of a right circular cone with a radius of 5 feet and a slant height of 13 feet.

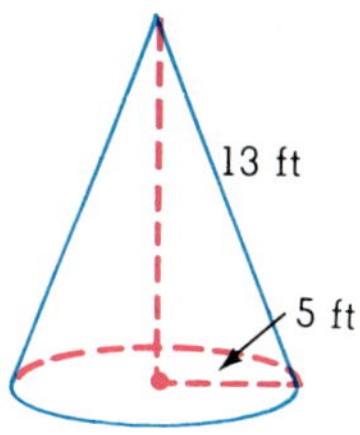

17. Find the surface area of the moon if its diameter is approximately 2160 miles.

11–7 For each of the following, use 3.14 for π and round to the nearest tenth.

18. Find the volume of a hexagonal prism if its radius is 10 centimeters and its height is 20 centimeters.

19. Find the volume of a right circular cone if its radius is 10 centimeters and its height is 20 centimeters.

11–8 For each of the following, use 3.14 for π and round to the nearest tenth.

20. Find the volume of a triangular prism if its base is an equilateral triangle with sides 9 centimeters long and its height is 15 centimeters.

21. Find the volume of a right circular cone if its height is 22 centimeters and its radius is 11 centimeters.

22. Find the volume of the sphere shown below.

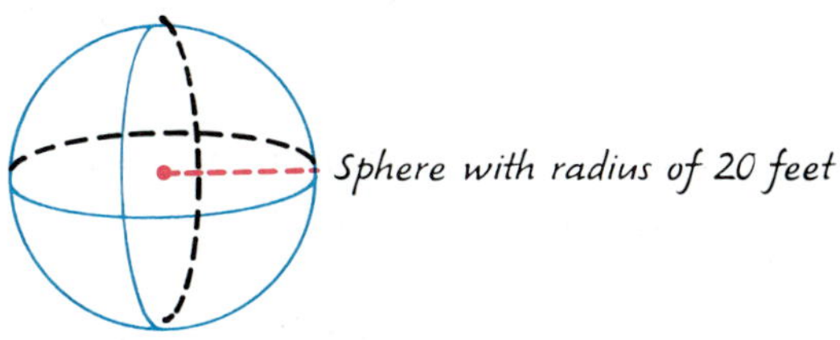

Chapter Test

Find the area for each of the following. Round each answer to the nearest tenth. Use 3.14 for π, if necessary.

1. a rectangle with length 7.1 millimeters and width 9.2 millimeters
2. a parallelogram with a height of 11 inches and a base of 12 inches
3. an equilateral triangle with each side 25 centimeters long
4. a right triangle with a hypotenuse of 13 centimeters and one leg 5 centimeters long
5. a regular hexagon with sides of 15 feet
6. a regular hexagon with an apothem of 2.5 inches
7. a circle with a diameter of 4 inches
8. a circle inscribed in a square with a diagonal of 21.5 centimeters
9. a sector of a circle, if its central angle measures 30° and the radius of the circle is 3 feet

Find the total surface area for each of the following. Round each answer to the nearest tenth. Use 3.14 for π, if necessary.

10. a regular hexagonal prism with base sides of 5 feet and a height of 10 feet
11. a right cylinder with base radius of 5 feet and a height of 10 feet
12. a regular triangular pyramid with base sides of 23 centimeters and a slant height of 40 centimeters
13. a right circular cone with a radius of 27 millimeters and a height of 30 millimeters
14. a sphere with radius of 300 feet
15. a regular square pyramid with base sides of 4 inches and a height of 6 inches

Find the volume for each of the following. Round each answer to the nearest tenth. Use 3.14 for π, if necessary.

16. a regular triangular prism with base sides of 8 yards and a height of 10 yards
17. a regular square pyramid with base sides of 4 inches and a height of 6 inches
18. a right cylinder with a diameter of 39 centimeters and a height of 50 centimeters
19. a right circular cone with a diameter of 39 centimeters and a height of 50 centimeters
20. a sphere with radius of 300 feet

Chapter

12

Coordinates

To prevent accidents, air traffic controllers must know the positions of various planes in the sky. Coordinates are used to identify these positions.

12–1 Graphing

The two axes do not lie in any quadrant

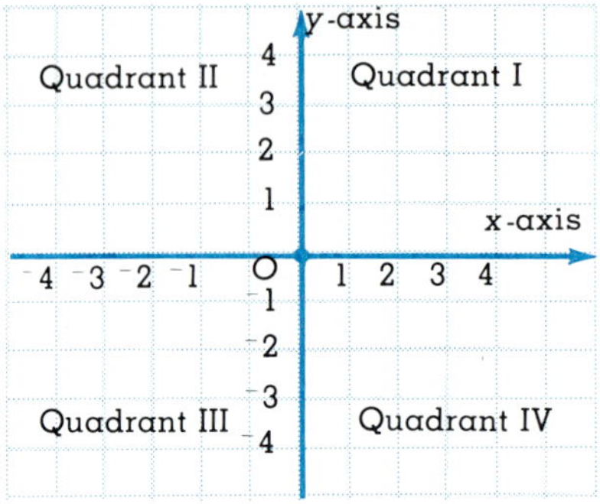

Ordered pairs of numbers can be used to locate points in a plane. Two perpendicular number lines separate the plane into four regions called quadrants. The horizontal number line is called the **x-axis.** The vertical number line is called the ***y*-axis.** Their point of intersection is called the origin and named O.

The figure at the right shows the graph of the ordered pair (4, 2). The first component, 4, is called the **x-coordinate.** It tells the number of units to the left or right of the origin. The second component, 2, is called the ***y*-coordinate.** It tells the number of units above or below the origin.

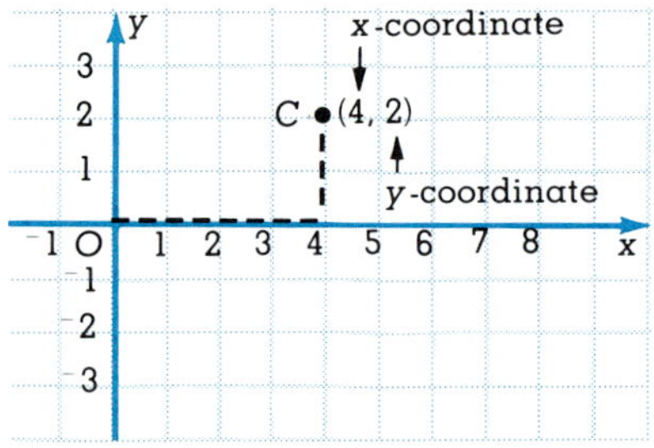

example 1

Graph *A* at (2, ⁻3).

Start at O. Move 2 units to the right. Then, move 3 units down.

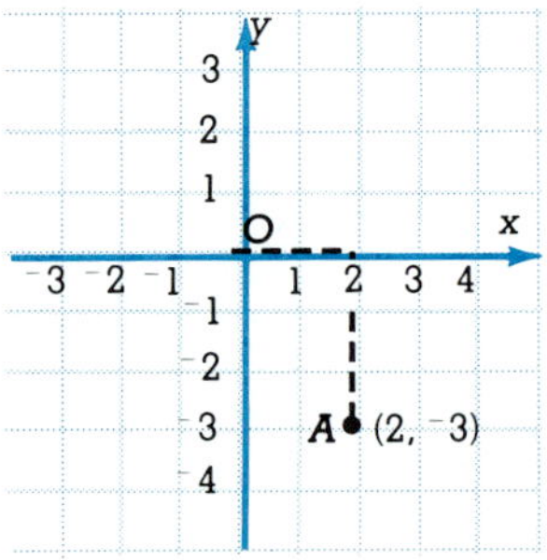

When plotting points, the following is true.

Postulate 12–1

Each point in a coordinate plane corresponds to exactly one ordered pair of numbers. Each ordered pair of numbers corresponds to exactly one point in a coordinate plane.

The solutions for $y = 2x - 1$ are ordered pairs. By substituting different values for x in the equation you can find the corresponding values for y. Some of the solutions are shown in the table below.

x	2x − 1	y	(x, y)
$^-2$	$2(^-2) - 1$	$^-5$	$(^-2, ^-5)$
$^-1$	$2(^-1) - 1$	$^-3$	$(^-1, ^-3)$
0	$2(0) - 1$	$^-1$	$(0, ^-1)$
1	$2(1) - 1$	1	(1, 1)
2	$2(2) - 1$	3	(2, 3)

For each value of x you choose there is a corresponding y value that satisfies the equation.

The solutions for $y = 2x - 1$ also can be shown by a graph. The figure below at the left shows some of the solutions to $y = 2x - 1$. The figure below at the right shows a line connecting the points.

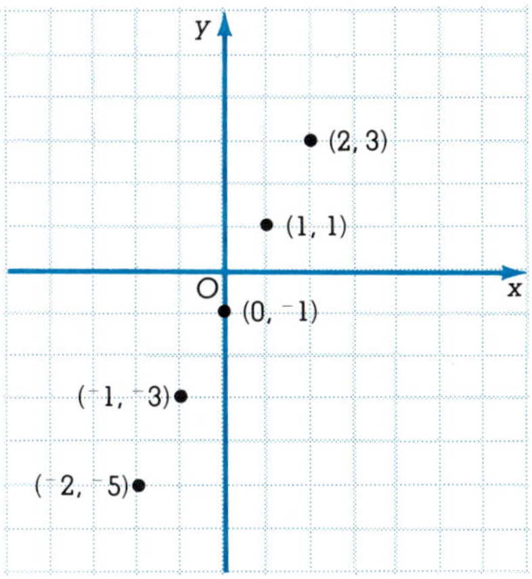

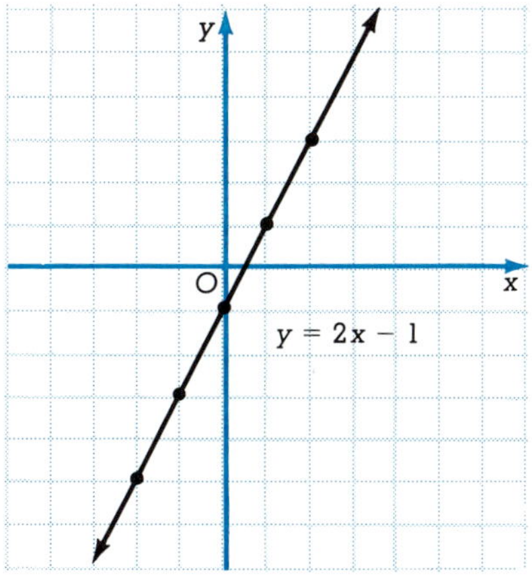

An equation whose graph is a straight line is called a **linear equation.**

An equation is linear if and only if it can be written in the form $Ax + By = C$, where A, B, and C are any numbers, and A and B are not both 0.

Definition of Linear Equation

The equations $5x + 6y = 8$, $3x = 4y + 9$, $5m - n = \frac{1}{2}$, and $x = 4$ are linear equations. Each can be written in the form $Ax + By = C$. The equations $4x + 5y^2 = 7$ and $\frac{1}{y} + x = 3$ are *not* linear equations.

example 2

Graph $y = 3x - 3$.

Make a table of values for x and y. Graph the ordered pairs and connect them with a line.

x	y
$^-2$	$^-9$
$^-1$	$^-6$
0	$^-3$
1	0
2	3

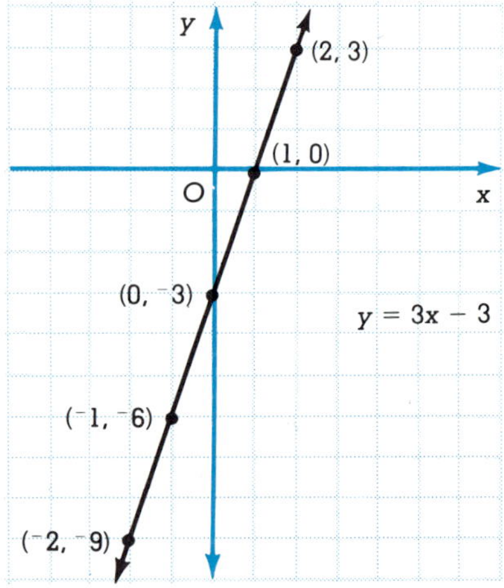

The equation $y = 3x - 3$ can be written in the form $3x - y = 3$. Therefore, it is a linear equation and the graph is a line.

According to Postulate 1–1, two points determine a line. Therefore, to graph any line, simply find two ordered pairs that satisfy the linear equation. Then, graph the ordered pairs and connect them with a straight line.

example 3

Graph $3x + 2y = 4$.

Transform the equation so that y is on one side by itself. Then, find two values for x and y.

$$3x + 2y = 4$$
$$2y = 4 - 3x$$
$$y = \frac{4 - 3x}{2}$$

x	y
0	2
2	$^-1$

Graph the ordered pairs and connect them with a line.

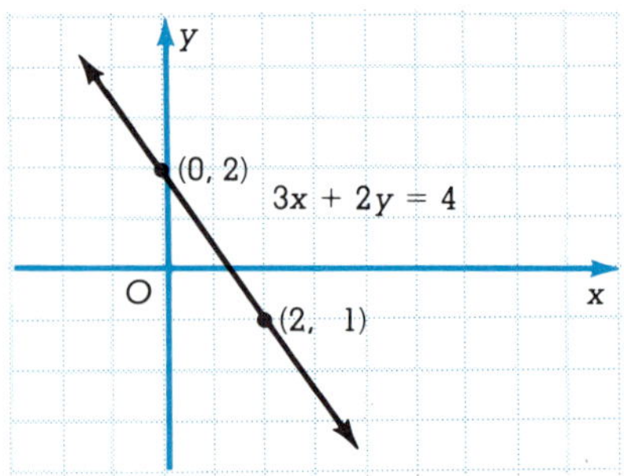

The equation $3x + 2y = 4$ is a linear equation. The graph is a line.

exercises

Exploratory Name the ordered pair for each point on the graph below.

1. A
2. I
3. F
4. W
5. P
6. X
7. Q
8. S
9. B
10. C
11. M
12. N

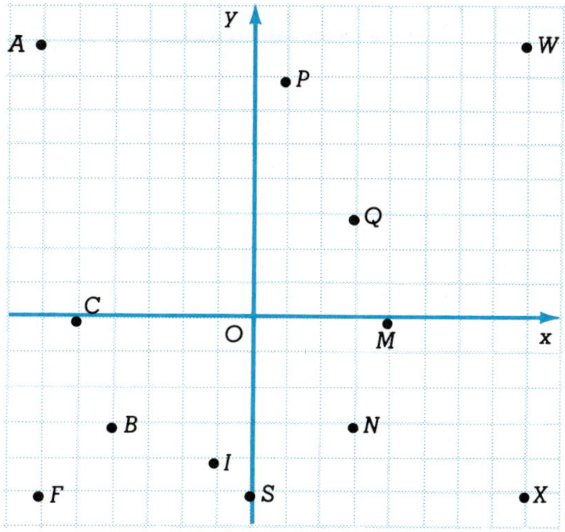

State which of the following are linear equations.

13. $3x + 2y = 6$
14. $y = 2x - 1$
15. $y = 8$
16. $3x + 4y^2 = 1$
17. $9m + 4n = {}^-2$
18. $\frac{1}{x} + \frac{3}{4}y = 7$
19. $5x + \frac{1}{2}y = 2$
20. $x^2 + y^2 = 16$
21. $2x + 3y = 4$

Written Graph each of the following linear equations.

1. $y = 2x$
2. $x + y = 5$
3. $x + 2y = 6$
4. $4x - 3y = 0$
5. $y = 4x$
6. $x + y = {}^-4$
7. $x - 4y = {}^-8$
8. $5x + 2y = 0$
9. $y = {}^-x + 2$
10. $2x + y = 6$
11. $4x + 2y = 6$
12. $2x - 3y = 4$
13. $y = {}^-2x + 3$
14. ${}^-3x + y = 2$
15. $6x - 3y = 9$

Each of the following are the coordinates of three vertices of a rectangle. Graph them. Then, find each fourth vertex.

16. $(3, 1), (3, {}^-3), ({}^-5, {}^-3)$
17. $(1, 0), (3, 0), (3, 3)$
18. $({}^-3, 4), (5, 4), (5, {}^-3)$
19. $(2, {}^-2), ({}^-2, {}^-2), ({}^-2, {}^-5)$
20. $(2, 0), (0, 2), ({}^-4, {}^-2)$
21. $({}^-1, 0), (1, 1), (0, 3)$
22. $(4, {}^-1), (1, {}^-4), (0, {}^-3)$

12–2 Distances and Midpoints

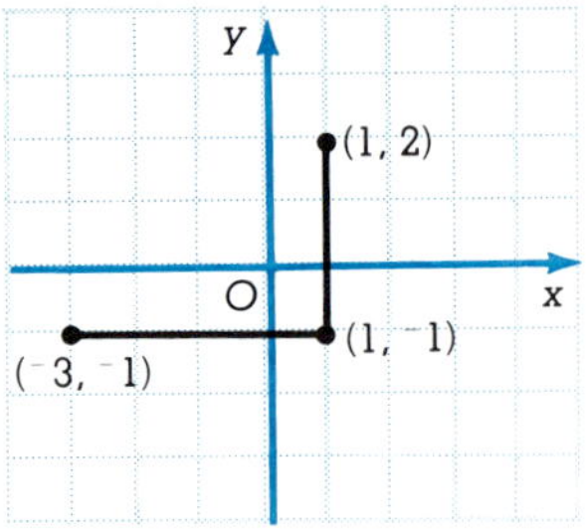

Consider two points in a coordinate plane with coordinates $(^-3, ^-1)$ and $(1, ^-1)$. These points lie on a horizontal line. You can use absolute value to find the distance between the points.

$|^-3 - 1| = |^-4|$ or 4 *The difference between the x-coordinates.*

The points with coordinates (1, 2) and $(1, ^-1)$ lie on a vertical line. The distance between these two points is 3 units.

$|2 - {^-1}| = |3|$ or 3 *The difference between the y-coordinates.*

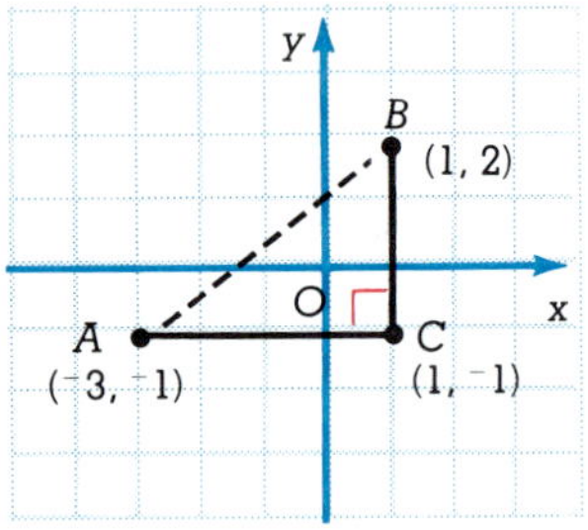

The figure at the left shows a right triangle, $\triangle ABC$. Notice that the line segment connecting A and B is neither horizontal nor vertical. The distance between A and B can be found by using the Pythagorean Theorem.

$$\begin{aligned} AB^2 &= AC^2 + BC^2 \\ &= |^-3 - 1|^2 + |2 - {^-1}|^2 \\ &= 4^2 + 3^2 \\ &= 16 + 9 \\ &= 25 \\ AB &= \sqrt{25} \text{ or } 5 \end{aligned}$$

Distance is always positive.

example 1

Find the distance between $(^-2, 3)$ and $(5, ^-3)$.

A vertical line through $(^-2, 3)$ and a horizontal line through $(5, ^-3)$ intersect at $(^-2, ^-3)$. These three points are the vertices of a right triangle, $\triangle ABC$. The measure of the hypotenuse is the distance between $(^-2, 3)$ and $(5, ^-3)$. Now, use the Pythagorean Theorem.

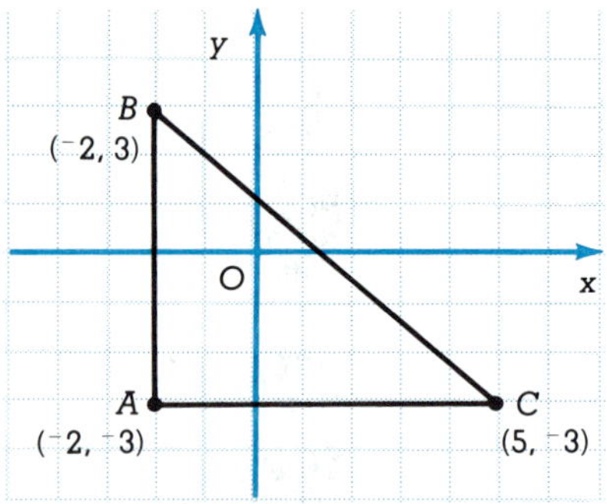

$$\begin{aligned} BC^2 &= AC^2 + AB^2 \\ BC^2 &= |5 - {^-2}|^2 + |3 - {^-3}|^2 \\ &= 7^2 + 6^2 \\ &= 49 + 36 \\ &= 85 \\ BC &= \sqrt{85} \end{aligned}$$

This is approximately 9.2195.

The Pythagorean Theorem can be used to develop a general formula for finding the distance between two points in a plane.

Suppose (x_1, y_1) and (x_2, y_2) are the endpoints of $\overline{BC}$ in the figure at the right. Form a right triangle, $\triangle ABC$, by drawing a vertical line through (x_1, y_1) and by drawing a horizontal line through (x_2, y_2). These lines intersect at (x_1, y_2).

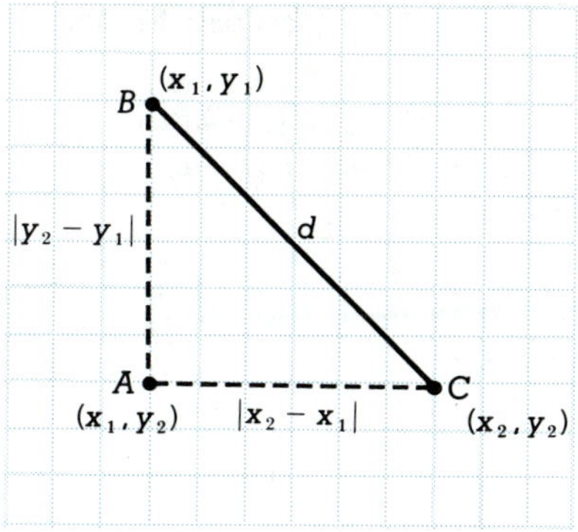

$$BC^2 = AC^2 + AB^2$$
$$d^2 = |x_2 - x_1|^2 + |y_2 - y_1|^2$$
$$d^2 = (x_2 - x_1)^2 + (y_2 - y_1)^2$$
$$d = \sqrt{(x_2 - x_1)^2 + (y_2 - y_1)^2}$$

x_1 is read x sub one.
y_1 is read y sub one.

The distance between two points with coordinates (x_1, y_1) and (x_2, y_2) is given by the following formula.

$$d = \sqrt{(x_2 - x_1)^2 + (y_2 - y_1)^2}$$

Theorem 12–1
Distance Formula

example 2

Use the Distance Formula to find the distance between ($^-1$, 6) and (5, $^-4$).

$$d = \sqrt{(x_2 - x_1)^2 + (y_2 - y_1)^2}$$
$$= \sqrt{(5 - {}^-1)^2 + ({}^-4 - 6)^2}$$
$$= \sqrt{(6)^2 + ({}^-10)^2}$$
$$= \sqrt{36 + 100}$$
$$= \sqrt{136}$$
$$= 2\sqrt{34}$$ *The distance is $2\sqrt{34}$ units.*

The Distance Formula can be used to show that a given point on a line segment is the midpoint of that line segment.

example 3

Show that the point represented by (3, 1) is the midpoint of a line segment having endpoints whose coordinates are (6, $^-1$) and (0, 3).

distance between (3, 1) and (6, $^-1$)

$$d = \sqrt{(6 - 3)^2 + ({}^-1 - 1)^2}$$
$$= \sqrt{(3)^2 + ({}^-2)^2}$$
$$= \sqrt{9 + 4}$$
$$= \sqrt{13}$$

distance between (3, 1) and (0, 3)

$$d = \sqrt{(0 - 3)^2 + (3 - 1)^2}$$
$$= \sqrt{({}^-3)^2 + (2)^2}$$
$$= \sqrt{9 + 4}$$
$$= \sqrt{13}$$

The Distance Formula also can be used to find the coordinates of the midpoint of a line segment.

example 4

Find the coordinates of M, the midpoint of a line segment having endpoints whose coordinates are (2, 3) and (6, 7).

Form trapezoid $ABQP$ by drawing vertical lines through A, through B, and through M to the x-axis. Since $\overline{MR}$ is the median of the trapezoid, R is the midpoint of $\overline{PQ}$. Use the Definition of Midpoint to find the x-coordinate of R and M.

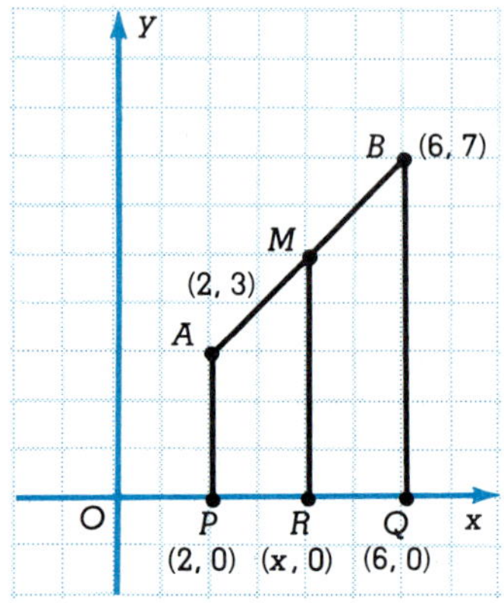

$$PR = RQ$$
$$x - 2 = 6 - x$$
$$2x = 2 + 6$$
$$x = \frac{2+6}{2}$$
$$x = \frac{8}{2} \text{ or } 4$$ *The x-coordinate of M is 4.*

To find the y-coordinate of M, form trapezoid $ABCD$ by drawing horizontal segments through A, through B, and through M to the y-axis. Since $\overline{MG}$ is the median of the trapezoid, G is the midpoint of $\overline{CD}$. Thus, the y-coordinate of G is the y-coordinate of M. Use the Definition of Midpoint to find the y-coordinate of G.

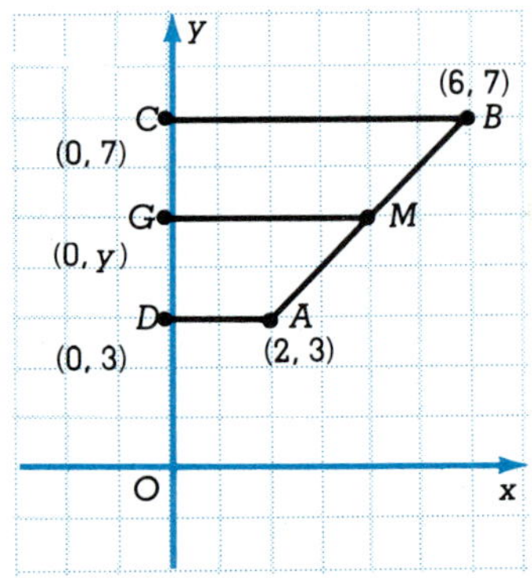

$$CG = GD$$
$$7 - y = y - 3$$
$$7 + 3 = 2y$$
$$\frac{7+3}{2} = y$$
$$y = \frac{10}{2} \text{ or } 5$$ *The y-coordinate of M is 5.*

The coordinates of M are (4, 5).

Example 4 suggests the following theorem.

Theorem 12–2
Midpoint Formula

If the coordinates of A and B are (x_1, y_1) and (x_2, y_2) respectively, then the midpoint M, of $\overline{AB}$, has coordinates $\left(\frac{x_1 + x_2}{2}, \frac{y_1 + y_2}{2}\right)$.

exercises

Exploratory State the distance between each of the following.

1. (5, 0), (12, 0)
2. (0, 3), (0, 6)
3. ($^-6$, 0), ($^-2$, 0)
4. (1, 2), (3, 4)
5. (2, 3), (5, 7)
6. ($^-1$, $^-2$), ($^-3$, $^-4$)

State the coordinates of the midpoint of each line segment which has endpoints with the following coordinates.

7. (0, 4), (0, 0)
8. (6, 0), (13, 0)
9. ($^-3$, 2), ($^-5$, 6)
10. ($^-1$, $^-2$), ($^-3$, $^-6$)
11. (4, 6), ($^-2$, $^-3$)
12. ($^-1$, $^-7$), (6, 1)

Written Find the distance between each of the following.

1. (2, 3), (8, 9)
2. (15, 0), (0, 15)
3. (1, 1), (8, 8)
4. ($^-7$, $^-5$), ($^-10$, $^-9$)
5. (0, 3), (5, 7)
6. ($^-2$, $^-6$), ($^-6$, $^-2$)
7. (4, 6), (0, $^-4$)
8. (6, 0), ($^-4$, $^-8$)
9. (0, 0), (3, 4)
10. (0, 0), (p, q)
11. (a, c), (c, a)
12. $(0, r)$, $(0, {}^-r)$
13. $(4, a)$, $({}^-8, b)$
14. $\left(\frac{2}{3}, \frac{4}{5}\right)$, $\left(\frac{1}{3}, \frac{2}{5}\right)$
15. (0.2, 0.7), (0.4, 0.9)

16–30. Find the coordinates of the midpoint of each line segment which has endpoints with the coordinates given in problems **1–15.**

31. The vertices of a rectangle have the coordinates (4, 1), ($^-5$, 1), ($^-5$, $^-3$), and (4, $^-3$). Find the perimeter of the rectangle.
32. Find the area of the rectangle in problem **31.**
33. Find the measures of the diagonals of the rectangle in problem **31.**
34. Find the coordinates of the center of a circle whose diameter has endpoints with coordinates ($^-6$, $^-8$), and (5, 7).
35. The center of a circle has coordinates ($^-8$, 3). One endpoint of a diameter has coordinates (5, 4). Find the coordinates of the other endpoint of the diameter.

Puzzle

Excursions in Geometry

The diagram shows 12 straws, each with the same length. The arrangement can be changed in several ways.

Exercises Show how to complete each of the following rearrangements.

1. Remove 4 straws so that 1 square remains.
2. Remove 4 straws so that 2 squares remain.
3. Remove 2 straws so that 2 squares remain.
4. Move 3 straws and have 3 squares.

Using Money

Prescriptions

The cost of a prescription may vary depending on the brand. In addition, unbranded products, called generic products, often cost less than the similar branded product. But, many doctors prescribe brand names because they are familiar with products through ads and company representatives.

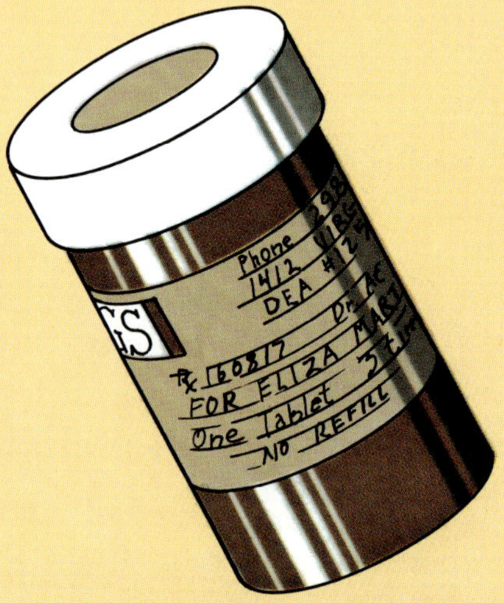

What can *you* do to lower the cost of a prescription?

Ask your doctor if there is a *generic equivalent* for the prescription.

If there is, ask the doctor to write *or generic equivalent* on the prescription.

When the prescription is filled, ask the pharmacist for the generic.

Exercises One well-known brand of tetracycline in 250 mg capsules sells for $4.05 per 50. The generic equivalent in 250 mg capsules sells for $1.50 per 25.

1. Find the cost of 100 capsules of tetracycline under the brand name.
2. Find the cost of the generic equivalent of 100 capsules of tetracycline.
3. Find the cost of 30 capsules of tetracycline under the brand name.
4. Find the cost of the generic equivalent of 30 capsules of tetracycline.

12–3 Slope

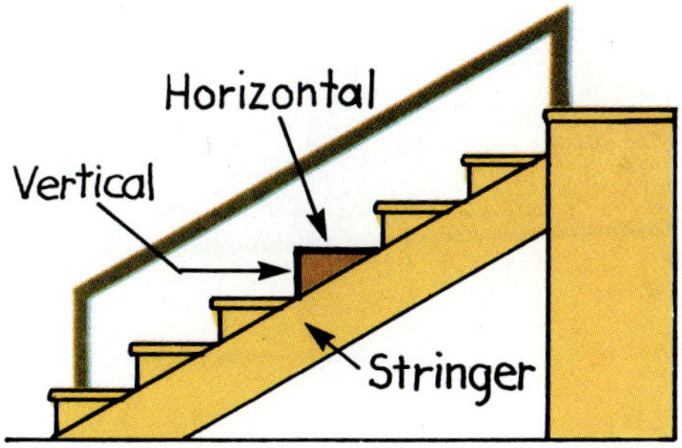

A staircase has two parts, the stringers and the steps. The stringers slope from one floor to the next and support the steps. The vertical and horizontal measurements of the stairs affect the steepness of the stringers.

The measure of the slope, or steepness, of a line is found by comparing the change in vertical units to the change in horizontal units. For example, in the graph of $y = 2x$ below, a vertical change of 2 units is accompanied by a horizontal change of 1 unit.

x	y
2	4
1	2
0	0
$^-1$	$^-2$

(Moving down the table: x changes by $^+1$ each step, y changes by $^+2$ each step.)

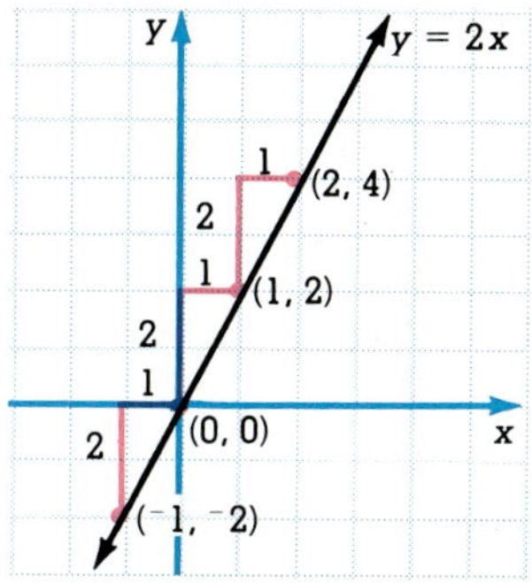

Notice that the vertical change is the difference of the y-coordinates. The horizontal change is the difference of the corresponding x-coordinates. You can use a ratio to compare the two differences.

$$\text{slope} = \frac{\text{difference of the } y\text{-coordinates}}{\text{difference of the corresponding } x\text{-coordinates}}$$

$$= \frac{2}{1} \text{ or } 2$$

example 1

Determine the slope of the line shown below.

Choose two points on the line. Find the ratio of the difference of the y-coordinates to the difference of the corresponding x-coordinates.

$$\text{slope} = \frac{\text{difference of } y\text{-coordinates}}{\text{difference of the corresponding } x\text{-coordinates}}$$

$$= \frac{5 - 2}{4 - 3}$$

$$= \frac{3}{1} \text{ or } 3$$ *The slope is 3.*

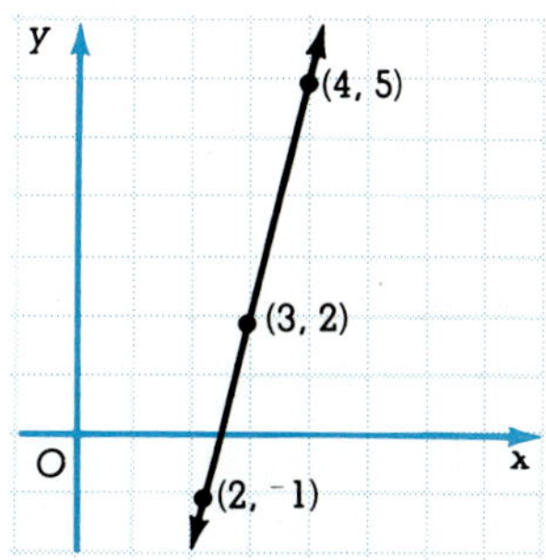

example 2

Determine the slope of each of the following lines.

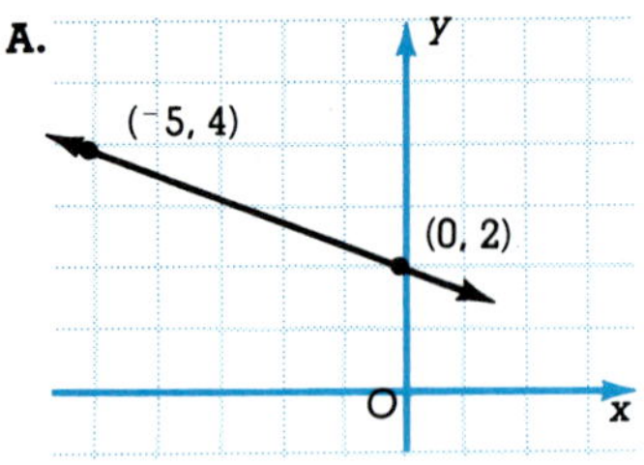

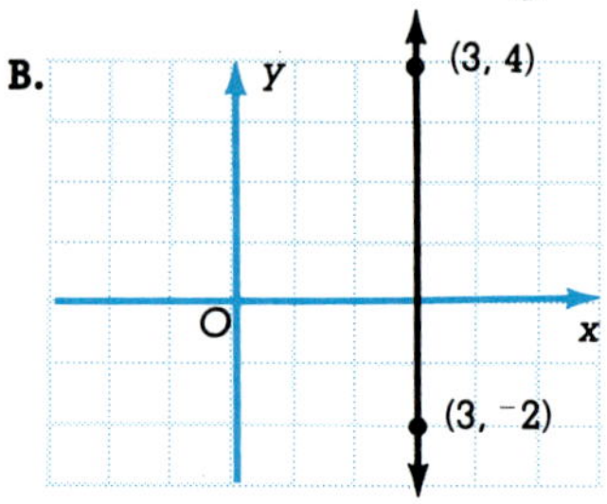

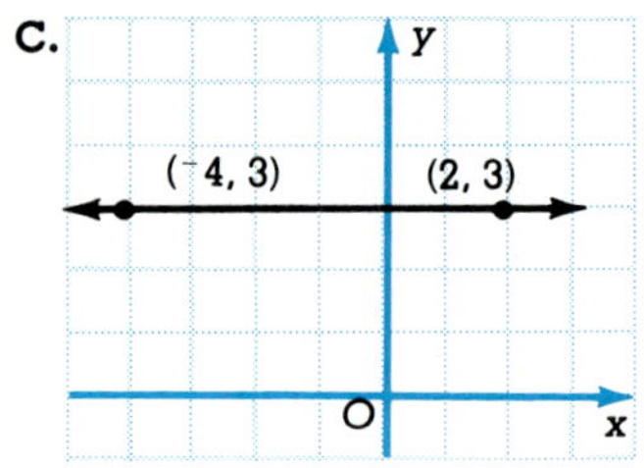

$$\text{slope} = \frac{\text{difference of the } y\text{-coordinates}}{\text{difference of the corresponding } x\text{-coordinates}}$$

A. $\text{slope} = \frac{4 - 2}{^-5 - 0}$

$= \frac{2}{^-5}$

$= -\frac{2}{5}$

The slope is $-\frac{2}{5}$.

B. $\text{slope} = \frac{4 - ^-2}{3 - 3}$

$= \frac{6}{0}$

The slope is undefined.

C. $\text{slope} = \frac{3 - 3}{^-4 - 2}$

$= \frac{0}{^-6}$

$= 0$

The line has zero slope.

The following conclusions about slope can be made.

If the line *rises* to the right, then the slope is *positive*.

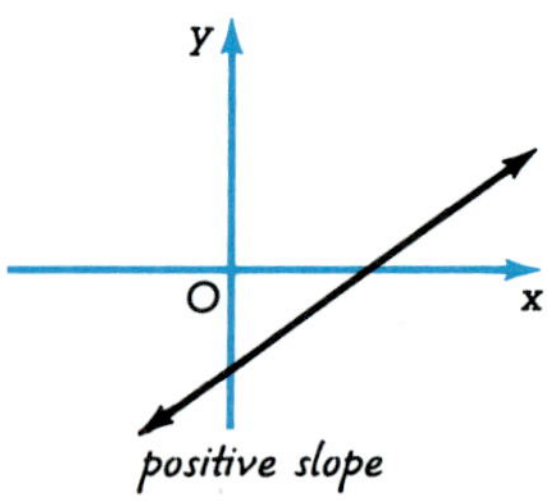

If the line *falls* to the right, then the slope is *negative*.

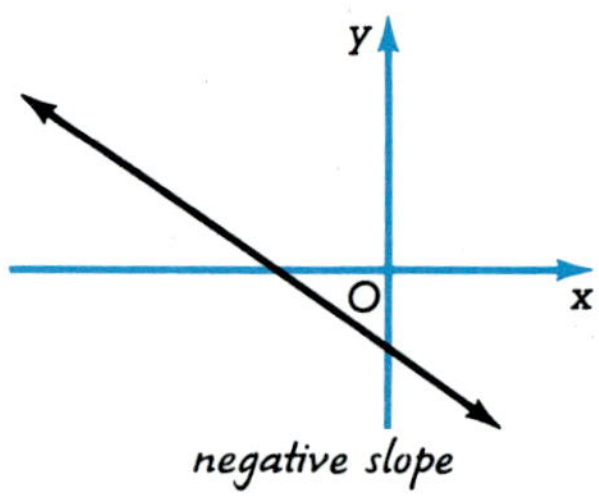

If the line is *vertical*, then the slope is *undefined*.

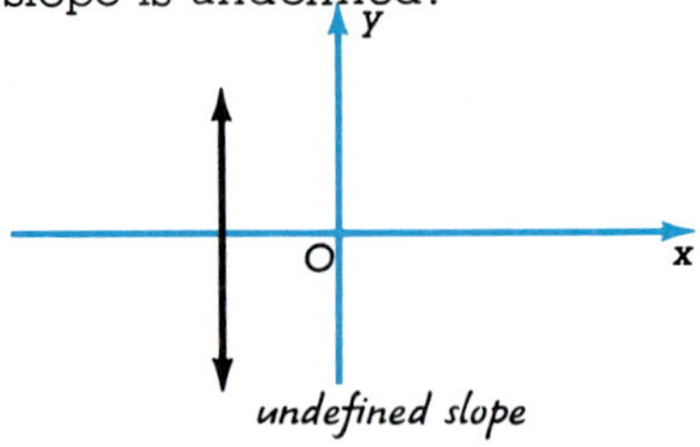

If the line is *horizontal*, then the slope is zero.

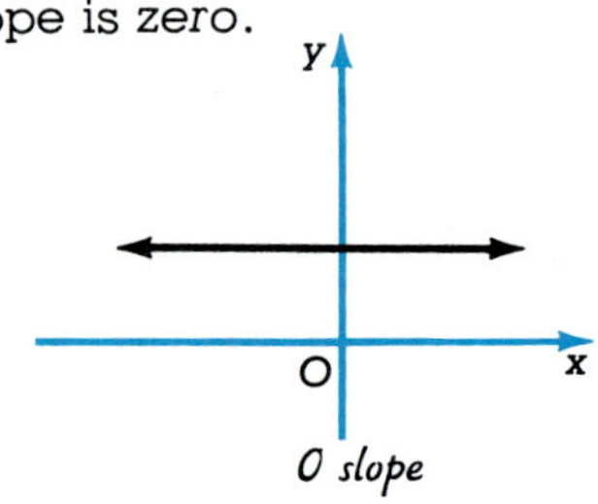

examples

3 **The following lines are parallel. Determine the slope of each line.**

$$\text{slope of } \ell = \frac{0 - ^{-}4}{^{-}2 - ^{-}3} = \frac{4}{1} \text{ or } 4$$

$$\text{slope of } m = \frac{4 - 0}{4 - 3} = \frac{4}{1} \text{ or } 4$$

The slopes are the same.

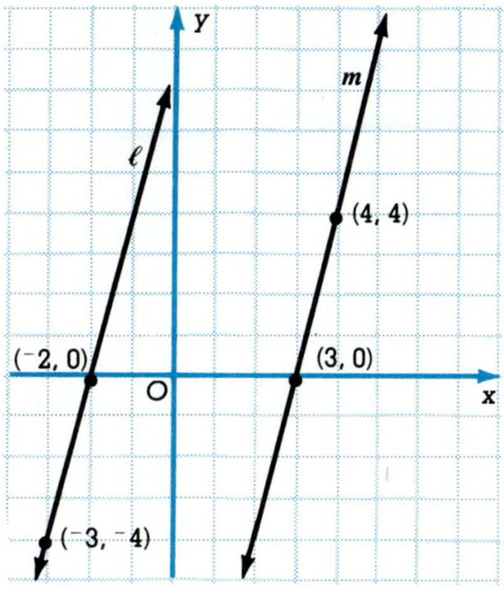

4 **The following lines are perpendicular. Determine the slope of each line.**

$$\text{slope of } a = \frac{4 - ^{-}3}{1 - ^{-}1} = \frac{7}{2}$$

$$\text{slope of } b = \frac{3 - 1}{^{-}4 - 3} = \frac{2}{^{-}7} \text{ or } -\frac{2}{7}$$

$$\frac{7}{2} \cdot -\frac{2}{7} = ^{-}1$$

The product of their slopes is $^{-}1$.

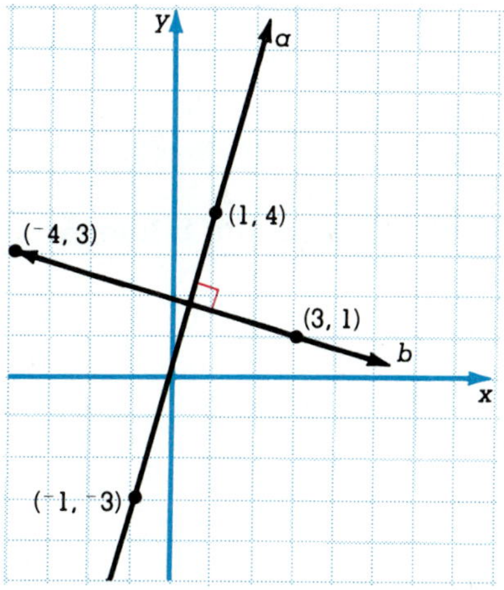

In general, the following theorems can be stated.

Two lines have the same slope if and only if they are parallel and nonvertical. Theorem 12–3

Two nonvertical lines are perpendicular if and only if the product of their slopes is $^{-}1$. Theorem 12–4

exercises

Exploratory Describe the slope of each of the following.

1. vertical lines
2. horizontal lines
3. nonvertical perpendicular lines
4. nonvertical parallel lines
5. lines which rise to the right
6. lines which fall to the right

For each table, state the change in y and the change in x. Then, determine the slope of the line passing through those points.

7.

x	y
0	0
1	2
2	4
3	6
4	8

8.

x	y
$^-2$	2
$^-1$	1
0	0
1	$^-1$
2	$^-2$

9.

x	y
6	$^-8$
3	$^-4$
0	0
$^-3$	4
$^-6$	8

10.

x	y
4	4
3	3
2	2
1	1
0	0

Written Find the slope of the lines passing through the following pairs of points whose coordinates are listed below.

1. (4, 6), (3, 4)
2. ($^-7$, 4), (2, 9)
3. (6, 3), ($^-7$, 3)
4. ($^-4$, 11), ($^-6$, 3)
5. (6, 9), (4, 6)
6. (4, 8), (4, 6)
7. ($^-8$, 1), ($^-5$, $^-8$)
8. (3, $^-2$), (5, $^-9$)
9. (4, 7), (10, $^-5$)

Determine the value of r so that a line through the points with the given coordinates has the slope listed.

10. (r, $^-5$), (5, 3); slope $= \frac{2}{3}$
11. (9, r), (6, 3); slope $= -\frac{1}{3}$
12. (8, r), (12, 6); slope $= \frac{1}{2}$
13. (r, 3), (5, 9); slope $= 2$
14. (r, 7), (11, r); slope $= -\frac{1}{5}$
15. (4, $^-7$), ($^-2$, r); slope $= \frac{8}{3}$
16. (r, 3), (7, r); slope $= 1$
17. (5, 1), (r, $^-3$); slope $= {}^-4$
18. Find the value of y if the points $A(1, 4)$, $B(3, 2)$, and $C(8, y)$ are collinear. *(Hint: Use slope.)*
19. Are points $A(1, 3)$, $B(5, 7)$, and $C(^-3, ^-1)$ collinear? Write *yes* or *no*.
20. Find the slope of each side of quadrilateral $ABCD$ to determine if the figure is a parallelogram. The vertices are $A(^-1, 1)$, $B(3, 1)$, $C(1, ^-2)$, and $D(^-3, ^-2)$.

Challenge Answer each of the following for $\triangle ABC$ whose vertices are $A(3, 2)$, $B(6, ^-3)$, and $C(1, ^-2)$.

1. Find the slope of the median to $\overline{AC}$.
2. Find the slope of the median to $\overline{BC}$.
3. Find the slope of the median to $\overline{AB}$.
4. Find coordinates for B so that $\triangle ABC$ is a right triangle.

Using Geometry

Angle of Inclination

If a line intersects the x-axis, the angle formed by the axis and the line is called the **angle of inclination.** The angle of inclination is measured from the positive direction at the x-axis counterclockwise to the line.

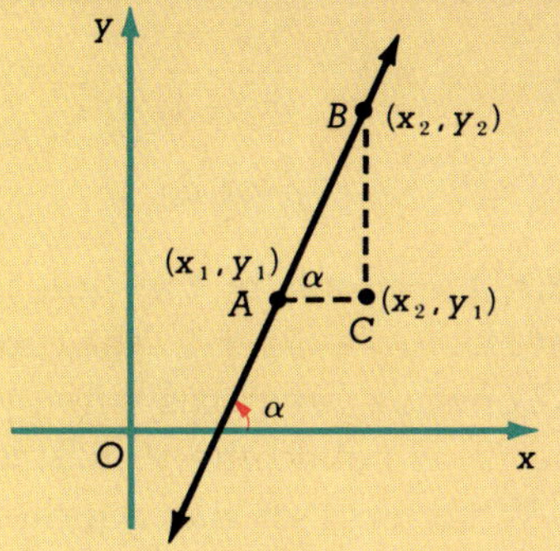

At the left, the coordinates of two points on $\overleftrightarrow{AB}$ are (x_1, y_1) and (x_2, y_2). The angle of inclination has a degree measure of α. Right triangle ABC is determined by constructing $\overline{AC}$ parallel to the x-axis and $\overline{BC}$ parallel to the y-axis. The $\tan \alpha = \dfrac{y_2 - y_1}{x_2 - x_1}$. Notice, the $\tan \alpha$ also is equal to the slope of $\overline{AB}$.

The slope (m) and the angle of inclination can be used to describe a line.

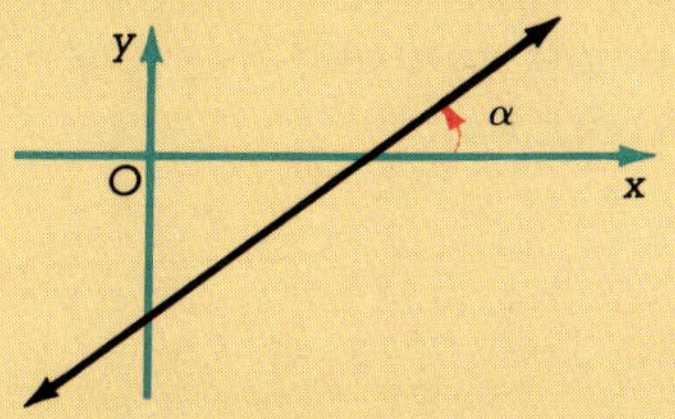

If $m > 0$, and $0 < \alpha < 90$, then the line rises to the right.

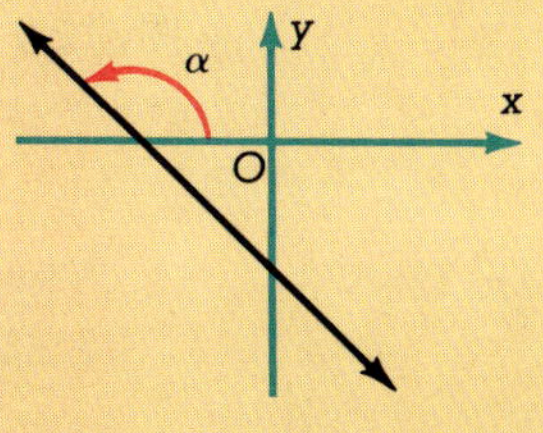

If $m < 0$, and $90 < \alpha < 180$, then the line falls to the right.

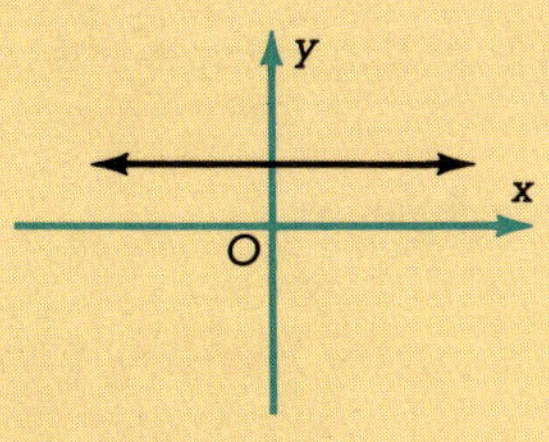

If $m = 0$ and $\alpha = 0$, then the line is horizontal.

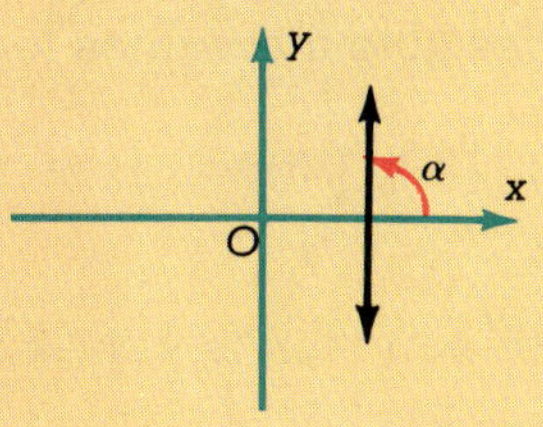

If m is undefined, and $\alpha = 90$, then the line is vertical.

Exercises For each of the following angles of inclination, tell whether the line rises to the right, falls to the right, is horizontal, or is vertical.

1. 55° **2.** 0° **3.** 135° **4.** 90°

5. 91° **6.** 3° **7.** 88° **8.** 100°

12–4 Equations for a Line

The graph of $y = 3x + 2$ is shown below.

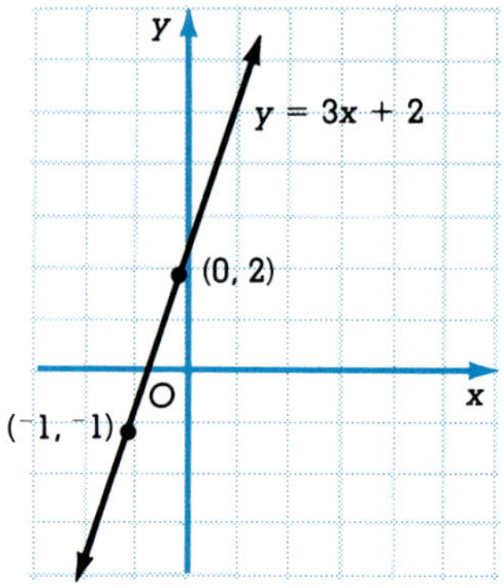

Notice that (0, 2) and ($^-1$, $^-1$) represent points on the line. The coordinates can be used to find the slope of the line.

$$\text{slope} = \frac{2 - {}^-1}{0 - {}^-1}$$
$$= \frac{3}{1} \text{ or } 3$$

Compare the slope of the line to the coefficient of x in $y = 3x + 2$. The graph of $y = 3x + 2$ crosses the y-axis when y is 2. This value, 2, is called the **y-intercept** of the line. What part of the equation involves 2?

In general, any linear equation can be written in the form $y = mx + b$. The slope of the line is m and its y-intercept is b. This form is called the slope-intercept form.

Theorem 12–5

The equation of a line having a slope m and a y-intercept b is $y = mx + b$.

examples

1 **Write the equation of a line having a slope of 4 and a y-intercept of $^-3$.**

Substitute the values for slope and y-intercept into the slope-intercept form.

$$\begin{aligned} y &= mx + b \\ &= 4x + {}^-3 \quad \textit{Substitute 4 for m.} \\ &= 4x - 3 \end{aligned}$$

The equation is $y = 4x - 3$.

2 **Name the slope and y-intercept of $5x + 3y = 6$.**

Rewrite the equation in slope-intercept form.

$$\begin{aligned} 5x + 3y &= 6 \\ 3y &= {}^-5x + 6 \quad \textit{Subtraction Property of Equality} \\ y &= -\tfrac{5}{3}x + 2 \quad \textit{Division Property of Equality} \end{aligned}$$

The slope is $-\frac{5}{3}$. The y-intercept is 2.

For example, suppose the slope and the coordinates of one point on a line are known. Then, the equation of the line can be determined. The coordinates of a point on a line whose slope is 4 are (2, 3). Let (x, y) be the coordinates of another point on the line.

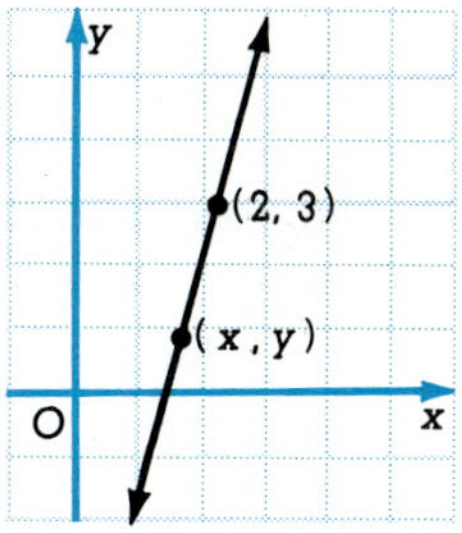

$$\text{slope} = \frac{\text{difference of the } y\text{-coordinates}}{\text{difference of the corresponding } x\text{-coordinates}}$$

$m = \dfrac{y - 3}{x - 2}$ *Simplify.*

$y - 3 = 4(x - 2)$ *Substitute 4 for m.*

This equation is said to be in *point-slope* form. *Could this equation be written in either standard form or slope-intercept form?*

The equation of a line passing through a point whose coordinates are (x_1, y_1) and having a slope, m, is $y - y_1 = m(x - x_1)$. This is called the point-slope form.

Theorem 12–6

example 3

Find the equation of a line passing through the point whose coordinates are ($^-3$, 2) and having a slope of 5. Write the equation in the point-slope form.

Substitute $^-3$, 2, and 5 into the point-slope form.

$y - y_1 = m(x - x_1)$ *Point-slope form*

$y - 2 = 5(x - {}^-3)$ *Substitute*

exercises

Exploratory State the slope and y-intercept of the graph of each of the following.

1. $y = -\frac{3}{4}x + 2$
2. $y = {}^-3x + 5$
3. $y = 4x - 3$
4. $y = \frac{1}{2}x + 6$
5. $y = 3x - 5$
6. $y = mx + b$
7. $y = 6$
8. $x = {}^-3$

State the equation of the line given the following information.

9. $m = 6$, y-intercept is $^-5$
10. $m = {}^-2$, y-intercept is 3
11. $m = 4$, through a point at (1, 3)
12. $m = {}^-7$, through a point at (2, 4)

Written Write the equation of the line given the following information.

1. $m = 2$, y-intercept is $^{-}3$
2. $m = -\frac{1}{3}$, y-intercept is 0.4
3. $m = {}^{-}4$, y-intercept is 7
4. $m = 0$, y-intercept is $^{-}5$
5. $m = 0$, through a point at $(^{-}4, 8)$
6. $m = 7$, through a point at $(10, {}^{-}3)$
7. $m = -\frac{2}{3}$, through a point at $(^{-}2, {}^{-}2)$
8. $m = \frac{1}{5}$, through a point at $\left(\frac{1}{2}, -\frac{1}{3}\right)$
9. parallel to $y = 3x + 4$, through a point at $(3, 7)$
10. parallel to the y-axis, through a point at $(3, 9)$
11. perpendicular to $3x - 5y = 6$, through a point at $(^{-}4, {}^{-}5)$
12. parallel to the x-axis, through a point at $(^{-}3, {}^{-}6)$
13. passing through points at $(^{-}7, 4)$ and $(^{-}7, {}^{-}5)$
14. parallel to the y-axis, through a point at $(2, 6)$

Find the slope and y-intercept of each of the following lines.

15. $y = 5x - 3$
16. $y = \frac{1}{2}x + 6$
17. $4x - 3y = 5$
18. $^{-}6x + 5y = 7$

Answer each of the following.

19. Write the equation of a line passing through points whose coordinates are $(^{-}3, {}^{-}7)$ and $(6, {}^{-}1)$.
20. Write the equation of a line which bisects a line segment whose endpoints have coordinates $(^{-}4, 10)$ and $(5, {}^{-}7)$. The line has a slope of $^{-}5$.
21. Write the equation of a line that is the perpendicular bisector of a line segment whose endpoints have coordinates $(^{-}3, {}^{-}7)$ and $(12, 14)$.

Using Computers

Collinear Points

The following computer program tests to see if any 3 points are on the same line. Suppose the coordinates of the points to be tested are (A, B), (C, D), and (E, F).

```
10 REM THIS PROGRAM DETERMINES IF 3 POINTS FALL ON THE SAME LINE.
20 INPUT A, B, C, D, E, F
30 IF (D – B)/(C – A) = (F – D)/(E – C) THEN 60
40 PRINT A; B, C; D, E; F, "POINTS NOT ON SAME LINE"
50 GO TO 70
60 PRINT A; B, C; D, E; F, "POINTS ARE ON THE SAME LINE"
70 END
```

Exercises Explain what each of the following statements tell the computer.

1. statement 20
2. statement 30
3. statement 40
4. statement 60

Using Geometry

Cross Country Skiing

Courses for cross country skiing run up and down hills as well as on flat ground. There are several types of races for men and for women. For each of the types of races, the climb must *not* have a slope greater than 0.33.

The figure below shows a cross-sectional view of a cross country course. To decide if this course can be used for a race, determine the slope.

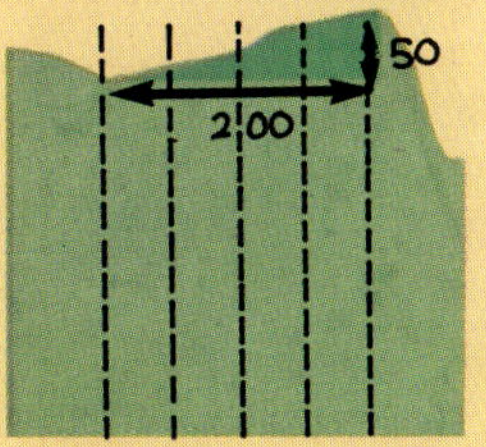

The steepest hill rises 50 meters over a distance of 200 meters.

$$\text{slope} = \frac{\text{change in vertical}}{\text{corresponding change in horizontal}}$$
$$= \frac{50}{200}$$
$$= 0.25$$

Since 0.25 is less than 0.33, the course can be used for a race.

Exercises Solve each problem.

1. A 10 kilometer course rises 75 meters over a distance of 300 meters. Can the course be used for a race? Write *yes* or *no*.
2. A 15 kilometer course rises 30 meters over a distance of 150 meters. Can the course be used for a race? Write *yes* or *no*.
3. A 30 kilometer course rises 100 meters over a distance of 1000 meters. Can the course be used for a race? Write *yes* or *no*.
4. A 50 kilometer course rises 100 meters over a distance of 1500 meters. Can the course be used for a race? Write *yes* or *no*.

12–5 Systems of Equations

The graphs of $y = 2x + 1$ and $y = {}^-2x + 5$ are shown below.

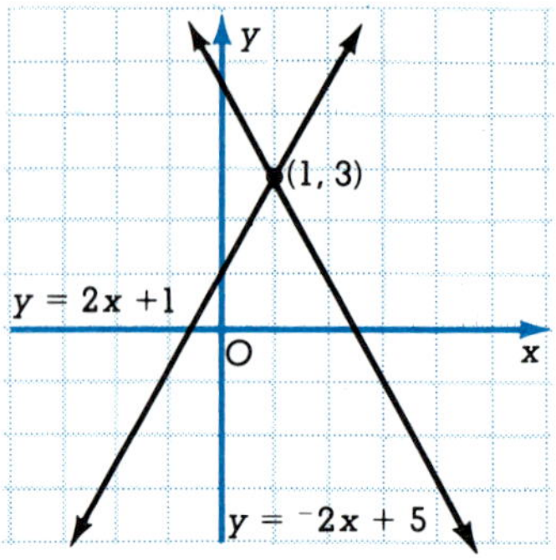

Notice that the graphs intersect at the point with coordinates (1, 3). Since this point lies on the graph of each equation, its coordinates satisfy both $y = 2x + 1$ and $y = {}^-2x + 5$. You can check this as follows.

$y = 2x + 1$	$y = {}^-2x + 5$
$3 \stackrel{?}{=} 2(1) + 1$	$3 \stackrel{?}{=} {}^-2(1) + 5$
$3 = 3$	$3 = 3$

The equations $y = 2x + 1$ and $y = {}^-2x + 5$ together are called a **system of equations**. The solution of this system of equations is (1, 3).

example 1

Graph the equations $x + y = 6$ and $y = 2x$. Then, find the solution of the system of equations.

The graphs intersect at (2, 4). Therefore, (2, 4) is the solution of the system of equations $x + y = 6$ and $y = 2x$.

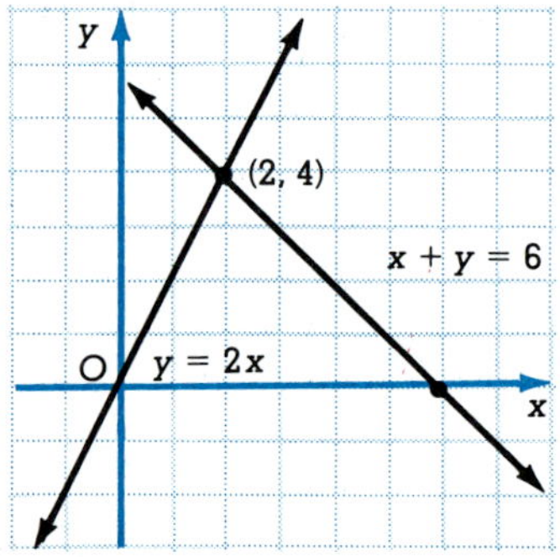

A system of equations can be solved by algebraic methods as well as by graphing. Two such methods are the *substitution method* and the *elimination method*.

examples 2

Use the substitution method to solve the following system of equations.

$y = x - 3$
$3x + 5y = 9$

By the first equation, y is equal to $x - 3$. Therefore, $x - 3$ can be *substituted* for y in the second equation.

$3x + 5y = 9$
$3x + 5(x - 3) = 9$ *Substitute $x - 3$ for y.*

The resulting equation has only one variable, x. Solve the equation.

$$\begin{aligned} 3x + 5x - 15 &= 9 \\ 8x - 15 &= 9 \\ 8x &= 24 \\ x &= 3 \end{aligned}$$

Now, find y by substituting 3 for x in the first equation, $y = x - 3$.

$y = x - 3$
$y = 3 - 3$ *Substitute 3 for x.*
$y = 0$

The solution is (3, 0).

3 **Use the elimination method to solve the system of equations.**

$3x + 4y = 6$
$2x + 3y = 5$

Sometimes adding or subtracting two equations will eliminate a variable. In this case, adding or subtracting the two equations will not eliminate a variable. However, suppose both sides of the first equation are multiplied by 2, and both sides of the second equation are multiplied by $^-3$. Then, the system can be solved by adding the equations.

$3x + 4y = 6$ *Multiply by 2* → $6x + 8y = 12$

$2x + 3y = 5$ *Multiply by $^-3$* → $^-6x - 9y = {}^-15$

Now, add to eliminate x. Then, solve for y.

$$\begin{aligned} 6x + 8y &= 12 \\ ^-6x - 9y &= {}^-15 \\ \hline ^-y &= {}^-3 \\ y &= 3 \end{aligned}$$

Add.

The variable x is eliminated.

Finally, substitute 3 for y in the first equation. Then, solve for x.

$$\begin{aligned} 3x + 4y &= 6 \\ 3x + 4(3) &= 6 \\ 3x + 12 &= 6 \\ 3x &= {}^-6 \\ x &= {}^-2 \end{aligned}$$

The solution is $(^-2, 3)$.

exercises

Exploratory State the coordinates of the point of intersection of each pair of lines.

1. a and b
2. a and c
3. a and d
4. b and d
5. b and c

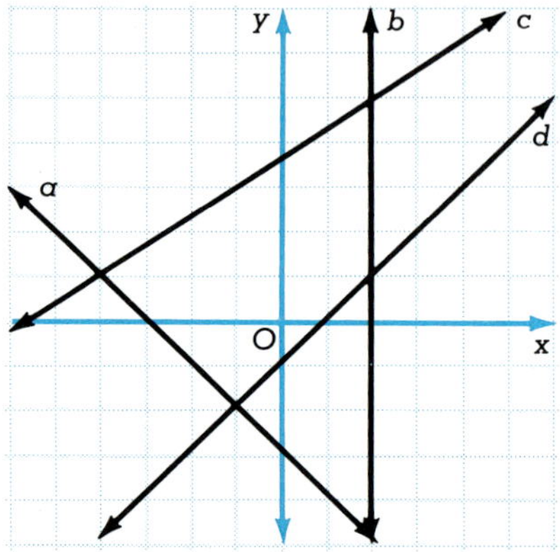

State the letter of the ordered pair that satisfies each equation.

6. $x + 3y = 6$	a. (0, 2)	b. $(^-1, 4)$	c. (6, 0)	d. $(^-3, 3)$
7. $2x - 5y = {}^-1$	a. (0, 5)	b. (2, 1)	c. $\left(-\frac{1}{2}, 0\right)$	d. $(^-2, {}^-1)$
8. $3x = 15$	a. (5, 1)	b. (5, 0)	c. (0, 5)	d. (5, 8)
9. $x - 7y = 0$	a. (7, 1)	b. (7, 0)	c. $(^-7, 1)$	d. (1, 7)
10. $y = {}^-2x + 4$	a. (3, 2)	b. (1, 2)	c. (4, 1)	d. (0, 4)

Written Graph each pair of equations. Then, state the solution of each system of equations.

1. $x + y = 6$
   $x - y = 2$
2. $y = x - 1$
   $x + y = 11$
3. $3x - 2y = 10$
   $x + y = 0$
4. $x + 2y = 7$
   $y = 2x + 1$
5. $y = x + 3$
   $3y + x = 5$
6. $y = 4x$
   $x + y = 5$

Solve each system of equations by the substitution method. Show your work.

7. $r - s = {}^-5$
   $r + s = 25$
8. $x - y = 6$
   $x + y = 5$
9. $x + 2y = 5$
   $2x + y = 7$
10. $y = 3x$
    $x + 2y = {}^-21$
11. $3x + 4y = {}^-7$
    $2x + y = {}^-3$
12. $y = x - 1$
    $4x - y = 19$
13. $y = x - 4$
    $2x + y = 5$
14. $5s + 4t = 12$
    $3s - 4t = 4$
15. $3x + y = 5$
    $2x + 3y = 8$
16. $x = y + 10$
    $2y = x - 6$
17. $12 - 3y = {}^-4x$
    $40 + 4x = 10y$
18. $9x + y = 20$
    $3x + 3y = 12$
19. $y = 8$
    $7x = 1 - y$
20. $2x - 5y = {}^-1$
    $5x + 3y = 13$
21. $y = {}^-2x + 4$
    $x = {}^-2y + 4$

22–36. Solve each system of equations in problems **7–21** by using the elimination method.

12–6 Circles

Circles are used to locate points on a radar scope.

The distance formula can be used to develop an equation for the graph of the circle.

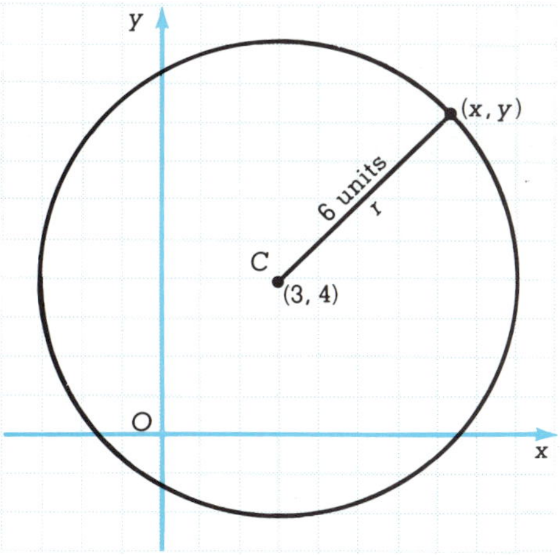

The circle at the left has its center (3, 4). The length of the radius is 6 units. Suppose (x, y) represents any point on the circle. This point must be 6 units from the center.

The distance between (x, y) and (3, 4) is 6 units.

$$\sqrt{(x-3)^2+(y-4)^2}=6$$

$$(x-3)^2+(y-4)^2=6^2 \quad \textit{Square both sides.}$$

$$(x-3)^2+(y-4)^2=36$$

The equation of a circle with center (3, 4) and radius measuring 6 units is $(x-3)^2+(y-4)^2=36$.

The equation of a circle can be written in the form given below.

The equation of a circle with center (h, k) and radius measuring r units is $(x-h)^2+(y-k)^2=r^2$. Theorem 12–7

examples

1 **Find the equation of a circle with center (0, 0) and a radius with a length of 4 units.**

The distance between (0, 0) and a point on the circle (x, y) is 4 units.

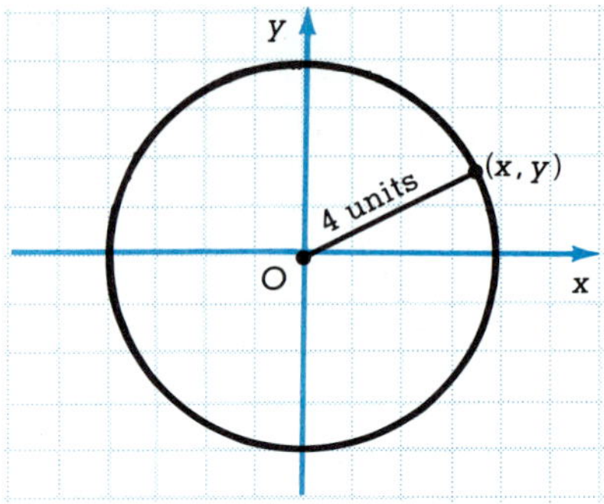

$$\sqrt{(x-0)^2+(y-0)^2} = 4$$
$$x^2+y^2 = 4^2$$
$$x^2+y^2 = 16$$

The equation of the circle is $x^2+y^2 = 16$.

2 **Graph the circle with equation $(x+1)^2+(y-4)^2 = 25$.**

First, rewrite the equation in the form $(x-h)^2+(y-k)^2 = r^2$.

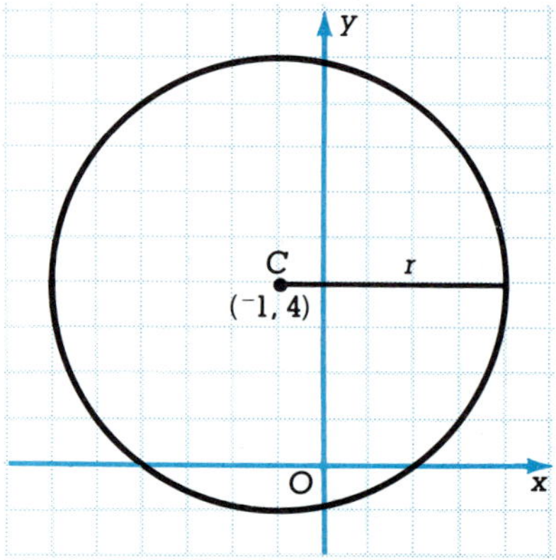

$$(x+1)^2+(y-4)^2 = 25$$
$$(x-{}^-1)^2+(y-4)^2 = 5^2$$

Therefore, $h = {}^-1$, $k = 4$, and $r = 5$.
The center is $({}^-1, 4)$.
The radius is 5 units long.

exercises

Exploratory Name the center and the measure of the radius for each circle whose equation is given below.

1. $x^2+y^2 = 16$
2. $(x+2)^2+(y+7)^2 = 81$
3. $x^2+y^2-25 = 0$
4. $\left(x+\frac{1}{2}\right)^2+\left(y+\frac{1}{3}\right)^2 = \frac{16}{25}$
5. $(x-4)^2+(y-6)^2 = 9$
6. $(x-2)^2+(y-5)^2 = 49$
7. $(x-3)^2+(y-12)^2-36 = 0$
8. $(x+5)^2+(y-7)^2 = 100$
9. $(x+8)^2+(y-9)^2 = 81$
10. $(x+4)^2+y^2-121 = 0$

Written For each of the following, state the equation of the circle. The center and measure of the radius are given.

1. (0, 0), 5
2. (3, 4), 6
3. $(^-2, 8)$, $\sqrt{2}$
4. (0, 0), 7
5. $(^-1, ^-1)$, $\frac{1}{4}$
6. $(^-5, 9)$, $\sqrt{20}$

Find the center and the measure of the radius for each circle whose equation is given below.

7. $x^2 + y^2 = 121$
8. $(x - 9)^2 + (y - 10)^2 = 1$
9. $x^2 + (y - 3)^2 = 4$
10. $0 = {^-y^2} - x^2 + 10$
11. $(x + 3)^2 + (y - 4)^2 = 20$
12. $(x + 7)^2 + (y + 3)^2 = 50$
13. $(x - 7)^2 + (y + 5)^2 = 4$
14. $x^2 + y^2 = 64$
15. $y^2 = 16 - x^2$
16. $(x + 1)^2 + y^2 = 11$
17. $(x + 7)^2 + (y + 3)^2 = 3$
18. $x^2 + (y - 3)^2 = 25$
19. $(x + 4)^2 + \left(y - \frac{1}{2}\right)^2 = 6$
20. $(x - 4)^2 + y^2 = \frac{16}{25}$
21. $(x + 5)^2 + (y - 2)^2 = \frac{3}{4}$
22. $x^2 + (y + 5)^2 = \frac{81}{64}$

For each of the following, write the equation of the circle. The center and measure of the radius are given.

23. (1, 2), 3
24. $(^-6, ^-10)$, 25
25. (0, 0), $\sqrt{14}$
26. (0, 0), 0.5
27. $(^-4, 3)$, $\frac{4}{3}$
28. (6, 0), 12
29. $\left(3, \frac{1}{2}\right)$, $\frac{4}{5}$
30. $\left(-\frac{3}{4}, 6\right)$, $\sqrt{18}$

Solve each problem.

31. Find the equation of the circle that has a diameter whose endpoints are (2, 7) and $(^-6, 15)$.
32. Find the equation of the circle that has a diameter of 12 units and a center at $(^-4, ^-7)$.

Contest Problem

Excursions in Geometry

The following problem appeared in a high school mathematics contest sponsored by the Mathematical Association of America.

Three circles are drawn on the same side of a line. One circle has a radius of 4 inches and is tangent to the line. The other two circles are congruent and each is tangent to the line and to the other circles. Find the radius of the congruent circles.

12–7 Coordinate Proofs

Theorems can be proved using coordinate geometry.

example 1

Prove that the diagonals of a square are perpendicular.

Given: Square $PQRS$

Prove: $\overline{PR} \perp \overline{QS}$

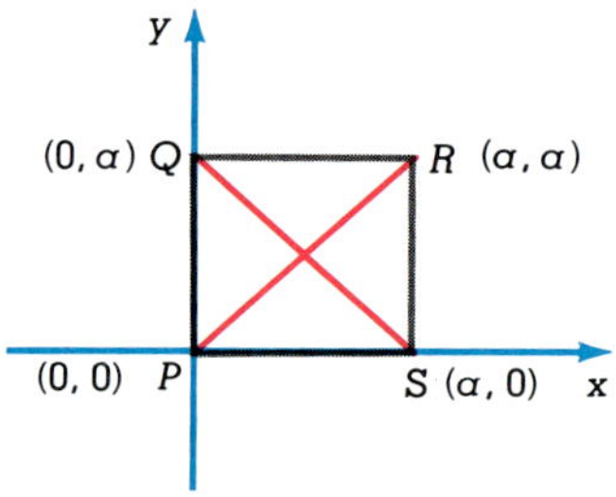

Proof:

By Theorem 12–4, the product of the slopes of perpendicular lines is $^-1$.

First, find the slope of $\overline{PR}$ and of $\overline{QS}$.

$$\text{slope of } \overline{PR} = \frac{a - 0}{a - 0}$$

$$= \frac{a}{a} \text{ or } 1$$

$$\text{slope of } \overline{QS} = \frac{0 - a}{a - 0}$$

$$= \frac{^-a}{a} \text{ or } ^-1$$

Then, find the product of the slopes.

$$\text{slope of } \overline{PR} \cdot \text{slope of } \overline{QS} = 1 \cdot {}^-1 \text{ or } ^-1$$

Therefore, the diagonals are perpendicular.

Placing the given geometric figure in a coordinate plane is an important part of planning a coordinate proof. In most cases, the following suggestions will help you place figures for your proofs.

Placing Figures on a Coordinate Plane.

1. Use the origin as a vertex or center.
2. Place at least one side of a polygon on a coordinate axis.
3. Keep the figure within the first quadrant.

example 2 **Position and label a right triangle on the coordinate plane.**

Use the origin as the vertex of the right angle. Place the legs of the triangle on the positive axes.

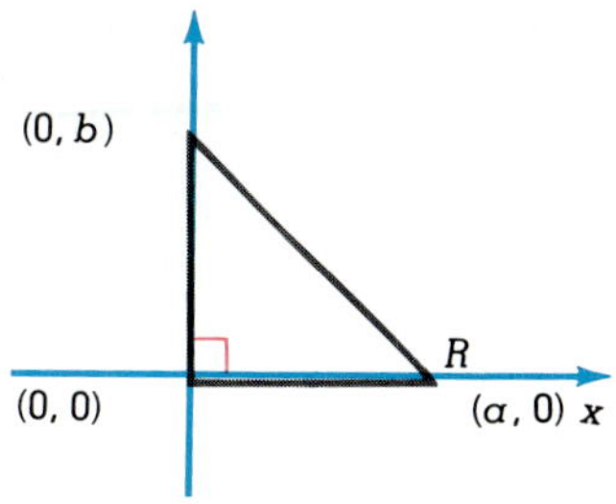

Label the vertices P, Q, and R. Since Q is on the y-axis, its x-coordinate is 0. Since R is on the x-axis, its y-coordinate is 0. Since P is at the origin, both coordinates are 0.

The most common ways of positioning and labeling several figures on the coordinate plane are shown below.

isosceles triangle

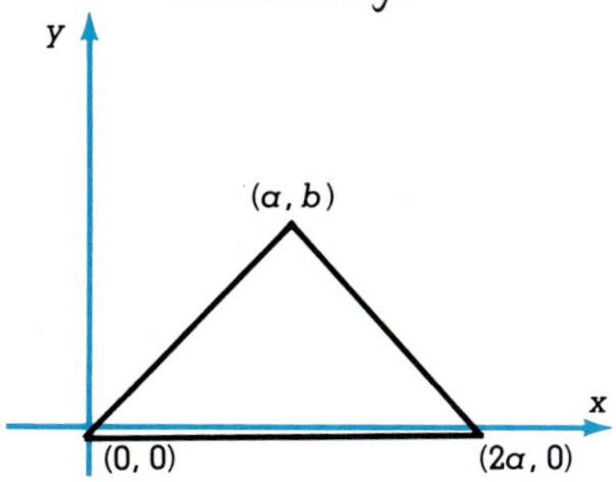

isosceles triangle

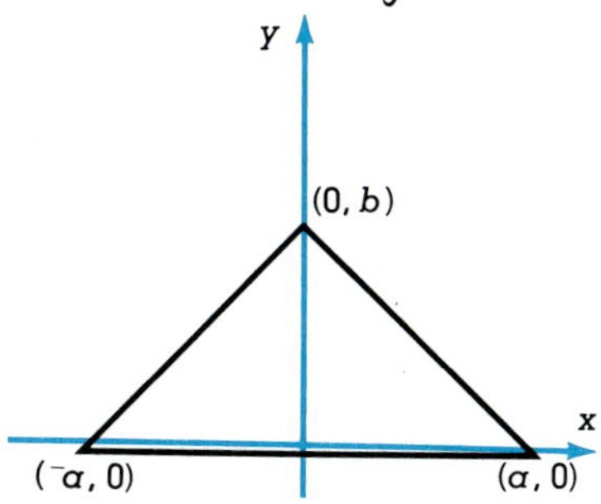

right triangle

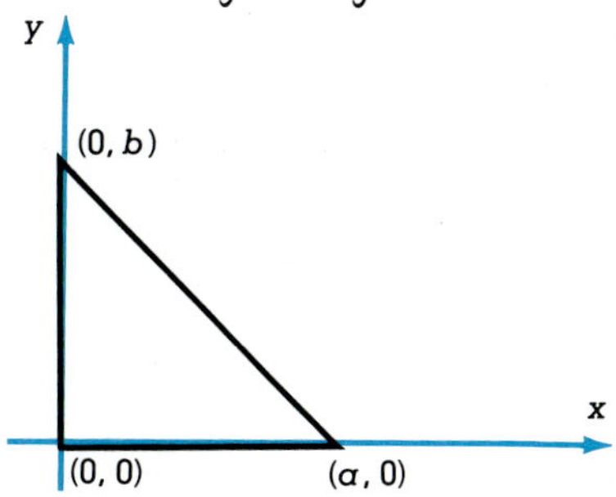

rectangle

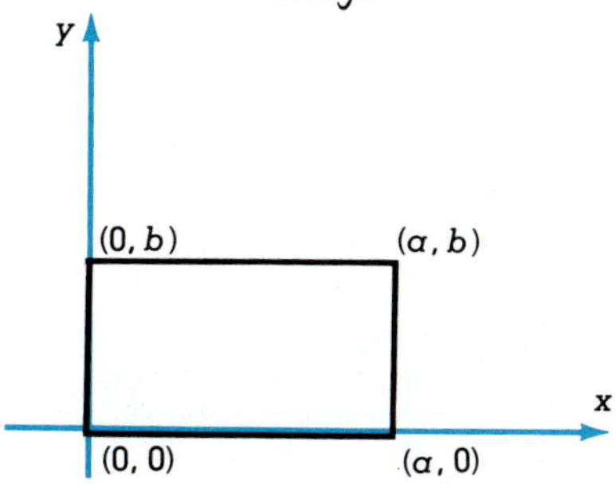

parallelogram

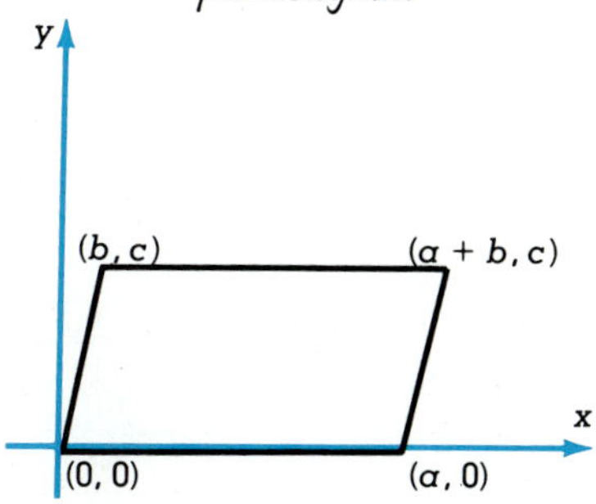

circle

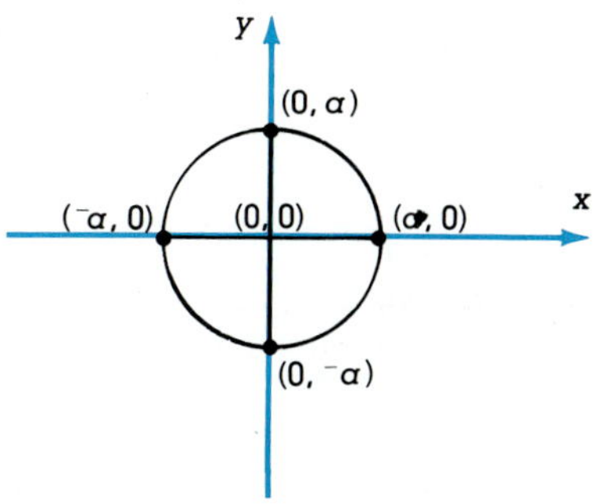

example 3

Prove that the midpoint of the hypotenuse of a right triangle is equidistant from the vertices.

Given: $\angle QPR$ is a right angle
M is the midpoint of $\overline{QR}$

Prove: M is equidistant from Q, P, and R

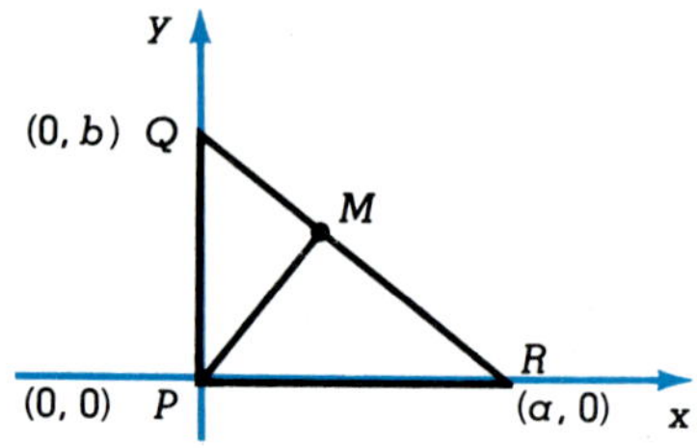

Proof:

By the Midpoint Formula, the coordinates of M are $\left(\frac{a}{2}, \frac{b}{2}\right)$.

Use the Distance Formula to find MR and PM.

$$MR = \sqrt{\left(\frac{a}{2} - a\right)^2 + \left(\frac{b}{2} - 0\right)^2} \qquad PM = \sqrt{\left(\frac{a}{2} - 0\right)^2 + \left(\frac{b}{2} - 0\right)^2}$$

$$= \sqrt{\left(\frac{^-a}{2}\right)^2 + \left(\frac{b}{2}\right)^2} \qquad = \sqrt{\left(\frac{a}{2}\right)^2 + \left(\frac{b}{2}\right)^2}$$

$$= \sqrt{\frac{a^2 + b^2}{4}} \qquad = \sqrt{\frac{a^2 + b^2}{4}}$$

Therefore, $MR = PM$. Also, by the definition of midpoint, $QM = MR$. Thus, $QM = MR = PM$ and M is equidistant from Q, P, and R.

exercises

Exploratory For each of the following, name the missing coordinates without introducing new letters.

1.

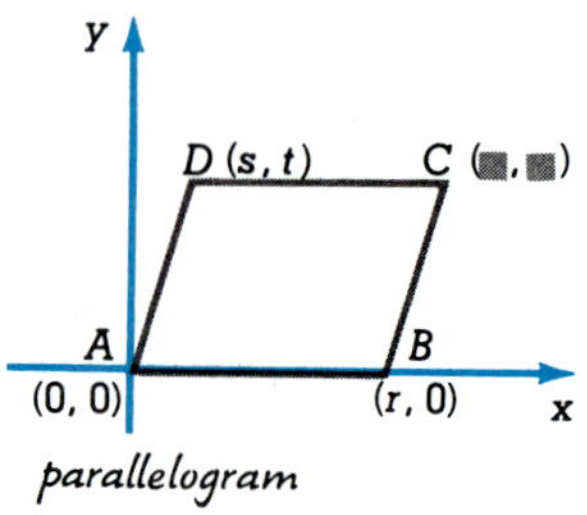

2.

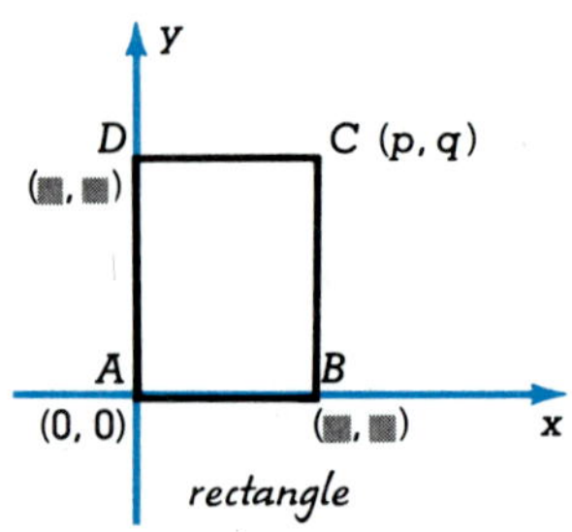

3.

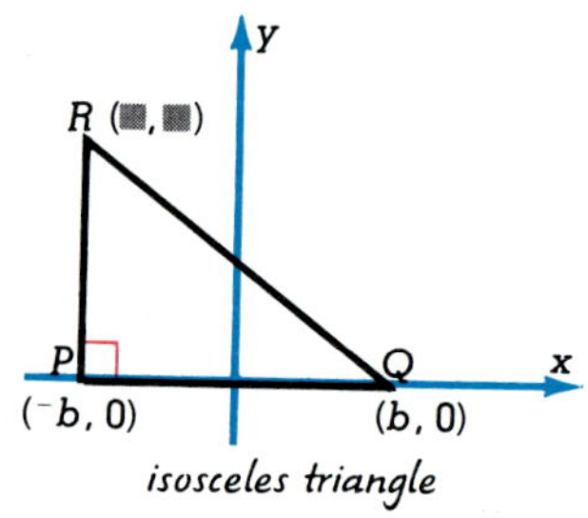

4.

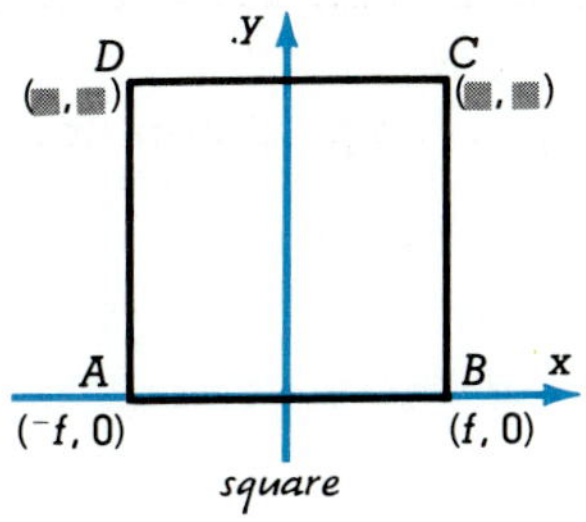

5.

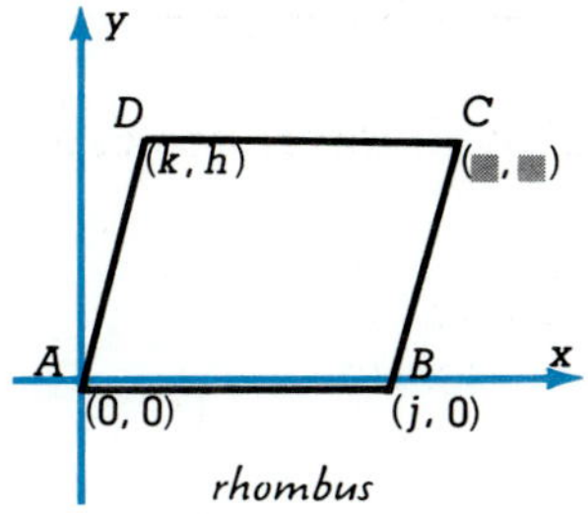

6.

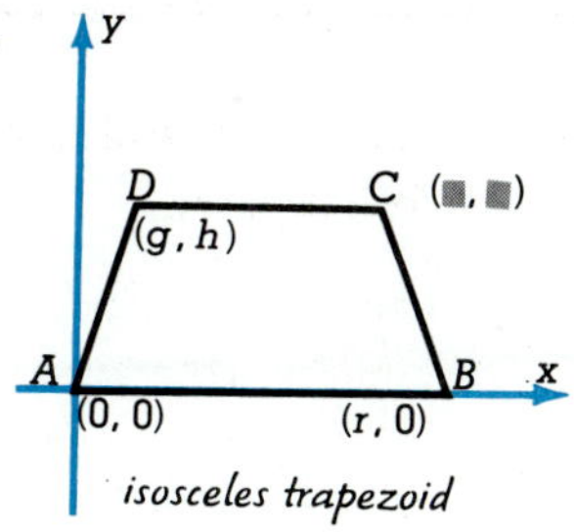

Written For each of the following, name the missing coordinates without introducing new variables.

1.

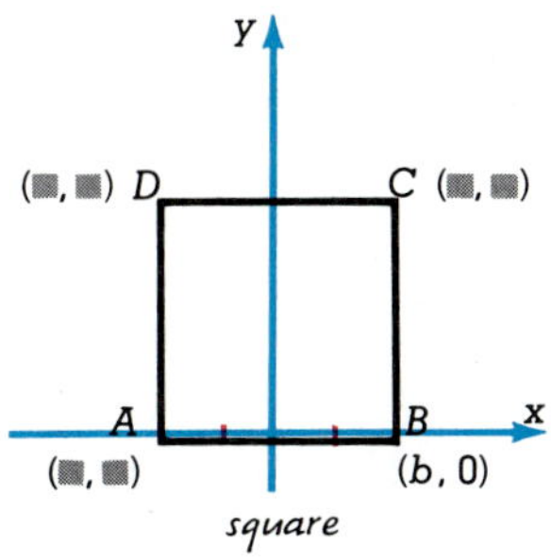

2.

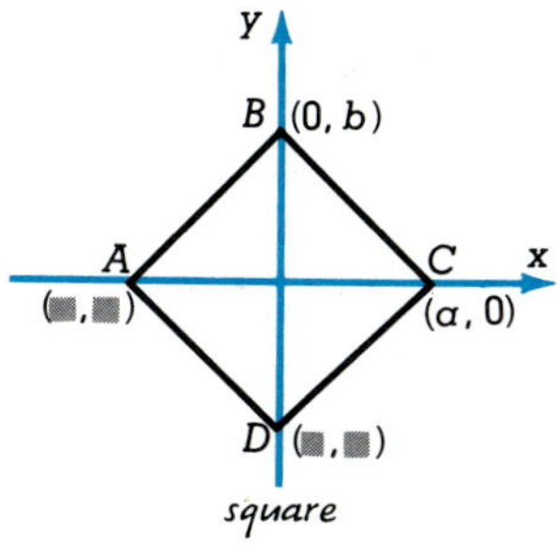

3.

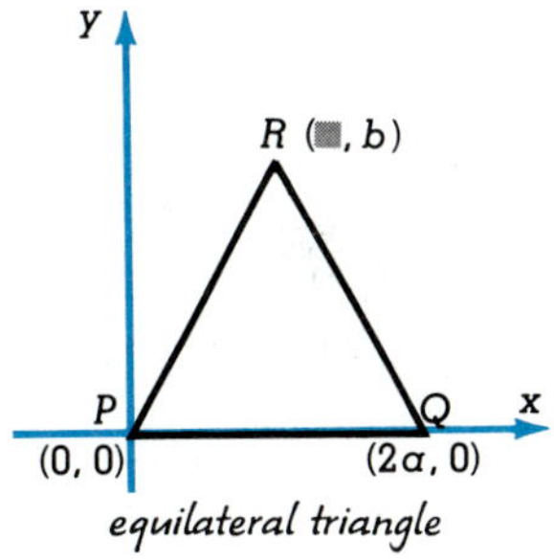

4.

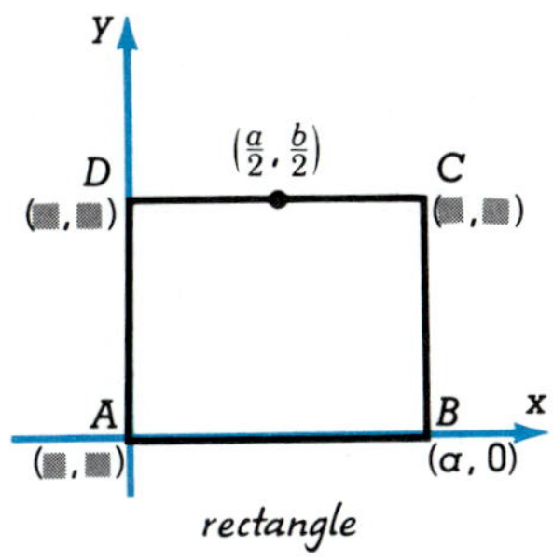

5.

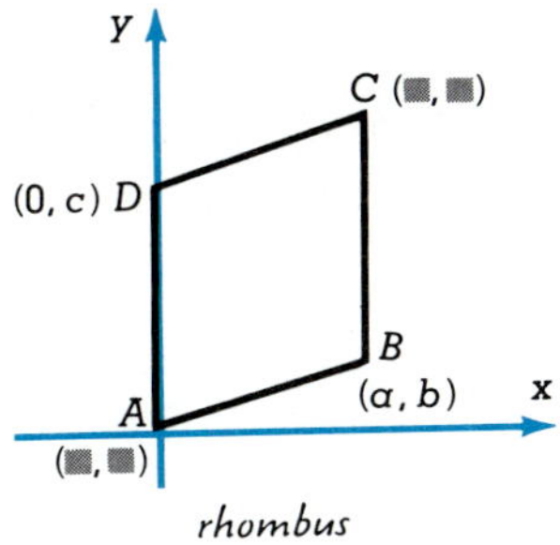

6.

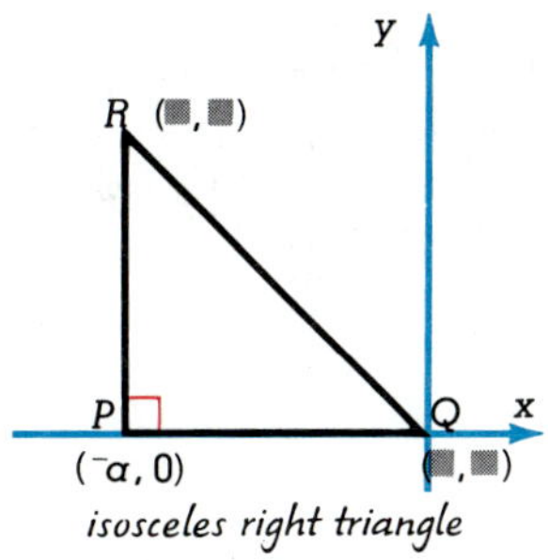

For each of the following theorems name the *given*, *prove*, and draw a diagram you would use in a formal proof.

7. The diagonals of a rectangle are congruent.
8. The diagonals of a rhombus are perpendicular.
9. The diagonals of an isosceles trapezoid are congruent.
10. The diagonals of a parallelogram bisect each other.
11. Opposite sides of a parallelogram are congruent.
12. Segments joining the midpoints of consecutive sides of a rectangle form a rhombus.

13–18. Prove each of the theorems in problems 7–12.

12–8 Coordinates in Space

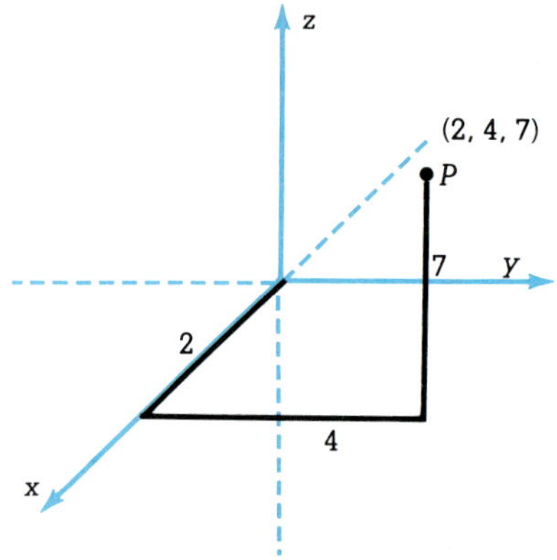

2 *is the* x*-coordinate.*
4 *is the* y*-coordinate.*
7 *is the* z*-coordinate.*

The two dimensional coordinate system can be extended to a three dimensional coordinate system in space. The x-axis, the y-axis, and the z-axis are perpendicular to each other. A point in space is represented by an ordered triple of real numbers, (x, y, z). In the figure at the left, the ordered triple (2, 4, 7) represents P.

The Pythagorean Theorem can be used to find the distance between two points in space.

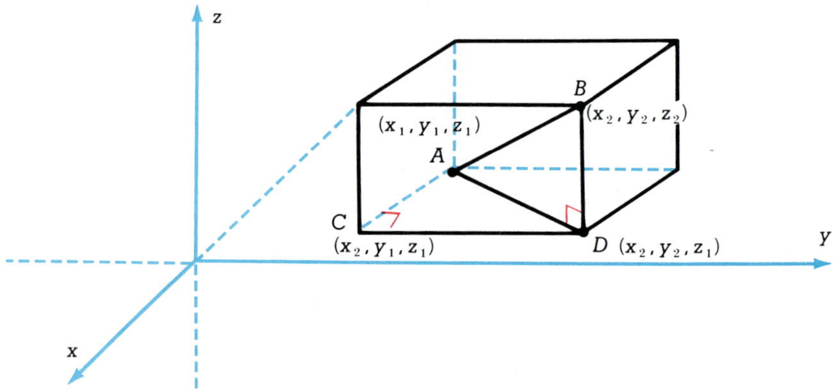

The distance between points A and B on $\triangle ABD$ is found as follows.

$AB^2 = AD^2 + BD^2$	*Pythagorean Theorem*
$= (AC^2 + CD^2) + BD^2$	*Substitute* $AC^2 + CD^2$ *for* AD^2.
$AB = \sqrt{AC^2 + CD^2 + BD^2}$	*Find the square root of both sides.*

But, $AC = (x_2 - x_1)$, $CD = (y_2 - y_1)$, and $BD = (z_2 - z_1)$.

$AB = \sqrt{(x_2 - x_1)^2 + (y_2 - y_1)^2 + (z_2 - z_1)^2}$ *Substitution.*

Theorem 12–8

This formula is an extension of the Distance Formula in the two-dimensional coordinate system.

Given two points $A(x_1, y_1, z_1)$ and $B(x_2, y_2, z_2)$ in space, the distance between A and B is given by the following equation.

$$AB = \sqrt{(x_2 - x_1)^2 + (y_2 - y_1)^2 + (z_2 - z_1)^2}$$

example 1

Find the distance between $P(6, {}^-1, 3)$ and $Q(2, 3, 5)$.

Let $(6, {}^-1, 3)$ be (x_1, y_1, z_1) and $(2, 3, 5)$ be (x_2, y_2, z_2).

$PQ = \sqrt{(x_2 - x_1)^2 + (y_2 - y_1)^2 + (z_2 - z_1)^2}$
$PQ = \sqrt{(6 - 2)^2 + ({}^-1 - 3)^2 + (3 - 5)^2}$
$PQ = \sqrt{4^2 + ({}^-4)^2 + ({}^-2)^2}$
$PQ = \sqrt{16 + 16 + 4}$
$PQ = \sqrt{36}$
$PQ = 6$

Suppose M in the figure at the right, is the midpoint of $\overline{PQ}$, a line segment in space. The midpoint is represented by the following.

$$\left(\frac{x_1 + x_2}{2}, \frac{y_1 + y_2}{2}, \frac{z_1 + z_2}{2}\right)$$

This formula is an extension of the Midpoint Formula in the two-dimensional coordinate system.

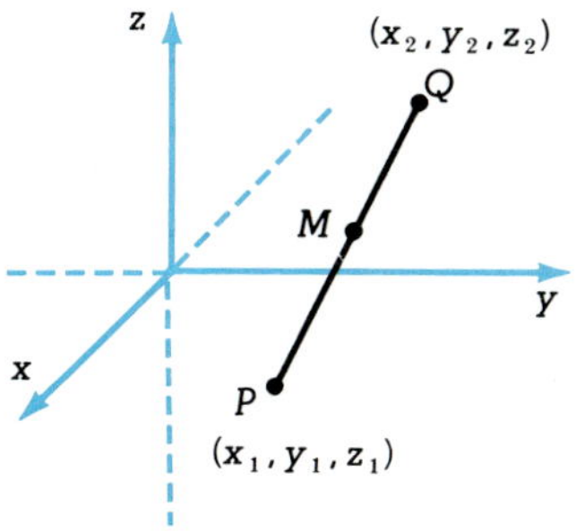

example 2

Find the coordinates of the midpoint of a line segment in space whose endpoints have coordinates $P(3, {}^-7, 0)$ and $Q(5, 1, 7)$.

Let $(3, {}^-7, 0)$ be (x_1, y_1, z_1) and $(5, 1, 7)$ be (x_2, y_2, z_2).

$$\left(\frac{x_1 + x_2}{2}, \frac{y_1 + y_2}{2}, \frac{z_1 + z_2}{2}\right)$$
$$\left(\frac{3 + 5}{2}, \frac{{}^-7 + 1}{2}, \frac{0 + 7}{2}\right)$$
$$\left(\frac{8}{2}, \frac{{}^-6}{2}, \frac{7}{2}\right)$$
$$\left(4, {}^-3, \frac{7}{2}\right)$$

The midpoint of $\overline{PQ}$ is represented by $\left(4, {}^-3, \frac{7}{2}\right)$

The formula for the equation of a sphere is an extension of the formula for the equation of a circle.

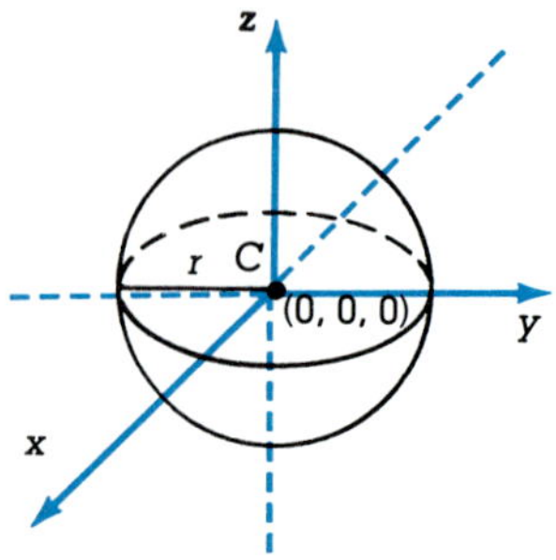

The equation of a sphere whose center is at (0, 0, 0) and has a radius measuring r units is as follows.

$$x^2 + y^2 + z^2 = r^2$$

The equation of a sphere whose center is at (i, j, k) and has a radius measuring r units is as follows.

$$(x - i)^2 + (y - j)^2 + (z - k)^2 = r^2$$

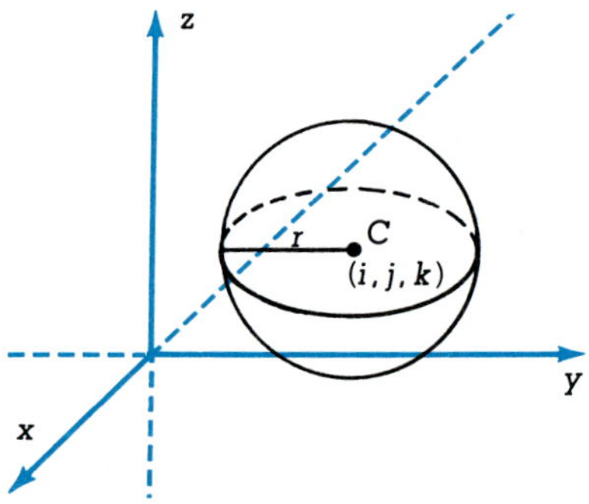

example 3

Write the equation of a sphere whose center is at (3, $^-2$, 4) and has a radius measuring 6 units.

Let $i = 3$, $j = {}^-2$, $k = 4$, and $r = 6$.

$(x - i)^2 + (y - j)^2 + (z - k)^2 = r^2$
$(x - 3)^2 + (y - {}^-2) + (z - 4)^2 = 6^2$
$(x - 3)^2 + (y + 2)^2 + (z - 4)^2 = 36$ *The equation of the sphere*

exercises

Exploratory **Name the distance between the following pairs of points.**

1. $A(0, 0, 0)$, $B(0, 4, 0)$
2. $P(0, 0, 0)$, $Q(3, 4, 0)$
3. $R(0, 0, 0)$, $S(2, 1, 1)$
4. $C(2, 4, 5)$, $D(2, 4, 7)$

Name the midpoint of the line segment whose endpoints are given.

5. $A(1, 3, {}^-2)$, $B(7, {}^-3, 2)$
6. $C({}^-5, 4, {}^-2)$, $D(5, {}^-4, 2)$
7. $P(0, 0, 6)$, $Q({}^-1, 8, 10)$
8. $R(14, {}^-10, {}^-8)$, $S({}^-4, 2, 6)$

Name the center and the measure of the radius for each sphere whose equation is given below.

9. $(x - 3)^2 + (y - 8)^2 + (z - 2)^2 = 49$
10. $x^2 + y^2 + z^2 = 7$
11. $(x + 4)^2 + (y - 2)^2 + (z + 1)^2 = 25$
12. $x^2 + y^2 + (z - 2)^2 = 2$

Written Name the distance between the following pairs of points.

1. $A(0, 0, 0)$, $B(3, 0, 2)$
2. $S(0, 4, 0)$, $T(2, 0, 0)$
3. $C(0, 0, 0)$, $D(0, {}^-4, 2)$
4. $L(1, 2, 3)$, $M(0, 0, 0)$
5. $P(3, 4, \sqrt{11})$, $Q(0, 0, 0)$
6. $E(3, 7, {}^-1)$, $F(4, 8, 9)$
7. $R({}^-1, {}^-4, 3)$, $S(2, {}^-5, 1)$
8. $H({}^-2, {}^-4, {}^-3)$, $J({}^-3, {}^-4, {}^-2)$
9. $M(2, 2, 2)$, $N({}^-5, 1, 7)$
10. $P(6, 1, 3)$, $Q(10, 8, 6)$

Name the midpoint of the line segment whose endpoints are given.

11. $A(0, {}^-4, 2)$, $B(3, 0, 2)$
12. $S({}^-6, 3, {}^-1)$, $T(6, 3, 1)$
13. $C(2, {}^-5, 1)$, $D(3, 2, 4)$
14. $R(2, 7, 4)$, $S(0, 8, {}^-4)$
15. $P({}^-1, 8, 10)$, $Q(0, 0, 0)$
16. $H(4, {}^-4, 2)$, $J(8, {}^-4, 6)$

Name the center and the measure of the radius for each sphere whose equation is given below.

17. $(x - 6)^2 + (y - 5)^2 + (z + 1)^2 = 81$
18. $(x + 1)^2 + (y - 8)^2 + z^2 = 11$
19. $x^2 + (y - 3)^2 + z^2 = 4$
20. $(x + 2)^2 + (y + 3)^2 + (z - 2)^2 = 100$

For each of the following, write the equation of the sphere. The center and the measure of the radius are given.

21. $(0, 0, 0)$, 3
22. $(0, 3, 1)$, 1
23. $({}^-1, 2, 4)$, 4
24. $(6, {}^-2, 3)$, 12
25. $\left(2, \frac{1}{2}, 1\right)$, $\frac{1}{3}$
26. $\left({}^-5, 0, \frac{2}{3}\right)$, $\frac{3}{5}$

Solve each of the following.

27. Find the perimeter of a triangle whose vertices are $A(6, 4, 1)$, $B(4, 6, 0)$, and $C(3, {}^-2, 3)$.
28. Find the perimeter of a triangle whose vertices are $P(0, 0, 0)$, $Q(3, 4, \sqrt{11})$ and $R(0, 5, 0)$.

The diameter of a sphere has endpoints $A({}^-3, 5, 7)$ and $B(5, {}^-1, 5)$.

29. Find the center.
30. Find the measure of the radius.
31. Write the equation of the sphere.
32. Sketch the graph of the sphere.

Solve the following.

33. Find the distance between $A(7, 5, 6)$ and $B(3, 1, 4)$ using the Pythagorean Theorem. Show your work. (Hint: See page 396 of lesson 12–8.)

Vocabulary

ordered pair (368)
quadrant (368)
axis (368)
origin (368)
coordinate (368)
linear equation (369)
distance formula (373)
midpoint formula (374)
slope (377)
slope-intercept form (382)
point-slope form (383)
system of equations (386)
equation of a circle (389)
coordinate system (396)
equation of a sphere (398)

Chapter Summary

1. Postulate 12–1: Each point in a coordinate plane corresponds to exactly one ordered pair of numbers. Each ordered pair of numbers corresponds to exactly one point in a coordinate plane. (368)
2. Definition of Linear Equation: An equation is linear if and only if it can be written in the form $Ax + By = C$, where A, B, and C are any numbers and A and B are not both 0. (369)
3. Theorem 12–1, Distance formula: The distance between two points with coordinates (x_1, y_1) and (x_2, y_2) is given by the following formula: $d = \sqrt{(x_2 - x_1)^2 + (y_2 - y_1)^2}$ (373)
4. Theorem 12–2, Midpoint Formula: If the coordinates of A and B are (x_1, y_1) and (x_2, y_2) respectively, then the midpoint M, of $\overline{AB}$, has coordinates $\left(\frac{x_1 + x_2}{2}, \frac{y_1 + y_2}{2}\right)$. (374)
5. Slope $= \dfrac{\text{difference of the } y\text{-coordinates}}{\text{difference of the corresponding } x\text{-coordinates}}$ (377)
6. If the line is vertical, then the slope is undefined; if the line is horizontal, the slope is zero. (378)
7. Theorem 12–3: Two lines have the same slope if and only if they are parallel and nonvertical. (379)
8. Theorem 12–4: Two nonvertical lines are perpendicular if and only if the product of their slopes is $^{-}1$. (379)
9. Theorem 12–5: The equation of a line having a slope m and a y-intercept b is $y = mx + b$. (382)
10. Theorem 12–6: The equation of a line passing through a point whose coordinates are (x_1, y_1) and having a slope, m, is $y - y_1 = m(x - x_1)$. This is called the point-slope form. (383)
11. A system of equations can be solved by 3 methods: graphing, the substitution method, and the elimination method. (386)

12. Theorem 12–7: The equation of a circle with center (h, k) and radius measuring r units is $(x - h)^2 + (y - k)^2 = r^2$. (389)
13. Placing figures in a coordinate plane:
    1. Use the origin as a vertex or center.
    2. Place at least one side of a polygon on a coordinate axis.
    3. Keep the figure within the first quadrant. (392)
14. In a three dimensional coordinate system in space, the x-axis, the y-axis, and the z-axis are perpendicular to each other. A point in space is represented by an ordered triple of real numbers, (x, y, z). (396)
15. Theorem 12–8: Given two points $A(x_1, y_1, z_1)$ and $B(x_2, y_2, z_2)$ in space, the distance between A and B is given by the following equation.
$$AB = \sqrt{(x_2 - x_1)^2 + (y_2 - y_1)^2 + (z_2 - z_1)^2} \quad (396)$$
16. Given two points $P(x_1, y_1, z_1)$ and $Q(x_2, y_2, z_2)$ in space, the midpoint of $\overline{PQ}$ is represented by the following.
$$\left(\frac{x_1 + x_2}{2}, \frac{y_1 + y_2}{2}, \frac{z_1 + z_2}{2}\right) \quad (397)$$
17. The equation of a sphere whose center is at (i, j, k) and has a radius measuring r units is as follows.
$$(x - i)^2 + (y - j)^2 + (z - k)^2 = r^2 \quad (398)$$

Chapter Review

12–1 Graph each of the following linear equations.

1. $x = 4$
2. $3x + y = 5$
3. $^-5x + 2y = 0$
4. $3x - 6y = 9$
5. $y = {}^-x + 1$
6. $2x = y - 1$

The following are the coordinates of three vertices of a rectangle. Graph them. Then, find each fourth vertex.

7. $(^-2, 2), (1, 2), (1, ^-4)$
8. $(0, 0), (1, ^-1), (^-3, ^-5)$
9. $(^-6, 4), (^-4, 6), (^-2, 4)$
10. $(1, 0), (5, 0), (5, 6)$

12–2 Find the distance between each of the following.

11. $(0, 3), (4, 0)$
12. $(^-3, ^-5), (3, 5)$
13. $(6, r), (4, r)$
14. $\left(\frac{2}{5}, \frac{3}{5}\right), \left(\frac{1}{10}, \frac{3}{10}\right)$
15. $(a, b), (b, a)$
16. $(0.1, 0.3), (0.2, 0.4)$

Find the coordinates of the midpoints of each line segment which has endpoints with the following coordinates.

17. $(^-4, 2), (^-4, 5)$
18. $(5, 0), (12, 0)$
19. $(^-2, ^-3), (^-4, ^-7)$
20. $(0, 0), (0, 5)$
21. $\left(\frac{1}{2}, \frac{1}{3}\right), \left(1, \frac{2}{3}\right)$
22. $(4, 6), (5, 7)$

23. The center of a circle has coordinates $(^-6, ^-8)$. One endpoint of a diameter has coordinates $(0, 0)$. Find the coordinates of the other endpoint of the diameter.

12–3 **Find the slope of the lines passing through the following pairs of points whose coordinates are listed below.**

24. $(5, 2), (^-6, 2)$
25. $(^-2, 1), (^-5, 3)$
26. $(3, 6), (9, ^-6)$
27. $(3, 8), (3, 6)$
28. $(2, 6), (5, 2)$
29. $(^-3, 10), (^-7, 2)$

Determine the value of r so that a line through the points with the given coordinates has the slope listed.

30. $(r, 6), (1, 5)$; slope $= \frac{1}{5}$
31. $(4, r), (^-2, 4)$; slope $= \frac{1}{2}$
32. $(5, 3), (8, r)$; slope $= 0$
33. $(r, ^-2), (5, r)$; slope $= 1$
34. Find the value of x if the points $A(x, 4)$, $B(0, 2)$, and $C(2, 0)$ are collinear.

12–4 **State the slope and y-intercept of the graphs of each of the following.**

35. $y = 4x + 6$
36. $y = ^-8$
37. $x = 7$
38. $4x - 6y = 8$

Write the equation of the line given the following information.

39. $m = 3$, y-intercept is $^-1$
40. $m = 0$, through a point at $\left(\frac{1}{2}, \frac{1}{3}\right)$
41. m is undefined; through a point at $(4, 6)$
42. $m = \frac{2}{3}$, y-intercept is $^-5$

43. Write the equation of a line passing through points whose coordinates are $(^-2, ^-6)$ and $(5, 0)$.

12–5 **Graph each pair of equations. Then, state the solution of each system of equations.**

44. $x + y = 4$
    $x - y = 2$
45. $y = x - 2$
    $2x + y = 13$

Solve each system of equations by the substitution method. Show your work.

46. $y = 2x - 1$
    $x + y = 7$
47. $3x - 4y = ^-1$
    $^-2x + y = ^-1$

Solve each system of equations by the elimination method.

48. $3x + y = 5$
    $2x + 3y = 8$
49. $4x + 5y = 29$
    $3x - 2y = 16$

12–6 Find the center and the measure of the radius for each circle whose equation is given below.

50. $x^2 + y^2 = 36$
51. $(x + 2)^2 + y^2 = 12$
52. $(x + 3)^2 + (y - 6)^2 = 16$
53. $x^2 + (y - 5)^2 = \frac{64}{121}$

For each of the following, write the equation of the circle. The center and measure of the radius are given.

54. (2, 3), 8
55. (0, 0), 0.3
56. $(^-7, 8)$, 5
57. $\left(-\frac{2}{3}, 5\right), \frac{6}{7}$

58. Find the equation of the circle that has a diameter of 12 units and a center at $(^-5, {^-6})$.

12–7 For each of the following, name the missing coordinates without introducing new letters.

59.

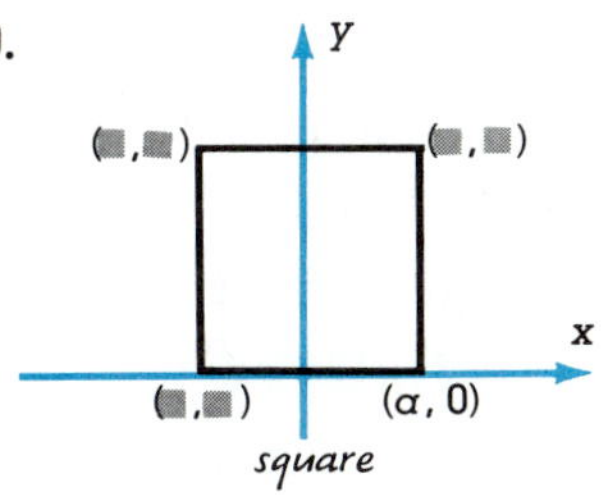

60.

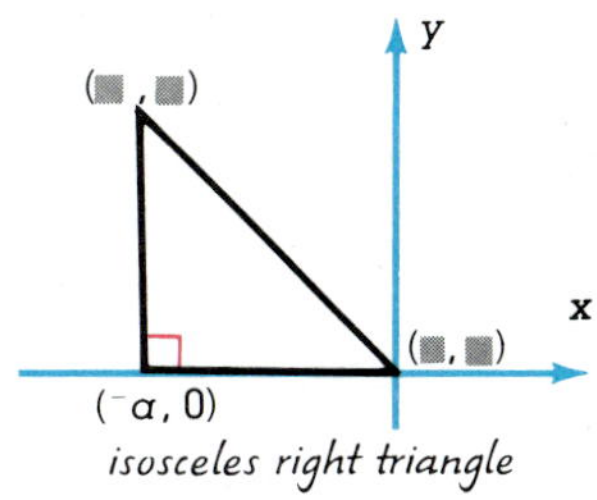

61. Name the *given* and the *prove* and then draw a diagram you would use to prove that the diagonals of a rhombus are perpendicular.

12–8 Name the distance between the following pairs of points.

62. (0, 0, 0), $(0, {^-3}, 7)$
63. (3, 3, 3), $(^-6, 0, 6)$
64. (0, 8, 0), (4, 0, 0)
65. (5, 0, 2), (9, 7, 5)

Name the midpoint of the line segment whose endpoints are given.

66. $(^-5, 2, 0)$, (5, 2, 0)
67. (1, 6, 4), $(0, 7, {^-3})$
68. $(3, {^-3}, 1)$, $(7, {^-3}, 5)$
69. (2, 4, 6), (0, 2, 4)

Name the center and the measure of the radius for each sphere whose equation is given below.

70. $(x - 4)^2 + (y - 1)^2 + (z + 2)^2 = 100$
71. $x^2 + y^2 + z^2 = 144$
72. $(x + 1)^2 + y^2 + (z - 5)^2 = 4$
73. $(x + 3)^2 + (y + 4)^2 + (z - 1)^2 = 13$

For each of the following, write the equation of the sphere. The center and the measure of the radius are given.

74. (0, 0, 0), 5
75. $(9, {^-6}, 4)$, 8
76. $(^-1, 2, {^-3})$, 4
77. $\left(^-2, 1, \frac{2}{5}\right), \frac{2}{7}$

Chapter Test

Graph each of the following linear equations.

1. $x + 2y = 6$
2. $x = {}^-3$

Find the distance between each of the following.

3. (2, 3), (6, 7)
4. $\left(\frac{2}{3}, \frac{4}{5}\right), \left(\frac{1}{3}, \frac{2}{5}\right)$

Find the coordinates of the midpoint of each line segment which has endpoints with the coordinates given below.

5. (0, 4), (6, 8)
6. ($^-5$, 7), ($^-5$, 9)

Find the slope of the lines passing through the following pairs of points whose coordinates are given below.

7. ($^-4$, 11), ($^-6$, 3)
8. (5, 2), (8, 2)
9. (3, 7), (3, $^-6$)

10. Determine the value of r so that a line through (r, 3) and (5, 9) has a slope of 2.
11. Write the equation of a line passing through points whose coordinates are ($^-2$, $^-6$) and (8, $^-2$).

Solve each system of equations. Show your work.

12. $y = 4x$
    $x + y = 5$
13. $3x - 2y = 10$
    $x + y = 0$
14. $3x + 4y = {}^-7$
    $2x + y = {}^-3$

Find the center and the measure of the radius for each circle whose equation is given below.

15. $x^2 + y^2 = 49$
16. $(x + 7)^2 + (y - 3)^2 = 3$

For each of the following, write the equation of the circle. The center and the measure of the radius are given.

17. (0, 0), 2
18. ($^-5$, $^-6$), $\sqrt{8}$
19. (2, $^-3$), 12

Solve each of the following problems.

20. Name the *given*, the *prove*, and draw a diagram you would use to prove that the diagonals of a rectangle are congruent.
21. Find the distance between (2, 4, 5) and (2, 4, 7).
22. Name the midpoint of the line segment whose endpoints are (0, $^-4$, 2) and (3, 0, 2).

Name the center and the measure of the radius for each sphere whose equation is given below.

23. $(x - 4)^2 + (y - 5)^2 + (z + 2)^2 = 81$
24. $x^2 + y^2 + z^2 = 7$

25. Write the equation of the sphere whose diameter has endpoints at ($^-3$, 5, 7) and (5, $^-1$, 5).

Chapter

13

Loci and Constructions

Among other instruments, artists use compasses and protractors to construct design layouts for engineering ideas. Their work involves the basic construction ideas included in this chapter.

13–1 Basic Constructions

It is possible to construct many geometric figures without making measurements. Two instruments are used, a **straightedge** and a **compass.**

A ruler may be used as a straightedge for constructions, as long as the measurement marks are not used.

A straightedge is like an unmarked ruler. It is used to draw lines and parts of lines. Two points lie on exactly one line. A straightedge is used to draw such a line.

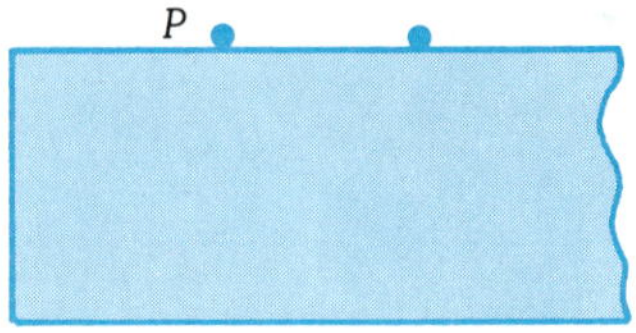

A compass is used to draw circles and arcs of circles. There is exactly one circle with a given center and radius. A compass is used to draw such a circle.

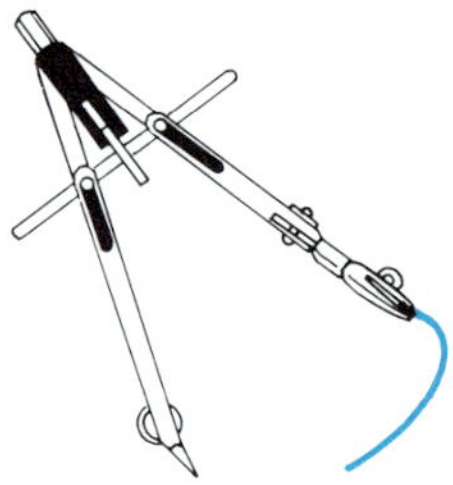

The following construction shows how to construct a line segment congruent to a given line segment. *Notice the parts of the presentation, given, construct, a diagram, method, and justification.*

construction

CONSTRUCTION 1: Construct a line segment congruent to a given line segment.

Given: $\overline{AB}$

Construct: $\overline{CD}$ so that $\overline{CD} \cong \overline{AB}$

Method:

1. Use a straightedge to draw a line. Call it ℓ.
2. Choose any point on ℓ. Label it C.
3. Place the compass point on A and the pencil on B. The compass radius now has measure AB.
4. With the same compass setting, place the compass point on C. Mark an arc intersecting ℓ. Label the point D. Then, $\overline{CD} \cong \overline{AB}$.

Justification: The radius for a circle with center C has the same measure as a circle with center A and radius $\overline{AB}$. Thus, $\overline{CD} \cong \overline{AB}$.

CONSTRUCTION 2: Construct an angle congruent to a given angle.

Given: $\angle ABC$

Construct: $\angle DEF$ so that $\angle DEF \cong \angle ABC$

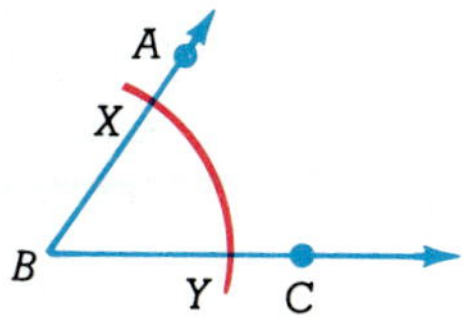

Method:

1. Draw $\overrightarrow{EF}$.
2. Using B as center, draw an arc that intersects both sides of $\angle ABC$. Call the intersection points X and Y.
3. Keep the compass at the same setting. Using E as a center, draw an arc intersecting $\overrightarrow{EF}$. Call the intersection R.
4. Using R as a center and radius setting XY, draw an arc that intersects the arc in Step 3. Call the intersection D.
5. Draw $\overrightarrow{ED}$. Then, $\angle DEF \cong \angle ABC$.

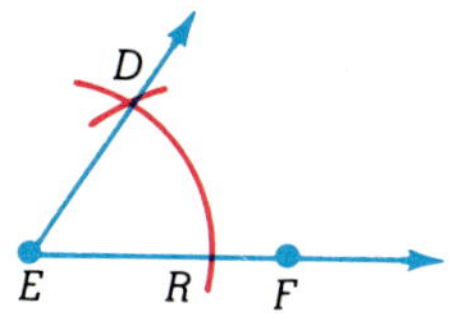

Justification: If $\overline{DR}$ and $\overline{XY}$ are drawn, then $\overline{DR} \cong \overline{XY}$. Also, by construction, $\overline{ED} \cong \overline{BX}$ and $\overline{ER} \cong \overline{BY}$. Thus, $\triangle DER \cong \triangle XBY$ by SSS. Then, $\angle DEF \cong \angle ABC$ by the definition of congruent triangles. *CPCTC*

CONSTRUCTION 3: Construct the bisector of a given angle.

Given: $\angle ABC$

Construct: $\overrightarrow{BD}$ so that $\overrightarrow{BD}$ bisects $\angle ABC$

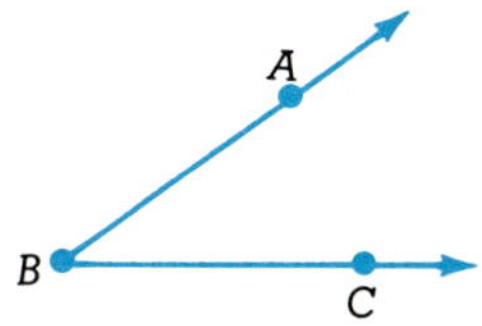

Method:

1. Using B as a center, draw an arc that intersects both sides of $\angle ABC$. Call the intersection points X and Y.
2. Keep the same compass setting. Using X as a center, draw an arc in the interior of $\angle ABC$.
3. Keep the same compass setting. Using Y as a center, draw an arc that intersects the arc in Step 2. Call the intersection D.
4. Draw $\overrightarrow{BD}$. Then, $\overrightarrow{BD}$ bisects $\angle ABC$.

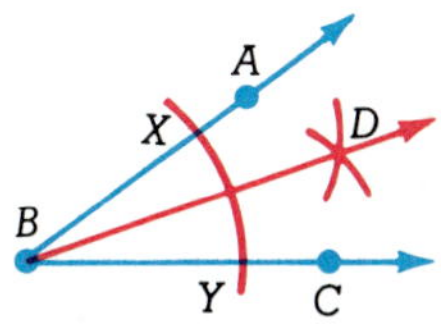

Justification: Draw $\overline{XD}$ and $\overline{YD}$. By construction, $\overline{XD} \cong \overline{YD}$ and $\overline{BX} \cong \overline{BY}$. Also, $\overline{BD} \cong \overline{BD}$. Thus, $\triangle BXD \cong \triangle BYD$ by SSS. By the definition of congruent triangles, $\angle XBD \cong \angle YBD$. Thus $\overrightarrow{BD}$ bisects $\angle ABC$.

exercises

Exploratory For each of the following, state the *given*, *construct*, and draw a diagram.

1. Construct an equilateral triangle given one side.
2. Construct a triangle congruent to a given triangle.
3. Construct a segment whose measure is twice the measure of a given segment.
4. Construct a segment whose measure is 3 times the measure of a given segment.
5. Construct a segment whose measure is the sum of the measures of two given segments.
6. Construct an angle whose measure is twice the measure of a given angle.
7. Construct an angle whose measure is one-half the measure of a given angle.
8. Construct an angle whose measure is the sum of the measures of two given angles.

Written On a piece of paper, draw and label two segments like the segments shown below. Then, construct segments for each of the following measures.

1. $AB + CD$
2. $CD - AB$
3. $2AB$
4. $3CD$
5. $4AB - CD$
6. $2AB - CD$
7. $3(CD - AB)$
8. $2(AB + CD)$

On a piece of paper, draw and label two angles like the angles shown below. Then, construct angles for each of the following degree measures.

9. $x + y$
10. $x - y$
11. $\frac{1}{2}x$
12. $\frac{1}{2}y$
13. $\frac{1}{2}(x + y)$
14. $\frac{1}{2}x + y$
15. $\frac{3}{4}x$
16. $x + \frac{1}{2}y$

Solve each problem.

17. Draw an acute triangle. Then, construct a triangle congruent to the triangle drawn.
18. Draw an obtuse triangle. Then, construct a triangle congruent to the triangle drawn.
19. Draw a line segment. Then, construct an equilateral triangle with the segment as one of its sides.
20. Construct a 60° angle.
21. Construct a 30° angle.
22. Construct a 15° angle.
23. Construct a 75° angle.
24. Construct a 45° angle.
25. Draw an angle and two line segments. Then, construct a triangle whose sides and an included angle are congruent to the figures drawn.
26. Draw two angles and a line segment. Then, construct a triangle whose angles and included side are congruent to the figures drawn.
27. Draw three angles with a total degree measure of 180 and a line segment. Then, inscribe a triangle with angles congruent to these angles in a circle with radius congruent to the segment.

13–2 Perpendiculars and Parallels

Notice how the following constructions are alike.

CONSTRUCTION 4: Construct a line perpendicular to a given line and through a given point *not* on the line.

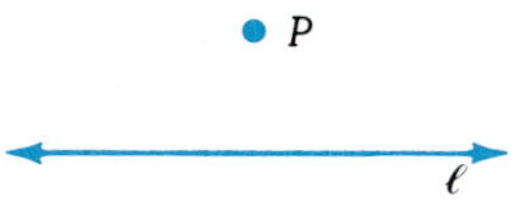

Given: P is *not* on ℓ

Construct: $\overleftrightarrow{PQ}$ so that $\overleftrightarrow{PQ} \perp \ell$

Method:

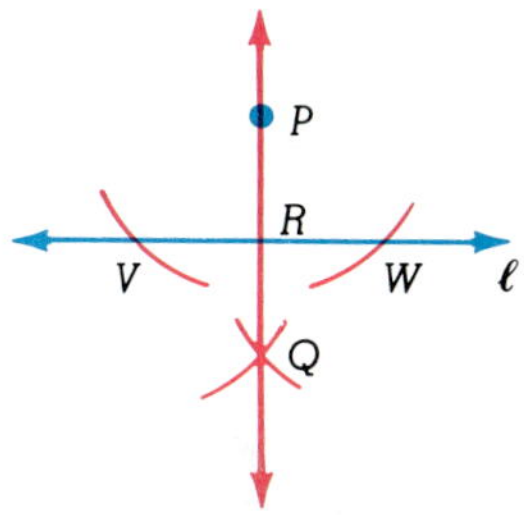

1. Using P as a center, draw two arcs that intersect ℓ. Call the intersection points V and W.
2. Using V as a center, and then W as a center, draw arcs that intersect at a point other than P. Call the intersection Q.
3. Draw $\overleftrightarrow{PQ}$. Then, $\overleftrightarrow{PQ} \perp \ell$.

Justification: Draw $\overline{PV}$, $\overline{PW}$, $\overline{VQ}$, and $\overline{WQ}$. By construction, $\overline{PV} \cong \overline{PW}$ and $\overline{VQ} \cong \overline{WQ}$. Also, $\overline{PQ} \cong \overline{PQ}$. Thus, $\triangle PVQ \cong \triangle PWQ$. By the definition of congruent triangles, $\angle VPR \cong \angle WPR$. Also, $\overline{PR} \cong \overline{PR}$, so $\triangle VPR \cong \triangle WPR$ by SAS. The linear pair, $\angle PRV$ and $\angle PRW$ are right angles because they are congruent, supplementary angles. Thus, $\overleftrightarrow{PQ} \perp \ell$.

CONSTRUCTION 5: Construct the perpendicular bisector of a given line segment.

Given: $\overline{AB}$

Construct: $\overleftrightarrow{PQ}$ so that $\overleftrightarrow{PQ} \perp \overline{AB}$ and $\overleftrightarrow{PQ}$ bisects $\overline{AB}$

Method:

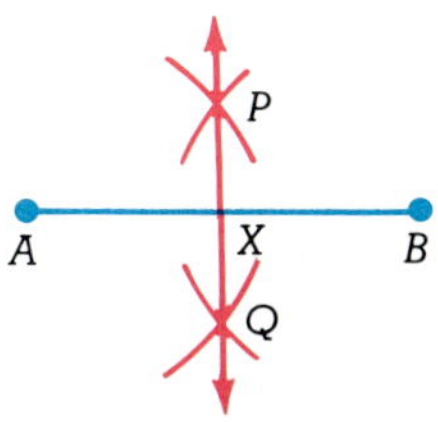

1. Set the compass with measure greater than $\frac{1}{2}AB$. Using A as a center, draw arcs on both sides of $\overline{AB}$.
2. Keep the same compass setting. Using B as a center, draw arcs intersecting the arcs in Step 1. Label the points P and Q.
3. Draw $\overleftrightarrow{PQ}$. Then $\overleftrightarrow{PQ} \perp \overline{AB}$ and $\overleftrightarrow{PQ}$ bisects $\overline{AB}$.

Justification: Draw $\overline{PA}$, $\overline{PB}$, $\overline{QA}$, and $\overline{QB}$. By construction, $\overline{PA} \cong \overline{PB}$ and $\overline{QA} \cong \overline{QB}$. Also, $\overline{PQ} \cong \overline{PQ}$. Thus, $\triangle PAQ \cong \triangle PBQ$. By the definition of congruent triangles, $\angle APX \cong \angle BPX$. Also, $\overline{PX} \cong \overline{PX}$, so $\triangle APX \cong \triangle BPX$ by SAS. It follows that $\overline{AX} \cong \overline{XB}$ and $\angle PXA \cong \angle PXB$. Thus, it is possible to conclude that $\overleftrightarrow{PQ} \perp \overline{AB}$ and $\overleftrightarrow{PQ}$ bisects $\overline{AB}$.

construction

CONSTRUCTION 6: Construct a line perpendicular to a given line through a given point on the line.

Given: A is on ℓ

Construct: $\overleftrightarrow{AB}$ so that $\overleftrightarrow{AB} \perp \ell$

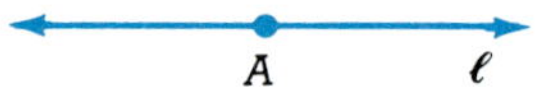

Method:

1. Using A as a center, draw two arcs with the same radius that intersect ℓ. Call the intersection points X and Y.
2. Increase the compass radius. Using X as a center, draw an arc on one side of the line.
3. Keep the same compass radius. Using Y as a center, draw an arc that intersects the arc in Step 2. Call the intersection B.
4. Draw $\overleftrightarrow{AB}$. Then, $\overleftrightarrow{AB} \perp \ell$.

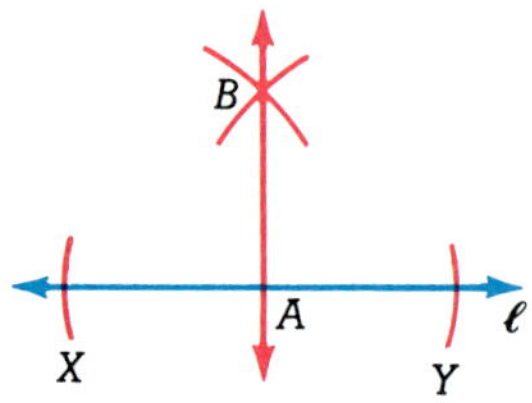

Justification: Draw $\overline{XB}$ and $\overline{YB}$. By construction, $\overline{XB} \cong \overline{YB}$ and $\overline{AX} \cong \overline{AY}$. Also, $\overline{AB} \cong \overline{AB}$. Thus, $\triangle AXB \cong \triangle AYB$ by SSS. By the definition of congruent triangles, $\angle BAX \cong \angle BAY$. Since $\angle BAX$ and $\angle BAY$ form a linear pair, by the Supplement Postulate they are supplementary. Congruent, supplementary angles are right angles. Thus, $\overline{AB} \perp \ell$.

According to Theorem 6–7, if two lines are perpendicular to the same line, then the two lines are parallel.

construction

CONSTRUCTION 7: Construct a line parallel to a given line and through a point *not* on the line.

Given: P is *not* on ℓ

Construct: n so that $n \parallel \ell$ and n contains P

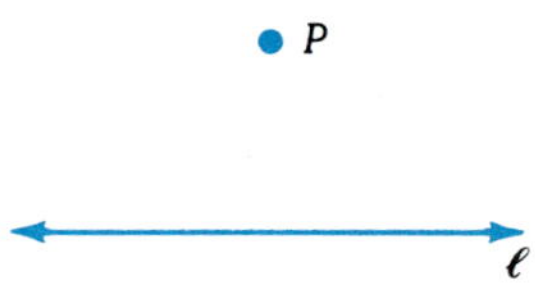

Method:

1. Draw a line through ℓ and through P. Call it $\overleftrightarrow{PX}$.
2. Copy one of the angles formed at X using P as a vertex and one side on $\overleftrightarrow{PX}$. Draw a line through P to form an angle congruent to the angle at X. Call it n. Then, $n \parallel \ell$ and n contains P.

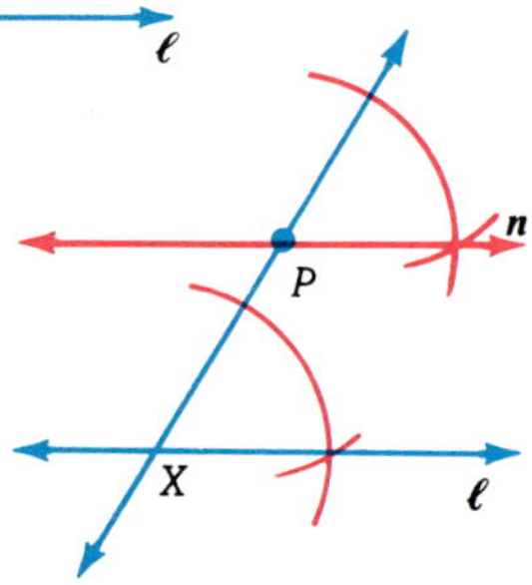

Justification: By construction, n contains P. Also, by construction, corresponding angles are congruent. Thus, $n \parallel \ell$.

exercises

Exploratory The figure at the right shows the construction of the perpendicular bisector of $\overline{AB}$. Supply reasons for each of the following statements in the justification for the construction.

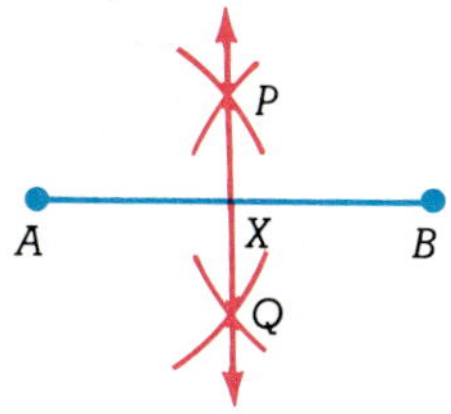

1. Draw $\overline{PA}$, $\overline{PB}$, $\overline{QA}$, and $\overline{QB}$.
2. $\overline{PA} \cong \overline{PB}$ and $\overline{QA} \cong \overline{QB}$.
3. $\overline{PQ} \cong \overline{PQ}$.
4. $\triangle PAQ \cong \triangle PBQ$.
5. $\angle APX \cong \angle BPX$.
6. $\overline{PX} \cong \overline{PX}$.
7. $\triangle APX \cong \triangle BPX$.
8. $\overline{AX} \cong \overline{BX}$ and $\angle PXA \cong \angle PXB$.
9. $\overleftrightarrow{PQ}$ bisects $\overline{AB}$.
10. $\overleftrightarrow{PQ} \perp \overline{AB}$.

Written Complete each of the following.

1. Draw a line and a point on the line. Label them ℓ and A. Construct a line perpendicular to ℓ and containing A.
2. Draw a line and a point *not* on the line. Label them ℓ and B. Construct a line perpendicular to ℓ and containing B.
3. Draw a line segment. Label it $\overline{PQ}$. Construct a line perpendicular to $\overline{PQ}$.
4. Draw a line segment. Label it $\overline{RS}$. Construct a bisector of $\overline{RS}$.
5. Draw a line and a point *not* on the line. Label them m and P. Construct a line parallel to m and containing P.
6. Describe a method for constructing a line parallel to a given line and through a point *not* on the line that is based on perpendicular lines.
7. Draw a line segment. Construct a square with sides congruent to the given segment.
8. Draw an acute triangle. Construct the medians of the triangle.
9. Draw an acute triangle. Construct the altitudes of the triangle.
10. Draw an obtuse triangle. Construct the altitudes of the triangle.
11. Draw a line segment. Construct an isosceles triangle whose congruent sides are congruent to the given segment.
12. Draw an acute angle and a line segment. Construct an isosceles triangle with base angles congruent to the given angle, and congruent sides congruent to the given segment.
13. Construct an isosceles right triangle.
14. Construct a $30°—60°—90°$ triangle.
15. Draw two line segments and an angle. Construct a parallelogram with adjacent sides congruent to the given segments, and included angle congruent to the given angle.

Draw two line segments. Label the shorter segment $\overline{AB}$, and the other $\overline{CD}$. Draw an acute angle. Label it $x°$. Use the figures to complete each of the following.

16. Construct a rectangle with sides measuring AB and CD.
17. Construct a rhombus with diagonals measuring AB and CD.
18. Construct a right triangle with hypotenuse measuring AB and containing an $x°$ angle.
19. Construct an isosceles triangle with congruent sides measuring CD and vertex angle $x°$.

13–3 Circles

The following show how to construct tangents to circles.

constructions

CONSTRUCTION 8: Construct a line tangent to a given circle at a given point on the circle.

Given: $\odot P$ with A on the circle

Construct: ℓ so that ℓ is tangent to $\odot P$ at A

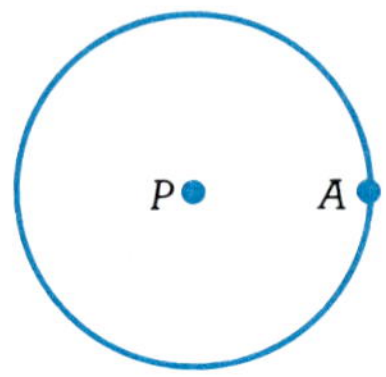

Method:

1. Draw $\overrightarrow{PA}$.
2. Construct ℓ through A and perpendicular to $\overrightarrow{PA}$. Line ℓ is tangent to $\odot P$ at A.

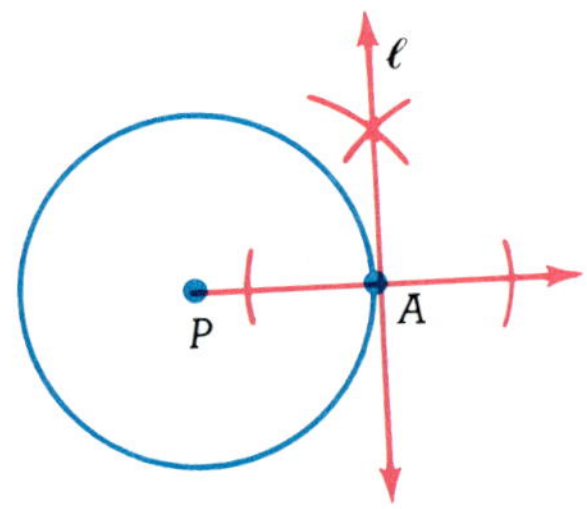

Justification: In a plane, if a line is perpendicular to a radius at a point on the circle, then the line is tangent to the circle. Thus, ℓ is tangent to $\odot P$ at A.

CONSTRUCTION 9: Construct a line tangent to a given circle and through a point *not* on the circle.

Given: A *not* on $\odot C$

Construct: $\overleftrightarrow{AD}$ so that $\overleftrightarrow{AD}$ is tangent to $\odot C$

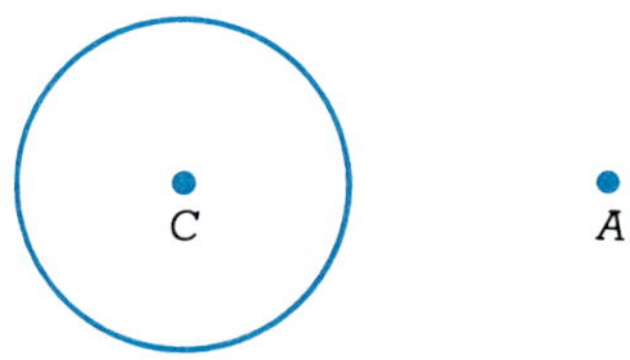

Method:

1. Draw $\overline{AC}$.
2. Construct the perpendicular bisector of $\overline{AC}$. Call it ℓ. Call X the intersection of ℓ and $\overline{AC}$.
3. Using X as a center, draw a circle with radius measuring XC. Call D and E the intersection points of the two circles.
4. Draw $\overleftrightarrow{AD}$. Then, $\overleftrightarrow{AD}$ is tangent to $\odot C$.

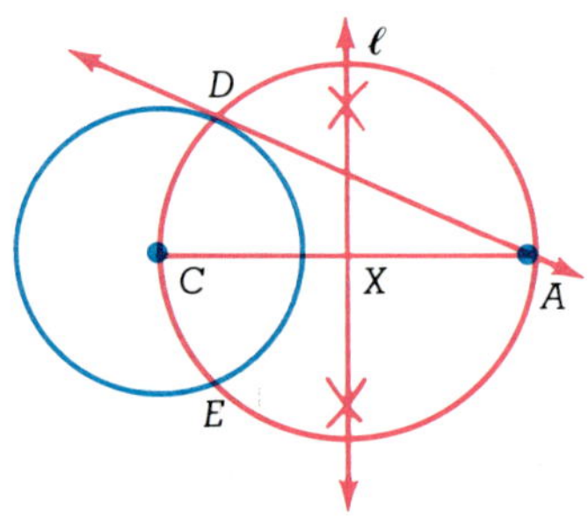

Justification: $\overline{AC}$ is a diameter of $\odot X$. Thus, $\widehat{CDA}$ is a semicircle and $\angle CDA$ is a right angle. $\overline{CD} \perp \overleftrightarrow{AD}$ implies that $\overleftrightarrow{AD}$ is tangent to $\odot C$ at D.

The justifications for each of the following are based on the same theorem.

constructions

CONSTRUCTION 10: Locate the center of a given circle.

Given: a circle

Construct: C so that C is the center of the circle

Method:

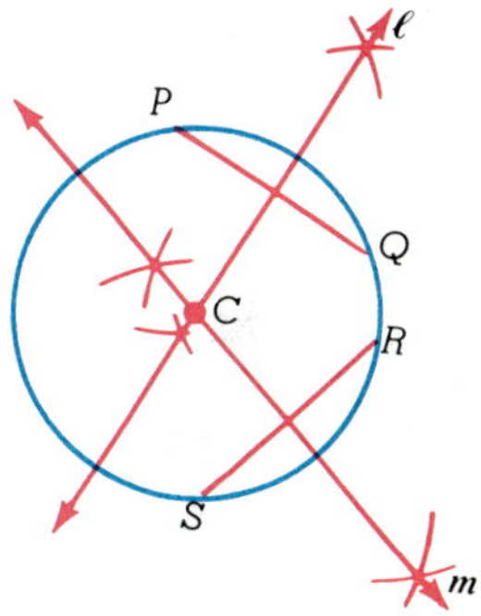

1. Draw two chords. Call them $\overline{PQ}$ and $\overline{RS}$.
2. Construct the perpendicular bisectors for each chord. Call them ℓ and m. Call C the intersection of ℓ and m. Then, C is the center of the circle.

Justification: The perpendicular bisector of a chord contains the center of the circle. Also, two lines intersect in exactly one point. Thus, ℓ and m must intersect in the center of the circle.

CONSTRUCTION 11: Construct a circle circumscribed about a given triangle.

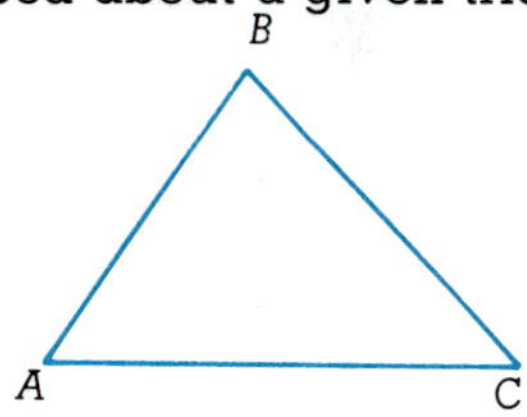

Given: $\triangle ABC$

Construct: $\odot Q$ so that $\odot Q$ is circumscribed about $\triangle ABC$

Method:

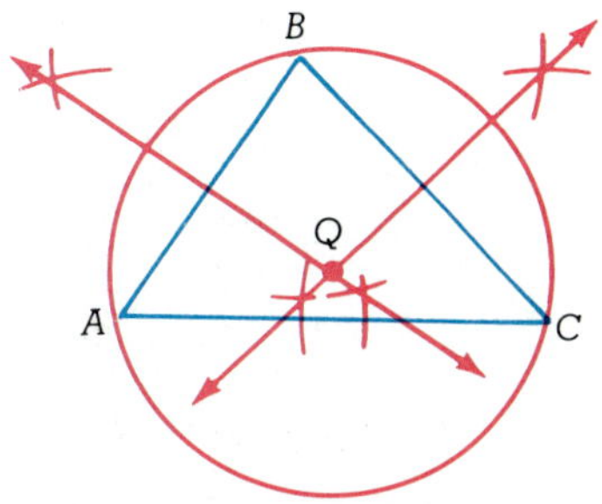

1. Construct perpendicular bisectors for two sides, $\overline{AB}$ and $\overline{BC}$. Call their intersection Q.
2. Using Q as a center, and QA as the measure of the radius, draw $\odot Q$. Then, $\odot Q$ is circumscribed about $\triangle ABC$.

Justification: The perpendicular bisector of a chord contains the center of a circle. Also, each side of a triangle is a chord of a circle circumscribed about the triangle. Thus, Q must be the center of the circle circumscribed about $\triangle ABC$.

exercises

Exploratory The figure at the right shows the construction of a line tangent to $\odot C$ through A. Supply reasons for each of the following statements in the justification for the construction.

1. $\overline{AC}$ is a diameter of $\odot X$.
2. $\overparen{CDA}$ is a semicircle.
3. $\angle CDA$ is a right angle.
4. $\overline{CD} \perp \overleftrightarrow{AD}$.
5. $\overleftrightarrow{AD}$ is tangent to $\odot C$ at D.

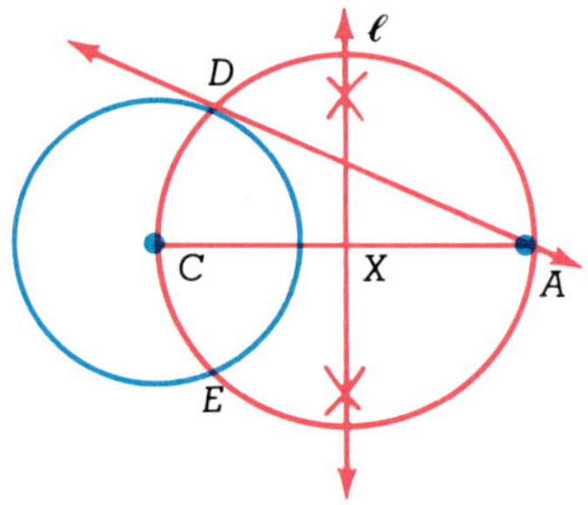

State *yes* or *no* for each of the following.

6. Suppose a circle is circumscribed about an acute triangle. Must the center of the circle be in the interior of the triangle?
7. Suppose a circle is circumscribed about an obtuse triangle. Must the center of the circle be in the interior of the triangle?

Written Complete each of the following.

1. Draw a circle. Label the center P and a point on the circle A. Construct a tangent to $\odot P$ at A.
2. Draw a circle. Label the center Q. Draw a point exterior to the circle. Label it B. Construct a tangent to $\odot Q$ containing B.
3. Draw an acute triangle. Circumscribe a circle about the triangle.
4. Draw an obtuse triangle. Circumscribe a circle about the triangle.
5. Construct a right triangle. Circumscribe a circle about the triangle.
6. Construct a square. Circumscribe a circle about the square.
7. Construct a rectangle. Circumscribe a circle about the rectangle.
8. Construct an equilateral triangle. Circumscribe a circle about the triangle.

The following method tells how to inscribe a circle in any triangle ABC.

Method:

1. Bisect $\angle A$. Bisect $\angle C$. Call X the intersection of the bisectors.
2. Construct a line through X and perpendicular to $\overline{AC}$. Call the intersection of the perpendicular with $\overline{AC}$, Y.
3. Using X as a center, draw a circle with radius measuring XY. Then, $\odot X$ is inscribed within $\triangle ABC$.

9. Draw a diagram to show the construction.
10. Complete a justification for the construction.

Using Symbols

Reading Geometry

The meaning of a mathematical expression can be changed by using different symbols.

Expression	Meaning
AB	measure of segment with endpoints A and B
$\overline{AB}$	line segment with endpoints A and B
$\overleftrightarrow{AB}$	line through points A and B
$\overrightarrow{AB}$	ray with endpoint A and going through B

Sometimes the meaning of one symbol is changed by adding another symbol.

Symbol	Meaning
$=$	is equal to
$\stackrel{?}{=}$	might be equal to
$\cong$	is congruent to

The position of a symbol with respect to other terms or letters affects the symbol's meaning. Notice the position of $<$ in the following expressions.

Expression	Meaning of $<$
$\angle A$	angle
$a < b$	is less than

The meaning of some symbols depend on the context in which they are used.

Expression	Meaning of °
sin 30°	degrees (angle measure)
30°C	degrees (temperature)
$30^0 = 1$	zero power

Exercises State the meaning for each of the following.

1. $\angle PQR$ **2.** $\triangle PQR$ **3.** $\widehat{PQR}$

4. $<$ **5.** $\not<$ **6.** $\leq$

7. $\sim$ **8.** $\approx$ **9.** $\cong$

State the meaning of $-$ in each of the following.

10. $7 - x = 3$ **11.** $^-8$ **12.** $\overline{RS}$

State the meaning of m in each of the following.

13. $m\angle C$ **14.** $9 = 2m + 3$ **15.** 4 m long

13–4 Proportions

It is possible to separate a line segment into congruent parts. The following construction shows the method for three congruent parts.

construction

CONSTRUCTION 12: Separate a line segment into three congruent parts.
A similar method can be used for any given number of congruent parts.

Given: $\overline{AB}$

Construct: P and Q on $\overline{AB}$ so that $\overline{AP} \cong \overline{PQ} \cong \overline{QB}$

Method:

1. Draw a ray from A.
2. With the compass point at A, mark off an arc on the ray at X. Construct $\overline{XY}$ and $\overline{YZ}$ so that $\overline{AX} \cong \overline{XY} \cong \overline{YZ}$.
3. Draw $\overline{ZB}$. Then, construct lines through Y and X that are parallel to $\overline{ZB}$. Call P and Q the intersection points on $\overline{AB}$. Then $\overline{AP} \cong \overline{PQ} \cong \overline{QB}$.

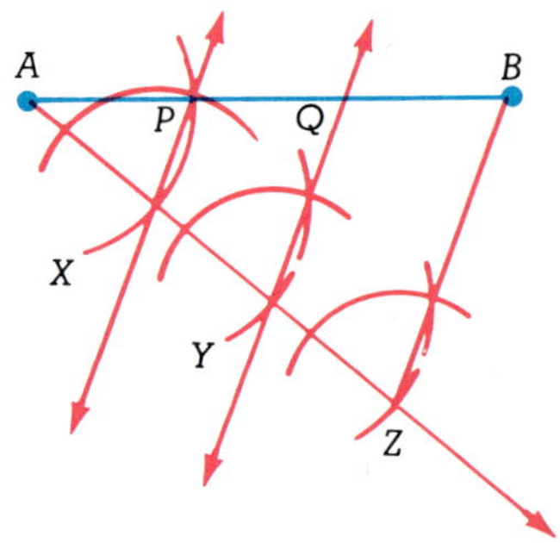

Justification: Since $\overleftrightarrow{PX} \parallel \overleftrightarrow{QY} \parallel \overline{BZ}$ and the lines cut off congruent segments on $\overrightarrow{AZ}$, they cut off congruent segments on $\overline{AB}$.

The following show how to construct proportional segments.

constructions

CONSTRUCTION 13: Given three segments, construct a fourth segment so the measures are proportional.

Given: segments with measures x, y, and z

Construct: a segment with measure a so that $\frac{x}{y} = \frac{z}{a}$.

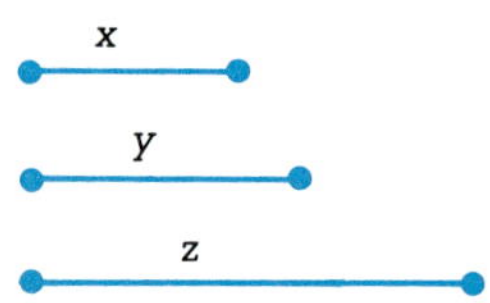

Method:

1. Draw an angle. Call it $\angle P$.
2. With compass setting at x and compass point at P, mark an arc on one side of the angle. Call the intersection point Q.

3. With compass setting at y and point at Q, mark an arc on $\overrightarrow{PQ}$ at R.
4. With compass setting at z and point at P, mark an arc on the other side of the angle at S.
5. Draw $\overline{QS}$. Then, construct a ray parallel to $\overline{QS}$ with endpoint R. The parallel intersects $\overrightarrow{PS}$ in a point. Call it T and say $ST = a$. Then, $\frac{x}{y} = \frac{z}{a}$ and $\overline{ST}$ is the required segment.

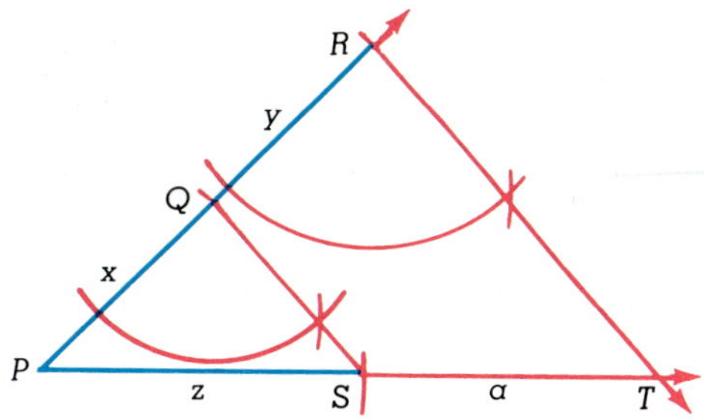

Justification: By Theorem 8–10, if a line is parallel to one side of a triangle and intersects the other two sides, then it divides them proportionally.

CONSTRUCTION 14: Construct the geometric mean of two given segments.

Given: $AB = x$ and $CD = y$

Construct: $\overline{PQ}$ so that PQ is the geometric mean of AB and CD

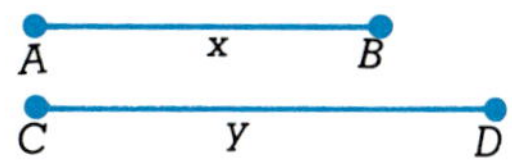

Method:

1. Draw a line. From a point P on the line, mark off segments measuring x and y. Label the endpoints L and M. 4
2. Construct the perpendicular bisector of $\overline{LM}$. Label the point of intersection S.
3. Using S as a center and $\overline{LS}$ as a radius, construct a semicircle with endpoints L and M.
4. Construct a line perpendicular to $\overline{LM}$ and containing P. Label the intersection of the line with the semicircle, Q. Then, PQ is the geometric mean of AB and CD.

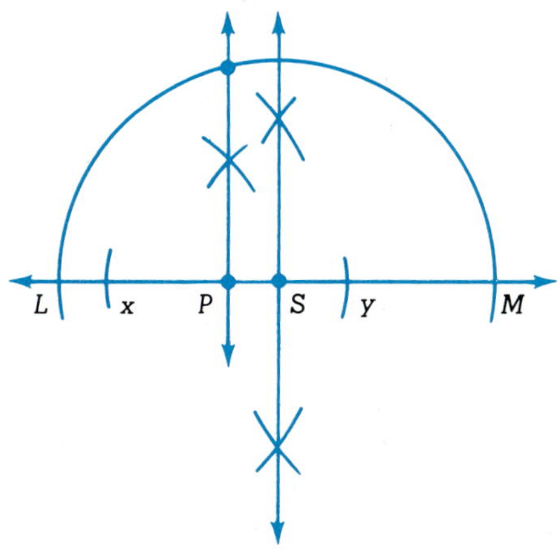

Justification: Draw $\overline{LQ}$ and $\overline{MQ}$. Since $\triangle LMQ$ is inscribed in a semicircle, $\angle LQM$ is a right angle and the triangle is a right triangle. By construction $\overline{PQ}$ is an altitude of $\triangle LQM$. Thus, PQ is the geometric mean between LP and PM. But, $\overline{LP} \cong \overline{AB}$ and $\overline{PM} \cong \overline{CD}$, so PQ is the geometric mean between AB and CD.

exercises

Exploratory The figure at the right shows the construction of the geometric mean of *PL* and *PM*. Supply reasons for each of the following statements in the justification for the construction.

1. Draw $\overline{LQ}$ and $\overline{MQ}$.
2. $\widehat{LQM}$ is a semicircle.
3. $\angle LQM$ is a right angle.
4. $\overline{PQ}$ is an altitude of $\triangle LQM$.
5. PQ is the geometric mean between PL and PM.

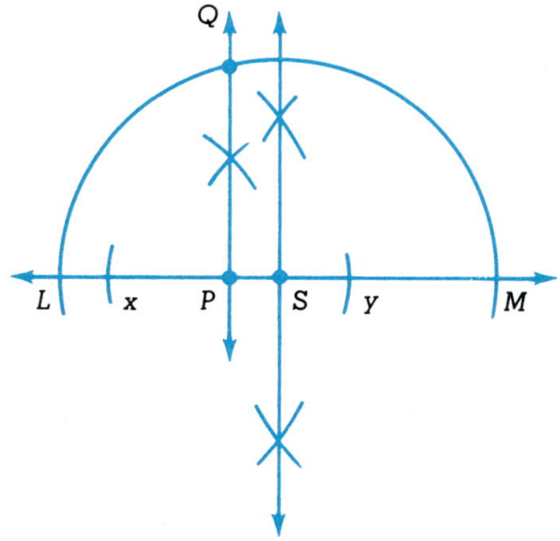

Written For each of the following, draw a segment that is about 10 centimeters long on a piece of paper. By construction, separate the segment into the given number of congruent parts.

1. 2	**2.** 3	**3.** 4
4. 5	**5.** 6	**6.** 7
7. 8	**8.** 9	**9.** 10

On a piece of paper, draw and label three segments like the segments shown below. Then, for each of the following, construct a segment with measure x so that the given proportion holds.

10. $\frac{a}{b} = \frac{c}{x}$

11. $\frac{a}{b} = \frac{x}{c}$

12. $\frac{a}{x} = \frac{b}{c}$

13. $\frac{x}{a} = \frac{c}{b}$

14. $\frac{a}{2b} = \frac{c}{x}$

15. $\frac{x}{2b} = \frac{c}{a}$

16. $\frac{x}{a} = \frac{a}{b}$

17. $\frac{x}{b} = \frac{b}{c}$

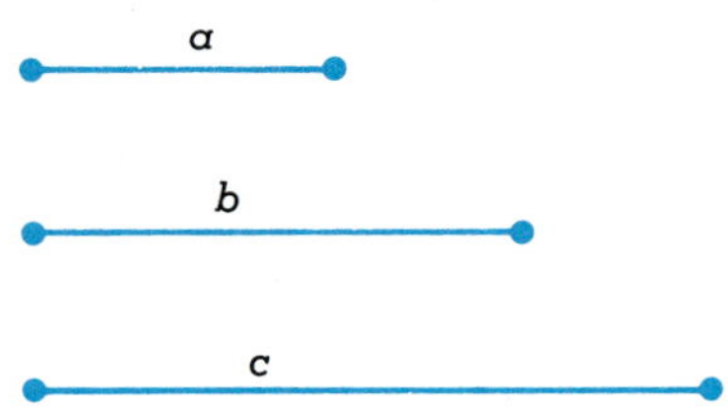

Complete each of the following.

18. Draw a line segment. Then, construct an equilateral triangle whose perimeter has the same measure as the segment.

19. Draw a line segment. Then, construct a square whose perimeter has the same measure as the segment.

20. Draw a line segment. Then, separate the segment into segments whose measures are in the ratio 2 to 3.

21. Draw a line segment. Then, separate the segment into segments whose measures are in the ratio 1 to 2.

CAREERS

Astronomer

Jim Herrick is an astronomer. He uses a telescope to study objects in space. Jim must calculate the distance from the earth of the objects he studies.

Because the earth rotates, at different times of the day, an object in space appears to be in a different part of the sky. If the object is very far away, the apparent change in position is unnoticeable without astronomical instruments. If the object is as close as the moon, the change may be as great as 2° in 10 hours. Since the radius of the earth is known, trigonometry can be used to find the object's distance from the earth. The equation used is shown below.

$$\sin 1° = \frac{6400}{d}$$

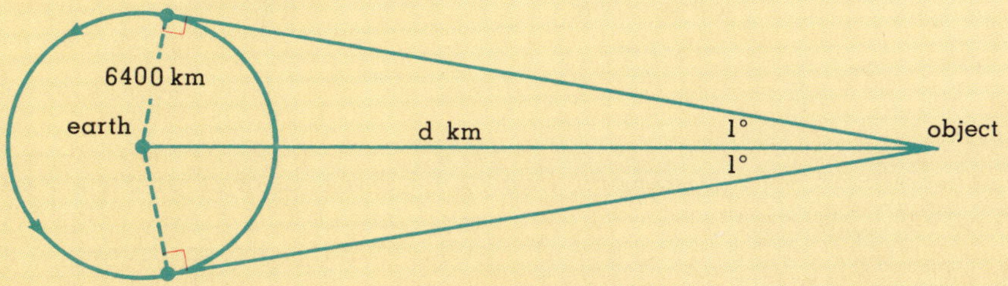

Exercises Solve each problem about the figure and equation.

1. What does d represent?
2. What does 6400 km represent?
3. Find sin 1°.
4. Solve for d.
5. The radius of the earth is about 4000 miles. The object is how many miles from the earth?

13–5 Loci

In geometry, a figure is a **locus** if and only if it is the set of *all* points and only those points that satisfy a given condition.

The circle at the right is the locus of all points that are 2 centimeters from P. *If a point is in the interior of the circle, it is less than 2 centimeters from P. If a point is in the exterior of the circle, it is more than 2 centimeters from P.*

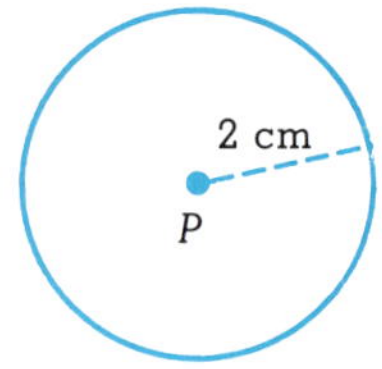

Suppose you are asked to describe a certain locus. First, you should read the problem carefully and draw the given figure.

Find the locus of all points in a plane that are 15 millimeters from a given point, C.

Draw the given figure. The given figure is point C.

After drawing the given figure, locate several points that satisfy the given conditions.

Locate several points. Draw points that are 15 millimeters from C. Locate enough points to suggest the shape of the locus.

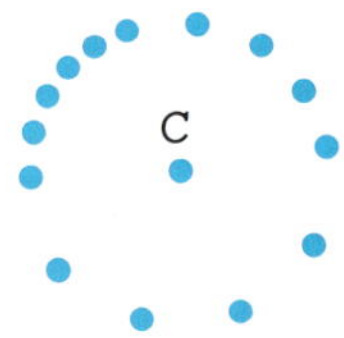

As soon as the shape of a geometric figure begins to appear, draw a smooth curve or line that contains the points. Then, describe the locus in words.

Draw a smooth curve or line. The points suggest a circle.

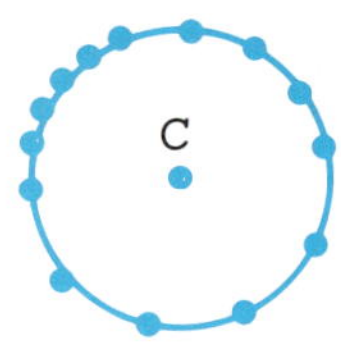

Describe the locus. The locus of all points in a plane that are 15 millimeters from a given point, C, is a circle with center C and radius measuring 15 millimeters.

Procedure for Determining Locus

1. Draw the given figure.
2. Locate several points.
3. Draw a smooth curve or line.
4. Describe the locus.

example 1

Determine the locus of all points in a plane that are 2 centimeters from a given line.

Draw the given figure. The given figure is a line.

Locate several points.

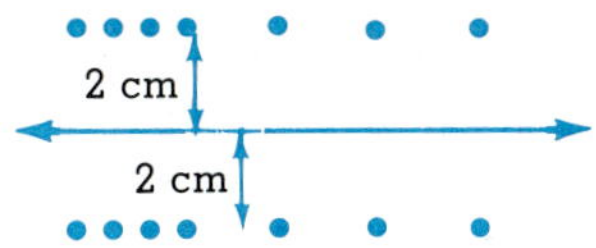

Draw a smooth curve or line.

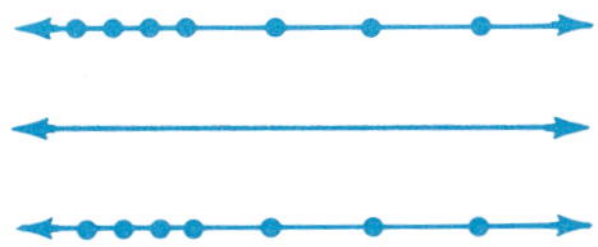

Describe the locus. The locus of all points in a plane 2 centimeters from a given line is a pair of lines parallel to the given line and 2 centimeters from the line.

A locus may differ depending on whether it is *in a plane* or *in space*.

The locus of all points *in a plane* that are equidistant from two parallel lines is a line that is parallel to the given lines and midway between them.

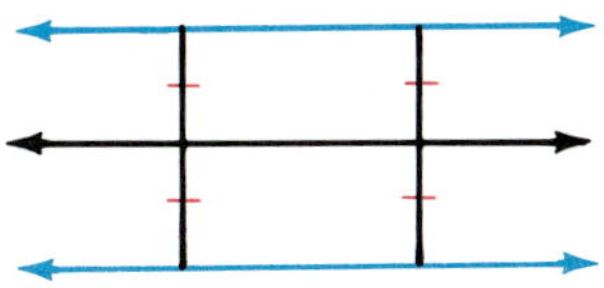

The locus of all points *in space* that are equidistant from two parallel lines is a plane that contains a line parallel to the given lines and midway between them, and is perpendicular to the plane containing the given lines.

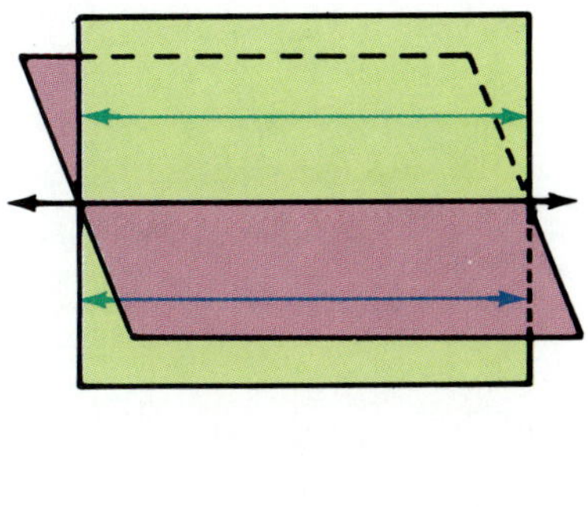

exercises

Exploratory For each of the following, describe the locus of all points *in a plane* that are the given distance from a given point C.

1. 1 mm
2. 2 mm
3. 3 mm
4. 4 mm
5. 40 mm
6. 100 mm
7. 1000 mm
8. x mm

9–16. Solve problems 1–8 for points *in space*.

For each of the following, describe the locus of all points *in a plane* that are the given distance from a given line ℓ.

17. 1 in.
18. 2 in.
19. 3 in.
20. 4 in.
21. 40 in.
22. 100 in.
23. 1000 in.
24. x in.

25–32. Solve problems 17–24 for points *in space*.

Written Describe the locus for each of the following.

1. All the midpoints of the radii of a circle with radius measuring 10 centimeters.
2. All the midpoints of the radii of a circle with radius measuring 12 inches.
3. All points in the interior of a 5-inch square that are 2 inches from a side of the square.
4. All points in the exterior of a 5-inch square that are 2 inches from a side of the square.
5. All points in a plane that are equidistant from the sides of a given angle.
6. All points in a plane that are equidistant from two intersecting lines in that plane.
7. All the points in a plane that are equidistant from two given points.
8. All the points in a plane that are equidistant from the centers of two given circles.
9. All the points in a plane that are 5 centimeters from a circle with radius measuring 3 centimeters.
10. All the points in a plane that are 3 centimeters from a circle with radius measuring 5 centimeters.
11. All the points in a plane that are centers of circles with radii measuring 7 inches and pass through a given point.
12. All points in a plane that are centers of circles tangent to a given line at a given point on the line.
13. All the points in a plane that are centers of circles which are tangent to both sides of a given angle.
14. All points in a plane that are centers of circles with radii measuring 15 millimeters and tangent to a given line.
15. All the points in a plane that are the third vertex of isosceles triangles with a given line segment for a base.
16. All points in a plane that are the right angle vertex of right triangles with a given line segment for a hypotenuse.
17. All the points in space that are a given distance from a given line.
18. All the points in space that are equidistant from two given points.
19. All the points in space that are the centers of spheres with radii measuring r units and tangent to a given plane.
20. All the points in space that are the centers of spheres that are tangent to a given plane at a given point in the plane.

13–6 Intersection of Loci

Some loci must satisfy several conditions. Such loci often can be determined by finding the intersection of loci that meet each condition.

example 1

Determine the locus of all points in a plane that are equidistant from two given points and equidistant from two given parallel lines.

The locus of all points in a plane that are equidistant from two given points is the perpendicular bisector of the segment joining the two points.

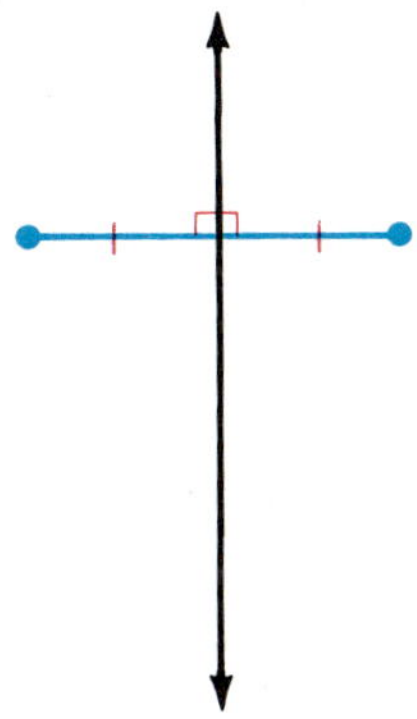

The locus of all points in a plane that are equidistant from two given parallel lines is a line parallel to the given lines and midway between them.

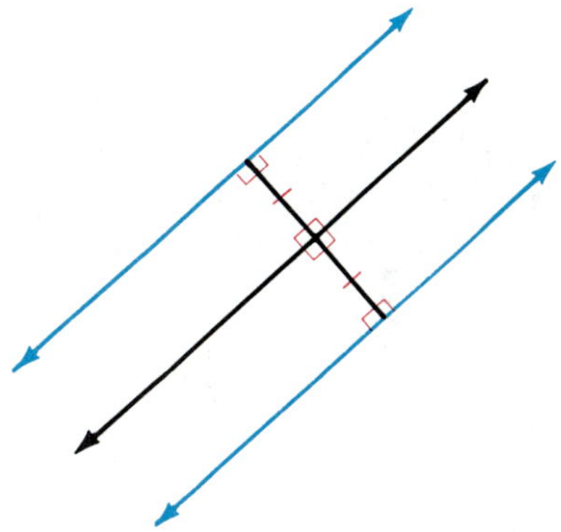

If A and B are the given points, and ℓ and m are the given lines, then C is the only point that is equidistant from A and B, *and* equidistant from ℓ and m.

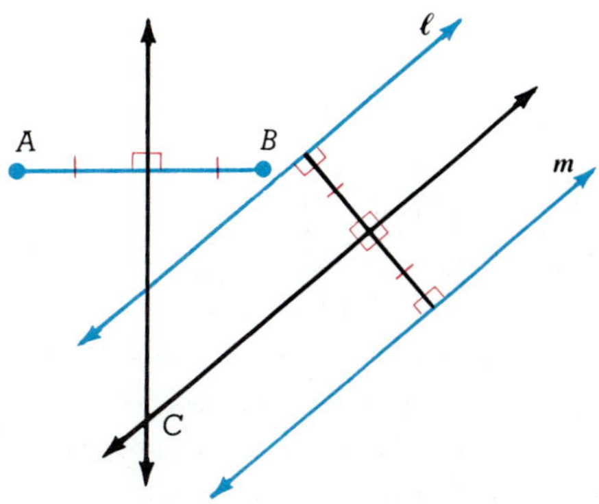

Some construction problems involve finding a point that satisfies several conditions.

example 2

Given AB units, AC units, and CD units, construct $\triangle ABC$ so that $\overline{CD}$ is an altitude.

The locus of all points that are CD units from $\overleftrightarrow{AB}$ are two lines on either side of $\overleftrightarrow{AB}$, parallel to $\overleftrightarrow{AB}$ and CD units away.

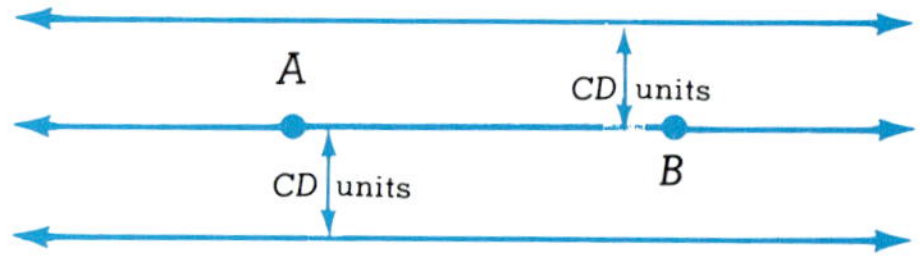

The locus of all points that are AC units from A is a circle with center A and radius measuring AC units.

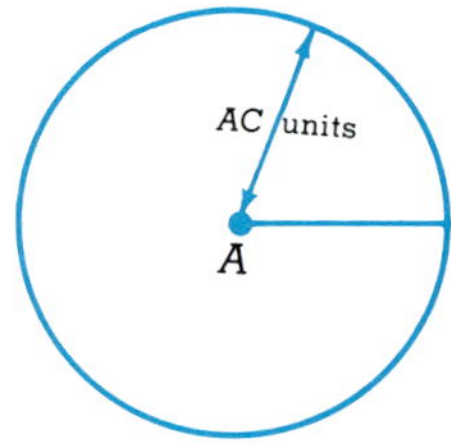

Since $\overline{CD}$ is an altitude of the triangle, C must be CD units from $\overleftrightarrow{AB}$ and lie on one of the two parallel lines. Since C is the endpoint of $\overline{AC}$, it must be AC units from A and lie on $\odot A$. Only four points satisfy these conditions. Thus, there are four possible ways to draw $\triangle ABC$.

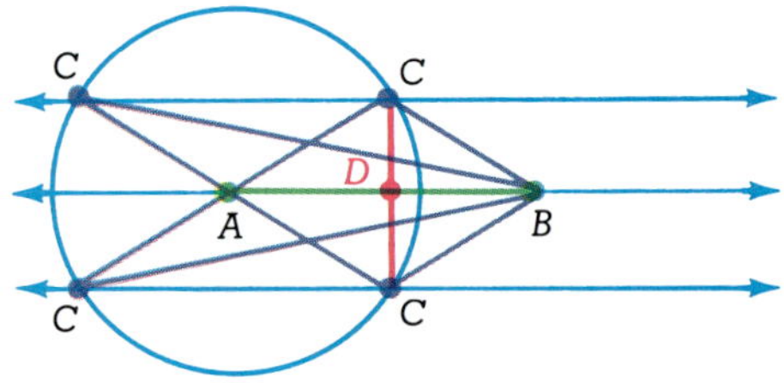

exercises

Exploratory **For each of the following, choose the best answer.**

1. The locus of points in a plane equidistant from the ends of a line segment is ______.
   a. a midpoint of the segment
   b. a bisector of the segment
   c. a perpendicular bisector of the segment
   d. two lines parallel to the segment

2. The locus of points in a plane equidistant from the sides of an angle is ______.
   a. a line joining the midpoints of the sides
   b. a bisector of the angle
   c. a circle about the vertex
   d. the angle vertical to the given angle

3. The locus of points in a plane equidistant from two parallel lines is ______.
   a. a line parallel to the lines and midway between them
   b. a line perpendicular to the given lines
   c. a pair of lines parallel to the given lines and at a given distance from them
   d. a circle between the lines and tangent to them

4. The locus of points in a plane equidistant from two intersecting lines is ______.
   a. the intersection of each pair of lines and at a given distance from them
   b. a circle with center at the intersection point and at a given distance from it
   c. a pair of lines perpendicular to the given lines
   d. a pair of lines which bisect the angles formed by the two intersecting lines

5. The locus of points in a plane at a given distance from a given point is ______.
   a. a pair of perpendicular lines that intersect at the given point
   b. a set of circles containing the point with radii of the given distance
   c. a circle whose center is the given point and whose radius measures the given distance
   d. a square whose diagonals intersect at the given point and sides measure the given distance

6. The locus of points in a plane at a given distance from a given line is ______.
   a. a pair of lines parallel to the given line and at a given distance from it
   b. a set of circles whose centers lie on the line and radii measure the given distance
   c. a line perpendicular to the given line
   d. a pair of planes parallel to each other and at the given distance from the line

Written **For each of the following, describe all the possible ways the figures can intersect.**

1. two parallel lines and a circle
2. two concentric circles and a line
3. two concentric circles and two parallel lines
4. a sphere and a plane
5. a sphere and two parallel lines
6. two circles
7. two spheres
8. a sphere and two parallel planes
9. a circle and a plane
10. a circle and a sphere

Describe the locus for each of the following.

11. all the points in a plane that are 2 centimeters from a given line and 5 centimeters from a given point on the line
12. all the points in a plane that are 3 centimeters from a given line and 1.5 centimeters from a given point on the line
13. all the points in space that are 2 inches from a given line and 5 inches from a given point on the line
14. all the points in space that are 3 inches from a given line and 2 inches from a given point on the line

15. all the points in a plane that are equidistant from three given points

16. all the points in a plane that are equidistant from a given line and equidistant from two points on the line

17. all the points in a plane that are equidistant from the sides of an angle and 4 inches from the vertex of the angle

18. all the points in a plane that are equidistant from the sides of an angle and a given distance from the vertex of the angle

19. all the points in a plane equidistant from two given parallel lines and a given distance from another line

20. all the points in a plane that are equidistant from two given parallel lines and a given distance from a point *not* on the lines

Complete each of the following.

21. Draw a line, m, and two points, C and D, on one side of the line. Then, construct the locus of points in the plane that are equidistant from C and D.

22. Draw two line segments, $\overline{PQ}$ and $\overline{RS}$. Choose a point between P and Q, and label it A. Then, construct a circle tangent to $\overline{PQ}$ at A and whose center lies on $\overline{RS}$.

23. Draw a line, ℓ, and a point on the line, P. Draw Q *not* on ℓ. Then, construct the locus of the centers of all circles in the plane that contain P and Q.

24. Draw two parallel lines, ℓ and m. Choose a point between ℓ and m, and label it B. Construct a circle tangent to the two lines and containing B.

Excursions in Geometry

Creative Blocks

The figure below shows a block of wood with three holes. One hole is circular, one hole is triangular, and one hole is square. The diameter of the circle, the altitude and base of the triangle, and each side of the square are congruent. The walls of each hole are perpendicular to the face of the block of wood.

Describe a single, solid object that will pass through each hole, and will entirely block any light during its passage.

Vocabulary

straightedge (406)
compass (406)
construction (406)
perpendicular bisector (409)
locus (420)

Chapter Summary

1. The parts of a construction are: *given*, *construct* a diagram, *method*, and *justification*. (406)
2. In geometry, a figure is a locus if and only if it is the set of all points and only those points that satisfies a given condition. (420)
3. Procedure for determining a locus:
   1. Draw the given figure.
   2. Locate several points.
   3. Draw a smooth curve or line.
   4. Describe the locus. (420)
4. The locus of all points *in a plane* that are equidistant from two parallel lines is a line that is parallel to the given lines and midway between them. (421)
5. The locus of all points *in space* that are equidistant from two parallel lines is a plane that contains a line parallel to the given lines and midway between them, and is perpendicular to the plane containing the given lines. (421)

Constructions:

Construction 1: Construct a line segment congruent to a given line segment. (406)
Construction 2: Construct an angle congruent to a given angle. (407)
Construction 3: Construct the bisector of a given angle. (407)
Construction 4: Construct a line perpendicular to a given line and through a given point not on the line. (409)
Construction 5: Construct the perpendicular bisector of a given line segment. (409)

Construction 6: Construct a line perpendicular to a given line and through a given point on the line. (410)

Construction 7: Construct a line parallel to a given line and through a point not on the line. (410)

Construction 8: Construct a line tangent to a given circle and through a point on the circle. (412)

Construction 9: Construct a line tangent to a given circle and through a point not on the circle. (412)

Construction 10: Locate the center of a given circle. (413)

Construction 11: Construct a circle circumscribed about a given triangle. (413)

Construction 12: Separate a line segment into three congruent parts. (416)

Construction 13: Given three segments, construct a fourth segment so the measures are proportional. (416)

Construction 14: Construct the geometric mean of two given segments. (417)

Chapter Review

13–1 On a piece of paper, draw and label two segments like the segments shown at the right. Then, construct segments for each of the following measures.

1. $XY + WZ$
2. $WZ - XY$
3. $4WZ$
4. $2XY - WZ$

On a piece of paper, draw and label two angles like the angles shown at the right. Then, construct angles for each of the following degree measures.

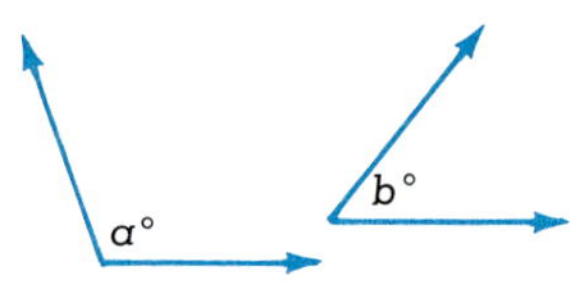

5. $a + b$
6. $a - b$
7. $\frac{1}{2}a$
8. $a - \frac{1}{2}b$

13–2 Complete each of the following.

9. Draw a line segment. Label it $\overline{AB}$. Construct a line perpendicular to $\overline{AB}$.

10. Draw a line and a point not on the line. Label them ℓ and Q. Construct a line parallel to ℓ and containing Q.

13–3 Complete each of the following.

11. Draw a circle. Label the center C. Draw a point exterior to the circle. Label it A. Construct a tangent to $\odot C$ containing A.

12. Draw a triangle. Circumscribe a circle about the triangle.

13–4 On a piece of paper, draw and label three segments like the segments shown at the right. Then, for each of the following, construct a segment with measure x so that the given proportion holds.

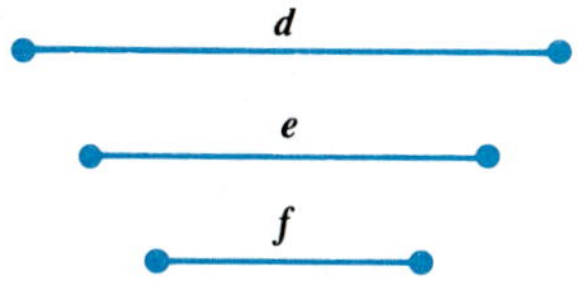

13. $\frac{d}{e} = \frac{f}{x}$

14. $\frac{d}{f} = \frac{x}{e}$

15. $\frac{x}{d} = \frac{d}{e}$

16. $\frac{f}{x} = \frac{2e}{d}$

17. Draw a line segment. Then, construct a rhombus whose perimeter has the same measure as the segment.

13–5 Describe the locus for each of the following.

18. all the points in a plane that are 1 foot from a given point C

19. all the points in space that are 1 inch from a given point A

20. all the midpoints of the diameters of a circle

21. all the points in a plane that are centers of circles with radii measuring 11 centimeters and pass through a given point

13–6 For each of the following, describe all the possible ways the figures can intersect.

22. a circle and a plane

23. a sphere and a line

24. two concentric circles and a plane

25. two parallel planes and a circle

Describe the locus for each of the following.

26. the perpendicular bisector of a segment

27. a plane parallel to two given planes and midway between them

Complete the following.

28. Draw a line, ℓ, and two points, E and F, on one side of the line. Then, construct the locus of points in the plane that are equidistant from E and F.

Chapter Test

On a piece of paper, draw and label two segments like the segments shown at the right. Then, construct segments for each of the following measures.

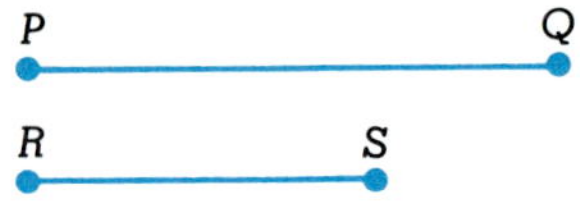

1. $PQ - RS$
2. $3PQ$
3. $3RS - PQ$
4. $2(RS + PQ)$
5. Draw two angles with degree measures c and d. Then, construct an angle having degree measure $c + d$.

Complete each of the following.

6. Draw a line and a point *not* on the line. Label them m and T. Construct a line perpendicular to m and containing T.
7. Draw an obtuse triangle. Construct the medians of the triangle.
8. Draw a circle. Label the center H and a point on the circle T. Construct a tangent to $\odot H$ at T.
9. Construct a 30°–60°–90° triangle. Circumscribe a circle about the triangle.
10. On a piece of paper, draw and label three segments with measures r, s, and t. Then, construct a segment with measure x so that the proportion $\frac{r}{s} = \frac{x}{t}$ holds.
11. Draw a line segment. Then, separate the segment into segments whose measures are in the ratio of 1 to 3.

Describe the locus for each of the following.

12. all the points in a plane that are 5 inches from a given ℓ
13. all the points in a plane that are 4 centimeters from a given point A
14. all the points in space that are 3 meters from a given point C
15. all the points in space that are 10 feet from a given line n

For each of the following, describe all the possible ways the figures can intersect.

16. two concentric spheres and a line
17. two parallel planes and a sphere

Describe the locus for each of the following.

18. the intersection of two planes
19. all the points in space that are 6 millimeters from a given line and 10 millimeters from a given point on the line

Complete the following.

20. Draw two line segments $\overline{AB}$ and $\overline{CD}$. Choose a point between A and B and label it P. Then, construct a circle tangent to $\overline{AB}$ at P and whose center lies on $\overline{CD}$.

Chapter

14

Transformations

Reflections are one kind of transformation. The photograph above was created by using a reflection in a pond.

14–1 Mappings

Figures can be changed in many ways. Some are shown below.

A figure can be *reflected*.

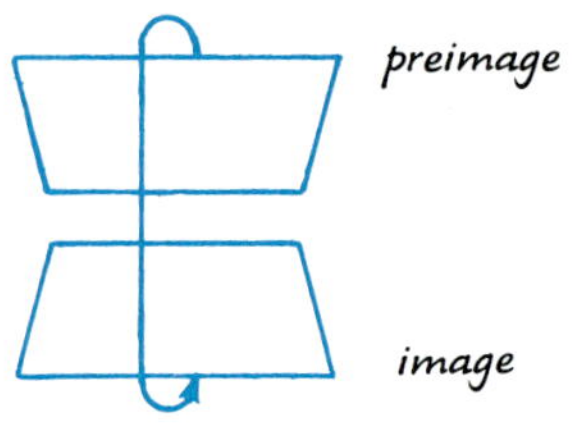

A figure can be *rotated*.

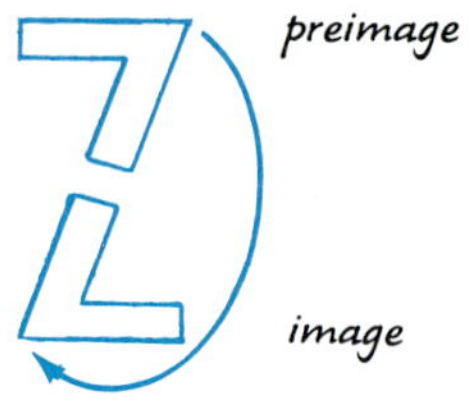

A figure can be *slid*.

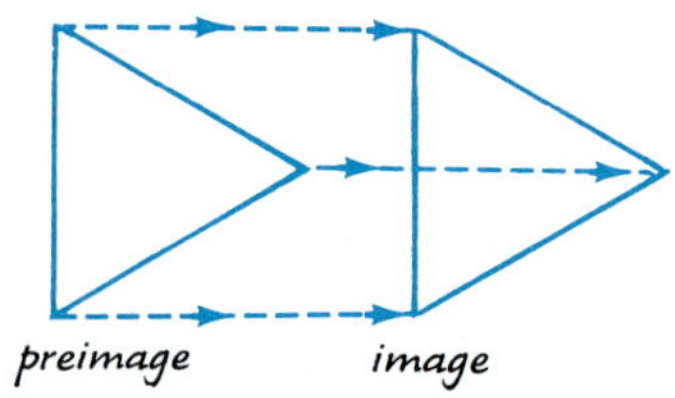

A figure can be *enlarged*.

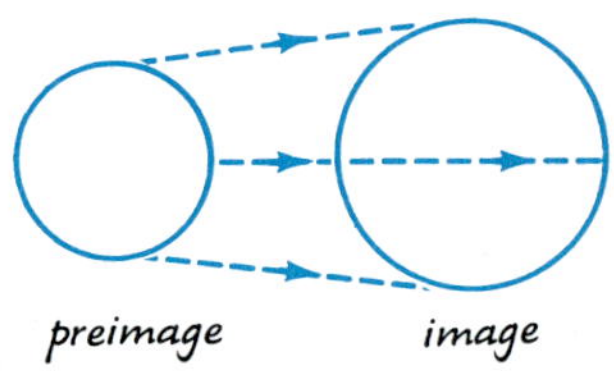

The mappings shown above are called **transformations.** A transformation maps a *preimage* onto an *image*.

Definition of Transformation

In a plane, a mapping is a transformation if and only if each point has exactly one image point and each image point has exactly one preimage point.

The symbol → is used to indicate mapping. For example, $\triangle ABC \rightarrow \triangle PQR$ means $\triangle ABC$ *is mapped onto* $\triangle PQR$. The order of the letters indicates the correspondence of the preimage to the image. The first vertices named, A and P, are corresponding vertices. The second vertices named, B and Q, are corresponding vertices. The third vertices named, C and R, are corresponding vertices.

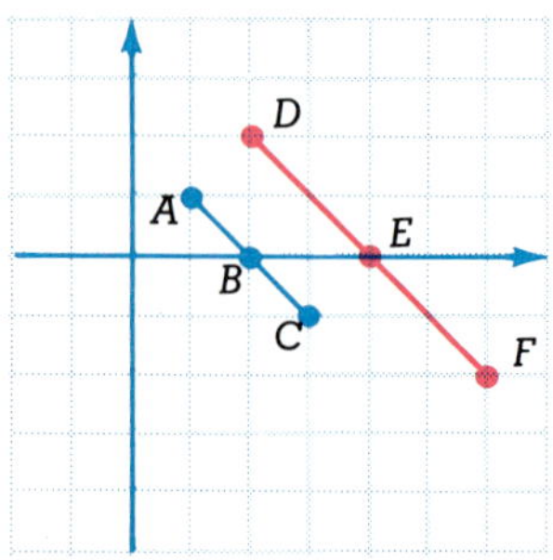

In the figure at the left, $\overline{AC} \rightarrow \overline{DF}$, A and D are corresponding points. C and F are corresponding points. B and E are corresponding points. Every point on $\overline{AC}$ corresponds to a point on $\overline{DF}$, and every point on $\overline{DF}$ corresponds to a point on $\overline{AC}$. Thus, the mapping is a transformation.

When a geometric figure and its transformation image are congruent, the mapping is called an *isometry* or a **congruence transformation.** Other mappings are called **similarity transformations.**

example 1

Suppose $\triangle ABC \rightarrow \triangle PQR$. Show that this mapping is an isometry.

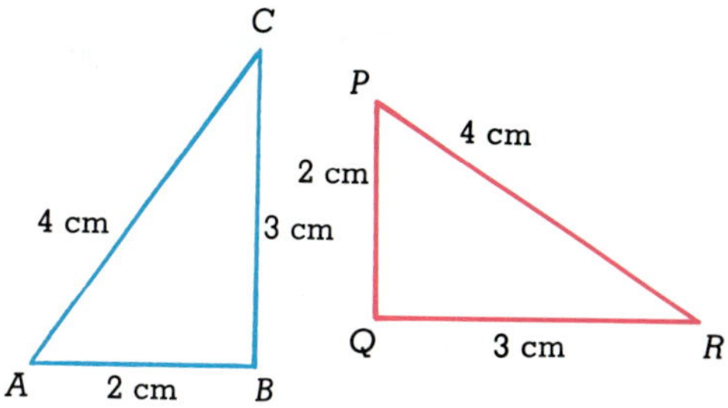

$\angle A \cong \angle P$, $\angle B \cong \angle Q$, $\angle C \cong \angle R$
$\overline{AB} \cong \overline{PQ}$, $\overline{BC} \cong \overline{QR}$, $\overline{AC} \cong \overline{PR}$
$\triangle ABC$ is the preimage. $\triangle PQR$ is the image.

exercises

Exploratory For each of the following, suppose quadrilateral $RSTU \rightarrow$ quadrilateral $ABCD$.

1. Name the image of R.
2. Name the image of $\overline{UT}$.
3. Name the preimage of D.
4. Name the preimage of B.
5. Name the image of $\overline{ST}$.
6. Name the preimage of $\overline{DC}$.
7. Name the image of $\angle U$.
8. Name the preimage of $\angle D$.
9. Name the image of $\angle T$.
10. Name the preimage of $\overline{AB}$.

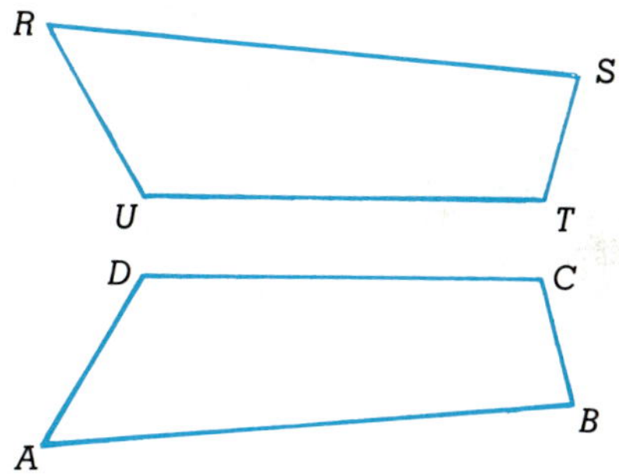

Written In the figure at the right, $\triangle ABC \rightarrow \triangle EBD$. For each of the following, name the image.

1. A 2. B 3. C
4. $\angle CAB$ 5. $\angle BCA$ 6. $\overline{AC}$

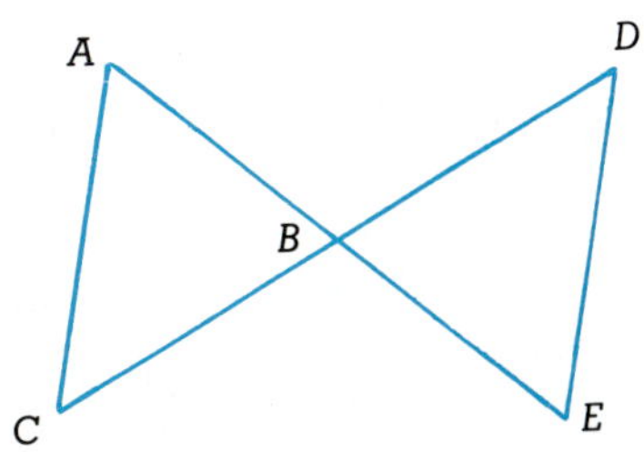

For each of the following, name the preimage if $\triangle ABC \rightarrow \triangle EBD$.

7. E 8. B 9. D
10. $\overline{BE}$ 11. $\overline{DE}$ 12. $\angle DBE$
13. List the information needed to show that the mapping of $\triangle ABC$ onto $\triangle EBD$ is an isometry.

Use the figures below to name the image of the preimage given in problems **14–21,** if the mapping is an isometry.

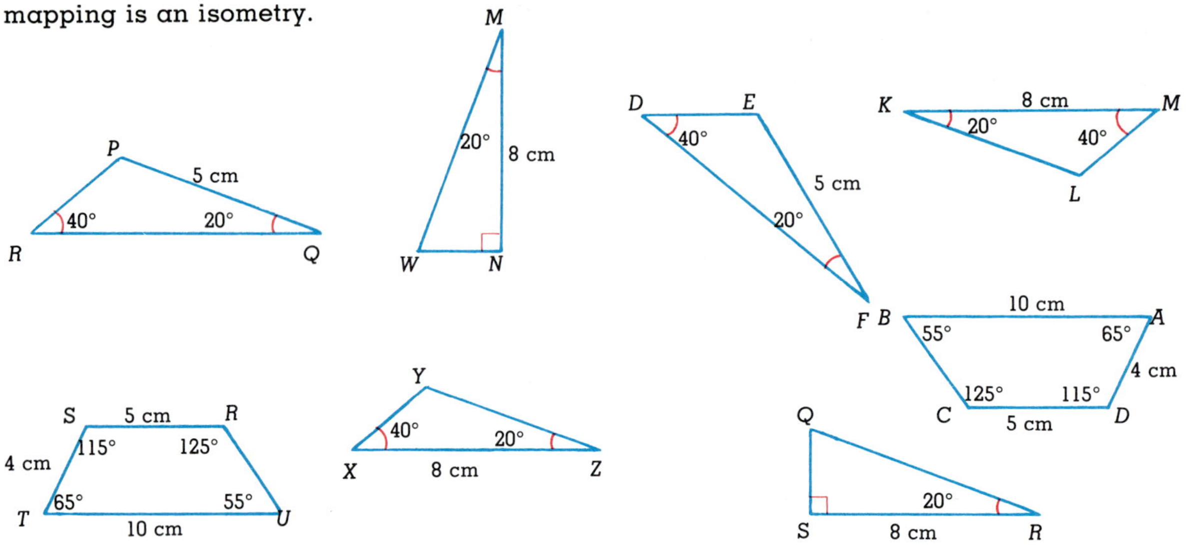

14. $\triangle MWN$
15. $\triangle PQR$
16. $\triangle SRQ$
17. $\triangle MKL$
18. $\triangle LMK$
19. quadrilateral $RSTU$
20. quadrilateral $BADC$
21. $\triangle ZYX$

In the figure at the right, $\triangle XYW \cong \triangle ZYW$.

22. Name a segment that is its own preimage.
23. Name two points that are their own preimages.
24. Name a mapping that describes the congruence.

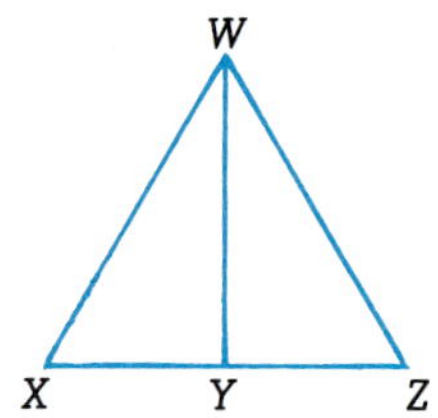

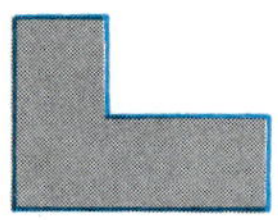

The L-shaped tile at the left is gray on top and blue underneath. Suppose the L-shaped tile is used to form each of the following. Name the colors for figure A and figure B.

25.

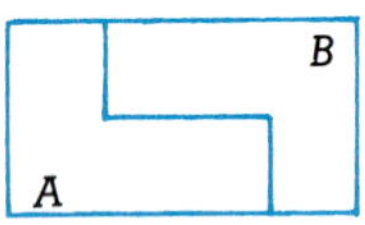

26.

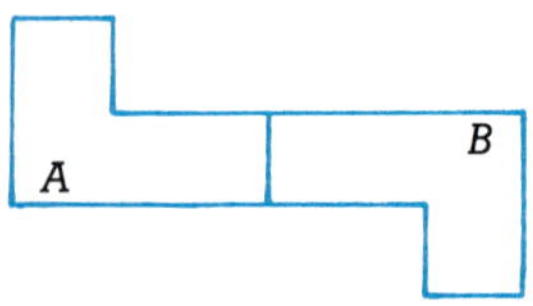

27.

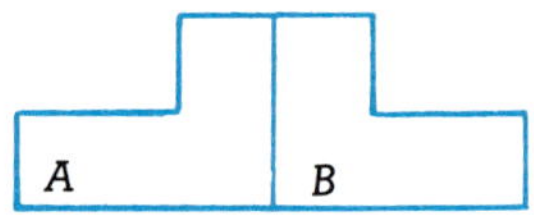

28.

29.

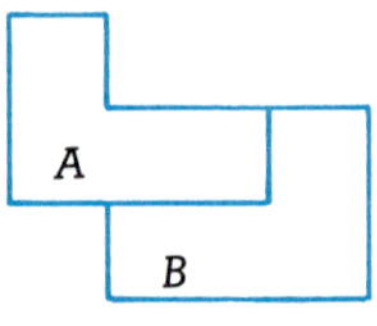

30.

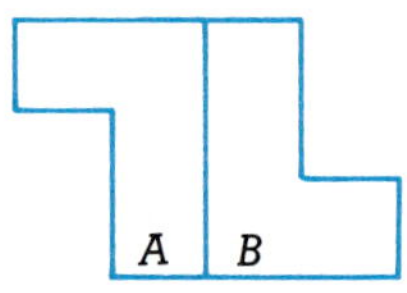

14–2 Reflections

One type of transformation is a **reflection.** For example, in the figure at the right, A is the reflection image of X with respect to ℓ. The *line of reflection*, ℓ, is a perpendicular bisector of the segment drawn from X to A. Since P is on the line of reflection, its image is P itself.

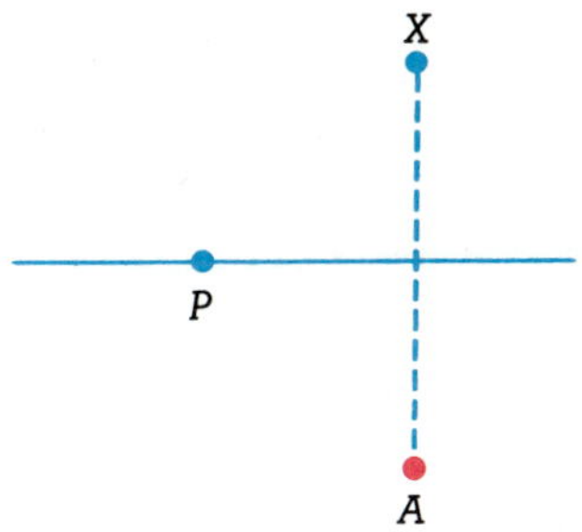

A line that is perpendicular to a segment and bisects the segment is a perpendicular bisector.

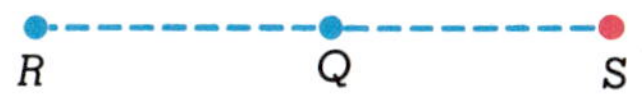

It also is possible to have a reflection image with respect to a *point of reflection*. S is the reflection of R with respect to Q, the point of reflection Q is the midpoint of the segment drawn from R to S.

The reflection images of collinear points are collinear. The images of collinear points A, B, and C are the collinear points P, Q, and R. Therefore, it is said that *reflections preserve collinearity*.

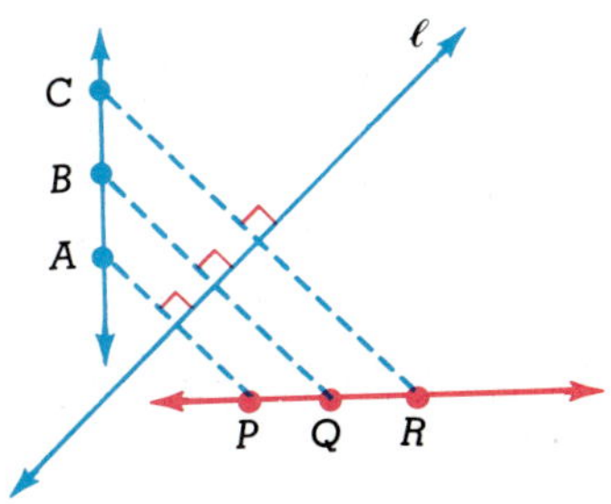

Points are collinear if and only if they lie on the same line.

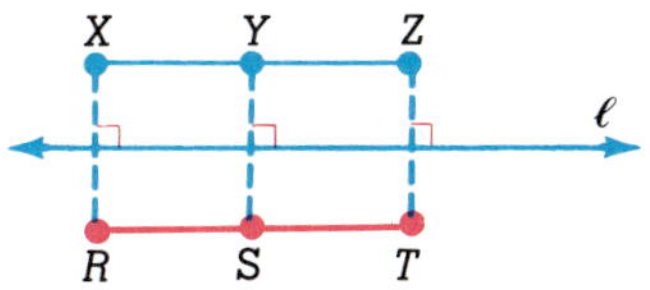

The reflection image of Y is between the image of X and Z if and only if Y is between X and Z. Thus, *reflections preserve betweenness of points*.

Reflections also preserve angle measure and distance measure. $\triangle XYZ$ is the reflection image of $\triangle ABC$. By measuring the corresponding parts of $\triangle ABC$ and $\triangle XYZ$ it can be shown that $\triangle ABC$ is congruent to $\triangle XYZ$. Because only the orientation of a geometric figure is changed, *a reflection is an isometry*.

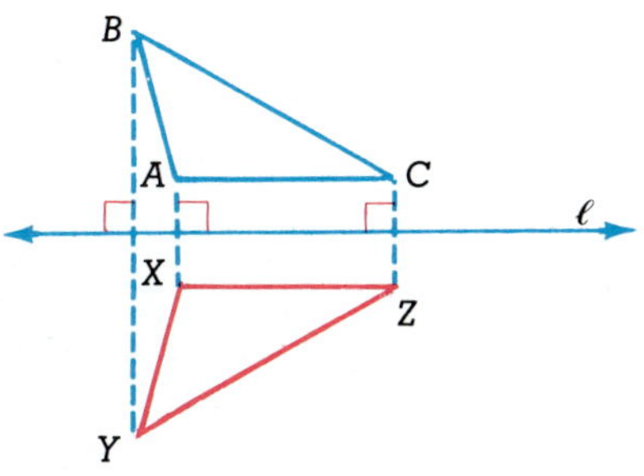

example 1

Name the reflection image of $\triangle ABC$ with respect to ℓ.

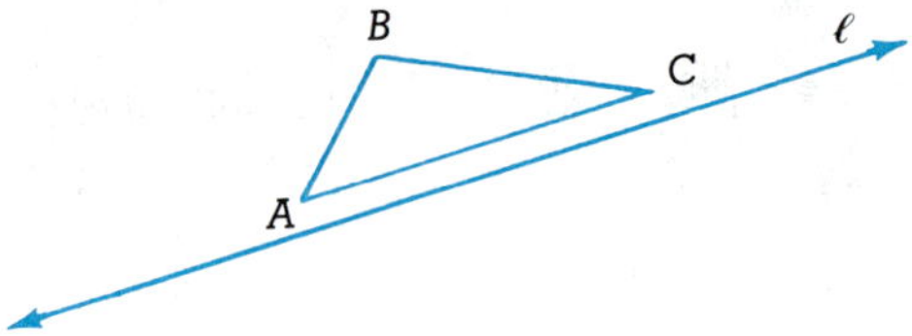

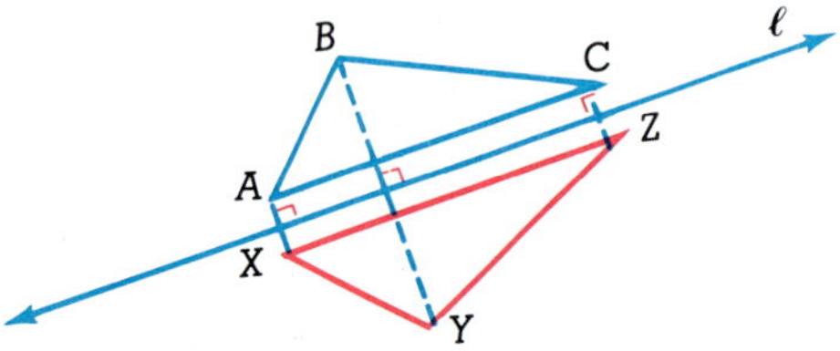

Construct perpendiculars from A, B, and C through ℓ. Locate X, Y, and Z so that ℓ is the perpendicular bisector of $\overline{AX}$, $\overline{BY}$, and $\overline{CZ}$. X, Y, and Z are the corresponding vertices of A, B, and C. The reflection image of $\triangle ABC$ is found by connecting the vertices X, Y, and Z.
$\triangle ABC \rightarrow \triangle XYZ$.

A line can be drawn through many plane figures so that the figure on one side is a reflection image of the figure on the opposite side. In such a case, the line of reflection is called a **line of symmetry.**

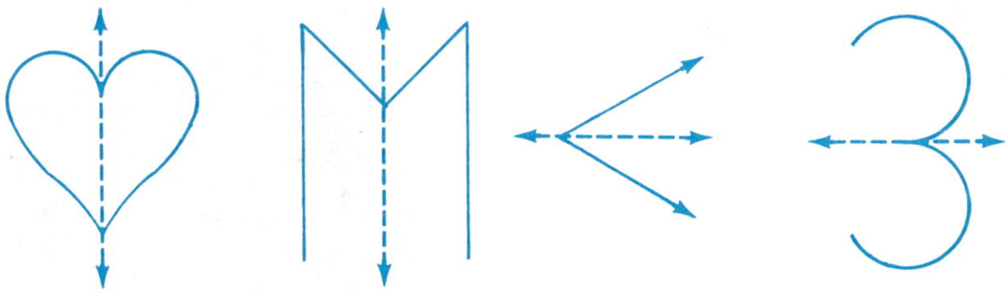

example 2

Test the given lines to see if they are lines of symmetry for rectangle *ABCD*.

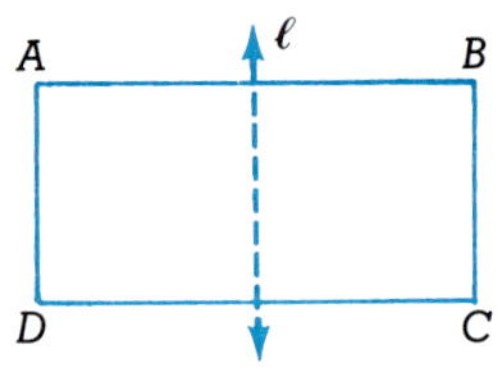

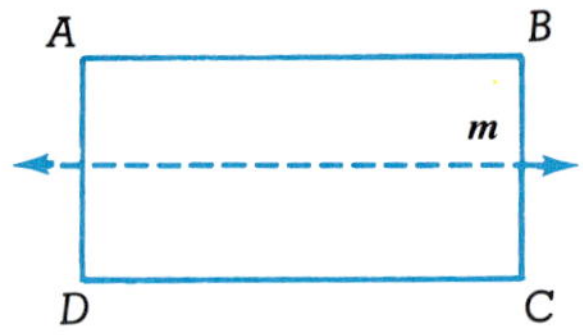

By measuring, it can be seen that ℓ is the perpendicular bisector of both $\overline{AB}$ and $\overline{CD}$. Any point to the left of ℓ has its reflection image to the right of ℓ. Likewise, by measuring it can be seen that m is the perpendicular bisector of $\overline{AD}$ and $\overline{BC}$. Also, any point above m has its reflection image below m. Therefore, ℓ and m are lines of symmetry for the rectangle $ABCD$.

exercises

Exploratory For each of the following figures name the reflection image with respect to ℓ.

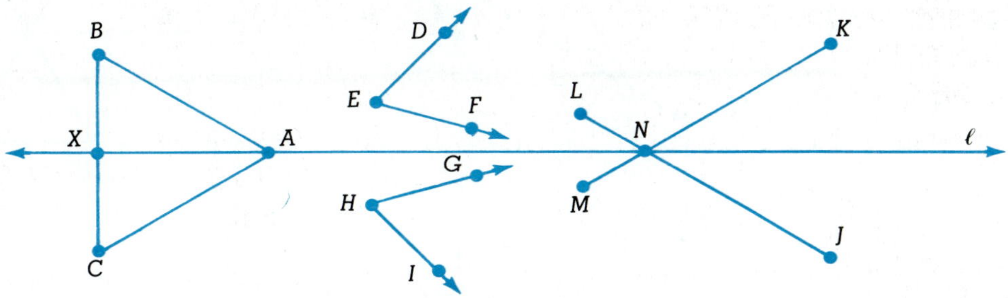

1. A
2. B
3. $\overline{AB}$
4. F
5. M
6. K
7. $\overline{KM}$
8. $\angle DEF$
9. $\overline{EF}$
10. $\triangle BXA$
11. N
12. $\overline{DE}$

Written Copy each figure below. Then draw the reflection image of each figure with respect to m. Show your work.

1.

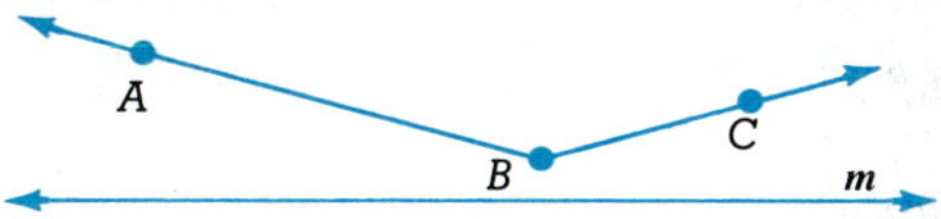

2.

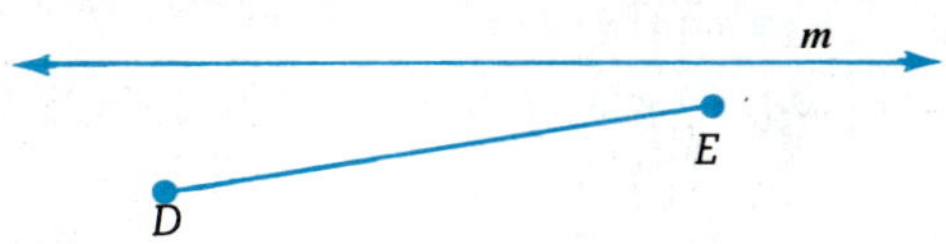

3.

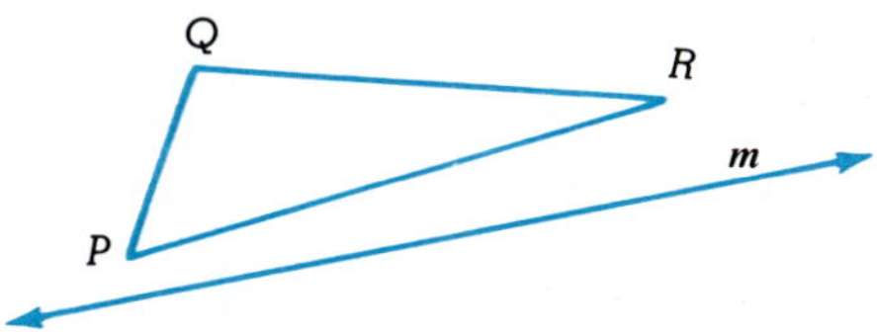

4.

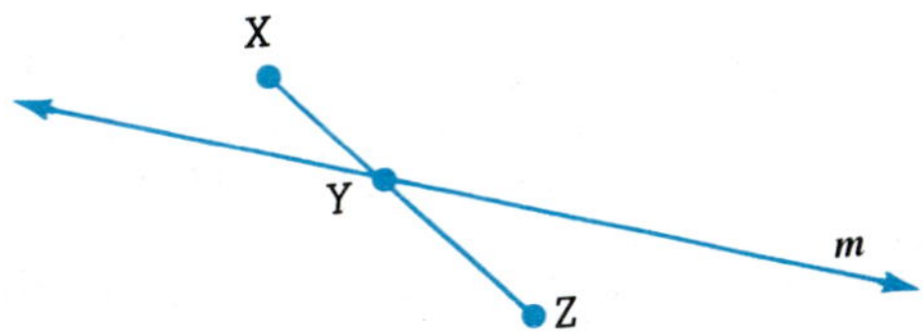

5.

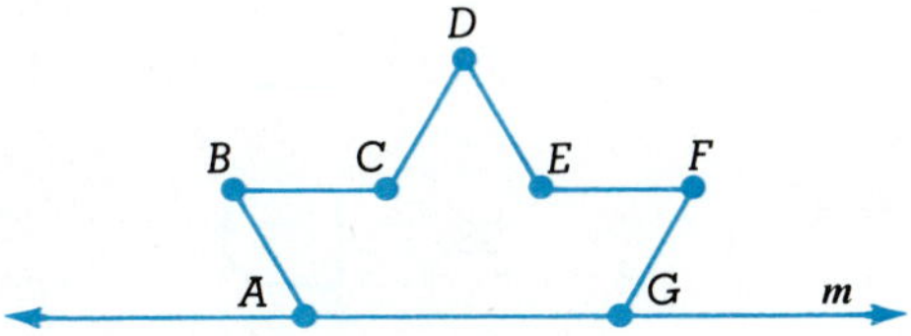

6.

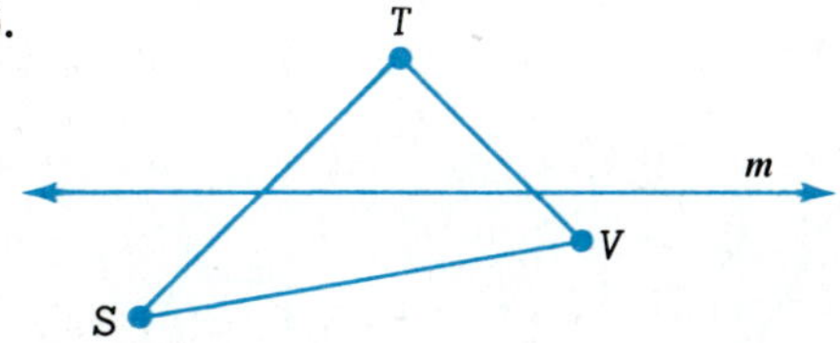

For each of the following, tell whether or not ℓ is a line of symmetry. Write *yes* or *no*. Then, explain your answer.

7. 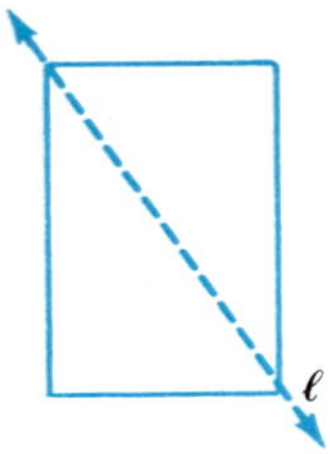

8. 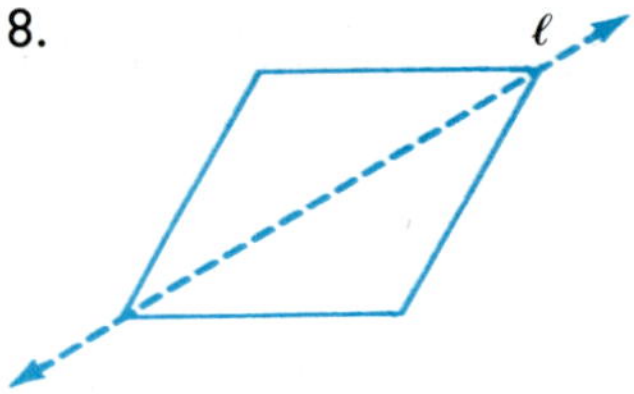

9. 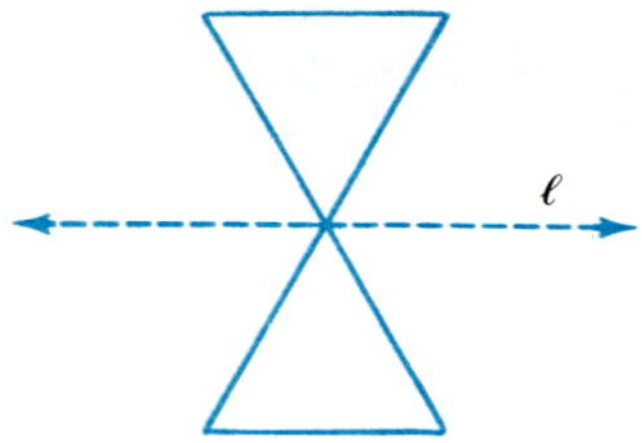

10. 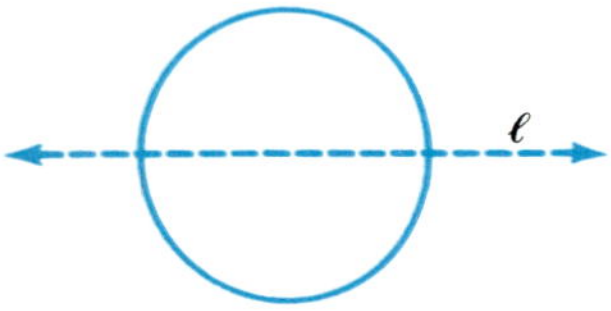

11. 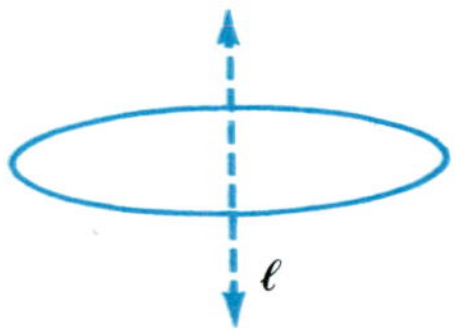

12. 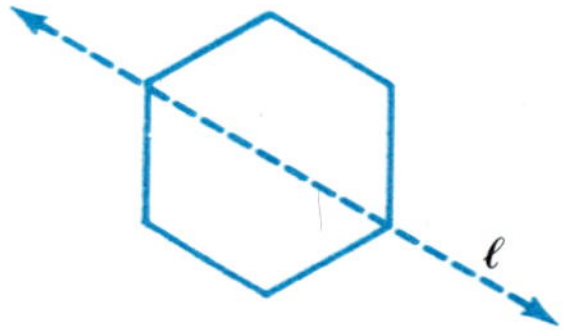

13–18. Copy each of the figures in problems 7–12. Draw a line of symmetry other than the one given. If there is no other line of symmetry, write *none*.

Copy each of the following. Draw the reflection image if R is the point of reflection.

19. 

20. 

21. 

22. 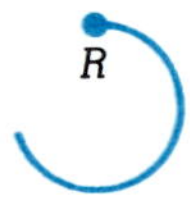

Copy the following letters. Draw all possible lines of symmetry for each. If none exist, write *none*.

23. A
24. B
25. C
26. D
27. E
28. F
29. G
30. H
31. I
32. J
33. K
34. L
35. N
36. P
37. Q
38. R

14–3 Translations

Two successive reflections together are called a composite of reflections.

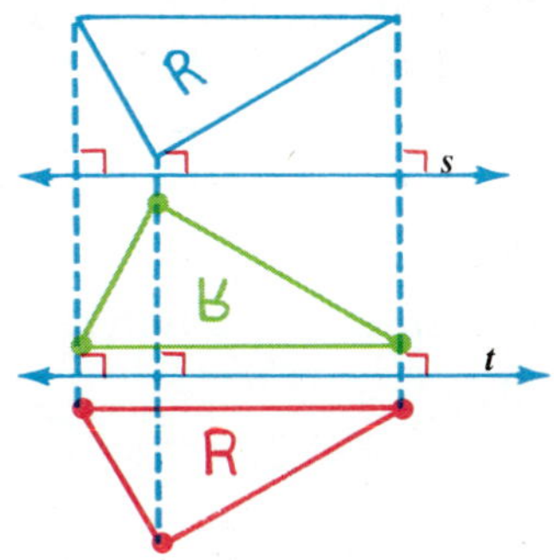

A racing car speeds along a track to the finish line. The result of a movement in one direction is a transformation called a **translation.**

To find a translation image, perform two reflections in a row with respect to two parallel lines. For example, the translation image of the blue figure with respect to the parallel lines, s and t, is the red figure. First the blue figure is reflected onto the green figure with respect to s. Then, the green figure is reflected onto the red figure with respect to t.

example 1

Draw the translation image of $\triangle ABC$ with respect to the parallel lines s and t.

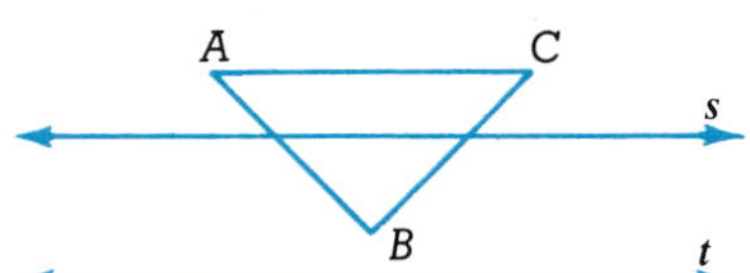

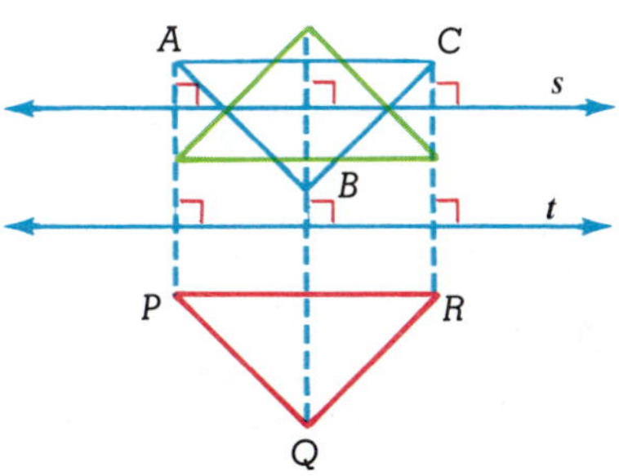

First draw the reflection image of $\triangle ABC$ with respect to s. Then draw the reflection image of that figure with respect to t. $\triangle PQR$ is the translation image of $\triangle ABC$.

A translation is an isometry

Since translations are composites of two reflections, all properties preserved by reflections are preserved by translations. The properties preserved are collinearity, angle and distance measure, and betweenness of points.

exercises

Exploratory In the figure below m and n are parallel. For each of the following name the reflection image with respect to the given line.

1. A, m 2. F, m 3. E, m
4. B, n 5. H, n 6. G, n
7. A, n 8. F, n 9. E, n

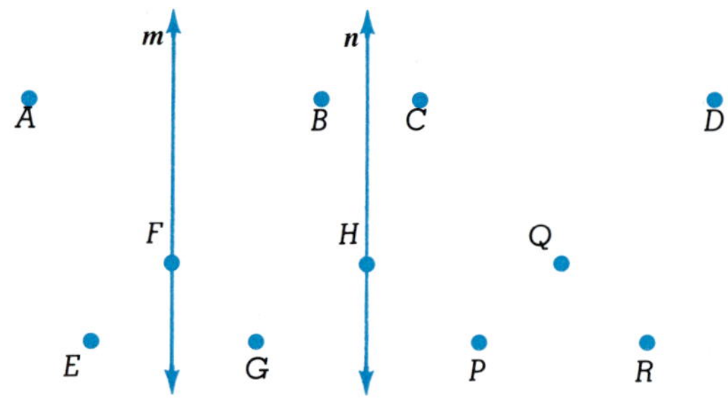

For each of the following name the translation image with respect to m, then n.

10. A 11. F 12. E

Written For each of the following ℓ and m are parallel. State whether or *not* each red figure is a translation image of the blue figure. Write *yes* or *no*. Then, explain your answer.

1.
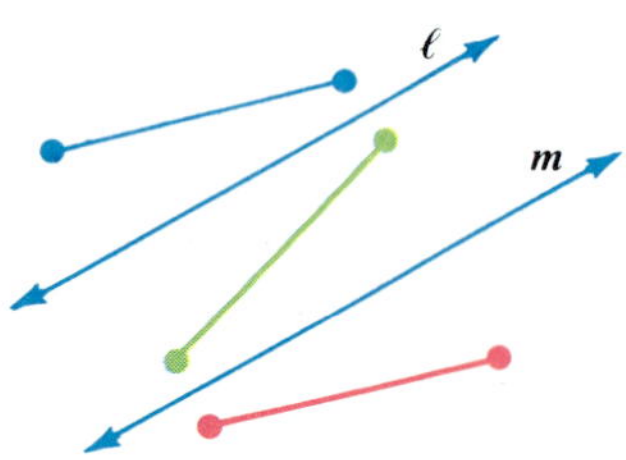

2.
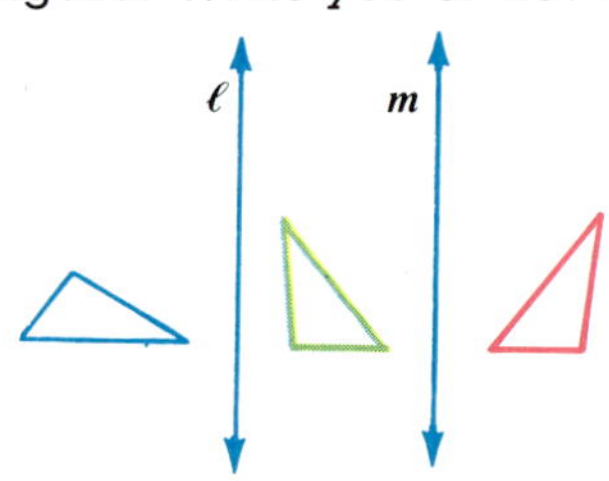

3.
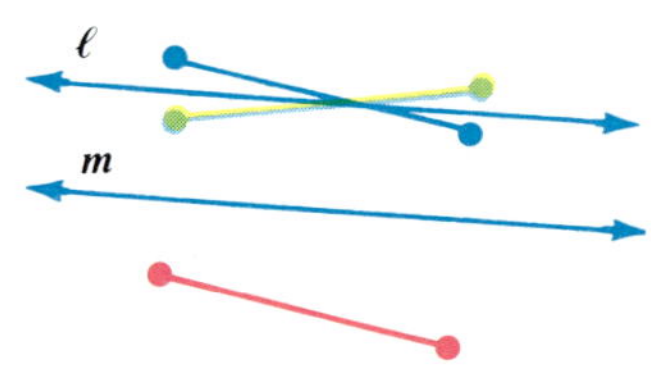

4.
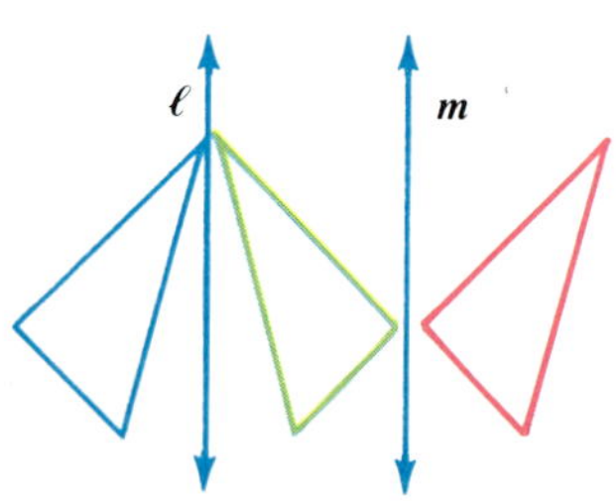

5.
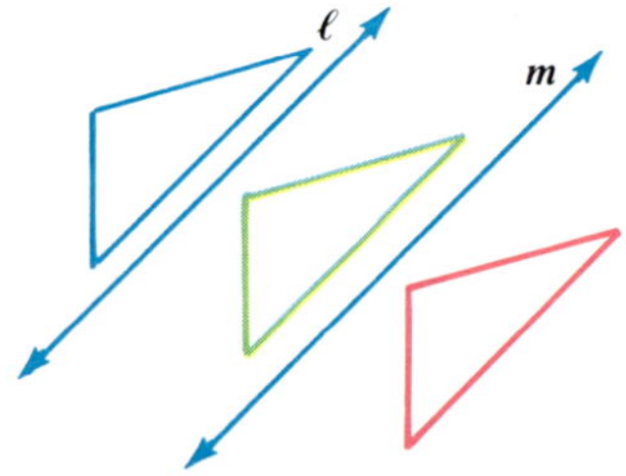

6.
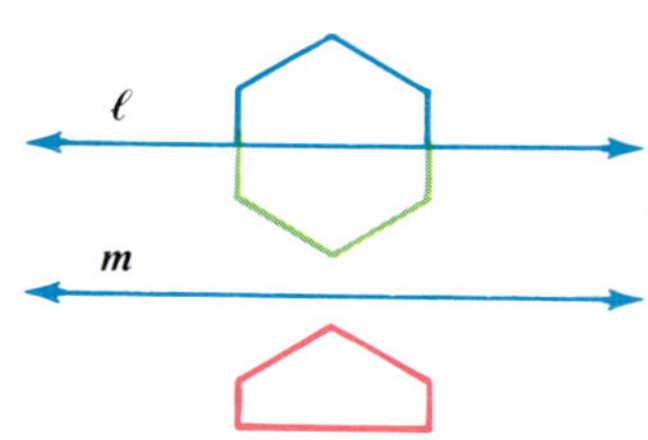

7–12. Name the reflections with respect to ℓ, in problems 1–6. If there are none, write *none*.

13–18. Name the reflections with respect to m, in problems 1–6. If there are none, write *none*.

Copy each of the following. Then find the translation image of each geometric figure with respect to the parallel lines *s* and *t*.

19. 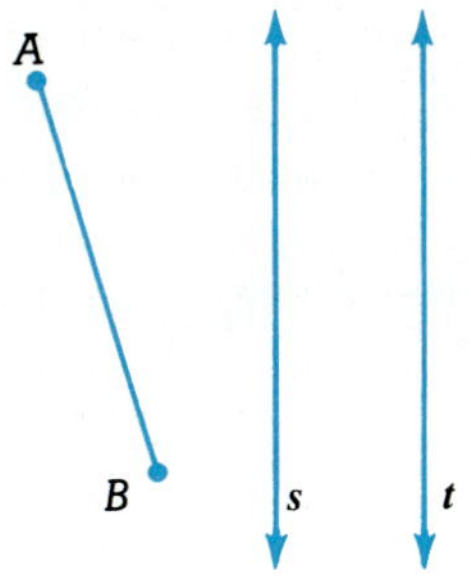

20. 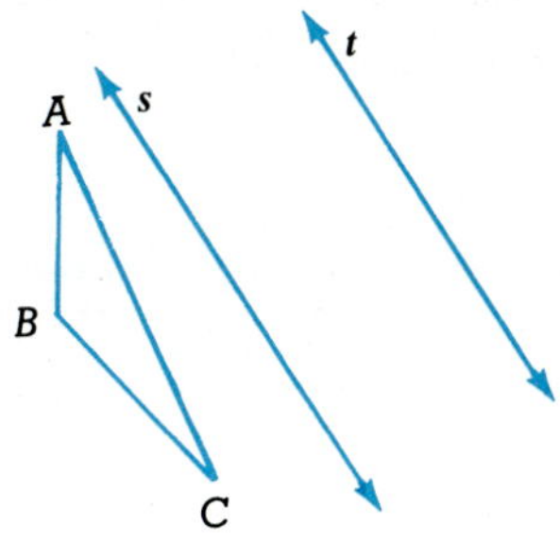

21. 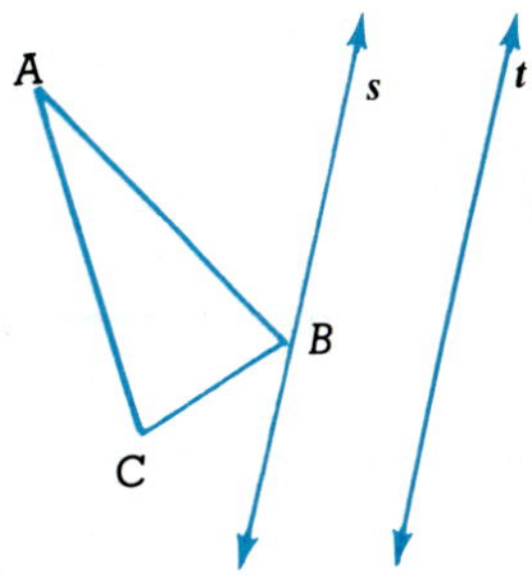

22. 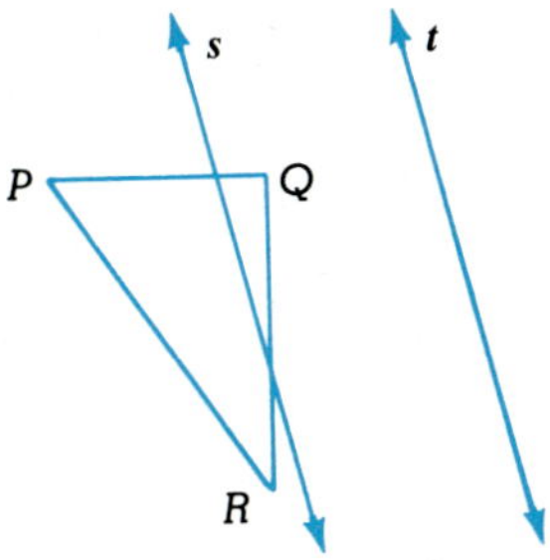

23. 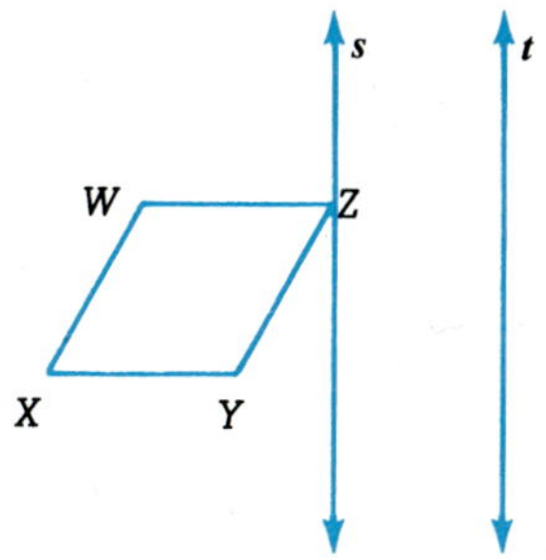

24. 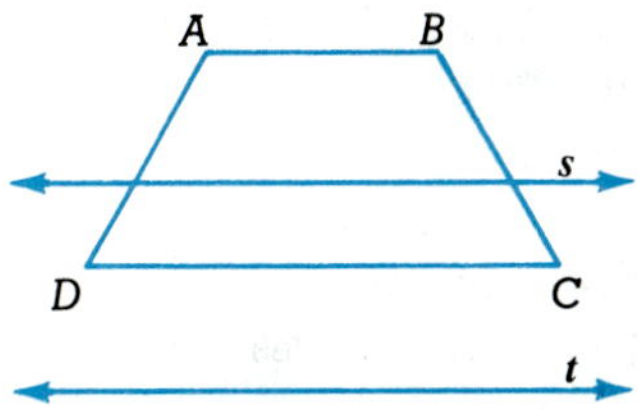

Use the figures below to name each of the following triangles. Assume $s \parallel t$.

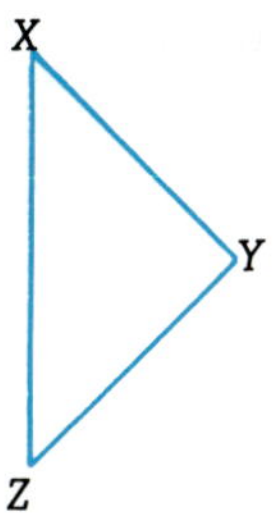

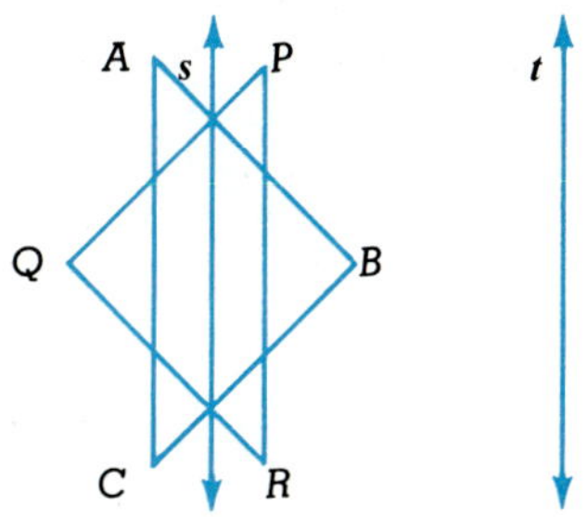

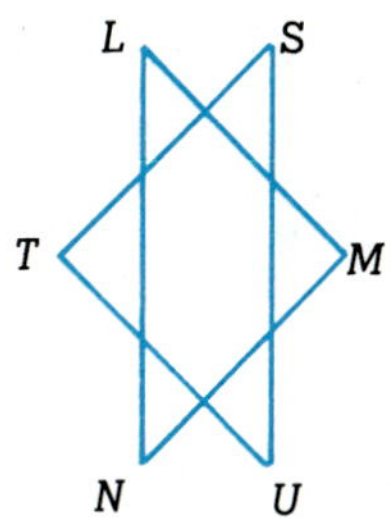

25. Reflection image of $\triangle ABC$ with respect to *s*.
26. Reflection image of $\triangle XYZ$ with respect to *s*.
27. Reflection image of $\triangle PQR$ with respect to *t*.
28. Translation image of $\triangle ABC$ with respect to *s* and *t*.
29. Translation image of $\triangle PQR$ with respect to *s* and *t*.

30. Plan a proof to show that the translation image of $\triangle ABC$ with respect to the parallel lines ℓ and *m* preserves collinearity.
31. Plan a proof to show that the translation image of $\triangle ABC$ with respect to the parallel lines ℓ and *m* preserves betweenness of points.
32. Plan a proof to show that the translation image of $\triangle ABC$ with respect to the parallel lines ℓ and *m* preserves angle and distance measure.

14–4 Rotations

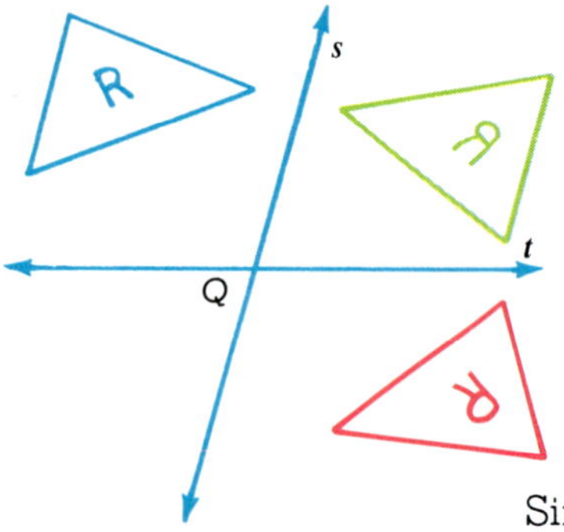

The composite of two reflections with respect to two intersecting lines is a transformation called a **rotation.** The reflection of the blue figure with respect to s is the green figure. The reflection of the green figure with respect to t is the red figure. Since s and t intersect, the red figure is the rotation image of the blue figure. Q, the intersection of the two lines, is called the **center of rotation.**

A rotation is an isometry.

Since rotations are composites of two reflections, all properties preserved by reflections are preserved by rotations.

example 1

Suppose t and s intersect and $\triangle ABC$ is on one side of t and s. Construct the rotation image of $\triangle ABC$ with respect to s and then t.

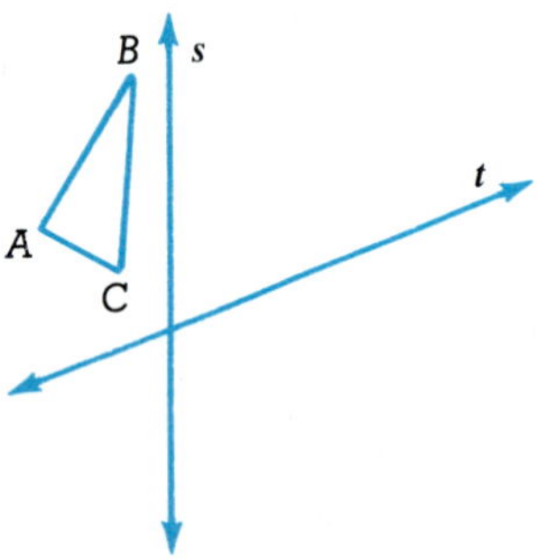

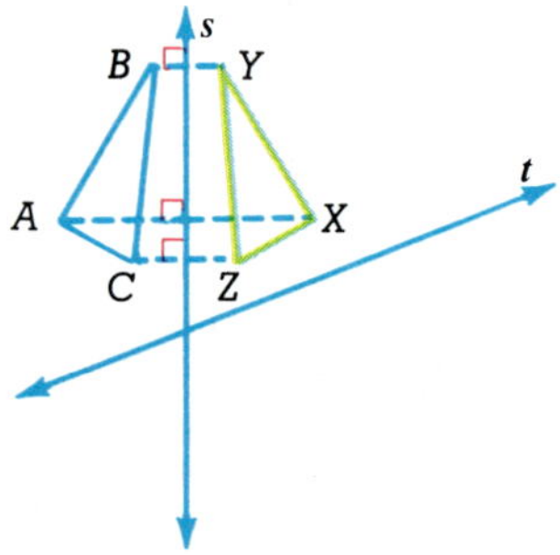

First, reflect $\triangle ABC$ with respect to s. The image is $\triangle XYZ$.

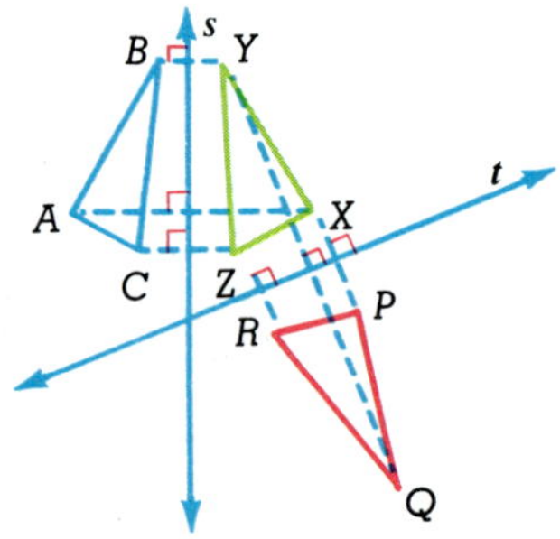

Then, reflect $\triangle XYZ$ with respect to t. The image is $\triangle PQR$.

The figure at the right shows how two reflections can be used to find the rotation image of $\overline{AB}$ with respect to ℓ and m. The image is $\overline{PQ}$.

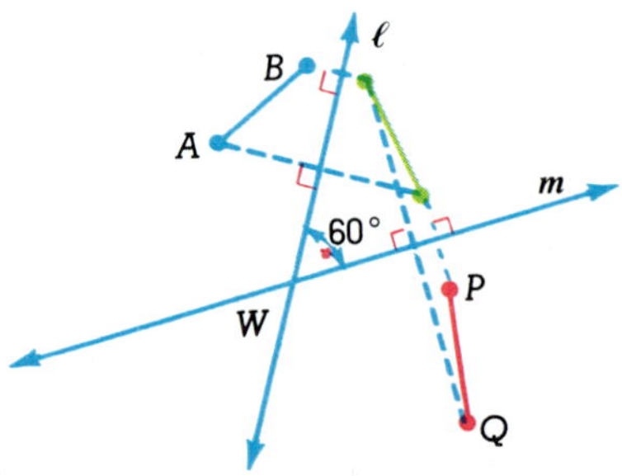

The same rotation image can be determined by using angles. Notice that ℓ and m form a 60-degree angle. It can be shown that $\angle AWP$ measures 2(60) or 120. Also, $\overline{AW} \cong \overline{WP}$. Likewise, it can be shown that if segments are drawn from B to W and from Q to W, then $m\angle BWQ = 120$ and $\overline{BW} \cong \overline{WQ}$.

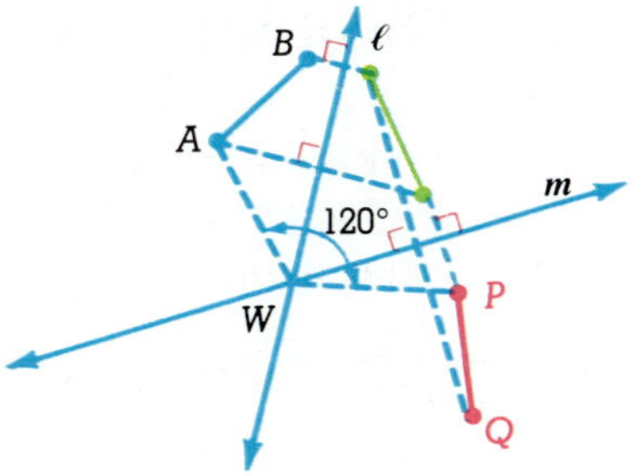

The angles, $\angle AWP$ and $\angle BWQ$, are called **angles of rotation.** In both cases, the measure of the angles is 2(60) or 120.

$\angle AWP$ is determined by A, the center of rotation, W, and the rotation image, P.

In a given rotation, if A is the preimage, P is the image, and W is the center of rotation, then the measure of the angle of rotation, $\angle AWP$, equals twice the measure of the angle between the intersecting lines of reflection.

Postulate 14–1

The following example shows how to find a rotation image using the angles of rotation.

example 2

The intersection of ℓ and m at P forms a 40-degree angle. Use the angles of rotation to find the rotation image of $\overline{XY}$.

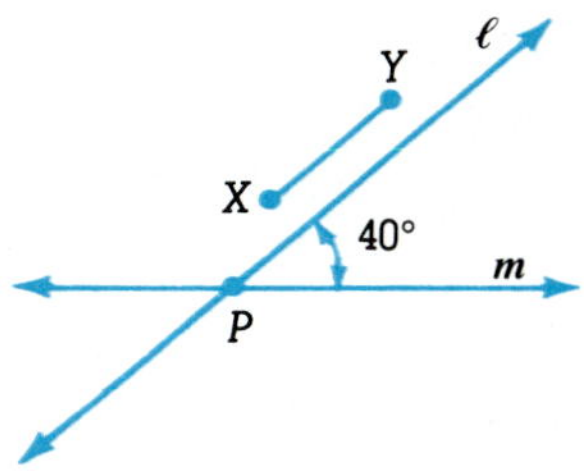

The angle of rotation has degree measure 2(40), or 80.

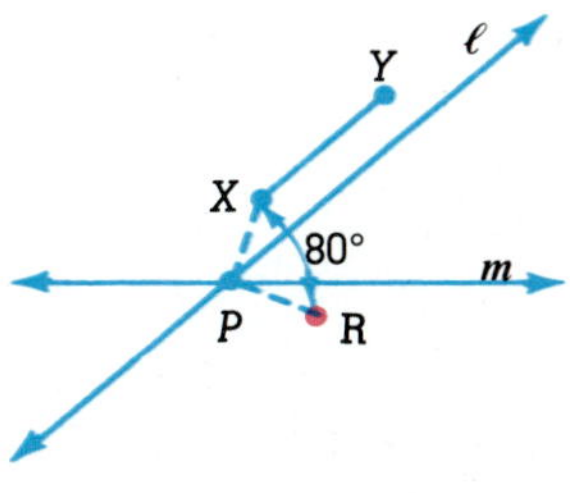

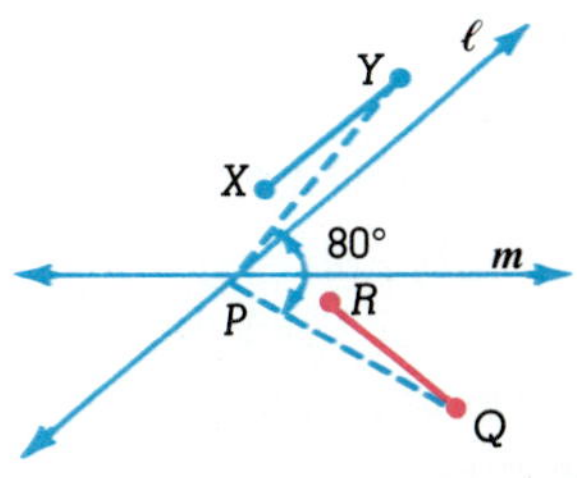

Construct $\angle XPR$ so that its degree measure is 80, and $\overline{XP} \cong \overline{PR}$. Then, construct $\angle YPQ$ so that its degree measure is 80, and $\overline{YP} \cong \overline{PQ}$. The rotation image of $\overline{XY}$ is $\overline{RQ}$.

exercises

Exploratory For each of the following state whether or not the indicated composition of reflections is a rotation. Explain your answer.

1.

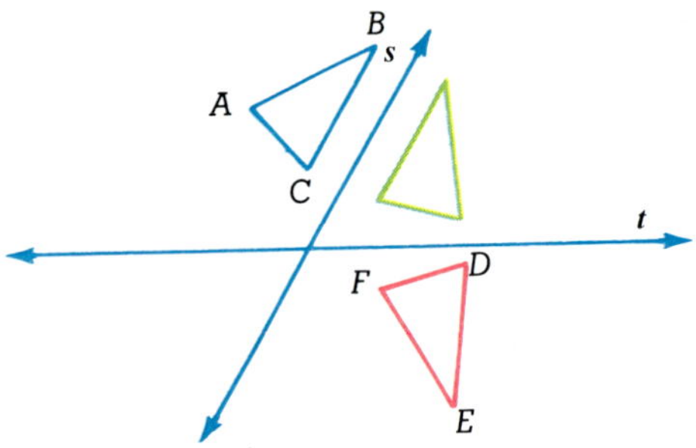

2.

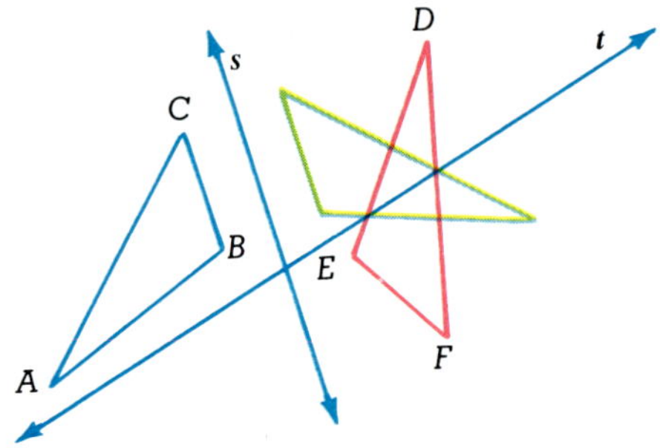

Use the figure below to solve each problem.

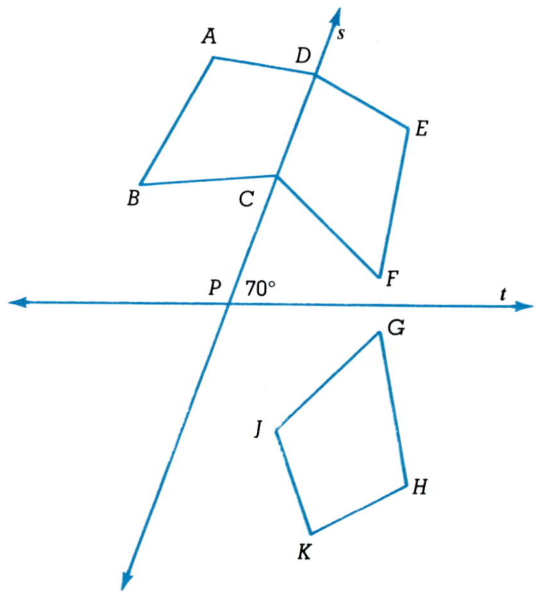

3. Find the reflection image of quadrilateral $ABCD$ with respect to s.
4. Find the reflection image of quadrilateral $CDEF$ with respect to t.
5. Find the rotation image of quadrilateral $ABCD$ with respect to s and t.
6. Find the reflection image of quadrilateral $JKHG$ with respect to t.
7. Find the rotation image of quadrilateral $JKHG$ with respect to t and s.
8. Find the measure of $\angle APH$.
9. Find the measure of $\angle BPG$.
10. Find the measure of $\angle CPK$.
11. Find the measure of $\angle DPJ$.
12. Find the reflection image of C with respect to s.
13. Find the rotation image of $\overline{BD}$ with respect to s and t.

Written Solve each of the following.

1. Draw a line segment and two intersecting lines. Find the rotation image of the line segment with respect to the two intersecting lines.
2. Draw a triangle and two intersecting lines. Find the rotation image of the triangle with respect to the two intersecting lines.

Two lines intersect to form the given angle measurements. Find the measurement of the angle of rotation.

3. 30°
4. 45°
5. 60°
6. 37°

Copy each of the following. Then, use the angles of rotation to find the rotation image of each geometric figure with respect to s and t.

7.

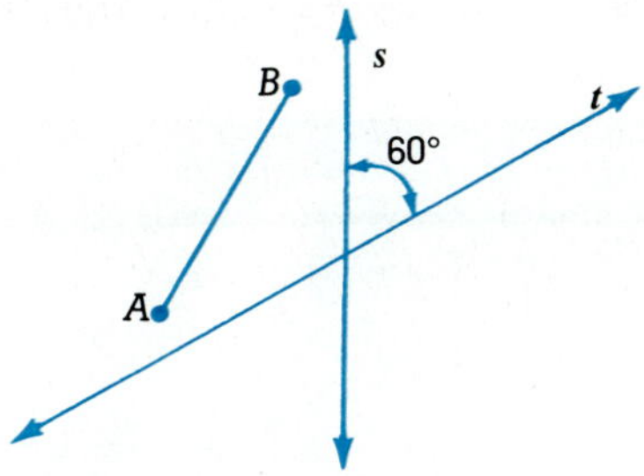

8.

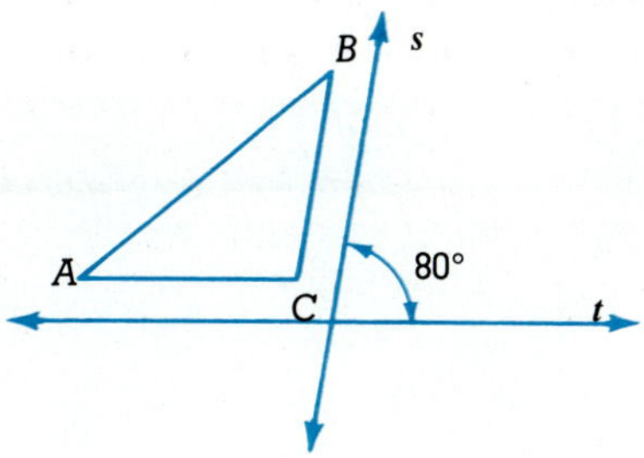

9.

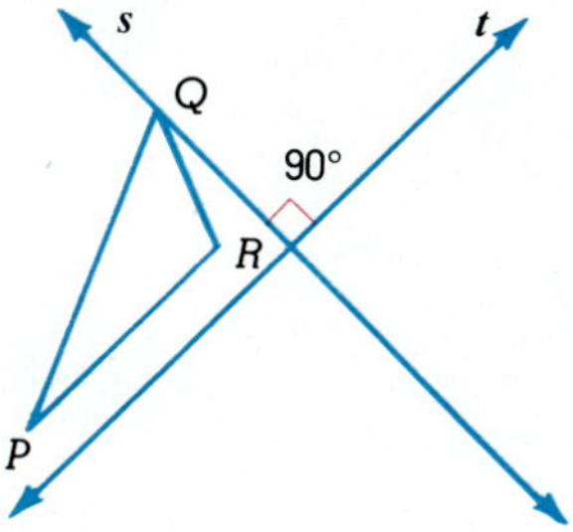

10.

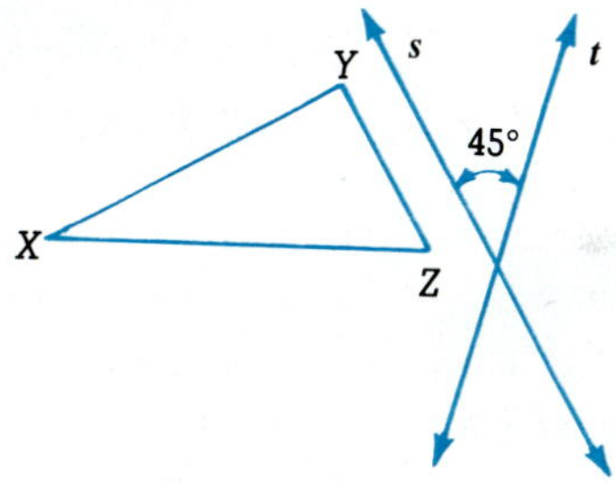

11.

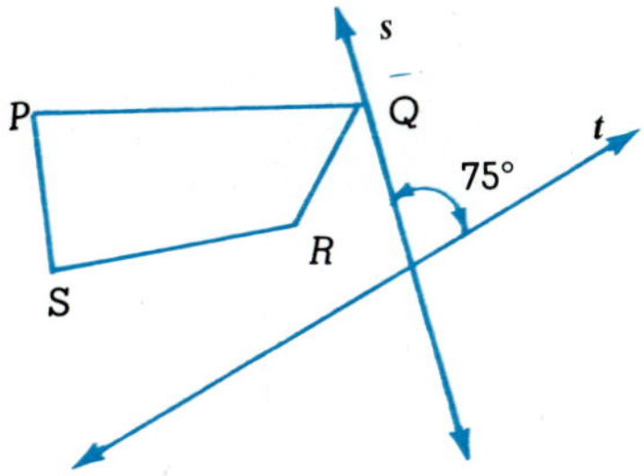

12.

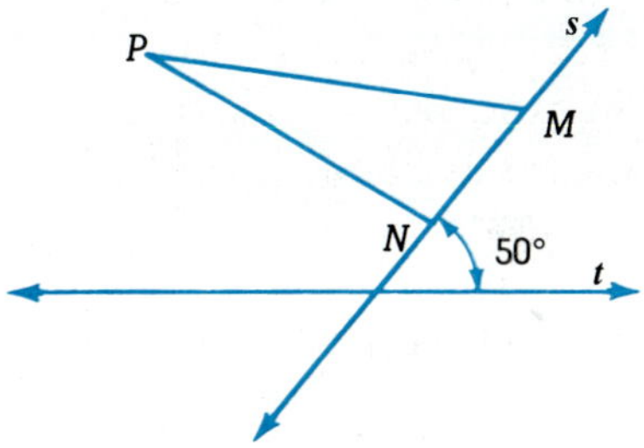

Tell whether or *not* each of the following is preserved by rotation. Write *yes* or *no*.

13. collinearity 14. betweenness of points 15. angle measure 16. distance measure

Challenge Plan a proof for each of the following.

1. The rotation image of $\triangle PQR$ with respect to the intersecting lines ℓ and m preserves collinearity.
2. The rotation image of $\triangle PQR$ with respect to the intersecting lines ℓ and m preserves betweenness of points.
3. The rotation image of $\triangle PQR$ with respect to the intersecting lines ℓ and m preserves angle and distance measure.

CAREERS

Physicist

Physicists use symmetry to describe various subatomic particles. The following diagrams are arrangements used to describe the interrelationships among various groups of these particles.

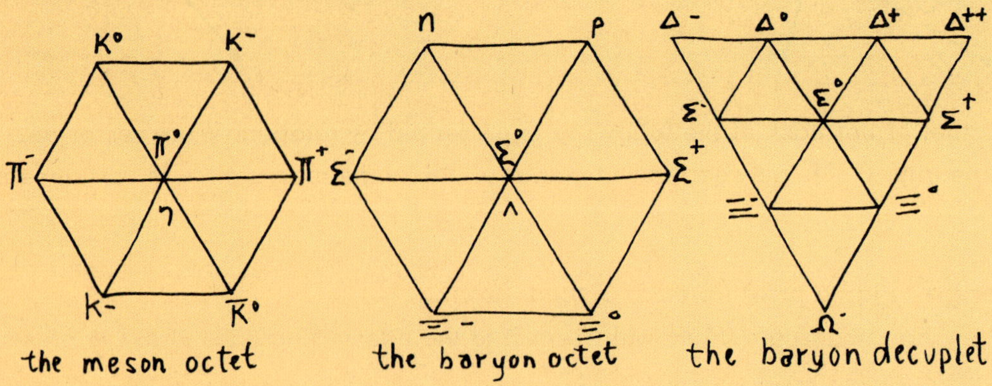

Exercises How would you use symmetry to describe each of the following?

1. the meson octet
2. the baryon octet
3. the baryon decuplet

14–5 Composites

The picture at the left was taken using mirror reflections. How many reflections do you see?

Recall that composites of two reflections, rotations and translations, are isometries. Also, other composites of reflections also can be described as isometries. The number of reflections and the relationship of the lines of reflections determines the kind of isometry.

Consider three lines intersecting at one point. The composite of three reflections with respect to these lines actually is a composite of a rotation and a reflection. But, this rotation and reflection is the same as a single reflection with respect to a line.

In the figure at the right, $\triangle ABC$ is transformed to $\triangle GHF$. This transformation can be described in the following three ways.

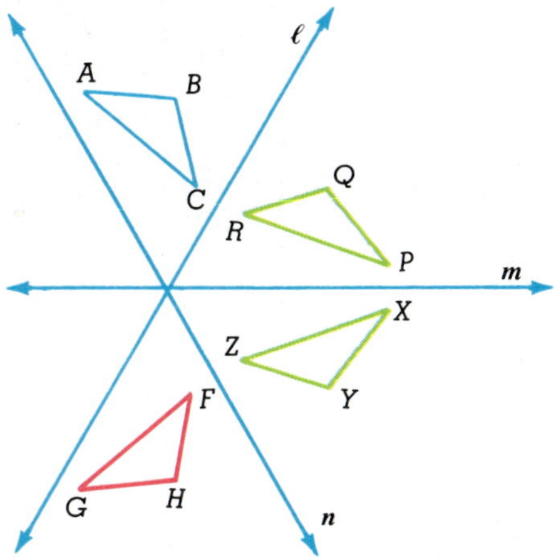

$$\triangle ABC \xrightarrow{\text{reflection}} \triangle PQR \xrightarrow{\text{reflection}} \triangle XYZ \xrightarrow{\text{reflection}} \triangle GHF$$

$$\triangle ABC \xrightarrow{\text{rotation}} \triangle XYZ \xrightarrow{\text{reflection}} \triangle GHF$$

$$\triangle ABC \xrightarrow{\text{reflection}} \triangle GHF$$

examples

1 **Construct the composite of three reflections of $\overline{AB}$ with respect to the three parallel lines ℓ, m and n.**

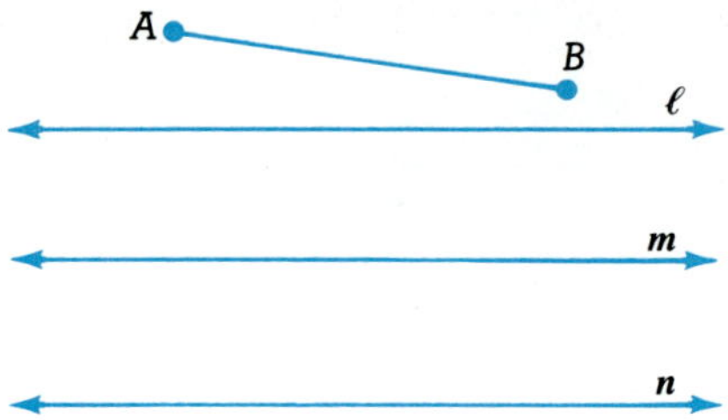

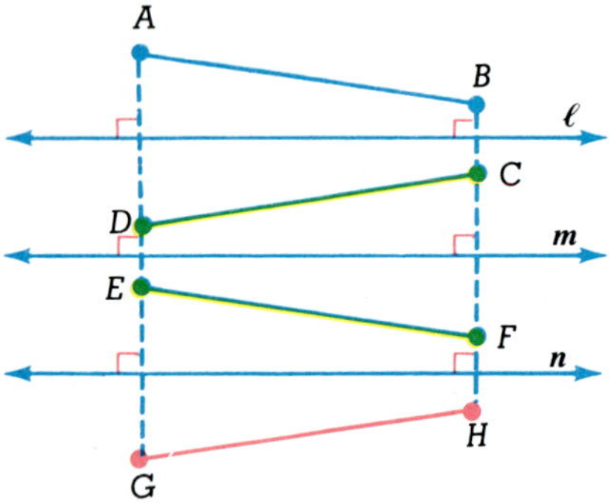

Construct perpendicular lines from A and B through ℓ, m, and n. Locate D and C so that ℓ bisects $\overline{AD}$ and $\overline{BC}$. Likewise, locate E, F, G and H. $\overline{GH}$ is the reflection image of $\overline{AB}$ with respect to the parallel lines ℓ, m, and n.

2 **Describe the single isometry that can be used to map $\overline{PQ}$ onto $\overline{VW}$.**

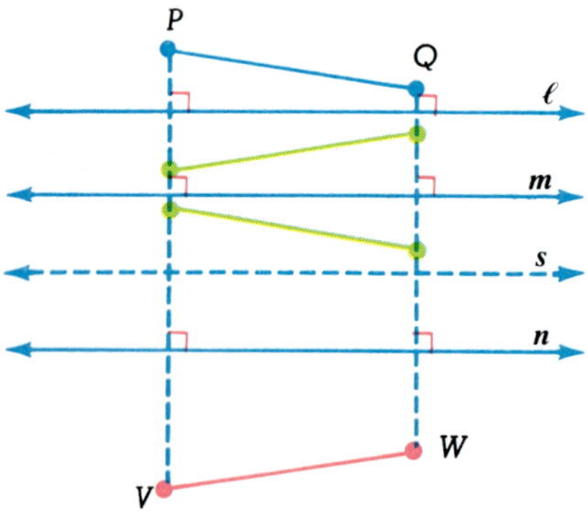

Draw $\overline{PV}$ and $\overline{QW}$ perpendicular to ℓ, m and n. Locate s, the perpendicular bisector of $\overline{PV}$ and $\overline{QW}$. Thus, $\overline{PQ}$ maps onto $\overline{VW}$ by a single reflection of $\overline{PQ}$ with respect to s.

The figure at the right shows a composite of four reflections with respect to the parallel lines ℓ, m, n, and t. A composite of two translations with respect to ℓ, m, n and t will also map $\triangle ABC$ onto $\triangle PQR$. But a single translation also can map $\triangle ABC$ onto $\triangle PQR$.

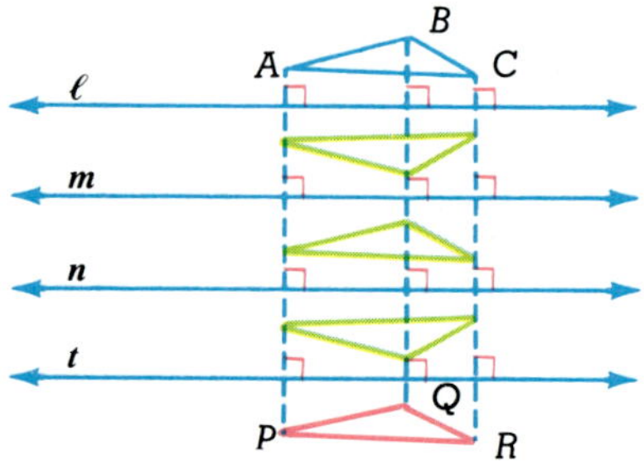

Using only ℓ, m, n, or t as the possible lines of reflection, there are two choices for finding the translation.

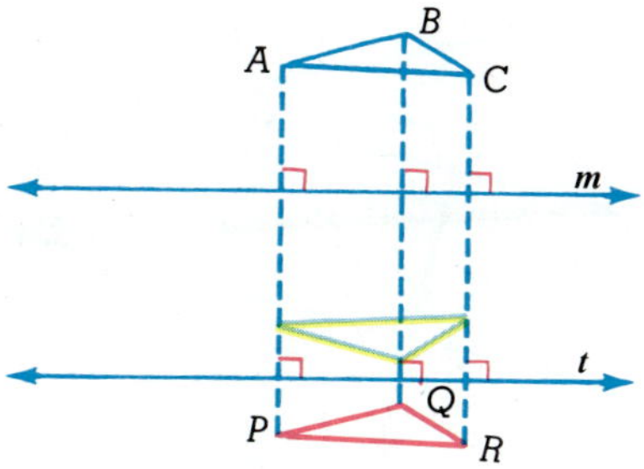

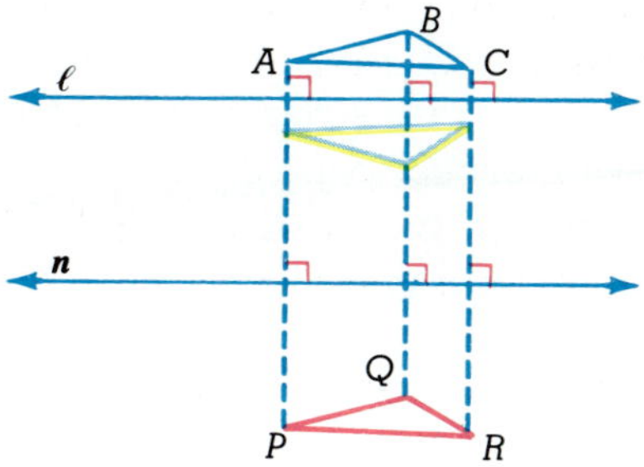

$\triangle ABC$ is mapped onto $\triangle PQR$ by a single translation with respect to either m parallel to t, or ℓ parallel to n.

example 3

Suppose $\triangle PQR \cong \triangle ABC$. Name an isometry that maps $\triangle PQR$ onto $\triangle ABC$. Construct the necessary lines of reflection.

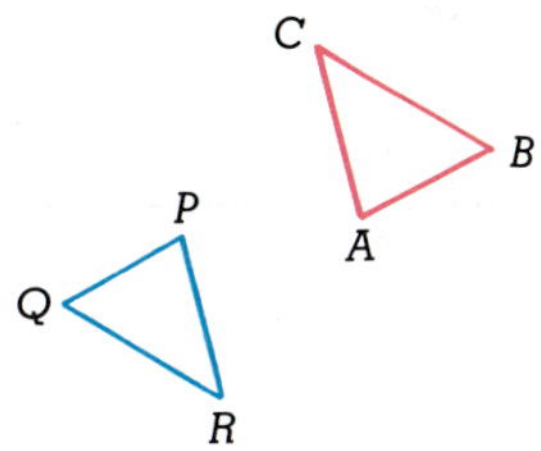

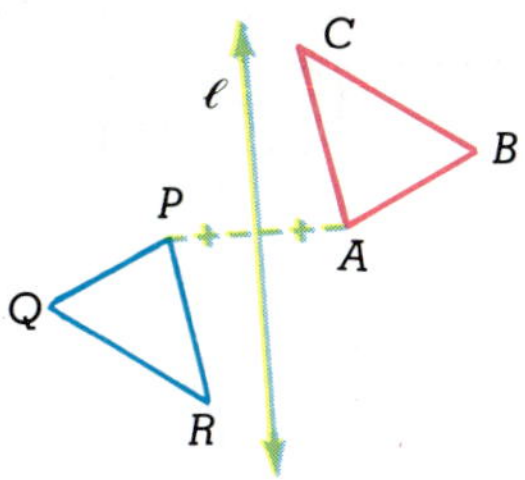

Construct ℓ, the perpendicular bisector of $\overline{PA}$.

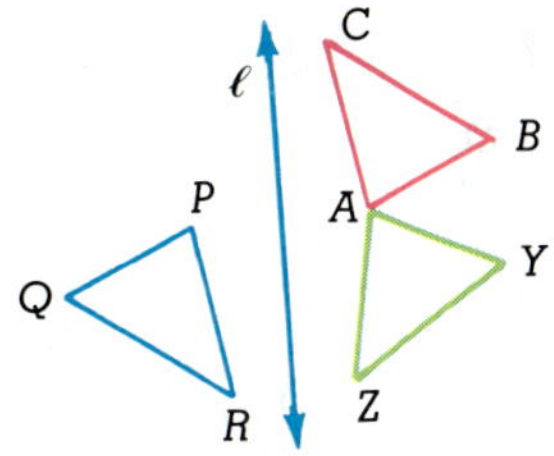

Reflect $\triangle PQR$ with respect to ℓ. Call this image $\triangle AYZ$.

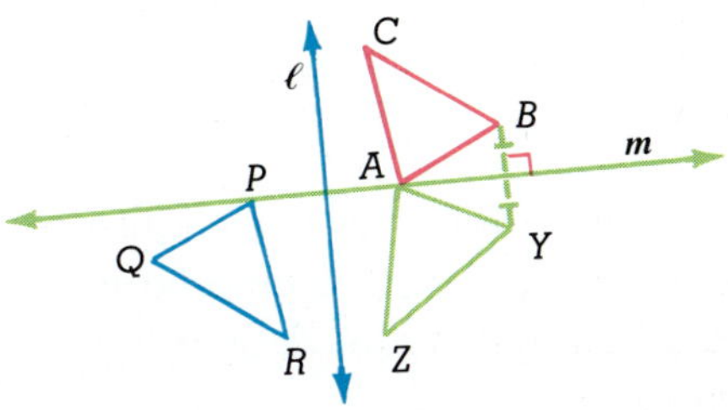

Construct m, the perpendicular bisector of $\overline{YB}$. The reflection of $\triangle AYZ$ with respect to m is $\triangle ABC$.
Since ℓ intersects m, and $\triangle ABC$ is a composite of two reflections, $\triangle PQR$ maps onto $\triangle ABC$ by a rotation transformation.

Sometimes, the result of a composition of three reflections is not a single reflection, translation, or rotation. The result is an isometry called a **glide reflection.**

In the figure at the right, $\triangle ABC$ is reflected onto $\triangle GHK$ with respect to the parallel lines ℓ and m and the transversal n. The result is a glide reflection.

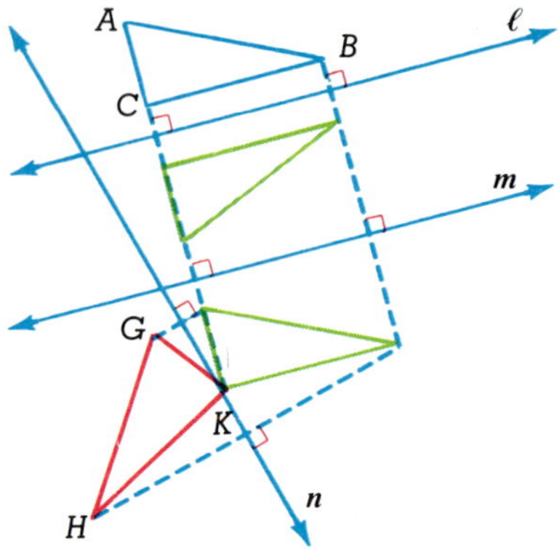

exercises

Exploratory **Solve each of the following for the figures labeled *A*, *B*, *C*, *D*, *E*, *F*, *G*, and *H*. Assume, ℓ, m, n, p, q, r, and s are parallel.**

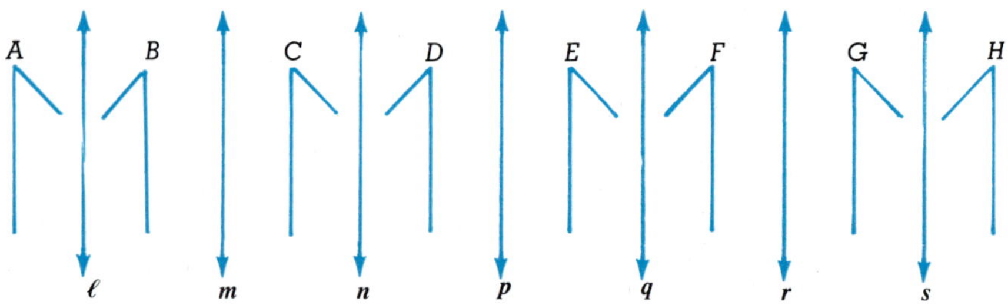

1. Name the reflection image of figure *A* with respect to n.
2. Name the image of figure *A* as a composite of three reflections with respect to ℓ, n, and q.
3. Name the reflection image of figure *C* as a composite of five reflections with respect to n, p, q, r, and s.
4. Name the reflection image of figure *G* as a composite of two reflections with respect to q and m.
5. Name the one reflection line that can be used to map figure *A* onto figure *H*.
6. Name the two reflection lines that can be used to map figure *A* onto figure *G*.
7. Name the three reflection lines that can be used to map figure *A* onto figure *F*.
8. Name the single isometry that maps figure *A* onto figure *E*.
9. Name the single isometry that maps figure *A* onto figure *F*.

Written **For each of the following copy the diagram. Then, solve each problem.**

1. Construct the composite of five reflections with respect to the parallel lines ℓ, m, n, p, and q.
2. Describe the single isometry that can be used to map $\overline{XY}$ onto its image.

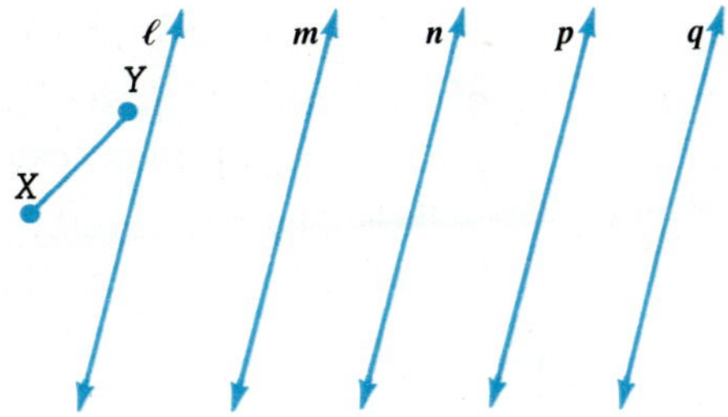

3. Construct the composite of six reflections of $\triangle ABC$ with respect to the parallel lines ℓ, m, n, p, q, and r.
4. Describe the single isometry that can be used to map $\triangle ABC$ onto its image with respect to n and r.
5. Find another pair of lines that can be used to form a single isometry that maps $\triangle ABC$ onto its image.
6. Name the different ways a composite of two or more reflections, with respect to parallel lines, can be described.

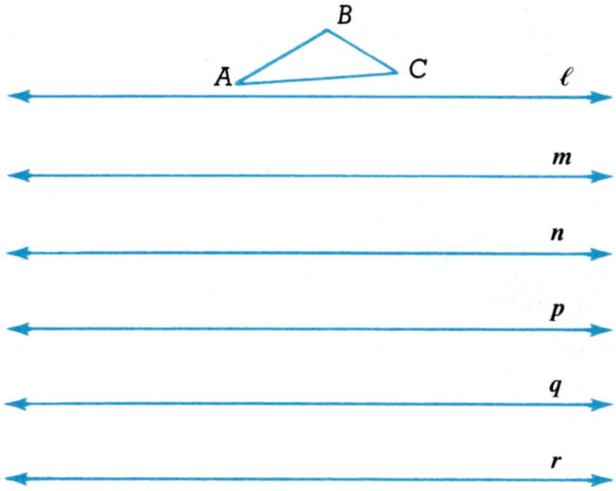

Name the kind of isometry that maps one figure onto the other.

7. 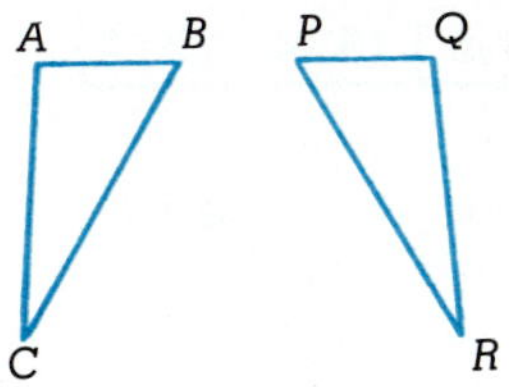

8. 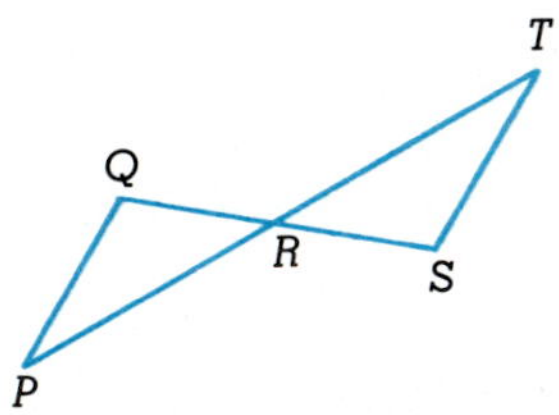

9. 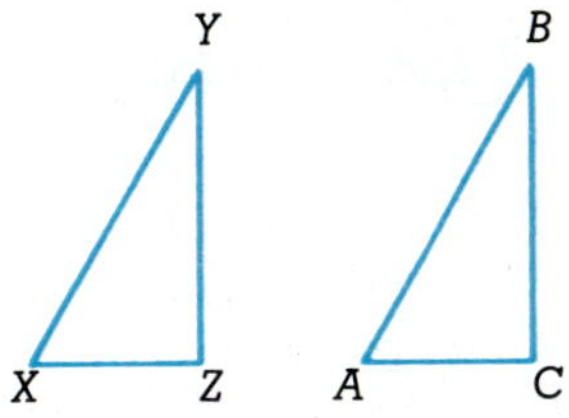

10. 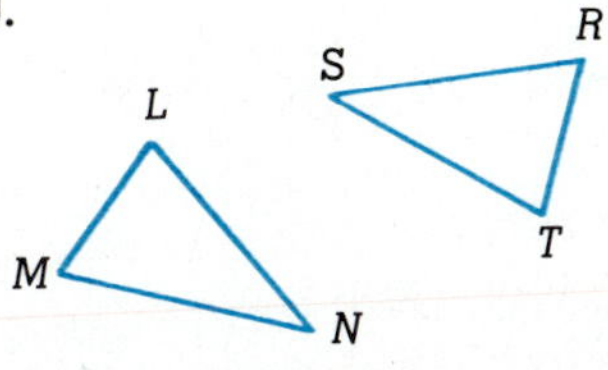

11. 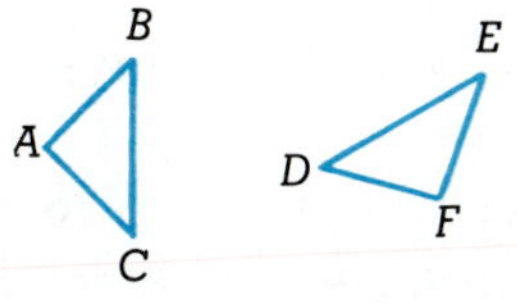

12. 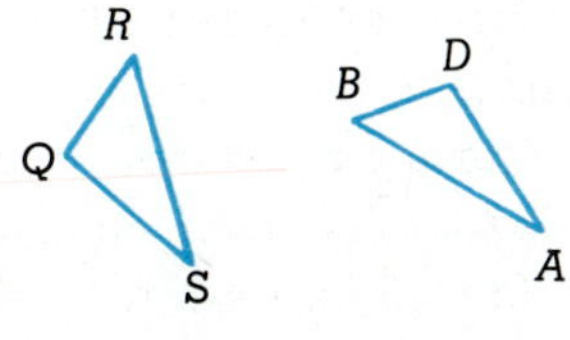

13–18. Copy each of the figures in problems **7–12.** Then, construct the lines of reflection.

A period of inflation is a time of rising prices. A given amount of money buys less than it did before. Inflation usually is described as a monthly or annual rate of price increases.

INFLATION 15%
GOVERNMENT WARNS

WASHINGTON (CEM) The typical family is paying 15% more than it did last year. According to a report issued by the Bureau of Labor Statistics, the consumer price index showed an annual rise of 15% in the cost of living, based on figures from the first two quarters of this year. John Jacobson

Suppose the rate of inflation for the year is 15%. This means that prices at the beginning of the year are about 15% higher than at the end of the year. *Not* every price rises 15%. Some prices go up, some go down, and some stay the same. On the average, the prices rise about 15%.

The government publishes the *consumer price index* which shows price changes of certain items purchased by large numbers of people. It considers items such as food, housing, cars, fuel, clothing, and so on.

CONSUMER PRICE INDEXES, 1970–1979
Source: Department of Labor, Bureau of Labor Statistics

Year	1970	1971	1972	1973	1974	1975	1976	1977	1978	1979
Index	116.3	121.3	125.3	133.1	147.7	161.2	170.5	181.5	195.4	220.4

Changes in the consumer price index can be used as a measure of the rate of inflation.

Example Find the rate of inflation for 1971.

$$\textit{rate of inflation for 1971} = \frac{\textit{change in index for year}}{\textit{index for previous year}}$$

$$= \frac{121.3 - 116.3}{116.3}$$

$$= \frac{5}{116.3} \text{ or about } 4.3\%$$

Exercises **1–8.** Find the annual rate of inflation for each year, 1972–1979.

14–6 Dilations

Seen through a magnifying glass, a figure is an enlargement of the original figure.

A geometric figure can be altered in size. Enlarging or reducing the figure will *not* change its shape. This type of transformation is called a **dilation** or a *similarity transformation*.

In the figure at the right, $\triangle XYZ$ is the dilation image of $\triangle PQR$. The measure of the distance from C to a point on $\triangle XYZ$ is twice the measure of the distance from C to a point on $\triangle PQR$. For example, the following equations hold.

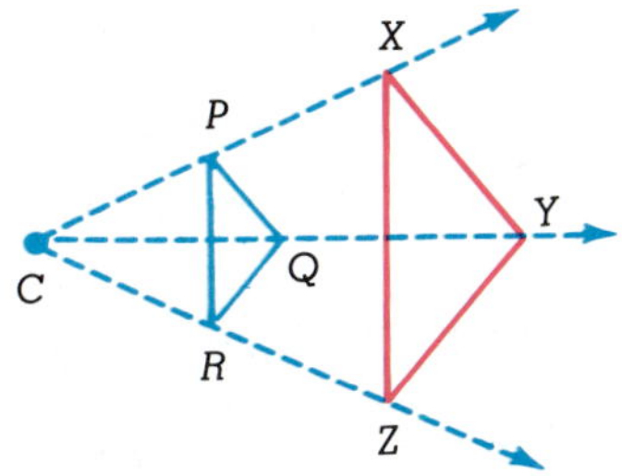

$$CX = 2(CP)$$
$$CY = 2(CQ)$$
$$CZ = 2(CR)$$

In this transformation, $\triangle PQR$, with **center** C and a **scale factor** of 2, is enlarged to $\triangle XYZ$.

The figure at the right shows a dilation where the preimage $\overline{AB}$ is reduced to $\overline{ED}$ by a scale factor of $\frac{1}{3}$. Thus, $CE = \frac{1}{3}(CA)$. Therefore, $\frac{CE}{CA} = \frac{1}{3}$ and $\frac{CD}{CB} = \frac{1}{3}$. By proving $\triangle CAB \sim \triangle CED$, it can be shown that $ED = \frac{1}{3}(AB)$.

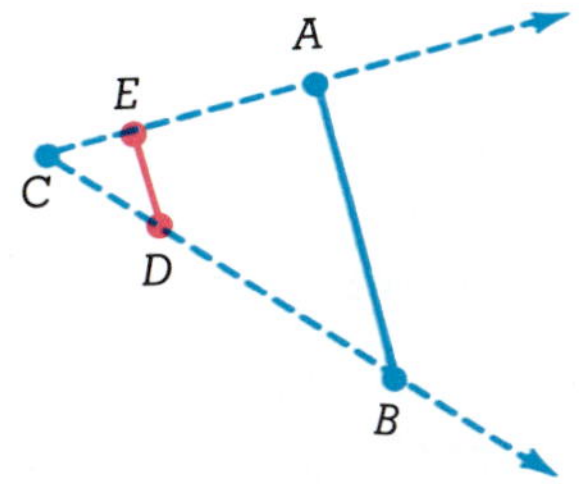

If a dilation with center C and a scale factor k maps A onto E and B onto D, then $ED = k(AB)$. Theorem 14–1

Notice that when the scale factor is 3, the figure is enlarged. When the scale factor is $\frac{1}{3}$, the figure is reduced. In general, if k is the scale factor for a dilation, then the following is true.

If $k > 1$, the dilation is an *enlargement*.
If $0 < k < 1$, the dilation is a *reduction*.
If $k = 1$, the dilation is a *congruence transformation*.

example 1

Given center C and a scale factor of $\frac{3}{4}$, find the dilation image of $\triangle PQR$.

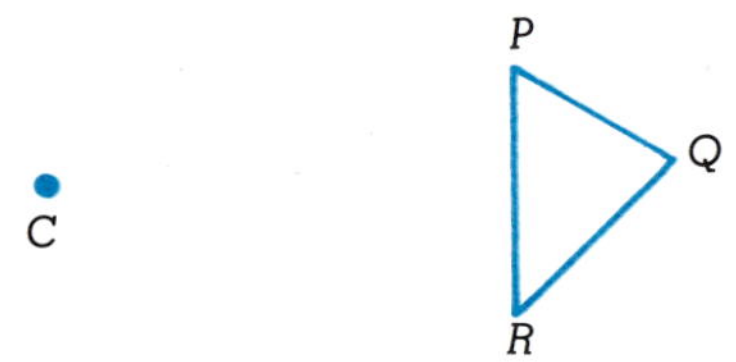

Since the scale factor is less than 1, the dilation is a reduction.

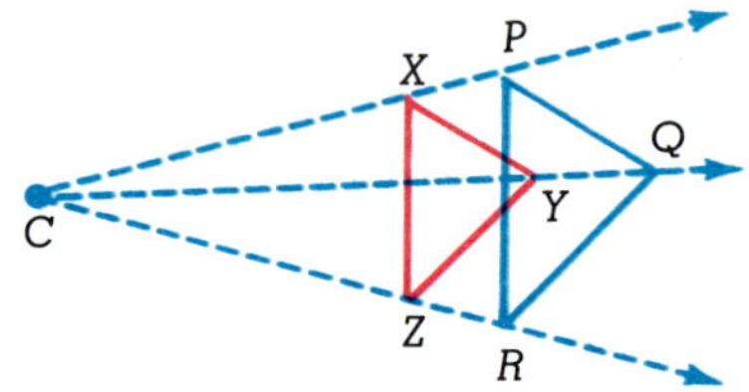

Draw $\overrightarrow{CP}$, $\overrightarrow{CQ}$, and $\overrightarrow{CR}$. Find X, Y, and Z so that $CX = \frac{3}{4}(CP)$, $CZ = \frac{3}{4}(CR)$, and $CY = \frac{3}{4}(CQ)$. $\triangle XYZ$ is the dilation image of $\triangle PQR$.

exercises

Exploratory For each of the following scale factors, tell whether dilation is an *enlargement*, *reduction*, or a *congruence transformation*.

1. $4\frac{2}{5}$
2. $\frac{3}{8}$
3. 1
4. 2.5
5. 0.61
6. 7
7. $\frac{1}{6}$
8. $\frac{3}{2}$
9. Tell whether or not it is possible to have a scale factor of 0. Explain your answer.

Written A dilation with center C and a scale factor k maps A onto D, and B onto E. Find k for each of the following conditions.

1. $CD = 10$, $CA = 5$
2. $CE = 18$, $CB = 9$
3. $CD = 6$, $CA = 4$
4. $CA = 2$, $CD = 10$
5. $AB = 3$, $DE = 1$
6. $AB = 3$, $DE = 4$

7–12. State whether the dilation is an *enlargement*, *reduction*, or *congruence transformation* for each of the conditions in problems **1–6.**

Draw and label a figure like the one shown below. Then, find the dilation image of $\triangle ABC$ for the given scale factor and center.

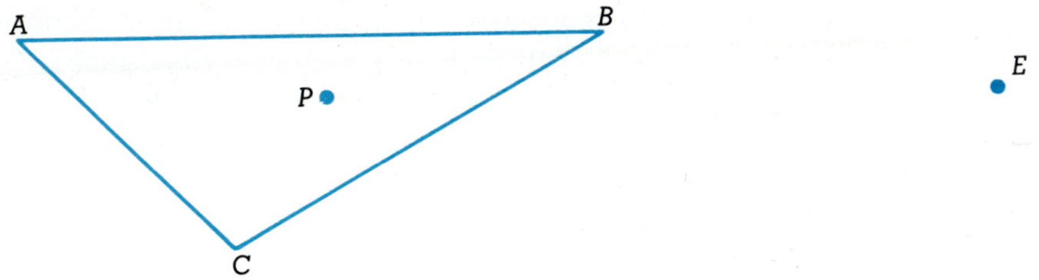

13. 3, center E

14. $\frac{1}{3}$, center E

15. 2, center P

16. $\frac{1}{2}$, center P

For each of the following scale factors, find the image of A with respect to a dilation with center P.

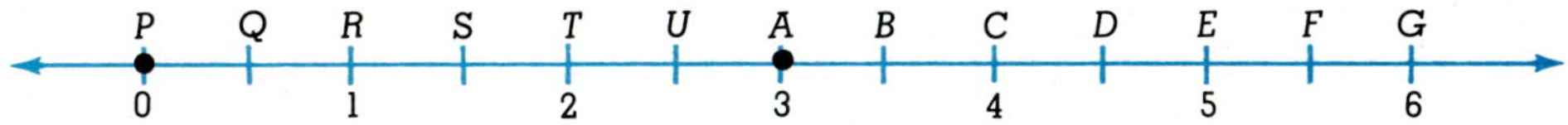

17. 1

18. 2

19. $\frac{2}{3}$

20. $1\frac{1}{3}$

21. $\frac{1}{2}$

22. $1\frac{1}{6}$

23. $1\frac{5}{6}$

24. $1\frac{1}{2}$

Graph each of the following points. Then, connect the points in order. Using (0, 0) as the center of dilation and a scale factor of 2, draw the dilation image.

25. (0, 2), (4, 0)

26. (3, $^-3$), ($^-2$, $^-2$)

27. ($^-2$, $^-1$), ($^-2$, $^-2$)

28. (3, 4), (6, 2)

29. (3, 4), (6, 10), ($^-3$, 5)

30. (6, 5), (4, 5), (3, 7)

31. ($^-1$, 4), (0, 1), (2, 3)

32. (1, $^-2$), (4, $^-3$), (6, $^-1$)

33. (1, 2), (3, 3), (3, 5), (1, 4)

34. (4, 2), ($^-4$, 6), ($^-6$, $^-8$), (6, $^-10$)

35–44. Draw the dilation images of the geometric figures in problems **25–34** using center (0, 0) and a scale factor of $\frac{1}{2}$.

Find the measure of the image of $\overline{AB}$ with respect to a dilation with the given scale factors.

45. $AB = 5$ in., $k = 6$

46. $AB = \frac{2}{3}$ cm, $k = \frac{1}{2}$

47. $AB = 16$ ft, $k = 1\frac{1}{2}$

48. $AB = 3.1$ m, $k = 5$

49. $AB = 12$ cm, $k = \frac{1}{4}$

50. $AB = 3\frac{1}{3}$ in., $k = 9$

51. Plan a proof to show that a dilation with center C and scale factor $\frac{1}{2}$ maps $\overline{AB}$ onto $\overline{ED}$ such that $ED = \frac{1}{2}(AB)$. (Hint: See page 453 of this text.)

52. Plan a proof to show that a dilation with center C and scale factor k maps $\overline{AB}$ onto $\overline{ED}$ such that $ED = k(AB)$.

Using Geometry

Total Probable Error

Accurate measurements are important for many occupations. For example, the construction of a safe building depends upon accurate measurements. However, all measurements involve errors. Errors may occur because of human mistakes, weather conditions, or flaws in measuring instruments.

Builders must plan for these types of errors when they order materials and erect a building. Estimating *total probable error* helps builders plan around errors. Total probable error is based upon the error in one measurement and the total number of measurements to be made.

The following formula is used to find total probable error.

$$T = E\sqrt{n}$$

T is the total probable error.
E is the error in one measurement.
n is the number of measurements.

Example Tia Kanawa uses a tape measure which is 0.005 centimeters shorter than the standard. To measure the boundary between two houses, she uses the tape 36 times. A maximum error of 0.40 centimeters is allowed.

$T = E\sqrt{n}$ *E is 0.005*
$= (0.005)(\sqrt{36})$ *n is 36*
$= (0.005)(6)$
$= 0.03$ The *total probable error* is 0.03 centimeters, which is within the 0.40 centimeter maximum.

Exercises Solve each problem.

1. A 100-foot long tape is 0.02 feet longer than standard. The tape will be used 49 times. A maximum error of 0.19 feet is allowed. Find the total probable error and tell whether or *not* it is within the maximum.
2. A worker makes consistent errors of 1.3 millimeters each time for 81 times. A maximum error of 15 millimeters is allowed. Find the total probable error and tell whether or *not* it is within the maximum.
3. One end of a tape is held 1.5 centimeters too low. This error occurs consistently 25 times. A maximum error of 4 centimeters is allowed. Find the total probable error and tell whether or *not* it is within the maximum.
4. A tape measure is consistently held 2 feet too high. The tape is used 49 times. A maximum error of 5 feet is allowed. Find the total probable error and tell whether or *not* it is within the maximum.
5. A machine makes consistent errors of 1.3 millimeters each time for 300 times. A maximum error of 15 millimeters is allowed. Find the total probable error and tell whether or *not* it is within the maximum.
6. A machine makes consistent errors of 0.4 millimeters each time for 5000 times. A maximum error of 30 millimeters is allowed. Find the total probable error and tell whether or *not* it is within the maximum.

Vocabulary

transformation (432)
image (432)
preimage (432)
isometry (433)
congruence transformation (433)
similarity transformation (433)
reflection (435)
line of reflection (435)
point of reflection (435)
line of symmetry (436)
composite of reflections (439)
translation (439)
rotation (442)
center of rotation (442)
angle of rotation (443)
glide reflection (450)
dilation (453)
scale factor (453)
enlargement (454)
reduction (454)

Chapter Summary

1. Moving a geometric figure to show that it coincides with another geometric figure is called a mapping. The second figure is called the image of the first figure. The first figure is called the preimage of the second figure. (432)
2. Definition of a transformation: In a plane, a mapping is a transformation if and only if each point has exactly one image point and each image point has exactly one preimage point. (432)
3. When a geometric figure and its transformation image are congruent, the mapping is called an isometry. (433)
4. A reflection with respect to ℓ is a transformation such that if A is on ℓ then A and its image B are the same point. If A is not on ℓ, then ℓ is the perpendicular bisector of $\overline{AB}$. (435)
5. Reflections preserve collinearity, betweenness of points, angle measure, and distance measure. (435)
6. A line of symmetry is a line of reflection. (436)
7. A translation is a composite of two reflections with respect to two parallel lines. (439)
8. All properties preserved by reflections are preserved by translations. (439)
9. A rotation is the composite of two reflections with respect to two intersecting lines. The point of intersection is called the center of rotation. (442)

10. All properties preserved by reflections are preserved by rotations. (442)
11. Postulate 14–1: In a given rotation, if A is the preimage, P is the image, and W is the center of rotation, then the measure of the angle of rotation, $\angle AWP$ equals twice the measure of the angle between the intersecting lines of reflection. (443)
12. A composite of more than two reflections can be described as a single isometry or a congruence transformation. (447)
13. A composite of three reflections that does *not* result in a reflection, translation, or a rotation is an isometry called a glide reflection. (450)
14. A dilation is a similarity transformation. (453)
15. Theorem 14–1: If a dilation with center C and a scale factor k maps A onto E and B onto D then $ED = k(AB)$. (453)
16. A dilation with center C and a scale factor k changes the size of the figure but not its shape. If $k > 1$, then the dilation is an enlargement. If $0 < k < 1$, then the dilation is a reduction. If $k = 1$, then the dilation is a congruence transformation. (454)

Chapter Review

14–1 **Suppose $\triangle ABE \rightarrow \triangle CBD$ and is an isometry. Name the preimage for each of the following.**

1. D 2. $\angle CBD$ 3. B 4. $\triangle CBD$

For each of the following, name a part congruent to the given part.

5. $\angle A$ 6. $\overline{BE}$ 7. $\overline{AB}$ 8. $\triangle BDC$

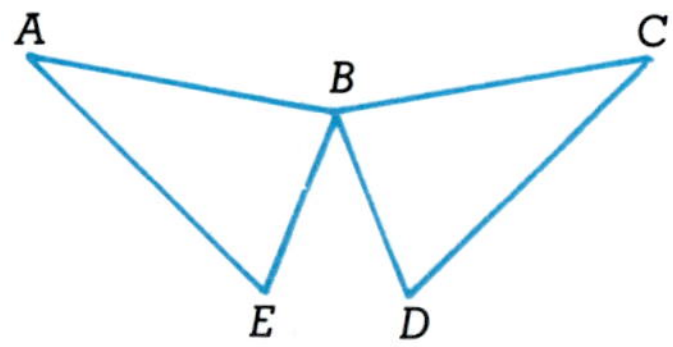

14–2 9. Draw an equilateral triangle. Then, draw all the lines of symmetry.

Copy each of the following. Then draw the reflection image of each figure with respect to ℓ.

10.

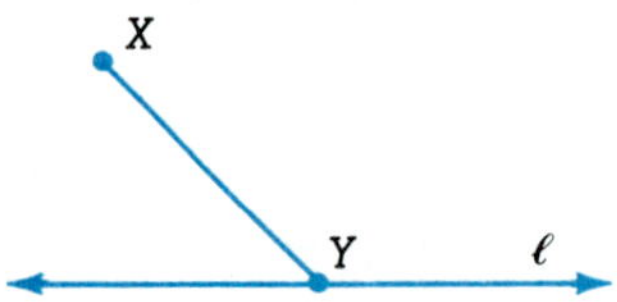

11.

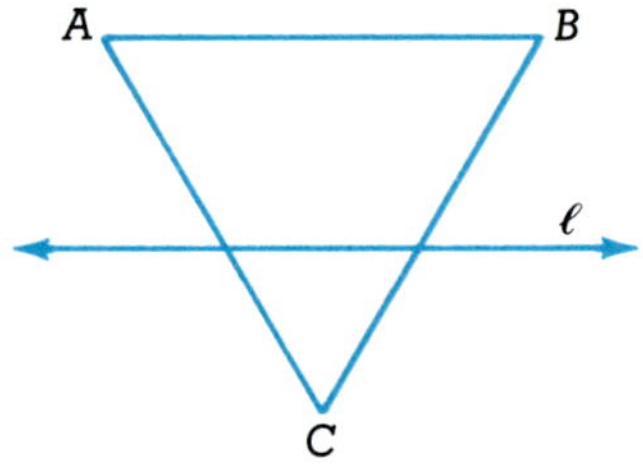

14–3 Copy each of the following. Then, draw the translation image of each figure with respect to the parallel lines s and t.

12. 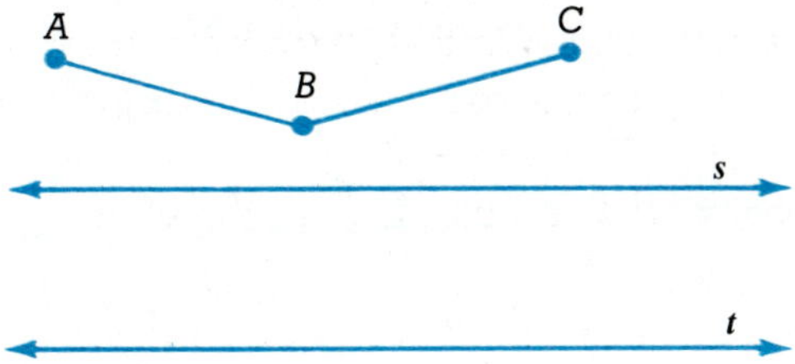

13. 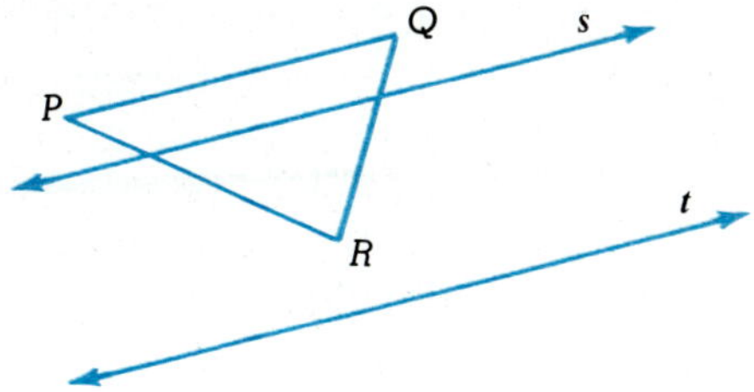

14–4 Use the figure at the right to solve each of the following.

14. Find the measure of the angle of rotation.
15. Copy the figure at the right. Then, draw the rotation image with respect to ℓ and m.

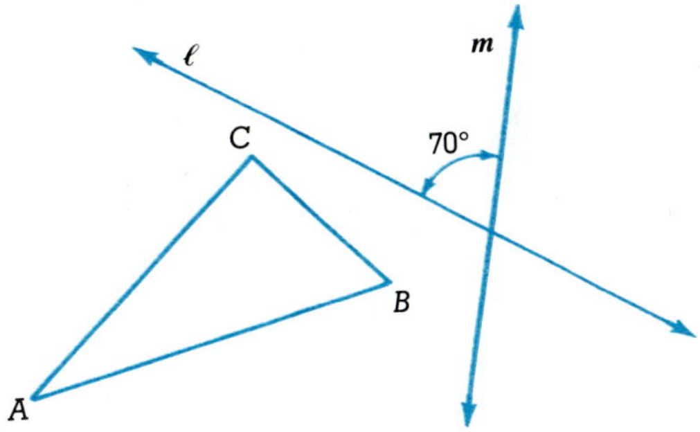

14–5 For each of the following, name a single isometry that can be used to map $\triangle ABC$ onto $\triangle PQR$.

16. 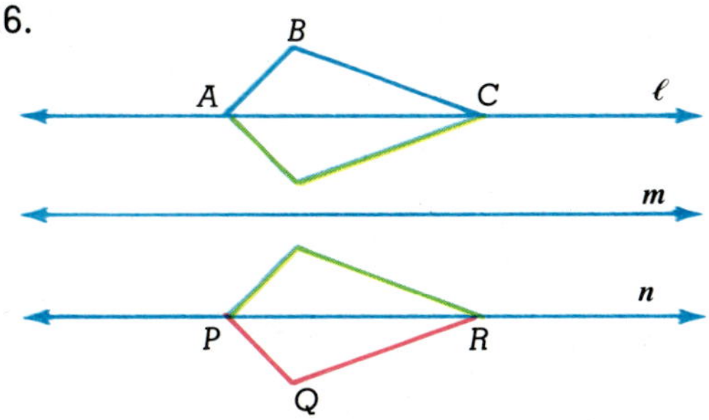

17. 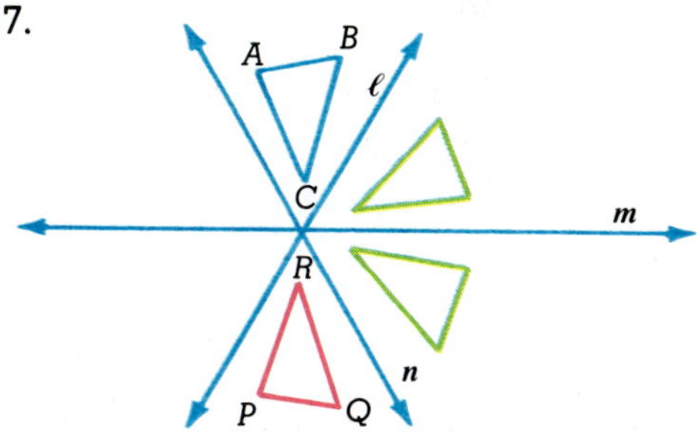

14–6 For each of the following scale factors tell whether the dilation is an *enlargement*, *reduction*, or *congruence transformation*.

18. 5   19. $\frac{1}{4}$ 20. 1

A dilation image with center C and a scale factor k maps A onto D, and B onto E. Find k for each of the following conditions.

21. $CD = 8$; $CA = 2$ 22. $CB = 3$; $CE = 9$

Chapter Test

Tell whether or *not* each of the following are isometries. Write *yes* or *no*.

1. reflection
2. translation
3. rotation
4. glide reflection

5–8. Name the properties preserved by each of the transformations in problems 1–4.

Describe each of the following as a *reflection*, a *rotation*, a *translation*, or a *glide reflection*.

9.

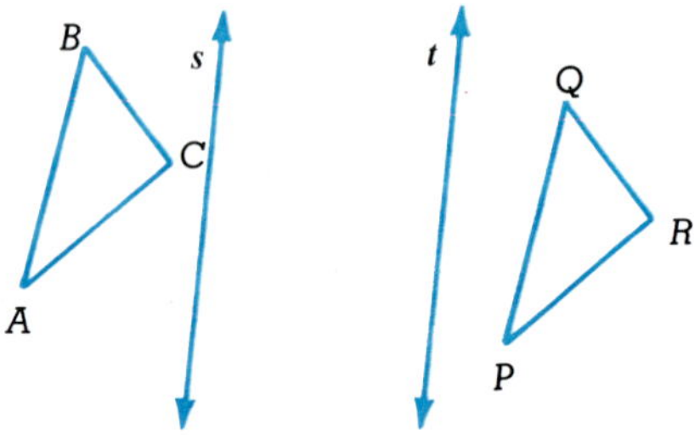

10.

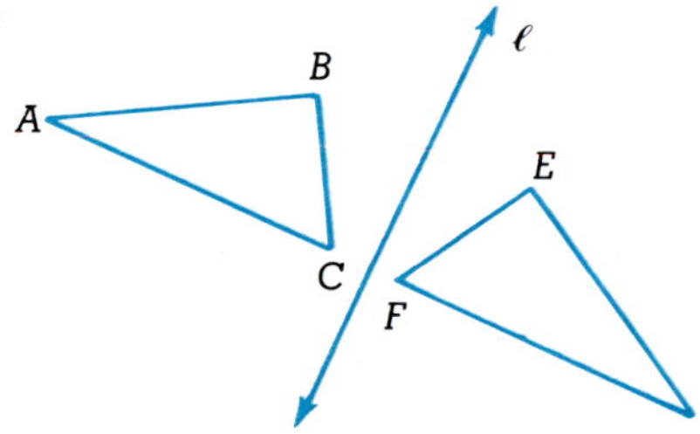

11.

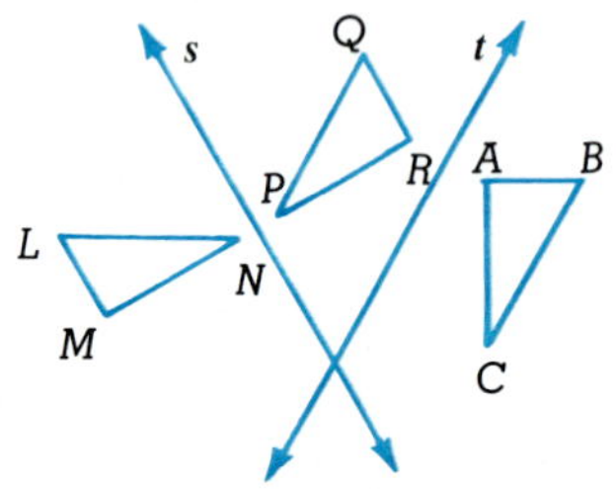

12.

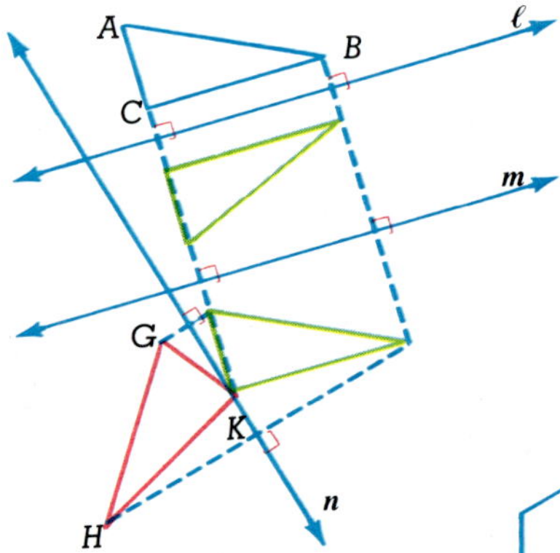

13. Copy, then draw all possible lines of symmetry for the figure at the right.

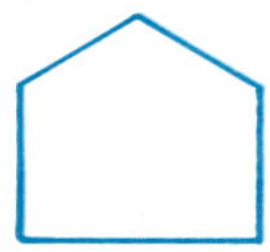

14. Draw a triangle and four parallel lines. Reflect the triangle with respect to the four lines.

15. Name the single isometry which can be used in problem **14.**

State whether the dilation is an *enlargement*, *reduction*, or *congruence transformation* for each of the following scale factors.

16. $1\frac{1}{3}$
17. 1
18. 0.3
19. 8

Copy, then draw the dilation image of the figure at the right for the given conditions.

20. center C, $k = 3$
21. center C, $k = \frac{1}{3}$
22. center C, $k = 1$

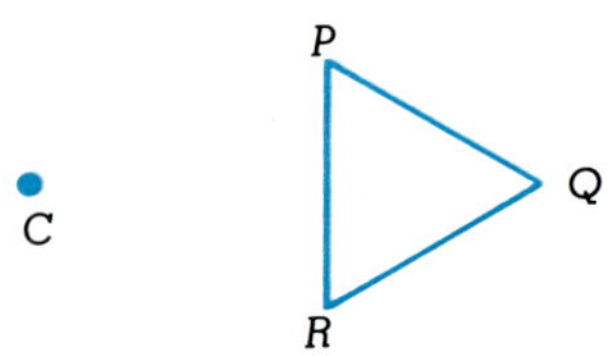

Appendix

BASIC

The Language of BASIC

BASIC is a frequently used computer language. BASIC is an abbreviation for **B**eginner's **A**ll-purpose **S**ymbolic **I**nstruction **C**ode. In BASIC, some operation symbols are similar to those in algebra.

In BASIC, raising to a power is indicated by ↑.

Algebra	BASIC	Algebra	BASIC
+	+	÷	/
−	−	=	=
×	*	5^2	5 ↑ 2

Variables are represented by a letter or by a letter and a numeral. A, B, X, C1, D2, and M5 are examples of variables.

In BASIC an operation symbol *cannot* be omitted. For example, A times B must be written A*B, *not* AB.

The computer performs operations in the same order as in algebra.

Order of Operations in BASIC

1. **Do all operations in parentheses, from the innermost parentheses outward.**
2. **Evaluate all powers from left to right.**
3. **Do all multiplications and/or division from left to right.**
4. **Do all additions and/or subtractions from left to right.**

The order of operations is shown in the following example.

example 1

Evaluate 18/3 − (6 + 2) * 3 ↑ 2 + 4.

18/3 − (6 + 2) * 3 ↑ 2 + 4 = 18/3 − 8 * 3 ↑ 2 + 4 *Do operations in parentheses.*
= 18/3 − 8 * 9 + 4 *Evaluate powers.*
= 6 − 72 + 4 *Do the division and multiplication.*
= ⁻62 *Do the addition and subtraction.*

The value of the expression is ⁻62.

A computer program is a series of statements which give directions to the computer. The purpose of a computer program is to put information into the computer *(input)*, have the computer do calculations *(execution)*, and get results out of the computer *(output)*. Here is a sample program.

```
line    ——→ 10  PRINT  7.8 + 2.5  ←
numbers ——→ 20  END  ←——————————————  statements
```

In a BASIC program each statement must have a line number. Integers from 1 to 9999 are usually used as line numbers. The computer performs the instructions in the order in which the statements are numbered. The last statement must be an END statement. Other words used to begin statements are READ and DATA.

example 2 **Write a program to find the sum and product of two numbers.**

```
10  READ A, B
20  DATA 16, 5
30  PRINT A+B
40  PRINT A*B
50  END
```

The computer assigns the numbers from the DATA statement in line 20 to the variables A and B in order.

A = 16   B = 5

The sum and product of any two numbers may be found by changing the numbers in the DATA statement.

```
10  READ A, B
20  DATA 6, 4
30  PRINT A+B
40  PRINT A*B
50  END
    RUN
    10
    24
```

The computer assigns A = 6, B = 4 from the data in line 20.

RUN tells the computer to execute the program.

10 = A+B

24 = A∗B

example 3 **Write a BASIC program to compute and print the perimeter and area of a rectangle with length 2.6 centimeters and width 1.7 centimeters.**

```
10  READ L, W
20  DATA 2.6, 1.7
30  PRINT 2*L+2*W, L*W
40  END
```

The comma in line 30 tells the computer to print the output on one line.

exercises

Written Write an expression in BASIC for each of the following.

1. $5a + 2$
2. $xyz + 7$
3. $9x + 4y - 3$
4. $(3x - 4y)^2$
5. $\dfrac{m + 6}{b}$
6. $\dfrac{2x + 5}{3y}$
7. $4a^2 + 6a - 7$
8. 7^{y+4}
9. $\dfrac{5x + 2}{3y - 7}$
10. $\left(\dfrac{x}{y} + w\right)^2$

Evaluate each of the following expressions.

11. 7*3 ↑ 2
12. (4+78)/2+5
13. ((13+11)/4)/6
14. (4*(4+8))/16
15. 4*3 ↑ 2
16. (4*3) ↑ 2

Evaluate each of the following. Use A = 10, B = 5, and C = 2.

17. A*B+3
18. A*(B+3)
19. C−B+5
20. A*B ↑ C
21. (A/B) ↑ 2
22. A/B ↑ 2

Write programs using BASIC to compute and print the following. Use READ and DATA statements.

23. Find the value of $A^2 - B^2$, $(A - B)^2$, $A^2 + B^2$, and $(A + B)^2$ when $A = 5$ and $B = 2$.
24. Find the circumference and area of a circle with radius 2.58 cm. Use $\pi = 3.14159$.
25. Find the area of a triangle with base 7.3 cm and height 4.8 cm.
26. Find the volume and surface area of a cube with each edge 5.5 cm.
27. Find the complement and supplement of an angle which has degree measure 68.
28. Two angles of a triangle have degree measures 48 and 30. Find the degree measure of the third angle.

Assignment of Variables

In the BASIC language, the left side of an equation can have *only one* variable. The equals sign tells the computer to assign the value of the expression on the right to the variable on the left. An equals sign is used in a LET statement.

Program	Comment
10 LET X=5	*5 is assigned to X.*
20 LET Y=2	*2 is assigned to Y.*
30 LET A=5*Y+20/X	*The value of A is computed and assigned.*
40 PRINT A	*The value of A is printed.*
50 END	
RUN	
14	*The output is 14.*

A different program may be written to solve the same problem.

Program	Comment
10 READ X, Y	*In this program, the data are*
20 DATA 5, 2	*assigned as follows. X = 5, Y = 2.*
30 LET A=5*Y+20/X	*The value of 5*Y+20/X is then*
40 PRINT A	*computed and assigned to A in*
50 END	*line 30. This value is then printed.*
RUN	
14	

LET statements also can be used to assign a new value to a variable as shown in the following example.

example 1

Write a program using BASIC to compute the volumes of three cubes with sides measuring 2 centimeters, 3 centimeters, and 4 centimeters.

Program	Comment
10 LET M=2	*The value of M is 2.*
20 PRINT M, M ↑ 3	*Output 2 8*
30 LET M=M+1	*Now the value of M is 3.*
40 PRINT M, M ↑ 3	*Output 3 27*
50 LET M=M+1	*Now what value is assigned to M?*
60 PRINT M, M ↑ 3	*What is the output?*
70 END	

When the output is more than six significant digits, the computer will use **E notation.** This is the computer equivalent of scientific notation. For example, 5.15076E+07 means 5.15076×10^7 or 51,507,600.

Output	Scientific Notation	E Notation
17161308	1.7161308×10^7	1.71613E+07
0.009514	9.514×10^{-3}	9.514E−03

exercises

Written Write each of the following using scientific notation. Then, write each in E notation.

1. 7,437,600
2. 0.000002827381
3. 0.002987
4. 14,630,040,000

Write the value of each expression *without* using scientific notation.

5. 7.063E+10
6. 3.33127E−05
7. 2.322E−07
8. 2.8062E+06

Find the value of X in each BASIC statement. Let A = 5, B = 2, and C1 = 9.

9. LET X=B*A+C1
10. LET X=A*B+6
11. LET X=A/B+9
12. LET X=C1+A+6

Correct the error in each expression or statement below.

13. (7+8/2
14. 4*X↑2−2Y
15. LET 2*Y=4+A
16. LET A+B=C+D

Each mathematical expression is followed by an incorrect BASIC expression. Write the correct BASIC expression.

17. $\frac{m+2}{r-3}$, M+2/(R−3)
18. $\frac{ab}{y+3}$, (A*B)/Y+3
19. $\frac{(x+a)^2}{2z}$, X+A↑2/2Z
20. $\left(\frac{x}{y}\right)^{n-1}$, X/Y↑N−1

For each of the following problems, write a BASIC program using the given values of A, B, and C to compute and print the value of X. Use LET statements.

21. A = 5, B = 10, C = 4
    $X = 3A + 5B - 6C$
22. A = 10, B = 2, C = 13
    $X = 3A^2 - 4(B + C)$

23–24. Adapt each program above so it has a READ statement and a DATA statement.

Write programs using BASIC to compute and print each of the following problems.

25. Convert 1 yard, 2 yards, and 3 yards to meters.
26. Find the average of the numbers 12, 254, and 73.
27. Find the volume of a rectangular solid whose length is 4 cm, width is 7.5 cm, and height is 2.5 cm.
28. Find the sum of the measures of the angles of a triangle, convex quadrilateral, and convex pentagon.

FOR-NEXT Loops

Sometimes an operation needs to be used on different sets of data. FOR-NEXT statements allow the computer to loop through the same operation several times.

The following program uses LET statements to print multiples of two.

```
10  LET M=1              Output
20  PRINT M, 2*M         1    2
30  LET M=M+1
40  PRINT M, 2*M         2    4
50  LET M=M+1
60  PRINT M, 2*M         3    6
70  END
```

A simpler way to write the program uses the FOR-NEXT statement.

```
10  FOR M=1 to 3 STEP 1
20  PRINT M, 2*M
30  NEXT M
40  END
```

The phrase STEP 1 tells the computer to increase M by one each time. The computer continues until there are no values of M remaining.

The general form of the FOR statement is as follows.

FOR <u>Variable</u> = <u>Number</u> TO <u>Number</u> STEP <u>Number</u>

If the step is not indicated, most computers will automatically use a step of one.

Every time a FOR statement is used there must be a NEXT statement.

The PRINT statement can be used to print words as well as the results of calculations. The computer will print the symbols that are placed between quotation marks.

example 1

Write a program using BASIC to compute and print the sum of the squares of all even integers from 2 to 100.

```
10  LET S=0
20  FOR N=2 TO 100 STEP 2
30  LET Y=N ↑ 2
40  LET S=S+Y
50  NEXT N
60  PRINT "SUM = ", S
70  END
```

The variable S represents the sum of all values of Y up to this point. What would be different about the output if line 60 were given the line number 45?

The following program shows how headings may be included in the output.

```
10  PRINT "INTEGER", "SQUARE", "CUBE"
20  FOR N=1 TO 4
30  PRINT N, N↑2, N↑3
40  NEXT N
50  END
```

Output

INTEGER	*SQUARE*	*CUBE*
1	*1*	*1*
2	*4*	*8*
3	*9*	*27*
4	*16*	*64*

It often is useful to have more than one loop. There are only two ways that loops can appear in a program.

Nested Loops	**Independent Loops**	***Not* Acceptable**
FOR X FOR Y NEXT Y NEXT X	FOR X NEXT X FOR Y NEXT Y	FOR X FOR Y NEXT X NEXT Y
The loops do <u>not</u> cross. They are nested.	*The loops do <u>not</u> cross.*	*The loops cross.*

example 2

Write a program to find the possible areas of a rectangle whose width may be 4 cm, 5 cm, 6 cm, or 7 cm and length may be 10 cm or 12 cm.

```
10  PRINT "WIDTH", "LENGTH", "AREA"
20  FOR W=4 TO 7
30  FOR L=10 TO 12 STEP 2
40  PRINT W, L, W*L
50  NEXT L
60  NEXT W
70  END
    RUN
```

This is example of nested loops. W has the same value while L goes through its loop.

WIDTH	LENGTH	AREA
4	10	40
4	12	48
5	10	50
5	12	60
6	10	60
6	12	72
7	10	70
7	12	84

exercises

Written Find the error in each of the following programs and write a correct program.

1.
```
10 FOR K=1 TO 16 STEP 0
20 PRINT K
30 NEXT K
40 END
```

2.
```
10 FOR M=3 TO 25, STEP 2
20 PRINT M
30 NEXT M
40 END
```

3.
```
10 LET X=4
20 FOR T=X TO 40 STEP 3
30 PRINT T
40 NEXT X
50 END
```

4.
```
10 FOR X=1 TO 6 STEP 3
20 FOR Y=2 TO 4 STEP 2
30 PRINT X*Y
40 NEXT X
50 NEXT Y
60 END
```

Write the output of each program.

5.
```
10 FOR N=4 TO 31 STEP 3
20 PRINT N
30 NEXT N
40 END
```

6.
```
10 FOR X=1 TO 10
20 LET Y=11-X
30 PRINT Y
40 NEXT X
50 END
```

7.
```
10 FOR I=1 TO 3 STEP 1
20 LET A=8
30 PRINT A
40 NEXT I
50 END
```

8.
```
10 FOR L=10 TO 100 STEP 10
20 LET M=L↑2
30 PRINT "INTEGER", L, "SQUARE", M
40 NEXT L
50 END
```

Write a program using BASIC to solve each of the following problems.

9. Find and print the sum of the cubes of the integers from 1 to 20 inclusive.

10. Find and print the complements of angles with degree measures from 0 to 90 in steps of five.

11. Find and print the values of y for the equation $y = x^2 - x + 6$ when $x = {}^-3, {}^-2, {}^-1, 0, 1, 2$, and 3.

12. The formula for Fahrenheit to Celsius conversion is $C = \frac{5}{9}(F - 32)$. Find the Celsius temperatures for Fahrenheit temperatures from 0° to 100° in steps of two.

13. The measures of the lengths and widths of three rectangles are 8 by 5, 9.2 by 3.4, and 17 by 18. Compute and print both the perimeter and area for each rectangle. Arrange the headings *length*, *width*, *perimeter*, and *area* at the top.

14. Given the ordered pairs (2, 5), (4, ⁻6); (7, 8), (⁻9, 5); and (4, 2.8), (3.6, ⁻2.1), find the slope of each line joining each pair of points.

IF-THEN Statements

The IF-THEN statement is used to compare two numbers. It tells the computer what to do based on the results of the comparison.

The general form of the IF-THEN statement is as follows.

IF *algebraic sentence* THEN *line number*

If the algebraic sentence is true, then the computer goes to the line whose number follows THEN. If the algebraic sentence is false, then the computer proceeds in normal sequence to the next line.

The algebraic sentence in the IF-THEN statement uses one of the following symbols.

BASIC symbol	Meaning
=	is equal to
<	is less than
<=	is less than or equal to
>	is greater than
>=	is greater than or equal to
<>	is *not* equal to

example 1

Write a program to compute and print integers whose fourth powers are less than 1000.

```
10  LET N=0                          Assign a value of zero to N.
20  LET N=N+1                        Increase the value of N by one.
30  LET F=N↑4                        Assign the value of N↑4 to F.
40  IF F>=1000 THEN 70 --+           Test line.
       |                 |
       No               Yes
       ↓                 |
50  PRINT N, F           |
60  GO TO 20             |           Return to line 20 of the program.
70  END <----------------+
```

When the computer comes to line 40, it tests the value of F. If F is less than 1000, then the computer goes to line 50. Next, line 60 sends it to line 20 to compute the next integer. If F is greater than or equal to 1000, the computer goes to line 70.

Sometimes a programmer needs to count how many times an operation is performed. In the program at the top of the next page, line 30 keeps the count. Each time through the loop the count is increased by one.

example 2

Write a program to count the number of powers of five between 100 and 1,000,000 and print each one. Express each one both as a number with no exponent, and as a power of five.

Program	
`10 LET K=1`	*The counter K starts at 1.*
`20 LET X=125`	*$5^3 = 125$ is the first power of 5 greater than 100.*
`30 PRINT K, X, "5 ↑" (K+2)`	
`40 LET K=K+1`	*The counter increases by one.*
`50 LET X=X*5`	*The next power of 5 is calculated.*
`60 IF X>=1000000 THEN 80`	*The test line checks to see if that power*
`70 GO TO 30`	*of 5 is greater than or equal to 1,000,000.*
`80 END`	*In line 30, why is the (K+2) power printed?*

exercises

Written Let A = 5, B = 8, and X = 10. Tell the line number of the statement that the computer will do after it executes the IF-THEN statement.

1.
```
10 IF A>=-4 THEN 90
20 PRINT 2*A
```

2.
```
10 IF A<>B THEN 50
20 PRINT X
```

3.
```
41 IF A-B>X THEN 75
44 PRINT B-A
```

4.
```
45 IF B*X<A↑3 THEN 10
50 PRINT B*X
```

Use the program at the right to tell whether A, B, or both A and B will be printed. Give the value of any variable printed.

5. A = 12, B = 12
6. A = 14, B = 19
7. A = 9, B = 21
8. A = 21, B = 5
9. A = 28, B = 38

```
10 IF A>B THEN 40
20 LET A=A+10
30 IF A>=B THEN 50
40 PRINT A
50 PRINT B
60 END
```

Write programs using BASIC to solve each of the following problems. Use IF-THEN statements.

10. Print the integers from 10 to 1 in descending order.

11. Read three numbers and print them from least to greatest.

12. Determine if a triangle exists given the length of its sides. Use data 4, 4, 4; 2, 5, 1; 4, 7, 9; 6, 7, 20.

13. Print the degree measure of a given angle and state whether the angle is acute, obtuse, or right.

14. Classify a triangle as acute, obtuse, or right given the degree measure of two angles of the triangle.

15. Count the number of powers of six less than 900,000. Express each power as a number with no exponent.

BASIC Functions

BASIC functions are subprograms stored in the computer which may be used to simplify a program. For example, programmers may need to find the absolute value of a number. Instead of writing a program to find the absolute value, they can use a code to have this done automatically by the computer.

Some common BASIC functions are listed below.

ABS(X)	*Absolute value of X*
SQR(X)	*Square root of X*
INT(X)	*Greatest integer less than or equal to X*

The absolute value function ABS(X) gives the absolute value of X as illustrated in the following examples.

ABS(3) = 3 ABS(0) = 0 ABS(−4.8) = 4.8

example 1

Write a program using BASIC to find the measure of the distance between points A and B, A and C, A and D, B and C, B and D, C and D on the number line.

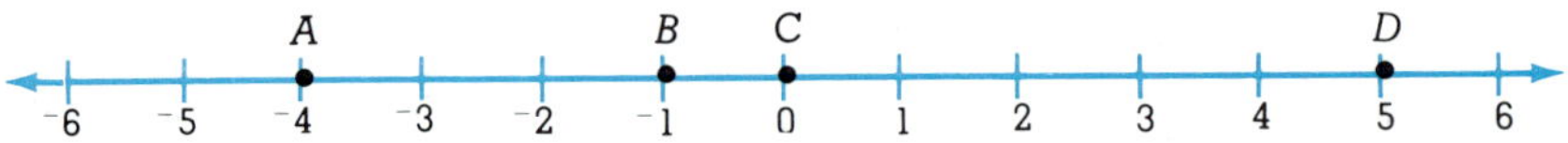


```
10 PRINT "POINT ONE", "POINT TWO", "MEASURE OF DISTANCE"
20 PRINT
30 FOR L=1 TO 6
40 READ X, Y
50 LET D=ABS(X−Y)
60 PRINT X, Y, D
70 NEXT L
80 DATA −4,−1,−4,0,−4,5,−1,0,−1,5,0,5
90 END
```

A blank line will appear in the output. (line 20)

The computer will assign the absolute value of (X−Y) to D. (lines 50–60)

The SQR(X) function tells the computer to calculate the square root of X. Some examples of valid BASIC statements using the SQR(X) function follow.

```
10 LET A=SQR(144)
10 IF B=SQR(X) THEN 80
10 FOR I=1 TO SQR(X) STEP 4
```

Since SQR(X) represents a number, it can be used in IF-THEN and FOR-NEXT statements.

example 2

Write a program using BASIC to find the square roots of the integers from 1 to 10.

```
10  FOR I=1 TO 10
20  PRINT "THE SQUARE ROOT OF", I, "IS", SQR(I)
30  NEXT I
40  END
```

The INT(X) function is called the greatest integer function. After computing or reading the value of X, it finds the greatest integer less than or equal to X. Here are some examples.

INT(8.02) = 8 INT(−5.6) = −6
INT(15) = 15 INT(72.99) = 72

example 3

Write a program using BASIC to find the factors of a given positive integer.

```
10  READ N
20  DATA 20, 48, 90
30  FOR X=1 TO N
40  IF N/X<>INT(N/X) THEN 60
50  PRINT X, "IS A FACTOR OF", N
60  NEXT X
70  GO TO 10
80  END
```

If N/X = INT(N/X), then N/X must be an integer. Therefore, X divides N evenly or X is a factor of N.

Pythagorean triples are sets of positive integers {A, B, C} such that $A^2 + B^2 = C^2$. A, B, and C represent the measures of the sides of a right triangle. The following program finds and prints all Pythagorean triples when A and B are less than or equal to 15.

```
10  FOR A=1 TO 15
20  FOR B=A TO 15
30  LET C=SQR(A↑2+B↑2)
40  IF C<>INT(C) THEN 60
50  PRINT A, B, C
60  NEXT B
70  NEXT A
80  END
```

Why does B go from A to 15?

Output

3	4	5
5	12	13
6	8	10
8	15	17
9	12	15

exercises

Written Write the output of each of the following PRINT statements. Let A = ⁻3, B = 16, C = 4.2, and D = 1.

1. PRINT ABS(A)
2. PRINT ABS(B ↑ D)
3. PRINT
4. PRINT INT(C)
5. PRINT SQR(B)−SQR(D)
6. PRINT INT(SQR(C))
7. PRINT 12*INT(C)
8. PRINT INT(ABS(A))

Write programs using BASIC to solve each of the following problems. Use BASIC functions.

9. Find the absolute value of a given number. Print both the number and its absolute value. Use data ⁻2, 4.761, 0, ⁻0.011, 7.999.
10. Find the square root of a number. Print both the number and its square root if it exists. Use data 2, 4.761, 0, ⁻3, 8.
11. Find and print the sum of the greatest integers of the square roots of integers from 1 to 10.
12. Find and print the greatest integer of the sum of the square roots of integers from 1 to 10.
13. Print an integer and state whether it is odd or even. Use the INT(X) function in your program. Use data ⁻6, ⁻5, ⁻1, 0, 2, 9.
14. Use the Pythagorean theorem to find the measure of the hypotenuse of a right triangle given the measures of the legs of the right triangle. Print the measures of all sides of the triangle. Use data 3, 4; 8, 15; 4, 4; 9, 10.
15. Find and print the distance between the points (1, 5) and (⁻7, 2); (2, ⁻2.5) and (⁻11.7, ⁻1). Recall that the distance between two points (x_1, y_1) and (x_2, y_2) is $\sqrt{(x_2 - x_1)^2 + (y_2 - y_1)^2}$.
16. Given three measures of the sides of a triangle, state if the triangle is a right triangle. Use data 4, 5, 7; 9, 40, 41; 20, 21, 29.

DIAGNOSTIC SKILLS REVIEW

Integer Equations: Addition and Subtraction

Solve each equation.

1. $7 + 5 = x$
2. $6 + 4 = x$
3. $8 + 9 = x$
4. $19 + 20 = x$
5. $13 + 15 = x$
6. $45 + 24 = x$
7. $38 + 16 = x$
8. $96 + 52 = x$
9. $77 + 9 = x$
10. $104 + 47 = x$
11. $83 + 86 = x$
12. $27 + 34 = x$
13. $636 + 104 = x$
14. $108 + 597 = x$
15. $302 + 969 = x$
16. $47 + 38 + 76 = x$
17. $29 + 62 + 35 = x$
18. $51 + 89 + 96 = x$
19. $20 + 63 + 57 = x$
20. $34 + 49 + 72 = x$
21. $66 + 28 + 92 = x$
22. ${}^{-}8 + {}^{-}5 = x$
23. ${}^{-}7 + {}^{-}9 = x$
24. ${}^{-}10 + {}^{-}12 = x$
25. $0 + {}^{-}6 = x$
26. ${}^{-}4 + {}^{-}3 = x$
27. ${}^{-}10 + 0 = x$
28. ${}^{-}40 + {}^{-}30 = x$
29. ${}^{-}25 + {}^{-}71 = x$
30. $0 + {}^{-}62 = x$
31. ${}^{-}66 + {}^{-}33 = x$
32. ${}^{-}36 + {}^{-}42 = x$
33. ${}^{-}86 + {}^{-}200 = x$
34. ${}^{-}152 + {}^{-}18 = x$
35. ${}^{-}206 + {}^{-}606 = x$
36. ${}^{-}829 + {}^{-}91 = x$
37. ${}^{-}709 + {}^{-}196 = x$
38. ${}^{-}643 + {}^{-}167 = x$
39. ${}^{-}229 + {}^{-}280 = x$
40. ${}^{-}7468 + {}^{-}4923 = x$
41. ${}^{-}25{,}406 + {}^{-}9{,}329 = x$
42. ${}^{-}2397 + {}^{-}4945 = x$
43. ${}^{-}11 + 11 = x$
44. ${}^{-}1 + 12 = x$
45. $5 + {}^{-}5 = x$
46. ${}^{-}10 + 7 = x$
47. $10 + {}^{-}4 = x$
48. $56 + {}^{-}10 = x$
49. ${}^{-}8 + 13 = x$
50. $12 + {}^{-}19 = x$
51. ${}^{-}15 + 14 = x$
52. $24 + {}^{-}36 = x$
53. $52 + {}^{-}48 = x$
54. ${}^{-}24 + 30 = x$
55. $40 + {}^{-}27 = x$
56. ${}^{-}54 + 46 = x$
57. $38 + {}^{-}50 = x$
58. ${}^{-}16 + 31 = x$
59. $75 + {}^{-}92 = x$
60. $123 + {}^{-}106 = x$
61. ${}^{-}2046 + 8752 = x$
62. ${}^{-}104 + 96 = x$
63. $728 + {}^{-}469 = x$

Solve each equation.

64. $47 - 41 = x$
65. $13 - 5 = x$
66. $21 - 17 = x$
67. $39 - 0 = x$
68. $85 - 72 = x$
69. $65 - 0 = x$
70. $165 - 141 = x$
71. $236 - 125 = x$
72. $387 - 281 = x$
73. $603 - 591 = x$
74. $409 - 268 = x$
75. $506 - 152 = x$
76. $792 - 237 = x$
77. $176 - 59 = x$
78. $321 - 218 = x$
79. $226 - 177 = x$
80. $833 - 546 = x$
81. $920 - 639 = x$
82. $1072 - 295 = x$
83. $2510 - 88 = x$
84. $6302 - 2567 = x$
85. $9 - 5 = x$
86. $5 - 9 = x$
87. $6 - 10 = x$
88. $13 - 10 = x$
89. $10 - 13 = x$
90. $2 - 7 = x$
91. $21 - 38 = x$
92. $15 - 11 = x$
93. $42 - 50 = x$
94. $9 - {}^{-}6 = x$
95. $18 - {}^{-}2 = x$
96. $4 - {}^{-}5 = x$
97. $22 - {}^{-}11 = x$
98. $40 - {}^{-}32 = x$
99. $25 - {}^{-}8 = x$
100. $34 - {}^{-}6 = x$
101. $75 - {}^{-}24 = x$
102. $67 - {}^{-}40 = x$
103. ${}^{-}6 - {}^{-}2 = x$
104. ${}^{-}8 - {}^{-}7 = x$
105. ${}^{-}10 - {}^{-}6 = x$
106. ${}^{-}31 - {}^{-}40 = x$
107. ${}^{-}23 - {}^{-}28 = x$
108. ${}^{-}15 - {}^{-}35 = x$
109. ${}^{-}23 - 5 = x$
110. ${}^{-}31 - 18 = x$
111. ${}^{-}68 - 30 = x$
112. ${}^{-}5 - 7 = x$
113. ${}^{-}84 - 6 = x$
114. ${}^{-}92 - 7 = x$

Integer Equations: Multiplication and Division

Solve each equation.

1. $20 \cdot 5 = x$
2. $10 \cdot 100 = x$
3. $6 \cdot 40 = x$
4. $100 \cdot 1000 = x$
5. $204 \cdot 7 = x$
6. $505 \cdot 6 = x$
7. $32 \cdot 9 = x$
8. $68 \cdot 8 = x$
9. $43 \cdot 4 = x$
10. $28 \cdot 10 = x$
11. $46 \cdot 100 = x$
12. $1000 \cdot 97 = x$
13. $43 \cdot 36 = x$
14. $86 \cdot 57 = x$
15. $62 \cdot 79 = x$
16. $709 \cdot 16 = x$
17. $32 \cdot 603 = x$
18. $204 \cdot 38 = x$
19. $840 \cdot 603 = x$
20. $958 \cdot 643 = x$
21. $657 \cdot 399 = x$
22. $^-5 \cdot {}^-3 = x$
23. $^-6 \cdot {}^-7 = x$
24. $^-9 \cdot {}^-6 = x$
25. $^-11 \cdot {}^-12 = x$
26. $^-2 \cdot {}^-1000 = x$
27. $^-1 \cdot {}^-576 = x$
28. $^-100 \cdot {}^-341 = x$
29. $^-907 \cdot {}^-700 = x$
30. $^-87 \cdot {}^-10 = x$
31. $^-74 \cdot {}^-6 = x$
32. $^-89 \cdot {}^-5 = x$
33. $^-21 \cdot {}^-9 = x$
34. $^-308 \cdot {}^-92 = x$
35. $^-403 \cdot {}^-37 = x$
36. $^-36 \cdot {}^-605 = x$
37. $^-58 \cdot {}^-91 = x$
38. $^-26 \cdot {}^-19 = x$
39. $^-49 \cdot {}^-68 = x$
40. $^-6072 \cdot {}^-52 = x$
41. $^-5416 \cdot {}^-31 = x$
42. $^-62 \cdot {}^-9557 = x$
43. $^-8 \cdot 0 = x$
44. $1 \cdot {}^-17 = x$
45. $0 \cdot {}^-6 = x$
46. $100 \cdot {}^-9 = x$
47. $45 \cdot {}^-1000 = x$
48. $2 \cdot {}^-1 = x$
49. $^-106 \cdot 5 = x$
50. $^-3 \cdot 406 = x$
51. $^-309 \cdot 7 = x$
52. $^-26 \cdot 32 = x$
53. $85 \cdot {}^-19 = x$
54. $66 \cdot {}^-58 = x$
55. $^-7 \cdot {}^-6 \cdot 4 = x$
56. $^-8 \cdot 6 \cdot {}^-7 = x$
57. $10 \cdot {}^-11 \cdot {}^-9 = x$
58. $^-5 \cdot {}^-3 \cdot {}^-2 = x$
59. $^-9 \cdot 0 \cdot {}^-11 = x$
60. $^-5 \cdot {}^-6 \cdot {}^-4 = x$
61. $^-2 \cdot {}^-1 \cdot 5 \cdot {}^-4 = x$
62. $3 \cdot 4 \cdot {}^-7 \cdot 8 = x$
63. $^-3 \cdot {}^-1 \cdot {}^-8 \cdot {}^-6 = x$

Solve each equation.

64. $40 \div 5 = x$
65. $24 \div 6 = x$
66. $38 \div 1 = x$
67. $144 \div 12 = x$
68. $169 \div 13 = x$
69. $56 \div 7 = x$
70. $912 \div 16 = x$
71. $928 \div 29 = x$
72. $5644 \div 83 = x$
73. $1441 \div 131 = x$
74. $4080 \div 255 = x$
75. $6408 \div 712 = x$
76. $35{,}288 \div 88 = x$
77. $11{,}495 \div 55 = x$
78. $77{,}478 \div 37 = x$
79. $^-55 \div {}^-11 = x$
80. $^-84 \div {}^-12 = x$
81. $^-51 \div {}^-17 = x$
82. $^-206 \div {}^-1 = x$
83. $^-175 \div {}^-25 = x$
84. $^-78 \div {}^-13 = x$
85. $^-720 \div {}^-18 = x$
86. $^-3600 \div {}^-60 = x$
87. $^-1950 \div {}^-65 = x$
88. $^-6570 \div {}^-365 = x$
89. $^-5280 \div {}^-220 = x$
90. $^-1728 \div {}^-12 = x$
91. $48 \div {}^-6 = x$
92. $^-28 \div 4 = x$
93. $^-18 \div 9 = x$
94. $63 \div {}^-9 = x$
95. $64 \div {}^-16 = x$
96. $^-42 \div 14 = x$
97. $^-968 \div 8 = x$
98. $^-6054 \div 6 = x$
99. $5810 \div {}^-83 = x$
100. $1950 \div {}^-65 = x$
101. $1116 \div {}^-31 = x$
102. $^-1441 \div 131 = x$
103. $^-3784 \div 172 = x$
104. $4500 \div {}^-50 = x$
105. $920 \div {}^-23 = x$

Solve each equation.

106. $8 + 5 + 6 \div 3 - 2 = x$
107. $7 + 4 + 3 - 1 = x$
108. $7 + 6 - 4 + 3 = x$
109. $8 - 5 + 3 - 2 = x$
110. $12 \cdot 8 + 12 \div 2 = x$
111. $8 \div 2 + 5 \cdot 6 - 20 \div 4 = x$

Fraction Equations: Addition and Subtraction

Solve each equation and express answers in simplest form.

1. $\frac{1}{3} + \frac{1}{3} = x$
2. $\frac{2}{9} + \frac{5}{9} = x$
3. $\frac{4}{11} + \frac{6}{11} = x$
4. $\frac{8}{15} + \frac{11}{15} = x$
5. $\frac{2}{3} + \frac{2}{3} = x$
6. $\frac{11}{18} + \frac{17}{18} = x$
7. $1\frac{1}{5} + \frac{2}{5} = x$
8. $4\frac{1}{4} + 3\frac{3}{4} = x$
9. $5\frac{1}{12} + 11\frac{7}{12} = x$
10. $\frac{1}{2} + \frac{3}{4} = x$
11. $\frac{1}{3} + \frac{2}{9} = x$
12. $\frac{5}{12} + \frac{1}{6} = x$
13. $7\frac{5}{6} + 4\frac{5}{9} = x$
14. $2\frac{9}{10} + 1\frac{11}{12} = x$
15. $3\frac{7}{20} + 1\frac{11}{12} = x$
16. $-\frac{1}{6} + -\frac{2}{6} = x$
17. $-\frac{3}{4} + -\frac{1}{4} = x$
18. $-\frac{5}{7} + -\frac{1}{7} = x$
19. $-\frac{5}{6} + -\frac{5}{6} = x$
20. $-\frac{3}{5} + -\frac{4}{5} = x$
21. $-\frac{2}{13} + -\frac{12}{13} = x$
22. ${}^{-}1\frac{1}{4} + {}^{-}3\frac{1}{4} = x$
23. ${}^{-}2\frac{10}{21} + {}^{-}8\frac{17}{21} = x$
24. ${}^{-}5\frac{7}{10} + {}^{-}3\frac{3}{10} = x$
25. ${}^{-}6\frac{5}{8} + -\frac{1}{4} = x$
26. ${}^{-}3\frac{1}{4} + {}^{-}2\frac{1}{2} = x$
27. ${}^{-}4\frac{1}{7} + {}^{-}2\frac{2}{21} = x$
28. $\frac{5}{7} + -\frac{1}{7} = x$
29. $-\frac{14}{15} + \frac{1}{15} = x$
30. $-\frac{4}{5} + \frac{3}{5} = x$
31. $-\frac{2}{3} + \frac{1}{3} = x$
32. $\frac{3}{8} + -\frac{1}{8} = x$
33. $-\frac{3}{7} + \frac{5}{7} = x$
34. $\frac{1}{4} + -\frac{2}{3} = x$
35. $-\frac{4}{5} + \frac{5}{6} = x$
36. $-\frac{3}{8} + \frac{4}{7} = x$
37. ${}^{-}2\frac{1}{10} + 3\frac{4}{5} = x$
38. $1\frac{4}{9} + {}^{-}2\frac{5}{6} = x$
39. ${}^{-}4\frac{1}{4} + 5\frac{3}{20} = x$

Solve each equation and express answers in simplest form.

40. $\frac{3}{5} - \frac{2}{5} = x$
41. $\frac{7}{9} - \frac{2}{9} = x$
42. $\frac{9}{11} - \frac{4}{11} = x$
43. $\frac{1}{2} - \frac{3}{8} = x$
44. $\frac{8}{9} - \frac{1}{6} = x$
45. $\frac{11}{12} - \frac{2}{3} = x$
46. $4\frac{3}{5} - 3\frac{1}{5} = x$
47. $1 - \frac{2}{5} = x$
48. $1\frac{1}{9} - \frac{4}{9} = x$
49. $4 - \frac{7}{8} = x$
50. $5\frac{1}{2} - 3\frac{1}{4} = x$
51. $3\frac{11}{12} - 1\frac{2}{3} = x$
52. $-\frac{2}{3} - -\frac{1}{3} = x$
53. $-\frac{4}{11} - -\frac{3}{11} = x$
54. $-\frac{8}{15} - -\frac{4}{15} = x$
55. $-\frac{15}{17} - -\frac{7}{17} = x$
56. $-\frac{1}{2} - -\frac{1}{4} = x$
57. $-\frac{3}{4} - -\frac{3}{8} = x$
58. ${}^{-}5\frac{6}{7} - {}^{-}2\frac{4}{7} = x$
59. ${}^{-}5\frac{3}{5} - {}^{-}2\frac{3}{10} = x$
60. ${}^{-}8\frac{7}{8} - {}^{-}4\frac{5}{12} = x$
61. ${}^{-}1\frac{7}{8} - -\frac{5}{8} = x$
62. ${}^{-}2\frac{3}{10} - {}^{-}1\frac{1}{10} = x$
63. ${}^{-}3\frac{5}{6} - {}^{-}1\frac{1}{6} = x$
64. $2 - -\frac{3}{4} = x$
65. ${}^{-}3 - \frac{5}{6} = x$
66. $1\frac{1}{5} - -\frac{3}{5} = x$
67. ${}^{-}1\frac{1}{6} - \frac{5}{6} = x$
68. ${}^{-}8\frac{1}{12} - 5\frac{5}{12} = x$
69. $2\frac{7}{15} - {}^{-}6\frac{13}{15} = x$
70. $-\frac{5}{6} - \frac{1}{3} = x$
71. $\frac{3}{5} - -\frac{9}{10} = x$
72. $-\frac{5}{8} - \frac{17}{24} = x$
73. $\frac{3}{5} - {}^{-}3\frac{1}{4} = x$
74. ${}^{-}4\frac{2}{3} - 8\frac{1}{5} = x$
75. ${}^{-}5\frac{1}{2} - 2\frac{1}{3} = x$
76. $1\frac{1}{2} - -\frac{3}{4} = x$
77. $1\frac{1}{12} - -\frac{2}{3} = x$
78. $7\frac{1}{6} - {}^{-}5\frac{2}{3} = x$

Fraction Equations: Multiplication and Division

Solve each equation and express answers in simplest form.

1. $\frac{1}{3} \cdot \frac{2}{3} = x$
2. $\frac{3}{4} \cdot \frac{4}{3} = x$
3. $\frac{5}{6} \cdot \frac{3}{6} = x$
4. $\frac{3}{5} \cdot \frac{3}{7} = x$
5. $\frac{5}{7} \cdot \frac{2}{13} = x$
6. $\frac{8}{11} \cdot \frac{9}{10} = x$
7. $\frac{2}{3} \cdot \frac{9}{16} = x$
8. $\frac{10}{9} \cdot \frac{15}{8} = x$
9. $\frac{25}{24} \cdot \frac{21}{10} = x$
10. $3 \cdot \frac{1}{2} = x$
11. $\frac{2}{3} \cdot 9 = x$
12. $\frac{3}{5} \cdot 5\frac{1}{6} = x$
13. $4\frac{1}{11} \cdot \frac{3}{5} = x$
14. $2\frac{1}{2} \cdot 3\frac{2}{7} = x$
15. $3\frac{1}{8} \cdot 2\frac{4}{5} \cdot \frac{5}{7} = x$
16. $-\frac{3}{4} \cdot -\frac{1}{2} = x$
17. $-\frac{1}{4} \cdot -\frac{3}{5} = x$
18. $-\frac{1}{7} \cdot -\frac{5}{8} = x$
19. $-\frac{9}{15} \cdot -\frac{15}{9} = x$
20. $-\frac{9}{16} \cdot {}^{-}64 = x$
21. ${}^{-}16 \cdot -\frac{5}{24} = x$
22. ${}^{-}3\frac{2}{3} \cdot -\frac{1}{8} = x$
23. $-\frac{4}{9} \cdot {}^{-}3\frac{1}{8} = x$
24. ${}^{-}3\frac{2}{3} \cdot {}^{-}2\frac{3}{5} = x$
25. ${}^{-}5\frac{1}{7} \cdot {}^{-}4\frac{4}{5} = x$
26. ${}^{-}1\frac{3}{8} \cdot {}^{-}2\frac{1}{7} \cdot -\frac{1}{11} = x$
27. ${}^{-}2\frac{1}{9} \cdot -\frac{5}{7} \cdot {}^{-}4\frac{2}{3} = x$
28. $-\frac{1}{2} \cdot \frac{3}{5} = x$
29. $\frac{6}{7} \cdot -\frac{5}{8} = x$
30. $\frac{2}{3} \cdot -\frac{4}{9} = x$
31. $\frac{7}{10} \cdot -\frac{5}{28} = x$
32. $-\frac{4}{5} \cdot 30 = x$
33. ${}^{-}6 \cdot \frac{5}{11} = x$
34. $-\frac{5}{9} \cdot 4\frac{2}{5} = x$
35. $\frac{3}{10} \cdot {}^{-}2\frac{3}{5} = x$
36. $-\frac{1}{3} \cdot 3\frac{3}{7} = x$
37. $3\frac{1}{3} \cdot {}^{-}1\frac{3}{5} = x$
38. $2\frac{1}{6} \cdot {}^{-}2\frac{4}{7} = x$
39. $1\frac{5}{6} \cdot {}^{-}5\frac{1}{2} = x$

Solve each equation and express answers in simplest form.

40. $\frac{2}{5} \div \frac{3}{7} = x$
41. $\frac{4}{9} \div \frac{5}{7} = x$
42. $\frac{6}{13} \div \frac{5}{7} = x$
43. $\frac{8}{17} \div \frac{15}{13} = x$
44. $\frac{1}{2} \div \frac{4}{5} = x$
45. $\frac{8}{15} \div \frac{4}{5} = x$
46. $3 \div 2\frac{3}{4} = x$
47. $2\frac{3}{5} \div 2 = x$
48. $\frac{2}{5} \div 2\frac{1}{3} = x$
49. $7 \div \frac{7}{8} = x$
50. $3\frac{1}{8} \div 3\frac{3}{4} = x$
51. $5\frac{3}{5} \div 4\frac{1}{5} = x$
52. $-\frac{5}{8} \div -\frac{8}{5} = x$
53. $-\frac{7}{16} \div -\frac{7}{11} = x$
54. $-\frac{9}{14} \div -\frac{3}{7} = x$
55. ${}^{-}8 \div {}^{-}3\frac{2}{3} = x$
56. ${}^{-}1\frac{4}{7} \div {}^{-}3 = x$
57. ${}^{-}3\frac{2}{5} \div -\frac{3}{4} = x$
58. $-\frac{5}{2} \div {}^{-}2\frac{1}{3} = x$
59. ${}^{-}4\frac{5}{8} \div {}^{-}3\frac{4}{7} = x$
60. ${}^{-}4\frac{4}{7} \div -\frac{8}{9} = x$
61. $-\frac{6}{7} \div {}^{-}5\frac{2}{5} = x$
62. ${}^{-}5\frac{1}{5} \div {}^{-}2\frac{1}{6} = x$
63. ${}^{-}2\frac{4}{7} \div {}^{-}1\frac{1}{2} = x$
64. $\frac{3}{8} \div -\frac{5}{7} = x$
65. $-\frac{2}{3} \div \frac{1}{2} = x$
66. $\frac{1}{9} \div -\frac{1}{9} = x$
67. $-\frac{2}{5} \div 7 = x$
68. $4 \div -\frac{2}{3} = x$
69. ${}^{-}7 \div 4 = x$
70. $2 \div {}^{-}2\frac{2}{3} = x$
71. ${}^{-}2\frac{8}{9} \div 4 = x$
72. $1\frac{3}{4} \div -\frac{7}{12} = x$
73. $4\frac{2}{5} \div -\frac{11}{15} = x$
74. ${}^{-}2 \div 2\frac{1}{7} = x$
75. ${}^{-}2\frac{5}{8} \div 7\frac{1}{2} = x$
76. ${}^{-}1\frac{3}{5} \div 2\frac{2}{5} = x$
77. $3\frac{1}{8} \div {}^{-}1\frac{3}{8} = x$
78. ${}^{-}3\frac{1}{21} \div 1\frac{21}{35} = x$

Decimals: Addition and Subtraction

Find each sum.

1. $\begin{array}{r}\$0.17\\0.29\\\underline{0.54}\end{array}$

2. $\begin{array}{l}7\\8.49\\\underline{0.78}\end{array}$

3. $\begin{array}{r}4.03\\0.95\\\underline{13.98}\end{array}$

4. $\begin{array}{l}\ \ 0.95\\\ \ 4.1\\\underline{15\quad\ }\end{array}$

5. $\begin{array}{r}1.68\\0.93\\\underline{2.75}\end{array}$

6. $\begin{array}{l}9.74\\0.05\\\underline{3.764}\end{array}$

7. $\begin{array}{r}1.5\\70.081\\\underline{68.499}\end{array}$

8. $\begin{array}{r}142.7\\0.02\\82.6\\\underline{1.584}\end{array}$

9. $\begin{array}{r}58.27\\3.1\\\underline{0.045}\end{array}$

10. $\begin{array}{r}\$18.93\\45.27\\\underline{23.84}\end{array}$

11. $\begin{array}{r}752.95\\17.461\\0.3\\\underline{8.77}\end{array}$

12. $\begin{array}{r}\$1375.34\\503.19\\594.19\\\underline{65.04}\end{array}$

13. 6 + 4.2 + 9
14. 0.42 + 0.06 + 1.11
15. 0.2 + 6.51 + 2.03
16. 0.006 + 2 + 10.01
17. 4.4 + 30.6 + 11.2
18. 6.501 + 1.1 + 1
19. ⁻28 + ⁻45.50
20. ⁻22.49 + ⁻83
21. ⁻10.04 + ⁻0.18
22. ⁻5.9 + 18.1
23. 7.11 + ⁻1.02
24. ⁻6.51 + 2.03
25. 3.75 + ⁻6.8
26. ⁻65 + 15.64
27. 0.046 + ⁻0.567

Find each difference.

28. $\begin{array}{r}7.8\\\underline{-\ 5.3}\end{array}$

29. $\begin{array}{r}57.96\\\underline{-\ 23.71}\end{array}$

30. $\begin{array}{r}4.01\\\underline{-\ 3.92}\end{array}$

31. $\begin{array}{r}6.0437\\\underline{-\ 1.9864}\end{array}$

32. $\begin{array}{r}8.00\\\underline{-\ 7.87}\end{array}$

33. $\begin{array}{r}\$4\\\underline{-\ \ 1.34}\end{array}$

34. $\begin{array}{r}11.000\\\underline{-\ \ 9.741}\end{array}$

35. $\begin{array}{r}5\\\underline{-\ 2.896}\end{array}$

36. $\begin{array}{r}20.48\\\underline{-\ 16.8}\end{array}$

37. $\begin{array}{r}28.05\\\underline{-\ \ 9.95}\end{array}$

38. $\begin{array}{r}8\\\underline{-\ 3.49}\end{array}$

39. $\begin{array}{r}932\\\underline{-\ \ \ \ 0.003}\end{array}$

40. $\begin{array}{r}26.706\\\underline{-\ \ 6.897}\end{array}$

41. $\begin{array}{r}1.7\\\underline{-\ 0.846}\end{array}$

42. $\begin{array}{r}86.4\\\underline{-\ 75.92}\end{array}$

43. $\begin{array}{r}86.4\\\underline{-\ 75.92}\end{array}$

44. ⁻8.4 − ⁻2.3
45. ⁻0.41 − ⁻65
46. ⁻3.1 − ⁻7.5
47. ⁻11.0 − ⁻6.9
48. ⁻5 − ⁻1.3
49. ⁻6.8 − ⁻9.9
50. ⁻7.25 − ⁻15.86
51. ⁻27.2 − ⁻6.4
52. ⁻6 − ⁻1.4
53. 0.8 − ⁻0.76
54. ⁻6.51 − 4.3
55. 4.5 − ⁻2.1
56. ⁻29.3 − 14.22
57. 16 − ⁻2.9
58. ⁻7 − 35.8
59. 101 − ⁻76.4
60. ⁻3.2 − 18.42
61. ⁻13.0 − 6.8

Decimal Equations: Multiplication and Division

Solve each equation.

1. $x = 68 \cdot 0.7$
2. $406 \cdot 0.9 = x$
3. $4065 \cdot 0.7 = y$
4. $y = 11.5 \cdot 22$
5. $a = 4.68 \cdot 47$
6. $b = 0.09 \cdot 18$
7. $12.5 \cdot 0.5 = a$
8. $11.54 \cdot 4.3 = c$
9. $a = 32.1 \cdot 2.5$
10. $c = {}^{-}22.6 \cdot 7.4$
11. $7.52 \cdot {}^{-}1.8 = x$
12. $20.07 \cdot {}^{-}9.5 = y$
13. ${}^{-}16.8 \cdot 0.55 = a$
14. ${}^{-}4.65 \cdot 1.58 = c$
15. $x = {}^{-}25.06 \cdot 1.05$
16. ${}^{-}47.5 \cdot {}^{-}3.19 = m$
17. $y = {}^{-}24.2 \cdot {}^{-}0.25$
18. ${}^{-}7.24 \cdot {}^{-}0.36 = p$
19. $y = {}^{-}6.15 \cdot {}^{-}0.51$
20. ${}^{-}31.24 \cdot {}^{-}8.61 = c$
21. ${}^{-}452.2 \cdot {}^{-}0.08 = a$
22. $9.06 \cdot {}^{-}0.53 = y$
23. $a = {}^{-}12.08 \cdot 4.25$
24. $8.94 \cdot 0.005 = y$
25. $m = 32.7 \cdot 2.29$
26. ${}^{-}4.88 \cdot 1.25 = r$
27. ${}^{-}4.48 \cdot {}^{-}6.25 = m$
28. $a = 216.17 \cdot 9.63$
29. $m = {}^{-}0.4082 \cdot 11.85$
30. ${}^{-}140.15 \cdot {}^{-}0.186 = t$
31. $q = 19.042 \cdot 8.54$
32. $4.007 \cdot 1.95 = n$
33. $0.0645 \cdot 0.81 = x$
34. $a = 8.265 \cdot {}^{-}3.32$
35. $x = {}^{-}2.508 \cdot {}^{-}0.975$
36. $m = 43.616 \cdot 2.405$
37. $0.0182 \cdot 0.007 = y$
38. $r = 0.0615 \cdot 0.13$
39. $z = {}^{-}0.0036 \cdot {}^{-}0.28$
40. $y = {}^{-}723 \cdot {}^{-}0.0068$
41. $462.1 \cdot 0.0094 = q$
42. $a = 87.061 \cdot 0.0016$
43. $0.0074 \cdot {}^{-}61.3 = a$
44. $0.0076 \cdot 0.00821 = b$
45. $0.0086 \cdot 0.00909 = c$
46. $x = 1249.867 \cdot 0.069$
47. $n = 187.411 \cdot 0.0098$
48. $t = 9.16739 \cdot 8.721$
49. $0.00886 \cdot 0.0708 = b$
50. $y = 0.007201 \cdot 0.00899$
51. $0.06121 \cdot 0.000619 = x$
52. $m = 2.7 \div 3$
53. $z = 4.8 \div 8$
54. $11.4 \div 6 = q$
55. $z = 68 \div 8$
56. $41 \div 5 = n$
57. $79 \div 5 = p$
58. $m = {}^{-}8 \div 20$
59. $12 \div {}^{-}30 = a$
60. $b = 98 \div {}^{-}8$
61. $p = 2.92 \div {}^{-}4$
62. $52.8 \div {}^{-}24 = b$
63. ${}^{-}164.97 \div 47 = z$
64. ${}^{-}0.451 \div {}^{-}11 = q$
65. $r = {}^{-}1.881 \div {}^{-}19$
66. ${}^{-}1.681 \div {}^{-}41 = h$
67. $m = {}^{-}3.286 \div {}^{-}62$
68. ${}^{-}1.953 \div {}^{-}63 = a$
69. ${}^{-}1.892 \div {}^{-}43 = b$
70. $63 \div {}^{-}0.9 = p$
71. $c = 86 \div {}^{-}4.3$
72. $d = 105 \div 2.1$
73. $81 \div 0.27 = q$
74. $d = 54 \div 0.18$
75. $m = {}^{-}33 \div 0.66$
76. $x = 30.1 \div 0.7$
77. $82.8 \div 0.4 = m$
78. ${}^{-}565.6 \div {}^{-}0.7 = y$
79. $19.5 \div 1.5 = q$
80. ${}^{-}18.53 \div {}^{-}1.7 = d$
81. $0.0418 \div 0.19 = m$
82. $1.781 \div 0.13 = r$
83. $0.1804 \div 4.4 = m$
84. $0.1696 \div 3.2 = s$
85. $b = 0.2808 \div 0.078$
86. $x = 0.1001 \div {}^{-}0.77$
87. $p = 0.5805 \div 2.15$
88. $c = 19.317 \div 0.94$
89. $y = 9.557 \div 1.9$
90. $m = 15.33 \div {}^{-}14.6$
91. $2.006 \div 0.118 = p$
92. $0.4484 \div 1.18 = n$
93. $a = 29.45 \div 6.2$
94. $x = 0.24102 \div 0.117$
95. $q = {}^{-}9.8049 \div {}^{-}0.966$
96. $0.1107 \div 2.25 = r$
97. $m = 0.32539 \div 5.006$
98. $0.046626 \div 8.18 = z$
99. $n = 21.99256 \div 3.14$
100. $8.9238 \div {}^{-}83.4 = a$
101. $1.573 \div {}^{-}6.05 = b$
102. ${}^{-}2800 \div {}^{-}625 = d$
103. $21.5025 \div 7.05 = p$

Forms of Real Numbers

Write each fraction as a decimal.

1. $\frac{2}{5}$ 2. $-\frac{1}{4}$ 3. $-\frac{7}{10}$ 4. $\frac{3}{20}$
5. $-\frac{4}{5}$ 6. $-\frac{17}{20}$ 7. $\frac{1}{3}$ 8. $-\frac{1}{6}$
9. $-\frac{2}{3}$ 10. $\frac{1}{9}$ 11. $-\frac{5}{6}$ 12. $\frac{2}{9}$
13. $\frac{3}{7}$ 14. $-\frac{1}{12}$ 15. $\frac{5}{7}$ 16. $-\frac{7}{12}$
17. $-\frac{8}{9}$ 18. $\frac{4}{25}$ 19. $\frac{3}{11}$ 20. $-\frac{9}{16}$

Write each mixed numeral as a decimal.

21. $1\frac{3}{4}$ 22. $2\frac{1}{2}$ 23. $^-5\frac{9}{10}$ 24. $^-6\frac{3}{5}$
25. $4\frac{1}{3}$ 26. $^-1\frac{5}{6}$ 27. $7\frac{3}{16}$ 28. $2\frac{5}{7}$
29. $^-2\frac{7}{11}$ 30. $8\frac{2}{3}$ 31. $^-9\frac{5}{9}$ 32. $11\frac{1}{6}$
33. $75\frac{1}{4}$ 34. $^-101\frac{4}{5}$ 35. $3\frac{5}{8}$ 36. $^-52\frac{7}{12}$
37. $32\frac{1}{8}$ 38. $73\frac{3}{10}$ 39. $^-18\frac{11}{12}$ 40. $16\frac{6}{7}$

For each of the following, write an equivalent fraction with a denominator of 100.

41. $-\frac{1}{2}$ 42. $-\frac{1}{5}$ 43. $\frac{1}{4}$ 44. $-\frac{1}{10}$
45. $\frac{1}{20}$ 46. $-\frac{3}{4}$ 47. $-\frac{4}{5}$ 48. $\frac{7}{25}$
49. $-\frac{7}{4}$ 50. $-\frac{9}{20}$ 51. $\frac{9}{50}$ 52. $-\frac{67}{10}$
53. $\frac{49}{25}$ 54. $-\frac{27}{5}$ 55. $\frac{19}{4}$ 56. $-\frac{25}{20}$
57. $\frac{13}{10}$ 58. $-\frac{4320}{1000}$ 59. $\frac{14}{5}$ 60. $\frac{21}{20}$

Write each decimal as a fraction in simplest form.

61. 0.7 62. $^-0.25$ 63. $^-0.6$ 64. $^-0.15$
65. $^-0.625$ 66. 0.85 67. $^-0.875$ 68. $0.\overline{3}$
69. $^-9.50$ 70. $^-2.75$ 71. 3.125 72. $^-1.34$
73. $11.\overline{6}$ 74. $^-9.\overline{3}$ 75. 25.25 76. $^-2.75$
77. $^-13.125$ 78. 4.15 79. $^-8.\overline{3}$ 80. 22.875
81. 1.375 82. $^-2.25$ 83. $9.\overline{6}$ 84. 7.24
85. $0.\overline{4}$ 86. $0.\overline{37}$ 87. $0.\overline{524}$ 88. $0.\overline{163}$
89. $2.\overline{3}$ 90. $^-1.\overline{26}$ 91. $^-0.3\overline{27}$ 92. $5.\overline{371}$

Scientific Notation

Express each of the following using scientific notation.

1. 40
2. 400
3. 4,000
4. 40,000
5. 20,691
6. 2069.1
7. 206.91
8. 20.691
9. $^{-}42.3$
10. $^{-}8{,}200{,}000$
11. 36.241
12. 95,236
13. 300,000
14. 0.00015
15. 0.0015
16. 0.015
17. 0.15
18. 0.0692
19. 0.00000308
20. 0.007
21. $^{-}0.029$
22. 0.0009
23. $^{-}325{,}000{,}000$
24. 1,200,000

Express each of the following using decimal notation.

25. 5×10^2
26. 5×10^3
27. 5×10^4
28. 5×10^5
29. $^{-}2.1 \times 10^3$
30. 1.05×10^5
31. $^{-}5.704 \times 10^3$
32. 3.157×10^2
33. $^{-}9.5 \times 10^6$
34. 6.33×10^8
35. $^{-}7.2 \times 10^7$
36. $^{-}5.402 \times 10^3$
37. $^{-}62.1 \times 10^7$
38. 9×10^{-1}
39. 9×10^{-2}
40. 9×10^{-3}
41. 9×10^{-4}
42. 4.8×10^{-1}
43. $^{-}3.6 \times 10^{-3}$
44. $^{-}7.7 \times 10^{-4}$
45. $^{-}8 \times 10^{-7}$
46. $^{-}6.4 \times 10^{-8}$
47. $^{-}1.002 \times 10^{-3}$
48. $^{-}1.6 \times 10^{-4}$

Evaluate. Express each answer in scientific notation.

49. $(4.3 \times 10^3)(2.0 \times 10^2)$
50. $(3.6 \times 10^5)(7.5 \times 10^3)$
51. $(1.5 \times 10)(3.4 \times 10^6)$
52. $(7.22 \times 10^4)(5.1 \times 10^7)$
53. $(^{-}45 \times 10^2)(^{-}5 \times 10^4)$
54. $(^{-}4.8 \times 10)(^{-}9.6 \times 10^8)$
55. $(^{-}6 \times 10^7)(^{-}8.2 \times 10^5)$
56. $(^{-}12.1 \times 10^3)(^{-}3.4 \times 10^2)$
57. $(8 \times 10^3)(^{-}5 \times 10^9)$
58. $(^{-}24 \times 10^6)(3 \times 10^9)$
59. $(7.01 \times 10^8)(^{-}9.1 \times 10)$
60. $(^{-}3 \times 10^4)(7 \times 10^4)$
61. $(5 \times 10^{-4})(6 \times 10^{-6})$
62. $(2 \times 10^{-3})(4 \times 10^{-2})$
63. $(2.3 \times 10^{-1})(9.2 \times 10^{-5})$
64. $(34.01 \times 10^{-7})(6 \times 10^{-8})$
65. $(7 \times 10^2)(6 \times 10^{-2})$
66. $(4 \times 10^{-3})(8 \times 10^5)$
67. $(8.8 \times 10^{-7})(1.1 \times 10^4)$
68. $(12 \times 10^{10})(12 \times 10^{-6})$
69. $(^{-}2 \times 10^{-5})(9 \times 10^3)$
70. $(7 \times 10^{-8})(^{-}4 \times 10^6)$
71. $\dfrac{24 \times 10^6}{3 \times 10^2}$
72. $\dfrac{45 \times 10^3}{9 \times 10^5}$
73. $\dfrac{12 \times 10^{12}}{4 \times 10^6}$
74. $\dfrac{8 \times 10^{-5}}{2 \times 10^{-3}}$
75. $\dfrac{33 \times 10^{-4}}{11 \times 10^{-5}}$
76. $\dfrac{63 \times 10^{-4}}{7 \times 10^{-4}}$
77. $\dfrac{10 \times 10^8}{5 \times 10^{-3}}$
78. $\dfrac{169 \times 10^{-4}}{13 \times 10^6}$
79. $\dfrac{6 \times 10}{2 \times 10^{-2}}$
80. $\dfrac{^{-}2.5 \times 10^2}{5 \times 10^{-3}}$
81. $\dfrac{^{-}48 \times 10^5}{^{-}2 \times 10^{-4}}$
82. $\dfrac{9 \times 10^{-1}}{^{-}3 \times 10^{-1}}$

Evaluating Expressions

Evaluate.

1. $t \div 4$ if $t = {}^-16$
2. $k \div 3$ if $k = 15$
3. ${}^-96 \div q$ if $q = 12$
4. ${}^-76 \div m$ if $m = {}^-19$
5. $\frac{x}{15}$ if $x = 60$
6. $\frac{y}{{}^-9}$ if $y = 36$
7. $\frac{{}^-54}{p}$ if $p = {}^-9$
8. $\frac{8}{n}$ if $n = {}^-40$
9. $\frac{z}{32}$ if $z = {}^-8$
10. $\frac{t}{{}^-7}$ if $t = {}^-84$
11. $\frac{2x}{{}^-15}$ if $x = {}^-30$
12. $\frac{{}^-112}{{}^-4n}$ if $n = {}^-7$
13. $2y + 3$ if $y = {}^-5$
14. $3(x - 9)$ if $x = 7$
15. $t(t + 4)$ if $t = {}^-8$
16. $r^2 - 2r$ if $r = {}^-3$
17. $q^2(q - 8)$ if $q = 4$
18. $\frac{5a^2 + a}{2a}$ if $a = {}^-2$

Evaluate if $a = 6$, $b = 4$, and $c = {}^-3$.

19. $a + (b + c)$
20. $a + (b - c)$
21. $6(a + b)$
22. $a - (b - c)$
23. $a - b - c$
24. $a(b + |c|)$
25. $ab|c|$
26. $ab + ac$
27. $a^2 + b^2 + c^2$
28. $7a - (2b + c)$
29. $a^2 - c^2$
30. $36 - 4c^2$
31. $\frac{4(a - b)}{|c - 1|}$
32. $\frac{3(4b + 5a)}{23}$
33. $\frac{(a - b)(a + b)}{2c}$
34. $\frac{3ab}{2a + |c|}$
35. $\frac{5a + 3c}{3b}$
36. $\frac{3ab^2 - c^3}{a}$

The relationship between Celsius temperature (C) and Fahrenheit temperature (F) is given by $C = \frac{5(F - 32)}{9}$. Find the Celsius temperature for the given temperatures.

37. 68°F
38. 86°F
39. ${}^-22$°F
40. 32°F
41. 23°F
42. ${}^-4$°F

The formula for the area of a trapezoid is $A = \frac{h}{2}(b + B)$

A stands for the area, h stands for the altitude, and b and B stand for the bases. Calculate the area of the trapezoid given the following.

43. $b = 16$, $B = 30$, and $h = 18$
44. $b = 7$, $B = 12$, and $h = 9$
45. $b = 20$, $B = 25$, and $h = 22$
46. $b = 5$, $B = 11$, and $h = 8$

The formula for the volume of a sphere is $V = \frac{4}{3}\pi r^3$. V stands for the volume and r for the radius. Calculate the volume of the sphere given the following.

47. $\pi = \frac{22}{7}$, $r = \frac{1}{4}$
48. $\pi = \frac{22}{7}$, $r = 2\frac{1}{2}$
49. $\pi = 3.14$, $r = 3$
50. $\pi = 3.14$, $r = 6$

Solving Linear Equations

Solve each equation.

1. $5 + x = 9$
2. $^{-}14 + y = 10$
3. $56 + z = {}^{-}3$
4. $38 = t + 21$
5. $72 = p + {}^{-}8$
6. $^{-}19 = q + {}^{-}4$
7. $^{-}25 + h = {}^{-}33$
8. $m + {}^{-}14 = 6$
9. $d + 40 = 27$
10. $^{-}x + 9 = 15$
11. $^{-}y + {}^{-}2 = {}^{-}85$
12. $^{-}f + {}^{-}7 = 1$
13. $a - 4 = 72$
14. $b - 9 = {}^{-}23$
15. $c - 11 = 6$
16. $^{-}102 = w - 99$
17. $61 = s - 71$
18. $^{-}28 = k - 80$
19. $z - {}^{-}15 = 36$
20. $x - {}^{-}40 = {}^{-}58$
21. $y - {}^{-}16 = 59$
22. $^{-}66 = a - {}^{-}1$
23. $524 = m - {}^{-}300$
24. $^{-}75 = q - {}^{-}24$
25. $6x = 18$
26. $8y = 64$
27. $10z = 5$
28. $^{-}20 = {}^{-}4a$
29. $^{-}8 = 8c$
30. $35 = {}^{-}5d$
31. $\frac{1y}{2} = 12$
32. $\frac{x}{7} = 9$
33. $\frac{z}{^{-}5} = 3$
34. $^{-}13 = \frac{h}{7}$
35. $^{-}32 = \frac{t}{^{-}14}$
36. $25 = \frac{n}{^{-}11}$
37. $\frac{2r}{3} = {}^{-}24$
38. $\frac{^{-}5s}{7} = {}^{-}20$
39. $\frac{6p}{^{-}11} = 54$
40. $^{-}21 = \frac{^{-}7y}{10}$
41. $36 = \frac{9x}{^{-}15}$
42. $^{-}42 = \frac{^{-}4z}{^{-}5}$
43. $4x + 8 = 20$
44. $6y - 8 = 22$
45. $9z - 30 = 6$
46. $3m + 4 = {}^{-}11$
47. $^{-}5r + 6 = {}^{-}14$
48. $3 + {}^{-}4m = 7$
49. $13 = 2x - 7$
50. $47 = 3k - 7$
51. $^{-}50 = 6m - 8$
52. $8a - 10 = {}^{-}90$
53. $25 = 4g + 5$
54. $^{-}3z - 18 = 9$
55. $\frac{x}{5} + 2 = 7$
56. $\frac{m}{3} + 6 = 14$
57. $\frac{y}{3} + 6 = {}^{-}9$
58. $\frac{b}{7} - 8 = {}^{-}12$
59. $\frac{d}{10} - 3 = {}^{-}5$
60. $\frac{r}{^{-}6} - 4 = 7$
61. $\frac{d + 5}{3} = {}^{-}9$
62. $\frac{3 + f}{7} = {}^{-}5$
63. $\frac{m - 5}{4} = 5$
64. $\frac{4r + 8}{16} = 7$
65. $\frac{3d - 4}{5} = 4$
66. $\frac{7n - {}^{-}1}{8} = 8$
67. $5y + 9 = 8y$
68. $6a - 10 = 4a$
69. $4c + 15 = 9c$
70. $7p - 12 = 4p + 6$
71. $7m + 56 = 3m + 36$
72. $^{-}3x + 6 = 12 - 2x$
73. $2(x + 2) = {}^{-}32$
74. $^{-}52 = 2(3z - 8)$
75. $^{-}4(x + 1) = {}^{-}12$
76. $2(n + 4) = n + 10$
77. $7x - 2 = 3(3x + 3)$
78. $4y - 3(y - 2) = 41$

Solve each equation for the variable indicated.

79. $P = 2l + 2w$, for l
80. $A = bh$, for h
81. $d = rt$, for r
82. $C = 2\pi r$, for r
83. $a = (b - c)p$, for c
84. $x(y + z) = t$, for y
85. $\frac{e}{f} = g + h$, for f
86. $A = \frac{1}{2}bh$ for b
87. $A = \frac{h}{2}(b + B)$, for B

Solving Linear Inequalities

Solve each inequality.

1. $y + 3 < 6$
2. $n + 2 < 5$
3. $x + 6 \geq {}^{-}14$
4. $a + 4 > 2$
5. $b - 3 > 8$
6. $c - 5 < 3$
7. $4y \leq 3y + 4$
8. $3y < 2y + 6$
9. $6a > 5a - 2$
10. $5n > 4n - 3$
11. $11 \geq y - 5$
12. $16 > n - 5$
13. $13 < b + 6$
14. $4 < c + 16$
15. $f - {}^{-}3 \leq {}^{-}6$
16. $r - {}^{-}4 \geq 12$
17. ${}^{-}20 + 5y \geq 6y$
18. $12d - 18 < 11d$
19. ${}^{-}7 + 6x < 7x + 2$
20. $14z - {}^{-}3 \leq 15z + 6$
21. $9y - 4 > 10y + {}^{-}12$
22. $\frac{x}{4} < 9$
23. $\frac{y}{5} < 3$
24. $\frac{z}{4} > {}^{-}5$
25. $\frac{r}{{}^{-}2} < 4$
26. $\frac{s}{{}^{-}8} \geq {}^{-}7$
27. $\frac{k}{{}^{-}9} < {}^{-}6$
28. $\frac{h}{12} \leq {}^{-}40$
29. $\frac{m}{{}^{-}10} > 11$
30. $\frac{y}{15} < 12$
31. $2 < \frac{x}{8}$
32. ${}^{-}3 \leq \frac{d}{7}$
33. $20 > \frac{g}{4}$
34. $38 > \frac{t}{{}^{-}2}$
35. $9 < \frac{z}{{}^{-}6}$
36. ${}^{-}13 \geq \frac{s}{{}^{-}12}$
37. $5x \geq 20$
38. $6y < {}^{-}12$
39. $14z > {}^{-}70$
40. ${}^{-}21 \leq 7a$
41. $46 > 2c$
42. ${}^{-}84 < 4b$
43. ${}^{-}2f > 26$
44. ${}^{-}15x > {}^{-}105$
45. ${}^{-}3z \geq 27$
46. ${}^{-}8r \leq {}^{-}48$
47. ${}^{-}9s < 63$
48. ${}^{-}31t < {}^{-}651$
49. $1600 > {}^{-}50k$
50. ${}^{-}925 < 37p$
51. $72 \geq {}^{-}8y$
52. ${}^{-}85x \leq {}^{-}255$
53. ${}^{-}33w > 297$
54. ${}^{-}18 < 3z$
55. $3a + 4 > 16$
56. $5t + 40 \leq 55$
57. $7 + 6z > 19$
58. $12b - 32 < {}^{-}6$
59. $9y - 5 \geq {}^{-}50$
60. $8m - 13 < {}^{-}37$
61. $16 - 5x > 31$
62. ${}^{-}4c + 3 < {}^{-}21$
63. ${}^{-}24 \leq 8 - 16q$
64. ${}^{-}7k - 12 \geq 37$
65. ${}^{-}20 - 3w > {}^{-}32$
66. ${}^{-}13y - 11 < 54$
67. $13x - 5 \leq 10x + 4$
68. $2y + 8 > 5y - 7$
69. $z - 6 > 3z + 2$
70. $16s - 15 \geq 12s - 3$
71. $9r + 3 < 6r + 3$
72. $4t - 18 < t + 9$
73. $2(n + 4) > n + 10$
74. $3(x + 6) > 2x + 12$
75. $3y + 7 \leq 4(y + 2)$
76. $7(m + 8) \leq 3m + 36$
77. ${}^{-}7({}^{-}x + 4) > 14$
78. $9 > {}^{-}3b + 6(b - 4)$
79. $\frac{3a - 5}{2} > 2a + 1$
80. $\frac{3c}{4} - \frac{5}{8} < 0$
81. $\frac{5 - 2k}{3} \leq {}^{-}7$

Find the set of all integers, x, satisfying the given conditions.

82. $x > {}^{-}5$ and $x < 4$
83. $x \geq {}^{-}2$ and $x \leq 2$
84. $x \leq 10$ and $x < {}^{-}1$
85. $x < 9$ and $x \geq {}^{-}3$
86. $x > 0$ or $x \geq 3$
87. $x > 25$ or $x < 30$
88. $2 < 3x + 2$ and $3x + 2 < 14$
89. $9 - 2x > 11$ and $5x < 2x + 9$
90. $x + 4 > {}^{-}2$ and $x - 5 < {}^{-}1$
91. $2x < 30$ or $3x > 60$

Proportions and Percents

Cross multiply to determine whether or not the ratios are equivalent. Write yes or no.

1. $\frac{1}{2}, \frac{2}{4}$
2. $\frac{1}{3}, \frac{2}{6}$
3. $\frac{1}{5}, \frac{2}{11}$
4. $\frac{1}{10}, \frac{2}{5}$
5. $\frac{1}{8}, \frac{3}{24}$
6. $\frac{1}{7}, \frac{4}{28}$
7. $\frac{2}{5}, \frac{4}{9}$
8. $\frac{3}{4}, \frac{6}{8}$
9. $\frac{2}{3}, \frac{3}{9}$
10. $\frac{2}{7}, \frac{6}{21}$
11. $\frac{4}{5}, \frac{15}{20}$
12. $\frac{5}{6}, \frac{15}{18}$
13. $\frac{4}{9}, \frac{12}{27}$
14. $\frac{6}{15}, \frac{12}{45}$
15. $\frac{8}{21}, \frac{4}{10}$
16. $\frac{11}{24}, \frac{33}{72}$

Solve each proportion.

17. $\frac{2}{5} = \frac{4}{c}$
18. $\frac{5}{12} = \frac{10}{y}$
19. $\frac{8}{12} = \frac{2}{e}$
20. $\frac{18}{24} = \frac{3}{k}$
21. $\frac{3}{4} = \frac{m}{12}$
22. $\frac{4}{7} = \frac{x}{14}$
23. $\frac{2}{9} = \frac{z}{27}$
24. $\frac{3}{10} = \frac{b}{30}$
25. $\frac{h}{6} = \frac{2}{12}$
26. $\frac{u}{12} = \frac{3}{36}$
27. $\frac{x}{22} = \frac{15}{11}$
28. $\frac{h}{30} = \frac{102}{60}$
29. $\frac{1}{v} = \frac{9}{27}$
30. $\frac{84}{d} = \frac{28}{5}$
31. $\frac{7}{r} = \frac{21}{28}$
32. $\frac{7}{q} = \frac{28}{16}$
33. $\frac{p}{84} = \frac{25}{100}$
34. $\frac{p}{50} = \frac{115}{100}$
35. $\frac{11}{44} = \frac{r}{100}$
36. $\frac{12}{60} = \frac{r}{100}$
37. $\frac{5}{B} = \frac{10}{100}$
38. $\frac{75}{B} = \frac{50}{100}$
39. $\frac{135}{B} = \frac{675}{100}$
40. $\frac{90}{40} = \frac{r}{100}$

Solve each of the following.

41. 10% of 60 is ________.
42. 25% of 84 is ________.
43. 6% of 150 is ________.
44. 485% of 180 is ________.
45. 205% of 22 is ________.
46. 110% of 40 is ________.
47. ________ is 4.1% of 200.
48. ________ is 16.5% of 300.
49. ________ is 42.8% of 150.
50. ________ is $16\frac{2}{3}$% of 48.
51. ________ is $112\frac{1}{2}$% of 536.
52. ________ is $11\frac{1}{4}$% of 800.
53. ________ is 19% of 170.
54. 10.8% of 20 is ________.
55. 40.5% of 120 is ________.
56. $13\frac{3}{4}$% of 4000 is ________.
57. $83\frac{1}{3}$% of 150 is ________.
58. ________ is 20.8% of 220.
59. 8 is ________% of 32.
60. 9 is ________% of 45.
61. 11 is ________% of 22.
62. 120 is ________% of 800.
63. 57 is ________% of 95.
64. 713 is ________% of 3100.
65. ________% of 40 is 16.
66. ________% of 500 is 55.
67. ________% of 71 is 17.04.
68. ________% of 85 is 102.
69. ________% of 705 is 2538.
70. ________% of 45 is 90.
71. 12 is ________% of 96.
72. 25 is ________% of 40.
73. 56 is ________% of 16.
74. 20% of ________ is 12.
75. 30% of ________ is 21.
76. 25% of ________ is 75.
77. 3% of ________ is 18.
78. 65% of ________ is 26.
79. 44% of ________ is 37.4.
80. 85 is 68% of ________.
81. 117 is 26% of ________.
82. 738 is 72% of ________.
83. 770 is $87\frac{1}{2}$% of ________.
84. 32 is $66\frac{2}{3}$% of ________.
85. 135 is 675% of ________.

Absolute Value

Solve each equation.

1. $|{}^{-}5| = x$
2. $|{}^{-}9| = x$
3. $|{}^{-}3.7| = y$
4. $y = |105.2|$
5. $z = |9.7|$
6. $r = |0.18|$
7. $t = |5 - 12|$
8. $m = |{}^{-}3 + {}^{-}4|$
9. $k = |51 + {}^{-}48|$
10. $|{}^{-}30 - {}^{-}5| = s$
11. $|{}^{-}60 - 25| = q$
12. $|24 + {}^{-}16| = x$
13. $|2 \cdot 3 - 7| = f$
14. $|32 - 4 \cdot 8| = k$
15. $|{}^{-}11 \cdot 9 + {}^{-}5 \cdot 10| = y$
16. $|{}^{-}5| + 4 = x$
17. ${}^{-}8 + |12| = z$
18. $7 - |{}^{-}24| = l$
19. $a = |11.4| - |{}^{-}12|$
20. $r = {}^{-}|3| + {}^{-}29$
21. $c = |45 - 10| - 37$
22. $x + |{}^{-}72| = 91 - |38|$
23. ${}^{-}62 + |56| = b - |47|$
24. ${}^{-}|84| - |100| = 12 - y$

Evaluate if $a = {}^{-}3$.

25. $|a|$
26. $|{}^{-}7a|$
27. $|11a|$
28. $|a - 4|$
29. $|3a + 6|$
30. $|3 - a|$
31. $a - |a|$
32. $a + |a|$
33. $|{}^{-}5a - 9|$
34. $25 - |2a + 7|$
35. $2|4a - 9|$
36. ${}^{-}3|a + 1| + 2|a - 6|$
37. $5|a + 4| - |6a|$
38. $|{}^{-}8a| + {}^{-}|12a + 3|$
39. ${}^{-}4|7a| + 3|a|$
40. $|a - 8| - 10|3a + 10|$
41. $|{}^{-}a + 56| + 53$
42. ${}^{-}6|7 - a| + |20a|$

Solve each of the following open sentences.

43. $|y + 3| = 2$
44. $|x + 10| = 6$
45. $|m - 7| = 8$
46. $|t - 2| = 5$
47. $|r - 6| = {}^{-}3$
48. $|k - 9| = 4$
49. $|2x + 1| = 7$
50. $|3x + 7| = {}^{-}21$
51. $|4a - 6| = 10$
52. $|5s - 5| = 20$
53. $8|p - 3| = 88$
54. $5|y + 4| = 45$
55. $2|3d + 3| = 18$
56. $4|2b + 6| = 32$
57. ${}^{-}6|2x - 14| = {}^{-}42$
58. $4|6r - 1| = 29$
59. $3|p + 9| + 6 = 0$
60. $9|3 - 2a| = 15$
61. $2|7 - 3f| = 3$
62. $3|y + 5| = y - 7$
63. $|4r + 6| = r - 5$
64. $|7 + 4s| = 12 - s$
65. $5|3t - 4| = t + 1$
66. $42 + |3z - 8| = 10$
67. $|a| > 3$
68. $|b| > 4$
69. $|z| \leq 9$
70. $|f| \geq 2$
71. $|r| \leq 6$
72. $|s| \leq 7$
73. $|m| > {}^{-}5$
74. $|t| < 0$
75. $|x + 1| \geq 3$
76. $|r + 2| \geq 0$
77. $|p - 5| < 0$
78. $|c| - 5 < 0$
79. $|6 - s| \leq 2$
80. $|2 - y| \geq 1$
81. $|a + 4| > {}^{-}2$
82. $|8 + b| < 2$
83. $|7y| \geq 21$
84. $|5x| < 35$
85. $|2m| < 26$
86. $|2s - 9| \leq 27$
87. $|3x + 11| > 42$
88. $|4x - 3| \geq 12$

Factoring

Find the prime factorization of each integer. Write each negative integer as the product of $^-1$ and its prime factors.

1. 21 **2.** 18 **3.** 36 **4.** 60
5. $^-24$ **6.** $^-36$ **7.** $^-95$ **8.** $^-48$
9. 16 **10.** $^-27$ **11.** $^-55$ **12.** 72
13. $^-450$ **14.** 200 **15.** 372 **16.** $^-163$

Factor each expression.

17. $16y - 8$
18. $3z + 9$
19. $13a^2 + 39$
20. $42r^2 - 6$
21. $^-2x + 5ax$
22. $z - 2az$
23. $qx - 5x$
24. $^-20xy - 16xz$
25. $x + 7x^2$
26. $85m - 6m^2$
27. $^-17y^2 + 11y^4$
28. $5c^5 - 3c^9$
29. $2x^6 + 8x^7$
30. $^-7p^3 + 56p^4$
31. $y^2 - 8y + 16$
32. $a^2 - 16a + 64$
33. $p^2 - 2p + 1$
34. $n^2 + 10n + 25$
35. $9a^2 - 12ab + 4b^2$
36. $9x^2 - 24xy + 16y^2$
37. $x^2 - 2xy + y^2$
38. $49p^2 - 14p + 1$
39. $64b^2 + 48bc + 9c^2$
40. $36 - 24x + 4x^2$
41. $9x^2 + 48xy + 64y^2$
42. $4m^2 + 12m + 9$
43. $4a^4 + 4a^2 + 1$
44. $49 - 28x^2 + 4x^4$
45. $1 - 14p^4 + 49p^8$
46. $4x^6 - 12x^3y^3 + 9y^6$
47. $y^2 - 81$
48. $m^2 - 100$
49. $r^2 - 4x^2$
50. $c^2 - 49b^2$
51. $25a^2 - b^2$
52. $4x^2 - 9$
53. $36s^2 - 100$
54. $1 - 64y^2$
55. $^-49 + 16t^2$
56. $^-121 + 9x^2$
57. $4x^2 - 9y^2$
58. $a^8 - b^{10}$
59. $12c^2 - 12$
60. $81 - 9x^2$
61. $(a + b)^2 - m^2$
62. $(x - y)^2 - z^2$
63. $y^4 - x^4$
64. $4y^6 - 9x^4$
65. $x^2 + 9x + 18$
66. $a^2 + 20a + 36$
67. $r^2 + 11r + 10$
68. $p^2 + 13p + 30$
69. $y^2 - 10y + 16$
70. $t^2 - 11t + 28$
71. $k^2 - 16k + 15$
72. $m^2 - 17m + 72$
73. $d^2 + 4d - 21$
74. $x^2 + 8x - 20$
75. $a^2 + 12a - 45$
76. $h^2 + 6h - 16$
77. $c^2 - 5c - 36$
78. $y^2 - 2y - 48$
79. $2s^2 + 7s + 3$
80. $b^2 - 2b - 63$
81. $3f^2 + 13f + 12$
82. $2h^3 - 8h^2 - 42h$

Simplifying Radicals

Simplify.

1. $-\sqrt{4}$
2. $\sqrt{25}$
3. $\pm\sqrt{9}$
4. $-\sqrt{36}$
5. $\sqrt{81}$
6. $\pm\sqrt{49}$
7. $-\sqrt{64}$
8. $\sqrt{121}$
9. $\pm\sqrt{169}$
10. $-\sqrt{225}$
11. $\sqrt{144}$
12. $\pm\sqrt{100}$
13. $\sqrt{0.0036}$
14. $\sqrt{0.16}$
15. $\sqrt{0.64}$
16. $\sqrt{1.44}$
17. $\sqrt{8}$
18. $\sqrt{27}$
19. $\sqrt{12}$
20. $\sqrt{32}$
21. $-\sqrt{75}$
22. $\pm\sqrt{72}$
23. $\sqrt{240}$
24. $-\sqrt{180}$
25. $\sqrt{128}$
26. $-\sqrt{125}$
27. $\sqrt{147}$
28. $\pm\sqrt{108}$
29. $-\sqrt{112}$
30. $-\sqrt{44}$
31. $\pm\sqrt{45}$
32. $\sqrt{162}$
33. $\sqrt{x^2}$
34. $\sqrt{y^4}$
35. $\sqrt{a^6}$
36. $\sqrt{t^{10}}$
37. $\sqrt{m^8}$
38. $\sqrt{a^2}$
39. $\sqrt{x}$
40. $\sqrt{z^{12}}$
41. $\sqrt{x^{30}}$
42. $\sqrt{y^{64}}$
43. $\sqrt{a^{150}}$
44. $\sqrt{c^{90}}$
45. $\sqrt{t^5}$
46. $\sqrt{k^7}$
47. $\sqrt{p^{11}}$
48. $\sqrt{s^{21}}$
49. $\sqrt{ab^2}$
50. $\sqrt{r^2t^4z}$
51. $\sqrt{3c^2}$
52. $\sqrt{5a^4}$
53. $\sqrt{4m^2}$
54. $\sqrt{9x^2z}$
55. $\sqrt{40k^8}$
56. $\sqrt{11x^6}$
57. $\sqrt{1600x^3}$
58. $\sqrt{4x^5y^7}$
59. $\sqrt{0.16a^{10}b^8}$
60. $\sqrt{m^4r^3t}$
61. $\frac{1}{5}\sqrt{54x^2y^4}$
62. $\frac{2}{3}\sqrt{24a^2b^2}$
63. $\sqrt{3ab^5}$
64. $\frac{1}{6}\sqrt{45x^2z}$
65. $-\sqrt{\frac{4}{9}}$
66. $\pm\sqrt{\frac{25a^2}{4b^2}}$
67. $\sqrt{\frac{49x^4}{36y^2}}$
68. $\sqrt{\frac{100a^2b^6}{81c^{10}}}$
69. $\sqrt{\frac{27}{x^3}}$
70. $\sqrt{\frac{y^4}{16y^2}}$
71. $\sqrt{\frac{64r^2t^{14}}{25s^6}}$
72. $\sqrt{\frac{9k^{12}}{25l^{10}}}$

Rationalize the denominators of the following.

73. $\frac{1}{\sqrt{5}}$
74. $\frac{3}{\sqrt{7}}$
75. $\frac{^-8}{\sqrt{11}}$
76. $\frac{^-8y}{\sqrt{15}}$
77. $\frac{3}{2\sqrt{5}}$
78. $\sqrt{\frac{a}{3}}$
79. $\frac{5\sqrt{21}}{\sqrt{5}}$
80. $\sqrt{\frac{10}{7}}$
81. $\sqrt{\frac{45}{2}}$
82. $-\sqrt{\frac{50}{3}}$
83. $-\frac{\sqrt{48}}{\sqrt{3}}$
84. $\sqrt{\frac{7}{8}}$
85. $-\sqrt{\frac{54}{r}}$
86. $\frac{\sqrt{8}}{\sqrt{9x}}$
87. $\pm\sqrt{\frac{7x^2}{4y}}$
88. $\frac{\sqrt{22a}}{\sqrt{8y^2}}$
89. $\sqrt{\frac{20m}{3}}$
90. $\pm\sqrt{\frac{2k}{11m}}$
91. $\frac{\sqrt{18s}}{\sqrt{3}}$
92. $-\sqrt{\frac{b}{6}}$
93. $\frac{^-5\sqrt{21}}{\sqrt{5}}$
94. $\frac{3\sqrt{7}}{\sqrt{2}}$
95. $\frac{6\sqrt{36}}{\sqrt{2}}$
96. $\frac{^-2\sqrt{27}}{\sqrt{3}}$
97. $\frac{11}{4\sqrt{6}}$
98. $\frac{^-3}{10\sqrt{15}}$
99. $\frac{20}{^-5\sqrt{2}}$
100. $\frac{4}{7\sqrt{8}}$
101. $\frac{^-6\sqrt{56}}{2\sqrt{10}}$
102. $\frac{9\sqrt{20}}{3\sqrt{11}}$
103. $\frac{^-4\sqrt{24}}{12\sqrt{27}}$
104. $\frac{16\sqrt{40}}{8\sqrt{32}}$

Operations With Radicals

Simplify.

1. $3\sqrt{5} + 2\sqrt{5}$
2. $8\sqrt{3} + {}^{-}2\sqrt{3}$
3. ${}^{-}6\sqrt{7} + 4\sqrt{7}$
4. ${}^{-}8\sqrt{6} + {}^{-}2\sqrt{6}$
5. $4\sqrt{x} + {}^{-}9\sqrt{x}$
6. ${}^{-}2\sqrt{y} + 11\sqrt{y}$
7. $\frac{1}{4}\sqrt{xy} + \frac{3}{4}\sqrt{xy}$
8. $\frac{2}{3}\sqrt{11} + -\frac{1}{3}\sqrt{11}$
9. $-\frac{5}{8}\sqrt{3} + -\frac{1}{8}\sqrt{3}$
10. $3\sqrt{21} - 11\sqrt{21}$
11. $18\sqrt{z} - 5\sqrt{z}$
12. $4\sqrt{t} - 10\sqrt{t}$
13. $84\sqrt{m} - 56\sqrt{m}$
14. ${}^{-}7\sqrt{3} - 11\sqrt{3}$
15. ${}^{-}2\sqrt{k} - {}^{-}5\sqrt{k}$
16. $2\sqrt{3} + \sqrt{27}$
17. $\sqrt{8} + 3\sqrt{2}$
18. $5\sqrt{5} + \sqrt{20}$
19. $\sqrt{8} + \sqrt{50}$
20. $\sqrt{27} - \sqrt{12}$
21. $\sqrt{32} - \sqrt{18}$
22. ${}^{-}\sqrt{44a} + \sqrt{99a}$
23. ${}^{-}\sqrt{50c} + {}^{-}\sqrt{32c}$
24. $\sqrt{20} - \sqrt{80}$
25. $\sqrt{24x} + {}^{-}\sqrt{54x}$
26. ${}^{-}\sqrt{18y} + \sqrt{50y}$
27. ${}^{-}\sqrt{45t} + {}^{-}\sqrt{125t}$
28. $\sqrt{18} + \sqrt{72} - \sqrt{50}$
29. $2\sqrt{40} - \sqrt{160} + \sqrt{72}$
30. $\sqrt{3} - 2\sqrt{12} - \sqrt{27}$
31. $3\sqrt{18} - \sqrt{8} + 2\sqrt{32}$
32. $3\sqrt{5} + 2\sqrt{20} - \sqrt{125}$
33. $2\sqrt{20} - 3\sqrt{80} + \sqrt{45}$
34. $7\sqrt{5x^2} + x\sqrt{20}$
35. $\sqrt{a^2b} - 2|a|\sqrt{b} + 9\sqrt{a^2b}$
36. $|y|\sqrt{x^2yz} - \sqrt{x^2y^3z}$

37. $\sqrt{3}\sqrt{2}$
38. $\sqrt{12}\sqrt{3}$
39. $\sqrt{18}\sqrt{6}$
40. $\sqrt{2}\sqrt{24}\sqrt{3}$
41. $\sqrt{a}\sqrt{a^3}$
42. $\sqrt{y^5}\sqrt{y^7}$
43. $\sqrt{5x}\sqrt{8x}$
44. $4\sqrt{2}\sqrt{10}$
45. $3\sqrt{5}\sqrt{18}$
46. $(\sqrt{7})^2$
47. $(2\sqrt{3})^2$
48. $(2\sqrt{32})^2$
49. $\sqrt{8}\sqrt{\frac{1}{2}}$
50. $\sqrt{\frac{1}{3}}\sqrt{9}$
51. $\sqrt{\frac{10}{4}}\sqrt{\frac{8}{5}}$
52. $2\sqrt{2} \cdot 4\sqrt{8}$
53. $5\sqrt{3} \cdot 7\sqrt{2}$
54. $2\sqrt{5} \cdot \sqrt{125}$
55. $\sqrt{5}(\sqrt{5} + \sqrt{3})$
56. $\sqrt{3}(\sqrt{3} + \sqrt{5})$
57. $\sqrt{2}(\sqrt{2} - \sqrt{6})$
58. $(3 - \sqrt{7})(3 + \sqrt{7})$
59. $(5 - \sqrt{5})(5 + \sqrt{5})$
60. $(\sqrt{2} - 3)(\sqrt{2} + 3)$
61. $(\sqrt{10} + \sqrt{3})(\sqrt{10} - \sqrt{3})$
62. $(\sqrt{7} - \sqrt{2})(\sqrt{7} + \sqrt{2})$
63. $(\sqrt{3} + 2\sqrt{5})(\sqrt{3} - 2\sqrt{5})$
64. $(\sqrt{a+b})(\sqrt{a+b})$
65. $(\sqrt{a} - \sqrt{b})(\sqrt{a} + \sqrt{b})$
66. $\sqrt{x-y} \cdot \sqrt{4x - 4y}$
67. $\frac{\sqrt{15}}{\sqrt{3}}$
68. $\frac{\sqrt{21}}{\sqrt{7}}$
69. $\frac{\sqrt{93}}{\sqrt{3}}$
70. $\frac{\sqrt{54}}{\sqrt{6}}$
71. $\frac{\sqrt{72}}{\sqrt{8}}$
72. $\frac{\sqrt{144}}{\sqrt{8}}$
73. $\frac{\sqrt{100}}{\sqrt{4}}$
74. $\frac{\sqrt{36}}{\sqrt{9}}$
75. $\frac{\sqrt{16x^4}}{\sqrt{4x^2}}$
76. $\frac{8\sqrt{x^5}}{3\sqrt{x}}$
77. $\frac{\sqrt{28}}{\sqrt{7}}$
78. $\frac{\sqrt{7}}{\sqrt{3}}$
79. $\frac{5\sqrt{21}}{\sqrt{5}}$
80. $\sqrt{\frac{3}{7}}$
81. $\sqrt{\frac{b}{6}}$
82. $\sqrt{\frac{x^4}{11}}$
83. $\sqrt{\frac{27}{y}}$
84. $\sqrt{\frac{54}{z}}$
85. $\frac{\sqrt{18y^2}}{\sqrt{3}}$
86. $\frac{\sqrt{24x}}{8\sqrt{x^3}}$
87. $\frac{a^3\sqrt{ab^3}}{\sqrt{a^3b}}$

Solving Radical and Quadratic Equations

Solve each equation.

1. $\sqrt{x} = 2$
2. $\sqrt{y} = 9$
3. $-\sqrt{r} = ^-6$
4. $-\sqrt{m} = ^-5$
5. $\sqrt{5y} = 5$
6. $\sqrt{x} - 9 = 0$
7. $\sqrt{4a} = 1$
8. $3 + \sqrt{m} = 5$
9. $2 + \sqrt{r} = 10$
10. $\sqrt{4b + 1} = 3$
11. $\sqrt{2c + 7} = 5$
12. $\sqrt{8y + 1} = 5$
13. $\sqrt{3b - 5} - 4 = 0$
14. $\sqrt{3x + 7} - 7 = 0$
15. $\sqrt{2x + 7} - 3 = 0$
16. $\sqrt{r} = 3\sqrt{5}$
17. $3\sqrt{7} = \sqrt{x}$
18. $\sqrt{y} = 2\sqrt{8}$
19. $\sqrt{m} = 4\sqrt{7}$
20. $5\sqrt{10} = \sqrt{x}$
21. $5\sqrt{5} = \sqrt{c}$
22. $\sqrt{\frac{p}{3}} = 5$
23. $\sqrt{\frac{4a}{3}} - 2 = 0$
24. $\sqrt{\frac{2x}{3}} - 4 = 0$
25. $\frac{8}{\sqrt{x + 3}} = 2$
26. $\frac{\sqrt{x}}{\sqrt{2}} = \sqrt{5}$
27. $\frac{\sqrt{2x + 1}}{3} + 2 = 5$
28. $\sqrt{2x^2 - 121} = x$
29. $\sqrt{x + 2} = x - 4$
30. $\sqrt{1 - 2y} = 1 + y$
31. $\sqrt{5a + 1} + 6 = 10$
32. $n + \sqrt{n^2 + 3} = 3n$
33. $\sqrt{10 + x} - \sqrt{10 - x} = 2$

Solve each equation.

34. $(x - 4)(x + 5) = 0$
35. $y(y - 6) = 0$
36. $(3x + 4)(2x - 5) = 0$
37. $(x - 7)^2 = 0$
38. $x^2 + 2x + 1 = 0$
39. $x^2 - 36 = 0$
40. $x^2 + 2x - 48 = 0$
41. $x^2 - 12x + 35 = 0$
42. $y^2 - 7y = 0$
43. $y^2 + 17y + 72 = 0$
44. $2x^2 + 12x + 16 = 0$
45. $3y^2 - 21y + 36 = 0$
46. $9a^2 - 64 = 0$
47. $25b^2 - 16 = 0$
48. $3y^2 - 15y = 0$
49. $9x^2 + 6x + 1 = 0$
50. $4x^2 - 12x + 9 = 0$
51. $x^2 = 121$
52. $17x^2 = 68x$
53. $2y^2 + 24y = ^-40$
54. $y^2 + 3y = 40$
55. $2a^2 - 3a = 9$
56. $10b^2 + 33b = 7$
57. $18y^2 - 3y = 15$

Use the quadratic formula to solve each equation.

58. $x^2 + 7x + 6 = 0$
59. $a^2 + 8a + 15 = 0$
60. $2a^2 + a - 15 = 0$
61. $2y^2 + 7y + 3 = 0$
62. $3m^2 + 5m + 2 = 0$
63. $3x^2 + 23x + 14 = 0$
64. $3x^2 + 14x = 5$
65. $m^2 - 2m = 8$
66. $r^2 + 9 = 10r$
67. $m^2 + 36 = 13m$
68. $4x^2 - 8x = ^-3$
69. $4b^2 = 20b$
70. $2x^2 = 3 + x$
71. $3y^2 + 2 = 5y$
72. $b^2 - b - 3 = 0$
73. $y^2 - 27 = 0$
74. $x^2 - 125 = 0$
75. $m^2 = 32$
76. $x^2 - 6x - 2 = 0$
77. $y^2 - 2y = 1$
78. $y^2 + 4y = 3$
79. $2a^2 - 5a + 1 = 0$
80. $3b^2 - 5b + 1 = 0$
81. $x^2 + 3x = 1$
82. $3y^2 + y - 1 = 0$
83. $x^2 + 2x + 5 = 0$

SYMBOLS

Symbol	Meaning
h	altitude (333)
$\angle$	angle (65)
a	apothem (338)
$\approx$	approximately equal to (274)
$\widehat{AB}$	arc with endpoints A and B (297)
A	area (329)
B	area of base of a prism (347)
b	base (333)
BASIC	**B**eginner's **A**ll-purpose **S**ymbolic **I**nstructional **C**ode (462)
cm	centimeter (328)
$\odot P$	circle with center P (294)
C	circumference (342)
$\cong$	congruent, is congruent to (51)
cos	cosine (281)
cm^3	cubic centimeters (354)
m^3	cubic meters (354)
$^\circ$	degree (68)
d	diagonal of a square (276); diameter (296); distance (373)
AB	distance between points A and B (39)
km	kilometer (331)
L	lateral area (347)
$\overleftrightarrow{DE}$	line containing points D and E (2)
$\overline{PQ}$	line segment with endpoints P and Q (42)
$\rightarrow$	mapping, is mapped onto (432)
$m\angle A$	measure of angle A (68)
$m\widehat{AB}$	measure of arc AB (297)
m	meter (329)
mm	millimeter (335)
$\sqrt{\ }$	nonnegative square root (264)
(4, 2)	ordered pair (368)
(x, y, z)	ordered triple (396)
$\parallel$	parallel, is parallel to (160)
▱	parallelogram (202)
P	perimeter (218)
$\perp$	perpendicular, is perpendicular to (82)
π	pi (36)
n-gon	polygon with n sides (195)
r	radius (296)
$\overrightarrow{PQ}$	ray with endpoint P passing through Q (64)
k	scale factor for a dilation (453)
s	side of a regular polygon (219)
$\sim$	similar, is similar to (238)
sin	sine (281)
l	slant height (350)
m	slope (382)
cm^2	square centimeter (328)
m^2	square meter (329)
S	sum of the degree measure of a convex polygon (198)
tan	tangent (281)
T	total surface area (347)
$\triangle$	triangle (96)
V	volume (355)

POSTULATES & THEOREMS

Chapter 1: Points, Lines, and Planes

Postulate 1–1 Through any two points there is exactly one line.

Postulate 1–2 Through any three points *not* on the same line there is exactly one plane.

Postulate 1–3 A line contains at least two points.

Postulate 1–4 A plane contains at least three points *not* on the same line.

Postulate 1–5 If two points lie in a plane, then the entire line containing those two points lies in that plane.

Postulate 1–6 If two planes intersect, then their intersection is a line.

Theorem 1–1 If there is a line and a point *not* on the line, then there is exactly one plane that contains them.

Theorem 1–2 If two lines intersect, then exactly one plane contains both lines.

Chapter 2: Measure

Postulate 2–1 Each real number corresponds to exactly one point on a number line. Each point on a number line corresponds to exactly one real number.

Postulate 2–2 Ruler Postulate The points on a number line can be paired with the real numbers so that, given any two points P and Q on the line, P corresponds to zero, and Q corresponds to a positive number.

Postulate 2–3 Distance Postulate For any two points on a line and a given unit of measure, there is a unique positive number called the measure of the distance between the two points.

Properties of equality:

	For any numbers a, b, and c	
Postulate 2–4	reflexive	$a = a$
Postulate 2–5	symmetric	If $a = b$, then $b = a$
Postulate 2–6	transitive	If $a = b$ and $b = c$, then $a = c$
Postulate 2–7	addition and subtraction	If $a = b$, then $a + c = b + c$, and $a - c = b - c$.
Postulate 2–8	multiplication and division	If $a = b$, then $a \cdot c = b \cdot c$, and if c is not zero, $\frac{a}{c} = \frac{b}{c}$.
Postulate 2–9	substitution	If $a = b$, then a may be replaced by b.

Properties of operations:

	For any numbers a, b, and c		
		Addition	Multiplication
Postulate 2–10	Commutative	$a + b = b + a$	$a \cdot b = b \cdot a$
Postulate 2–11	Associative	$(a + b) + c = a + (b + c)$	$(a \cdot b) \cdot c = a \cdot (b \cdot c)$
Postulate 2–12	Identity	$a + 0 = a = 0 + a$	$a \cdot 1 = a = 1 \cdot a$
Postulate 2–13	Inverse	$a + {}^{-}a = 0 = {}^{-}a + a$	If a is *not* zero, then $a \cdot \frac{1}{a} = 1 = \frac{1}{a} \cdot a$
Postulate 2–14	Distributive of multiplication over addition: $a(b + c) = ab + ac$ and $(b + c)a = ba + ca$		

Properties of inequality:

	For any number a, b, and c	
Postulate 2–15	Comparison	$a < b$, or $a = b$, or $a > b$
Postulate 2–16	addition and subtraction	1. If $a < b$, then $a + c < b + c$ and $a - c < b - c$. 2. If $a > b$, then $a + c > b + c$ and $a - c > b - c$.
Postulate 2–17	multiplication and division	1. If $c > 0$ and $a < b$, then $ac < bc$ and $\frac{a}{c} < \frac{b}{c}$. 2. If $c > 0$ and $a > b$, then $ac > bc$ and $\frac{a}{c} > \frac{b}{c}$. 3. If $c < 0$ and $a < b$, then $ac > bc$ and $\frac{a}{c} > \frac{b}{c}$. 4. If $c < 0$ and $a > b$, then $ac < bc$ and $\frac{a}{c} < \frac{b}{c}$.

Theorem 2–1 A line segment has exactly one midpoint.

Theorem 2–2 Congruence of line segments is reflexive.

Theorem 2–3 Congruence of line segments is symmetric.

Theorem 2–4 Congruence of line segments is transitive.

Theorem 2–5 Midpoint Theorem — If M is the midpoint of $\overline{PQ}$, then $\overline{PM} \cong \overline{MQ}$.

Theorem 2–6 Bisector Theorem — If $\overline{PQ}$ is bisected at M, then $\overline{PM} \cong \overline{MQ}$.

Chapter 3: Angles and Perpendiculars

Postulate 3–1 Angle Measure Postulate	For every angle there is a unique positive number between 0 and 180 called the degree measure of the angle.
Postulate 3–2 Protractor Postulate	Given a ray $\overrightarrow{QR}$ on the edge of a half plane, for every positive number r between 0 and 180, there is exactly one ray $\overrightarrow{QP}$ in the half plane so that $m\angle PQR = r$.
Postulate 3–3 Angle Addition Postulate	If a point R is in the interior of a given angle, $\angle PQS$, then $m\angle PQS = m\angle PQR + m\angle RQS$.
Postulate 3–4 Supplement Postulate	The angles in a linear pair are supplementary angles.
Theorem 3–1	Congruence of angles is reflexive, symmetric, and transitive.
Theorem 3–2	If two angles are supplementary to the same angle, then they are congruent.
Theorem 3–3	If two angles are supplementary to two congruent angles, then the two angles are congruent to each other.
Theorem 3–4	If two angles are complementary to the same angle, then they are congruent to each other.
Theorem 3–5	If two angles are complementary to two congruent angles, then the two angles are congruent to each other.
Theorem 3–6	If two angles are right angles, then the angles are congruent.
Theorem 3–7	If one angle in a linear pair is a right angle, then the other angle also is a right angle.
Theorem 3–8	If two angles are congruent and supplementary, then each is a right angle.
Theorem 3–9	If two intersecting lines form one right angle, then they form four right angles.
Theorem 3–10	Vertical angles are congruent.
Theorem 3–11	If two lines are perpendicular, then they form four right angles.
Theorem 3–12	In a plane, through a given point on a line, there is exactly one line perpendicular to the given line.
Theorem 3–13	If two lines are perpendicular, then they form congruent adjacent angles.
Theorem 3–14	If two intersecting lines form congruent adjacent angles, then they are perpendicular.
Theorem 3–15	Two planes are perpendicular if and only if they intersect to form a right dihedral angle.

Chapter 4: Congruent Triangles

Postulate 4–1 SSS	If each side of one triangle is congruent to the corresponding side of another triangle, then the triangles are congruent.

Postulate 4–2 SAS	If two sides and the included angle of one triangle are congruent to the corresponding sides and included angle of another triangle, then the triangles are congruent.
Postulate 4–3 ASA	If two angles and the included side of one triangle are congruent to the corresponding angles and included side of another triangle, then the triangles are congruent.
Postulate 4–4 HL	If the hypotenuse and a leg of one right triangle are congruent to the corresponding sides of another right triangle, then the triangles are congruent.
Theorem 4–1 Angle Sum Theorem	The sum of the degree measures of the angles of a triangle is 180.
Theorem 4–2	The degree measure of each angle of an equiangular triangle is 60.
Theorem 4–3	The acute angles of a right triangle are complementary.
Theorem 4–4	Congruence of triangles is reflexive, symmetric, and transitive.
Theorem 4–5 AAS	If two angles and a nonincluded side of one triangle are congruent to the corresponding angles and nonincluded side of another triangle, then the triangles are congruent.
Theorem 4–6 Isosceles Triangle Theorem	If two sides of a triangle are congruent, then the angles opposite those sides are congruent.
Theorem 4–7	If a triangle is equilateral, then the triangle is equiangular.
Theorem 4–8	Each angle of an equilateral triangle has degree measure 60.
Theorem 4–9	If two angles of a triangle are congruent, then the sides opposite those angles are congruent.
Theorem 4–10	If a triangle is equiangular, then the triangle is equilateral.
Theorem 4–11 HA	If the hypotenuse and an acute angle of one right triangle are congruent to the corresponding hypotenuse and angle of another right triangle, then the triangles are congruent.
Theorem 4–12 LL	If two legs of one right triangle are congruent to the corresponding legs of another right triangle, then the triangles are congruent.
Theorem 4–13 LA	If one leg and an acute angle of one right triangle are congruent to the corresponding leg and angle of another right triangle, then the triangles are congruent.

Chapter 5: Triangle Inequalities

Theorem 5–1	The measure of an exterior angle of a triangle is equal to the sum of the measures of the two remote interior angles.
Theorem 5–2 Inequality Theorem	For any numbers a and b, $a > b$ if and only if there is a positive number c such that $a = b + c$.

Theorem 5–3 The measure of an exterior angle of a triangle is greater than the measure of either remote interior angle.

Theorem 5–4 If the measures of two sides of a triangle are unequal, then the measures of the angles opposite those sides are unequal in the same order.

Theorem 5–5 If the measures of two angles of a triangle are unequal, then the measures of the sides opposite those angles are unequal in the same order.

Theorem 5–6 The shortest segment from a point to a line is the perpendicular segment.

Theorem 5–7 The shortest segment from a point to a plane is the perpendicular segment.

Theorem 5–8 Triangle Inequality — The sum of the measures of any two sides of a triangle is greater than the measure of the third side.

Theorem 5–9 Hinge Theorem — If two sides of one triangle are congruent to two sides of another triangle, but the measures of the included angles are unequal, then the measures of the third sides are unequal in the same order.

Theorem 5–10 Converse of the Hinge Theorem — If two sides of one triangle are congruent to two sides of another triangle, but the measures of the third sides are unequal, then the measures of the angles included between the pairs of congruent sides are unequal in the same order.

Chapter 6: Parallels

Postulate 6–1 Parallel Postulate — If there is a line and a point *not* on the line, then there is exactly one line through the point and parallel to the given line.

Theorem 6–1 If two lines are cut by a transversal and one pair of alternate interior angles are congruent, then the other pair of angles also are congruent.

Theorem 6–2 If two lines are cut by a transversal and one pair of corresponding angles are congruent, then all pairs of corresponding angles are congruent.

Theorem 6–3 A right triangle has no more than one right angle.

Theorem 6–4 In a plane, if two lines are cut by a transversal so that a pair of corresponding angles are congruent, then the two lines are parallel.

Theorem 6–5 In a plane, if two lines are cut by a transversal so that a pair of alternate interior angles are congruent, then the two lines are parallel.

Theorem 6–6 In a plane, if two lines are cut by a transversal so that a pair of consecutive interior angles are supplementary, then the lines are parallel.

Theorem 6–7 In a plane, if two lines are perpendicular to the same line, then the two lines are parallel.

Theorem 6–8 If two parallel lines are cut by a transversal, then each pair of corresponding angles are congruent.

Theorem 6–9 If two parallel lines are cut by a transversal, then each pair of alternate interior angles are congruent.

Theorem 6–10 If two parallel lines are cut by a transversal, then each pair of consecutive interior angles are supplementary.

Theorem 6–11 If a line is perpendicular to one of two parallel lines, then it is perpendicular to the other.

Theorem 6–12 Given a line and a point *not* on the line, there is exactly one line through the point and perpendicular to the given line.

Theorem 6–13 Two parallel lines are everywhere equidistant.

Chapter 7: Polygons

Theorem 7–1 If a convex polygon has n sides, for $n \geq 3$, then the sum of the degree measures of its angles is $(n - 2)\,180$.

Theorem 7–2 The sum of the degree measures of the exterior angles of a convex polygon, one at each vertex, is 360.

Theorem 7–3 A diagonal of a parallelogram separates the parallelogram into two congruent triangles.

Theorem 7–4 Opposite angles of a parallelogram are congruent.

Theorem 7–5 Opposite sides of a parallelogram are congruent.

Theorem 7–6 The diagonals of a parallelogram bisect each other.

Theorem 7–7 If both pairs of opposite sides of a quadrilateral are congruent, then the quadrilateral is a parallelogram.

Theorem 7–8 If two sides of a quadrilateral are parallel and congruent, then the quadrilateral is a parallelogram.

Theorem 7–9 If the diagonals of a quadrilateral bisect each other, then the quadrilateral is a parallelogram.

Theorem 7–10 The diagonals of a rectangle are congruent.

Theorem 7–11 The diagonals of a rhombus bisect the angles of the rhombus.

Theorem 7–12 The diagonals of a rhombus are perpendicular.

Theorem 7–13 The diagonals of an isosceles trapezoid are congruent.

Theorem 7–14 Each pair of base angles of an isosceles trapezoid is congruent.

Theorem 7–15 The median of a trapezoid is parallel to the bases and its measure is one-half the sum of the measures of the bases.

Theorem 7–16 If a rectangle has perimeter P units, length l units, and width w units, then $P = 2(l + w)$.

Theorem 7–17 If a regular n-gon has perimeter P units, and a side measures s units, then $P = ns$.

Chapter 8: Similarity

Postulate 8–1
AA Similarity
If two angles of one triangle are congruent to the corresponding angles of another triangle, then the triangles are similar.

Theorem 8–1
Equality of Cross Products
For any numbers a and c, and any nonzero numbers b and d, $\frac{a}{b} = \frac{c}{d}$ if and only if $ad = bc$.

For any numbers a, b, c, and d, the following properties hold whenever all terms are defined.

Theorem 8–2 If $\frac{a}{b} = \frac{c}{d}$, then $\frac{a}{c} = \frac{b}{d}$.

Theorem 8–3 If $\frac{a}{b} = \frac{c}{d}$, then $\frac{b}{a} = \frac{d}{c}$.

Theorem 8–4 Addition Property of Proportions: $\frac{a}{b} = \frac{c}{d}$ if and only if $\frac{a+b}{b} = \frac{c+d}{d}$

Theorem 8–5 Subtraction Property of Proportions: $\frac{a}{b} = \frac{c}{d}$ if and only if $\frac{a-b}{b} = \frac{c-d}{d}$.

Theorem 8–6 Summation Property of Proportions: $\frac{a}{b} = \frac{c}{d}$ if and only if $\frac{a}{b} = \frac{a+c}{b+d}$ or $\frac{c}{d} = \frac{a+c}{b+d}$.

Theorem 8–7 SSS Similarity: If there is a correspondence between two triangles so that the measures of their corresponding sides are proportional, then the two triangles are similar.

Theorem 8–8 SAS Similarity: If the measures of two sides of a triangle are proportional to the measures of two corresponding sides of another triangle and the included angles are congruent, then the triangles are similar.

Theorem 8–9 If a line intersects two sides of a triangle, cutting off segments with measures proportional to the measures of these two sides, then it is parallel to the third side.

Theorem 8–10 If a line is parallel to one side of a triangle and intersects the other two sides, then it divides them proportionally.

Theorem 8–11 If a segment joins the midpoints of two sides of a triangle, then it is parallel to the third side and its length is one-half the length of the third side.

Theorem 8–12 If three parallel lines intersect two transversals, then they divide the transversals proportionally.

Theorem 8–13 If three parallel lines cut off congruent segments on one transversal, then they cut off congruent segments on any transversal.

Theorem 8–14 If two triangles are similar, then the measures of corresponding perimeters are proportional to the measures of corresponding sides.

Theorem 8–15 If two triangles are similar, then the measures of corresponding altitudes are proportional to the measures of corresponding sides.

Theorem 8–16 If two triangles are similar, then the measures of corresponding angle bisectors of the triangles are proportional to the measures of corresponding sides.

Theorem 8–17 If two triangles are similar, then the measures of corresponding medians are proportional to the measures of corresponding sides.

Chapter 9: Right Triangles

Postulate 9–1 Product Property of Square Roots
For any nonnegative numbers a and b, $\sqrt{ab} = \sqrt{a} \cdot \sqrt{b}$.

Postulate 9–2 Quotient Property of Square Roots
For any nonnegative numbers a and b, $\sqrt{\frac{a}{b}} = \frac{\sqrt{a}}{\sqrt{b}}$ as long as b is *not* zero.

Theorem 9–1
If the altitude is drawn from the right angle to the hypotenuse of a right triangle, then the two triangles formed are similar to the given triangle and to each other.

Theorem 9–2
The measure of the altitude drawn from the right angle to the hypotenuse of a right triangle is the geometric mean between the measure of the two segments of the hypotenuse.

Theorem 9–3
If the altitude is drawn to the hypotenuse of a right triangle, then the measure of a leg of the triangle is the geometric mean between the measure of the hypotenuse and the measure of the segment of the hypotenuse adjacent to that leg.

Theorem 9–4 Pythagorean Theorem
For any right triangle, the square of the measure of the hypotenuse equals the sum of the squares of the measures of the legs.

Theorem 9–5 Converse of Pythagorean Theorem
If the square of the measure of one side of a triangle equals the sum of the squares of the other two sides, then the triangle is a right triangle.

Theorem 9–6
If each acute angle of a right triangle has degree measure 45, then the hypotenuse measures $\sqrt{2}$ times the measure of a leg.

Theorem 9–7
If the acute angles of a right triangle have degree measures 30 and 60, then the hypotenuse measures 2 times the measure of the shorter leg and the longer leg measures $\sqrt{3}$ times the measure of the shorter leg.

Chapter 10: Circles and Spheres

Postulate 10–1 Arc Addition Postulate
If Q is a point on $\widehat{PQR}$, then $m\widehat{PQ} + m\widehat{QR} = m\widehat{PQR}$.

Theorem 10–1
In a plane, if a line contains a point in the interior of a circle, then the line intersects the circle in exactly two points.

Theorem 10–2 In a circle or in congruent circles, two central angles are congruent if and only if their minor arcs are congruent.

Theorem 10–3 In a circle or in congruent circles, two minor arcs are congruent if and only if their corresponding chords are congruent.

Theorem 10–4 In a circle, if a diameter is perpendicular to a chord, then it bisects the chord and its arcs.

Theorem 10–5 In a circle or in congruent circles, two chords are congruent if and only if they are equidistant from the center.

Theorem 10–6 The measure of an inscribed angle of a circle equals one-half the measure of its intercepted arc.

Theorem 10–7 If two inscribed angles intercept congruent arcs, then the angles are congruent.

Theorem 10–8 If an angle is inscribed in a semicircle, then the angle is a right angle.

Theorem 10–9 If the angles of a quadrilateral are inscribed in a circle, then each pair of opposite angles are supplementary.

Theorem 10–10 A tangent to a circle is perpendicular to the radius drawn to the point of tangency.

Theorem 10–11 In a plane, if a line is perpendicular to a radius of a circle at its endpoint on the circle, then the line is a tangent.

Theorem 10–12 The two tangent segments from the same exterior point of a circle are congruent.

Theorem 10–13 If two secants intersect in the interior of a circle, then the measure of an angle formed is one-half the sum of the measures of the arcs intercepted by the angle and its vertical angle.

Theorem 10–14 If two secants intersect in the exterior of a circle, then the measure of an angle formed is one-half the positive difference of the measures of the intercepted arcs.

Theorem 10–15 If a secant and a tangent intersect at the point of tangency, then the measure of the angles formed is one-half the measure of its intercepted arc.

Theorem 10–16 If a secant and a tangent, or two tangents, intersect in the exterior of a circle, then the measure of the angle formed is one-half the positive difference of the measures of the intercepted arcs.

Theorem 10–17 If two chords intersect in a circle, then the product of the measures of the segments of one chord equals the product of the measures of the segments of the other chord.

Theorem 10–18 If two secant segments are drawn to a circle from an exterior point, then the product of the measures of one secant segment and its external secant segment equals the product of the measures of the other secant segment and its external secant segment.

Theorem 10–19 If a tangent segment and a secant segment are drawn to a circle from an exterior point, then the square of the measure of the tangent segment equals the product of the measures of the secant segment and its exterior secant segment.

Theorem 10–20 If a plane intersects a sphere in more than one point, then the intersection is a circle.

Chapter 11: Area and Volume

Postulate 11–1 For any polygonal region and a given unit of measure, there is a unique positive number called the measure of the area of the region.

Postulate 11–2 If two regions are congruent, then they have exactly the same area.

Postulate 11–3 Area Addition Postulate — If a region is separated into nonoverlapping regions, then the sum of the measures of the areas of these regions equals the measure of the area of the given region.

Postulate 11–4 Volume Postulate — For any solid region and a given unit of measure, there is a unique positive number called the measure of the volume of the region.

Postulate 11–5 If two solids are congruent, then they have exactly the same volume.

Postulate 11–6 Volume Addition Postulate — If a solid region is separated into nonoverlapping regions, then the sum of the measures of the volumes of these regions equals the volume of the given region.

Chapter 12: Coordinates

Postulate 12–1 Each point in a coordinate plane corresponds to exactly one ordered pair of numbers. Each ordered pair of numbers corresponds to exactly one point in a coordinate plane.

Theorem 12–1 Distance Formula — The distance between two points with coordinates (x_1, y_1) and (x_2, y_2) is given by the following formula.

$$d = \sqrt{(x_2 - x_1)^2 + (y_2 - y_1)^2}$$

Theorem 12–2 Midpoint Formula — If the coordinates of A and B are (x_1, y_1) and (x_2, y_2) respectively, then the midpoint M of $\overline{AB}$ has coordinates $\left(\frac{x_1 + x_2}{2}, \frac{y_1 + y_2}{2}\right)$.

Theorem 12–3 Two lines have the same slope if and only if they are parallel and nonvertical.

Theorem 12–4 Two nonvertical lines are perpendicular if and only if the product of their slopes is $^{-}1$.

Theorem 12–5 The equation of a line having a slope m and a y-intercept b is $y = mx + b$.

Theorem 12–6 The equation of a line passing through a point whose coordinates are (x_1, y_1) and having a slope, m, is $y - y_1 = m(x - x_1)$. This is called the point-slope form.

Theorem 12–7 The equation of a circle with center (h, k) and radius measuring r units is $(x - h)^2 + (y - k)^2 = r^2$.

Theorem 12–8 Given two points $A(x_1, y_1, z_1)$ and $B(x_2, y_2, z_2)$ in space, the measure of the distance between A and B is given by the following equation.

$$AB = \sqrt{(x_2 - x_1)^2 + (y_2 - y_1)^2 + (z_2 - z_1)^2}$$

Chapter 14: Transformations

Postulate 14–1 In a given rotation, if A is the preimage, P is the image, and W is the center of rotation, then the measure of the angle of rotation, $\angle AWP$, equals twice the measure of the angle between the intersecting lines of reflection.

Theorem 14–1 If a dilation with center C and a scale factor k maps A onto E and B onto D, then $ED = k(AB)$.

SQUARES AND APPROXIMATE SQUARE ROOTS

n	n^2	$\sqrt{n}$	n	n^2	$\sqrt{n}$
1	1	1.000	51	2601	7.141
2	4	1.414	52	2704	7.211
3	9	1.732	53	2809	7.280
4	16	2.000	54	2916	7.348
5	25	2.236	55	3025	7.416
6	36	2.449	56	3136	7.483
7	49	2.646	57	3249	7.550
8	64	2.828	58	3364	7.616
9	81	3.000	59	3481	7.681
10	100	3.162	60	3600	7.746
11	121	3.317	61	3721	7.810
12	144	3.464	62	3844	7.874
13	169	3.606	63	3969	7.937
14	196	3.742	64	4096	8.000
15	225	3.873	65	4225	8.062
16	256	4.000	66	4356	8.124
17	289	4.123	67	4489	8.185
18	324	4.243	68	4624	8.246
19	361	4.359	69	4761	8.307
20	400	4.472	70	4900	8.367
21	441	4.583	71	5041	8.426
22	484	4.690	72	5184	8.485
23	529	4.796	73	5329	8.544
24	576	4.899	74	5476	8.602
25	625	5.000	75	5625	8.660
26	676	5.099	76	5776	8.718
27	729	5.196	77	5929	8.775
28	784	5.292	78	6084	8.832
29	841	5.385	79	6241	8.888
30	900	5.477	80	6400	8.944
31	961	5.568	81	6561	9.000
32	1024	5.657	82	6724	9.055
33	1089	5.745	83	6889	9.110
34	1156	5.831	84	7056	9.165
35	1225	5.916	85	7225	9.220
36	1296	6.000	86	7396	9.274
37	1369	6.083	87	7569	9.327
38	1444	6.164	88	7744	9.381
39	1521	6.245	89	7921	9.434
40	1600	6.325	90	8100	9.487
41	1681	6.403	91	8281	9.539
42	1764	6.481	92	8464	9.592
43	1849	6.557	93	8649	9.644
44	1936	6.633	94	8836	9.695
45	2025	6.708	95	9025	9.747
46	2116	6.782	96	9216	9.798
47	2209	6.856	97	9409	9.849
48	2304	6.928	98	9604	9.899
49	2401	7.000	99	9801	9.950
50	2500	7.071	100	10000	10.000

TRIGONOMETRIC RATIOS

Angle	sin	cos	tan	Angle	sin	cos	tan
0°	0.0000	1.0000	0.0000	45°	0.7071	0.7071	1.0000
1°	0.0175	0.9998	0.0175	46°	0.7193	0.6947	1.0355
2°	0.0349	0.9994	0.0349	47°	0.7314	0.6820	1.0724
3°	0.0523	0.9986	0.0524	48°	0.7431	0.6691	1.1106
4°	0.0698	0.9976	0.0699	49°	0.7547	0.6561	1.1504
5°	0.0872	0.9962	0.0875	50°	0.7660	0.6428	1.1918
6°	0.1045	0.9945	0.1051	51°	0.7771	0.6293	1.2349
7°	0.1219	0.9925	0.1228	52°	0.7880	0.6157	1.2799
8°	0.1392	0.9903	0.1405	53°	0.7986	0.6018	1.3270
9°	0.1564	0.9877	0.1584	54°	0.8090	0.5878	1.3764
10°	0.1736	0.9848	0.1763	55°	0.8192	0.5736	1.4281
11°	0.1908	0.9816	0.1944	56°	0.8290	0.5592	1.4826
12°	0.2079	0.9781	0.2126	57°	0.8387	0.5446	1.5399
13°	0.2250	0.9744	0.2309	58°	0.8480	0.5299	1.6003
14°	0.2419	0.9703	0.2493	59°	0.8572	0.5150	1.6643
15°	0.2588	0.9659	0.2679	60°	0.8660	0.5000	1.7321
16°	0.2756	0.9613	0.2867	61°	0.8746	0.4848	1.8040
17°	0.2924	0.9563	0.3057	62°	0.8829	0.4695	1.8807
18°	0.3090	0.9511	0.3249	63°	0.8910	0.4540	1.9626
19°	0.3256	0.9455	0.3443	64°	0.8988	0.4384	2.0503
20°	0.3420	0.9397	0.3640	65°	0.9063	0.4226	2.1445
21°	0.3584	0.9336	0.3839	66°	0.9135	0.4067	2.2460
22°	0.3746	0.9272	0.4040	67°	0.9205	0.3907	2.3559
23°	0.3907	0.9205	0.4245	68°	0.9272	0.3746	2.4751
24°	0.4067	0.9135	0.4452	69°	0.9336	0.3584	2.6051
25°	0.4226	0.9063	0.4663	70°	0.9397	0.3420	2.7475
26°	0.4384	0.8988	0.4877	71°	0.9455	0.3256	2.9042
27°	0.4540	0.8910	0.5095	72°	0.9511	0.3090	3.0777
28°	0.4695	0.8829	0.5317	73°	0.9563	0.2924	3.2709
29°	0.4848	0.8746	0.5543	74°	0.9613	0.2756	3.4874
30°	0.5000	0.8660	0.5774	75°	0.9659	0.2588	3.7321
31°	0.5150	0.8572	0.6009	76°	0.9703	0.2419	4.0108
32°	0.5299	0.8480	0.6249	77°	0.9744	0.2250	4.3315
33°	0.5446	0.8387	0.6494	78°	0.9781	0.2079	4.7046
34°	0.5592	0.8290	0.6745	79°	0.9816	0.1908	5.1446
35°	0.5736	0.8192	0.7002	80°	0.9848	0.1736	5.6713
36°	0.5878	0.8090	0.7265	81°	0.9877	0.1564	6.3138
37°	0.6018	0.7986	0.7536	82°	0.9903	0.1392	7.1154
38°	0.6157	0.7880	0.7813	83°	0.9925	0.1219	8.1443
39°	0.6293	0.7771	0.8098	84°	0.9945	0.1045	9.5144
40°	0.6428	0.7660	0.8391	85°	0.9962	0.0872	11.4301
41°	0.6561	0.7547	0.8693	86°	0.9976	0.0698	14.3007
42°	0.6691	0.7431	0.9004	87°	0.9986	0.0523	19.0811
43°	0.6820	0.7314	0.9325	88°	0.9994	0.0349	28.6363
44°	0.6947	0.7193	0.9657	89°	0.9998	0.0175	57.2900
45°	0.7071	0.7071	1.0000	90°	1.0000	0.0000	∞

GLOSSARY

acute angle (78) An acute angle is one whose degree measure is less than 90.

acute triangle (97) An acute triangle is a triangle with all acute angles.

adjacent angles (70) Two angles are adjacent when they have a common side and vertex, but *no* points in the interior of one angle are in the interior of the other.

alternate interior angles (161) In the figure, transversal t intersects lines ℓ and m. $\angle 1$ and $\angle 7$, $\angle 2$ and $\angle 8$ are alternate interior angles.

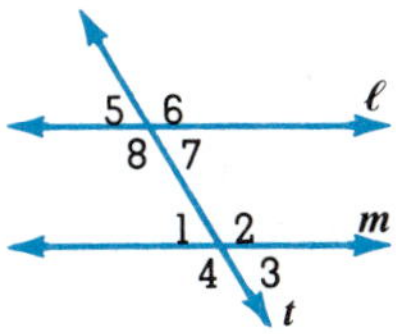

altitude (115) A perpendicular segment to the base from a vertex is an altitude.

angle (65) An angle is a pair of rays that have the same endpoint but do *not* lie on the same line. The rays are *sides* of the angle. The endpoint is the *vertex* of the angle. An angle separates a plane into three parts, the angle, its *interior*, and its *exterior*.

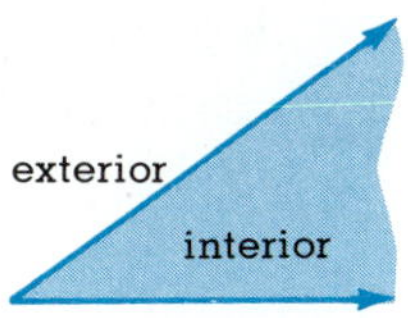

angle of elevation (depression) (287) An angle, called the angle of elevation, is formed by the line of sight and a horizontal line.

angle of rotation (443) The angle of rotation, angle ABC is determined by A, the preimage, B, the center of rotation, and C, the rotation image.

angles of a triangle (100) See triangle.

apothems (337) A segment of a regular polygon that is a radius of a circle inscribed in the polygon is an apothem.

arc (300) See major arc; minor arc.

arc measure (297) The degree measure of a minor arc is the degree measure of its central angle. The degree measure of a major arc is 360 minus the degree measure of its central angle. The degree measure of a semicircle is 180.

arc of a chord (300) When a minor arc and a chord have the same endpoints, we call the arc, the arc of the chord.

auxiliary figure (118) An auxiliary figure is included on a geometric figure in order to prove a given theorem.

axis of a cylinder (348) The segment whose endpoints are centers of the congruent circles that are called the bases is the axis. of a cylinder.

base of a trapezoid (214) See trapezoid.

BASIC (462) BASIC is an abbreviation for **B**eginner's **A**ll-purpose **S**ymbolic **I**nstruction **C**ode; a frequently used computer language.

between (42) A point Q is said to be between points P and R if and only if the following holds.

1. P, Q, and R are collinear, and
2. $PQ + QR = PR$

biconditional statement (10) A biconditional statement is a statement that can be written in *if and only if* form. All definitions are biconditional statements because they are reversible.

bisect (52) If a plane, a line, or a line segment intersects a line segment at its midpoint, it is said to bisect the segment.

bisector of an angle (116) A ray that separates an angle into two congruent angles is called the bisector of an angle.

center of circle (294) See circle.

center of rotation (442) See rotation.

central angle (297) A central angle of a circle is an angle formed by two rays drawn from the center of the circle.

chord of a circle (294) A chord of a circle is a line segment whose endpoints are points on the circle.

chord of a sphere (319) A chord of a sphere is a line segment whose endpoints are points on the sphere.

circle (293) In a plane, a figure is a circle if and only if it is the set of all points that are a given distance from a given point in the plane, called the center of the circle. Usually, circles are named by their centers. The symbol ⊙ means *circle*. Thus, $\odot P$ means circle P. If the measure of the segment joining a point to the center of a circle is less than the measure of the radius, the point is in the interior of the circle. If the measure of the segment joining a point to the center of the circle is greater than the measure of the radius, the point is in the exterior of the circle.

circumference (341) The measure of a circle that is the limit of the measure of the perimeters of the inscribed regular polygons as the number of sides increases is the circumference.

circumscribed polygon (309) A polygon is circumscribed about a circle if and only if each side of the polygon is tangent to the circle.

collinear points (5) Points are collinear if and only if they lie on the same line.

compass (406) An instrument used to draw circles and arcs of circles is a compass.

complementary angles (73) Two angles are complementary if and only if the sum of their degree measures is 90.

composite of reflections (439) Two successive reflections are called a composite of reflections.

concave polygon (nonconvex) (195) A polygon is concave if and only if it is *not* a convex polygon. See convex polygon.

conclusion (9) See conditional statement.

conditional statement (8) A conditional statement is a statement that can be written in *if-then* form. The part following *if* is called the *hypothesis*. The part following *then* is called the *conclusion*. Conditional statements may be true or false.

congruence transformation (433) If k is the scale factor for a dilation, and $k = 1$, then the dilation is a congruence transformation.

congruent angles (69) Angles having the same measure are called congruent angles.

congruent circles (298) Two circles are congruent if and only if their radii are congruent.

congruent segments (51) Segments having exactly the same length are called congruent segments. The symbol $\cong$ means *congruent* or *is congruent to*.

congruent triangles (104) Two triangles are congruent if and only if there is a correspondence so that all their corresponding parts are congruent.

construction (406) The process of drawing a figure that will satisfy certain given conditions, using only a compass and straightedge is a construction.

contrapositive (27) The contrapositive of a conditional statement is another conditional statement formed by interchanging the hypothesis and conclusion, and negating both. The contrapositive of a true statement is always a true statement. The contrapositive of a false statement is always a false statement.

converse (9) The converse of a conditional statement is another conditional statement formed by interchanging the hypothesis and conclusion.

convex polygon (195) A polygon is convex if and only if any line containing a side of the polygon does not contain a point interior to the polygon.

coordinate (370) In an ordered pair, the first component is called the x-coordinate and the second component is called the y-coordinate.

coordinate of a point (36) A number associated with a point on a number line is called the coordinate of the point.

coordinate system (368) It is made up by the x-axis, y-axis and z-axis. A point in space is represented by an ordered triple of real numbers, (x, y, z).

coplanar points (5) Points are coplanar if and only if they lie on the same plane.

corresponding angles (104) In the figure, transversal t intersects lines ℓ and m. $\angle 1$ and $\angle 5$, $\angle 2$ and $\angle 6$, $\angle 3$ and $\angle 7$, and $\angle 4$ and $\angle 8$ are corresponding angles.

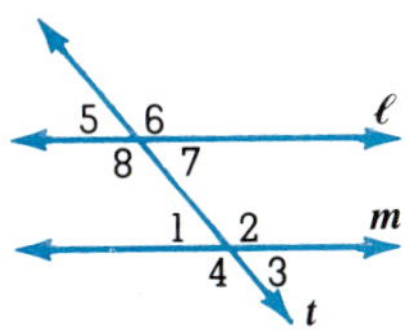

cosine (280) A ratio is the cosine of an acute angle of a right triangle if and only if it is the ratio of the measure of the leg adjacent to the acute angle to the measure of the hypotenuse.

cross products (231) Every proportion has two cross products. In the proportion $\frac{a}{b} = \frac{c}{d}$ where $b \neq 0$ and $d \neq 0$ the cross products are ad and bc. The cross products of a proportion are equal.

cross section (360) The intersection of a solid with a plane that is parallel to the base of the solid is a cross section.

cylinder (348) A solid figure with two circular bases and a rectangular face is a cylinder.

DATA statement (463) DATA tells the computer what numbers to assign to previously listed variables.

definition (5) A definition is an explanation of how a word is to be used.

degrees (68) The unit of measurement used to measure angles is called degrees.

diagonal (198) A line segment joining two nonconsecutive vertices of a convex polygon is called a diagonal of the polygon.

diameter of a circle (294) A diameter of a circle is a line segment whose endpoints are points on the circle and which contains the center of the circle.

diameter of a sphere (319) A diameter of a sphere is a line segment which contains the center and whose endpoints are points on the sphere.

dihedral angle (87) Two noncoplanar half planes with a common edge are called a dihedral angle.

dilation (453) Dilation is a transformation in which size is altered based on a center, C, and a scale factor, k; also called a similarity transformation.

distance (38) The absolute value of the difference between two points on the number line represents the measure of the distance between their corresponding points.

distance formula (372) The distance between any two points (x_1, y_1) and (x_2, y_2) is given by the formula

$$d = \sqrt{(x_2 - x_1)^2 + (y_2 - y_1)^2}.$$

E notation (465) E is the computer equivalent of scientific notation. Used when the output is more than six significant digits.

equation of a circle (389) The equation of a circle with center (h, k) and radius measuring r units is $(x - h)^2 + (y - k)^2 = r^2$.

equation of a sphere (398) The equation of a sphere whose center is at (i, j, k) and has a radius measuring r units is as follows.
$$(x - i)^2 + (y - j)^2 + (g - k)^2 = r^2$$

equiangular triangle (97) An equiangular triangle is a triangle with all three angles congruent.

equidistant (185) Equidistant means at the same distance.

equilateral triangle (97) An equilateral triangle is a triangle with all sides congruent.

edges of polyhedra (221) See polyhedron.

enlargement (454) If k is the scale factor for a dilation, and $k > 1$, then the dilation is an enlargement.

equality of cross products (231) For any numbers a and c, and any nonzero numbers b and d, $\frac{a}{b} = \frac{c}{d}$ if and only if $ad = bc$.

exactly one (166) Exactly one means at least one and no more than one.

exterior angle (131) The angle between a polygon and an extended side of that polygon is called an exterior angle. Angle 2 is an exterior angle.

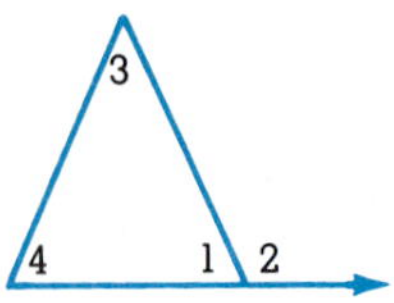

exterior angles (159) In the figure, transversal t intersects lines ℓ and m. $\angle 3$, $\angle 4$, $\angle 5$, and $\angle 6$ are exterior angles.

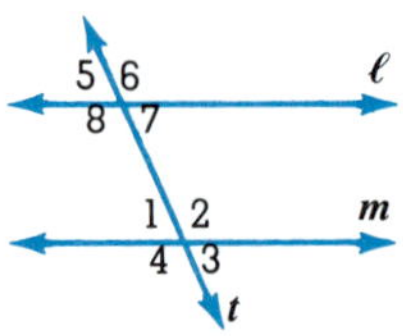

exterior of an angle (66) See angle.

exterior of a circle (295) See circle.

exterior of a triangle (98) See triangle.

external secant segment (316) The part of the secant segment that is exterior to the circle is the external secant segment.

face of a polyhedron (221) See polyhedron.

FOR-NEXT statements (467) FOR-NEXT is used when an operation needs to be used on different sets of data. It allows the computer to loop through the same operation several times.

geometric mean (269) For any positive numbers a and b, x is the geometric mean between a and b if and only if $\frac{a}{x} = \frac{x}{b}$ and x is positive.

glide reflection (459) The composition of a translation and a line reflection is a glide reflection.

golden ratio (236) A ratio of about 1 to 1.618 is called a golden ratio.

golden rectangles (236) Golden rectangles are rectangles whose measures of their sides are always in the ratio of about 1 to 1.618.

great circle (320) If a plane intersects a sphere in more than one point and contains the center of the sphere, the intersection of the plane and the sphere is called a great circle.

half plane (68) Part of a plane on one side of a straight line drawn in the plane is a half plane.

hypotenuse (123) The hypotenuse of a right triangle is the side opposite its right angle.

hypothesis (9) See conditional statement.

IF-THEN statements (470) IF-THEN is used to compare two numbers. It tells the computer what to do based on the results of the comparison.

image (432) If A is mapped onto A', then A' is called the image of A. A is the preimage of A'.

indirect reasoning (164) In indirect reasoning you assume the alternative of what you want to prove. Then, show that your assumption leads to a contradiction.

input (462) Input is putting information into the computer.

inscribed angle (304) An angle ABC is called an inscribed angle if $\angle ABC$ intercepts an arc of the circle and if the vertex of the angle lies on the circle.

inscribed polygon (306) A polygon is inscribed in a circle if each of its vertices lies on the circle.

integer (34) The integers consist of the set of positive and negative numbers including zero.

$\{\cdots {}^-3, {}^-2, {}^-1, 0, 1, 2, 3, \cdots\}$

intercepted arc (364) An angle intercepts an arc if and only if each of the following characteristics hold.

1. The endpoints of the arc lie on the angle.
2. All points of the arc, except the endpoints, are in the interior of the angle.
3. Each side of the angle contains an endpoint of the arc.

interior angles (161) In the figure, transversal t intersects lines ℓ and m. $\angle 1$, $\angle 2$, $\angle 7$, and $\angle 8$ are interior angles.

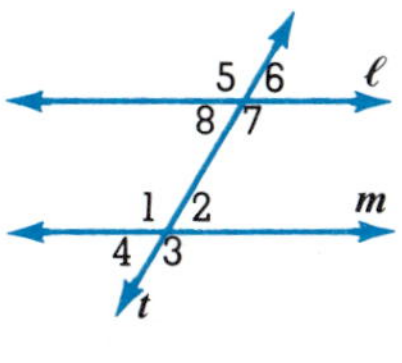

interior of an angle (66) See angle.

interior of a circle (295) See circle.

interior of a triangle (98) See triangle.

inverse (26) The inverse of a conditional statement is another conditional statement formed by negating both the hypothesis and conclusion.

irrational number (36) An irrational number is a real number named by a nonterminating and nonrepeating decimal. π is an irrational number.

$\pi \sim 3.14159265358979323846264338 \cdots$.

isometry or congruence transformation (433) When a geometric figure and its transformation image are congruent, the mapping is called an isometry or congruence transformation.

isosceles triangle (97) An isosceles triangle is a triangle with two sides congruent.

lateral area (347) The area of all the lateral faces is the lateral area.

leg of a right triangle (123) The leg of a right triangle is any side *not* opposite the right angle (. . . any side contained in the right angle).

leg of a trapezoid (214) See trapezoid.

LET statement (465) LET uses an equals sign to assign the value of the expression on the right to the variable on the left. It can also be used to assign a new value to a variable.

line (2) A line is one of the basic undefined terms of geometry. It extends indefinitely, having no thickness or width, and has only one dimension. A line is represented by a double arrow and named either by a lower case script letter or two points of the line.

linear equation (369) An equation is linear if and only if it can be written in the form $Ax + By = C$, where A, B and C are any numbers, and A and B are not both 0.

linear pair (73) If the noncommon sides of adjacent angles together form a straight angle, the angles are called a linear pair.

line of reflection (435) Line ℓ is a line of reflection if it is the perpendicular bisector of the segment drawn from point X to its reflection image point A.

line of symmetry (436) A line of symmetry is a line which can be drawn through a plane figure so that the figure on one side is a reflection image of the figure on the opposite side.

line segment (42) A line segment is a part of a line that consists of two endpoints and all the points between them.

locus (420) A figure is a locus if and only if it is the set of all points and only those points that satisfy a given condition.

major arc (297) If $\angle APB$ is a central angle of circle P and C is any point on the circle not in the interior of the angle, then the points A and B and all points of the circle exterior to $\angle APB$ form a major arc called $\widehat{ACB}$. Three letters are needed to name a major arc.

mapping (433) See transformation.

measure (38) See measurement.

measurement (38) The positive number which corresponds to the *measure* of the distance between two points is called the measurement. This measurement is a unique distance when given a certain *unit of measure*.

median of a trapezoid (215) See trapezoid.

meter (45) See metric system.

metric system (45) The metric system is a decimal system of weights and measures based on three basic units, the meter for length, liter for volume, and the gram for mass (weight). Each unit increases or decreases in size by multiples of ten, which is indicated by the prefixes of which milli-, centi-, deci-, deka-, hecto-, and kilo- are the most common.

midpoint (42) A midpoint separates a line segment into two segments of equal measure. A line segment has exactly one midpoint.

midpoint formula (374) If the coordinates of A and B are (x_1, y_1) and (x_2, y_2) respectively, then the midpoint, M, of $\overline{AB}$, has coordinates $\left(\frac{x_1 + x_2}{2}, \frac{y_1 + y_2}{2}\right)$.

minor arc (297) If $\angle APB$ is a central angle of circle P, the points A and B and all points of the circle interior to the angle form a minor arc called $\widehat{AB}$.

natural number (34) The natural numbers consist of the set of counting numbers.

$$\{1, 2, 3, 4, 5, \cdots\}$$

negation (26) A negation is a sentence formed by denying another sentence. Often you can form a negation by using *not*.

***n*-gon** (195) A polygon with n sides is called an n-gon.

noncollinear points (6) Points are noncollinear if and only if they do *not* all lie on the same line.

nonconvex polygon (195) See concave polygon.

noncoplanar points (7) Points are noncoplanar if and only if they do *not* all lie on the same plane.

number line (34) A number line is a line of infinite extent in both the positive and negative directions whose points correspond to the set of integers.

obtuse angle (78) An obtuse angle is one whose degree measure is greater than 90.

obtuse triangle (97) An obtuse triangle is a triangle with one obtuse angle.

opposite rays (64) Opposite rays are two rays that head in opposite directions from a common endpoint. $\overrightarrow{BA}$ is opposite $\overrightarrow{BC}$.

ordered pair (368) It is a pair of numbers in which the order is specified. An ordered pair is used to locate points in a plane.

origin (368) The point of intersection of the two axes of the coordinate plane is the origin.

output (463) Output are the results given by a computer.

parallel lines (160) Two lines that lie in the same plane and they do *not* intersect are called parallel lines.

parallelogram (333) A polygon is a parallelogram if and only if it is a quadrilateral with both pairs of opposite sides parallel.

perimeter (218) The sum of the measures of the sides of a polygon is the measure of the perimeter of the polygon.

perpendicular bisector (409) For a line segment in a plane, the line perpendicular to the segment at its midpoint is the perpendicular bisector.

perpendicular lines (82) Two lines that intersect to form a right angle are called perpendicular lines. The symbol $\perp$ means "is perpendicular to."

perpendicular planes (86) Two planes are perpendicular if and only if any line in one of them that is perpendicular to their line of intersection, is perpendicular to the other plane.

plane (2) A plane is one of the basic undefined terms of geometry. It extends indefinitely in two directions, has no thickness, and has two dimensions. A plane is represented by a four-sided figure and named by a capital script letter or three points of the plane.

plane angle (88) The intersection of a dihedral angle and a plane perpendicular to its edge is called a plane angle of a dihedral angle.

platonic solid (222) A platonic solid is any one of the five regular polyhedra: tetrahedron, hexahedron, octahedron, dodecahedron, or icosohedron.

point (2) A point is one of the basic undefined terms of geometry. It has no size or dimension, is represented by dots, and is named by capital letters.

point of reflection (435) Point S is the reflection of point R with respect to Q, the point of reflection, if Q is the midpoint of the segment drawn from R to S.

point-slope form (383) The equation of a line passing through a point whose coordinates are (x_1, y_1) and having a slope, m, is $y - y_1 = m(x - x_1)$.

polygon (194) A figure is a polygon if and only if it is formed by three or more coplanar line segments that are also noncollinear and intersect exactly two other segments only at their endpoints. The line segments are called sides and the endpoints are called vertices.

polygonal region (328) A polygon and its interior is the polygonal region.

polyhedron (221) A solid with flat surfaces that form polygons is called a polyhedron. The flat surfaces formed by polygons and their interiors are called *faces*. Pairs of faces intersect at *edges*. Three or more edges intersect at a vertex.

postulate (494) A postulate is a conditional statement that describes a fundamental property of basic terms.

preimage (432) See image.

principal square root (264) The principal square root is the nonnegative square root.

prism (346) A solid with the following characteristics is a prism.

1. Two faces are formed by congruent polygons that lie in parallel planes (bases).
2. The faces that are *not* bases are formed by parallelograms (lateral sides).
3. The intersection of two adjacent lateral faces are parallel segments (lateral edges).

proportion (231) A proportion is an equation of the form $\frac{a}{b} = \frac{c}{d}$ which states that two ratios are equivalent.

protractor (68) A protractor is a tool used to find the degree measure of a given angle.

pyramid (350) A solid with the following characteristics is a pyramid.

1. All the faces, except one face, intercepts at a point called the vertex.
2. The face that does not intersect at the vertex is a polygon and called the base.
3. The faces meeting at the vertex, called lateral faces, are triangles.

Pythagorean Theorem (272) For any right triangle, the square of the measure of the hypotenuse equals the sum of the squares of the measures of the legs.

quadrant (368) One of the four regions into which two perpendicular number lines separate the plane is a quadrant.

quadrilateral (195) A polygon with four sides is called a quadrilateral.

radical sign (264) The symbol $\sqrt{}$ called a radical sign, indicates a nonnegative square root. The expression under the radical sign is called the radicand.

radicand (264) See radical sign.

radius of a circle (294) A radius of a circle is a line segment whose endpoints are the center of the circle and a point on the circle.

radius of a sphere (319) A radius of a sphere is a line segment whose endpoints are the center of the sphere and a point on the sphere.

ratio (230) A ratio is a comparison of the numbers by division.

rational number (34) A rational number is any real number which can be written as the quotient of two integers.
$\left\{\cdots -\frac{4}{3}, 1, -\frac{2}{3}, -\frac{1}{3}, 0, \frac{1}{3}, \frac{2}{3}, 1, \frac{4}{3}, \cdots\right\}$

rationalizing the denominator (266) The method used to simplify a radical fraction so that the denominator of the fraction becomes a rational number is called rationalizing the denominator.

ray (64) A ray is part of a line and it has one endpoint. A never ending straight path in one direction is represented by a ray.

real number (36) The rational and the irrational numbers make up the real numbers. Each point on the number line can be paired with a real number.

rectangle (210) A quadrilateral is a rectangle if and only if it is a parallelogram with four right angles.

reduction (454) If k is the scale factor for a dilation, and $0 < k < 1$, then the dilation is a reduction.

reflection (435) See line of reflection or point of reflection.

regular polygon (195) A polygon in which all sides are congruent and all angles are congruent is a regular polygon.

regular polyhedron (222) Each face of a regular polyhedron forms a regular polygon and each edge of a regular polyhedron has the same measure.

regular pyramid (350) A pyramid whose base is a regular polygon and the segment joining the base to the vertex is perpendicular to the base is a regular pyramid.

remote interior angles (132) The angles in a triangle that are *not* adjacent to a given exterior angle are called remote interior angles.

repeating decimal (35) A repeating decimal is a decimal number in which a digit or a group of digits repeat without end. A bar over the repeating digit or digits indicates a repeating decimal. Examples are $0.\overline{3}$ and $0.4\overline{21}$.

rhombus (210) A quadrilateral is a rhombus if and only if it is a parallelogram with all four sides congruent.

right angle (78) A right angle is an angle whose degree measure is 90.

right circular cone (351) A solid figure which has a circular base and its axis from the vertex is perpendicular to the base is a right circular cone.

right cylinder (348) A cylinder whose axis is also the altitude of the figure is a right cylinder.

right prism (346) A prism in which lateral faces are rectangles is a right prism.

right triangle (97) A right triangle is a triangle with one right angle.

rotation (442) A rotation is the composite of two reflections with respect to two intersecting lines. The intersection of the two lines is called the *center of rotation*.

scalene triangle (97) A scalene triangle is a triangle with no sides congruent.

secant (312) A secant is a line which intersects a circle in exactly two points. A secant to a circle contains a chord of the circle.

sector (343) A region bounded by a central angle and the intercepted arc is a sector.

semicircle (297) A straight angle with its vertex at the center of a circle forms two semicircles with the circle.

sides of an angle (65) See angle.

sides of a polygon (194) See polygon.

sides of a triangle (96) See triangle.

significant digits (50) The degree of accuracy of a measurement can be shown by the number of significant digits.

similar figures (238) Figures that have the same shape but which may differ in size are called similar figures.

similar polygons (238) Two polygons are similar if and only if there is a correspondence such that their corresponding angles are congruent and the measures of their corresponding sides are proportional.

similarity transformation (433) See dilation.

simplified square roots (266) A square root is said to be simplified if

1. the radicand has no factor other than 1 that is a perfect square,
2. the radicand does not contain a fraction, and
3. no radical appears in the denominator of a fraction.

sine (280) A ratio is a sine of an acute angle of a right triangle if and only if it is the ratio of the measure of the leg opposite the acute angle to the measure of the hypotenuse.

skew lines (160) Two lines are skew lines if they do *not* intersect and are *not* in the same plane.

slant height (350) The height of each lateral face of the pyramid is the slant height.

slope (377) The slope of a line is the ratio of the change in y to the corresponding change in x.

$$\frac{\text{difference of the } y\text{-coordinates}}{\text{difference of the corresponding } x\text{-coordinates}}$$

slope-intercept form (382) The slope-intercept form of the equation of a line is $y = mx + b$. The slope of the line is m and the y-intercept is b.

sphere (319) In space, a figure is a sphere if and only if it is the set of all points that are a given distance from a given point, called the center.

square (211) A quadrilateral that is *both* a rectangle and a rhombus is called a square.

square root (264) For any numbers a and b if $a^2 = b$, then a is the square root of b.

straight angle (64) A straight angle is a pair of opposite rays.

straightedge (406) Any instrument used to draw a straight line is a straightedge.

supplementary angles (73) Two angles are supplementary if and only if the sum of their degree measures is 180.

surface area (346) The sum of the areas of the faces of a solid figure is the surface area.

system of equations (386) A set of equations with the same variables is a system of equations.

tangent (281) A ratio is the tangent of an acute angle of a right triangle if and only if it is the ratio of the measure of the leg opposite the acute angle to the measure of the leg adjacent to the acute angle.

tangent to a circle (295) A tangent is a line which intersects a circle in exactly one point. The point of intersection is called the *point of tangency*.

terminating decimal (35) A terminating decimal is a decimal number in which the division process stops or terminates because a remainder of zero has been reached. An example is 0.35.

theorem (16) A theorem is a conditional statement that must be proved.

transformations (432) In a plane, a mapping is a transformation if and only if each point has exactly one image point and each image point has exactly one preimage point.

translation (439) A composite of two reflections over two parallel lines is a translation.

transversal (161) In a plane, a transversal is a line that intersects two other lines in two different points.

trapezoid (214) A quadrilateral is a trapezoid if and only if it has exactly one pair of parallel sides. The parallel sides of a trapezoid are called *bases*. The nonparallel sides are called *legs*. The line segment joining the midpoints of the legs of the trapezoid is called the *median*.

triangle (96) A triangle is the union of three noncollinear line segments that intersect only at all three endpoints. The line segments are *sides* of the triangle. The endpoints are the *vertices* of the triangle. The angles formed at each vertex are called the *angles* of the triangle. A triangle separates a plane into three parts, the triangle, its *interior*, and its *exterior*.

trigonometric ratio (280) A ratio of two measures of sides of a right triangle is called a trigonometric ratio.

trigonometric table (283) A table which lists decimal approximations for values of trigonometric ratios is called a trigonometric table. Such a table is found on page **506**.

trigonometry (280) See trigonometric ratio.

undefined term (5) Undefined terms are basic terms from which other terms are defined. *Point*, *line*, and *plane* are undefined terms of geometry.

unit of measure (38) See measurement.

vertex of an angle (65) See angle.

vertical angles (79) Vertical angles are two nonadjacent angles formed by two intersecting lines.

vertices of a polygon (194) See polygon.

vertices of a polyhedron (221) See polyhedron.

vertices of a triangle (96) See triangle.

volume (355) The measure of the amount of space a figure encloses.

x-axis (368) The horizontal line of the two perpendicular number lines used to separate the plane into quadrants is the x-axis.

y-axis (368) The vertical line of the two perpendicular number lines used to separate the plane into quadrants is the y-axis.

INDEX

A

B

C

D

E

n

o

p

Q

R

S

T

U

V

X

Y

Z

SELECTED ANSWERS

CHAPTER 1 POINTS, LINES, AND PLANES

Exploratory Exercises Page 4 **1.** point **3.** line **5.** line **7.** line **9.** plane **11.** point **13.** line

Written Exercises Page 4 **15.** $\overleftrightarrow{AF}$, $\overrightarrow{FB}$, ***n*** **17.** none **19.** *A*, *E*, *D* **21.** $\overleftrightarrow{AF}$, $\overleftrightarrow{FB}$, $\overleftrightarrow{AB}$ **23.** one **25.** three

Exploratory Exercises Page 7 **1.** false **3.** false **5.** true **7.** false **9.** false

Written Exercises Page 7 **1.** collinear **3.** noncollinear **5.** collinear **7.** coplanar **9.** noncoplanar **11.** coplanar **13.** good **15.** not good; violates guidelines 3, 4, and 5 **17.** not good; violates guideline 3 **19.** good **21.** answers will vary

Exploratory Exercises Page 10 **1.** hypothesis: it rains; conclusion: the grass gets wet **3.** hypothesis: you live in Texas; conclusion: you are an American **5.** hypothesis: two lines are perpendicular; conclusion: the two lines intersect **7.** hypothesis: n is even; conclusion: n^2 is even **9.** hypothesis: an animal is a chimpanzee; conclusion: it is a mammal

Written Exercises Page 10 **1.** If lines are parallel, then they do not intersect. **3.** If a figure is a square, then it is a rectangle. **5.** If $x < 0$, then $5x > 6x$. **7.** If it rains, then it pours. **9.** If two lines intersect, then they are contained in exactly one plane. **11.** If lines do not intersect, then they are parallel. **13.** If a figure is a rectangle, then it is a square. **15.** If $5x > 6x$, then $x < 0$. **17.** If it pours, then it rains. **19.** If two lines are contained in exactly one plane, then they intersect. **21.** yes **23.** yes **25.** no; an imperative statement **27.** yes **29.** no; a question **31.** true **33.** true **35.** neither **37.** true **39.** neither **41–44.** answers will vary **45.** If points are collinear, then they lie on the same line. If points lie on the same line, then they are collinear. **47.** If two line segments are congruent, then they have the same length. If two line segments have the same length, then they are congruent. **49.** If two angles are supplementary, then the sum of their measures is 180. If the sum of the measures is 180, then two angles are supplementary. **51.** Two lines are perpendicular if and only if they intersect at right angles. **53.** When two lines are intersected by a transversal, alternate interior angles are congruent if and only if the lines are parallel.

Exploratory Exercises Page 14 **1.** Postulate 1 **3.** Postulate 4 **5.** Postulate 4 **7.** Postulate 1 **9.** Postulate 6

Written Exercises Page 14 **1.** true **3.** false **5.** true **7.** false **9.** true **11.** none **13.** three **15.** four **17.** infinitely many **19.** none **21.** one **23.** one **25.** one

Exploratory Exercises Page 18 **1.** theorem, given, to prove, diagram, proof **3.** conclusion of the theorem **5.** yes **7.** yes **9.** statements, reasons **11.** to justify the statements **13.** two lines intersect **15.** a line and a plane intersect, and the plane does not contain the line **17.** exactly one plane contains both lines **19.** their intersection is a point

Written Exercises Page 19 **1.** two intersecting lines **3.** a line and plane which intersect but the plane does not contain the line **5.** exactly one plane contains both lines **7.** the intersection of the line and the plane is a point **19.** a line and a point not on the line **21.** two intersecting planes **23.** two lines $\overleftrightarrow{PQ}$ and $\overleftrightarrow{QR}$ which intersect at Q **25.** a line and a plane which intersect at *P*, and the plane does not contain the line **27.** two intersecting lines, one of which, $\overleftrightarrow{AB}$, is contained in $\mathscr{M}$, while the other line, $\overleftrightarrow{BC}$, is not in $\mathscr{M}$

Exploratory Exercises Page 24 **1.** Theorem 1–2 **3.** Postulate 1–4 **5.** Postulate 1–2 **7.** Theorem 1–2 **9.** none **11.** none

Written Exercises Page 24 **1.** Postulate 1–1 **3.** Postulate 1–2 **5.** Postulate 1–5 **7.** Theorem 1–1 **9.** definition of noncollinear points **11.** $\mathscr{P}$ contains both ℓ and $\boldsymbol{m}$ **13.** The intersection of $\mathscr{M}$ and $\mathscr{N}$ is ℓ. **15.** *A* and *B* are on ℓ. **17.** $\mathscr{M}$ is the only plane that contains ℓ and *P*. **19–22.** See the given, prove, and diagram in **23–26.**

23. Given: $\mathcal{M}$ and $\mathcal{N}$ intersecting at ℓ
Prove: the intersection of $\mathcal{M}$ and $\mathcal{N}$ contains at least two points, A and B

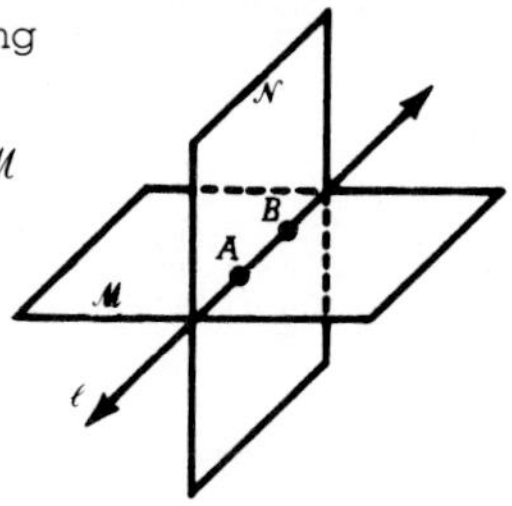

Proof:

STATEMENTS	REASONS
1. $\mathcal{M}$ and $\mathcal{N}$ intersecting at ℓ	1. Given
2. A and B are in ℓ	2. Postulate 1–3
3. the intersection of $\mathcal{M}$ and $\mathcal{N}$ contain at least two points	3. Conclusion

25. Given: $\boldsymbol{m}$ and $\boldsymbol{n}$ intersecting at P
Prove: $\boldsymbol{m}$ and $\boldsymbol{n}$ contain P, R, and S

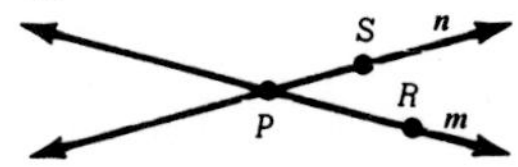

Proof:

STATEMENTS	REASONS
1. $\boldsymbol{m}$ and $\boldsymbol{n}$ intersect at P	1. Given
2. R and P lie on $\boldsymbol{m}$	2. Postulate 1–3
3. S and P lie on $\boldsymbol{n}$	3. Postulate 1–3
4. $\boldsymbol{m}$ and $\boldsymbol{n}$ contain R, P, and S	4. Statements 2 and 3

Exploratory Exercises Page 27 **1.** a figure is not a triangle **3.** three points are not collinear **5.** $\mathcal{M}$ does not intersect $\mathcal{N}$ **7.** ℓ lies in $\mathcal{M}$ **9.** P, Q, and R are coplanar **11.** two points do not lie in different planes

Written Exercises Page 28 **1.** yes **3.** no **5.** yes **7.** no **9.** no **11.** no **13.** yes **15.** no **17.** yes **19.** no **21.** no **23.** yes **25.** If it does not rain, then the grass does not get wet. **27.** If you do not live in Texas, then you are not an American. **29.** If two lines are not perpendicular, then the two lines do not intersect. **31.** If n is not even, then n^2 is not even. **33.** If an animal is not a chimpanzee, then it is not a mammal. **35.** Lines that are not parallel do intersect. **37.** If a figure is not a square, then it is not a rectangle. **39.** $x \nless 0$ implies $5x \ngtr 6x$. **41.** When it does not rain; it does not pour. **45.** If the grass does not get wet, then it does not rain. **47.** If you are not an American, then you do not live in Texas. **49.** If two lines do not intersect, then they are not perpendicular. **51.** If n^2 is not even, then n is not even. **53.** If it is not a mammal, then an animal is not a chimpanzee. **55.** Lines which intersect are not parallel. **57.** If a figure is not a rectangle, then it is not a square. **59.** $5x \ngtr 6x$ implies $x \nless 0$. **61.** If it does not pour, then it does not rain. **63.** none

Chapter Review Pages 30–31

1.

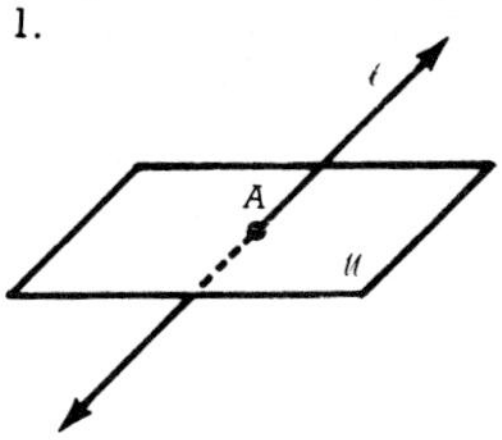

3.

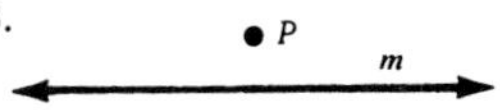

5. Noncoplanar points are points which do not all lie in the same plane **7.** If two points are given, then there is exactly one line through those points. **9.** If exactly one line passes through points, then there are two points. **11.** Lines are skew if and only if they are not coplanar. **13.** four **19.** exactly one plane contains ℓ and P **21.** If you do not want something, do not ask for it. **23.** If $2x \neq 8$, then $x \neq 4$ **25.** If you do not ask for something, then you do not want it. **27.** If $x \neq 4$, then $2x \neq 8$

CHAPTER 2 MEASURE

Exploratory Exercises Page 36 **1.** 1 **3.** $^{-}8$ **5.** 0 **7.** 5 **9.** $^{-}6$ **11.** $^{-}4$ **13.** $\frac{2}{3}$ **15.** $-\frac{4}{3}$ **17.** $^{-}3$ **19.** $\frac{7}{3}$ **21.** $-\frac{1}{3}$ **23.** 0

Written Exercises Page 37 **1.** 0.3 **3.** $^{-}0.8$ **5.** $^{-}0.75$ **7.** $0.\overline{6}$ **9.** $0.\overline{3}$ **11.** $0.\overline{4}$ **13.** 1.2 **15.** $^{-}2.7$ **17.** $2.\overline{6}$ **19.** $^{-}7.\overline{142857}$ **21.** $\frac{7}{9}$ **23.** $\frac{5}{9}$ **25.** $4\frac{4}{15}$ **27.** $1\frac{26}{99}$ **29.** no **31.** no **33.** no **35.** no **37.** no **39.** no **41.** yes **43.** no **45.** no **47.** no **49.** no **51.** no **53.** yes **55.** yes **57.** no **59.** yes **61.** yes **63.** no **65.** no **67.** no **69.** no **71.** no **73.** yes **75.** yes **77.** yes

Exploratory Exercises Page 40 **1.** 12 **3.** 30 **5.** 15 **7.** 11 **9.** 3 **11.** 89 **13.** 44 **15.** 52 **17.** 2.3 **19.** $\frac{3}{4}$ **21.** $12\frac{7}{8}$ **23.** 10.02

Written Exercises Page 40 **1.** 9 **3.** $^{-}12$ **5.** $^{-}69$ **7.** $^{-}153$ **9.** 11 **11.** $^{-}87$ **13.** $^{-}59$ **15.** $^{-}5$ **17.** 3 **19.** 9 **21.** 13.2

23. $2\frac{3}{4}$ 25. $11\frac{3}{4}$ 27. $4\frac{5}{6}$ 29. $\frac{1}{3}$ 31. $\frac{11}{3}$ 33. $\frac{5}{6}$ 35. $\frac{7}{6}$ 37. 2 39. 2 41. $\frac{1}{3}$ 43. 0.3 45. 1.1 47. 2.3 49. 0.7 51. 0 53. 1.5 55. 0 57. 0.7

Exploratory Exercises Page 43 1. $^-1, 0$ 3. $^-8, 9$ 5. $^-4, {}^-8$ 7. 6, 3 9. $^-8, 0$ 11. $^-1, 6$ 13. $6, {}^-8$ 15. $^-4, 0$ 17. 3, 6 19. $^-4, 9$

Written Exercises Page 44 1. 1 3. 17 5. 4 7. 3 9. 8 11. 7 13. 14 15. 4 17. 3 19. 13

21.
$RT + TS \overset{?}{=} RS$
$|7 - 1| + |7 - 3| \overset{?}{=} |3 - 1|$
$6 + 4 \overset{?}{=} 2$
$10 \neq 2$
Thus, $RT + TS \neq RS$

23. N does not lie between P and S.
$PN + NS \overset{?}{=} PS$
$|^-7 - {}^-9| + |^-9 - 3| \overset{?}{=} |^-7 - 3|$
$2 + 12 \overset{?}{=} 10$
$14 \neq 10$

25. Show that Q is between P and S
$PQ + QS \overset{?}{=} PS$
$|^-7 - {}^-2| + |^-2 - 3| \overset{?}{=} |^-7 - 3|$
$5 + 5 \overset{?}{=} 10$
$10 = 10$
Show that $PQ = QS$
$PQ \overset{?}{=} QS$
$|^-7 - {}^-2| \overset{?}{=} |^-2 - 3|$
$5 = 5$

27. 0

29.

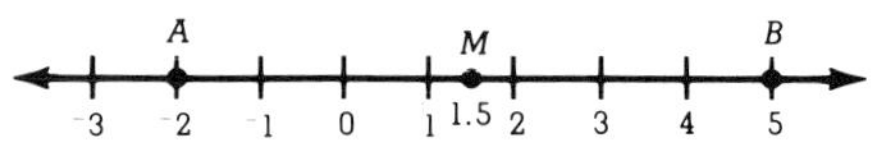

31.

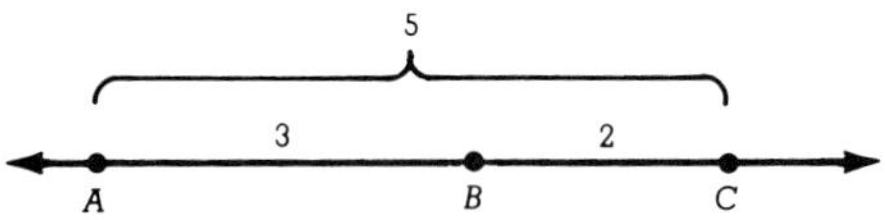

33.

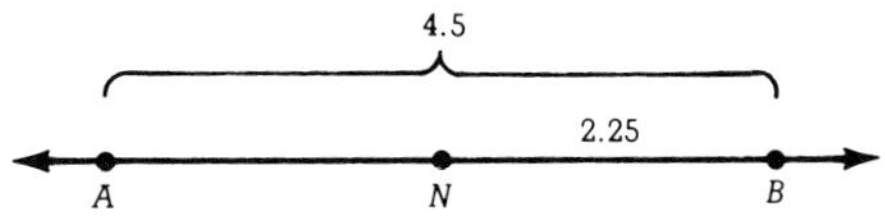

35. Definition of Betweenness, Definition of a Midpoint 37. Definition of Betweenness 39. Definition of a Midpoint 41. 1 or $^-11$ 43. $^-2$ or $^-8$

Exploratory Exercises Page 48 1. Reflexive 3. Symmetric 5. Transitive 7. Reflexive 9. Addition property of equality 11. Reflexive 13. Transitive 15. Addition property of equality 17. Symmetric 19. Addition property of equality

Written Exercises Page 48 1. Postulate 2–7 3. Postulate 2–9 5. Postulate 2–7

7.

STATEMENTS	REASONS
1. $MB = MQ$	1. Given
2. $AB = AM + MB$	2. Definition of Betweenness
3. $AB = AM + MQ$	3. Postulate 2–9

Exploratory Exercises Page 53 1. Reflexive 3. Symmetric 5. Transitive 7. Transitive

Written Exercises Page 53 1. true 3. true 5. true 7. true 9. true 11. M is the midpoint of $\overline{PQ}$ 13. If $\overline{PM} \cong \overline{MQ}$, then M is the midpoint of $\overline{PQ}$. 15. $\overline{PQ}$ is bisected at M 17. $\overline{PQ}$ is bisected at M 19. $\overline{PM} \cong \overline{MQ}$ 21. A midpoint separates a line segment into two segments of equal measure. 23. Show that M is the midpoint of $\overline{PQ}$ and conclude that $\overline{PM} \cong \overline{MQ}$ by the Midpoint Theorem.

25. Given: $\overline{AB}$
Prove: $\overline{AB} \cong \overline{AB}$

Proof:

STATEMENTS	REASONS
1. $\overline{AB}$	1. Given
2. $AB = AB$	2. Postulate 2–4
3. $\overline{AB} \cong \overline{AB}$	3. Definition of congruence

27. Given: M between P and Q; $\overline{PM} \cong \overline{MQ}$
Prove: M is the midpoint of $\overline{PQ}$

Proof:

STATEMENTS	REASONS
1. M between P and Q; $\overline{PM} \cong \overline{MQ}$	1. Given
2. $PM = MQ$	2. Definition of congruent line segments
3. M is the midpoint of $\overline{PQ}$	3. Definition of a midpoint

Exploratory Exercises Page 55 1. $<$ 3. $>$ 5. $>$ 7. $=$ 9. $<$ 11. $<$ 13. $>$ 15. $=$

Written Exercises Page 56 1. $=$ 3. $>$ 5. $=$ 7. $>$ 9. $<$ 11. $<$

13. Given: $3x + 7 > 43$
Prove: $x > 12$

Proof:

STATEMENTS	REASONS
1. $3x + 7 > 43$	1. Given
2. $3x > 36$	2. Postulate 2–16, Postulate 2–9
3. $x > 12$	3. Postulate 2–17, Postulate 2–9

15. Given: $11 - 5y < {}^{-}77$
Prove: $y > 17.6$

Proof:

STATEMENTS	REASONS
1. $11 - 5y < {}^{-}77$	1. Given
2. ${}^{-}5y < {}^{-}88$	2. Postulate 2–16, Postulate 2–9
3. $y > 17.6$	3. Postulate 2–17, Postulate 2–9

17. Given: $9(x + 2) < 72$
Prove: $x < 6$

Proof:

STATEMENTS	REASONS
1. $9(x + 2) < 72$	1. Given
2. $x + 2 < 8$	2. Postulate 2–17, Postulate 2–9
3. $x < 6$	3. Postulate 2–16, Postulate 2–9

19. $PQ > QR$ 21. $AB < EF$

Chapter Review Pages 60–61 1. ${}^{-}1\frac{3}{4}$ 3. 0 5. 0.7 7. ${}^{-}0.\overline{6}$ 9. $\frac{1}{3}$ 11. $\frac{4}{33}$ 13. $\frac{1}{2}$ 15. $4\frac{3}{4}$ 17. $3\frac{1}{2}$ 19. $\frac{1}{2}$

21.
$$PQ + QR \stackrel{?}{\geq} PR$$
$$|{}^{-}1 - {}^{-}2\tfrac{1}{2}| + |{}^{-}1 - 0| \stackrel{?}{\geq} |{}^{-}2\tfrac{1}{2} - 0|$$
$$1\tfrac{1}{2} + 1 \stackrel{?}{\geq} 2\tfrac{1}{2}$$
$$2\tfrac{1}{2} = 2\tfrac{1}{2}$$

23.
$$QS \stackrel{?}{\geq} ST$$
$$|1 - {}^{-}1| \stackrel{?}{\geq} |2\tfrac{1}{2} - 1|$$
$$2 \neq 1\tfrac{1}{2}$$

CHAPTER 3 ANGLES AND PERPENDICULARS

Exploratory Exercises Page 66 1. yes 3. no 5. yes 7. no 9. no 11. no 13. yes 15. yes 17. no 19. yes

Written Exercises Page 67 1. one 3. a point 5. negative

7.

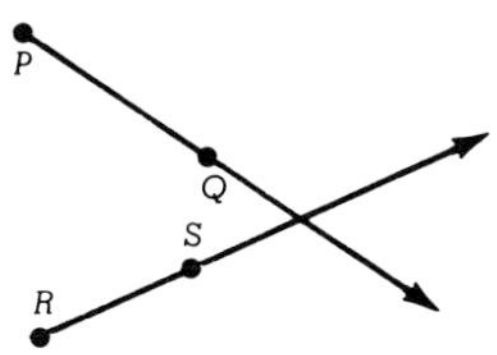

9.

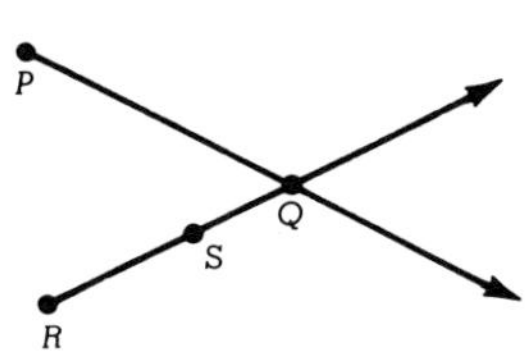

11. yes 13. Q 15. $\overrightarrow{QT}$, $\overrightarrow{QP}$ 17. $\overrightarrow{QT}$ 19. $\angle RQV$, $\angle TQV$, $\angle SQV$, $\angle PQV$ 21. V, Q, or R 23. V 25. V, R, or P

Exploratory Exercises Page 71 1. 170 3. 120 5. 75 7. 25 9. 80 11. 60 13. 65 15. 65

Written Exercises Page 71 17. oil, 10.8°; natural gas, 14.4°; uranium, 10.8° 19. true 21. true 23. true 25. true

29. Given: $\angle ABC$
Prove: $\angle ABC \cong \angle ABC$

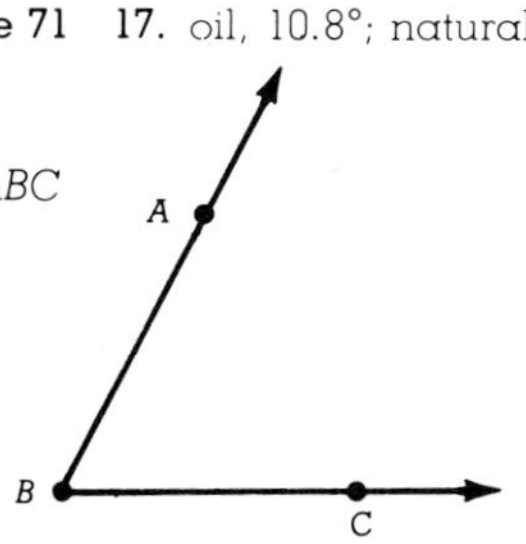

31. Given: $\angle ABC \cong \angle RPQ$,
$\angle RPQ \cong \angle STV$
Prove: $\angle ABC \cong \angle STV$

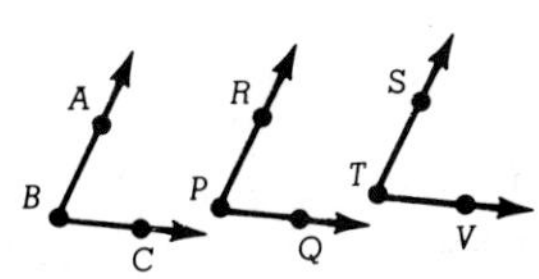

Proof:

STATEMENTS	REASONS
1. $m\angle ABC = m\angle ABC$	1. Postulate 2–4
2. $\angle ABC \cong \angle ABC$	2. Definition of congruent angles

33. 30 **35.** 60 **37.** 90

Proof:

STATEMENTS	REASONS
1. $\angle ABC \cong \angle RPQ$, $\angle RPQ \cong \angle STV$	1. Given
2. $m\angle ABC = m\angle RPQ$, $m\angle RPQ = m\angle STV$	2. Definition of congruent angles
3. $m\angle ABC = m\angle STV$	3. Postulate 2–6
4. $\angle ABC \cong \angle STV$	4. Definition of congruent angles

Exploratory Exercises Page 76 **1.** two sides of adjacent angles lie in a line **3.** $\angle ABC$ and $\angle CBD$ are adjacent angles such that their noncommon sides form a straight line **5.** $m\angle ABC + m\angle CBD = 180$ **7.** Postulate 3–4

Written Exercises Page 76 **1.** 45 **3.** 80 **5.** 45 **7.** 19 **9.** 91 **11.** 110 **13.** 55 **15.** 27.5 **17.** 87.5 **19.** 115
21. Given: $\angle A$ is the complement of $\angle C$
$\angle B$ is the complement of $\angle C$
Prove: $\angle A \cong \angle B$

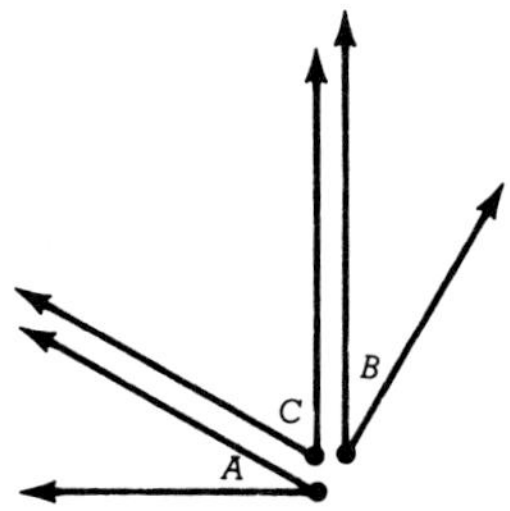

Proof:

STATEMENTS	REASONS
1. $\angle A$ is the complement of $\angle C$ $\angle B$ is the complement of $\angle C$	1. Given
2. $m\angle A + m\angle C = 90$ $m\angle B + m\angle C = 90$	2. Definition of complementary angles
3. $m\angle A + m\angle C = m\angle B + m\angle C$	3. Postulate 2–9
4. $m\angle A = m\angle B$	4. Postulate 2–7
5. $\angle A \cong \angle B$	5. Definition of congruent angles

Exploratory Exercises Page 80 **1.** acute **3.** acute **5.** obtuse **7.** right **9.** acute **11.** obtuse **13.** An angle measuring 90° and a right angle **15.** two angles are right angles **17.** one angle in a linear pair is a right angle **19.** the angles are congruent **21.** the angles are supplementary **23.** the other angle also is a right angle

Exploratory Exercises Page 84 **1.** no **3.** yes **5.** yes **7.** yes **9.** yes **11.** yes **13.** no **15.** yes **17.** no **19.** no

Written Exercises Page 84 **1.** Definition of a linear pair **3.** Definition of supplementary angles **5.** Definition of congruent angles **7.** Postulate 2–14 **9.** Definition of perpendicular lines
11. Given: $\overrightarrow{BA} \perp \overrightarrow{BD}$
Prove: $\angle ABC$ is complementary to $\angle CBD$

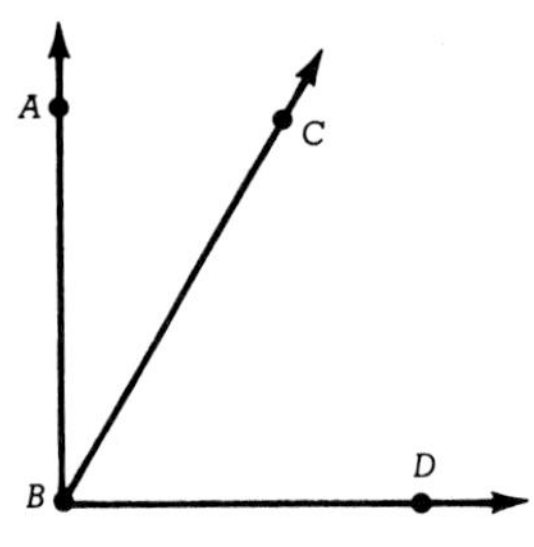

Proof:

STATEMENTS	REASONS
1. $\overrightarrow{BA} \perp \overrightarrow{BD}$	1. Given
2. $\angle ABD$ is a right angle	2. Definition of perpendicular lines
3. $m\angle ABD = 90$	3. Definition of right angles
4. $m\angle ABD = m\angle ABC + m\angle CBD$	4. Postulate 3–3
5. $m\angle ABC + m\angle CBD = 90$	5. Postulate 2–9
6. $\angle ABC$ and $\angle CBD$ are complementary	6. Definition of complementary angles

Exploratory Exercises Page 88 **1.** $\angle S - \overrightarrow{PQ} - R$ **3.** $\angle S - \overleftrightarrow{QR} - P$ **5.** $\angle Q - \overleftrightarrow{SR} - P$ **7.** $\angle PQR$ **9.** $\angle PQS$

Written Exercises Page 89 **1.** $\angle Q - \overleftrightarrow{XT} - V$, $\angle T - \overleftrightarrow{QR} - P$; answers may vary **3.** $\angle RXW$ or $\angle VTQ$ **5.** yes
7. no

9. Given: $\mathscr{P} \perp \mathscr{M}$
Prove: $\angle A - \overleftrightarrow{BX} - C$ is a right dihedral angle

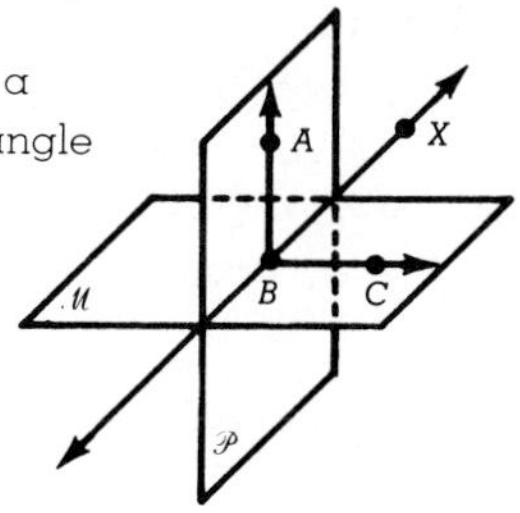

Proof:

STATEMENTS	REASONS
1. $\mathscr{P} \perp \mathscr{M}$	1. Given
2. $\overrightarrow{BA} \perp \overleftrightarrow{BX}$	2. Theorem 3–12
3. $\overrightarrow{BC} \perp \overleftrightarrow{BX}$	3. Theorem 3–12
4. $\overrightarrow{BA} \perp \mathscr{M}$	4. Definition of perpendicular planes
5. $\overrightarrow{BC} \perp \mathscr{P}$	5. Definition of perpendicular planes
6. $\angle A - \overleftrightarrow{BX} - C$ is a right dihedral angle	6. Definition of right dihedral angle

11. Given: $\mathscr{P}$ intersects $\mathscr{M}$, $\angle V - \overleftrightarrow{AB} - R$ is a right dihedral angle
Prove: $\mathscr{P} \perp \mathscr{M}$

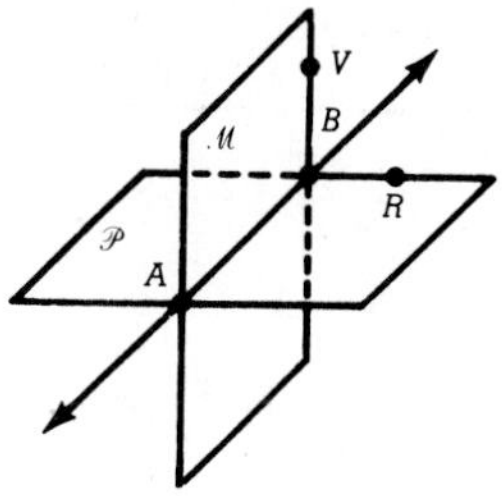

Proof:

STATEMENTS	REASONS
1. $\mathscr{P}$ intersects $\mathscr{M}$, $\angle V - \overleftrightarrow{AB} - R$ is a right dihedral angle	1. Given
2. $\overrightarrow{BR} \perp \overleftrightarrow{AB}$, $\overrightarrow{BR} \perp \mathscr{M}$	2. Definition of right dihedral angles
3. $\mathscr{P} \perp \mathscr{M}$	3. Definition of perpendicular planes

Chapter Review Page 93 **1.** Q **3.** T **5.** $\overrightarrow{RS}$ and $\overrightarrow{RP}$, or $\overrightarrow{QR}$ and $\overrightarrow{QP}$ **7.** 95 **9.** 21 **11.** 86

13. Given: $\angle A \cong \angle B$, $\angle A$ and $\angle B$ are supplementary
Prove: $m\angle A = 90$, $m\angle B = 90$

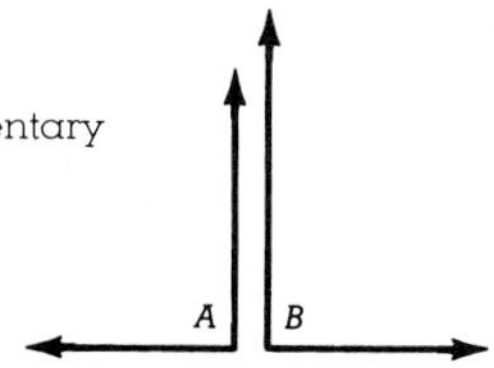

Proof:

STATEMENTS	REASONS
1. $\angle A \cong \angle B$, $\angle A$ and $\angle B$ are supplementary	1. Given
2. $m\angle A = m\angle B$	2. Definition of congruent angles
3. $m\angle A + m\angle B = 180$	3. Definition of supplementary angles
4. $m\angle A + m\angle A = 180$	4. Postulate 2–9
5. $2m\angle A = 180$	5. Postulate 2–14
6. $m\angle A = 90$	6. Postulate 2–8
7. $m\angle B = 90$	7. Postulate 2–9

15. Given: $\angle 1$ and $\angle 2$ are vertical angles, $\angle 3$ and $\angle 4$ are vertical angles, $\angle 1$ and $\angle 2$ are supplementary
Prove: $\angle 3$ and $\angle 4$ are supplementary

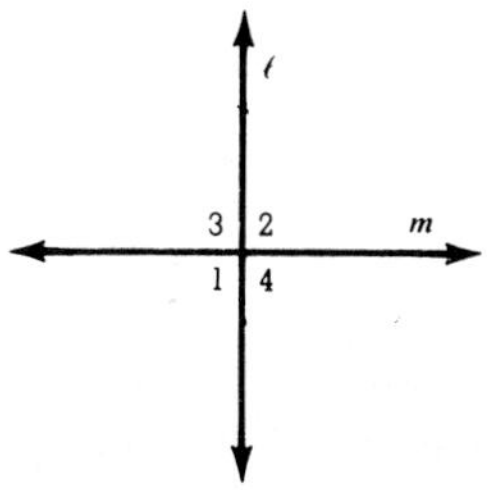

Proof:

STATEMENTS	REASONS
1. $\angle 1$ and $\angle 2$ are vertical angles, $\angle 3$ and $\angle 4$ are vertical angles, $\angle 1$ and $\angle 2$ are supplementary	1. Given
2. $\angle 1 \cong \angle 2$	2. Theorem 3–10
3. $m\angle 1 = m\angle 2$	3. Definition of congruent angles
4. $m\angle 1 + m\angle 2 = 180$	4. Definition of supplementary angles
5. $m\angle 1 + m\angle 1 = 180$	5. Postulate 2–9
6. $2m\angle 1 = 180$	6. Postulate 2–9, 2–14
7. $m\angle 1 = 90$, $m\angle 2 = 90$	7. Postulate 2–9, Postulate 2–8
8. $\angle 1$ and $\angle 3$ form a linear pair	8. Given
9. $\angle 1$ and $\angle 3$ are supplementary	9. Postulate 3–4

STATEMENTS	REASONS
10. $m\angle 1 + m\angle 3 = 180$	10. Definition of supplementary angles
11. $90 + m\angle 3 = 180$	11. Postulate 2–9
12. $m\angle 3 = 90$	12. Postulate 2–7
13. $\angle 3 \cong \angle 4$	13. Theorem 3–10
14. $m\angle 3 = m\angle 4$	14. Definition of congruent angles
15. $m\angle 4 = 90$	15. Postulate 2–9
16. $m\angle 3 + m\angle 4 = 180$	16. Postulate 2–7 Postulate 2–9
17. $\angle 3$ and $\angle 4$ are supplementary	17. Definition of supplementary angles

17. Given: $\angle ABC$ and $\angle CBD$ are a linear pair, $\angle ABC \cong \angle CBD$
Prove: $\overrightarrow{BC} \perp \overleftrightarrow{AD}$

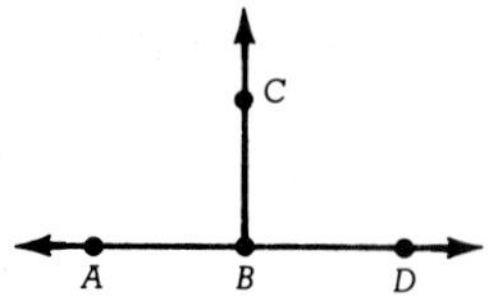

Proof:

STATEMENTS	REASONS
1. $\angle ABC$ and $\angle CBD$ are a linear pair	1. Given
2. $\angle ABC$ and $\angle CBD$ are supplementary	2. Postulate 3–4
3. $m\angle ABC + m\angle CBD = 180$	3. Definition of supplementary angles
4. $\angle ABC \cong \angle CBD$	4. Given
5. $m\angle ABC = m\angle CBD$	5. Definition of congruent angles
6. $m\angle ABC + m\angle ABC = 180$	6. Postulate 2–9
7. $2m\angle ABC = 180$	7. Postulate 2–14
8. $m\angle ABC = 90$	8. Postulate 2–8
9. $\angle ABC$ is a right angle	9. Definition of a right angle
10. $\overrightarrow{BC} \perp \overleftrightarrow{AD}$	10. Definition of perpendicular lines

CHAPTER 4 CONGRUENT TRIANGLES

Exploratory Exercises Page 98 1. D, E, F 3. $\angle D, \angle E, \angle F$ 5. $\overline{DE}, \overline{EF}, \overline{DF}$ 7. right triangle 9. scalene

Written Exercises Page 99 1–8. $\triangle RST, \triangle QRT, \triangle PST, \triangle PQS, \triangle QRS, \triangle PRS, \triangle PQR, \triangle PQT$ **9.** S, T, R **11.** $\angle TSR, \angle SRT, \angle RTS$ **13.** $\overline{QS}, \overline{QR}, \overline{RS}$ **15.** A figure is a triangle if and only if it is the union of three noncollinear line segments that intersect only at all three endpoints. **17.** A triangle is an obtuse triangle if and only if one of its angles is an obtuse angle. **19.** A triangle is equiangular if and only if all of its angles are congruent. **21.** A triangle is isosceles if and only if two of its sides are congruent. **23.** A point lies in the interior of a triangle if and only if it lies between two points, each of which lie on different sides of the triangle. **31.** none **35.** none **37.** $x = 18, PQ = 34, RP = 34, QR = 34$ **39.** $x = 20, PQ = 19, PR = 19$

Exploratory Exercises Page 102 1. 85 **3.** 81 **5.** 25 **7.** 60 **9.** 56 **11.** 85 **13.** 53 **15.** 66 **17.** 54

Written Exercises Page 102 1. No; otherwise the sum of two angles would equal 180, while in reality it is the sum of three angles which equals 180.

3. Given: $\triangle ABC$ is equiangular
Prove: the measure of each angle is 60

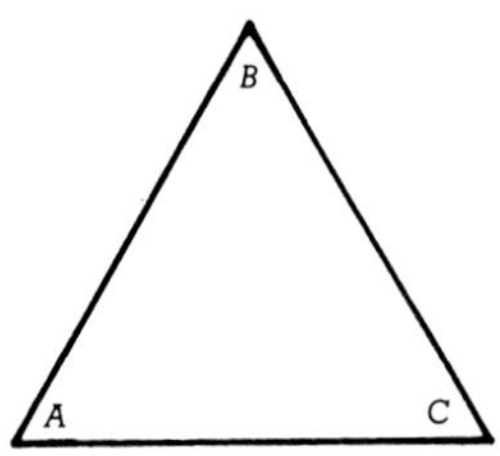

Proof:

STATEMENTS	REASONS
1. $\triangle ABC$ is equiangular	1. Given
2. $m\angle A = m\angle B = m\angle C$	2. Definition of equiangular triangle
3. $m\angle A + m\angle B + m\angle C = 180$	3. Theorem 4–1
4. $m\angle A + m\angle A + m\angle A = 180$	4. Postulate 2–9
5. $3m\angle A = 180$	5. Postulate 2–14
6. $m\angle A = 60$	6. Postulate 2–8
7. $m\angle B = 60, m\angle C = 60$	7. Postulate 2–9

5. 55 **7.** 55 **9.** 27.5

11. Given: $\angle Q$ and $\angle T$ are right angles
Prove: $m\angle P = m\angle S$

Proof:

STATEMENTS	REASONS
1. $\angle Q$ and $\angle T$ are right angles	1. Given
2. $\angle P$ is complementary to $\angle QRP$, $\angle S$ is complementary to $\angle SRT$	2. Theorem 4–3
3. $\angle QRP \cong \angle SRT$	3. Theorem 3–10
4. $\angle P \cong \angle S$	4. Theorem 3–4, 3–5
5. $m\angle P = m\angle S$	5. Definition of congruent angles

Exploratory Exercises Page 106 **1.** $\angle S$ **3.** $\angle V$ **5.** $\angle Q$ **7.** $\overline{ST}$ **9.** $\overline{VS}$ **11.** $\overline{QR}$ **13.** $\overline{AB} \cong \overline{FG}$, $\overline{BC} \cong \overline{GH}$, $\overline{CA} \cong \overline{HF}$, $\angle A \cong \angle F$, $\angle B \cong \angle G$, $\angle C \cong \angle H$

Written Exercises Page 106 **1.** no **3.** yes **5.** no **7.** no **9.** yes **11.** no **13.** no **15.** no **17.** no **19.** yes **21.** no **23.** $\angle S$ **25.** $\overline{QP}$ **27.** $\overline{RS}$ **29.** yes **31.** no

33. Given: $\triangle ABC \cong \triangle SRT$
Prove: $\triangle SRT \cong \triangle ABC$

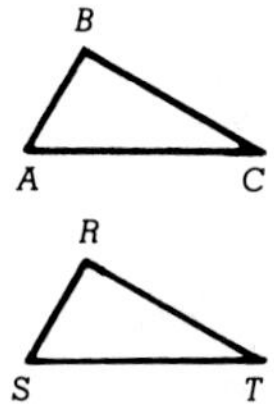

Proof:

STATEMENTS	REASONS
1. $\triangle ABC \cong \triangle SRT$	1. Given
2. $\overline{AB} \cong \overline{SR}$, $\overline{BC} \cong \overline{RT}$, $\overline{AC} \cong \overline{ST}$, $\angle A \cong \angle S$, $\angle B \cong \angle R$, $\angle C \cong \angle T$	2. Definition of congruent triangles
3. $\overline{SR} \cong \overline{AB}$, $\overline{RT} \cong \overline{BC}$, $\overline{ST} \cong \overline{AC}$, $\angle S \cong \angle A$, $\angle R \cong \angle B$, $\angle T \cong \angle C$	3. Theorem 2–3, Theorem 3–1
4. $\triangle SRT \cong \triangle ABC$	4. Definition of congruent triangles

Exploratory Exercises Page 109 **1.** no **3.** no **5.** no **7.** no **9.** no **11.** no **13.** yes **15.** no **17.** no **19.** no **21.** no **23.** no **25.** yes **27.** yes

Written Exercises Page 109

1.

STATEMENTS	REASONS
1. $\angle 1 \cong \angle 2$, R is the midpoint of $\overline{PT}$	1. Given
2. $PR = RT$	2. Definition of a midpoint
3. $\overline{PR} \cong \overline{RT}$	3. Definition of congruent line segments
4. $\angle P$ and $\angle T$ are right angles	4. Given
5. $\angle P \cong \angle T$	5. Theorem 3–6
6. $\triangle QRP \cong \triangle SRT$	6. ASA
7. $\overline{QP} \cong \overline{ST}$	7. Definition of congruent triangles

7.

STATEMENTS	REASONS
1. $\overline{AC} \cong \overline{CE}$, $\angle A \cong \angle E$	1. Given
2. $\angle 1 \cong \angle 2$	2. Theorem 3–10
3. $\triangle CAB \cong \triangle CED$	3. ASA
4. $\overline{BC} \cong \overline{CD}$	4. Definition of congruent triangles

13. no **15.** yes **17.** yes **19.** no **21.** yes **23.** no **25.** yes **27.** no **29.** no **31.** yes

Exploratory Exercises Page 113 **1.** AAS **3.** ASA **5.** none **7.** none **9.** AAS **11.** SAS **13.** AAS **15.** none **17.** none

Written Exercises Page 113

1.

STATEMENTS	REASONS
1. $\angle A \cong \angle C$, $\angle 1 \cong \angle 2$	1. Given
2. $\overline{BD} \cong \overline{BD}$	2. Theorem 2–2
3. $\triangle ABC \cong \triangle CBD$	3. AAS
4. $\overline{AB} \cong \overline{CB}$	4. Definition of congruent triangles

Exploratory Exercises Page 117 1. 90 3. 90 5. 45 7. 45 9. 45 11. 90

Written Exercises Page 117

11.

STATEMENTS	REASONS
1. $\overline{BC} \cong \overline{AD}$ $\overline{AB} \cong \overline{CD}$ $\overline{CD} \cong \overline{AD}$	1. Given
2. $\overline{DB} \cong \overline{DB}$	2. Theorem 2–2
3. $\triangle DAB \cong \triangle DCB$	3. SSS
4. $\angle CBD \cong \angle ABD$	4. Definition of congruent triangle
5. $m\angle CBD = m\angle ABD$	5. Definition of congruent angles
6. $\overline{BD}$ bisects $\angle ABC$	6. Definition of bisector of an angle

13. Given: $\triangle ABC$ is an isosceles triangle, $\overline{BD}$ is a median
Prove: $\overline{BD}$ bisects $\angle ABC$

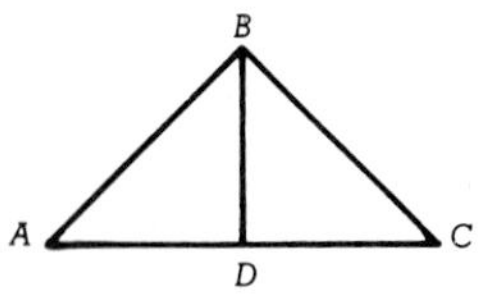

Proof:

STATEMENTS	REASONS
1. $\overline{BD}$ is median of isosceles triangle ABC	1. Given
2. $\overline{AB} \cong \overline{BC}$	2. Definition of isosceles triangle
3. $\overline{BD} \cong \overline{BD}$	3. Theorem 2–2
4. D is midpoint of $\overline{AC}$	4. Definition of median of a triangle
5. $\overline{AD} \cong \overline{DC}$	5. Theorem 2–5
6. $\triangle ABD \cong \triangle CBD$	6. SSS
7. $\angle ABD \cong \angle CBD$	7. Definition of congruent triangles
8. $m\angle ABD = m\angle CBD$	8. Definition of congruent angles
9. $\overline{BD}$ bisects $\angle ABC$	9. Definition of bisector of an angle

9.

STATEMENTS	REASONS
1. $\overline{PS}$ and $\overline{SR}$ are in $\mathcal{M}$, $\overline{QS} \perp \mathcal{M}$	1. Given
2. $\overline{QS} \perp \overline{PS}$, $\overline{QS} \perp \overline{SR}$	2. Definition of a line perpendicular to a plane
3. $\angle QSP$ and $\angle QSR$ are right angles	3. Theorem 3–11
4. $\angle QSP \cong \angle QSR$	4. Theorem 3–6
5. $\angle P \cong \angle R$	5. Given
6. $\overline{QS} \cong \overline{QS}$	6. Theorem 2–2
7. $\triangle QSP \cong \triangle QSR$	7. AAS
8. $\overline{PS} \cong \overline{SR}$	8. Definition of congruent triangles

15. Given: BD is an altitude and a median of $\triangle ABC$
Prove: $\triangle ABC$ is isosceles

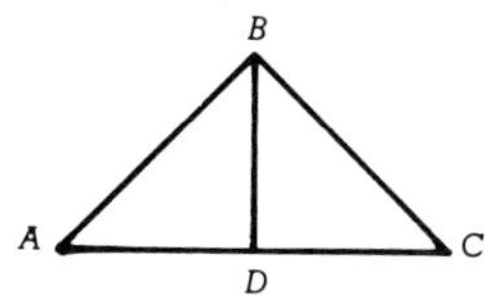

Proof:

STATEMENTS	REASONS
1. $\overline{BD}$ is an altitude and a median of $\triangle ABC$	1. Given
2. D is the midpoint of $\overline{AC}$	2. Definition of median of a triangle
3. $\overline{AD} \cong \overline{DC}$	3. Theorem 2–5
4. $\overline{BD} \cong \overline{BD}$	4. Theorem 2–2
5. $\overline{BD} \perp \overline{AC}$	5. Definition of altitude of a triangle
6. $\angle ADB \cong \angle CDB$	6. Theorem 3–13
7. $m\angle ADB = m\angle CDB$	7. Definition of congruent angles
8. $\triangle ADB \cong \triangle CDB$	8. SAS
9. $\overline{AB} \cong \overline{CB}$	9. Definition of congruent triangles
10. $\triangle ABC$ is isosceles	10. Definition of isosceles triangle

Exploratory Exercises Page 120 **1.** yes **3.** no **5.** no **7.** yes **9.** no

Written Exercises Page 120 **3.** isosceles **9.** isosceles **11.** No, the bisector of $\overline{PQ}$ may not necessarily be perpendicular to $\overline{PQ}$. **13.** yes **15.** No, not possible for $\overline{XY}$ to be perpendicular to both segments; violates theorem for degrees of a triangle.

17. Given: $\overline{PQ} \cong \overline{RS}$
$\overline{QR} \cong \overline{SP}$
Prove: $\angle Q \cong \angle S$

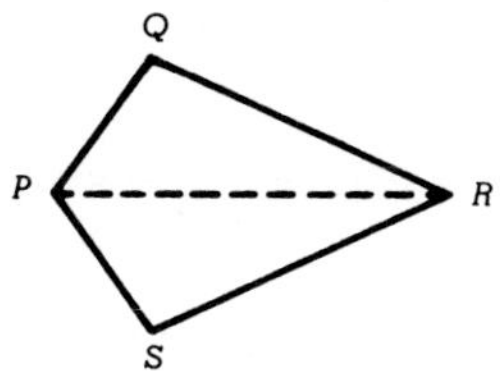

Proof:

STATEMENTS	REASONS
1. Draw $\overline{PR}$	1. Postulate 1–1
2. $\overline{PQ} \cong \overline{RS}$ $\overline{QR} \cong \overline{SP}$	2. Given
3. $\overline{PR} \cong \overline{PR}$	3. Theorem 2–2
4. $\triangle PQR \cong \triangle PSR$	4. SSS
5. $\angle Q \cong \angle S$	5. Definition of congruent triangles

21. Given: $\overline{CB} \cong \overline{AD}$
$\overline{CD} \cong \overline{AB}$
Prove: $\angle A \cong \angle C$

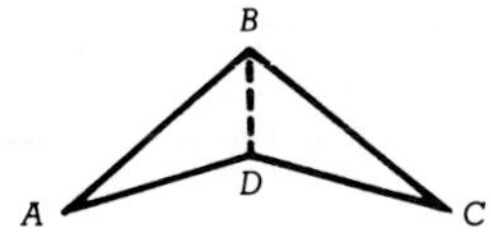

Proof:

STATEMENTS	REASONS
1. Draw $\overline{BD}$	1. Postulate 1–1
2. $\overline{CB} \cong \overline{AD}$ $\overline{CD} \cong \overline{AB}$	2. Given
3. $\overline{BD} \cong \overline{BD}$	3. Theorem 2–2
4. $\triangle ABD \cong \triangle CDB$	4. SSS
5. $\angle A \cong \angle C$	5. Definition of congruent triangles

25. Given: $\overline{AB} \cong \overline{BC}$, $\overline{BC} \cong \overline{AC}$,
$\overline{AC} \cong \overline{AB}$
Prove: $m\angle A = 60$,
$m\angle B = 60$,
$m\angle C = 60$

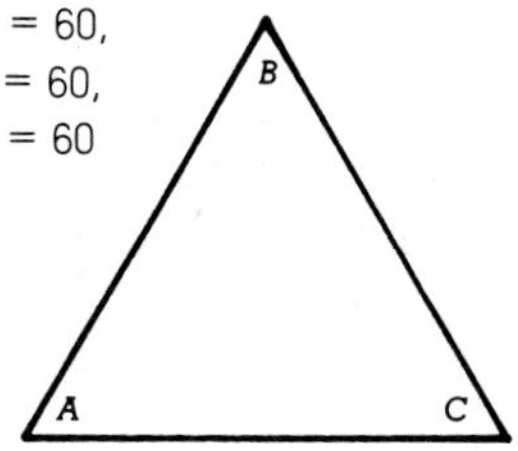

Proof:

STATEMENTS	REASONS
1. $\overline{AB} \cong \overline{BC}$, $\overline{BC} \cong \overline{AC}$, $\overline{AC} \cong \overline{AB}$	1. Given
2. $\angle A \cong \angle B$, $\angle B \cong \angle C$, $\angle C \cong \angle A$	2. Theorem 4–7
3. $m\angle A = m\angle B$, $m\angle B = m\angle C$, $m\angle C = m\angle A$	3. Definition of congruence
4. $m\angle A + m\angle B + m\angle C = 180$	4. Theorem 4–1
5. $m\angle A + m\angle A + m\angle A = 180$	5. Postulate 2–9
6. $3m\angle A = 180$	6. Postulate 2–14
7. $m\angle A = 60$	7. Postulate 2–8
8. $m\angle B = 60$, $m\angle C = 60$	8. Postulate 2–9

Exploratory Exercises Page 125 **9.** HA **11.** HA; Theorem 4–3 **13.** HL **15.** HL

Written Exercises Page 125

1.

STATEMENTS	REASONS
1. $\angle Q$ and $\angle S$ are right angles	1. Given
2. $\triangle PQR$ and $\triangle RSP$ are right triangles	2. Definition of a right triangle
3. $\angle 1 \cong \angle 2$	3. Given
4. $\overline{PR} \cong \overline{PR}$	4. Theorem 2–2
5. $\triangle PQR \cong \triangle RSP$	5. HA
6. $\overline{QP} \cong \overline{SR}$	6. Definition of congruent triangles

5.

STATEMENTS	REASONS
1. $\angle Q$ and $\angle S$ are right angles	1. Given
2. $\triangle PQR$ and $\triangle RSP$ are right triangles	2. Definition of a right triangle
3. $\overline{QP} \cong \overline{SR}$	3. Given
4. $\overline{PR} \cong \overline{PR}$	4. Theorem 2–2
5. $\triangle PQR \cong \triangle RSP$	5. HL
6. $\overline{QR} \cong \overline{PS}$	6. Definition of congruent triangles

9. Given: $\triangle ABC$ and $\triangle RPQ$ are right triangles, $\angle A$ and $\angle R$ are right angles, $\overline{BC} \cong \overline{PQ}$, $\angle B \cong \angle P$

Prove: $\triangle ABC \cong \triangle RPQ$

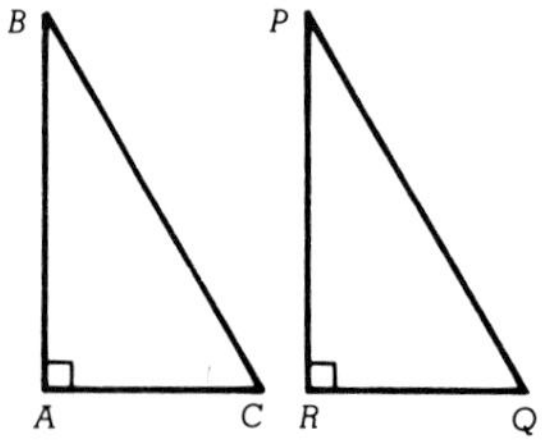

Proof:

STATEMENTS	REASONS
1. $\triangle ABC$ and $\triangle RPQ$ are right triangles, $\angle A$ and $\angle R$ are right angles, $\overline{BC} \cong \overline{PQ}$, $\angle B \cong \angle P$	1. Given
2. $\angle A \cong \angle R$	2. Theorem 3–6
3. $\triangle ABC \cong \triangle RPQ$	3. AAS

11. Given: $\triangle BDF$ and $\triangle TVZ$ are right triangles, $\angle B$ and $\angle T$ are right angles, $\overline{BF} \cong \overline{TZ}$, $\angle F \cong \angle Z$

Prove: $\triangle BDF \cong \triangle TVZ$

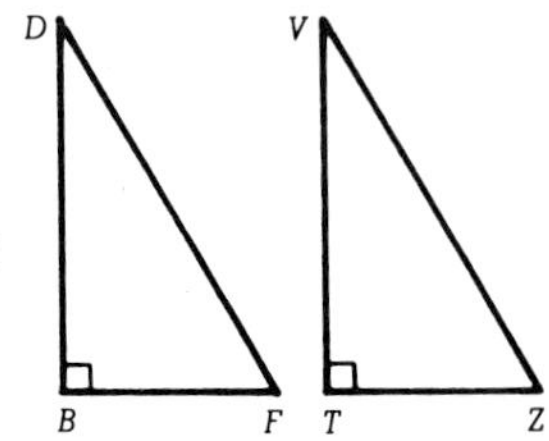

Proof:

STATEMENTS	REASONS
1. $\triangle BDF$ and $\triangle TVZ$ are right triangles, $\angle B$ and $\angle T$ are right angles, $\overline{BF} \cong \overline{TZ}$, $\angle F \cong \angle Z$	1. Given
2. $\angle B \cong \angle T$	2. Theorem 3–6
3. $\triangle BDF \cong \triangle TVZ$	3. ASA

Chapter Review Pages 128–129 1. $\triangle WTV$ 3. W, T, V 5. 28 7. 62 9. $\overline{RQ}$ 11. $\overline{QS}$ 13. $\overline{SR}$

15.

STATEMENTS	REASONS
1. $\overline{AC}$ and $\overline{BD}$ bisect each other	1. Given
2. E is the midpoint of $\overline{AC}$ and $\overline{BD}$	2. Definition of a segment bisector
3. $\overline{BE} \cong \overline{ED}$, $\overline{AE} \cong \overline{EC}$	3. Definition of a bisector
4. $\angle BEA \cong \angle CED$	4. Theorem 3–10
5. $\triangle BEA \cong \triangle CED$	5. SAS
6. $\overline{AB} \cong \overline{CD}$	6. Definition of congruent triangles

17.

STATEMENTS	REASONS
1. $\overline{PR} \perp \overline{QS}$	1. Given
2. $\angle 3 \cong \angle 4$	2. Theorem 3–13
3. $\angle Q \cong \angle S$	3. Given
4. $\overline{PR} \cong \overline{PR}$	4. Theorem 2–2
5. $\triangle PRQ \cong \triangle PRS$	5. AAS
6. $\overline{QR} \cong \overline{SR}$	6. Definition of congruent triangles
7. $QR = SR$	7. Definition of congruence
8. R is the midpoint of $\overline{QS}$	8. Definition of a midpoint

21.

STATEMENTS	REASONS
1. $\angle A \cong \angle D$, $\overline{AB} \cong \overline{CD}$, E is the midpoint of $\overline{AD}$	1. Given
2. $\overline{AE} \cong \overline{DE}$	2. Theorem 2–5
3. $\triangle ABE \cong \triangle DCE$	3. SAS
4. $\overline{BE} \cong \overline{CE}$	4. Definition of congruent triangles
5. $\angle 3 \cong \angle 4$	5. Theorem 4–6

Exploratory Exercises Page 134 **1.** $x > y$ **3.** $x > y$ **5.** $x < y$ **7.** $x < y$ **9.** $PQ < PR$ **11.** $PQ < PR$ **13.** $PQ > PR$ **15.** $m\angle 1 > m\angle 2$ **17.** $m\angle 1 < m\angle 2$ **19.** $m\angle 1 > m\angle 2$ **21.** $m\angle 1 > m\angle 2$

Written Exercises Page 134 **1.** 33 **3.** 30 **5.** 120 **7.** 53 **9.** 40 **11.** 100 **13.** 20 **15.** 30 **17.** 55 **19.** 70 **21.** 125

23.

STATEMENTS	REASONS
1. $\angle 1 \cong \angle 2$	1. Given
2. $\angle 3$ is supplementary to $\angle 1$, $\angle 4$ is supplementary to $\angle 2$	2. Postulate 3–4
3. $\angle 3 \cong \angle 4$	3. Theorem 3–3
4. $\overline{QR} \cong \overline{QP}$	4. Theorem 4–9
5. $\triangle PQR$ is isosceles.	5. Definition of isosceles triangle

25. Given: $\triangle PQR$, $\angle 1$ is an exterior angle of $\triangle PQR$
Prove: $m\angle 1 = m\angle 3 + m\angle 4$

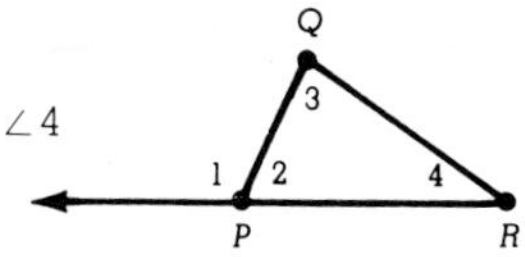

Proof:

STATEMENTS	REASONS
1. $\angle 1$ is an exterior angle of $\triangle PQR$	1. Given
2. $\angle 1$ and $\angle 2$ for a linear pair	2. Definition of an exterior angle
3. $m\angle 1 + m\angle 2 = 180$	3. Postulate 3–4
4. $m\angle 2 + m\angle 3 + m\angle 4 = 180$	4. Theorem 4–1
5. $m\angle 1 + m\angle 2 = m\angle 2 + m\angle 3 + m\angle 4$	5. Postulate 2–9
6. $m\angle 1 = m\angle 3 + m\angle 4$	6. Postulate 2–7

27. 1. Given 2. Postulate 2–16 3. Postulate 2–12 4. Postulate 2–13 and Postulate 2–9 5. Postulate 2–11 6. Postulate 2–10 7. Postulate 2–11 **29.** Show $m\angle y = m\angle c + m\angle b$. Replace $m\angle b$ with $m\angle x + m\angle c$. This yields: $m\angle y = m\angle c + m\angle x + m\angle c$. Then by the inequality theorem $m\angle y > m\angle x$.

Exploratory Exercises Page 140 **1.** yes **3.** no **5.** no **7.** yes **9.** yes **11.** yes **13.** yes **15.** no **17.** yes **19.** sometimes **21.** no **23.** no **25.** yes **27.** yes

Written Exercises Page 140 **1.** yes **3.** yes **5.** no **7.** yes **9.** no **11.** no **13.** 5 **15.** 3 **17.** 4 **19.** $\triangle SQR$, $\triangle SPQ$

Exploratory Exercises Page 145 **1.** $\angle B$, $\angle A$, $\angle C$ **3.** $\angle I$, $\angle G$, $\angle H$ **5.** $\overline{ML}$, $\overline{LN}$, $\overline{MN}$ **7.** $\overline{ST}$, $\overline{TV}$, $\overline{SV}$

Written Exercises Page 146 **1.** Show that $m\angle A = m\angle C + m\angle B$. Thus, $m\angle A > m\angle C$, which implies that $BC > AB$ by Theorem 5–5. **3.** Since $\overline{PR} \cong \overline{PQ}$, show that $PR = PQ$ by definition of congruent line segments. Also, since $QR > PQ$, then by substitution $QR > PR$. **9.** The measures of two sides of a triangle are unequal if and only if the measures of the angles opposite those sides are unequal in the same order.

Exploratory Exercises Page 149 **1.** yes **3.** yes **5.** no **7.** yes **9.** no **11.** no **13.** no **15.** yes **17.** yes **19.** no **21.** yes **23.** yes **25.** yes **27.** yes **29.** yes

Written Exercises Page 150 **1.** no **3.** yes **5.** yes **7.** no **9.** no **11.** no **13.** 4, 10 **15.** 3, 27 **17.** By triangle inequality it is known that $AD + AC > CD$. Since it is given that $\angle ABC \cong \angle ACB$, it can be shown that $\overline{AB} \cong \overline{AC}$ and $AB = AC$. By substitution, $AD + AB > CD$. **19.** By triangle inequality it is known that $BE + AE > BA$. Use congruent triangles to show that $AE = EC$, and substitute. Thus $BE + EC > BA$. By property of betweenness, $BE + EC = BC$. Substitute again, $BC > BA$.

23. Given: $\triangle PQR$
Prove: $PQ + PR > RQ$

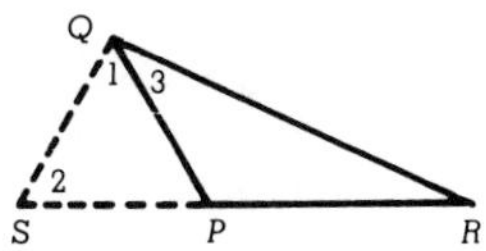

Proof:

STATEMENTS	REASONS
1. Draw $\overline{PS}$ so that P is between S and R and $PQ = PS$	1. Postulate 2–2
2. $PS + PR = RS$	2. Definition of betweenness
3. $\overline{PQ} \cong \overline{PS}$	3. Definition of congruent line segments
4. $\angle 1 \cong \angle 2$	4. Theorem 4–6
5. $m\angle 1 = m\angle 2$	5. Definition of congruent angles
6. $m\angle SQR = m\angle 1 + m\angle 3$	6. Postulate 3–3
7. $m\angle SQR = m\angle 2 + m\angle 3$	7. Postulate 2–9
8. $m\angle SQR > m\angle 2$	8. Theorem 5–3
9. $RS > RQ$	9. Theorem 5–5
10. $PS + PR > RQ$	10. Postulate 2–9
11. $QP + PR > RQ$	11. Postulate 2–9

Exploratory Exercises Page 154 **1.** Theorem 5–9 **3.** Theorem 5–9 **5.** Theorem 5–9

Written Exercises Page 154

1.

STATEMENTS	REASONS
1. $\overline{PQ} \cong \overline{SQ}$	1. Given
2. $\overline{QR} \cong \overline{QR}$	2. Theorem 2–2
3. $m\angle PQR = m\angle PQT + m\angle SQR$	3. Postulate 3–3
4. $m\angle PQR > m\angle SQR$	4. Theorem 5–2
5. $PR > SR$	5. Theorem 5–9

3.

STATEMENTS	REASONS
1. $\overline{AB} \cong \overline{BC}$, $m\angle 4 > m\angle 3$	1. Given
2. $\overline{BD} \cong \overline{BD}$	2. Theorem 2–2
3. $AD > DC$	3. Theorem 5–9

11. If $m\angle P = m\angle A$, then by definition of congruent angles $\angle P \cong \angle A$. Show by SAS that $\triangle PQR \cong \triangle ABC$. Conclude $\overline{QR} \cong \overline{BC}$. Thus by definition of congruent line segments $QR = BC$, which contradicts $QR > BC$.

Chapter Review Page 157 **1.** 120 **3.** 30 **5.** 60 **7.** 55 **9.** yes **11.** no **13.** yes **15.** 10; $\triangle BCD$, $\triangle AEB$, $\triangle ABD$, $\triangle BED$, $\triangle BEF$, $\triangle FED$, $\triangle BEC$, $\triangle BFC$, $\triangle FDC$, $\triangle ECD$ **17.** $\overline{WZ}$, $\overline{WY}$, $\overline{ZY}$ **19.** no **21.** yes **23.** no

25.

STATEMENTS	REASONS
1. $\overline{AB} \cong \overline{CD}$, $AD > BC$	1. Given
2. $\overline{AC} \cong \overline{AC}$	2. Theorem 2–2
3. $m\angle ACD > m\angle BAC$	3. Theorem 5–10

CHAPTER 6 PARALLELS

Exploratory Exercises Page 162 **1.** parallel **3.** skew or parallel **5.** parallel, intersecting **7.** parallel **9.** intersecting **11.** parallel **13.** intersecting **15.** parallel **17.** parallel **19.** parallel, intersecting

Written Exercises Page 163 **1.** ℓ and m, m and n **3.** ℓ and n **5.** t and ℓ **7.** ℓ and n, transversal $\overleftrightarrow{FB}$ **9.** $\overrightarrow{BD}$ and $\overline{BC}$ **15.** no **17.** no **19.** no **21.** yes **23.** yes **25.** Line segments in the same plane are parallel if and only if the lines that contain them do not intersect. **27.** Angles are alternate interior angles if and only if they lie on opposite sides of a transversal and between the intersected lines.

29. Given: $\angle 1 \cong \angle 5$

Prove: $\angle 4 \cong \angle 8$, $\angle 2 \cong \angle 6$, $\angle 3 \cong \angle 7$

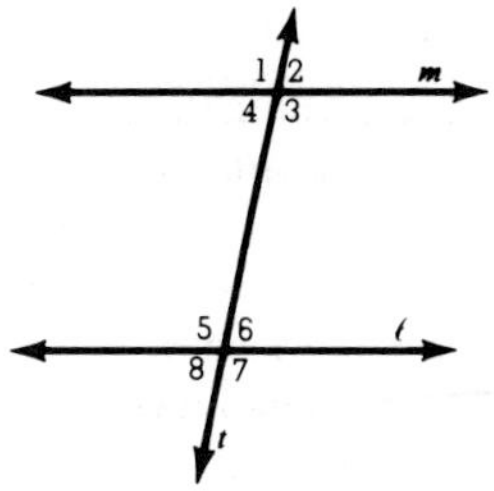

Proof:

STATEMENTS	REASONS
1. $\angle 1 \cong \angle 5$	1. Given
2. $\angle 1 \cong \angle 3$, $\angle 5 \cong \angle 7$	2. Theorem 3–10
3. $\angle 3 \cong \angle 7$	3. Postulate 2–9
4. $\angle 1$ and $\angle 2$ and $\angle 5$ and $\angle 6$ form linear pairs	4. Given
5. $\angle 1$ and $\angle 2$ are supplementary, $\angle 5$ and $\angle 6$ are supplementary	5. Postulate 3–4
6. $\angle 2 \cong \angle 6$	6. Theorem 3–3
7. $\angle 4 \cong \angle 2$, $\angle 8 \cong \angle 6$	7. Theorem 3–10
8. $\angle 4 \cong \angle 8$	8. Postulate 2–9

Exploratory Exercises Page 166 **1.** It is not poison ivy. **3.** It is not defective. **5.** They intersect in more than one point. **7.** The right triangle has no acute angles. **9.** They are not parallel to each other. **11.** The measures of the sides opposite those angles are equal in the same order.

Written Exercises Page 167 **1.** *b* **3.** *b* **5.** *c* **7.** *c* **9.** *b*

11. Given: $\triangle ABC$, $m\angle C = 90$

Prove: either $m\angle B < 90$ or $m\angle A < 90$

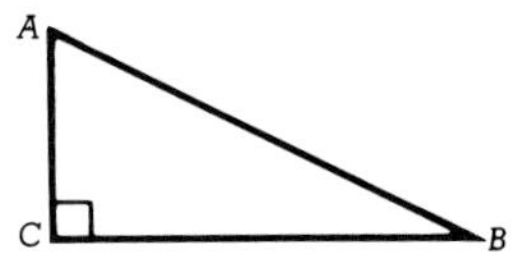

Proof:

Step 1. Suppose $\triangle ABC$ has no acute angles.

Step 2. Then, $m\angle A \geq 90$ and $m\angle B \geq 90$. Then by addition and substitution, $m\angle A + m\angle B + m\angle C \geq 270$. This contradicts the fact that the sum of the measures of the angles of a triangle equal 180.

Step 3. The original assumption that $\triangle ABC$ has no acute angles is false.

Step 4. Thus, $\triangle ABC$ has at least one acute angle.

13. Given: $a \parallel b$, $c \parallel b$

Prove: $a \parallel c$

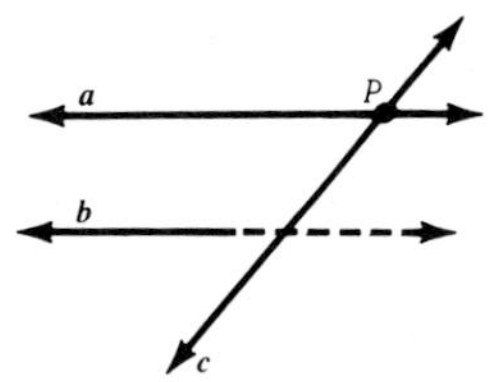

Proof:

Step 1. Suppose a is not parallel to c.

Step 2. Then, a and c must intersect at a point, say P. But, if c intersects with a, it must also intersect with b since $a \parallel b$. This contradicts the fact that $c \parallel b$ and they cannot intersect.

Step 3. Thus, the assumption that a is not parallel to c is false.

Step 4. Thus, $a \parallel c$.

15. Given: ℓ and P

Prove: There is exactly one line perpendicular to ℓ at P.

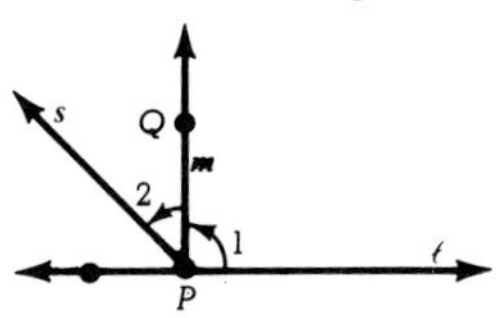

Proof:

Part I: There is at least one line, m, perpendicular to ℓ at P. By Postulate 3–2, given $\overrightarrow{PQ}$ on the edge of a half plane, for every positive number r between 0 and 180, there is one ray, $\overrightarrow{PQ}$, in the half plane so that $m\angle RPQ = r$.

Part II: There is no more than one line perpendicular to ℓ at P.

Step 1. Suppose there were more than one line perpendicular to ℓ at P, say s.

Step 2. Because $m \perp \ell$, $m\angle 1 = 90$. Thus, $m\angle 1 + m\angle 2 > 90$. This contradicts the fact that $s \perp \ell$.

Step 3. Therefore, the assumption that there is more than one line perpendicular to ℓ at P is false.

Step 4. There is no more than one line perpendicular to ℓ at P.

Conclusion: There is exactly one line perpendicular to ℓ at P.

17. Given: ℓ and P not on ℓ
Prove: Exactly one plane contains the point and the line.

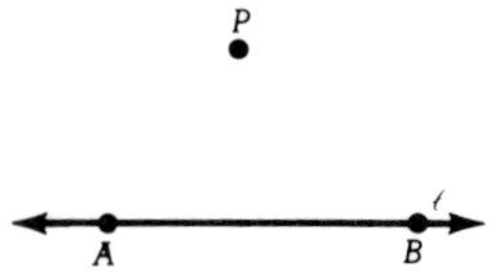

Proof:

By Postulate 1–3, a line contains at least two points, say A and B. P is given as not on ℓ. Then, by Postulate 1–2, through any three points not on the same line there is exactly one plane.

19. Indirect reasoning: either the radio or the outlet is defective. Assume that the radio is defective. However, the radio played well when it was purchased. This contradicts the assumption that the radio is defective. Therefore, the other alternative must be true, that is, the outlet is defective.

Exploratory Exercises Page 172 1. Theorem 6–5 3. Theorem 6–7 5. Theorem 6–5 7. Theorem 6–4

Written Exercises Page 173

1.

STATEMENTS	REASONS
1. $\angle R$ and $\angle S$ are supplementary, $\angle Q \cong \angle S$	1. Given
2. $m\angle Q = m\angle S$	2. Definition of angle congruence
3. $\angle R$ and $\angle Q$ are supplementary	3. Substitution
4. $\overline{QP} \parallel \overline{RS}$	4. Theorem 6–6

3.

STATEMENTS	REASONS
1. $\angle 1 \cong \angle 3$, $\overline{AD} \cong \overline{BC}$	1. Given
2. $\overline{AC} \cong \overline{AC}$	2. Theorem 2–2
3. $\triangle DAC \cong \triangle BCA$	3. SAS
4. $\angle 2 \cong \angle 4$	4. Definition of congruent triangles
5. $\overline{DC} \parallel \overline{AB}$	5. Theorem 6–5

5.

STATEMENTS	REASONS
1. $\overline{AD} \cong \overline{BC}$, $\overline{DC} \cong \overline{AB}$	1. Given
2. $\overline{AC} \cong \overline{AC}$	2. Theorem 2–2
3. $\triangle DAC \cong \triangle BCA$	3. SSS
4. $\angle 1 \cong \angle 3$	4. Definition of congruent triangles
5. $\overline{AD} \parallel \overline{BC}$	5. Theorem 6–5

7.

STATEMENTS	REASONS
1. $\overline{MP} \cong \overline{MQ}$, $\angle 1 \cong \angle N$	1. Given
2. $\angle 1 \cong \angle 2$	2. Theorem 4–6
3. $\angle 2 \cong \angle N$	3. Theorem 3–1
4. $\overline{PQ} \parallel \overline{LN}$	4. Theorem 6–4

9.

STATEMENTS	REASONS
1. $\angle 1 \cong \angle 2$	1. Given
2. $\overline{BC} \parallel \overline{FE}$	2. Theorem 6–5

11.

STATEMENTS	REASONS
1. $\angle F$ and $\angle 2$ are complementary, $\angle C$ and $\angle 1$ are complementary	1. Given
2. $m\angle F + m\angle 2 = 90$, $m\angle C + m\angle 1 = 90$	2. Definition of complementary angles
3. $m\angle F + m\angle 2 + m\angle A = 180$, $m\angle C + m\angle 1 + m\angle D = 180$	3. Theorem 4–1
4. $m\angle A = 90$, $m\angle D = 90$	4. Substitution and subtraction
5. $\overline{AF} \perp \overline{AD}$, $\overline{CD} \perp \overline{AD}$	5. Definition of right angles and perpendicular lines
6. $\overline{AF} \parallel \overline{CD}$	6. Theorem 6–7

13. Given: $\angle 2 \cong \angle 3$
Prove: $\ell \parallel m$

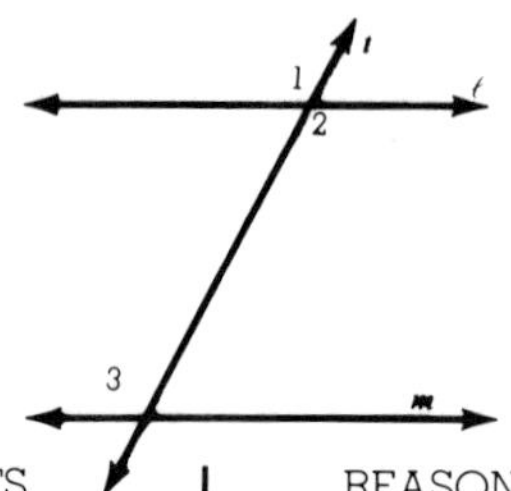

Proof:

STATEMENTS	REASONS
1. $\angle 2 \cong \angle 3$	1. Given
2. $\angle 1 \cong \angle 2$	2. Theorem 3–10
3. $\angle 1 \cong \angle 3$	3. Theorem 2–4
4. $\ell \parallel m$	4. Theorem 6–4

15. Given: $a \parallel b$, $a \parallel c$
Prove: $b \parallel c$

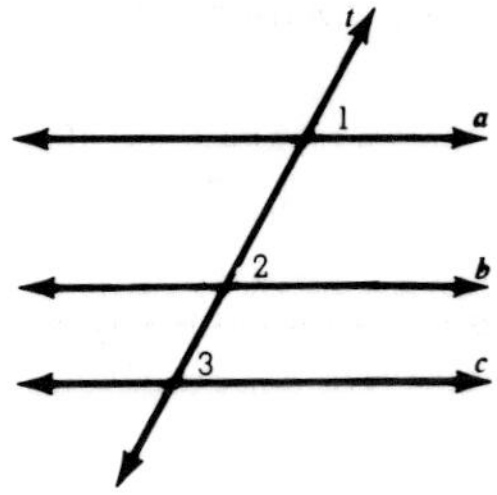

Proof:

STATEMENTS	REASONS
1. Draw t perpendicular to a	1. Auxiliary line
2. $a \parallel b$, $a \parallel c$	2. Given
3. $t \perp b$, $t \perp c$	3. Definition of perpendicular lines
4. $b \parallel c$	4. Theorem 6–7

Exploratory Exercises Page 177 **1.** $\angle 2$ and $\angle 12$, $\angle 3$ and $\angle 9$, $\angle 7$ and $\angle 13$, $\angle 6$ and $\angle 16$ **3.** $\angle 7$ and $\angle 16$, $\angle 6$ and $\angle 13$, $\angle 3$ and $\angle 12$, $\angle 2$ and $\angle 9$ **5.** $\angle 10$ and $\angle 14$, $\angle 11$ and $\angle 15$, $\angle 9$ and $\angle 13$, $\angle 12$ and $\angle 16$, $\angle 2$ and $\angle 6$, $\angle 3$ and $\angle 7$, $\angle 1$ and $\angle 5$, $\angle 4$ and $\angle 8$ **7.** $\angle 2$ and $\angle 4$, $\angle 1$ and $\angle 5$, $\angle 6$ and $\angle PQT$, $\angle 7$ and $\angle RQS$

Written Exercises Page 178 **1.** Postulate 6–1 **3.** Postulate 3–2 **5.** Theorem 2–1, Theorem 2–5 **7.** Theorem 2–1 **9.** Postulate 6–1 **11.** Theorem 3–12 **13.** 33 **15.** 57 **17.** 123 **19.** 90

21. Given: $\ell \parallel m$, with transversal t
Prove: $\angle 1 \cong \angle 2$

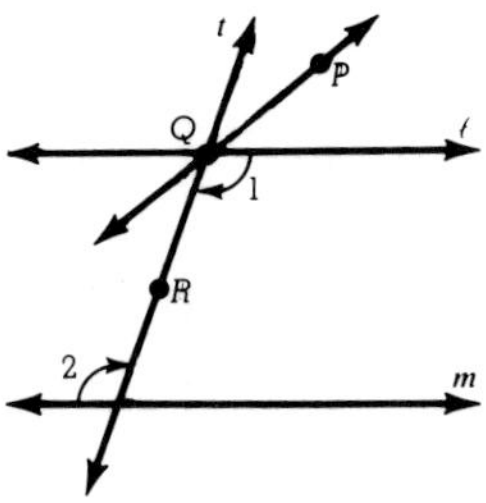

Proof:

Assume $\angle 1$ is not congruent to $\angle 2$. Then, draw $\overleftrightarrow{QP}$ so that $m\angle RQP = m\angle 2$. The protractor postulate justifies such a construction and guarantees that $\overleftrightarrow{QP}$ and ℓ are different lines. By definition of congruent angles, $\angle RQP \cong \angle 2$. Since they are alternate interior angles, $\overleftrightarrow{QP} \parallel m$. But $\ell \parallel m$ as given. Since $\overleftrightarrow{QP}$ and ℓ are different lines and both go through Q, the parallel postulate is contradicted. Thus, the assumption is false. Therefore, $\angle 1 \cong \angle 2$.

Exploratory Exercises Page 182 **1.** yes **3.** yes **5.** yes **7.** no **9.** no **11.** no **13.** yes **15.** yes **17.** yes **19.** yes **21.** yes

Written Exercises Page 182

1.

STATEMENTS	REASONS
1. $\overline{AB} \parallel \overline{CD}$, $\overline{AB} \perp \overline{AD}$, $\overline{BC} \perp \overline{CD}$	1. Given
2. $\overline{AD} \perp \overline{CD}$	2. Theorem 6–11
3. $\overline{BC} \parallel \overline{AD}$	3. Theorem 6–7

3.

STATEMENTS	REASONS
1. $\overline{QS} \parallel \overline{PT}$, $\angle T$ is a right angle	1. Given
2. $\overline{RT} \perp \overline{PT}$	2. Definition of perpendicular lines
3. $\overline{RT} \perp \overline{QS}$	3. Theorem 6–11
4. $\angle 2$ is a right angle	4. Theorem 3–11
5. $\triangle RQS$ is a right triangle	5. Definition of a right triangle

5.

STATEMENTS	REASONS
1. $\overline{BA} \parallel \overline{ED}$, $\overline{BA} \cong \overline{ED}$	1. Given
2. $\angle B \cong \angle D$, $\angle A \cong \angle E$	2. Theorem 6–9
3. $\triangle BAC \cong \triangle DEC$	3. ASA
4. $\overline{EC} \cong \overline{CA}$	4. Definition of congruent triangles

9.

STATEMENTS	REASONS
1. $\overline{BC} \parallel \overline{AE}$, $\angle 1 \cong \angle 2$, $\angle 3 \cong \angle 4$	1. Given
2. $\angle ABC$ and $\angle BAE$ are supplementary	2. Theorem 6–10
3. $m\angle ABC + m\angle BAE = 180$	3. Definition of supplementary angles
4. $m\angle ABC = m\angle 1 + m\angle 2$, $m\angle BAE = m\angle 3 + m\angle 4$	4. Postulate 3–3
5. $m\angle 1 + m\angle 2 + m\angle 3 + m\angle 4 = 180$	5. Substitution
6. $m\angle 1 = m\angle 2$, $m\angle 3 = m\angle 4$	6. Definition of angle congruence
7. $m\angle 2 + m\angle 2 + m\angle 3 + m\angle 3 = 180$	7. Substitution
8. $m\angle 2 + m\angle 3 = 90$	8. Distributive property, division property, and substitution
9. $m\angle 2 + m\angle 3 + m\angle BDA = 180$	9. Theorem 4–1
10. $90 + m\angle BDA = 180$	10. Substitution
11. $m\angle BDA = 90$	11. Division property
12. $\overline{BD} \perp \overline{AD}$	12. Definition of right angle definition of perpendicular lines

Exploratory Exercises Page 186 **1.** Given a line and a line not on the line. **3.** Line m, with point P not on m. **5.** There is one and only one line through P perpendicular to m. **7.** If two lines are parallel, then they are everywhere equidistant. **9.** then they are everywhere equidistant

Written Exercises Page 186 **1.** Direct **3.** Theorem 3–12 **5.** Indirect **7.** Yes, parallel. **9.** A point is on the bisector of an angle. **11.** Use congruent triangles to show that the distance to both legs are congruent. **13.** If a point is equidistant from the sides of an angle, then it lies on the bisector of the angle.

15. Given: $\angle ABC$, $PA = PC$,
$\overline{PA} \perp \overline{AB}$, $\overline{PC} \perp \overline{BC}$
Prove: $\angle PBA \cong \angle PBC$

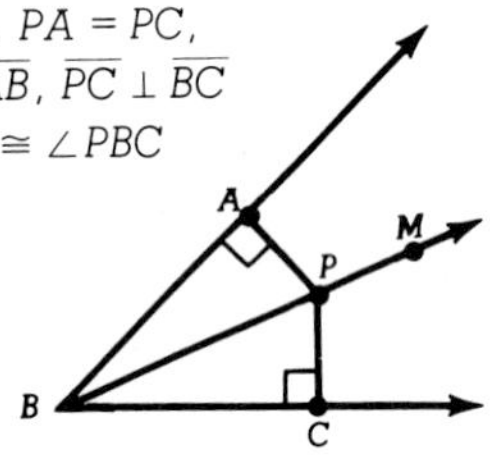

Proof:

STATEMENTS	REASONS
1. Draw m through P and B	1. Postulate 1–1
2. $PA = PC$	2. Given
3. $\overline{PA} \cong \overline{PC}$	3. Definition of congruence
4. $\overline{PA} \perp \overline{AB}$, $\overline{PC} \perp \overline{BC}$	4. Given
5. $\angle PAC$ and $\angle PCB$ are right angles	5. Theorem 3–11
6. $\overline{PB} \cong \overline{PB}$	6. Theorem 2–2
7. $\triangle PAB \cong \triangle PCB$	7. HL
8. $\angle PBA \cong \angle PBC$	8. Definition of congruent triangles
9. $\overline{PB}$ is the bisector of $\angle ABC$	9. Definition of angle bisector

Chapter Review Page 190 **1.** $n \parallel t$ **3.** t and m are skew **5.** $n \parallel t$ with ℓ as the transversal **7.** An isosceles triangle has two equal base angles. In indirect proof, assume that both such angles are obtuse. Then, the sum of the angles of the triangle would be greater than 180, which is a contradiction. Thus, the assumption is false. Therefore the isosceles triangle can have no obtuse base angles.

9.

STATEMENTS	REASONS
1. T is the midpoint of $\overline{QS}$ and $\overline{PR}$	1. Given
2. $\overline{QT} \cong \overline{TS}$, $\overline{PT} \cong \overline{TR}$	2. Theorem 2–5
3. $\angle 1 \cong \angle 3$	3. Theorem 3–10
4. $\triangle QTP \cong \triangle STR$	4. SAS
5. $\angle TSR \cong \angle TQP$	5. Definition of congruent triangles
6. $\overline{QP} \parallel \overline{RS}$	6. Theorem 6–5

11. auxiliary line; alternate interior angles

13. Given: $\overline{AC} \perp m$, $\overline{BD} \perp \ell$, $AC = BD$

Prove: $\ell \parallel m$

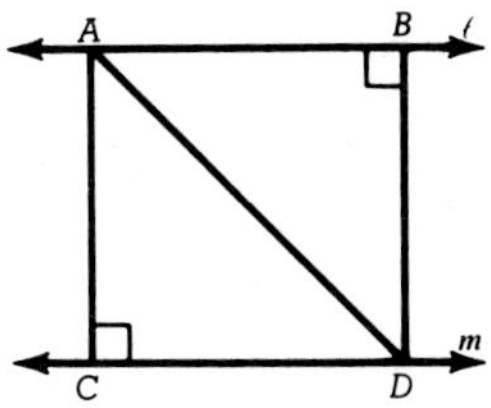

Proof:

STATEMENTS	REASONS
1. Draw $\overline{AD}$	1. Postulate 1–1
2. $\overline{AC} \perp m$, $\overline{BD} \perp \ell$	2. Given
3. $\angle ACD$ is a right angle, $\angle DBA$ is a right angle	3. Theorem 3–11
4. $\triangle ACD$ is a right triangle, $\triangle DBA$ is a right triangle	4. Definition of right triangle
5. $AC = BD$	5. Given
6. $\overline{AC} \cong \overline{BD}$	6. Definition of congruence of line segments
7. $\overline{AD} \cong \overline{AD}$	7. Theorem 2–2
8. $\triangle ACD \cong \triangle DBA$	8. HL
9. $\angle BAD \cong \angle CDA$	9. Definition of congruent triangles
10. $\ell \parallel m$	10. Theorem 6–5

CHAPTER 7 POLYGONS

Exploratory Exercises Page 196 1. no 3. no 5. yes 7. no

Written Exercises Page 197 1. triangle 3. decagon 5. pentagon 7. heptagon 9. 24-gon 11. n-gon 13. pentagon 15. hexagon 17. heptagon 19. convex 21. convex 23. non-convex 25. A, B, C, D, E 27. $\overline{AB}$, $\overline{BC}$, $\overline{CD}$, $\overline{DE}$, $\overline{EA}$ 29. pentagon 31. $\angle A$, $\angle B$ 33. Angles of a polygon are consecutive if and only if they have one side in common. 35. regular, it is both equilateral and equiangular 37. not regular, it is not equiangular

Exploratory Exercises Page 200 1. none 3. two 5. four 7. seven 9. $n - 3$ 11. two 13. four 15. six 17. thirteen 19. none 21. five 23. fourteen 25. thirty-five 27. $\frac{n(n-3)}{2}$

Written Exercises Page 201 1. 2700 3. 1800 5. 10,260 7. $(x - 2) \cdot 180$ 9. 12 11. 6 13. 24 15. 10 17. 90 19. 135 21. $\frac{1980}{13}$ 23. $\frac{(d-2) \cdot 180}{d}$ 25. 90 27. 45 29. $\frac{360}{13}$ 31. $\frac{360}{d}$ 33. 20 35. 12 37. 36 39. 90

41. (sum of measure of exterior angles) = (sum of measure of linear pairs) = (sum of measure of angles of quadrilateral)

If $n = 4$, sum $= 4 \cdot 180 - (4 - 2) \cdot 180$

sum $= 4 \cdot 180 - (4 \cdot 180) + (2 \cdot 180)$

sum $= 360$

Thus, the sum of the degree measures of the exterior angles, one at each vertex, of a convex quadrilateral is 360.

43. $\begin{pmatrix}\text{sum of} \\ \text{measure of two} \\ \text{sets of exterior} \\ \text{angles}\end{pmatrix} = 2 \cdot \begin{pmatrix}\text{sum of} & & \text{sum of} \\ \text{measure of} & - & \text{measure of} \\ \text{linear} & & \text{angles} \\ \text{pairs} & & \text{on } n\text{-gon}\end{pmatrix}$

sum $= 2 \cdot (n \cdot 180 - [(n - 2) \cdot 180])$

sum $= 2 \cdot [n \cdot 180 - (n \cdot 180) + 360]$

sum $= 2 \cdot 360$

sum $= 720$

Thus, the sum of the degree measure of the exterior angles, two at each vertex, of a convex n-gon is 720.

Exploratory Exercises Page 204 **1.** yes **3.** yes **5.** yes **7.** yes **9.** no **11.** yes **13.** no **15.** yes

Written Exercises Page 204 1. $x = 16$ 3. $x = 1.2$ 5. $x = 5$ 7. $x = 8$

9. Given: ▱$ABCD$

Prove: $\angle B \cong \angle D$
$\angle A \cong \angle C$

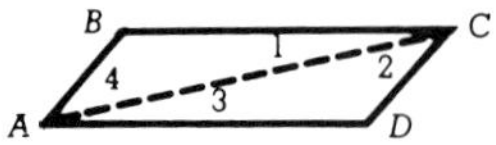

Proof:

STATEMENTS	REASONS
1. Draw diagonal $\overline{AC}$.	1. Postulate 1–1
2. $\overline{AC} \cong \overline{AC}$	2. Theorem 2–2
3. $\overline{AB} \parallel \overline{CD}$, $\overline{BC} \parallel \overline{AD}$	3. Definition of parallelogram
4. $\angle 1 \cong \angle 3$, $\angle 2 \cong \angle 4$	4. Theorem 6–9
5. $\triangle ABC \cong \triangle CDA$	5. Postulate 4–3
6. $\angle B = \angle D$	6. Corresponding parts of congruent triangles
7. $m\angle 1 = m\angle 3$, $m\angle 2 = m\angle 4$	7. Definition of angle congruence
8. $m\angle 1 + m\angle 2 = m\angle 3 + m\angle 4$	8. Postulate 2–7
9. $m\angle 1 + m\angle 2 = m\angle BCD$ $m\angle 3 + m\angle 4 = m\angle BAD$	9. Postulate 3–3
10. $m\angle BCD = m\angle BAD$	10. Postulate 2–9
11. $\angle BCD \cong \angle BAD$	11. Definition of angle congruence

11.

STATEMENTS	REASONS
1. ▱$PQST$ $\overline{RP}$ bisects $\angle QPT$ $\overline{VS}$ bisects $\angle QRT$	1. Given
2. $\angle Q \cong \angle T$ $\angle QST \cong \angle QPT$	2. Theorem 7–4
3. $m\angle QST = m\angle QPT$	3. Definition of angle congruence
4. $m\angle QST = m\angle QSV + m\angle VST$, $m\angle QPT = m\angle QPR + m\angle RPV$	4. Postulate 3–3
5. $m\angle QSV = m\angle VST$ $m\angle QPR = m\angle RPV$	5. Definition of angle bisector and angle congruence
6. $m\angle QST = 2m\angle VST$ $m\angle QPT = 2m\angle QPR$	6. Postulates 2–9 and 2–14.
7. $m\angle VST = m\angle QPR$	7. Postulates 2–9 and 2–8.
8. $\angle VST \cong \angle QPR$	8. Definition of angle congruence
9. $\overline{QP} \cong \overline{ST}$	9. Theorem 7–5
10. $\triangle QPR \cong \triangle TVS$	10. Postulate 4–3
11. $\angle QRP \cong \angle SVT$	11. Corresponding parts of congruent triangles
12. $\overline{QS} \parallel \overline{PT}$	12. Definition of parallelogram
13. $\angle QRP \cong \angle RPT$	13. Theorem 6–9
14. $\angle SVT \cong \angle RPT$	14. Theorem 3–1
15. $\overline{RP} \parallel \overline{VS}$	15. Theorem 6–4

13.

STATEMENTS	REASONS
1. ▱$ABCD$, $\overline{DE} \perp \overline{AC}$, $\overline{BF} \perp \overline{AC}$	1. Given
2. $\angle BFC$ is a right angle, $\angle DEA$ is a right angle	2. Theorem 3–11
3. $\angle BFC \cong \angle DEA$	3. Theorem 3–6
4. $\overline{AD} \parallel \overline{BC}$	4. Definition of parallelogram
5. $\angle DAE \cong \angle BCF$	5. Theorem 6–9
6. $\overline{AD} \cong \overline{BC}$	6. Theorem 7–4
7. $\triangle AED \cong \triangle CFB$	7. Theorem 4–5
8. $\overline{DE} \cong \overline{BF}$	8. Corresponding parts of congruent triangles

Exploratory Exercises Page 207 **1.** yes **3.** yes **5.** no **7.** No, congruent sides are adjacent not opposites. **9.** yes, by Theorem 7–9 **11.** No, diagonals should bisect each other. **13.** yes, by Theorem 6–6 **15.** yes, by Theorem 7–7

Written Exercises Page 208

1. Given: $\overline{RA} \cong \overline{TH}$, $\overline{AT} \cong \overline{RH}$
Prove: $\overline{RA} \parallel \overline{TH}$, $\overline{AT} \parallel \overline{RH}$

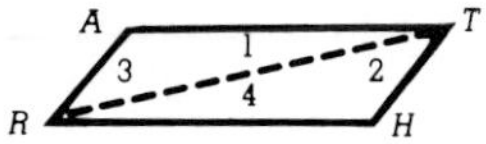

Proof:

STATEMENTS	REASONS
1. Draw diagonal $\overline{RT}$	1. Postulate 1–1
2. $\overline{RT} \cong \overline{RT}$	2. Theorem 2–2
3. $\overline{RA} \cong \overline{TH}$, $\overline{AT} \cong \overline{RH}$	3. Given
4. $\triangle ART \cong \triangle HTR$	4. Postulate 4–1
5. $\angle 1 \cong \angle 4$, $\angle 2 \cong \angle 3$	5. Corresponding parts of congruent triangles
6. $\overline{AT} \parallel \overline{RH}$, $\overline{RA} \parallel \overline{TH}$	6. Theorem 6–5
7. $RATH$ is a parallelogram	7. Definition of parallelogram

3. Given: $\angle A \cong \angle C$, $\angle B \cong \angle D$
Prove: $ABCD$ is a parallelogram

Proof:

STATEMENTS	REASONS
1. $\angle A \cong \angle C$, $\angle B \cong \angle D$	1. Given
2. $m\angle A = m\angle C$, $m\angle B = m\angle D$	2. Definition of congruent angles
3. $m\angle A + m\angle B = m\angle C + m\angle D$	3. Postulate 2–7
4. $m\angle A + m\angle B + m\angle C + m\angle D = 360$	4. Theorem 7–1
5. $m\angle A + m\angle B = 180$	5. Postulates 2–9, 2–14 and 2–8
6. $\overline{BC} \parallel \overline{AD}$	6. Theorem 6–6
7. $m\angle D + m\angle C = 180$	7. Postulates 2–9, 2–14 and 2–8
8. $m\angle B + m\angle C = 180$	8. Postulate 2–9
9. $\overline{AB} \parallel \overline{DC}$	9. Theorem 6–6
10. $ABCD$ is a parallelogram	10. Definition of a parallelogram

5.

STATEMENTS	REASONS
1. $\overline{AE} \cong \overline{EC}$, $\angle ECD \cong \angle EAB$	1. Given
2. $\angle AEB \cong \angle CED$	2. Theorem 3–10
3. $\triangle AEB \cong \triangle CED$	3. Postulate 4–3
4. $\overline{AB} \cong \overline{CD}$, $\angle BAE \cong \angle DCE$	4. Corresponding parts of congruent triangles
5. $\overline{AB} \parallel \overline{CD}$	5. Theorem 6–5
6. $ABCD$ is a parallelogram	6. Theorem 7–8

7.

STATEMENTS	REASONS
1. $\triangle PQR \cong \triangle STV$	1. Given
2. $\overline{PR} \cong \overline{SV}$	2. Corresponding parts of congruent triangles
3. $\overline{PR} \parallel \overline{VS}$	3. Given
4. $PRSV$ is a parallelogram	4. Theorem 7–8

9.

STATEMENTS	REASONS
1. $\square ABCD$	1. Given
2. $\overline{BC} \cong \overline{AD}$, $\overline{AB} \cong \overline{DC}$	2. Theorem 7–5
3. $BC = AD$, $AB = DC$	3. Definition of congruent line segments
4. E is the midpoint of $\overline{AB}$, F is the midpoint of $\overline{BC}$, G is the midpoint of $\overline{CD}$ H is the midpoint of $\overline{DA}$	4. Given
5. $\overline{BF} \cong \overline{FC}$, $\overline{AE} \cong \overline{EB}$ $\overline{CG} \cong \overline{GD}$, $\overline{DH} \cong \overline{HA}$	5. Theorem 2–5
6. $AE = EB$, $BF = FC$, $CG = GD$, $DH = HA$	6. Definition of congruent line segments
7. $AE = EB = \frac{1}{2}AB$ $BF = FC = \frac{1}{2}BC$ $CG = GD = \frac{1}{2}CD$ $DH = HA = \frac{1}{2}DA$	7. If E, F, G, and H are the midpoints of their respective segments, then each subsegment is one-half of the whole.
8. $BF = FC = DH = HA$ $AE = EB = CG = AH$	8. Postulate 2–9
9. $\overline{BF} \cong \overline{HD}$, $\overline{FC} \cong \overline{AH}$ $\overline{EB} \cong \overline{GD}$, $\overline{AE} \cong \overline{CG}$	9. Definition of congruent line segments
10. $\angle B \cong \angle D$, $\angle C \cong \angle A$	10. Theorem 7–4
11. $\triangle BFE \cong \triangle DHG$ $\triangle CGF \cong \triangle AEH$	11. Postulate 4–2
12. $\overline{EF} \cong \overline{GH}$, $\overline{FG} \cong \overline{HE}$	12. Corresponding parts of congruent triangles
13. $EFGH$ is a parallelogram	13. Theorem 7–7

11.

STATEMENTS	REASONS
1. D is the midpoint of $\overline{CF}$, D is the midpoint of $\overline{BE}$	1. Given
2. $\overline{CD} \cong \overline{FD}$, $\overline{ED} \cong \overline{BD}$	2. Theorem 2–5
3. $\angle CDB \cong \angle FDE$	3. Theorem 3–8
4. $\triangle CDB \cong \triangle FDE$	4. Postulate 4–2
5. $\angle DEF \cong \angle CBD$, $\overline{EF} \cong \overline{BC}$	5. Corresponding parts of congruent triangles
6. $\overline{AB} \cong \overline{BC}$	6. Given
7. $\overline{AB} \cong \overline{EF}$	7. Theorem 2–4
8. $\overline{AB} \parallel \overline{EF}$	8. Theorem 6–5
9. $ABEF$ is a parallelogram	9. Theorem 7–8

13.

STATEMENTS	REASONS
1. $\overline{VS} \parallel \overline{RP}$, $\angle PQV \cong \angle TSV$	1. Given
2. $\angle PQV \cong \angle QVS$	2. Theorem 6–9
3. $\angle TSV \cong \angle QVS$	3. Theorem 3–1
4. $\overline{RT} \parallel \overline{QV}$	4. Theorem 6–5
5. $VQRS$ is a parallelogram	5. Definition of parallelogram

15.

STATEMENTS	REASONS
1. $\square BCDG$, $\angle E \cong \angle A$ $\overline{CE} \parallel \overline{AF}$	1. Given
2. $\angle A$ is supplementary to $\angle C$	2. Theorem 6–10
3. $\angle E$ is supplementary to $\angle C$	3. Substitution
4. $\overline{AC} \parallel \overline{EF}$	4. Theorem 6–6
5. $ACEF$ is a parallelogram	5. Definition of parallelogram

Exploratory Exercises Page 212 **1.** rectangles **3.** parallelograms **5.** rhombuses **7.** parallelograms **9.** rectangles **11.** squares

Written Exercises Page 212 **1.** $10\frac{7}{8}$ **3.** $TP = 52.1$ **5.** $SV = 13\frac{1}{2}$ **7.** $m\angle RSV = 56.3$ **9.** $m\angle RVT = 51.8$
11. $m\angle SPR = 90$
13. Given: $SQRE$ is a square, $\overline{SR}$ and $\overline{QE}$ are diagonals
Prove: $\overline{SR}$ bisects $\overline{QE}$

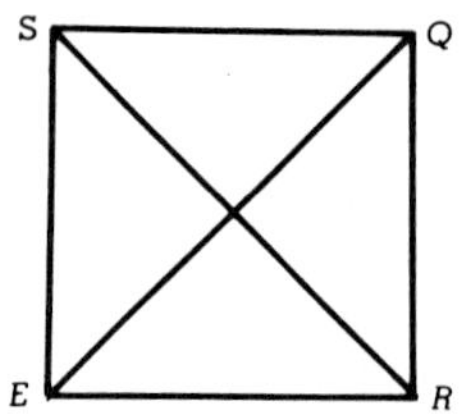

Proof:

STATEMENTS	REASONS
1. $SQRE$ is a square	1. Given
2. $\overline{SQ} \cong \overline{ER}$ and $\overline{QR} \cong \overline{SE}$	2. Definition of a square
3. $SQRE$ is a parallelogram	3. Theorem 7–7
4. $\overline{SR}$ bisects $\overline{QE}$	4. Theorem 7–6

15.

STATEMENTS	REASONS
1. $\square PRTW$ is a square Q is the midpoint of $\overline{PR}$ S is the midpoint of $\overline{RT}$ V is the midpoint of $\overline{TW}$ X is the midpoint of $\overline{PW}$	1. Given
2. $\overline{RT} \cong \overline{PW}$, $\overline{RP} \cong \overline{TW}$	2. Definition of a square
3. $\overline{RS} \cong \overline{ST}$, $\overline{PX} \cong \overline{XW}$ $\overline{RQ} \cong \overline{QP}$, $\overline{TV} \cong \overline{VW}$	3. Theorem 2–5
4. S lies between R and T, Q lies between R and P, V lies between T and W, X lies between P and W	4. Given
5. $RS + ST = RT$ $TV + VW = TW$ $PX + XW = PW$ $RQ + QP = RP$	5. Definition of betweenness
6. $RS = \frac{1}{2}RT$, $PX = \frac{1}{2}PW$ $RQ = \frac{1}{2}RP$, $TV = \frac{1}{2}TW$	6. Meaning of congruence, substitution, and division, property of equality
7. $\overline{RS} \cong \overline{PX}$, $\overline{RQ} \cong \overline{TV}$	7. Substitution, meaning of congruence
8. $\overline{RT} \parallel \overline{PW}$, $\overline{RP} \parallel \overline{TW}$	8. Definition of square
9. $RSXP$ is a parallelogram $RQVT$ is a parallelogram	9. Theorem 7–8
10. $\angle P$, $\angle R$, $\angle T$, and $\angle W$ are right angles	10. Definition of square
11. $\angle S \cong \angle P$, $\angle Q \cong \angle T$	11. Theorem 7–4
12. $\angle S$ is a right angle $\angle Q$ is a right angle	12. Substitution
13. $m\angle R = m\angle S = m\angle Q = 90$	13. Definition of right angles
14. $m\angle R + m\angle S + m\angle Q + m\angle QYS = 360$	14. Theorem 7–1
15. $m\angle QYS = 90$	15. Postulates 2–9 and 2–7
16. $\angle QYS$ is a right angle	16. Definition of right angle
17. $\overline{QV} \perp \overline{SX}$	17. Definition of perpendicular lines

17. Given: $\square MFPQ$, $\overline{MP} \cong \overline{FQ}$
Prove: $MFPQ$ is a rectangle

Proof:

STATEMENTS	REASONS
1. $\square MFPQ$, $\overline{MP} \cong \overline{FQ}$	1. Given
2. $\overline{FM} \cong \overline{PQ}$	2. Theorem 7–5
3. $\overline{FP} \cong \overline{FP}$	3. Theorem 2–2
4. $\triangle FMP \cong \triangle PFQ$	4. Postulate 4–1
5. $\angle F \cong \angle P$	5. Corresponding parts of congruent triangles
6. $m\angle F = m\angle P$	6. Definition of angle congruence
7. $FM \parallel PQ$	7. Definition of parallelogram
8. $\angle F$ is supplementary to $\angle P$	8. Theorem 6–6
9. $m\angle F + m\angle P = 180$	9. Definition of supplementary angles.
10. $m\angle F = m\angle P = 90$	10. Postulate 2–9 and 2–8
11. $\angle F$ and $\angle P$ are right angles	11. Definition of right angles
12. $\angle F \cong \angle Q$, $\angle P \cong \angle M$	12. Theorem 7–4
13. $\angle Q$ and $\angle M$ are right angles	13. Substitution
14. $MFPQ$ is a rectangle	14. Definition of rectangle

19. Given: $\square FSTR$, $\overline{SR}$ bisects $\angle S$ and $\angle R$
$\overline{FT}$ bisects $\angle F$ and $\angle T$
Prove: $FSTR$ is a rhombus

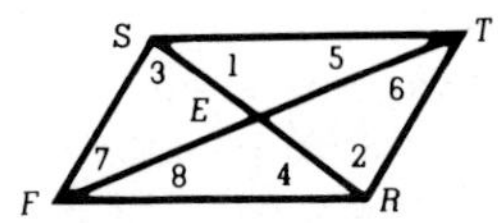

Proof:

STATEMENTS	REASONS
1. $\square FSTR$, $\overline{SR}$ bisects $\angle S$ and $\angle R$, $\overline{FT}$ bisects $\angle F$ and $\angle T$	1. Given

2. $\overline{SE} \cong \overline{ER}$	2. Theorem 7–6
3. $\angle S \cong \angle R$	3. Theorem 7–4
4. $m\angle 1 + m\angle 3 = m\angle S$ $m\angle 2 + m\angle 4 = m\angle R$	4. Postulate 3–3
5. $\angle 1 \cong \angle 3$, $\angle 2 \cong \angle 4$, $\angle 5 \cong \angle 6$	5. Definition of angle bisector
6. $m\angle 1 = m\angle 3$, $m\angle 2 = m\angle 4$, $m\angle S = m\angle R$	6. Definition of angle congruence
7. $m\angle 1 = \frac{1}{2}m\angle S$ $m\angle 2 = \frac{1}{2}m\angle R$	7. Substitution and division of equality
8. $m\angle 1 = m\angle 2$	8. Postulate 2–9
9. $\angle 1 \cong \angle 2$	9. Definition of angle congruence
10. $\triangle SET \cong \triangle RET$	10. Theorem 4–5
11. $\overline{ST} \cong \overline{TR}$	11. Corresponding parts of congruent triangles
12. $\overline{ST} \cong \overline{FR}$, $\overline{SF} \cong \overline{TR}$	12. Theorem 7–5
13. $\overline{ST} \cong \overline{TR} \cong \overline{SF} \cong \overline{FR}$	13. Theorem 7–4
14. *FSTR* is a rhombus	14. Definition of rhombus

21. Given: ▭*ABCD* is a rectangle
Prove: $\overline{AC} \cong \overline{BD}$

Proof:

STATEMENTS	REASONS
1. ▭*ABCD* is a rectangle	1. Given
2. $\overline{AB} \cong \overline{CD}$	2. Definition of rectangle
3. $\angle BAD$ and $\angle CDA$ are right angles	3. Definition of rectangle
4. $\angle BAD \cong \angle CDA$	4. Theorem 3–6
5. $\overline{AD} \cong \overline{AD}$	5. Theorem 2–2
6. $\triangle ABD \cong \triangle DCA$	6. Postulate 4–2
7. $\overline{AC} \cong \overline{BD}$	7. Corresponding parts of congruent triangles

Exploratory Exercises Page 216 1. yes 3. yes 5. yes 7. no 9. yes 11. no

Written Exercises Page 216 1. 12 cm 3. 3 ft 5. 7 mm 7. 89.5 cm **9.** $20\frac{3}{4}$ mi 11. 48.4 m **13.** 8.2 **15.** 13.5 17. $m\angle CDA = 115$ 19. $m\angle DAB = 132$ 21. $m\angle ECB = 25$ 23. $m\angle ACB = 27$ 25. $m\angle DCA = 23$ 27. $m\angle ECB = 27$

29.

STATEMENTS	REASONS
1. $\overline{BC} \parallel \overline{AD}$, $\overline{AB} \cong \overline{DC}$ $\overline{BF} \perp \overline{AD}$, $\overline{CE} \perp \overline{AD}$	1. Given
2. $\angle AFB$ and $\angle DEC$ are right angles	2. Theorem 3–11
3. *ABCD* is an isosceles trapezoid	3. Definition of isosceles trapezoid
4. $\angle A \cong \angle D$	4. Theorem 7–14
5. $\triangle AFB \cong \triangle DEC$	5. Theorem 4–11
6. $\overline{BF} \cong \overline{CE}$	6. Congruent parts of congruent triangles

31.

STATEMENTS	REASONS
1. $\overline{BC} \parallel \overline{AD}$, $\overline{AB} \cong \overline{DC}$ $\overline{BF} \perp \overline{AD}$, $\overline{CE} \perp \overline{AD}$	1. Given
2. $\angle AFB$ is a right angle $\angle DEC$ is a right angle	2. Theorem 3–11
3. *ABCD* is an isosceles trapezoid	3. Definition of isosceles trapezoid
4. $\angle A \cong \angle D$	4. Theorem 7–14
5. $\triangle ABF \cong \triangle DCE$	5. Theorem 4–11

33.

STATEMENTS	REASONS
1. $\overline{RT} \parallel \overline{PW}$, $\overline{TX} \parallel \overline{RP}$, *Q* is the midpoint of $\overline{PR}$, *V* is the midpoint of $\overline{SW}$	1. Given
2. $\overline{QV}$ is the median of trapezoid *PRSW*	2. Definition of median
3. $\overline{QV} \parallel \overline{PW}$, $\overline{QV} \parallel \overline{RT}$	3. Theorem 7–16
4. *QRTV* is a parallelogram *QPXV* is a parallelogram	4. Definition of parallelogram
5. $\overline{QR} \cong \overline{VT}$, $\overline{QP} \cong \overline{VX}$	5. Theorem 7–5
6. $\overline{QR} \cong \overline{QP}$	6. Theorem 2–5
7. $\overline{VT} \cong \overline{VX}$	7. Theorem 2–4
8. *V* is the midpoint of $\overline{TX}$	8. Definition of midpoint

35.

STATEMENTS	REASONS
1. $\overline{RT} \parallel \overline{PW}$, $\overline{TX} \parallel \overline{RP}$ *V* is the midpoint of $\overline{SW}$ *Q* is the midpoint of $\overline{PR}$	1. Given
2. $\overline{QV}$ is the median of trapezoid *PRSW*	2. Definition of median
3. $\overline{QV} \parallel \overline{RS}$	3. Theorem 7–16
4. *QRTV* is a parallelogram	4. Definition of parallelogram

37.

STATEMENTS	REASONS
1. $\overline{RT} \parallel \overline{PW}$, $\overline{TX} \parallel \overline{RP}$, Q is midpoint of $\overline{PR}$, V is the midpoint of $\overline{SW}$	1. Given
2. $\overline{QV}$ is the median of trapezoid $PRSW$	2. Definition of median
3. $\overline{QV} \parallel \overline{RS}$	3. Theorem 7–16

Exploratory Exercises Page 220 **1.** 46 ft **3.** 52 ft **5.** 59 ft **7.** 60 ft **9.** 53 ft

Written Exercises Page 220 **1.** 10 m **3.** $8\frac{3}{4}$ **5.** 163.6 mi **7.** 55.6 cm **9.** 20 in. **11.** 3.9 mi **13.** 2.5 in. **15.** $2\frac{1}{12}$ yd **17.** 72 ft **19.** 99.6 mm **21.** 30 yds **23.** 2.4 m **25.** 3.1 cm **27.** 0.7 km **29.** 214 ft

Explorafory Exercises Page 223 **1.** faces: $\triangle ABD$, $\triangle ADC$, $\triangle BDC$, $\triangle ABC$; edges: $\overline{AB}$, $\overline{BC}$, $\overline{BA}$, $\overline{BD}$, $\overline{AD}$, $\overline{DC}$; vertices: A, B, C, D **3.** faces: polygons $ABFE$, $BFGC$, $AEHD$, $EFGH$, $ABCD$, $DHGC$; edges: $\overline{AB}$, $\overline{BC}$, $\overline{CD}$, $\overline{AD}$; $\overline{EF}$, $\overline{FG}$, $\overline{GH}$, $\overline{HE}$, $\overline{AE}$, $\overline{BF}$, $\overline{GC}$, $\overline{HD}$; vertices: A, B, C, D, E, F, G, H **5.** 3 **7.** 3 **9.** 240

Written Exercises Page 223 **1.** triangles **3.** triangles **5.** 60 **9.** 240 **11.** 270 **13.** 450 **15.** 360 **17.** yes **19.** yes **21.** no **23.** no **25.** 4 faces **27.** 6 edges **29.** 8 vertices **31.** 8 faces **33.** 12 edges **35.** $6 + 8 = 12 + 2$

Chapter Review Page 226 **1.** Q, R, S, T, U **3.** $\angle Q$, $\angle R$, $\angle S$, $\angle T$, $\angle U$ **5.** non-convex **7.** 720 **9.** 1800 **11.** 108 **13.** 40 **15.** $BA = 7.2$ **17.** $x = 3$ **19.** $EC = 8.2$

21.

STATEMENTS	REASONS
1. $\square ABCD$, $\overline{AE} \cong \overline{CF}$	1. Given
2. $\overline{AB} \cong \overline{CD}$	2. Theorem 7–5
3. $\angle BAC \cong \angle ACD$	3. Theorem 6–9
4. $\triangle ABE \cong \triangle CDF$	4. Postulate 4–2
5. $\overline{BE} \cong \overline{FD}$, $\angle AEB \cong \angle CFD$	5. Congruent parts of congruent triangles
6. $\angle BEF$ is supplementary to $\angle AEB$; $\angle EFD$ is supplementary to $\angle CFD$	6. They form linear pairs
7. $\angle BEF \cong \angle EFD$	7. Theorem 3–3
8. $\overline{BE} \parallel \overline{FD}$	8. Theorem 6–5
9. $BFDE$ is a parallelogram	9. Theorem 7–8

23. no **25.** yes **27.** $47\frac{1}{2}$

29.

STATEMENTS	REASONS
1. $\overline{BC} \parallel \overline{AD}$, $\overline{AB} \cong \overline{DC}$, $\overline{AB} \nparallel \overline{CD}$	1. Given
2. $ABCD$ is an isosceles trapezoid	2. Definition of isosceles trapezoid
3. $\overline{BD} \cong \overline{AC}$	3. Theorem 7–13
4. $\overline{AD} \cong \overline{AD}$	4. Theorem 2–2
5. $\triangle ABD \cong \triangle ACD$	5. Postulate 4–1
6. $\angle CAD \cong \angle BDA$	6. Congruent parts of congruent triangles
7. $\overline{AE} \cong \overline{ED}$	7. Theorem 4–9
8. $\triangle AED$ is an isosceles triangle	8. Definition of isosceles triangle

31. 16 m **33.** 855 cm

CHAPTER 8 SIMILARITY

Exploratory Exercises Page 232 **1.** 0.517 **3.** 0.655 **5.** 0.286 **7.** 0.519 **9.** 0.769 **11.** 0.538 **13.** 0.50 **15.** 0.50 **17.** 0.33 **19.** 0.33 **21.** 0.40 **23.** 0.55 **25.** 0.57

Written Exercises Page 232 **1.** $11 \cdot 24 = 12x$ **3.** $6t = 18 \cdot 5$ **5.** $1.2 \cdot 4 = 1.6k$ **7.** $8b = 3.24 \cdot 1$ **9.** $3x = (30 - x) \cdot 2$ **11.** $(3 - n) = 2(n + 1)$ **13.** $x = 22$ **15.** $t = 15$ **17.** $k = 3$ **19.** $b = 0.405$ **21.** $x = 12$ **23.** $n = \frac{1}{3}$ **25.** $\frac{5}{8} = \frac{n}{100}$ **27.** $\frac{1}{3} = \frac{n}{100}$ **29.** $\frac{3}{8} = \frac{n}{100}$ **31.** $\frac{5}{12} = \frac{n}{100}$ **33.** $\frac{4}{5} = \frac{n}{100}$ **35.** $\frac{1}{10} = \frac{n}{100}$ **37.** 62.5% **39.** $33.\overline{3}\%$ **41.** 37.5% **43.** $41.\overline{6}\%$

45. 80% **47.** 10% **49.** 75% **51.** 6% **53.** 37.5 g

55. Given: $ad = bc$, b and d are nonzero

Prove: $\frac{a}{b} = \frac{c}{d}$

Proof:

STATEMENTS	REASONS
1. $ad = bc$, b and d are nonzero numbers.	1. Given
2. $\frac{ad}{bd} = \frac{bc}{bd}$	2. Postulate 2–8
3. $\frac{a}{b} \cdot \frac{d}{d} = \frac{b}{b} \cdot \frac{c}{d}$	3. Multiplication property of fractions
4. $\frac{a}{b} = \frac{c}{d}$	4. Postulates 2–9, 2–12

Exploratory Exercises Page 237 **1.** Subtraction property of proportions **3.** Addition property of proportions **5.** Theorem 8–3 **7.** Summation property of proportions **9.** Theorem 8–2 **11.** Theorem 8–3 **13.** Theorem 8–2 **15.** Theorem 8–3

Written Exercises Page 237

1. Given: $\frac{a}{b} = \frac{c}{d}$

Prove: $\frac{a}{c} = \frac{b}{d}$

Proof:

STATEMENTS	REASONS
1. $\frac{a}{b} = \frac{c}{d}$	1. Given
2. $ad = bc$	2. Theorem 8–1
3. $\frac{ad}{cd} = \frac{bc}{cd}$	3. Postulate 2–8
4. $\frac{a}{c} = \frac{b}{d}$	4. Postulates 2–9, 2–12

5. Given: $\frac{a}{b} = \frac{a+c}{b+d}$

Prove: $\frac{a}{b} = \frac{c}{d}$

Proof:

STATEMENTS	REASONS
1. $\frac{a}{b} = \frac{a+c}{b+d}$	1. Given
2. $a(b+d) = b(a+c)$	2. Theorem 8–1
3. $ab + ad = ba + bc$	3. Postulate 2–14
4. $(ab + ad) - ab = (ab + bc) - ab$	4. Postulates 2–10, 2–7
5. $ad = bc$	5. Postulates 2–11, 2–13, 2–12, 2–9
6. $\frac{a}{b} = \frac{c}{d}$	6. Theorem 8–1

3. Given: $\frac{a}{b} = \frac{c}{d}$

Prove: $\frac{a-b}{b} = \frac{c-d}{d}$

Proof:

STATEMENTS	REASONS
1. $\frac{a}{b} = \frac{c}{d}$	1. Given
2. $ad = cb$	2. Theorem 8–1
3. $ad - bd = cb - bd$	3. Postulate 2–7
4. $(a-b)d = b(c-d)$	4. Postulates 2–14, 2–9
5. $\frac{a-b}{b} = \frac{c-d}{d}$	5. Theorem 8–1

7. $\frac{6}{5}$ **9.** 5 **11.** 13

Exploratory Exercises Page 240 **1.** $\angle A \cong \angle D$, $\angle B \cong \angle E$, $\angle C \cong \angle F$, $\frac{AB}{DE} = \frac{BC}{EF} = \frac{AC}{DF}$ **3.** $\angle R \cong \angle L$, $\angle S \cong \angle M$, $\angle T \cong \angle N$, $\angle V \cong \angle O$, $\angle W \cong \angle P$, $\angle X \cong \angle Q$, $\frac{RS}{LM} = \frac{ST}{MN} = \frac{TV}{NO} = \frac{VW}{OP} = \frac{WX}{PQ} = \frac{XR}{QL}$ **5.** yes **7.** yes

Written Exercises Page 241 **1.** yes **3.** no **5.** no **7.** no **9.** no **11.** $\frac{5}{3}$ **13.** 15 **15.** $1\frac{1}{3}$ **17.** 35

19. Given: $\triangle ABC$
Prove: $\triangle ABC \sim \triangle ABC$

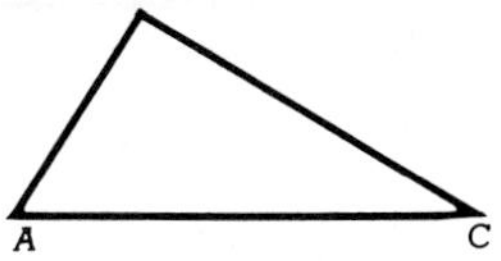

Proof:

STATEMENTS	REASONS
1. $\triangle ABC$	1. Given
2. $\angle A \cong \angle A$, $\angle B \cong \angle B$ $\angle C \cong \angle C$	2. Theorem 3–1
3. $\overline{AB} \cong \overline{AB}$, $\overline{BC} \cong \overline{BC}$, $\overline{AC} \cong \overline{AC}$	3. Theorem 2–2
4. $AB = AB$, $BC = BC$ $AC = AC$	4. Definition of congruent segments
5. $\frac{AB}{AB} = 1$, $\frac{BC}{BC} = 1$ $\frac{AC}{AC} = 1$	5. Postulate 2–8
6. $\triangle ABC \sim \triangle ABC$	6. Definition of similar triangles

21. Given: $\triangle ABC \sim \triangle RST$, $\triangle RST \sim \triangle PQM$
Prove: $\triangle ABC \sim \triangle PQM$

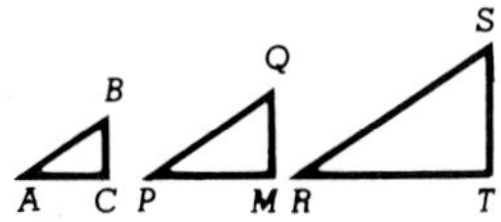

Proof:

STATEMENTS	REASONS
1. $\triangle ABC \sim \triangle RST$, $\triangle RST \sim \triangle PQM$	1. Given
2. $\angle A \cong \angle R$, $\angle B \cong \angle S$, $\angle C \cong \angle T$, $\angle R \cong \angle P$, $\angle S \cong \angle Q$, $\angle T \cong \angle M$	2. Definition of similar polygons
3. $\angle A \cong \angle P$, $\angle B \cong \angle Q$ $\angle C \cong \angle M$	3. Theorem 3–1
4. $\frac{AB}{RS} = \frac{BC}{ST} = \frac{CA}{TR}$, $\frac{RS}{PQ} = \frac{ST}{QM} = \frac{TR}{MP}$	4. Postulate 2–8
5. $\dfrac{\frac{AB}{RS}}{\frac{PQ}{RS}} = \dfrac{\frac{BC}{ST}}{\frac{QM}{ST}} = \dfrac{\frac{CA}{TR}}{\frac{MP}{TR}}$	5. Postulate 2–8
6. $\frac{AB}{PQ} = \frac{BC}{QM} = \frac{CA}{MP}$	6. Simplification
7. $\triangle ABC \sim \triangle PQM$	7. Definition of similar polygons

23. 8 m **25.** 240 m

Exploratory Exercises Page 245 **1.** yes **3.** no **5.** yes

Written Exercises Page 246 **1.** 11

3.

STATEMENTS	REASONS
1. $\angle D$ is a right angle, $\overline{BE} \perp \overline{AC}$	1. Given
2. $\angle ABE$ is a right angle.	2. Theorem 3–11
3. $\angle A \cong \angle A$	3. Theorem 3–1
4. $\angle D \cong \angle ABE$	4. Theorem 3–6
5. $\triangle ADC \sim \triangle ABE$	5. Postulate 8–1

5.

STATEMENTS	REASONS
1. $\angle Q$ is a right angle; $WSTV$ is a square.	1. Given
2. $m\angle VWS = m\angle WST = m\angle STV = m\angle TVW = 90$	2. Definition of a square
3. $\angle VWS$ and $\angle TVW$ are right angles.	3. Definition of right angles
4. $\angle PWS$ and $\angle RVT$ are right angles.	4. Theorem 3–9
5. In $\triangle PQR$, $\angle P$ is complementary to $\angle R$.	5. Theorem 4–3
6. In $\triangle PSW$, $\angle P$ is complementary to $\angle S$.	6. Theorem 4–3
7. $\angle S \cong \angle R$	7. Theorem 3–4
8. $\angle PWS \cong \angle RVT$	8. Theorem 3–6
9. $\triangle PWS \sim \triangle TVR$	9. AA

7. Given: $\frac{AB}{PQ} = \frac{BC}{QM}$,

$\angle B \cong \angle Q$

Prove: $\triangle ABC \sim \triangle PQM$

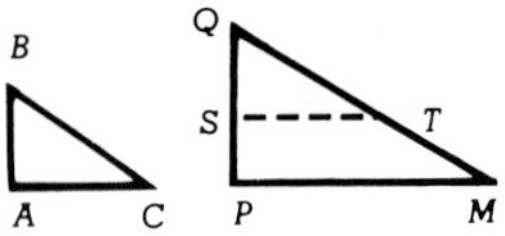

Proof:

STATEMENTS	REASONS
1. Draw $\overline{ST} \parallel \overline{PM}$ so that $\overline{SQ} \cong \overline{AB}$.	1. Postulate 6–1
2. $\angle Q \cong \angle Q$	2. Theorem 3–1
3. $\angle QST \cong \angle QPM$	3. Theorem 6–8
4. $\triangle SQT \sim \triangle PQM$	4. AA
5. $\frac{SQ}{PQ} = \frac{QT}{QM}$	5. Definition of similar triangles
6. $\frac{AB}{PQ} = \frac{BC}{QM}$	6. Given
7. $SQ = AB$	7. Definition of congruent segments
8. $\frac{SQ}{PQ} = \frac{BC}{QM}$	8. Postulate 2–9
9. $SQ \cdot QM = PQ \cdot QT$, $SQ \cdot QM = PQ \cdot BC$	9. Theorem 8–1
10. $PQ \cdot QT = PQ \cdot BC$	10. Postulate 2–9
11. $QT = BC$	11. Postulate 2–8
12. $\overline{QT} \cong \overline{BC}$	12. Definition of congruent segments
13. $\angle B \cong \angle Q$	13. Given
14. $\triangle ABC \cong \triangle SQT$	14. SAS
15. $\triangle ABC \sim \triangle PQM$	15. Substitution

Exploratory Exercises Page 250 1. yes 3. no 5. no 7. yes 9. no 11. yes 13. no 15. no

Written Exercises Page 250 1. no 3. no 5. 3 7. 6.84 9. 3

11. Given: $a \parallel b$, $b \parallel c$,

a, b and c intersect with t and m.

Prove: $\frac{FE}{ED} = \frac{GH}{AK}$

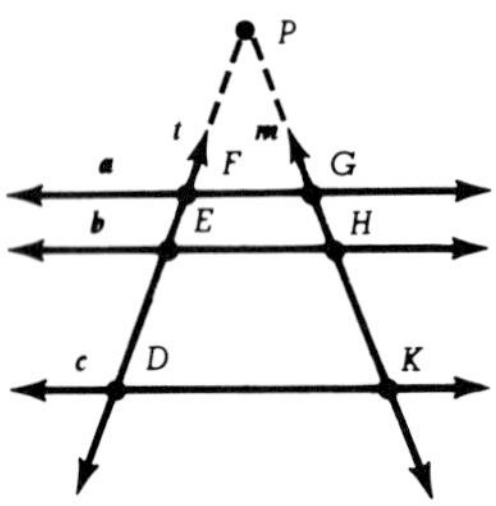

Proof:

STATEMENTS	REASONS
1. Extend t and m to meet at P, forming $\triangle PDK$.	1. Non-parallel lines meet at a point.
2. $a \parallel b$, $b \parallel c$	2. Given
3. $\frac{PE}{ED} = \frac{PH}{HK}$	3. Theorem 8–9
4. $\frac{PF}{FE} = \frac{PG}{GH}$	4. Theorem 8–9
5. $\frac{PF + FE}{FE} = \frac{PG + GH}{GH}$	5. Theorem 8–4
6. $PF + FE = PE$, $PG + GH = PH$	6. Definition of betweenness
7. $\frac{PE}{FE} = \frac{PH}{GH}$	7. Postulate 2–9
8. $\dfrac{\frac{PE}{ED}}{\frac{PE}{FE}} = \dfrac{\frac{PH}{HK}}{\frac{PH}{GH}}$	8. Postulate 2–9
9. $\frac{FE}{ED} = \frac{GH}{HK}$	9. Simplification

13. Since Q and S are midpoints, show that $\overline{QS}$ gives congruent segments on each side of the triangle. Therefore, it cuts off segments proportional to the two sides. Thus, $\overline{QS} \parallel \overline{PT}$ by Theorem 8–8. **15.** By extension, $\overline{QS} \cong \overline{SV}$. $\angle QRS \cong \angle VSR$ because they are vertical angles. $\overline{SR} \cong \overline{ST}$ since S is the midpoint. Congruence occurs by SAS. **17.** Since $\triangle RSQ \cong \triangle STV$ then $\overline{VT} \cong \overline{QR}$ and $\angle V \cong \angle RQS$. Therefore, by alternate interior angles $\overline{QP} \parallel \overline{VT}$. Since $\overline{QR} \cong \overline{QR}$ then $\overline{VT} \cong \overline{QP}$ by the transitive property. In polygon $PQVT$, opposite sides are parallel and equal. Thus, $PQVT$ is a parallelogram, and $\overline{QV} \cong \overline{PT}$.

Exploratory Exercises **Page 254** **1.** yes **3.** no **5.** yes **7.** no **9.** yes **11.** no

Written Exercises **Page 254** **17.** $x = 6$ **19.** $x = 44$ **21.** 5.1 **23.** 11.96 **25.** Show $\triangle ABD \sim \triangle PQR$. Since $\triangle ABC \sim \triangle PQM$, $\angle A \cong \angle P$. Let $\overline{BD}$ and $\overline{QR}$ be the altitudes, then $\angle D$ and $\angle R$ are right angles. Therefore, $\triangle ABD \sim \triangle PQR$ by AA. Thus, the sides are proportional. Then, use the similarity of $\triangle ABC$ and $\triangle PQM$ to extend the proportion to all sides of the original triangle. **27.** Show that $\triangle ADC \sim \triangle PMR$. Since $\triangle ABC \sim \triangle PQR$, $\angle C \cong \angle R$ and $\angle CAB \cong \angle RPQ$. Since the latter are bisected, $\angle CAD \cong \angle RPM$. Thus the triangles are similar. Therefore, $\frac{AD}{PM} = \frac{AC}{PR}$. Because $\triangle ABC \sim \triangle PQR$, the corresponding sides are also proportional.

Exploratory Exercises **Page 258** **1.** 4.5 cm **3.** 11 m

Written Exercises **Page 258** **1.** 48 ft **3.** 2.7 miles **5.** 0.78 cm

Chapter Review **Page 261** **1.** $15 = 2x$ **3.** $x + 5 = 5x$ **5.** $x = 7.5$ **7.** $x = \frac{5}{4}$

9. Given: $\frac{a}{b} = \frac{c}{d}$

Prove: $\frac{a-b}{c-d} = \frac{b}{d}$

Proof:

STATEMENTS	REASONS
1. $\frac{a}{b} = \frac{c}{d}$	1. Given
2. $\frac{a-b}{b} = \frac{c-d}{d}$	2. Theorem 8–5
3. $\frac{a-b}{c-d} = \frac{b}{d}$	3. Theorem 8–2

13.

STATEMENTS	REASONS
1. $\frac{PR}{QS} = \frac{RS}{QP}$; trapezoid $PQRS$	1. Given
2. $\angle QRP \cong \angle SPR$ and $\angle RQS \cong \angle PSQ$	2. Theorem 6–9
3. $\triangle PTS \sim \triangle RTQ$	3. Postulate 8–1 (AA Similarity)
4. $\frac{TS}{TQ} = \frac{TP}{TR}$	4. Definition of Similarity
5. $\angle QTP \cong \angle RTS$	5. Theorem 3–10
6. $\triangle QTP \sim \triangle RTS$	6. Theorem 8–8 (SAS Similarity)
7. $\angle QPR \cong \angle RSQ$	7. Definition of Similarity
8. $\triangle PQR \sim \triangle SRQ$	8. Theorem 8–8 (SAS Similarity)

15. $x = 4$ **17.** about 3.214 **19.** 20.125

CHAPTER 9 RIGHT TRIANGLES

Exploratory Exercises **Page 267** **1.** 64 **3.** 49 **5.** 0.04 **7.** $\frac{4}{25}$ **9.** 1.69 **11.** 8 **13.** $^-7$ **15.** 0.2 **17.** ± 1.3 **19.** $\frac{2}{5}$

Written Exercises **Page 267** **1.** 1 **3.** 0.5 **5.** 0.3 **7.** $\frac{7}{2}$ **9.** 13 **11.** 16 **13.** 30 **15.** $4\sqrt{5}$ **17.** $5\sqrt{3}$ **19.** $5\sqrt{3}$ **21.** $\frac{\sqrt{5}}{2}$ **23.** $\frac{\sqrt{2}}{4}$ **25.** $\frac{5\sqrt{3}}{3}$ **27.** $\frac{\sqrt{15}}{3}$ **29.** $\frac{\sqrt{6}}{2}$ **31.** rational **33.** rational **35.** irrational **37.** irrational **39.** irrational **41.** irrational **43.** irrational **45.** irrational **47.** irrational **49.** irrational **51.** 8.718 **53.** 4.123 **55.** $^-1.015$ **57.** 1.871 **59.** 14.142 **61.** ± 6.325 **63.** ± 6.928 **65.** ± 5 **67.** ± 6 **69.** ± 6.708

Exploratory Exercises **Page 270** **1.** 4 **3.** 6 **5.** 12 **7.** 6 **9.** 4 **11.** 12 **13.** 9 **15.** 8

Written Exercises **Page 270** **1.** $\sqrt{15}$ **3.** $2\sqrt{10}$ **5.** 12 **7.** $4\sqrt{3}$ **9.** $5\sqrt{3}$ **11.** $3\sqrt{6}$ **13.** $\frac{3}{2}$ **15.** 1 **17.** $\frac{\sqrt{3}}{4}$ **19.** $\frac{\sqrt{2}}{4}$ **21.** 5.477 **23.** 4.899 **25.** 5.292

27. Given: $\triangle PQR$ is a right triangle. $\overline{QS}$ is the altitude to $\overline{PR}$, the hypotenuse

Prove: $\frac{PS}{QS} = \frac{QS}{SR}$

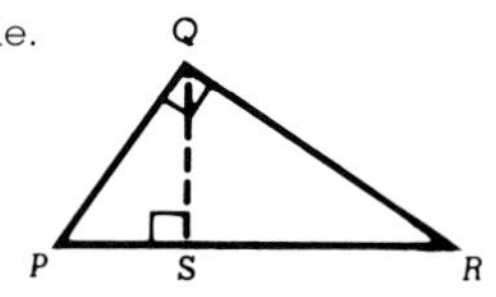

Proof:

STATEMENTS	REASONS
1. $\triangle PQR$ is a right triangle. $\overline{QS}$ is the altitude to $\overline{PR}$, the hypotenuse	1. Given
2. $\triangle PQS \sim \triangle QRS$	2. Theorem 9–1
3. $\frac{PS}{QS} = \frac{QS}{SR}$	3. Definition of similar triangles
4. QS is the geometric mean between segments of the hypotenuse.	4. Definition of geometric mean

Exploratory Exercises **Page 275** **1.** no **3.** yes **5.** yes **7.** yes **9.** yes **11.** yes **13.** yes **15.** no **17.** no **19.** yes **21.** yes **23.** yes **25.** yes **27.** no **29.** yes

Written Exercises **Page 275** **1.** yes **3.** no **5.** yes **7.** no **9.** yes **11.** yes **13.** 13 **15.** 6.7 **17.** 15 **19.** 2 **21.** 8.9 **23.** 13.6 mm **25.** 17 ft
27. Given: $\triangle ABC$ $(AB)^2 + (BC)^2 = (AC)^2$
Prove: $\angle B$ is a right angle

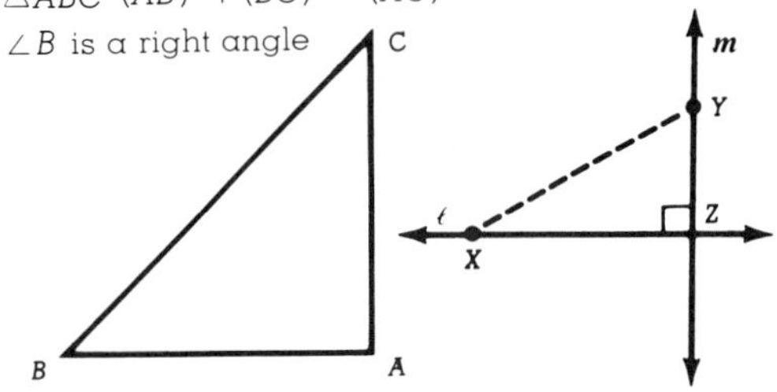

Proof:

STATEMENTS	REASONS
1. Let m be perpendicular to ℓ. On ℓ mark off $ZX = BA$ and on m mark off $ZY = BC$. Complete the triangle by drawing $\overline{XY}$.	1. Draw auxiliary figures
2. $\triangle ZYX$ is a right triangle	2. ℓ is given as perpendicular to m
3. $(XY)^2 = (ZX)^2 + (ZY)^2$	3. Theorem 9–4
4. $(AC)^2 = (AB)^2 + (BC)^2$	4. Given
5. $(XZ)^2 + (YZ)^2 = (AB)^2 + (BC)^2$	5. Postulate 2–9
6. $(XY)^2 = (AC)^2$	6. Postulate 2–9
7. $XY = AC$	7. If the squares of two line segments are equal, the line segments are equal.
8. $\triangle XYZ \cong \triangle ABC$	8. Postulate 4–1
9. $\angle Z \cong \angle B$	9. Corresponding parts of congruent triangles
10. $\angle Z$ is a right angle	10. ℓ is given as perpendicular to m
11. $\angle B$ is a right angle	11. Postulate 2–9

Exploratory Exercises **Page 277** **1.** $\sqrt{2}$ ft **3.** $2\sqrt{2}$ in. **5.** $45\sqrt{2}$ mm **7.** $1.8\sqrt{2}$ m **9.** $31.2\sqrt{2}$ cm **11.** $\frac{14}{3}\sqrt{2}$ yd **13.** $\frac{\sqrt{3}}{2}$ in. **15.** $2\sqrt{3}$ cm **17.** $\frac{25\sqrt{3}}{2}$ cm **19.** $1.25\sqrt{3}$ cm **21.** $\frac{\sqrt{3}}{3}$ yd **23.** $\frac{11\sqrt{3}}{8}$ ft

Written Exercises **Page 278** **1.** 7.1 cm **3.** 13 in. **5.** 4.9 m **7.** 1.0 m **9.** 18.0 cm **11.** 4.2 cm **13.** 18 **15.** $\sqrt{3}$ **17.** $\sqrt{5}$ **19.** $\sqrt{7}$

Exploratory Exercises **Page 282** **1.** $\frac{21}{29}$ **3.** $\frac{21}{20}$ **5.** $\frac{21}{29}$ **7.** $\frac{4}{5}$ **9.** $\frac{4}{3}$ **11.** $\frac{4}{5}$ **13.** $\frac{15}{17}$ **15.** $\frac{15}{8}$ **17.** $\frac{15}{17}$

Written Exercises **Page 282** **1.** $\sin A = 0.882$, $\cos A = 0.471$, $\tan A = 1.875$, $\sin B = 0.471$, $\cos B = 0.882$, $\tan B = 0.533$ **3.** $\sin A = 0.969$, $\cos A = 0.246$, $\tan A = 3.938$, $\sin B = 0.246$, $\cos B = 0.969$, $\tan B = 0.254$ **5.** $\sin A = 0.800$, $\cos A = 0.600$, $\tan A = 1.333$, $\sin B = 0.600$, $\cos B = 0.800$, $\tan B = 0.750$ **7.** $\sin 30° = \frac{1}{2}$ **9.** $\tan 30° = \frac{\sqrt{3}}{3}$ **11.** $\cos 60° = \frac{1}{2}$ **13.** They are equal.

Exploratory Exercises **Page 284** **1.** 0 **3.** 1.000 **5.** 0 **7.** 1.000 **9.** 0 **11.** 45°

Written Exercises **Page 285** **1.** 0.000 **3.** 1.0000 **5.** 0.3584 **7.** 0.8480 **9.** 0.7813 **11.** 1.000 **13.** 0.9272 **15.** 0.8090 **17.** 0.2250 **19.** 1.1918 **21.** 9.5144 **23.** 0.9613 **25.** 1.2349 **27.** 0.9998 **29.** 6° **31.** 26° **33.** 64° **35.** 75° **37.** 88° **39.** 26° **41.** 73° **43.** 39° **45.** 76° **47.** 17° **49.** 49° **51.** 14° **53.** 56° **55.** 87° **57.** 37° **59.** 0.3660 **61.** 1.8660 **63.** 1.000 **65.** 0.500 **67.** $^{-}0.5773$ **69.** 0.4142 **71.** $^{-}0.0670$ **73.** 2.7321 **75.** 0.6880

Exploratory Exercises **Page 288** **1.** $\sin 15° = \frac{QR}{37}$ **3.** $\sin 47° = \frac{10}{PQ}$ **5.** $\cos 16° = \frac{13.4}{PQ}$

Written Exercises **Page 288** **1.** $AC = 39.2$ **3.** $A = 65°$ **5.** $B = 23°$ **7.** $AC = 10.1$ **9.** $AB = 105.7$ **11.** 71.47 **13.** 132.63 m **15.** 2.80 mile

Chapter Review **Page 290** **1.** 5 **3.** $\frac{9}{4}$ **7.** 13.229 **9.** 1.528 **11.** 8.485 **13.** 0.441 **15.** yes **17.** yes **19.** yes **21.** 9.8 **23.** 4.2 in. **25.** 10 mm **27.** $\sin R = 0.471$, $\cos R = 0.882$, $\tan R = 0.533$, $\sin Q = 0.882$, $\cos Q = 0.471$, $\tan Q = 1.875$ **29.** 0.8387 **31.** $A = 30°$ **33.** $C = 44°$

CHAPTER 10 CIRCLES AND SPHERES

Exploratory Exercises Page 296 1. P 3. $\overline{PB}$, $\overline{PC}$, $\overline{PD}$ 5. $\overline{AE}$, $\overline{BD}$ 7. $\overleftrightarrow{AE}$ 9. F, H

Written Exercises Page 296 1. yes 3. no 5. no 7. yes 9. no 11. one 13. one 15. zero 17. zero, one, or two 19. A line segment is a radius of a circle if and only if its endpoints are the center of the circle and a point on the circle.

21. Given: $\odot P$ with diameter $\overline{AB}$

Prove: $AB = 2r$

Proof:

The diameter of a circle is a line segment which contains the center of the circle, P. By definition, $\overline{PB}$ is a radius whose length is r. $\overline{PA}$ is also a radius whose length is r. By betweenness, $AB = AP + PB$. By substitution, $AB = r + r$. Thus, $AB = 2r$.

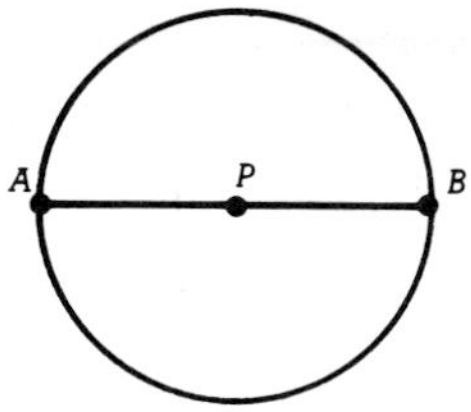

Exploratory Exercises Page 299 1. minor 3. semicircle 5. major 7. major 9. minor 11. major 13. semicircle 15. major 17. minor 19. minor 21. major 23. semicircle

Written Exercises Page 299 1. 38 3. 28 5. 180 7. 180 9. 114 11. 246 13. 152 15. 40 17. 140 19. 140 21. 85 23. 40 25. 135 27. yes 29. yes 31. no 33. yes

35. Given: $\odot P \cong \odot M$, $\widehat{AB} \cong \widehat{RS}$

Prove: $\angle APB \cong \angle RMS$

Proof:

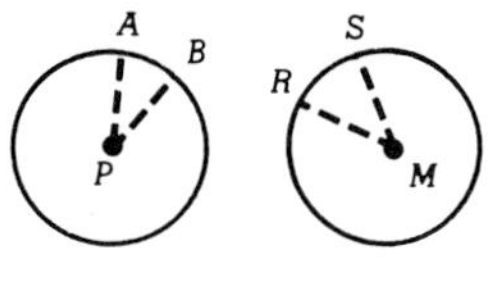

STATEMENTS	REASONS
1. $\widehat{AB} \cong \widehat{RS}$	1. Given
2. $m\widehat{AB} = m\widehat{RS}$	2. Definition of congruent arcs
3. $m\angle APB = m\angle RMS$	3. Arcs are measured by their corresponding central angles.
4. $\angle APB \cong \angle RMS$	4. Definition of congruent angles

37. 156

Exploratory Exercises Page 302 1. Theorem 10–3 3. Theorem 10–5 5. Theorem 10–4

Written Exercises Page 302 1. $\overline{VQ}$ 3. V 5. $\widehat{SQ}$ 7. $\overline{WP}$ 9. no 11. 13 inches 13. 6 cm

15. Given: $\overline{AB} \cong \overline{RS}$

Prove: $\widehat{AB} \cong \widehat{RS}$

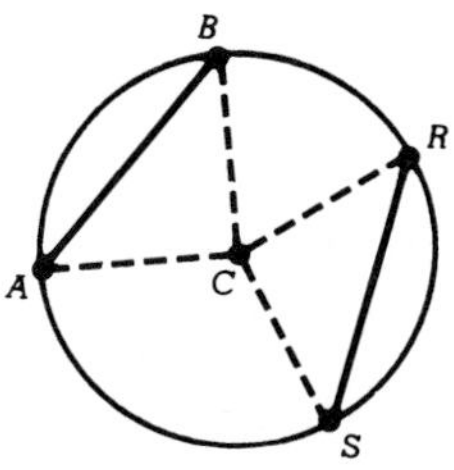

Proof:

STATEMENTS	REASONS
1. Draw radii $\overline{AC}$, $\overline{BC}$, $\overline{RC}$, $\overline{SC}$	1. Postulate 1–1
2. $\overline{AC} \cong \overline{RC}$, $\overline{BC} \cong \overline{SC}$	2. Radii of a circle are congruent.
3. $\overline{AB} \cong \overline{RS}$	3. Given

17. Given: radius $\overline{PR}$ bisects $\widehat{AB}$

Prove: $\overline{PR}$ bisects $\overline{AB}$

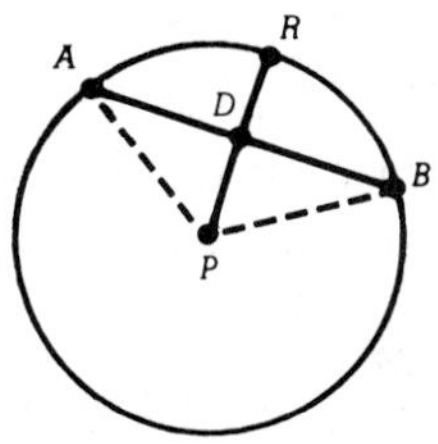

Proof:

STATEMENTS	REASONS
1. Draw radii $\overline{PA}$ and $\overline{PB}$.	1. Postulate 1–1
2. $\overline{PR}$ bisects $\widehat{AB}$	2. Given
3. $m\widehat{AR} = m\widehat{RB}$	3. Definition of bisector
4. $\widehat{AR} \cong \widehat{RB}$	4. Definition of congruence

STATEMENTS	REASONS
4. $\triangle CAB \cong \triangle CRS$	4. SSS
5. $\angle ACB \cong \angle RCS$	5. Definition of congruent triangles
6. $\overset{\frown}{AB} \cong \overset{\frown}{RS}$	6. Theorem 10–2

19. Given: $\overline{AB} \cong \overline{CD}$, PM and PQ are distances of the chords to the center.
Prove: $PM = PQ$

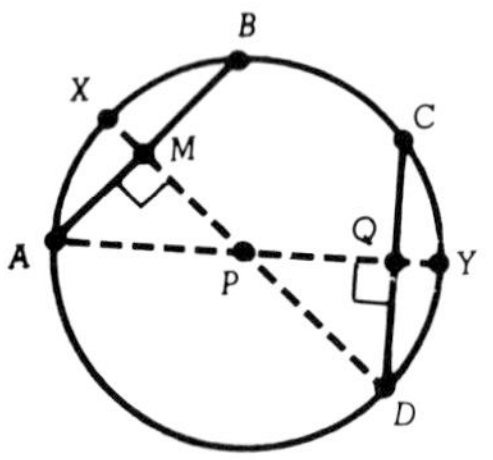

Proof:

STATEMENTS	REASONS
1. PM and PQ are the distances of the chords to the center.	1. Given
2. $\overline{PM} \perp \overline{AB}$, $\overline{PQ} \perp \overline{CD}$	2. The shortest distance from a point to a line is a perpendicular line segment.
3. $\overline{AB} \cong \overline{CD}$	3. Given
4. $\overset{\frown}{AB} \cong \overset{\frown}{CD}$	4. Theorem 10–3
5. $\overline{PX}$ bisects $\overset{\frown}{AB}$, $\overline{PY}$ bisects $\overset{\frown}{CD}$	5. Theorem 10–4
6. $\overset{\frown}{AX} \cong \overset{\frown}{DY}$	6. $\overset{\frown}{AB}$ and $\overset{\frown}{CD}$ are congruent and bisected
7. Draw radii $\overline{AP}$ and $\overline{PD}$	7. Postulate 1–1
8. $\overline{AP} \cong \overline{PD}$	8. Radii of a circle are congruent.
9. $\angle APM \cong \angle DPY$	9. Theorem 10–2
10. $\triangle APM \cong \triangle DPQ$	10. HA
11. $\overline{PM} \cong \overline{PQ}$	11. Definition of congruent triangles
12. $PM = PQ$	12. Definition of congruent line segments

STATEMENTS	REASONS
5. $\angle APD \cong \angle BPD$	5. Theorem 10–2
6. $\overline{AP} \cong \overline{BP}$	6. Radii of a circle are congruent.
7. $\overline{PD} \cong \overline{PD}$	7. Theorem 2–2
8. $\triangle APD \cong \triangle BPD$	8. SAS
9. $\overline{AD} \cong \overline{BD}$	9. Definition of congruent triangles
10. $AD = BD$	10. Definition of congruent line segments
11. $\overline{PR}$ bisects $\overline{AB}$	11. Definition of bisector

21. Given: $\overline{AB} \perp \overline{CD}$, $\overline{CM} \cong \overline{MD}$
Prove: $\overline{AB}$ is a diameter.

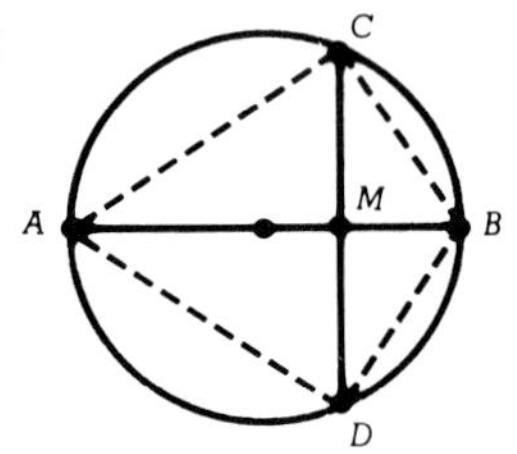

Proof:

STATEMENTS	REASONS
1. $\overline{AB} \perp \overline{CD}$, $\overline{CM} = \overline{MD}$	1. Given
2. $\angle AMC$, $\angle BMC$, $\angle BMD$, and $\angle AMD$ are right angles.	2. Theorem 3–11
3. $\overline{MB} \cong \overline{MB}$, $\overline{MA} \cong \overline{MA}$	3. Theorem 2–2
4. $\triangle BMC \cong \triangle BMD$, $\triangle AMC \cong \triangle AMD$	4. LL
5. $\overline{AC} \cong \overline{AD}$, $\overline{BC} \cong \overline{BD}$	5. Definition of congruent triangles
6. $\overset{\frown}{AC} \cong \overset{\frown}{AD}$, $\overset{\frown}{CB} \cong \overset{\frown}{DB}$	6. Theorem 10–3
7. $m\overset{\frown}{AC} + m\overset{\frown}{CB} = m\overset{\frown}{AD} + m\overset{\frown}{BD}$	7. Addition property of equality
8. $m\overset{\frown}{ACB} = m\overset{\frown}{ADB}$	8. Postulate 10–1
9. $m\overset{\frown}{ACB} = 180$	9. Definition of semicircle
10. $\overline{AB}$ is a diameter.	10. Definition of diameter

23.

STATEMENTS	REASONS
1. $\overline{AC} \perp \overline{BD}$, $\overline{AC}$ is a diameter.	1. Given
2. $\angle CEB$ and $\angle CED$ are right angles.	2. Theorem 3–11
3. $\overline{EC} \cong \overline{EC}$	3. Theorem 2–2
4. $\overline{BE} \cong \overline{ED}$	4. Theorem 10–4
5. $\triangle CEB \cong \triangle CED$	5. LL
6. $\overline{BC} \cong \overline{DC}$	6. Definition of congruent triangles
7. $\overset{\frown}{BC} \cong \overset{\frown}{DC}$	7. Theorem 10–3

Exploratory Exercises Page 307 1. $\widehat{BC}$ 3. $\widehat{HK}, \widehat{IJ}$ 5. $\widehat{QT}, \widehat{RS}$ 7. $\widehat{YZ}$ 9. yes 11. no 13. no 15. yes

Written Exercises Page 307 1. 104 3. 47 5. 52 7. 104 9. 38 11. 99 13. 94 15. 81 17. 94 19. 12 21. $12\sqrt{3}$

23. Given: $\angle QPR$; diameter $\overline{PR}$

Prove: $m \angle QPR = \frac{1}{2} m\widehat{QR}$

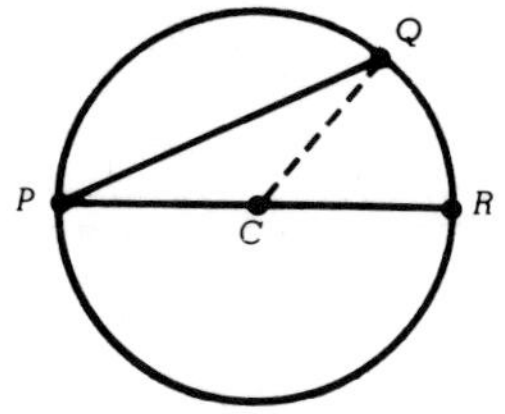

Proof:

STATEMENTS	REASONS
1. Draw radius $\overline{QC}$.	1. Postulate 1–1
2. $\overline{PC} \cong \overline{QC}$	2. Radii of a circle are congruent.
3. $\angle CPQ \cong \angle CQP$	3. Theorem 4–6
4. $m \angle CPQ \cong m \angle CQP$	4. Definition of angle congruence
5. $m \angle QCR = m\widehat{QR}$	5. Arcs are measured by their corresponding central angles.
6. $m \angle QCR = m \angle CPQ + m \angle CQP$	6. Theorem 5–1
7. $m \angle QCR = 2m \angle CPQ$	7. Substitution and addition
8. $m \angle CPQ = \frac{1}{2} m \angle QCR$	8. Postulate 2–8
9. $m \angle QPR = \frac{1}{2} m\widehat{QR}$	9. Substitution

25. Given: $\angle AMB$ intercepts $\widehat{AB}$, $\angle SPR$ intercepts $\widehat{RS}$, $\widehat{AB} \cong \widehat{RS}$

Prove: $\angle AMB \cong \angle SPR$

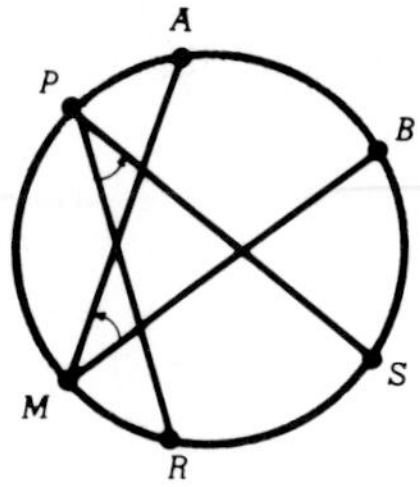

Proof:

STATEMENTS	REASONS
1. $\angle AMB$ intercepts $\widehat{AB}$, $\angle SPR$ intercepts $\widehat{RS}$, $\widehat{AB} = \widehat{RS}$	1. Given
2. $m\widehat{AB} = m\widehat{RS}$	2. Definition of congruent arcs
3. $\frac{1}{2}m\widehat{AB} = \frac{1}{2}m\widehat{RS}$	3. Division property of equality
4. $m \angle AMB = \frac{1}{2}m\widehat{AB}$, $m \angle SPR = \frac{1}{2}m\widehat{RS}$	4. Theorem 10–6
5. $m \angle AMB = m \angle SPR$	5. Postulate 2–6
6. $\angle AMB \cong \angle SPR$	6. Definition of angle congruence

Exploratory Exercises Page 310 1. 2 3. 1 5. tangents

Written Exercises Page 310 1. 12 3. 14 5. 8 7. 45 9. 45 11. 45 13. 45 15. 5 17. 13 19. $5\sqrt{2}$ 21. $\sqrt{329}$

23. Given: $\overline{PT}$ is radius of $\odot P$, $\overline{PT} \perp \overleftrightarrow{AB}$ at T

Prove: $\overleftrightarrow{AB}$ is a tangent to $\odot P$ at T

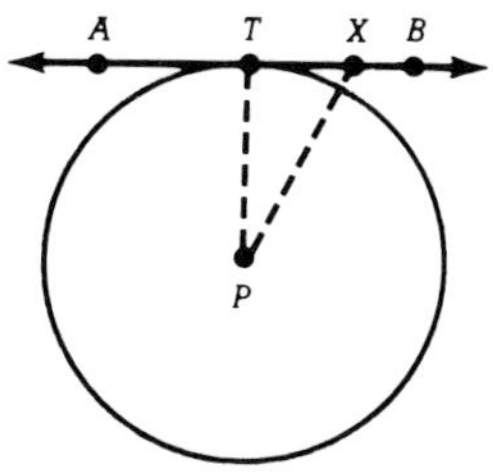

Proof:

STATEMENTS	REASONS
1. $\overline{PT}$ is radius of $\odot P$, $\overline{PT} \perp \overleftrightarrow{AB}$ at T	1. Given
2. Select any point X on $\overleftrightarrow{AB}$, draw $\overline{PX}$	2. Postulate 1–1
3. $PX > PT$	3. Theorem 5–6
4. Any point X, except T, on $\overleftrightarrow{AB}$ lies outside $\odot P$.	4. Step 3
5. $\overleftrightarrow{AB}$ is a tangent to $\odot P$ at T.	5. Definition of tangent

25. Given: hexagon $ABCDE$ circumscribed about $\odot P$ and inscribed in $\odot S$

Prove: $ABCDE$ is regular

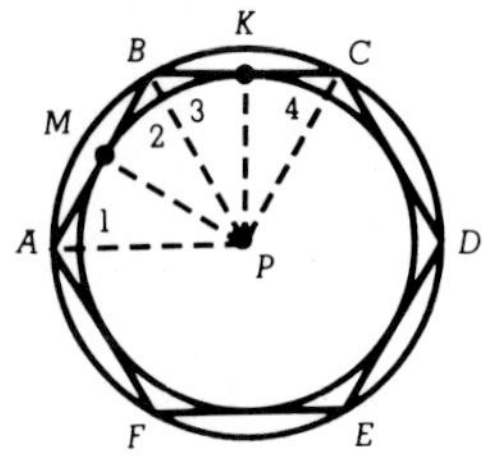

Proof:

STATEMENTS	REASONS
1. $ABCDE$ inscribed in $\odot S$ and circumscribed about $\odot P$	1. Given
2. Draw radii $\overline{PA}$, $\overline{PM}$, $\overline{PB}$, $\overline{PK}$ and $\overline{PC}$.	2. Postulate 1–1
3. $\overline{PM} \cong \overline{PK}$, $\overline{PA} \cong \overline{PB} \cong \overline{PC}$	3. Radii of a circle are congruent.
4. $\overline{PM} \perp \overline{AB}$, $\overline{PK} \perp \overline{BC}$	4. Theorem 10–10

5. $\triangle AMP \cong \triangle BMP$ $= \triangle BKP \cong \triangle CKP$	5. HL
6. $\overline{AM} \cong \overline{MB} \cong \overline{BK} \cong \overline{KC}$, $\angle 1 \cong \angle 2 \cong \angle 3 \cong \angle 4$	6. Definition of congruent triangles
7. $\overline{AB} \cong \overline{BC}$	7. Addition property of equality, definition of betweenness
8. $m\angle B = m\angle 2 + m\angle 3$	8. Angle addition
9. $m\angle B = 2m\angle 2$	9. Substitution

Repeat for all vertices and sides of the hexagon. Thus, hexagon $ABCDE$ is regular.

Exploratory Exercises Page 314 **1.** 63 **5.** $17\frac{1}{2}$

Written Exercises Page 314 **1.** 114 **3.** 66 **5.** 138 **7.** 174 **9.** 49 **11.** 198 **13.** 236 **15.** 38

17. Given: $\overline{AB}$ intersects $\overleftrightarrow{TB}$ at B, $\overleftrightarrow{TB}$ is tangent to $\odot P$ at B

Prove: $m\angle ABT = \frac{1}{2}m\widehat{BA}$

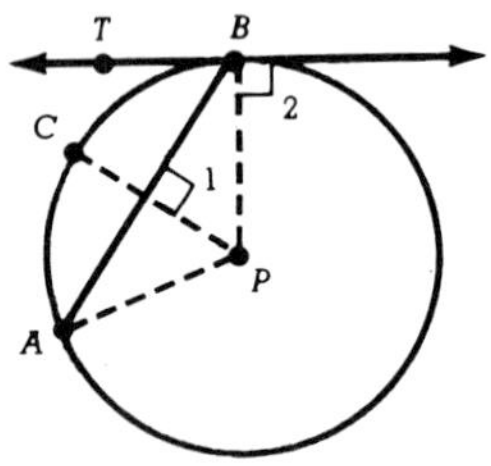

Proof:

STATEMENTS	REASONS
1. $\overline{AB}$ intersects $\overleftrightarrow{TB}$ at B, $\overleftrightarrow{TB}$ is tangent to $\odot P$ at B.	1. Given
2. Draw radii $\overline{PB}$ and $\overline{PA}$.	2. Postulate 1–1
3. $\overline{PB} \perp \overleftrightarrow{TB}$	3. Theorem 10–8
4. $m\angle 2 = 90$	4. Definition of perpendicular
5. Draw $\overline{PC} \perp \overline{AB}$.	5. Auxiliary line
6. $m\angle 1 = 90$	6. Definition of perpendicular
7. $m\angle APB \cong m\widehat{AB}$	7. Definition of arc measure
8. $m\widehat{AC} = m\widehat{CB}$	8. Theorem 10–2
9. $m\angle APC = m\angle CPB$	9. Substitution
10. $m\angle CPB = \frac{1}{2}mAB$	10. Substitution
11. $m\angle ABT + m\angle ABP + m\angle 2 = 180$	11. Definition of a line
12. $m\angle 1 + m\angle ABP + \frac{1}{2}m\angle CPB = 180$	12. Definition of triangle
13. $m\angle ABT + m\angle ABP + m\angle 2 = m\angle 1 + m\angle ABP + \frac{1}{2}m\angle CPB$	13. Substitution
14. $m\angle ABT = \frac{1}{2}m\angle CPB$	14. Subtraction
15. $m\angle ABT = \frac{1}{2}m\widehat{AB}$	15. Substitution

19. Given: $\overrightarrow{TA}$ and $\overrightarrow{TB}$ tangent to $\odot R$, at A and B respectively, $\overrightarrow{TA}$ and $\overrightarrow{TB}$ intersect at T

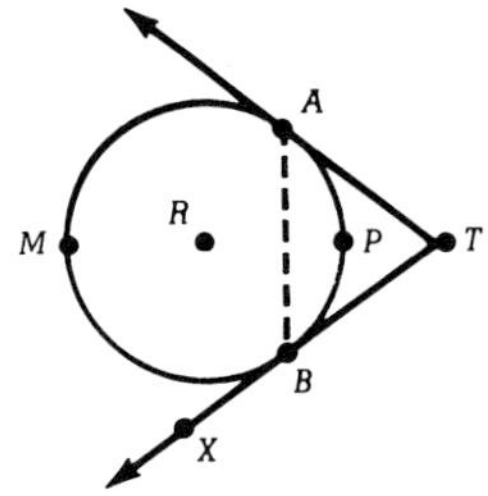

Prove: $m\angle ATB = \frac{1}{2}m\widehat{AMB} - \frac{1}{2}m\widehat{APB}$

Proof:

STATEMENTS	REASONS
1. $\overrightarrow{TA}$ and $\overrightarrow{TB}$ tangent to $\odot R$ at A and B respectively $\overrightarrow{TA}$ and $\overrightarrow{TB}$ intersect at T	1. Given
2. Draw secant $\overline{AB}$	2. Postulate 1–1
3. $m\angle TAB = \frac{1}{2}m\widehat{APB}$, $m\angle ABX = \frac{1}{2}m\widehat{AMB}$	3. Theorem 10–15
4. $m\angle ABX = m\angle TAB + m\angle ATB$	4. Theorem 5–1
5. $m\angle ATB = m\angle ABX - m\angle TAB$	5. Postulate 2–7
6. $m\angle ATB = \frac{1}{2}m\widehat{AMB} - \frac{1}{2}m\widehat{APB}$	6. Substitution
7. $m\angle ATB = \frac{1}{2}(m\widehat{AMB} - m\widehat{APB})$	7. Postulate 2–14

Exploratory Exercises Page 317 **1.** $3x = 9 \cdot 4$ **3.** $4 \cdot 12 = 5(5 + x)$ **5.** $10(10 + x) = 20^2$ **7.** $x(x + 5) = 3 \cdot 12$ **9.** $x(x + 12) = 8^2$

Written Exercises Page 318 **1.** 12 **3.** 0.3 **5.** 5 **7.** 5 **9.** 5 **11.** 2 **13.** 4 **15.** 6.1 **17.** 19.6

19. Given: secants $\overline{RP}$ and $\overline{RT}$
Prove: $RQ \cdot RP = RS \cdot RT$

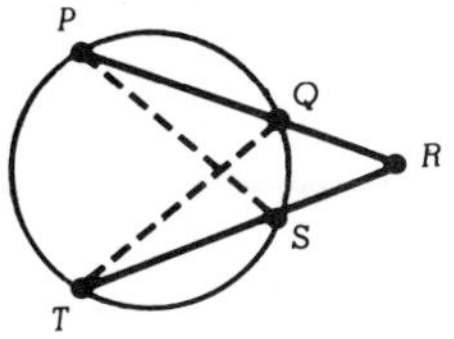

Proof:

STATEMENTS	REASONS
1. Secants $\overline{RP}$ and $\overline{RT}$	1. Given
2. Draw $\overline{PS}$ and $\overline{TQ}$	2. Postulate 1–1
3. $\angle R \cong \angle R$	3. Theorem 3–1
4. $\angle QTS \cong \angle QPS$	4. Theorem 10–7
5. $\triangle RPS \sim \triangle RQT$	5. AA
6. $\frac{RS}{RQ} = \frac{RP}{RT}$	6. Definition of similar polygons
7. $RS \cdot RT = RQ \cdot RP$	7. Theorem 8–1

21.

STATEMENTS	REASONS
1. $\overleftrightarrow{CF}$ is tangent to $\odot P$ at F, $\overleftrightarrow{CF}$ is tangent to $\odot Q$ at F	1. Given
2. $BC \cdot CA = CF^2$, $CD \cdot CE = CF^2$	2. Theorem 10–19
3. $CA \cdot CB = CE \cdot CD$	3. Substitution

Exploratory Exercises Page 320 **1.** sphere **3.** sphere **5.** circle **7.** circle **9.** circle **11.** neither

Written Exercises Page 321 **1.** yes **3.** yes **5.** yes **7.** yes **9.** no **11.** yes **13.** yes **15.** A line segment is a radius of a sphere if and only if its endpoints are the center of the sphere and a point on the sphere. **17.** The intersection of a sphere and a plane is a great circle if and only if the intersection contains the center of the sphere. **19.** A point lies in the interior of a sphere if and only if the measure of the segment joining the point to the center is less than the measure of the radius. The interior of the sphere is the set of all such points. **21.** A line is tangent to a sphere if and only if it intersects the sphere in exactly one point. **23.** A line is called a secant of a sphere if and only if it intersects the sphere in exactly two points. **25.** Spheres are congruent if and only if they have the same center and have radii which are congruent. **27.** 1 **29.** 1 **31.** 1 **33.** infinite

Chapter Review Page 324 **1.** no **3.** A line segment is a diameter of a circle if and only if its endpoints lie on the circle and it contains the center of the circle. **5.** 180 **7.** 117 **9.** 57 **11.** 123 **13.** 10 cm **15.** 106 **17.** 106 **19.** 36 **21.** 36 **23.** $3\sqrt{2}$ **25.** tangents **27.** 55 **29.** 18.5 **31.** 133 **33.** 0.806 **35.** no **37.** 2 **39.** 0

CHAPTER 11 AREA AND VOLUME

Exploratory Exercises Page 330 **1.** 1×12, 2×6, 3×4 **3.** 1×24, 2×12, 3×8, 4×6 **5.** 36, 24 **7.** 36, 30 **9.** 25, 20 **11.** 21, 20 **13.** no **15.** yes

Written Exercises Page 331 **1.** 20 cm^2 **3.** 120 ft^2 **5.** 2269.5 cm^2 **7.** 667 mi^2 **9.** $18\frac{1}{3}$ mi^2 **11.** $23\frac{3}{8}$ in^2 **13.** $(x^2 + 3x)$ ft^2 **15.** $(12a^2 + 10a + 2)$ in^2 **17.** 10 in. **19.** 21.2 m **21.** $4\frac{1}{3}$ yd **23.** $2\frac{1}{4}$ ft **25.** 448 **27.** 120 **29.** 12.5 **31.** 240 bags

Exploratory Exercises Page 335 **1.** 96 in^2 **3.** 18 cm^2 **5.** 252 mm^2 **7.** 84 ft^2 **9.** 48 in^2 **11.** 9 cm^2 **13.** 126 mm^2 **15.** 42 ft^2 **17.** 38.5 cm^2 **19.** 850 mi^2 **21.** 64 in^2

Written Exercises Page 336 **1.** 4 ft **3.** 13.16 m^2 **5.** 1.3 m **7.** $20\frac{1}{4}$ in^2 **9.** $(2x^2 + 8x)$ m^2 **11.** 8 ft **13.** 6.58 m^2 **15.** 2.6 m **17.** $10\frac{1}{8}$ in^2 **19.** $(x^2 + 4x)$ m^2 **21.** 4.68 m^2 **23.** 42.56 m^2 **25.** 31.5 ft^2

Exploratory Exercises **Page 339** 1. 120 3. 72 5. 45 7. 36 9. $25\frac{5}{7}$

Written Exercises **Page 340** 1. D 3. $\overline{DC}$, $\overline{DB}$, $\overline{DA}$ 5. $\overline{AC}$ 7. $\overline{DE}$ 9. $\overline{AB}$ 11. $\odot D$ with radius $\overline{DE}$ 13. 174 cm² 15. 17.6 ft² 17. 110 in² 19. 937.65 mm² 21. 186 ft² 23. 62.4 units² or $36\sqrt{3}$ units² 25. <498.8 cm² or $288\sqrt{3}$ cm² 27. 200 ft² 29. 33 in² 31. $4\sqrt{3}$ cm, $8\sqrt{3}$ cm, 72 cm, $144\sqrt{3}$ cm² or 249.4 cm²

33. Given: a regular polygon with an inscribed circle; $\overline{AP}$ is an apothem
Prove: $\overline{AP}$ is perpendicular to the side of the polygon ($\overline{BC}$) at the point of tangency

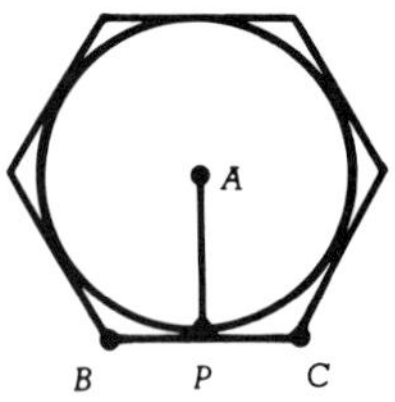

Proof:

STATEMENTS	REASONS
1. $\overline{AP}$ is an apothem of a regular polygon; circle A is inscribed in the polygon	1. Given
2. $\overline{BC}$ is tangent to $\odot A$	2. Definition of inscribed circle
3. $\overline{AP}$ is the radius of the inscribed circle	3. Definition of apothem
4. $\overline{AP} \perp \overline{BC}$	4. Theorem 10–10

35. Given: regular hexagon $ABCDEF$
Prove: each central angle is $\frac{360}{6}$

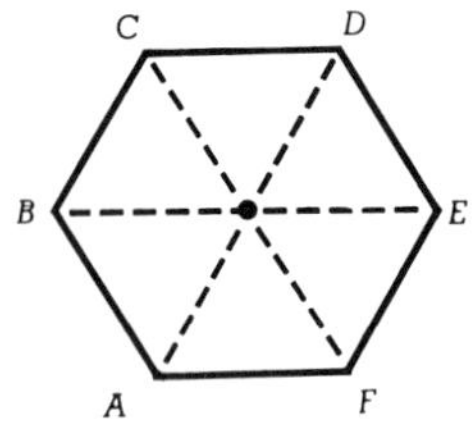

Proof: Let radii be drawn from the vertices of the polygon (hexagon) to its center. Since there are 6 sides of the hexagon, there are 6 central angles. Also, 6 triangles are formed by the radii. These 6 triangles are all congruent because of SSS (radii are congruent, so are the sides of the hexagon). Thus all central angles are equal by corresponding parts of congruent triangles. Since there are 6 central angles, and the sum total of degrees at the center of the polygon is 360, each central angle must measure $\frac{360}{6}$.

Exploratory Exercises **Page 344** 1. 14π cm 3. 8π in. 5. 14.8π cm 7. 4.8π km 9. π ft 11. $9\frac{1}{3}\pi$ yd 13. $10\pi\sqrt{2}$ cm 15. $6\pi\sqrt{2}$ in. 17. 100π m² 19. 64π ft² 21. 92.16π m² 23. 2.25π mi² 25. $11\frac{1}{9}\pi$ ft² 27. $33\frac{1}{16}\pi$ ft² 29. 48π in²

Written Exercises **Page 344** 1. 44.0 cm 3. 25.1 in. 5. 46.5 cm 7. 15.1 km 9. 3.1 ft 11. 29.3 yd 13. 44.4 cm 15. 26.6 in. 17. 314 m² 19. 200.96 ft² 21. 289.38 m² 23. 7.07 mi² 25. 34.89 yd² 27. 103.82 ft² 29. 150.72 in² 31. $\frac{1}{9}$ 33. $\frac{1}{12}$ 35. $\frac{1}{8}$ 37. $\frac{1}{20}$ 39. 7.4 cm² 41. 5.5 cm² 43. 8.3 cm² 45. 3.3 cm² 47. 40.7 m² 49. 9.1 ft²

Exploratory Exercises **Page 349** 1. polygon $ABCDEFG$, polygon $PQRSTUVW$ 3. 7 5. yes

Written Exercises **Page 349** 1. 60 cm² 3. 264 in² 5. 40.5 m² 7. 175.2 units² 9. 72 cm² 11. 312 cm² 13. 46.8 m² 15. 452.1 units² 17. 439.6 cm² 19. 384.3 ft² 21. 1704.4 m² 23. 747.3 cm² 25. 610.4 ft² 27. 2578.8 m² 29. 10,362 ft² 31. hexagonal crystal

Exploratory Exercises **Page 352** 1. 27 cm² 3. 36 cm² 5. 45 cm² 7. 54 cm² 9. 72 cm²

Written Exercises **Page 352** 1. 188.4 cm² 3. 132.6 m² 5. 149.5 cm² 7. 77.7 in² 9. 376,800 ft² 11. 301.4 cm² 13. 185.4 m² 15. 299.0 cm² 17. 141.3 in² 19. 408,200 ft² 21. 1256 cm² 23. 50,240,000 ft² 25. 1,738,103 miles² 27. approximately 922,069 ft²

Exploratory Exercises **Page 356** 1. 42 cm³ 3. 67.2 ft³ 5. 14,400 cm³

Written Exercises **Page 357** 1. 48 m³ 3. 2598.1 cm³ or $1500\sqrt{3}$ cm³ 5. 58.5 cm³ 7. 100.5 m³ 9. 337.6 cm³ 11. 28.3 ft³ 13. eight times larger 15. eight times larger 17. 36 people 19. 19.6 ft³

Exploratory Exercise **Page 361** 1. 35 ft³ 3. 10 ft³ 5. 120.6 cm³ 7. 446.6 cm³ 9. 6.3 in³

Written Exercise **Page 361** 1. 418.7 ft³ 3. 340.2 m³ 5. 928.9 cm³ 7. 1,360,666.7 ft³ 9. 1.097×10^{12} km³ 11. 268.0 cm³ 13. $\frac{\pi r^2 h}{3}$ 15. $\frac{3s^2\sqrt{3}}{2}\left(h + \frac{l^2}{3} - \frac{s^3}{3}\right)$

Chapter Review Page 364 1. 24 ft^2 3. 24 m^2 5. 130 ft^2 7. 1250 cm^2 9. 2660.4 cm^2 11. 289.4 in^2 13. 48 cm^2 15. 45 cm^2 17. 14,649,984 $miles^2$ 19. 2093.3 cm^3 21. 2786.2 cm^3

CHAPTER 12 COORDINATES

Exploratory Exercises Page 371 1. $(^-6, 8)$ 3. $(^-6, ^-5)$ 5. $(0, 7)$ 7. $(3, 3)$ 9. $(^-4, ^-3)$ 11. $(4, 0)$ 13. linear 15. linear 17. linear 19. linear 21. linear

Written Exercises Page 371 17. $(1, 3)$ 19. $(2, ^-5)$ 21. $(^-2, 2)$

Exploratory Exercises Page 375 1. 7 3. 4 5. 5 7. $(0, 2)$ 9. $(^-4, 4)$ 11. $\left(1, 1\frac{1}{2}\right)$

Written Exercises Page 375 1. $6\sqrt{2}$ 3. $7\sqrt{2}$ 5. $\sqrt{41}$ 7. $2\sqrt{29}$ 9. 5 11. $(a \cdot c)\sqrt{2}$ 13. $\sqrt{12^2 + (a - b)^2}$ 15. $\frac{1}{5}\sqrt{2}$ 17. $(7.5, 7.5)$ 19. $(^-8.5, ^-7)$ 21. $(^-4, ^-4)$ 23. $(1, ^-4)$ 25. $\left(\frac{p}{2}, \frac{q}{2}\right)$ 27. $(0, 0)$ 29. $\left(\frac{1}{2}, \frac{3}{5}\right)$ 31. 26 33. $\sqrt{97}$

Exploratory Exercises Page 380 1. no slope 3. product is $^-1$ 5. positive slope 7. change in y is 2 units, change in x is 1 unit, slope is 2 9. change in y is 4 units, change in x is $^-3$ units, slope is $-\frac{4}{3}$

Written Exercises Page 380 1. 2 3. 0 5. $\frac{3}{2}$ 7. $^-3$ 9. $^-2$ 11. 2 13. 2 15. $^-23$ 17. 6 19. yes

Exploratory Exercises Page 383 1. $-\frac{3}{4}, 2$ 3. $4, ^-3$ 5. $3, ^-5$ 7. $0, 6$ 9. $y = 6x - 5$ 11. $y - 3 = 4(x - 1)$

Written Exercises Page 384 1. $y = 2x - 3$ 3. $y = {}^-4x + 7$ 5. $y = 8$ 7. $y + 2 = -\frac{2}{3}(x + 2)$ 9. $y - 7 = 3(x + 3)$ 11. $y + 5 = -\frac{5}{3}(x + 4)$ 13. $x = {}^-7$ 15. $5, ^-3$ 17. $\frac{4}{3}, -\frac{5}{3}$ 19. $y + 1 = \frac{2}{3}(x - 6)$ 21. $y - \frac{7}{2} = -\frac{5}{7}\left(x - \frac{9}{2}\right)$

Exploratory Exercises Page 388 1. $(2, ^-5)$ 3. $(^-1, ^-2)$ 5. $(2, 5)$ 7. b, c 9. a

Written Exercises Page 388 1. $(4, 2)$ 3. $(2, ^-2)$ 5. $(^-1, 2)$ 7. $r = 10, s = 15$ 9. $x = 3, y = 1$ 11. $x = {}^-1, y = {}^-1$ 13. $x = 3, y = {}^-1$ 15. $x = 1, y = 2$ 17. $x = 0, y = 4$ 19. $x = {}^-1, y = 8$ 21. $x = \frac{4}{3}, y = \frac{4}{3}$

Exploratory Exercises Page 390 1. $(0, 0)$, 4 3. $(0, 0)$, 5 5. $(4, 6)$, 3 7. $(3, 12)$, 6 9. $(^-8, 9)$, 9

Written Exercises Page 391 1. $x^2 + y^2 = 25$ 3. $(x + 2)^2 + (y - 8)^2 = 2$ 5. $(x + 1)^2 + (y + 1)^2 = \frac{1}{16}$ 7. $(0, 0)$, 11 9. $(0, 3)$, 2 11. $(^-3, 4)$, $2\sqrt{5}$ 13. $(7, ^-5)$, 2 15. $(0, 0)$, 4 17. $(^-7, ^-3)$, $\sqrt{3}$ 19. $\left(^-4, \frac{1}{2}\right)$, $\sqrt{6}$ 21. $(^-5, 2)$, $\frac{1}{2}\sqrt{3}$ 23. $(x - 1)^2 + (y - 2)^2 = 9$ 25. $x^2 + y^2 = 14$ 27. $(x + 4)^2 + (y - 3)^2 = \frac{16}{9}$ 29. $(x - 3)^2 + \left(y - \frac{1}{2}\right)^2 = \frac{16}{25}$ 31. $(x + 2)^2 + (y - 11)^2 = 32$

Exploratory Exercises Page 394 1. $(r + s, t)$ 3. $(^-b, 2b)$ 5. $(j + k, k)$

Written Exercises Page 395 1. $A(^-b, 0)$, $C(b, 2b)$, $D(^-b, 2b)$ 3. $R(a, b)$ 5. $A(0, 0)$, $C(a, b + c)$

7. Given: rectangle $ABCD$
Prove: $\overline{DB} \cong \overline{AC}$

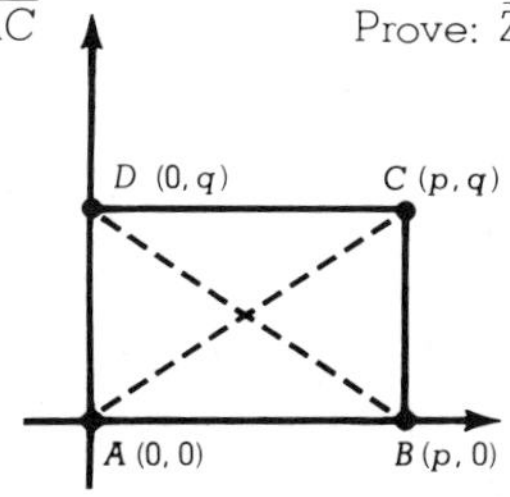

9. Given: isosceles trapezoid $TRPZ$
Prove: $\overline{ZR} \cong \overline{PT}$

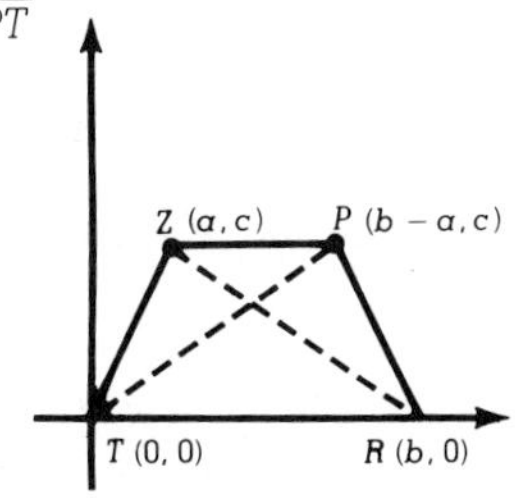

11. Given: parallelogram $PLGM$
Prove: $\overline{MG} \cong \overline{PL}$, $\overline{MP} \cong \overline{GL}$

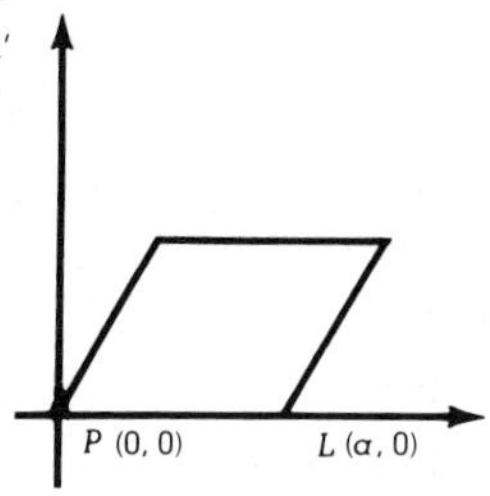

13. Proof: Use the distance formula to show that $DB = AC$.

$DB = \sqrt{(p - 0)^2 + (0 - q)^2}$
$= \sqrt{p^2 + (^-q)^2}$
$= \sqrt{p^2 + q^2}$

$AC = \sqrt{(p - 0)^2 + (q - 0)^2}$
$= \sqrt{p^2 + q^2}$

Therefore, $DB = AC$
Thus, $\overline{DB} \cong \overline{AC}$, by definition of congruent line segments.

15. Proof: Use the distance formula to show that $ZR = PT$.

$ZR = \sqrt{(b - a)^2 + (c - 0)^2}$
$= \sqrt{(b - a)^2 + c^2}$

$PT = \sqrt{[(b - a) - 0]^2 + (c - 0)^2}$
$= \sqrt{(b - a)^2 + c^2}$

Therefore, $ZR = PT$.
Thus, by definition of congruent line segments $\overline{ZR} \cong \overline{PT}$.

17. Proof: Use the distance formula to show that $MG = PL$, and $MP = GL$.

$$MG = \sqrt{[(a+b)-b]^2 + (c-c)^2} = \sqrt{a^2 + 0^2} = \sqrt{a^2} = a$$

$$PL = \sqrt{(a-0)^2 + (0-0)^2} = \sqrt{a^2 + 0^2} = \sqrt{a^2} = a$$

Thus, $MG = PL$.

$$MP = \sqrt{(b-0)^2 + (c-0)^2} = \sqrt{b^2 + c^2}$$

$$GL = \sqrt{[(a+b)-a]^2 + (c-0)^2} = \sqrt{b^2 + c^2}$$

Thus, $MP = GL$.

By definition of congruent line segments, $\overline{MG} \cong \overline{PL}$ and $\overline{MP} \cong \overline{GL}$.

Exploratory Exercises Page 398 **1.** 4 **3.** $\sqrt{6}$ **5.** (4, 0, 0) **7.** $\left(-\frac{1}{2}, 4, 8\right)$ **9.** (3, 8, 2), 7 **11.** $(^-4, 2, ^-1)$, 5

Written Exercises Page 399 **1.** $\sqrt{13}$ **3.** $2\sqrt{5}$ **5.** 6 **7.** $\sqrt{14}$ **9.** $5\sqrt{3}$ **11.** $\left(\frac{3}{2}, ^-2, 2\right)$ **13.** $\left(\frac{5}{2}, -\frac{3}{2}, \frac{5}{2}\right)$ **15.** $\left(-\frac{1}{2}, 4, 5\right)$ **17.** $(6, 5, ^-1)$, 9 **19.** (0, 3, 0), 2 **21.** $x^2 + y^2 + z^2 = 9$ **23.** $(x+1)^2 + (y-2)^2 + (z-4)^2 = 16$ **25.** $(x-2)^2 + \left(y - \frac{1}{2}\right)^2 + (z-1)^2 = \frac{1}{9}$ **27.** $10 + \sqrt{74}$ **29.** (1, 2, 6) **31.** $(x-1)^2 + (y-2)^2 + (z-6)^2 = 26$ **33.** 6

Chapter Review Page 401 **7.** $(^-2, ^-4)$ **9.** $(^-4, 2)$ **11.** 5 **13.** 2 **15.** $\sqrt{(a-b)^2 + (b-a)^2}$ **17.** $\left(^-4, \frac{7}{2}\right)$ **19.** (3, 5) **21.** $\left(\frac{3}{4}, \frac{1}{2}\right)$ **23.** $(^-12, ^-16)$ **25.** $-\frac{2}{3}$ **27.** undefined **29.** 2 **31.** 7 **33.** $\frac{3}{2}$ **35.** 4, 6 **37.** none, none **39.** $y = 3x - 1$ **41.** $x = 4$ **43.** $7y = 6x - 30$ **45.** $x = 5$, $y = 3$ **47.** $x = \frac{^-5}{11}$, $y = \frac{^-1}{11}$ **49.** $x = 6$, $y = 1$ **51.** (2, 0), $\sqrt{12}$ **53.** (0, 5), $\frac{8}{11}$ **55.** $x^2 + y^2 = 0.09$ **57.** $\left(x + \frac{2}{3}\right)^2 + (y-5)^2 = \frac{36}{49}$ **59.** $(^-a, 0)$, $(^-a, 2a)$, $(a, 2a)$ **61.** Given: rhombus $RHMB$
Prove: $\overline{RM} \perp \overline{BH}$

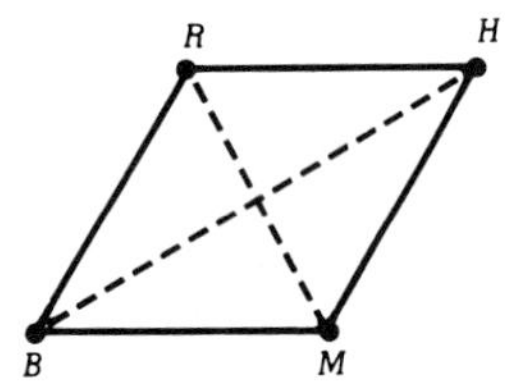

63. $\sqrt{99}$ **65.** $\sqrt{74}$ **67.** $\left(\frac{1}{2}, \frac{13}{2}, \frac{1}{2}\right)$ **69.** (1, 3, 5) **71.** (0, 0, 0), 12 **73.** $(^-3, ^-4, 1)$, $\sqrt{13}$ **75.** $(x-9)^2 + (y+6)^2 + (z-4)^2 = 64$ **77.** $(x+2)^2 + (y-1)^2 + \left(z - \frac{2}{5}\right)^2 = \frac{4}{49}$

CHAPTER 13 LOCI AND CONSTRUCTIONS

Exploratory Exercises Page 411 **1.** Postulate 1–1 **3.** Theorem 2–2 **5.** definition of congruent triangles **7.** SAS **9.** definition of midpoint

Exploratory Exercises Page 414 **1.** $\overline{AC}$ contains the center of $\odot X$. **3.** $\angle CDA$ intercepts $\widehat{CEA}$ which is a semicircle; $\angle CDA$ is inscribed in a semicircle $\widehat{CDA}$. **5.** Definition of a tangent: $\overleftrightarrow{AD}$ is perpendicular to radius $\overline{CD}$ at its endpoint. **7.** no

Exploratory Exercises Page 418 **1.** Postulate 1–1 **3.** An angle inscribed in a semi-circle is a right angle. **5.** Theorem 9–21.

Exploratory Exercises Page 422 **1.** A circle with a 1 mm radius. **3.** A circle with a 3 mm radius. **5.** A circle with a 40 mm radius. **7.** A circle with a 1000 mm radius. **9.** A sphere with a 1 mm radius. **11.** A sphere with a 3 mm radius. **13.** A sphere with a 40 mm radius. **15.** A sphere with a 1000 mm radius. **17.** A pair of parallel lines one 1 in. above and one 1 in. below ℓ. **19.** A pair of parallel lines, one 3 in. above and one 3 in. below ℓ. **21.** A pair of parallel lines, one 40 in. above and one 40 in. below ℓ. **23.** A pair of parallel lines, one 1000 in. above and one 1000 in. below ℓ. **25.** A cylinder drawn about ℓ with a 1 in. radius. **27.** A cylinder drawn about ℓ with a 3 in. radius. **29.** A cylinder drawn about ℓ with a 40 in. radius. **31.** A cylinder drawn about ℓ with a 1000 in. radius.

Written Exercises Page 422 **1.** The locus is another circle with the same center, but with a radius of 5 centimeters. **3.** The locus is another square in the interior of the original square whose sides measure 1 inch and whose sides lie

2 inches away from the original. **5.** The locus of all points in a plane that are equidistant from the sides of a given angle is the bisector of the angle. **7.** The locus of all points in a plane that are equidistant from two given points is the perpendicular bisector of the line segment connecting the two points. **9.** The locus of all points in a plane that are 5 centimeters from a circle with radius measuring 3 centimeters is a concentric circle whose radius is 8 centimeters. **11.** The locus of all the points in a plane that are center of circles with radii measuring 7 inches and pass through a given point is a circle with radius of 7 inches and whose center is the given point. **13.** The locus of all the points in a plane that are centers of circles which are tangent to both sides of a given angle is the bisector of the angle. **15.** The locus of all the points in a plane that are the third vertex of isosceles triangles with a given line segment for a base is the perpendicular bisector of the base segment. **17.** The locus of all the points in space that are a given distance from a given line is a cylinder having the given line as its center and the given distance as its radius. **19.** The locus of all the points in space that are the centers of spheres with radii measuring r units and tangent to a given plane is a pair of planes parallel to the original plane and each r units from the original.

Exploratory Exercises Page 424 **1.** c **3.** a **5.** c

Written Exercises Page 425 **1.** May have no points of intersection; one line may be tangent to the circle, the other not intersecting; one line may be secant to the circle, the other not intersecting; one line may be secant to the circle while the other is tangent; both lines may be secant to the circle. **3.** May have no points of intersection; one of the parallel lines may be tangent to the outer circle; one of the parallel lines may be secant to the outer circle; one of the parallel lines may be secant to the outer circle while the other is tangent to the outer circle; one parallel line may be tangent to the inner circle and secant to the outer while the other is tangent to the outer circle; one parallel line may be tangent to the inner circle and secant to the outer while the other is secant to the outer circle; one parallel may be secant to the inner circle and secant to the outer while the other is tangent to the outer circle; one parallel line may be secant to the inner circle and secant to the outer while the other is secant to the outer circle; one parallel line may be secant to the inner circle and secant to the outer while the other is tangent to the inner and secant to the outer; both parallel lines may be secant to the inner circle and both secant to the outer; one of the parallel lines could be tangent to the inner circle and secant to the outer while the other intersects neither circle; one of the parallel lines could be secant to the inner circle and secant to the outer while the other intersects neither. **5.** Could have no point of intersection; one line could be tangent to the sphere; one line could be secant to the sphere; one line could be secant to the sphere while the other is tangent; both lines could be secant to the sphere. **7.** Could have no point of intersection; could be tangent at a point; could intersect in a circle; could have all points in common. **9.** Could have no points in common; have one point in common; the circle would lie in the plane. **11.** The locus of the points 2 centimeters from the line is a pair of parallel lines each 2 centimeters away. The locus of the points 5 centimeters from the point on the line is a circle of radius 5 centimeters with the point as center. The locus of all points meeting both conditions is the set of four points where the circle and parallel lines intersect. **13.** The locus of points 2 inches from the line is a cylinder with the line as a center and a radius of 2 inches. The locus of points 5 inches from the point on the line is a sphere with the point as center and a radius of 5 inches. The locus of all points meeting both conditions are the two circles formed by the intersection of the cylinder and the sphere. **15.** The locus of all the points in a plane that are equidistant from three given points is the center of the circle containing the three points. **17.** The locus of points equidistant from the sides of the angle is the bisector of the angle. The locus of points 4 inches from the vertex is a circle with the vertex as center and a radius of 4 inches. The locus of all points meeting both conditions is the intersection of the circle and the bisector. **19.** The locus of points equidistant from two given parallel lines is a third line parallel to both and midway between both. The locus of points a given distance from another line is a pair of parallel lines each the given distance away. The locus of points meeting both conditions can be an infinite set of the loci, a parallel, and coincide, or none, if all lines above are parallel and do not coincide, or the two points of intersection of the midway line and the latter pair of parallel lines.

Chapter Review Page 428 **19.** The locus of all points in space that are one inch from a given point A is a sphere whose radius is one inch. **21.** The locus of all the points in a plane that are centers of circles with radii measuring 11 centimeters and pass through a given point is a circle with the given point as center and a radius of 11 centimeters. **23.** Two points of intersection if the line passes through the sphere; one point of intersection if the line is tangent to the sphere; no points of intersection. **25.** The circle intersects both parallel planes in two points each; the circle is tangent to one plane and intersects the other in two points; the circle is tangent to both planes; the circle intersects one plane only in two points; the circle is tangent to one plane only; no points of intersection. **27.** A plane parallel to two given planes and midway between them is the locus of points which lie halfway between the planes.

CHAPTER 14 TRANSFORMATIONS

Exploratory Exercises Page 433 1. A 3. U 5. $\overline{BC}$ 7. $\angle D$ 9. $\angle C$

Written Exercises Page 433 1. E 3. D 5. $\angle BDE$ 7. A 9. C 11. $\overline{CA}$ 13. $\angle A \cong \angle E$, $\angle ABC \cong \angle EBD$, $\angle C \cong \angle D$, $\overline{AB} \cong \overline{EB}$, $\overline{AC} \cong \overline{ED}$, $\overline{CB} \cong \overline{DB}$ 15. $\triangle EFD$ 17. $\triangle XZY$ 19. quadrilateral $DCBA$ 21. $\triangle KLM$ 23. Y, W 25. gray, gray 27. blue, gray 29. gray, blue

Exploratory Exercises Page 437 1. A 3. $\overline{AC}$ 5. L 7. $\overline{JL}$ 9. $\overline{HG}$ 11. N

Written Exercises Page 437 7. no 9. yes 11. yes 29. none 33. none 35. none 37. none

Exploratory Exercises Page 440 1. B 3. G 5. H 7. D 9. R 11. Q

Written Exercises Page 440 1. yes 3. yes 5. no 7. reflection 9. reflection 11. none 13. reflection 15. reflection 17. none 25. $\triangle PQR$ 27. $\triangle LMN$ 29. $\triangle STU$ 31. Plan: A translation is composed of two successive reflections. The first reflection with respect to ℓ preserves betweenness of points. The second reflection with respect to m preserves betweenness of points. Therefore, by transitivity, betweenness of points is preserved from preimage to image.

Exploratory Exercises Page 444 1. yes 3. quadrilateral $EFCD$ 5. quadrilateral $HGJK$ 7. quadrilateral $CDAB$ 9. 140 11. 140 13. $\overline{GK}$

Written Exercises Page 444 3. 60° 5. 120°

7.
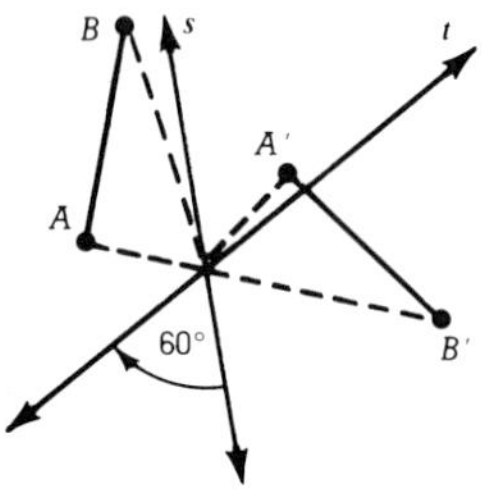

9.
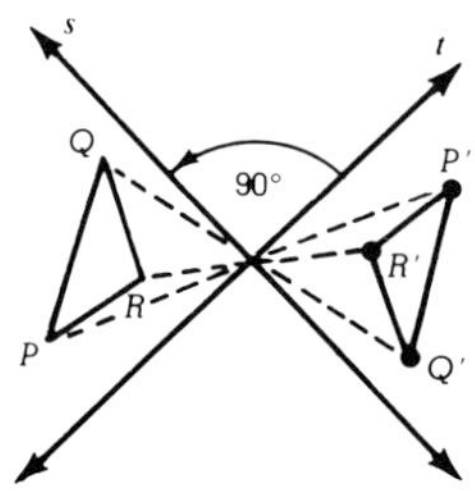

11.
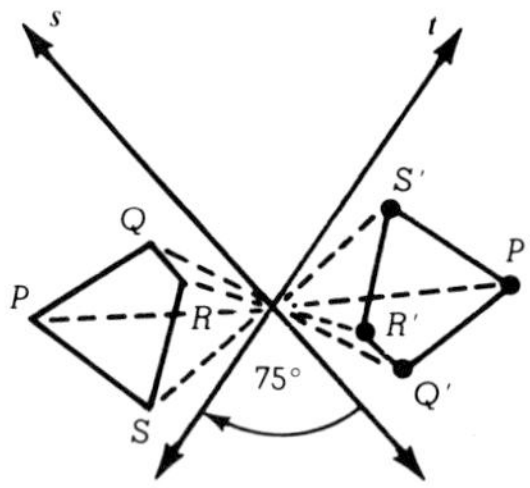

13. yes 15. yes

Exploratory Exercises Page 450 1. F 3. H 5. p 7. ℓ, m, p; m, p, q 9. reflection

Written Exercises Page 451 5. translation with respect to ℓ and p 7. reflection 9. translation 11. rotation

13.
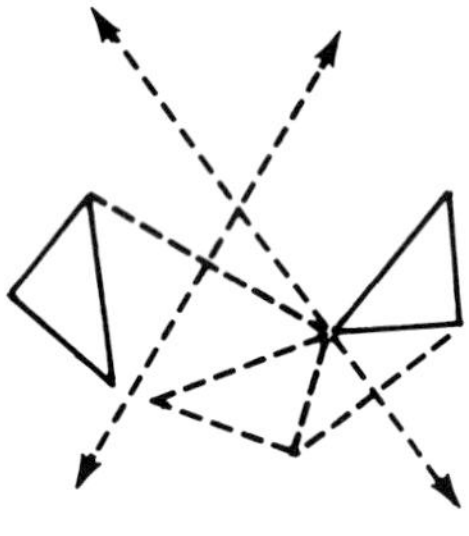

15.
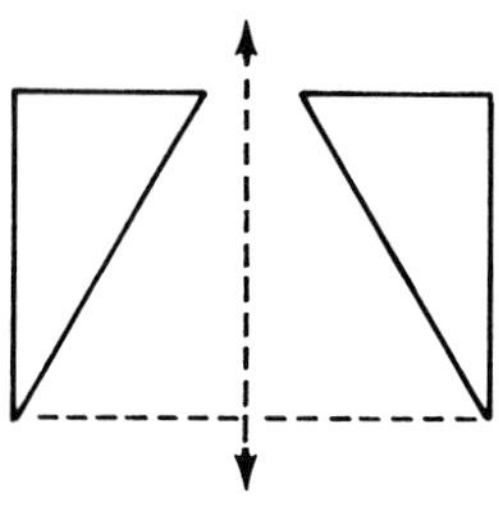

17.
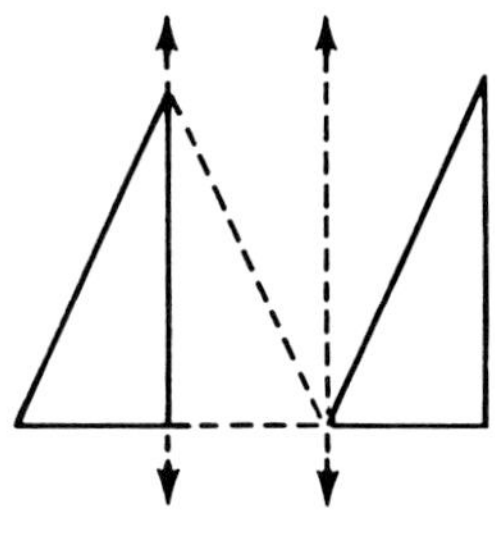

Exploratory Exercises Page 454 1. enlargement 3. congruence transformation 5. reduction 7. reduction 9. No. The figure would be reduced to a point.

Written Exercises Page 454 1. 2 3. $\frac{3}{2}$ 5. $\frac{1}{3}$ 7. enlargement 9. enlargement 11. reduction 17. A 19. T 21. S 23. F 45. 30 in. 47. 24 ft 49. 3 cm 51. Plan: By a scale factor of $\frac{1}{2}$, $CE = \frac{1}{2}CA$ and $CD = \frac{1}{2}CB$. That is, $\frac{CE}{CA} = \frac{1}{2}$ and $\frac{CD}{CB} = \frac{1}{2}$. By the reflexive property of congruence $\angle ECD \cong \angle ACD$ for $\triangle CED$ and $\triangle CAB$. Thus, $\triangle CED \sim \triangle CAB$. Therefore $ED = \frac{1}{2}AB$.

Chapter Review Page 458 1. E 3. B 5. $\angle C$ 7. $\overline{CB}$ 17. reflection with respect to m 19. reduction 21. 4

7 8 9 10 11 12 13 14 15 — 89 88 87 86 85